AF738195

Lecture Notes in Electrical Engineering

Volume 556

Series Editors

Leopoldo Angrisani, Department of Electrical and Information Technologies Engineering, University of Napoli Federico II, Napoli, Italy
Marco Arteaga, Departament de Control y Robótica, Universidad Nacional Autónoma de México, Coyoacán, Mexico
Bijaya Ketan Panigrahi, Electrical Engineering, Indian Institute of Technology Delhi, New Delhi, Delhi, India
Samarjit Chakraborty, Fakultät für Elektrotechnik und Informationstechnik, TU München, München, Germany
Jiming Chen, Zhejiang University, Hangzhou, Zhejiang, China
Shanben Chen, Materials Science & Engineering, Shanghai Jiao Tong University, Shanghai, China
Tan Kay Chen, Department of Electrical and Computer Engineering, National University of Singapore, Singapore, Singapore
Rüdiger Dillmann, Humanoids and Intelligent Systems Lab, Karlsruhe Institute for Technology, Karlsruhe, Baden-Württemberg, Germany
Haibin Duan, Beijing University of Aeronautics and Astronautics, Beijing, China
Gianluigi Ferrari, Università di Parma, Parma, Italy
Manuel Ferre, Centre for Automation and Robotics CAR (UPM-CSIC), Universidad Politécnica de Madrid, Madrid, Madrid, Spain
Sandra Hirche, Department of Electrical Engineering and Information Science, Technische Universität München, München, Germany
Faryar Jabbari, Department of Mechanical and Aerospace Engineering, University of California, Irvine, CA, USA
Limin Jia, State Key Laboratory of Rail Traffic Control and Safety, Beijing Jiaotong University, Beijing, China
Janusz Kacprzyk, Systems Research Institute, Polish Academy of Sciences, Warsaw, Poland
Alaa Khamis, German University in Egypt El Tagamoa El Khames, New Cairo City, Egypt
Torsten Kroeger, Stanford University, Stanford, CA, USA
Qilian Liang, Department of Electrical Engineering, University of Texas at Arlington, Arlington, TX, USA
Ferran Martin, Departament d'Enginyeria Electrònica, Universitat Autònoma de Barcelona, Bellaterra, Barcelona, Spain
Tan Cher Ming, College of Engineering, Nanyang Technological University, Singapore, Singapore
Wolfgang Minker, Institute of Information Technology, University of Ulm, Ulm, Germany
Pradeep Misra, Department of Electrical Engineering, Wright State University, Dayton, OH, USA
Sebastian Möller, Quality and Usability Lab, TU Berlin, Berlin, Germany
Subhas Mukhopadhyay, School of Engineering & Advanced Technology, Massey University, Palmerston North, Manawatu-Wanganui, New Zealand
Cun-Zheng Ning, Electrical Engineering, Arizona State University, Tempe, AZ, USA
Toyoaki Nishida, Graduate School of Informatics, Kyoto University, Kyoto, Kyoto, Japan
Federica Pascucci, Dipartimento di Ingegneria, Università degli Studi "Roma Tre", Rome, Italy
Yong Qin, State Key Laboratory of Rail Traffic Control and Safety, Beijing Jiaotong University, Beijing, China
Gan Woon Seng, School of Electrical & Electronic Engineering, Nanyang Technological University, Singapore, Singapore
Joachim Speidel, Institute of Telecommunications, Universität Stuttgart, Stuttgart, Baden-Württemberg, Germany
Germano Veiga, Campus da FEUP, INESC Porto, Porto, Portugal
Haitao Wu, Academy of Opto-electronics, Chinese Academy of Sciences, Beijing, China
Junjie James Zhang, Charlotte, NC, USA

The book series *Lecture Notes in Electrical Engineering* (LNEE) publishes the latest developments in Electrical Engineering - quickly, informally and in high quality. While original research reported in proceedings and monographs has traditionally formed the core of LNEE, we also encourage authors to submit books devoted to supporting student education and professional training in the various fields and applications areas of electrical engineering. The series cover classical and emerging topics concerning:

- Communication Engineering, Information Theory and Networks
- Electronics Engineering and Microelectronics
- Signal, Image and Speech Processing
- Wireless and Mobile Communication
- Circuits and Systems
- Energy Systems, Power Electronics and Electrical Machines
- Electro-optical Engineering
- Instrumentation Engineering
- Avionics Engineering
- Control Systems
- Internet-of-Things and Cybersecurity
- Biomedical Devices, MEMS and NEMS

For general information about this book series, comments or suggestions, please contact leontina. dicecco@springer.com.

To submit a proposal or request further information, please contact the Publishing Editor in your country:

China
Jasmine Dou, Associate Editor (jasmine.dou@springer.com)

India
Swati Meherishi, Executive Editor (swati.meherishi@springer.com)
Aninda Bose, Senior Editor (aninda.bose@springer.com)

Japan
Takeyuki Yonezawa, Editorial Director (takeyuki.yonezawa@springer.com)

South Korea
Smith (Ahram) Chae, Editor (smith.chae@springer.com)

Southeast Asia
Ramesh Nath Premnath, Editor (ramesh.premnath@springer.com)

USA, Canada:
Michael Luby, Senior Editor (michael.luby@springer.com)

All other Countries:
Leontina Di Cecco, Senior Editor (leontina.dicecco@springer.com)
Christoph Baumann, Executive Editor (christoph.baumann@springer.com)

**** Indexing: The books of this series are submitted to ISI Proceedings, EI-Compendex, SCOPUS, MetaPress, Web of Science and Springerlink ****

More information about this series at http://www.springer.com/series/7818

Vijay Nath · Jyotsna Kumar Mandal
Editors

Proceedings of the Third International Conference on Microelectronics, Computing and Communication Systems

MCCS 2018

Springer

Editors
Vijay Nath
Department of Electronics
and Communication Engineering
Birla Institute of Technology, Mesra
Ranchi, Jharkhand, India

Jyotsna Kumar Mandal
Department of Computer Science
and Engineering
University of Kalyani
Kalyani, West Bengal, India

ISSN 1876-1100 ISSN 1876-1119 (electronic)
Lecture Notes in Electrical Engineering
ISBN 978-981-13-7090-8 ISBN 978-981-13-7091-5 (eBook)
https://doi.org/10.1007/978-981-13-7091-5

Library of Congress Control Number: 2019934362

© Springer Nature Singapore Pte Ltd. 2019
This work is subject to copyright. All rights are reserved by the Publisher, whether the whole or part of the material is concerned, specifically the rights of translation, reprinting, reuse of illustrations, recitation, broadcasting, reproduction on microfilms or in any other physical way, and transmission or information storage and retrieval, electronic adaptation, computer software, or by similar or dissimilar methodology now known or hereafter developed.
The use of general descriptive names, registered names, trademarks, service marks, etc. in this publication does not imply, even in the absence of a specific statement, that such names are exempt from the relevant protective laws and regulations and therefore free for general use.
The publisher, the authors and the editors are safe to assume that the advice and information in this book are believed to be true and accurate at the date of publication. Neither the publisher nor the authors or the editors give a warranty, expressed or implied, with respect to the material contained herein or for any errors or omissions that may have been made. The publisher remains neutral with regard to jurisdictional claims in published maps and institutional affiliations.

This Springer imprint is published by the registered company Springer Nature Singapore Pte Ltd.
The registered company address is: 152 Beach Road, #21-01/04 Gateway East, Singapore 189721, Singapore

Committee Members

Patron

Dr. K. T. V. Reddy, President IETE New Delhi

Chairman Committee

Sh. K. K. Thakur, CGMT Jharkhand State Circle, ARTTC BSNL Ranchi
Sh. V. B. Pandey, DOT BSNL Term Cell1 and Chairman ISVE Ranchi
Sh. Ajay Kumar, AGM (HR) BSNL and Chairman IETE Ranchi
Prof. P. S. Neelakanta, C. Engg., Fellow IEE, Florida Atlantic University (FAU), USA
Prof. J. K. Mondal, Professor, Kalyani University, West Bengal
Prof. C. K. Sarkar, Chairman IEEE Kolkata Section, West Bengal
Prof. Vinay Gupta, Professor, Delhi University, Delhi
Sh. Sanjay Kumar Jha, Immediate Past Chairman IETE Ranchi and Chief Executive Engineer Government of Jharkhand
Prof. Subir Sarkar, Professor, Jadavpur University, West Bengal

Co-chairman Committee

Dr. Umesh Yadav, DDU GU
Sh. Rajan Kumar Ram, ARTTC BSNL Ranchi
Dr. R. K. Singh Founder Chairman IETE Ranchi
Dr. Anshuman Sarkar, Kalyani Government Engineering College, Kalyani, West Bengal

Organizing Secretary

Dr. Anand Kr. Thakur, Ranchi University and Treasurer IETE Ranchi

General Chair

Dr. Vijay Nath, BIT Mesra and Secretary IETE Ranchi

Convener

Dr. Raj Kumar Singh, Ranchi University and Executive Mem IETE Ranchi

International Advisory Committee

Prof. Bernd Michel, Micro Materials Centre (MMC) Berlin, Germany
Prof. Bharath Bhushan, Ohio Eminent Scholar and The Horward
D. Winbigler Professor, Director NBLL, The Ohio State University, Columbus, Ohio USA
Dr. A. K. S. Chandelle, Immediate Past President IETE New Delhi
Smt. Smriti Dagur, Former President IETE, New Delhi
Dr. A. A. Khan, Former VC, Ranchi University, Ranchi
Dr. M. K. Mishra, VC, BIT Mesra, Ranchi
Dr. Ramgopal Rao, Professor and Director IIT Delhi
Dr. P. K. Barhai, Former VC, BIT Mesra, Ranchi
Dr. S. Pal, Professor and SD, Satellite Navigation, ISRO Bangalore
Dr. M. S. Kori, Chairman IETE, TPC, New Delhi
Sh. R. K. Gupta, Former President, IETE, New Delhi
Dr. Rajendra Prasad, Professor, IIT Roorkee
Dr. Labh Singh, Former CGM, ARTTC, BSNL, Ranchi
Sh. R. Mishra, Former CMD, HEC, Ranchi
Dr. Yogesh Singh Chauhan, Associate Professor, Nano Lab, Department of Electrical Engineering, IIT Kanpur
Dr. S. N. Verma, Former CMD, EDC, Ltd. Jharkhand
Sh. S. C. Thakur, Chief Engineer, Rural Electrification Energy Distribution Corporation Limited Jh.
Sh. Ravindra Kr. Rakesh, Editor, Dainik Bhasker, Jharkhand
Sh. Gopal Jha, Journalist, New Delhi
Dr. A. N. Mishra, VC, Central University, Jharkhand

Dr. R. Pandey, VC, RU, Ranchi
Dr. A. Chakrabarty, Professor, IIT Kharagpur
Dr. S. Banerjee, Professor, IIT Kharagpur
Dr. Nandita Das Gupta, Professor, IIT Chennai
Dr. L. K. Singh, Former Professor, Dr. RML AU, Faizabad
Dr. B. S. Rai, Former Professor, MMMUT, Gorakhpur
Dr. D. Samathanam, Former Adviser and Head TDT, DST New Delhi
Dr. P. Chakrabarty, Professor, IIT BHU
Dr. G. A. Murthy, Scientist-G, DRDO Hyderabad
Dr. M. Srinivasa, Scientist-G, DRDO Hyderabad
Dr. S. C. Bose, Scientist-G, CEERI Pilani
Dr. Jamir Akhatar, Sr. Scientist, CEERI Pilani
Dr. Arokiaswami Alphones, Vice–Chairman, IEEE Singapore Section and Professor, NTU Singapore
Dr. K. Rajasekhar, Dy. Director General, NIC, DEIT, Mo CIT, Government of India, Hyderabad
Dr. N. V. Kalyankar, VC, Marathwara University (MS)
Dr. R. P. Panda, Professor, VSSUT, Burla, Odissa
Dr. Allen Klinger, Professor, University of California
Dr. Hisao Ishibuchi, Professor, Osaka Prefecture University, Japan
Dr. T. K. Bhattacharya, Professor, IIT Kharagpur
Dr. N. Gupta, Professor, BIT Mesra, Ranchi and Fellow Member IETE New Delhi
Dr. V. R. Gupta, Professor, BIT Mesra, Ranchi and Fellow Member IETE New Delhi
Dr. M. Chakrabarty, Professor, IIT Kharagpur
Dr. A. S. Dhar, Professor, IIT Kharagpur
Dr. D. K. Sharma, Professor, IIT Bombay
Dr. Nandita Das Gupta, Professor, IIT Chennai
Dr. B. Mishra, Professor, BIT Mesra, Ranchi
Dr. Swaroop Gosh, Assistant Professor, University of South Florida
Dr. S. P. Maity, Professor, IIEST, Shibpur
Dr. S. K. Ghorai, Professor, BIT Mesra, Ranchi and Executive Member IETE Ranchi Centre
Dr. M. Bhuyan, Professor, Tejpur University, Assam
Dr. S. Hosimin Thilangar, Professor, Anna University, Chennai
Dr. V. N. Mani, Senior Scientist, CMET, Hyderabad
Dr. V. Kumar, Professor, ITT-ISM Dhanbad
Dr. S. K. Paul, Professor, ISM Dhanbad
Dr. J. P. Gupta, Former Pro-VC, DDU Gorakhpur University
Dr. H. C. Prasad, Former Professor, DDU Gorakhpur University
Dr. S. Bhaumik, Associate Professor, NIT Tripura
Dr. P. D. Kashyap, Professor, NIT Arunanchal Pradesh
Dr. J. Akhatar, Senior Scientist, CEERI Pilani
Dr. S. Ahmad, Former Director, CEERI Pilani
Dr. P. Kapoor, Former Director, CSIO, Chandigarh

Dr. Uma Maheshwari, Professor, Anna University
Dr. Sandip Rakshit, Professor, Kaziranga University, Assam
Dr. Abhijit Biswas, Professor, Institute of Radio Physics and Electronics, Calcutta University
Dr. Vikash Patel, SAC ISRO Ahmedabad
Dr. Parul Patel, SAC ISRO Ahmedabad
Dr. S. Pal, Professor, BIT Mesra Ranchi
Dr. V. Bhattacharya, Professor BIT Mesra Ranchi

National Advisory Committee

Dr. Gaurav Trivedi, Assistant Professor, IIT Guwhati
Dr. B. K. Kaushik, Professor, IIT Roorkee
Dr. K. K. Khatua, Professor, NIT Rourkela
Dr. M. Bhaskar, Associate Professor, NIT Tirchi
Dr. P. Kumar, Associate Professor, IIT Patna
Dr. Soumya Pandit, Assistant Professor, Kolkata University
Dr. Soma Berman, Assistant Professor, University of Cacutta
Dr. K. B. Raja, Professor, Bangalore College of Engineering, Bangalore
Dr. R. P. Panda, Professor, VSSUT, Burla, Odisa
Dr. P. R. Thakura, Professor, BIT Mesra, Ranchi
Dr. S. S. Solanki, Professor, BIT Mesra, Ranchi
Dr. Mahesh Chandra, Professor, BIT Mesra, Ranchi
Dr. D. K. Malik, Associate Professor, BIT Mesra, Ranchi
Dr. Nutan Lata, Associate Professor, BIT Mesra, Ranchi
Dr. K. K. Senapati, Assistant Professor, BIT Mesra, Ranchi
Dr. K. K. Patnaik, Associate Professor, IIIT Gwalior
Dr. M. Goswami, Associate Professor, IIIT Allahabad
Dr. Sukalayam Chakraborty, Assistant Professor, BIT Mesra Ranchi
Dr. Lallan Yadav, Professor, DDU University Gorakhpur
Dr. S. Chakrabarty, Professor, BIT Mesra, Ranchi
Dr. D. Devaraj, Professor, Kalasalingam University, Tamilnadu
Dr. J. S. Roy, Professor, KIIT Bhubneshwar
Dr. N. K. Kamila, Professor, CV RCE, Bhubneshwar, Odisa
Dr. B. K. Ratha, Associate Professor, Utkal University, Odisa
Dr. A. Srinivasulu, Professor, Vignan University, Andhra Pradesh
Dr. Manish Prateek, Professor, University of Petroleum and Energy Studies, Dehradun
Dr. Vijay Laxmi, Professor, BIT Mesra, Ranchi
Dr. V. K. Jha, Associate Professor, BIT Mesra, Ranchi
Dr. R. K. Lal, Associate Professor, BIT Mesra, Ranchi
Dr. L. B. Singh, Professor, RPSIT Patna
Dr. H. S. Gupta, Senior Scientist ISRO, Bangalore

Dr. N. S. Rao, Associate Professor, MECS, Hyderabad
Dr. Usha Mehta, Professor, Nirma Institute of Technology Ahmedabad
Sh. Manish Mehta, Senior Scientist, SAC ISRO Ahmadabad
Sh. Vishvanath Vedam, Senior Scientist, SAC ISRO Ahmadabad

Technical Program Committee

Dr. Kota Solomon Raju, Scientist-F, CEERI Pilani
Dr. Amalin Prince, Associate Professor, BITS Pilani Goa Campus
Dr. M. Mishra, Assistant Professor, DDU University Gorakhpur
Dr. J. B. Sharma, Associate Professor, Rajasthan Technical University, Kota
Dr. Prabir Saha, Assistant Professor, NIT Meghalaya
Dr. S. P. Tiwari, Assistant Professor, IIT Jodhpur
Dr. Santosh Vishvakerma, Associate Professor, IIT Indore
Dr. S. N. Shukla, Professor, Dr. RML Avadh University, Faizabad
Dr. B. N. Sinha, Associate Professor, SSMC, Ranchi
Dr. V. S. Rathore, Assistant Professor, BIT Mesra, Ranchi
Dr. Manish Kumar, Assistant Professor, NERIST
Dr. A. N. Jadhav, Professor, Y.M. R.T. Marathwada University, Nanded
Dr. Sudip Sahana Assistant Professor, BIT Mesra Ranchi
Dr. S. Pushkar, Assistant Professor, BIT Mesra Ranchi
Dr. P. Pal, Assistant Professor, BIT Mesra Ranchi
Dr. Amritanjali, Assistant Professor, BIT Mesra Ranchi
Dr. Rishi Sharma, Assistant Professor, BIT Mesra Ranchi
Dr. K. Bose, Assistant Professor, BIT Mesra Ranchi

Joint Secretary

Prof. D. Acharya, President ISTM Kolkata
Prof. Rajeev Ranjan, ISM, Dhanbad
Prof. Amar Prakash Sinha, BIT Sindri
Prof. Jayant Pal, NIT, Agarpara, Kolkata
Prof. Adesh Kumar, Energy and Petroleum University, Dehradun
Prof. J. Dinesh Reddy, BMS College of Engineering, Bangalore
Prof. N. Srinivasa Rao, BMS College of Engineering, Bangalore
Prof. P. Kumar, CIT Ranchi
Sh. Ramkrishna Kundu, IBM, Bangalore
Sh. Dipayan Gosh, GM Aircel, Kolkata
Sh. S. Chakrabarty, IBM, Bangalore
Prof. Jyoti Kumari, RBS, Bangalore
Sh. Rahul Kumar Singh, ST Microelectronics, Noida

Sh. Suraj Kumar, NIT Agartala
Prof. Anand Kr. Signha, GNIT, Gaziabad
Prof. Abhishek Pandey, Usha Martin University Ranchi and Secretary ISVE Ranchi

Treasurer

Dr. Anand Kr. Thakur, IETE Ranchi
Sh. S. K. Saw, MCCS-2018
Smt. Saroj, ISVE Ranchi

Preface

The volume presents the high-quality, best-selected papers presented at the "Third International Conference on Microelectronics, Computing and Communication Systems (MCCS 2018)". The conference was organized by Indian Society of VLSI Education (ISVE), Ranchi, and Institution of Electronics and Telecommunication Engineers (IETE), Ranchi Centre, on 12–13 May 2018. MCCS 2018 gave a unique platform to young researchers, scientists, engineers and professionals to present their work. Out of 400 submissions, 70 papers were selected for the presentation. Each paper has undergone a process of double-blind peer review and was reviewed thrice before final acceptance. The papers are published in Springer book series "Lecture Notes in Electrical Engineering (LNEE)" and are in the areas of advancement in MEMS and nanoelectronics, wireless communications, optical communication, instrumentation, signal processing, Internet of things, image processing, bioengineering, green energy, hybrid vehicles, environmental science, weather forecasting, cloud computing, renewable energy, RFID, CMOS sensors, actuators, transducers, telemetry systems, embedded systems, sensor network applications in mines, etc.

The conference was conducted with the blessings of Honourable Governor of Jharkhand Smt. Draupadi Murmu, and the inaugural session of the conference was graced by Dr. K. T. V. Reddy, President, IETE, New Delhi. The other key dignitaries present were Sh. K. K. Thakur, CGMT BSNL, Ranchi; Dr. A. A. Khan, Former VC, Ranchi University, Ranchi; Dr. R. K. Singh, Former Chairman, IETE Ranchi; Sh. Sanjay Kumar Jha, Past Chairman, IETE Ranchi, and Chief Executive Engineer, Government of Jharkhand; Dr. P. R. Thakura, Professor BIT Mesra, and Executive Member, ISVE and IETE Ranchi; Dr. M. Chandra, Professor BIT Mesra, and Executive Member, ISVE and IETE Ranchi; Dr. S. S. Salanki, Professor, BIT Mesra; Sh. Ajay Kumar, Chairman, IETE Ranchi, and AGM (Admin), ARTTC BSNL, Ranchi; Dr. Anand Kumar Thakur, Assistant Professor, Ranchi University, and Organizing Secretary, MCCS 2018; Dr. Raj Kumar Singh, Assistant Professor, Ranchi University, and Convenor, MCCS 2018, Ranchi; Dr. Vijay Nath, Assistant Professor, BIT Mesra, and General Chair, MCCS 2018;

and Keynote Speaker Dr. J. K. Mandal, Professor, Kalyani University, West Bengal.

The inaugural programme commenced with the welcome address by Sh. Ajay Kumar, Chairman, IETE Ranchi Centre, and followed by technical details of the conference presented by Dr. Vijay Nath, General Chair of MCCS 2018. The Chief Guest of the function Dr. K. T. V. Reddy presented his viewpoint about the impact of research on education and how new technologies are changing society. He further explained the necessity of quality research for the growth of technical education in the country. All other dignitaries gave their views for the growth of society, quality of publications, research innovations and challenges.

The technical sessions were arranged into four tracks: Microelectronics and Signal Processing; VLSI and Communication Systems; Embedded System and Signal Processing; and Grid Energy and Robotics. In addition, another session was conducted for the model presentation. In total, ten models were presented and three were selected for IETE award.

We extend our thanks to all authors for their contribution to this book/proceedings by sharing their valuable research findings. We specially thank reviewers for reviewing the papers submitted to the conference. We are grateful to the volunteers, invited speakers, session chairs, sponsors, sub-committee members, members of International Advisory Committee, and members of National Advisory Committee, members of Technical Program Committee, members of Joint Secretary, and members of Scientific Advisory Committee for their support. Last but not least, we express our sincere gratitude towards the staff members of Springer Nature, India, for their continuous support in publishing this book.

Kalyani, India — Jyotsna Kumar Mandal
Ranchi, India — Vijay Nath

Editorial Acknowledgements

We extend our thanks to all the authors for contributing to this book/proceeding by sharing their valuable research findings. We specially thank a number of reviewers for promptly reviewing the papers submitted to the conference. We are grateful to the volunteers, invited speakers, session chairs, sponsors, sub-committee members, members of International advisory committee, members of national advisory committee, members of technical programme committee, members of joint secretary and members of scientific advisory committee for successful conduct of conference. The editors express their heartfelt gratitude towards Honourable Governor of Jharkhand, Smt. Droupadi Murmu. The editors also express their heartfelt gratitude towards Dr. K. T. V. Reddy, President, IETE, New Delhi; Dr. A. K. S. Chandelle Immediate, Past President, IETE, New Delhi; Smt. Srimati Dagur, Former President of IETE, New Delhi; Sh. Sanjay Kumar Jha, Immediate Past Chairman, IETE, Ranchi, and Executive Engineer, Government of Jharkhand; Prof. Bernd Michel, Micro Materials Centre (MMC), Berlin, Germany; Prof. Bharath Bhushan, Ohio Eminent Scholar and The Horward D. Winbigler Professor and Director, NBLL, The Ohio State University, Columbus, Ohio, USA; Prof. P. S. Neelakanta, C. Engg. Fellow, IEE, Florida Atlantic University (FAU), USA; Sh. Prasad Vijay Bhushan Pandey, DTO Term Cell1, BSNL Ranchi, and Chairman ISVE, Ranchi; Prof. A. A. Khan Former VC, Ranchi University; Prof. M. K. Mishra, VC BIT Mesra; Prof. Gopal Pathak, VC, Jharkhand Technical University, Ranchi; Dr. K. K. Thakur, CGMT, BSNL, Ranchi; Prof. R. K. Pandey, VC, Ranchi University; Prof. P. K. Barhai, Former VC, BIT Mesra; Sh. R. Mishra Former, CMD HEC, Ranchi; Dr. M. Chakraborty, Professor, IIT Kharagpur; Dr. Ramgopal Rao, Professor, IIT Bombay, and Director IIT Delhi; Dr. P. Chakraborty, Professor, IIT BHU; Dr. S. Jit, Professor, IIT BHU; Dr. J. K. Mandal, Professor, Kalyani University; Dr. Abhijit Biswas, Professor Kolkata University; Dr. Subir Kumar Sarkar, Professor Jadavpur University; Dr. Gaurav Trivedi, Associate Professor, IIT Guwahati; Dr. Y. S. Chauhan, Associate Professor, IIT Kanpur; Dr. B. K. Kaushik, Professor, IIT Roorkee; Dr. Shree Prakash Tiwari, Faculty, IIT Jodhpur; Dr. P. Kumar Associate, Professor, IIT Patna; Dr. M. Bhaskar, Professor, NIT Trichy; Dr. Adesh

Kumar, Assistant Professor, UPES University, Dehradun; Dr. Manish Kumar, Associate Professor, MMMUT, Gorakhpur; Dr. Manish Mishra, Associate Professor, DDU University, Gorakhpur; Dr. Umesh Yadav, Professor, DDU University, Gorakhpur; Prof. D. Acharjee, President, ISTM, Kolkata; Dr. N. Gupta, Professor, BIT Mesra, Ranchi; Dr. S. Pal, Professor, BIT Mesra; Dr. S. Konar, Professor, BIT Mesra; Dr. Vibha Rani Gupta, Professor, BIT Mesra; Dr. B. K. Mishra, Principal Jumeritelaya, Government of Jharkhand; Dr. V. K. Jha, BIT Mesra; Sh. Ajay Kumar AGM (admin), ARTTC BSNL, Ranchi, and Chairman, IETE, Ranchi; Dr. P. R. Thakura, Executive Member of IETE and ISVE, Ranchi, and Professor of BIT Mesra, Ranchi; Dr. M. Chandra, Executive Member of IETE, Ranchi, and Professor of BIT Mesra, Ranchi; Dr. S. K. Ghorai, Executive Member of IETE, Ranchi, and Professor of BIT Mesra, Ranchi; Dr. B. Chakraborty, Executive Member of IETE, Ranchi, and Executive Engineer, MECON, Ranchi; Dr. S. Chakraborty, Executive Member, IETE, Ranchi, and Professor, BIT Mesra, Ranchi; Dr. S. S. Solanki, Professor, BIT Mesra, Ranchi; Dr. S. Pal, Professor, BIT Mesra, Ranchi; Dr. S. Kumar, Executive Member, IETE, Ranchi, and Associate Professor, BIT Mesra, Ranchi; Dr. B. K. Bhattacharya, Professor, NIT Agartala; Dr. Anand Kumar Thakur, Treasurer of IETE, Ranchi, and Assistant Professor and Director, FM Ranchi University; Dr. Raj Kumar Singh, Executive Member, IETE, Ranchi, and Assistant Professor and Coordinator, UGC Refresher Course, Ranchi University; Dr. R. K. Lal, Associate Professor, BIT Mesra, Ranchi; Dr. Sudip Sahana, Mesra; Dr. P. Pal; Dr. Amritanjali; Dr. Rishi Sharma; Dr. K. K. Senapati; Dr. S. S. Sahu; Dr. M. K. Mukul; Dr. K. Bose of BIT Mesra; Smt. Saroj, Treasurer of ISVE, Ranchi; Prof. Jyoti Singh, Joint Secretary of ISVE, Ranchi; Prof. A. K. Pandey, Secretary, ISVE, Ranchi; Sh. Suraj Kumar Saw; Sh. Subro Chakraborty; Sh. Dipayan Ghosh; Sh. Ramkrishna Kundu, Executive Member of ISVE, Ranchi; Sh. Deepak Prasad; Sh. Sumit Singh; Sh. H. Kar; Sh. Rajanish Yadav; Sh. Anup Tirkey for their endless support, encouragement and motivation to organize such prestigious event that paved the way for this book on proceeding of Third International Conference on Microelectronics, Computing and Communication Systems (MCCS-2018). At last, we express our sincere gratitude towards the staff members of Springer, India, who helped in publishing this book.

Contents

About the Editors

Dr. Vijay Nath was born in Gorakhpur (U.P.) India in 1976. He received his Bachelor degree in Physics and Master degree in Electronics from DDU Gorakhpur University, India in 1998 and 2001. He also received PGDCN (GM) from MMMUT Gorakhpur in 1999. He received his Ph.D. degree in VLSI Design & Technology from Dr. RML Avadh University Faizabad in association with CEERI-Pilani in 2008. From 2000 to 2001 he served as project trainee in IC Design Group CEERI Pilani under the guidance of Dr. K.S.Yadav (Senior Scientist). From 2002 to 2006 he severed as faculty in the Department of Electronics, DDU University Gorakhpur. In 2006, he joined as a faculty in the Department of Electronics and Communication Engineering, Birla Institute of Technology Mesra Ranchi (JH) India. Presently he is Professor In-charge of Embedded System Design Lab, Advisor of NAPS (News & Publication Society), Member of BIT Mesra Brand Management, Faculty Advisor IEI Student Chapter (ECE) BIT Mesra and Honorary Secretary of IETE Ranchi Centre. He is the recipient of Viveka Nanda Techno Fiesta Award-2002, Young Scientist Award-2004, CCSN Best Paper Award-2013, First Prize in IEI Technical Contest-2013, and Cadence Design Contest-2014, 2016. His research interests include microelectronics, nanoelectronics, analog, digital, mixed CMOS VLSI circuits, low power VLSI circuits, ultra low power CMOS temperature sensor, ADC, DAC, PTAT, CMOS bandgap voltage reference, ASICs, Embedded Systems Designs, Smart Cardiac Pacemaker, Smart Grids, Internet of Things, and early stage cancer detection. He has to his credit around 150 publications in reputed Scopus & SCI journals, Scopus books and conferences. He has successfully completed two R&D projects funded by DST New Delhi jointly with DRDL Hyderabad and MHRD New Delhi, and the third project is in ongoing stage funded by RESPOND SAC ISRO Ahmadabad, Government of India. He has developed VLSI Design course under Developing Suitable Pedagogical Methods for Various Classes, Intellectual Calibres & Research in E-learning sponsored by Ministry of Human Resource Development Government of India. Now the complete course is available in IIT Kharagpur official website and open for all IITs, NITs, BITs and technical universities of India & abroad. He is the resource person of 16 Winter & Summer School on Digital VLSI Design. He is the program coordinator and resource person of one day seminar/workshop on Electronic System Design & Manufacturing (ESDM), East Zone of India funded by MHRD New Delhi under the aegis of IETE New Delhi. He has conducted 15 ESDM workshops in IITs NITs, BITs and other university/ college of India in FY2015-16. He is the Editor of four books Proceeding of International Conference on Nanoelectronics Circuits & Communication Systems (NCCS-2015) & NCCS-2017 and Proceeding of International Conference on Microelectronics Computing & Communication Systems (MCCS-2015) & MCCS-2018 published in Scopus book series: Lecture Notes of Electrical Engineering, Springer. He is a member of several professional societies and academic bodies including IETE, ISTE, ISVE and IEEE.

Prof. Jyotsna Kumar Mandal received his M.Sc. in Physics from Jadavpur University in 1986 and M.Tech. in Computer Science from the University of Calcutta, and was awarded a Ph.D. in Computer Science and Engineering by Jadavpur University in 2000. Currently, he is Professor of Computer Science and Engineering and was Dean of the Faculty of Engineering, Technology and Management, Kalyani University, West Bengal for two consecutive terms. He started his career as a lecturer at NERIST, Arunachal Pradesh in September 1988, and has 30 years' teaching and research experience. His areas of research include coding theory, data and network security, remote sensing and GIS based applications, data compression, error correction, visual cryptography, steganography, security in MANET, wireless networks and unify computing. He has been a life member of the Computer Society of India since 1992, CRSI since 2009, ACM since 2012, IEEE since 2013 and a fellow of IETE since 2012. He has chaired more than 60 sessions at various international conferences and delivered more than 60 expert/invited lectures during the last 5 years. He has acted as program chair of several international conferences and edited more than 30 proceedings volumes. He is a reviewer for various international journals and conferences and has published over 360 articles and 6 books. He is one of the editors for the Springer AISC and CCIS Series.

He has published more than 450 articles out of which 160 papers are published in the international journal. He has authored 7 books. 23 scholars were awarded Ph.D. degree under his supervision.

A Novel Approach for Pain Intensity Detection by KAZE Features

Ashutosh Vaish and Sagar Gupta

Abstract Successful pain evaluation is one of the most difficult problems of clinical practitioners. Continuous pain evaluation of the patient in the intensive care unit (ICU) is burdensome and expensive. An individual, generally a nurse, has to be present at all times to gage the level of pain of a person during as well as after an operation. This research paper is an attempt to automate the pain intensity detection through the scrutiny of the facial features. The intent of this research is to provide a new method of detecting the intensity of pain, i.e., by Kaze features. Kaze features differ from the previous features in the sense that the blurring made by it is locally adaptive. The building of Fisher vector is done through GMM. This paper also reviews the various features used for the purpose of pain detection. The high accuracy of 91.8% instills the confidence in using of Kaze for facial features and makes the realization of an actual state of the art closer than before.

Keywords Real time · Pain intensity detection · Facial expressions · Kaze features

1 Introduction

Managing and assessing pain is a difficult task. The most widely used methods for assessing pain is the patient self-report as no special skill set is needed and is convenient. Other methods which have been successfully used for pain assessment include a visual analog scale and clinical interviews [1]. The scale varies from zero pain (0) implying no pain to high pain, i.e., the worst imaginable pain (100) on a scale of 100 mm [2, 3].

A. Vaish
IIIT Delhi, New Delhi, India
e-mail: ashutoshv@iiitd.ac.in

S. Gupta (✉)
IIITD, Okhla Industrial Area, New Delhi, India
e-mail: absagargupta@gmail.com; sagar18174@iiitd.ac.in

© Springer Nature Singapore Pte Ltd. 2019
V. Nath and J. K. Mandal (eds.), *Proceedings of the Third International Conference on Microelectronics, Computing and Communication Systems*,
Lecture Notes in Electrical Engineering 556, https://doi.org/10.1007/978-981-13-7091-5_1

But as suggested by [4], self-reporting methods have several drawbacks and are thus unreliable. This method cannot be of much help with young children, with people with various types of mental disease. Many patients require the presence of a nurse to observe the level of pain in post-operative care and the ones who are under a transient state due to anesthesia [5].

The observer rating method is inefficient if an onlooker is needed for a large amount of time such as when the patient is in ICU [6]. The high subjectivity of the self-report adds to the inefficiency of the method.

A potential solution is to use facial cues. Several researchers have devised algorithms that are able to automatically distinguish pain and no- pain faces. Initially, the researchers focused on detecting the presence or absence pain on facial images. Later on, with improved methods they tried to detect the intensity of the pain in various facial images. A significant advantage in a patient's care at lower costs can be achieved by the realization of the pain detection systems [5, 7, 8]. The pain detection or pain intensity estimation is accomplished in two major steps: feature extraction and classification.

Against the background noise, the face has to be extracted for reducing the computation required and for making the system robust. A very popular method of face localization is Haar Cascade Classifier, proposed by Viola and Jones in [9].

Once the faces have been extracted, extracting the information from the shown face is the next step. It is a very arduous task for the machines to extract facial features because of the high variability in the types of faces. Changes in shape of eyebrows, mouth, and eyes are some of the information from facial features [10]. Along with edges, shape and appearance information also relies on texture features. Thus, researchers have used various model-based methods and pixel-based methods. The most popular model used for modeling the faces for pain feature extraction is active appearance model (AAM) and used by [5, 7, 8]. Ashraf et al. [5] extract S-PTS, S-APP, and their combination for the purpose of binary classification of pain and relies on OPI and PSPI scores as a measure of pain. They conclude that using the combination of both these features yield better results than when they were used alone.

The authors in [11] describe an AAM-based method for detecting. According to [6], canonical appearance features (CAPP) achieve the best results as they couple together both the appearance and shape representation. Moreover, the fusion of SPTS, CAPP, and SAPP performs better at detecting pain than any other combination or individual descriptors. They also take the head movements into account for better detection of pain by improving the face registration technique.

Hammal and Cohn [12] extracted CAPP using active appearance modeling and processed them via Log-Normal filters. Recently, Rathee and Ganotra [13] adopted thin plate spline to separate rigid and nonrigid deformations. Model-based methods have an inherent advantage of illumination and scaling invariance, but suffer from the problem of exhaustive computation due to the iterative fitting of models (Fig. 1).

Khan et al. [14] extracted PHOG and PLBP in order to get discriminative representation of face.

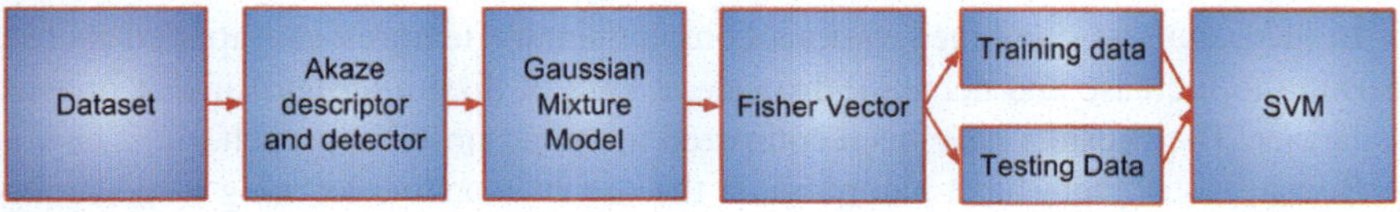

Fig. 1 Working of the proposed approach

Kaltwang et al. [15] use shape features as well as appearance features and then perform their late fusion. They employ facial landmarks as the shape features and DCT and LBP as appearance features. As shown in [16, 17], finding the LBP features is an efficient method for texture feature extraction. The computation simplicity and illumination invariance of LBP are an added advantage. They conclude that their proposed approach performs better when appearance-based features are used instead of shape features. They also show that fusing DCT and LBP features gives the best performance for pain intensity detection. Sikka et al. [18] propose a framework for utilizing multiple instance learning (MIL).

After the image frames are extracted from the image frame, they are tested for classification in different classifiers such as SVM, KNN, LLR, SVR, random forest, and decision trees.

A hyperplane is constructed between training classes during the training phase in the SVM approach. Then, during the test phase, SVM classifies the given expression into one of the given pain intensity. Ashraf et al. [5] and Lucey et al. [8] used SVM for binary classification of pain, i.e., Pain or No Pain. PSPI score 0 indicates No Pain, whereas PSPI 1 and above indicates Pain. They all use leave-one-subject-out-cross-validation. Ashraf et al. [5] achieved an accuracy of 82%, [8] scored an accuracy of 78.03, whereas [19] achieves an accuracy of 83.9. Hammal and Cohn [12] classified the pain intensity into four levels (PSPI 0, PSPI 1, PSPI 2, PSPI $\geq$ 3) and adopted both fivefold cross-validation as well as leave-one-subject-out-CV. They achieved an accuracy of 82 in fivefold CV and accuracy of 57% with leave-one-subject-out-CV.

Another method of classification, linear logistical regression (LLR) utilizes learning the weights via logistic regression where the calibration is improved compared to SVM using regularization. Lucey et al. [7] used SVM + LLR for binary classification of pain. They achieved the highest accuracy of 84.70 with a combination of all the three descriptors (SAPP + CAPP + SPTS).

RVR utilizes relevance vectors. Kaltwang et al. [15] use RVR for the task of estimation of 16 levels of pain. They also adopted leave-one-subject-out- cross-validation and achieved the mean square error (MSE) of 1.386 and Pearson's correlation coefficient (PCC) of 0.590.

Khan et al. [14] use four different classifiers, namely, SVM, 2NN, random forests, decision tree, and compare their performance for the binary classification of pain. They used tenfold cross-validation and scored an accuracy of 96.4 for 2NN classifier.

Rathee and Ganotra [13] use SVR for 16-level pain intensity estimation. SVR is unique in the sense that SVR minimizes the generalization error bound for achieving generalized performance.

In [13], authors combined various complementary features and applied metric learning to increase and decrease distance in case of dissimilar and similar images in dataset. Their results have been compared to the proposed approach.

Recent researchers have incorporated the use of machine learning in detecting pain. Rodriguez et al. [20] utilized CNN to learn the facial features. LSTM is utilized by them for making use of temporal relations among the frames.

Egede et al. [21] proposed a method combining both deep learned as well as handcrafted features. Shape and appearance information is encoded in the features learned by deep learning.

The novelty of the presented approach is the usage of accelerated Kaze features for facial analysis introduced by [22]. It preserves the boundary lines in the image. It does so by utilizing an algorithm which blurs adaptively. In this method, facial features are extracted. Preprocessing of the extracted image is undertaken. One against all techniques has been used in this paper to find the accuracy.

2 Dataset Availability

All of the proposed models for the purpose of recognition of pain expression intensities are made on UNBC-McMaster Shoulder Pain expression archive. The database contains face videos of 66 female and 63 male patients while they suffered shoulder pain during different ranges of motion tests of their arms. Videos of 25 subjects, i.e., 200 sequences, 48,398 frames are available openly to the research community. These videos are recorded at 320×240 pixel resolution, out of which the face area spanned an approximate 140×200 pixels.

Prkachin and Solomon pain intensity (PSPI) [23] score is calculated for each frame to assign pain intensity to 16 discrete levels (0–15) based on AUs as follows:

$$\text{Pain} = \text{AU4} + \max(\text{AU6}, \text{AU7}) + \max(\text{AU9}, \text{AU10}) + \text{AU43}$$

The PSPI ratings along with other measures of pain such as VAS and observer-dependent score (such as OPI) are provided with the dataset. PSPI is the only frame-based measure, rest all are sequence based.

3 Feature Descriptor

Kaze features [24] are a powerful tool for feature extraction. It preserves the natural boundary of images unlike the other similar feature descriptors. The other feature descriptors blurred the boundary lines of an image in a similar way that they used to blur the noise, thereby severely affecting the efficiency.

The object borderline conserved in the Gaussian blurring is provincially adaptive, thereby it reduces the noise along with conserving the frame boundary. On the basis of research in [25], two-dimensional features are obtained in a nonlinear scale space by the means of nonlinear diffusion filtering. The limitation of high computation is resolved by the use of additive operator splitting (AOS) techniques [26]. The classic nonlinear diffusion formulation is depicted in the equation given below:

$$\frac{\partial K}{\partial t} = (o(X, Y, T) \cdot \nabla K) \tag{1}$$

where div and ∇ are, respectively, the divergence and gradient operators. Conductivity function is represented by (o).

The scale space in logarithmic steps that are arranged in a series of D octaves and Q sublevels is discretized. Downsampling is not performed at every new octave as was the case with SIFT, thus the original image resolution is maintained. The sublevel indexes are usually identified by a discrete index d and a sublevel q and are mapped to their corresponding scale f by

$$\begin{aligned} f_j(d, q) = f_0 2^{d+q/Q}, d\epsilon[0 \ldots D-1] \\ q\epsilon[0 \ldots Q-1], j\epsilon[0 \ldots V] \end{aligned} \tag{2}$$

where f_0 represents the base scale level and V gives the total number of the filtered images. A set of evolution times is obtained and the scale space is transformed to time units through the mapping given below:

$$f_j \to t_j \tag{3}$$

$$t_j = \frac{1}{2} f_j^2, j = \{0 \ldots V\}, \tag{4}$$

The maxima is searched in all images except where j = 0 and j = V. The response is checked over a window of 3 × 3. We have used Akaze features as presented by [27].

In raw form, Kaze features cannot be used for training the classifier, thus the Fisher vectors were constructed using GMM. The number of Gaussians was kept fixed at 100.

4 Experimental Results

The proposed method is evaluated on UNBC MacMaster Shoulder Pain database detection of intensity of pain. The number of frames is selected randomly to reduce the data nonuniformity. The features were localized using the Viola Jones detector from the images.

This problem is considered as a classification problem. The cross-validation approach used in our work is leave-one-sequence-fivefold-out-cross-validation. The number of frames in each intensity is presented in Table 1.

Confusion matrix of the testing procedure on UNBC dataset is shown in Table 2.

We utilized LIBSVM toolkit [28]. The F1-score, precision, and recall are calculated and mentioned in Table 3.

The average accuracy for each PSPI level is given in Table 4.

It can be seen from the confusion matrix that this method scores more than 90% hit rate of detecting PSPI levels 0, 1, 2 and has comparable accuracy in PSPI ≥ 3.

The accuracy of 91.87% was achieved. Results from other researchers have been compiled in Table 5 for comparison.

Table 1 Frames taken for testing

PSPI level	No. of frames
PSPI 0	541
PSPI 1	606
PSPI 2	585
PSPI ≥ 3	582

Table 2 Confusion matrix of pain detection

Pain intensity	Classification of pain intensity while testing			
	PSPI 0	PSPI 1	PSPI 2	PSPI ≥ 3
PSPI 0	506	22	5	8
PSPI 1	29	546	24	7
PSPI 2	9	9	555	12
PSPI ≥ 3	56	5	3	518

Table 3 Percentage accuracy of pain intensity detection

Pain intensity	Precision	Recall	F1-score
PSPI 0	0.84	0.94	0.89
PSPI 1	0.94	0.90	0.92
PSPI 2	0.95	0.95	0.95
PSPI ≥ 3	0.95	0.89	0.92
Average	0.92	0.91	0.92

Table 4 Percentage accuracy of pain intensity detection

Pain intensity	% accuracy
PSPI 0	93.53
PSPI 1	90.09
PSPI 2	94.87
PSPI ≥ 3	89.00
Average	91.87

Table 5 Comparison with other approaches

Approach	% accuracy
Lucey et al. [19]	83.90
Hammal et al. [12]	82.00
Rathee et al. [13]	89.59
Proposed approach	91.87

5 Conclusion

In this paper, an approach for pain intensity recognition using facial features is presented. The Akaze features have been used *fivefolds leave-one-sequence-out-cross-validation* accuracy was computed. The presented approach results in 91.87% accuracy when evaluated on UNBC dataset. The number of Gaussians was varied between 10 and 200 and the optimum results are reported in the study. In future, this method may be revised for improvement and the number of Gaussians will be varied more widely.

References

1. McCormack HM, David JL, Sheather S (1988) Clinical applications of visual analogue scales: a critical review. Psychol Med 18(4):1007–1019
2. Jensen MP, Karoly P, Braver S (1986) The measurement of clinical pain intensity: a comparison of six methods. Pain 27(1):117–126
3. Burckhardt CS, Jones KD (2003) Adult measures of pain: the McGill pain questionnaire (MPQ), rheumatoid arthritis pain scale (RAPS), short-form mcgill pain questionnaire (SF-MPQ), verbal descriptive scale (VDS), visual analog scale (VAS), and west haven-yale multidisciplinary pain inventory (WHYMPI). Arthritis Care Res 49(S5)
4. Williams ACC, Davies HTO, Chadury Y (2000) Simple pain rating scales hide complex idiosyncratic meanings. Pain 85(3):457–463
5. Ashraf AB et al (2009) The painful face-pain expression recognition using active appearance models. Image Vision Comput 27(12):1788–1796
6. Lucey P et al (2011) Painful data: the UNBC-McMaster shoulder pain expression archive database. In: 2011 IEEE international conference on automatic face & gesture recognition and workshops (FG 2011). IEEE
7. Lucey P et al (2011) Automatically detecting pain in video through facial action units. IEEE Trans Syst Man Cybern Part B (Cybernetics) 41(3):664–674
8. Lucey P et al (2009) Automatically detecting pain using facial actions. In: 3rd international conference on affective computing and intelligent interaction and workshops, 2009. ACII 2009. IEEE
9. Viola P, Jones MJ (2004) Robust real-time face detection. Int J Comput Vision 57(2):137–154
10. Dongcheng S, Jieqing J (2010) The method of facial expression recognition based on DWT-PCA/LDA. In: 2010 3rd international congress on image and signal processing (CISP), vol 4. IEEE
11. Lucey P et al (2009) Automatically detecting action units from faces of pain: comparing shape and appearance features. In: IEEE computer society conference on computer vision and pattern recognition workshops, 2009. CVPR workshops 2009. IEEE

12. Hammal Z, Cohn JF (2012) Automatic detection of pain intensity. In: Proceedings of the 14th ACM international conference on multimodal interaction. ACM
13. Rathee N, Ganotra D (2015) A novel approach for pain intensity detection based on facial feature deformations. J Vis Commun Image Represent 33:247–254
14. Khan RA et al (2013) Pain detection through shape and appearance features. In: 2013 IEEE international conference on multimedia and expo (ICME). IEEE
15. Kaltwang S, Rudovic O, Pantic M (2012) Continuous pain intensity estimation from facial expressions. Adv Vis Comput 368–377
16. Feng X, Pietikainen M, Hadid A (2005) Facial expression recognition with local binary patterns and linear programming. Pattern Recogn Image Anal C/C of Raspoznavaniye Obrazov I Analiz Izobrazhenii 15(2):546
17. Shan C, Gong S, McOwan PW (2009) Facial expression recognition based on local binary patterns: a comprehensive study. Image Vis Comput 27(6):803–816
18. Sikka K, Dhall A, Bartlett M (2013) Weakly supervised pain localization using multiple instance learning. In: 2013 10th IEEE international conference and workshops on automatic face and gesture recognition (FG). IEEE
19. Lucey P et al (2012) Painful monitoring: automatic pain monitoring using the UNBC-McMaster shoulder pain expression archive database. Image Vis Comput 30(3):197–205
20. Rodriguez P, Cucurull G, Gonzàlez J, Gonfaus JM, Nasrollahi K, Moeslund TB, Roca FX (2017) Deep pain: exploiting long short-term memory networks for facial expression classification. IEEE Trans Cybern
21. Egede J, Valstar M, Martinez B (2017) Fusing deep learned and hand-crafted features of appearance, shape, and dynamics for automatic pain estimation. In: 2017 12th IEEE international conference on automatic face & gesture recognition (FG 2017), pp 689–696. IEEE
22. Alcantarilla PF, Oh SM, Mariottini GL, Bergasa LM, Dellaert F (2010) Learning visibility of landmarks for vision-based localization. In: 2010 IEEE international conference on robotics and automation (ICRA), pp 4881–4888. IEEE
23. Prkachin KM, Solomon PE (2008) The structure, reliability and validity of pain expression: evidence from patients with shoulder pain. Pain 139(2):267–274
24. Alcantarilla P, Bartoli A, Davison A (2012) Kaze features. Comput Vis ECCV 2012, 214–227
25. Weickert J (2001) Efficient image segmentation using partial differential equations and morphology. Pattern Recogn 34(9):1813–1824
26. Weickert J, Ter Haar Romeny BM, Viergever MA (1998) Efficient and reliable schemes for nonlinear diffusion filtering. IEEE Trans Image Process 7(3):398–410
27. Alcantarilla PF, Solutions T (2011) Fast explicit diffusion for accelerated features in nonlinear scale spaces. IEEE Trans Patt Anal Mach Intell 34(7):1281–1298
28. Chang CC, Lin CJ (2011) LIBSVM: a library for support vector machines. ACM Trans Intell Syst Technol (TIST) 2(3):27

Multimedia Encryption on Bitplanes of Image Using ECC on Koblitz Curves with Lopez–Dahab Projective Coordinates

Anil Pinapati and R. Padmavathy

Abstract The requirements of video communication make people to use the Internet daily in their life. The security also plays a vital role to achieve smooth communication in multimedia. Multimedia encryption provides elegant solution to fulfill all these requirements. Among all public key encryption algorithms, elliptic curve cryptography (ECC) is attracted toward encryption of multimedia for better security with less bandwidth. We proposed a method named as perceptual encryption, which performs encryption on selective bitplanes of a grayscale image. The proposed method uses ECC on Koblitz curves with Lopez–Dahab projective coordinates to improve the efficiency of multimedia encryption and decryption. Moreover, scalar multiplication dominates the performance of ECC over Koblitz curves. The performance of Koblitz curves can be further improved by reducing number of field operations required at the time of point addition using Lopez–Dahab projective coordinates, anyway doubling needs less cost due to Frobenius endomorphism. Two parameters such as peak signal-to-noise ratio (PSNR) and structural similarity measure (SSIM) are considered to measure the quality of an image. The requirements of multimedia are analyzed based on the results obtained.

Keywords Multimedia encryption · Koblitz curves · Elliptic curve cryptography (ECC) · Scalar multiplication (SM) · Lopez–Dahab projective coordinates · Perceptual encryption

---

A. Pinapati (✉) · R. Padmavathy
Department of Computer Science & Engineering, National Institute of Technology, Warangal, Warangal, Telangana, India
e-mail: ap@nitw.ac.in

R. Padmavathy
e-mail: rpadma@nitw.ac.in

© Springer Nature Singapore Pte Ltd. 2019

V. Nath and J. K. Mandal (eds.), *Proceedings of the Third International Conference on Microelectronics, Computing and Communication Systems*, Lecture Notes in Electrical Engineering 556, https://doi.org/10.1007/978-981-13-7091-5_2

1 Introduction

Performance of networking technologies helps to improve popularity of multimedia applications like Internet television, video technology, video on demand (VoD), and video conference. For commercial benefits, multimedia communications were made confidential in open networks. While sending information in video on demand, the service providers are willing to send high-quality data to legitimate users only [1]. Confidentiality is the main concern to communicate legitimate parties during business meets which were held through video conference. These issues are usually addressed by encryption of multimedia data to maintain confidentiality. Only authorized members have the encryption keys to access the multimedia data.

One of the old and basic methods is to protect the multimedia data called naive algorithm [2], which is resolved by encrypting multimedia after compression with conventional algorithms such as DES [3], AES [4–6]. DES, AES are basically used for text encryption to achieve confidentiality, but applying security to multimedia data is complex due to encrypting huge amount of multimedia data which is difficult even after compression. Conventional cryptographic algorithms require more mathematical computation, so it is difficult to process huge amount of multimedia data in real time. After encryption of multimedia data, it is difficult to show perceptual data for commercial purpose [7, 8].

Perceptual encryption requires special algorithms to encrypt the multimedia data [9]. For instance, in video on demand (VoD), in order to maintain the purchase of the high-quality products of multimedia, perceptual encryption makes the video perceptual even after encryption. A novel video encryption was proposed in [10, 11]. Selective video scrambling was proposed by Zeng et al. [12]. Wu et al. proposed multimedia compression and encryption [13]. Li proposed a perceptual video encryption in [9].

Existing methods give interesting direction between video compression and encryption. Figure 1 shows different ways of multimedia encryption, which are restricted to video compression algorithms, but other algorithms are works with independence of compression. Video encryptions are categorized based on encryption operation and compression, such as encryption done before or after compression is called as compression-independent algorithms and performing encryption during compression is named as joint video encryption and compression. A requirement of multimedia decides the performance and efficiency of an algorithm. Section 2 will explain in detail about requirements of multimedia encryption. Existing method [14] performs encryption on video with ECC over Galois Field GF(P). Proposed method is introducing efficient algorithms for encryption and decryption in ECC over Binary Field GF(2^m). An analyzed result with requirements of multimedia encryption, requirements do not affect the compression efficiency and codec compliance and remaining fields also.

The unique feature of ECC makes public key cryptography less computational with small key size. ECC provides security to each bit in data. The scalar multiplication consumes more computational cost during encryption and decryption of

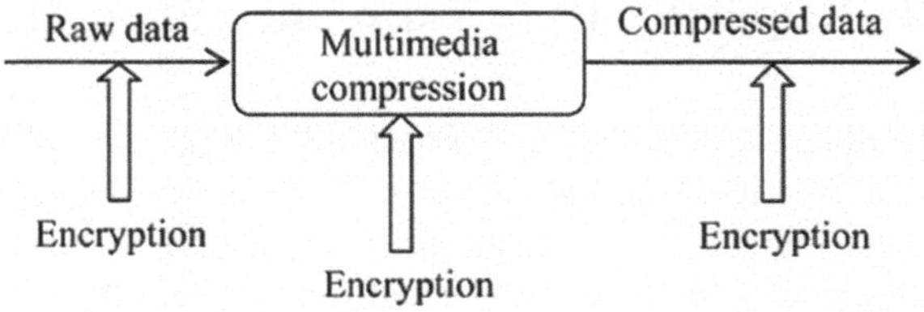

Fig. 1 Possible ways of encrypting multimedia data

ECC. Scalar multiplication (sP) is, adding P to P, s number of times, where s is a scalar (positive integer), P is point on the curve E_a. Different methods have been proposed to improve the efficiency of ECC by reducing number of field operations required to achieve scalar multiplication (i.e., LSB, MSB, NAF, Windows width-NAF method, Montgomery) [15]. Koblitz curves are special curves of binary curve, which is defined over GF(2), also named as anomalous binary curves (ABC) by Solinas [16]. τ accelerates scalar multiplication on Koblitz curves. Koblitz curves are very efficient because of Frobenius endomorphism (simple squaring with respect to $GF(2^m)$). Scalar multiplication is achieved by continuous Frobenius endomorphism followed by point addition if nonzero exists in τ-NAF (τ-NAF contains elements $\{0, \pm 1\}$).

In general, scalar multiplication is achieved with combinations of point addition and doubling. Point addition, doubling are achieved through field operations such as I (inverse), A (addition), M (multiplication), and S (squaring). Among all, inversion requires more computational power because one inversion is equal to six field multiplications. Projective coordinates reduce the number of inversions required at the time of point addition and doubling and improve the performance of scalar multiplication.

Image encryption with ECC on Koblitz curves increases the security of each bit in an image. Proposed method performs multimedia encryption using ECC on Koblitz curves using perceptual encryption. Performance of scalar multiplication is improved by reducing the number of field operations required during point addition on Koblitz curves, anyway doubling requires less cost because of Frobenius endomorphism. Performance of ECC on Koblitz curves is further improved by reducing field operations required in scalar multiplication using Lopez–Dahab projective coordinates. Proposed algorithms are three times efficient as compared with existing method [14].

The paper is scattered into six sections. Requirements of video encryption are discussed in Sect. 2. Section 3 explains preliminaries required to improve the performance of ECC on Koblitz curves with projective coordinates. The proposed method, perceptual-algorithm-based ECC on Koblitz curves, is discussed in Sect. 4. Section 5 discusses results and comparative study of proposed method in terms of requirements of multimedia. Section 6 concludes the paper.

2 Requirements to Video Encryption Algorithms

Multimedia communication contains huge amount of information to be transformed which includes formats, standard compression codec, and real-time payout [14]. All these parameters inculcate other parameters to be included; those are encryption efficiency which satisfies the real-time usage. When designing a cryptographic algorithm, these requirements of multimedia encryption are to be considered [1]. They are as follows:

- **Compression efficiency**: It reduces the size of multimedia data further, and after compression algorithms, the data size should be appropriate to the network bandwidth. These data rates should be intact even after applying encryption algorithms.
- **Encryption efficiency**: Multimedia data is huge even after compression. But applying conventional cryptography algorithm on multimedia data is not feasible, it creates overhead on CPU. We need an algorithm which requires efficient amount of time to encrypt and decrypt the multimedia data.
- **Codec compliance**: In compression and decompression, different standard codecs are used. Those codecs are intact even after multimedia encryption.
- **Security level**: Security of application varies based on the requirements. For instance, security of video on demand requires less whereas compared with security of sensitive applications such as video conference.

3 Preliminaries

3.1 Background of ECC Over GF(2^m)

Cryptosystems are categorized into two types based on security keys used at source and destination. They are symmetric/private and asymmetric/public key cryptosystems. Private key cryptography uses single key for encryption and decryption at both sides. Public key cryptography uses a key pair, one is public known to all members which are connected to system and another key is private known to owner only. Both keys are used to perform encryption and decryption based on application. ECC is a public key cryptography; prerequisites are required to find a key pair in order to establish data communication in ECC [15, 17].

An elliptic curve $E_{(a,b)} : y^2 + xy = \left(x^3 + ax^2 + b\right) mod\ f(x)$ which generates a group of points (x, y) $\in E_{(a,b)}$ including a special point named as point at infinity $\mathcal{O}$, where f(x) is an irreducible polynomial with highest degree m, m, $(a, b) \in GF(2^m)$. Altering values of (a, b) in the curve will generate different domains of ECC and it performs operations with respect to that domain. The ECC domain parameters are known to all members in the entities such as a, b, G, f(x). ECC working on GF(2^m), where G is a generator point which is also a point on the curve but it generates all remaining points on the curve. In ECC, public key is a point on the curve, private key is a positive integer scalar number known to owner only (i.e., s $\in$ [1, n], where n is order of P, nP $= \mathcal{O}$). Generation of key pair is discussed in Sect. 3.7. While calculating public key, G will be added s number of times known as scalar multiplication (sG) [15]. In general, scalar multiplication is achieved by combination of point add-and-double.

Let *P, Q, R* are points on the curve $E_{(a,b)}$ The point addition of *P, Q* is *R* which is generated by drawing a straight line passing through *P, Q* which intersects the curve at third point replication (w. r. t y coordinate) of that third point is *R* (*R* = *P* + *Q*). Point doubling is also achieved in the same way but we draw a tangent line which passes through the point intersect at third point replication of intersected point *R* (*R* = *P* + *P*). The group laws on non-super singular curves are as follows:

$$1. \mathcal{O} + P = P + \mathcal{O} = P$$

$$2. -P = (x, x + y)\, and\, P + (-P) = \mathcal{O}$$

$$3. R = P + Q \text{ where } P, Q \in E_a(2^m) \text{ and } P \neq \pm Q$$

$$4. R = P + P \text{ where } P \neq -P$$

$$x_3 : \begin{cases} \lambda^2 + \lambda + x_2 + x_1 + a\; where\; P \neq Q \\ \lambda^2 + \lambda + a + x_1^2 + \frac{2}{x_1^2} where\; P = Q \end{cases}$$

$$y_3 : \begin{cases} \lambda(x_1 + x_3) + x_3 + y_1\; where\; P \neq Q \\ x_1^2 + \lambda x_3 + x_3\; where\; P = Q \end{cases}$$

$$\lambda : \begin{cases} \frac{y_2+y_1}{x_2+x_1} where\; P \neq Q \\ x_1 + \frac{y_1}{x_1} where\; P = Q \end{cases}$$

3.2 *Koblitz Curves*

Koblitz curves are defined over GF(2) [18–20]. τ-NAF will accelerate the scalar multiplication on Koblitz curves [16]. Koblitz curves are similar to binary curves, and so they are also named as anomalous binary curves (ABC) by Solinas [16]. The representation of curve is as follows:

$$E_a : y^2 + xy = x^3 + ax^2 + 1\; where\, a \in \{0, 1\} \tag{1}$$

$$E_a\left(2^m\right) : \left\{(x, y) \in F_{2^m} X F_{2^m} : y^2 + xy = x^3 + ax^2 + 1\right\} U\{O\} where \\ a \in \{0, 1\} \tag{2}$$

The group of rational points on E_a is denoted by $E_a(2^m)$ and it has order $\#E_a(2^m)$. A Koblitz curve (E_a) is almost prime order if $\#E_a(2^m)= hn$, where *n* is a prime. Co-factor (*h*) depends on parameter *a*, $h = 2$ if $a = 1$, $h = 4$ if $a = 0$. $\tau : E_a(2^m) \rightarrow E_a(2^m)$ is Frobenius map, $\tau(\infty) = \infty$ and $\tau(x, y) = \left(x^2, y^2\right)$, where (*x, y*) is a point on the curve E_a. Using point addition formula, we can write $\tau^2+2 = \mu\tau$, where $\tau = \frac{\mu+\sqrt{-7}}{2}$ with $\mu = (-1)^{1-a}$.

3.3 τ-adic Nonadjacent Form (NAF)

τ accelerates scalar multiplication on Koblitz curves which were introduced by Solinas [16]. Initially, a scalar s is converted into a unique representation of τ-NAF (contains $\{0, \pm 1\}$), where $s = a_0 + a_1\tau \in Z[\tau]$ such that $s = \sum_{i=0}^{l-1} u_i\tau^i$, and satisfy $u_i u_{i+1} = 0$, where $Z[\tau]$ is a Euclidean domain with respect to the norm function (refer more about norm function [15, 16]). Point P involved in multiplication over $E_a(2^m)$ by elements of the ring $Z[\tau]$: if $u_{l-1}\tau^{l-1} + \cdots + u_1\tau + u_0 \in Z[\tau]$ then $(u_{l-1}\tau^{l-1}+\cdots+u_1\tau+u_0)P = u_{l-1}\tau^{l-1}(P)+\cdots+u_1\tau(P)+u_0(P)$, where $P \in E_a(2^m)$.

To assess the τ-NAF, divide scalar s with τ repeatedly and add remainders to the τ-NAF sequence and it maintains the property of NAF (no two nonzeros should come beside). In detail, discussion about generation of τ-NAF of a scalar was presented in [16]. $l(\alpha)$ be the length of τ-NAF representation of α. The length of τ-NAF should satisfy the bound [15, 21]:

$$\log_2(N(\alpha)) - 0.55 < l(\alpha) < \log_2(N(\alpha)) + 3.52 \tag{1}$$

where $\alpha \in Z[\tau]$, $N(\alpha)$ is norm of any element α. The Hamming weight (number of nonzeros) in the τ-NAF is $l = 3$ of its length l [16, 22]. The length of τ-NAF of a scalar is double of its binary length, to reduce its length, and Solinas proposed a method which reduces its length to approximately m over $GF(2^m)$ [16]. Recently, efficient hardware architectures for τ-NAF conversion have been proposed in [23, 24].

Algorithm 1: Probabilistic mapping of message to a point P where $P \in E_a(2^m)$

1: Select $x_1 \in_R GF(2^m)$ such that x_1 is an x-coordinate of a $P \in G$ where (G is prime subgroup)
2: Let $P_M = (x_1, M)$
3: Set $\gamma \leftarrow (\frac{M}{x_1})^2 + (\frac{M}{x_1}) + x1 + a2 + \frac{a6}{x_1^2}\{Then\ (x_1, M)\ belongs\ to\ (E_{(a+\gamma)}, b)\}$

3.4 Encoding and Decoding of Message Over $GF(2^m)$

In [25], King Brian proposed an algorithm for mapping an arbitrary data to a point on the curve with respect to binary field. Introduced a curve with $E_{(a+\gamma,b)} : y^2 + xy = x^3 + (a+\gamma)x^2 + b$ where $a, b \in GF(2^m)$ can be solved in quadratic roots over $GF(2^m)$. Algorithm 1 shows about message encoding from M to P_M. Here, γ does not reveal any information about message M because x_1 was picked randomly.

3.5 Point Addition Using Lopez–Dahab Coordinates on Koblitz Curves

Projective coordinates are used to reduce the inversion operations required during point addition, doubling. Different projective coordinates are available to perform point addition over binary fields such as Jacobian, standard, and Lopez–Dahab projective coordinates. Among all, minimal field operations are required with Lopez—Dahab projective coordinates. The affine coordinate uses (x, y) to represent a point on the curve E_a, whereas in Lopez–Dahab coordinates [24], the three coordinates (X, Y, Z) are represented in affine coordinates as follows (X/Z, Y/Z^2), where Z ≠ 0 and point at infinity O = (1, 0, 0). The domain of the curve with updated affine coordinates is $Y^2XYZ = X^3Z + aX^2Z^2 + bZ^4$, where a, b ∈ $GF(2^m)$. It requires field operation cost of 13M + 4S + 9A, 5M + 4S + 5A to achieve point addition, doubling, respectively, where M, S, A are multiplication, squaring, and addition, respectively.

Mixed coordinates are used in Lopez–Dahab coordinates during point addition. Mixed coordinate means, one of the point is represented in affine and other is in projective coordinate, i.e., $(X_3, Y_3, Z_3) = (X_1, Y_1, Z_1) + (x_2, y_2)$, field operations were reduced further 9M + 5S + 9A after applying Lopez–Dahab coordinates. The explicit formulation is given as follows [15]:

$$
\begin{aligned}
Z_3 &= [Z_1(X_1Z_2 + X_2Z_1)]^2 \\
X_3 &= (Y_1Z_1^2y_2)^2 + (Y_1Z_1^2y_2)(X_1 + Z_1x_2)Z_1 \\
&\quad + Z_1(X_1 + Z_1x_2)^3 + a[Z_1(X_1Z_2 + X_2Z_1)]^2 \\
&= (Y_1 + Z_1^2y_2)^2 + Z_1(X_1 + Z_1x_2)[(Y_1 + Z_1^2y_2) \\
&\quad + (X_1 + Z_1x_2)^2 + aZ_1(X_1Z_2 + X_2Z_1)] \\
Y_3 &= (Y_1 + Z_1^2y_2)^2 + (X_1 + Z_1x_2)(X_1Z_3 + X_3Z_1) \\
&\quad + X_3[Z_1(X_1Z_2 + X_2Z_1)]^2 + Y_1Z_3(X_1 + Z_1x_2)^2
\end{aligned}
$$

$$
Z_3 : \begin{cases} A = Y_1 + y_2Z_1^2, \quad B = X_1 + x_2Z_1 \\ C = BZ_1 \\ Z_3 = C^2 \end{cases}
$$

$$
X_3 : \begin{cases} D = x_3Z_3 \\ X_3 = A^2 + C(A + B^2 + aC) \end{cases}
$$

$$
Y_3 : \{Y_3 = (D + X_3)(AC + Z_3) + (y_2 + x_2)Z_3^2
$$

3.6 Scalar Multiplication on Koblitz Curves

ECC is secure due to elliptic curve discrete logarithm problem (ECDLP), that is, there is no polynomial solution to find s from Q = sP by knowing {Q, P}, where {Q, P} ∈ $GF(2^m)$, s is a scalar. Significant research has been carried out to improve the efficiency of scalar multiplication on hardware design using parallel prepossessing [26, 27]. Efficiency of scalar multiplication was improved on software design over binary fields [28]. Efficiency of scalar multiplication is improved using point halving and improvement of point halving over binary fields is explained in [29–31]. Koblitz curves are special type of curve under binary curves, τ-NAF accelerates the scalar multiplication on Koblitz curves and it contains elements $\{0, \pm 1\}$ [16]. Avanzi [32] further increased the performance of scalar multiplication on Koblitz curves by combining point halving and Frobenius endomorphism. The scalar multiplication in ECC is computationally costly operation during encryption and decryption techniques. The performance of scalar multiplication in ECC on Koblitz curves can be further increased by reducing the number of field operations required in point addition over $GF(2^m)$.

Koblitz curve needs point addition only, and it does not require point doubling due to Frobenius endomorphism. The point addition requires field operations such as I, M, S, and A. Among all inversion consumes more computational power. The projective coordinates are used to reduce the number of inversions required at the time of point addition, but one extra inversion is required to convert projective to affine coordinates at the end. Algorithm 2 for computing scalar multiplication, i.e., Q = kP on Koblitz curves, where conversion of scalar into τ-NAF is mentioned in [15, 16]. This algorithm requires on an average m − 1 Frobenius maps and (m/3) − 1 point subtractions and additions. The computation of point addition determines the efficiency of scalar multiplication in ECC on Koblitz curves.

Algorithm 2: Scalar multiplication on Koblitz curves [15].
Inputs: $P = (x, y) \in E_K(2^m)$ on curve and scalar k, $k = \sum_{i=0}^{l-1} k_i \tau^i$.
Output: $Q = kP$.

1: initialize
   a: if $k_{l-1} = 1$ then $Q \leftarrow (x, y, 1)$
   b: if $k_{l-1} = -1$ then $Q \leftarrow (x, x + y, 1)$
2: for i from $l - 2$ to 0 do
   $Q \leftarrow \tau(Q) = (X^2, Y^2, Z^2)$
   if $k_i \neq 0$ then
      $Q \leftarrow Q + k_i P$
   end if
   end for
3: $Q \leftarrow (X/Z, Y/Z^2)$

3.7 Encryption and Decryption of ECC on Koblitz Curves

Initially, choose a base point G (Generator point which generates all the points in a group) for encryption and decryption. After encoding M to P_M PM ($P_M \in E_a$), each entity has its own pair called private and public keys. For instance, A is an entity, choosing k_a as its private key and its public key becomes $P_a = k_a G$, where $k_a \in [1, \text{order(G)}]$. The entities in a group will generate unique key pairs and all public keys are kept at one shared location. Here, entity B is sending message point PM to entity A. Entity B collects public key P_a of entity A and choosing a scalar k_r randomly. Generates four cipher values or two points as follows [33]:

$$Ci_M = \{Ci_1 = k_r G, Ci_2 = P_M + k_r P_a\}$$

The cipher Ci_M is sent to entity A. Entity A can decrypt the cipher by using its private key k_a to assess plain text (P_M). Those steps are as follows:

$$P_M = P_M + k_r P_a - k_a(k_r G) = P_M + k_r(k_a G) - k_r(k_a G)$$

Field operations such as A, S, M, and I are done with respect to $GF(2^m)$(see Sect. 3.1). Frobenius operation is achieved by using point squaring with respect to $GF(2^m)$. Each time one inversion is required during point addition. The number of inversions is reduced using Lopez–Dahab coordinates (see Sect. 3.5 for projective coordinates). At the time of converting projective to affine coordinates, we used the extended Euclidean algorithm to find the multiplicative inverse of denominator over $GF(2^m)$[15].

Proposed method is implemented experimentally with system configuration: MATLAB R2015a, Intel(R) core(TM) i5-2450M CPU 2.5 GHz, 4 GB DDR3. We choose the curves K-163 recommended by NIST [5]. The parameters are as follows: m is the extension degree of the binary field $GF(2^m)$ f(z). The irreducible polynomial of degree m, a The coefficients of the elliptic curve. n The prime order of the base point P, h is the co-factor. (x, y) are the coordinates of point P.

m = 163, f(z) = z163 + z7 + z6 + z3 + 1, a = 1, b = 1, h = 2,
n = 0x 00000004 00000000 00000000 00020108 A2E0CC0D 99F8A5EF,
x = 0x 00000002 FE13C053 7BBC11AC AA07D793 DE4E6D5E 5C94EEE8,
y = 0x 00000002 89070FB0 5D38FF58 321F2E80 0536D538 CCDAA3D9

4 Multimedia Encryption Algorithms

4.1 Perceptual Encryption

We perform perceptual encryption by applying Ea over GF(2^{163}) (recommended by NIST [5]) before compression. Combinations of eight binary images are included in a grayscale image, named as bitplanes: $b_8b_7b_6b_5b_4b_3b_2b_1$. Each bit in bitplane contains elements either 0 or 1. The bitplane b_i where $i \in [1, 8]$ if i = 8, bitplane formed with MSB bits whereas i = 1, bitplane formed with LSB bits, and so on. Significant information holds by the higher order bitplanes like b_8, b_7, whereas bitplanes b_1, b_2 consist of subtle information as shown in Fig. 2. We will make use of this advantage by storing encrypted higher order bitplanes into lower order bitplanes and this is named as perceptual encryption. The perceptual encryption is independent of compression and usually it happens before compression. Figure 3 shows the block diagram of perceptual encryption.

First step in perceptual encryption is from a group of segments with 8 bits in a bitplane and then encrypts each one with E_a. Each segment is converted into a point P_M (using Algorithm 1) and then generates two points or four cipher values after encryption. Whereas each value except segment length is having 163 bits. The cipher values are hidden in the LSB bitplane because it contains only subtle details.

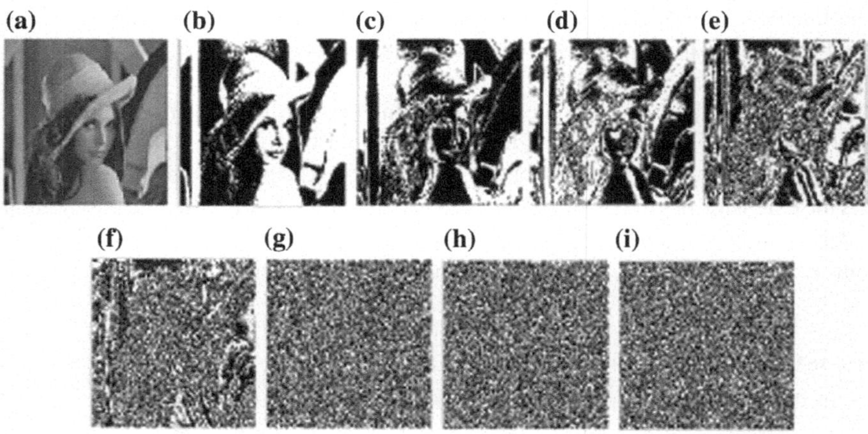

Fig. 2 **a** Lena grayscale image (256 × 256) and **b–i** are its 8 bitplanes from b_8 to b_1, respectively

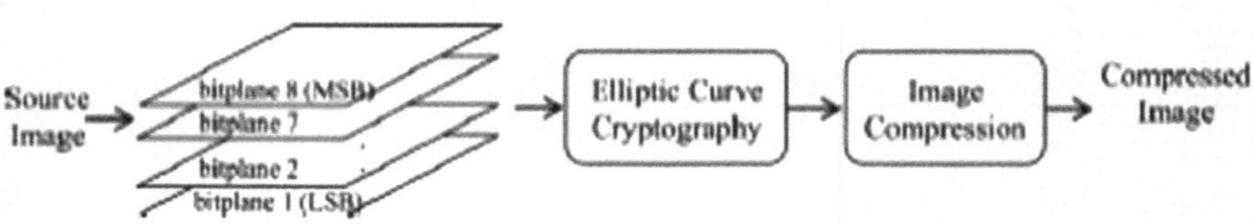

Fig. 3 Image bitplanes encryption and compression

Each encrypted segment is associated with a block of 497 bits (8-bit segment, remaining three points with 163 bits), where the original segment contains block number. 256 blocks are required at most to store the cipher values of all segments due to segment values ranging from 0 to 255. For instance, a bitplane of size 256 × 256 bits after encryption, approximately two images (256 × 497 bits) are required to store all the cipher values. Lower bitplanes b_1 and b_2 store encrypted segment values.

5 Results and Comparative Study

The results of perceptual encryption using ECC on Koblitz curves with Lopez–Dahab projective coordinates are discussed with respect to the requirements of multimedia encryption. The perceptual encryption is explained with three different encryption methods: (i) bitplane b7 encryption, (ii) bitplane b_8 encryption, and (iii) both bitplanes b_7 and b_8 encryption. Figure 4 shows the results obtained after applying E_a on bitplane b_7 of Lena image. Figure 5 shows the obtained results of applying E_a on bitplane b_8. Figure 6 depicts the perceptual quality of the Lena image with its higher order bitplanes b_7 and b_8 after encryption.

Figure 4b shows the case of encrypting only bitplane 8, whereas Fig. 5b shows the case of encrypting only bitplane 7. Figure 6b, c depicts the perceptual quality after encryption of both bitplanes 7 and 8 of Lena image. The visual distortion is evaluated after perceptual encryption, PSNR, SSIM are used to assess the image quality.

Encryption tightly effected on image quality that means quality reduction of bitplane b8 (PSNR = 9.4235, SSIM = 0.1343) compared with encryption of bitplane b7 (PSNR = 16.175, SSIM = 0.632). The parameters are analyzed with ECC-based

(a) (b)

Fig. 4 Bitplane 8 before and after applying perceptual encryption

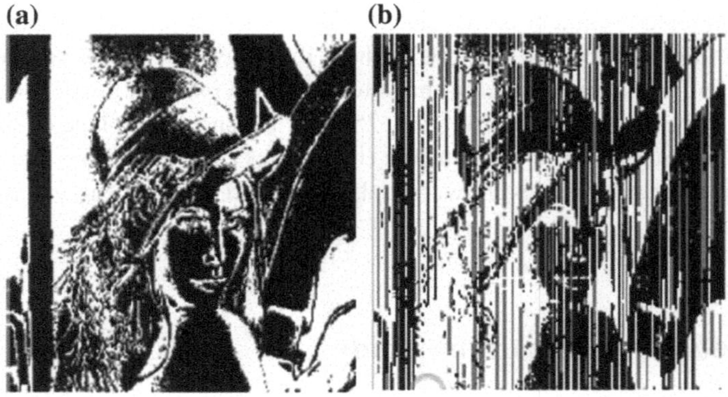

Fig. 5 Bitplane 8 before and after applying perceptual encryption

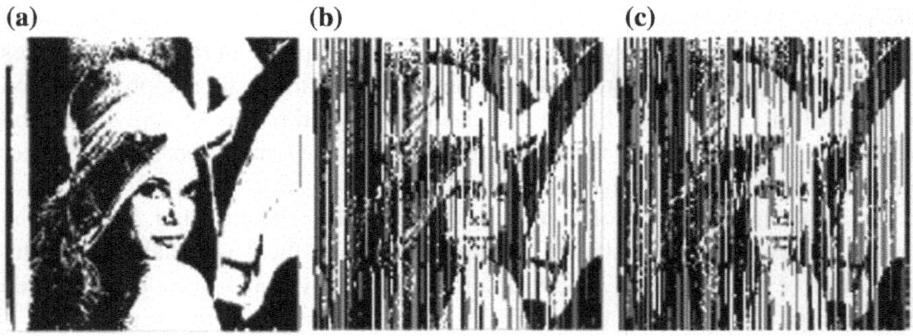

Fig. 6 Both bitplanes 7 and 8 before and after applying perceptual encryption

multimedia encryption as compared with requirements of multimedia. The ECC operations are fast and can meet the multimedia requirements, and streams are sync at the receiver end as compared with encryption efficiency.

ECC uses lightweight operations to perform field arithmetic as compared with remaining public key algorithms. The perceptual encryption performs encryption on each 8-bit segment of higher order bitplane. Extra data which were produced at the time of encryption are stored in LSB bitplanes because lower level bitplanes contain subtle information which does not increase the size of the compressed data. Whereas requirements of multimedia, codec compliance does not have effect in case of perceptual encryption because it performs encryption of bitplanes before compression.

6 Conclusion

The proposed method ECC on Koblitz curves with Lopez–Dahab coordinates accomplished multimedia encryption algorithm named as perceptual encryption. The perceptual encryption performs single or multiple bitplanes encryption before compression. The proposed method does not affect the requirements of multimedia and it meets the requirements of multimedia. The structure of the codec is intact if we apply ECC before compression, and structure of codec will alter if we apply ECC during compression. The proposed scalar multiplication with Lopez–Dahab projective coordinates consumes less computational cost due to Frobenius endomorphism and reduction of field operations. The encryption and decryption of ECC on Koblitz curves are better than existing method [14]. The future enhancement of proposed method is possible by encrypting reversible data hiding (RDH) image or video, and by this we can get the benefit of sending two images or videos with one encryption.

References

1. Liu F, Koenig H (2010) A survey of video encryption algorithms. Comput Sec 29:3–15
2. Agi I, Gong L (1996) An empirical study of MPEG video transmission. In: Proceedings of the internet society symposium on network and distributed system security. San Diego, CA, USA
3. FIPS 46–2, Data encryption standard, Nov 1993
4. NIST Data encryption standard. FIPS PUB 46, Jan 1977
5. NIST Advanced encryption standard. FIPS PUB 197, Nov 2001
6. FIPS 197, Advanced encryption standard (AES), Nov 2001
7. Fuhrt B, Kirovski D (2004) Multimedia security handbook. CRC Press
8. Liu X, Eskicioglu AM Selective encryption of multimedia content in distribution networks: challenges and new directions. In: IASTED international conference on communications, internet and information technology (CIIT). Scottsdale, AZ, USA, Nov 2003
9. Li S, Chen G, Cheung A, Bhargava B, Lo KT (2007) On the design of perceptual MPEG-video encryption algorithms. IEEE Trans Circuits Syst Video Technol 17:214–223
10. Liu F, Koenig H A novel encryption algorithm for high resolution video. In: Proceeding of ACM NOSSDAV05. ACM Press, New York, pp 69–74, June 2005
11. Liu F, Koenig H Puzzle-a novel video encryption algorithm. In: IFIP CMS 2005, LNCS 3677. Springer, Salzburg, Austria, pp 88–97, Sept 2005
12. Zeng W, Lei SE (2003) Efficient frequency domain selective scrambling of digital video. IEEE Trans Multimed 5:11–29
13. Wu P, Kuo CJ (2005) Design of integrated multimedia compression and encryption systems. IEEE Trans Multimed 7:828–839
14. Tawalbeh L, Mowafi M, Aljoby W (2012) Use of elliptic curve cryptography for multimedia encryption. IET Inf Sec 285:101–108
15. Hankerson D, Menezes A, Vanstone S (2004) Guide to elliptic curve cryptography. Springer
16. Solinas JA Efficient arithmetic on koblitz curves. Design Codes Cryptogr 19:195–249 (2000)
17. Menezes AJ, van Oorschot PC, Vanstone SA (1997) Hand-book of applied cryptography. In: CRC press series on discrete mathematics and its applications
18. Koblitz N (1987) Elliptic curve cryptosystems. Math Comput 48:203–209
19. Koblitz N (1998) An elliptic curve implementation of the finite field digital signature algorithm. In: Advances in cryptology CRYPTO 98, lecture notes in computer science, 1462, pp 327–337

20. Koblitz N (1991) CM-curves with good cryptographic properties. In: Proceedings of CRYPTO 1991, LNCS 576, pp 279–287. Springer
21. Roy SS, Rebeiro C, Mukhopadhyay D, Takahashi J, Fukunaga T (2011) Scalar multiplication on Koblitz curves using_2-NAF. IACR Cryptology ePrint Archive
22. Morai F, Olivos J (1990) Speeding up the computations on an elliptic curve using addition-subtraction chains. Inf Theory Appl 531–543
23. Adikari J, Dimitrov V, Jarvinen K (2012) A fast hardware architecture for integer to τ-NAF conversion for Koblitz curves. IEEE Trans Comput 61:732–737
24. Lopez J, Dahab R (1999) Fast multiplication on elliptic curves over without precomputation. In: Proceeding workshop CHES, pp 316–327
25. King B (2009) Mapping an arbitrary message to an elliptic curve when defined over GF(2^n). Int J Netw Sec 169–176
26. Lijuan L, Shuguo L (2016) High-performance pipelined architecture of elliptic curve scalar multiplication over GF(2^m). IEEE Trans VLSI Syst 24:1223–1232
27. Reza A, Arash R (2013) High-performance implementation of point multiplication on koblitz curves. IEEE Trans Circuits Syst II Exp Briefs 60:41–45
28. Hankerson D, Lopez-Hernandez J, Menezes A (2001) Software implementation of elliptic curve cryptography over binary fields. In: Proceedings of CHES 2000. LNCS 1965, pp 1–24
29. Schroeppel R (2000) Point halving wins big. Talks at: (i) midwest arithmetical geometry in cryptography workshop, November 1719, 2000, University of Illinois at Urbana Champaign; and (ii) ECC 2001 Workshop, October 2931, University of Waterloo, Ontario, Canada
30. Knudsen EW (1999) Elliptic scalar multiplication using point halving. In: Proceedings of ASIACRYPT 1999, LNCS 1716, pp 135–149
31. Fong K, Hankerson D, Lopez J, Menezes A (2004) Field inversion and point halving revisited. IEEE Trans Comput 53:1047–1059
32. Avanzi RM, Ciet M, Sica F (2004) Faster scalar multiplication on koblitz curves combining point halving with the frobenius endomorphism. In: Proceedings of PKC, 2004, LNCS 2947, pp 1–14
33. Washington Lawrence C (2008) Elliptic curves number theory and cryptography, 2nd edn. Chapman & Hall/CRC Taylor & Francis Group

Prediction of Dyslexia Using Machine Learning—A Research Travelogue

A. Jothi Prabha and R. Bhargavi

Abstract Dyslexia is regarded as a common learning disorder characterized by a persistent deficit in rapid word recognition and by spelling. It affects the individual's ability to decode letters and words fluently and accurately. The research community has worked on distinguishing dyslexic from non-dyslexic people by using various machine learning approaches, image processing techniques, design assessment, and assistive tools to support dyslexia. This survey paper looks at different dimensions of research toward dyslexia. This review identifies the research gaps, open issues, and challenges in this field. It also motivates for the application of ML techniques toward early prediction of dyslexia.

Keywords Dyslexia · Support vector machine · Image processing

1 Introduction

Dyslexia is a kind of reading disorder characterized through specific reading impairment. Dyslexia is not a problem related to visual acuity, schooling, or overall mental development. Dyslexic has an abnormal anatomy of the brain having damage in language processing control region. The symptoms of dyslexia are not static and can vary based on the language. It is not a disease and is not treatable. People with dyslexia have to live with it. It affects people of all age groups and is not related to the intelligence quotient or vision of an individual. It affects around 10% of the world's population. Detection of dyslexia is critical, expensive, and requires an able expert to handle dyslexic. Different approaches used toward prediction of dyslexia are Machine learning algorithms, Eye-tracking measures and Image processing techniques.

A. Jothi Prabha (✉) · R. Bhargavi
School of Computing Science and Engineering, Vellore Institute of Technology, Chennai, India
e-mail: jothiprabha@gmail.com

R. Bhargavi
e-mail: bhargavi.r@vit.ac.in

© Springer Nature Singapore Pte Ltd. 2019

V. Nath and J. K. Mandal (eds.), *Proceedings of the Third International Conference on Microelectronics, Computing and Communication Systems*, Lecture Notes in Electrical Engineering 556, https://doi.org/10.1007/978-981-13-7091-5_3

Machine learning is a type of artificial intelligence that learns and identifies new patterns from an enormous amount of data. It allows data scientists and analysts to effectively identify the possible opportunities, and design strategies/techniques to improve customer satisfaction, using relevant information hidden in enormous data sets. Machine learning is an emerging trend in the health domain that helps medical experts for better analysis, prevention, and treatment of individuals. There exist many machine learning models, each of which formulates prediction in various fashions. Choosing an appropriate machine learning algorithm is a key role because there are many to choose from.

Machine learning algorithms are primarily classified as supervised and unsupervised learning. Supervised learning is learning from already trained data whose class category is already known. Unsupervised learning categorizes the given data whose class category is unknown by learning the similarities in the given data. Most predominantly used machine learning methods are support vector machine (SVM), Neural Networks, Decision Tree, Bayesian Classifier, k-means clustering, and Logistic Regression.

Support vector machine is a classifier using kernel for pattern analysis, ranking raw data, clustering, and classification of data. It creates a model that can classify a given new dataset to some class using a set of training examples. It is a nonparametric supervised learning method suitable for data which has many features.

Neural networks mimics the functionality of the brain which understands information and comes up with various predictions based on the information. Artificial Neural Networks has three layers such as input, hidden, and output layer. The input layer has the raw features, hidden layer has the highly connected neurons, and the output layer has the prediction. The neurons are fine-tuned until we get a fine prediction. It finds its use in speech and object recognition, natural language processing, and image segmentation.

Decision trees is a type of supervised learning which use directed graphs to model decision-making by predictive modeling approach. This approach is to learn the decision tree from a given data. Random Forest is an ensemble of thousands of decision trees widely used in many applications.

K-means clustering is an unsupervised learning approach that finds distinct groups in a given data based on their similarities. It works on data whose class labels are not known. In the name, K refers to the number of distinct groups created. If finds it use in various business applications.

Logistic regression is from statistics. This machine learning algorithm is simple and works best when there is a possibility of only two outputs and they are highly dependent on explanatory variables. It estimates the probable outcome when given a set of observed variables.

2 Literature Survey

Research toward dyslexia has different dimensions. Few of them are: 1. Machine Learning Approaches and Image Processing Techniques, and 3. Design assistive tools to predict the severity of dyslexia.

2.1 Machine Learning Approaches Toward Dyslexia

Alex Frid and Zvia-Breznitz, developed an algorithm to classify individuals as Regular or Dyslexic Readers by using electroencephalogram (EEG) recorded channels. Support vector machine (SVM) is used for feature analysis and extraction. The ensemble of SVMs was used for classification and decision-making is done by majority voting. Features such as Maximal peak amplitude, Positive Area, and Spectral Flatness measure were extracted. The algorithm does not focus on the subtypes of dyslexia like dyscalculia and dysgraphia, and deals only with young adult dyslexics. The accuracy of this algorithm was not recorded [1].

Perera et al., worked toward prediction of dyslexia by electroencephalogram-based analysis (EEG). Electrodes were used to monitor the electrical activities of the brain. Support vector machine classifier model has been used for feature extraction and classification of dyslexics. They have identified the pros and cons of EEG approach and also suggested optimization techniques for better prediction of dyslexia [2].

Iwabuchi et al., explores to create an easy way for students to check their reading and writing difficulties. URAWSS (Understanding Reading and Writing skills of school children) assessment tool was used in this study. It uses rule-based decision technique. The URAWSS test data were analyzed by a professional. The results were compared with the analysis done by Decision tree and random forest. The result showed that using machine learning (ML) approach predicts better than the ordinary rule-based decision approach. It is suggested that prediction using ML can still be improvised using RTI (Response to Intervention) model-based approach [3].

Rello et al., designed a statistical model to differentiate dyslexic readers from control readers using eye-tracking mechanisms. Eye movements are tracked using eye tracker. Support Vector Machine binary classifier was used to build the model. It was witnessed that dyslexics show longer fixations, tinier saccades, and added regressions than non-dyslexic readers. The accuracy level of 80.18% was achieved using 10-fold cross-validation. They have suggested working on other types of classification algorithms such as Neural Networks and Perceptron learning [4].

Palacios et al., (2016) described a screening method for early identification of dyslexia to support children in school. Datasets are categorized through Fuzzy Unordered Rules Induction Algorithm (FURIA). The methods diagnose the dyslexic by considering the tests made by parents and teachers. The disadvantage of this approach is the time needed to classify dataset through FURIA is high [5].

Al-Barhamtoshy et al., proposed a computational analysis classifier with the help of dyslexia metrics. Gibson test of brain skills was taken into consideration for deriving dyslexia metrics. Initially, the given dataset is preprocessed, examined and analyzed by k-means classifier, ANN and Fuzzy based classifier. The results of these classifiers help in categorizing whether an individual is dyslexic or normal. An accuracy level of 96% in average was achieved using the three classifiers. They have suggested using EEG signals while reading to predict dyslexics [6].

Maitrei Kohli and T. V. Prasad, proposed a systematic approach for prediction of dyslexia using Artificial Neural Network (ANN). They also analyzed the potential causes of dyslexia more accurately by using ANN. The preliminary results obtained from the test data were reasonable and can act as a tool to classify dyslexics and non-dyslexics. The approach is easy to use but the accuracy of the model can be improved [7].

Zainuddin et al., described a classification technique for identifying dyslexic children from normal children using EEG signals. The EEG signals were extracted and then normalized for classification. The enhanced version of k-nearest neighbor (KNN) classifier is used in this approach that classifies the dyslexic and non-dyslexic better [8].

Jain et al., used the perceptron model of Artificial Neural Network (ANN) for identifying reading disabilities by conducting academics based assessments through remediation teachers. The model has an input layer with 11 units that correspond to various sections in the test and an output layer. The technique has less computational complexity and sophistication compared to other approaches. It also gives good and equivalent experimental outcomes on detection measures like sensitivity, accuracy, and specificity [9].

2.2 Image Processing Techniques Toward Dyslexia

Płonski et al., investigates the abnormal anatomies of the dyslexic brain by using multivariate classification technique to examine the gray matter disruption in brain images. The MR brain images of dyslexic children from three different countries (French, German, and Polish) were taken as input. The images are preprocessed to retrieve the features that are used for classification. Classification is done by Cross-validation classification framework includes the following steps in sequence. 1. Confounding factor corrections, 2. Feature selection, and 3. Classification. The results show that a typical curvature pattern with extra folds in the left hemisphere region of the brain characterizes dyslexia. Accuracy is better when compared to existing systems [10].

A. El-Baz et al., explored ways to differentiate dyslexic and control brain. Quantitative analysis of the shape of Cerebral White Matter gyrifications from 3D MRI is used for classification. An individual is classified as dyslexic or not by quantifying the volume of the extracted gyrifications. This quantitative classification result of this approach is efficient when compared to the level-segment-based segmentation approach. The research can be extended to investigate other brain structures that can help to understand the change of the dyslexic brain over a period of time. In addition to CWM, gray matter can also be investigated [11].

Tamboer et al., studied about voxel-based morphometry in dyslexia. Gray matter, white matter volume alterations. Group differences in local and total GM, white matter (WM) volume were analyzed. Classification and prediction of dyslexia is done using brain image datasets. Images were segmented and smoothed using Gaussian kernel. It was observed that multiple cognitive nature of dyslexia have a high influence on anatomical alterations based on the subtypes of dyslexia [12].

Feng et al., analyzed the cerebellar functional connectivity and cerebellar activation during phonological processing in dyslexics. Brain images used were taken using 3.0 T Siemens MRI (Magnetic Resonance Image) scanners at Beijing Normal University. The obtained input data is preprocessed using SUIT (toolbox for Software Process Management). Then the images are analyzed using a statistical technique. Children with dyslexia are observed to have abnormally high activation of bilateral cerebellum during orthographic activity. The results show that dyslexics had high bilateral activation of the cerebellum during reading. They have suggested studying the functional connectivity between the cerebral and cerebellum regions [13].

Cui et al., employed MRI along with diffusion tensor imaging to acquire multiple white matter features level-wise. This was carried out on 28 dyslexic school children with 33 age-matching attributes. Support vector machine has been employed to discriminate dyslexia kids from non-dyslexic using WMW, AD, FA, RD, and MD and achieved an accuracy of 83.61% [14].

Morken et al., studied the cortical connectivity of brain images in dyslexic children. The participants were across three different literacy groups. The dynamic causal model (DCM) is used to calculate the connectivity measures. These metrics are further analyzed through the ANOVA analysis technique. The results revealed that the dyslexics showed a delay during a move into preemergent and emergent literacy stages [15].

Mohamad et al., studied the correlation between the brain images from fMRI and 2D EEG topography, in dyslexic and non-dyslexic children while writing. The EEG signal acquired is filtered and feature extraction was done. It investigated whether EEG topography can be replaced in place of MRI and PET to make prediction less expensive and noninvasive. Left hemisphere of the brain activates for normal children, right hemisphere for dyslexic capable and both left and right hemispheres for dyslexic poor children. Hence variances in hemispheric activation were found between non-dyslexics and dyslexics [16].

2.3 Assistive Tools for Dyslexics

Martins et al., invented a mobile application for the diagnosis of dyslexia. It would help the dyslexic children to practice and improve their reading skill. It acts as an auxiliary tool to help health professionals to diagnose their patients. This can act as assistance for diagnosis of dyslexia and also support dyslexics to improve their reading abilities. The results show good usability aspects and good accuracy [17].

Alsobhi, proposed the ontology-based learning technique which can help the dyslexic children to make use of e-learning tools just like normal children. This work has inspected how learning objectives should be framed, by taking student's competencies and needs into consideration. It can be mapped with a suitable supportive technique in order to give better educational resources and e-learning support. The ontology used in this technique is developed through ontology web language. It mainly concentrates two types: Type of dyslexia, and the educational experience targeted to each dyslexia types. It does not offer proper evaluation and accuracy measure for dyslexia children's learning performance [18].

Wang et al., developed a theoretical motivation model for dyslexic people by designing supportive learning tools. This approach carried out an empirical study in real-time use cases. The key factors are collected by conducting interviews. This approach creates a motivation to dyslexic students and helps to improve learning experience and efficiency [19].

Sarpudin et al., came up with a prototype for mild dyslexics to read web pages easily. The prototype has special adjustments to the text environment to read easily using cascade style sheets. It was also tested on dyslexics in Titiwangsa center and results were acceptable [20].

Research on Dyslexia is carried out in different perspectives using different kinds of data like EEG signals, Imaging data such as MRI, fMRI, PET. Eye-tracking data is also used nowadays to predict and support dyslexia [21–25]. Table 1 shows the comparative literature survey on Dyslexia.

3 Research Dimensions Toward Dyslexia

The research community has been carried out in many dimensions 1. Prediction of dyslexia is done in two ways. a. by applying Machine Learning techniques b. image processing 2. Design assessment tests and assistive tools for supporting dyslexics. It has been observed that most of the research work is done using machine learning techniques only for predicting dyslexia.

Figure 1 shows different dimensions of research carried out toward the prediction and support of dyslexic people.

Table 1 Summary of literature survey

Author and year	Research objective	Proposed algorithm	Advantages	Disadvantages	Possible improvements over existing work
Alex Frid, Zvia-Breznitz	Use support vector machine algorithm for differentiating dyslexic readers from normal readers using ERP signals	Support vector machine	Easy procedure to detect dyslexia	Does not focus on the subtypes of dyslexia like dyscalculia and dysgraphia	Features related to subtypes of dyslexia can be extracted and used to diagnose the subtypes of dyslexia
Perera et al.	EEG-based analysis for classification for dyslexia	Support vector machine (SVM)	Identified the pros and cons of using EEG for prediction of dyslexia	Optimization techniques not considered which may contribute for better prediction	Optimization techniques like mirror prox algorithm can be used
Iwabuchi et al.	Machine learning-based evaluation of reading and writing difficulties	Decision tree and random forest	Prediction accuracy is better compared to rule-based approach	Prediction can be improvised by using a response to Intervention model	Rule-based decision approach along with machine learning can be used for prediction
Rello et al.	Eye-tracking measures to predict dyslexia	Support vector machine	Accuracy level (80.18%) is high compared to other systems	Accuracy level can be improved	Feature extraction can be improved
Palacios et al.	Diagnosis of dyslexia using low-quality data	Fuzzy unordered rules induction algorithm	Parents and teachers assessment helps to improve accuracy	Time needed to classify dataset through FURIA is high	Use of distributed approach can increase the execution speed

(continued)

Table 1 (continued)

Author and year	Research objective	Proposed algorithm	Advantages	Disadvantages	Possible improvements over existing work
Al-Barhamtoshy et al.	Diagnosis of dyslexia using computation analysis	K-means classifier, ANN and fuzzy based classifier	An accuracy level of 96% in average was achieved using the three classifiers	EEG signals during reading or writing would be more effective	Record EEG signals while reading to predict dyslexics
M. Kohli and T. V. Prasad	Application of artificial neural networks to predict dyslexia	Artificial neural networks	Screening tool for dyslexia	Accuracy needs to be improved	Accuracy of the model can be improved
Zainuddin et al.	EEG for discriminating capable dyslexic from normal children	The K-nearest neighbor (KNN) classifier	Classification error is less compared to existing systems	Standardization of data can be done before applying KNN	Check other clustering models for improved accuracy
Jain et al.	Computational diagnosis of learning disability	Perceptron based artificial neural network	Less complexity and sophistication	Assessment by a special teacher and machine learning can be ensembled	The ensemble of machine learning algorithms
Płonski	Neuroanatomical analysis to identify developmental dyslexia	Multi-parameter ML approach	Typical curvature pattern with extra folds in the left hemisphere region of the brain was identified as a characteristic of dyslexia	Other parts of the brain are not considered	Cerebral white matter and gray matter volume can be used as features to classify dyslexics

(continued)

Table 1 (continued)

Author and year	Research objective	Proposed algorithm	Advantages	Disadvantages	Possible improvements over existing work
A. El-Baz et al.	CAD system for early detection of dyslexic brains using MRI	Quantitative shape analysis	This approach is better than the level-segment-based approach	Gray matter volume is not considered	Extend the work to understand the change of the dyslexic brain over a period of time
Tamboer et al.	Classification of neuro-imaging scans to identify whether it dyslexic or control by using machine learning	Gaussian Kernel for classification	Subtypes of dyslexia are analyzed	Levels and types of dyslexia are not considered during classification	More features can be extracted for better classification of subtypes
Cui et al.	Analysis of white matter connectivity to developmental dyslexia using machine learning approaches	SVM classifier	Accuracy of 83.61% is achieved	Only white matter regions were analyzed	Gray matter and cerebral white matter can be analyzed
Morken et al.	Reading in dyslexics across literacy groups and development: a longitudinal study	ANOVA analysis technique	Identified delay in moves of dyslexics	Less no of subjects were analyzed	Check for classification accuracy on large datasets
Mohamad et al.	Study of neurological techniques to diagnose dyslexia in kids	Statistical technique	Variance in hemispheric activation was found	Only a few features were identified	The ensemble of machine learning algorithms

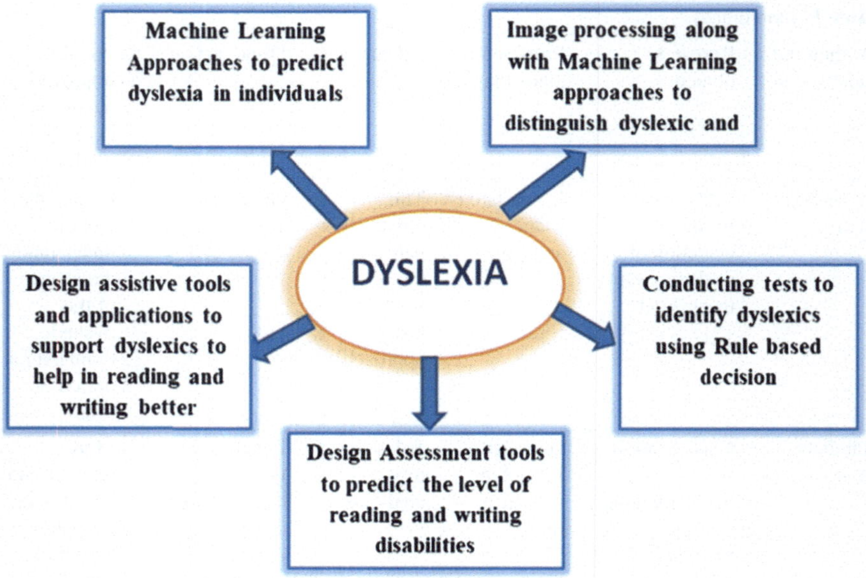

Fig. 1 Different dimensions of research toward support for dyslexia

4 Dyslexia Open Issues and Challenges

Dyslexia disease symptoms and characteristics can vary based on the different languages so language-based classification need to be considered. Several soft computing methods used have less accuracy level which can be improved. Many cases while using image datasets of the brain for prediction of dyslexia, not all parts of the brain are analyzed. They analyzed only a particular part of the brain for dyslexia prediction which is not efficient, because dyslexia may evolve through various part of the brain. Many assistive tools can be designed for helping dyslexics to improve their reading and writing skills [26–29]. Early and better prediction of dyslexia is also a challenging issue.

5 Conclusion and Future Scope

In this work, the application of machine learning techniques, image processing, and assistive tools used for prediction/assistance of dyslexia has been discussed. Most of the machine learning algorithms used for prediction of dyslexia focuses only on the few characteristics or symptoms of dyslexia. Dyslexia characteristics may vary depending on the language, hence language-based classification need to be done for better prediction. Mostly, SVM and ANN are used for classification whose accuracy level achieved is less. The efficiency of ML algorithm can be improved to predict dyslexia better by changing the input parameters used and also by improvising the feature extraction methods.

References

1. Frid A, Breznitz Z An SVM based algorithm for analysis and discrimination of dyslexic readers from regular readers using ERPs. In: 2012, IEEE 27th convention of electrical & electronics engineers in Israel, pp 1–4
2. Perera H, Fairuz Md, Wong KW (2016) A review of electroencephalogram-based analysis and classification frameworks for dyslexia. In: International conference on neural information processing. Springer International Publishing
3. Iwabuchi M, Hirabayashi R, Nakamura K, Dim NK (2017) Machine learning based evaluation of reading and writing difficulties. Stud Health Technol Info 242:1001
4. Rello L, Ballesteros Detecting readers with dyslexia using machine learning with eye tracking measures. In: Proceedings of the 12th web for all conference. ACM, p 16
5. Palacios A, Sánchez C, Destercke An extension of the FURIA classification algorithm to low quality data through fuzzy rankings and its application to the early diagnosis of dyslexia. Neuro Comput 176:60–71
6. Al-Barhamtoshy HM, Motaweh DM (2017) Diagnosis of dyslexia using computation analysis. In: Informatics, health & technology (ICIHT), pp 1–7. IEEE
7. Kohli M, Prasad TV (2010) Identifying dyslexic students by using artificial neural networks. In: Proceedings of the world congress on engineering, vol 1. London, UK
8. Zainuddin L, Mansor, Mahmoodin Optimized KNN classify rule for EEG based differentiation between capable dyslexic and normal children. In: Biomedical engineering and sciences (IECBES), 2016 IEEE, pp 685–688
9. Jain M, Dongardive, Abraham Computational diagnosis of learning disability. Int J Recent Trends Eng 2(3)
10. Płoński G, Altarelli M, Marbach VE, Grande, Jednoróg (2017) Multi-parameter machine learning approach to the neuroanatomical basis of developmental dyslexia. Human Brain Map 38(2):900–908
11. El-Baz A et al (2008) A new CAD system for early diagnosis of dyslexic brains. In: 15th IEEE international conference on image processing, 2008. ICIP 2008. IEEE
12. Tamboer V, Ghebreab, Scholte (2016) Machine learning and dyslexia: classification of individual structural neuro-imaging scans of students with and without dyslexia. NeuroImage Clin 11:508–514
13. Feng Z, Yang T, Xie, Ding (2017) Dyslexic children show atypical cerebellar activation and cerebro-cerebellar functional connectivity in orthographic and phonological processing. Cerebellum 16(2):496–507
14. Cui X, Gong G (2016) Disrupted white matter connectivity underlying developmental dyslexia: a machine learning approach. Hum Brain Mapp 37(4):1443–1458
15. Morken H, Hugdahl, Specht (2017) Reading in dyslexia across literacy development: a longitudinal study of effective connectivity. NeuroImage 144:92–100
16. Mohamad, Mansor, Lee Review of neurological techniques of diagnosing dyslexia in children. In: 2013, 3rd international conference on system engineering and technology (ICSET), pp 389–393. IEEE
17. Martins, Lima, Sampaio M Mobile application to support dyslexia diagnostic and reading practice
18. Alsobhi, Yaquob A, Khan N, Rahanu H (2015) Personalised learning materials based on dyslexia types: ontological approach. Proc Comput Sci 60:113–121

19. Wang R, Chen L, Solheim I, Schulz T, Ayesh A (2017) Conceptual motivation modeling for students with dyslexia for enhanced assistive learning. In: Proceedings of the 2017 ACM workshop on "intelligent interfaces for ubiquitous and smart learning", pp 11–18. ACM
20. Sarpudin SNS, Zambri S (2014) Web readability for students with dyslexia: Malaysian case study. In: 3rd international conference on user science and engineering (i-USEr). IEEE, pp 192–197
21. Manghirmalani PZ, Jain K (2011) Learning disability diagnosis and classification—a soft computing approach. In: 2011 world congress on information and communication technologies (WICT). IEEE, pp 479–484
22. Chen A et al (2017) Individualized early prediction of familial risk of dyslexia: a study of infant vocabulary development. Front Psychol 8
23. Loizou, Laouris (2011) Developing prognosis tools to identify learning difficulties in children using machine learning technologies. Cogn Comput 3(3):490–500
24. Georgopoulos, Stylios (2009) Diagnosis support using fuzzy cognitive maps combined with genetic algorithms. In: Engineering in medicine and biology society EMBC 2009. "Annual international conference of the IEEE", pp 6226–6229. IEEE
25. Mahmoodin Z et al Selection of symlets wavelet function order for EEG signal feature extraction in children with dyslexia. In: Biomedical engineering & sciences (ISSBES) in student symposium in IEEE
26. Dimitriadis, Stavros (2016) Classifying children with reading difficulties from non-impaired readers via symbolic dynamics and complexity analysis of MEG resting-state data. In: Signal processing and information technology (ISSPIT), international symposium on IEEE
27. Atakan VH et al (2009) Early prediction of reading disability using machine learning. In: Annual symposium proceedings, vol 2009. American medical informatics association
28. Schmalz X, Altoè, G Mulatti C (2016) Statistical learning and dyslexia: a systematic review. Annals Dyslexia, 1–16
29. Mahmoodin et al (2016) Electroencephalogram electrode localization in the support vector machine classification of dyslexic children. In: EMBS conference on biomedical engineering and sciences (IECBES). IEEE

Comparative Assessment of Passive Islanding Detection Techniques for Distributed Generations

Bhatraj Anudeep, Aman Verma and Paresh Kumar Nayak

Abstract In this paper, a comparative study of the available important passive islanding techniques used for distributed generation is carried out. The following passive methods are compared and analyzed: under/over voltage, under/over frequency, rate of change of frequency, voltage unbalance and total harmonic distortion, rate of change of active and reactive power, rate of change of phase angle difference, and rate of change of negative- and positive-sequence current methods. The above methods are tested and compared for numerous islanding and non-islanding cases by generating data on the IEEE 399-1997 system through PSCAD/EMTDC software.

Keywords Distributed generation · Islanding detection · Non-detection zone · Passive islanding detection technique

1 Introduction

During the last decade, there has been an urge to integrate distributed generation (DG) into the electrical power distribution system, since it provides several benefits over the conventional distribution system. However, the integration of DGs into distribution system brings great challenges to the existing protection schemes. Islanding detection is one of them. Islanding occurs when the utility grid is disconnected, but the power supply is maintained to the load by the DGs. The unintentional islanding is the fundamental concern as it not only damages the utility and consumer equipment, but it also risks the life of working personnel.

B. Anudeep (✉) · A. Verma · P. K. Nayak
Department of Electrical Engineering, Indian Institute of Technology (ISM) Dhanbad, Dhanbad 826004, India
e-mail: anu17deepu@gmail.com

A. Verma
e-mail: amanverma0496@gmail.com

P. K. Nayak
e-mail: pareshkumar.nayak@gmail.com

© Springer Nature Singapore Pte Ltd. 2019

V. Nath and J. K. Mandal (eds.), *Proceedings of the Third International Conference on Microelectronics, Computing and Communication Systems*,
Lecture Notes in Electrical Engineering 556, https://doi.org/10.1007/978-981-13-7091-5_4

Several methods are available in the literature on unintentional islanding. In [1], an up-to-date review on islanding detection is carried out. Passive islanding detection is performed either using local or remote information. Detection of islanding condition and hence, the disconnection of DG in remote passive islanding technique is carried out at the grid substation. Thus, it necessitates communication medium between the DG and the grid substation [2]. The performance point of view remote methods are superior over the local methods. However, the economic point of view local methods are preferable over remote methods. Local passive islanding detection methods use measurements at the DG terminal and are of two types; active- and passive islanding methods. In active islanding method, a disturbance is introduced into the system which forces the power system to give a certain response that cannot be kept after islanding [3]. However, the introduction of a disturbance in active method affects the power quality. Also, the economic point of view active methods are not attractive. On the other hand, passive techniques based on local information are simple and rigid. Passive methods utilize the changes in the voltage, current, frequency, harmonic distortion, voltage unbalance, etc., at the DG point for accomplishing the islanding detection task.

In the passive method, a relay monitors and one/more of the above system parameters trips if the parameters deviate from the preset threshold. However, the selection of a proper threshold is very difficult. If the threshold value is low, then there are chances of false tripping during non-fault transients (switching on/off of large loads and capacitor banks, etc.). Again, if the threshold value is set at a higher value, then there will be large non-detection zones (NDZ), where the system exhibits an inability in the detection of islanding condition. The IEEE Standard 1547-2003 recommends the disconnection of DGs within 2 s after the isolation of the grid substation [4]. Thus, it is very important to select an ideal protection scheme, which can detect the islanding condition quickly and should have the capability to avoid the false tripping.

The under/over voltage and under/over frequency are the two most widely used traditional methods of passive islanding detection techniques [5, 6]. Implementation point of view these traditional passive islanding methods are simple and cost-effective. However, these simple methods have larger NDZ and have limited performance during low power mismatch islanding.

To overcome the limitations of the traditional passive islanding techniques, several improved techniques are also proposed subsequently. In [7], the voltage unbalance and total harmonic distortion of current are used together for islanding detection. Further, in [8], the rate of change of reactive power and total harmonic distortion of current are used together for islanding detection. The phase angle between voltage and current are used as a criterion for islanding detection [9]. The methods based on proportional power spectral density [10], capacitor insertion technique [11], the rate of change of frequency dependent impedance [12], harmonic grid impedance [13], and the sequence components of currents [14] are some of the advanced passive islanding techniques reported in the literature for accomplishing fast islanding detection.

For further improvement in the speed and accuracy of the islanding detection methods, the wavelet singular entropy-based method [15], wavelet and S-transform-based method [16], probabilistic neural network-based method [17], and data mining-based intelligence approach [18] are also reported in the literature.

This paper presents a comparative assessment of the following available seven important passive islanding detection techniques. The working principle of each of the methods is provided first. Then, the comparative assessment is carried out for different islanding and non-islanding cases generated on the IEEE standard 399-1997 system through PSCAD/EMTDC.

2 A Review on the Available Important Passive Islanding Methods

2.1 Under/Over Voltage (UV/OV) [5]

This is one of the conventional methods used for islanding detection. In the case of grid interruption, there is a sudden mismatch in the load and generation in the island. Sudden power mismatch in the island causes changes in the voltage profile at the DG terminals. These changes are monitored using a relay and with suitable threshold, islanding condition can be detected. Consider the system in Fig. 1, in which the direction of power flow and point of common coupling (PCC) are shown. At the PCC, the active- and reactive power supplied from the grid are

$$\Delta P = P_{load} - P_{PV} \quad (1)$$

$$\Delta Q = Q_{load} - Q_{inv} \quad (2)$$

The system's behavior can be identified by the values of ΔP and ΔQ just before the islanding. If the value of $\Delta P \neq 0$, then voltage magnitude at the PCC will change and it will be detected by the UV/OV relays. However, this method is generally effective only in the case of large power mismatch and it is very difficult to set a threshold value of voltage for small power mismatch. The disadvantage of this method is that it has a large NDZ and hence has a slow response time.

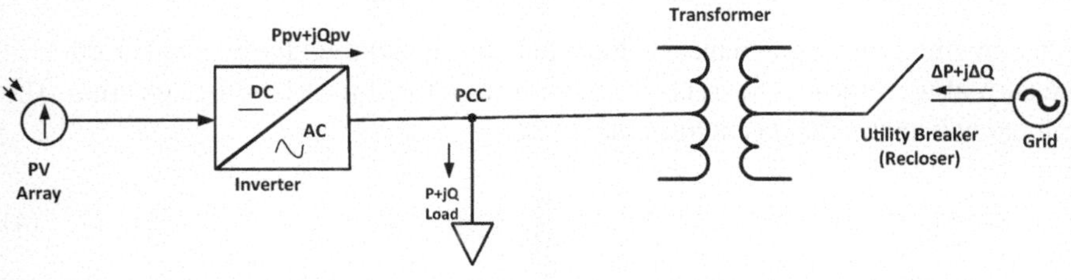

Fig. 1 Inverter interfaced photovoltaic (PV) DG source is connected to the utility grid

2.2 Under/Over Frequency (UF/OF) [6]

This is also one of the oldest techniques used for islanding detection. In the case of $\Delta Q \neq 0$ (Fig. 1), a sudden change in the phase of the DG terminal voltage will occur. As a result, the frequency of the inverter output will also change. Such a change in the frequency at the DG terminal can be detected using UF/OF relays. Thus, with a suitable threshold, islanding condition can be detected. However, this method has also large NDZ and slow detection time.

2.3 Rate of Change of Frequency (ROCOF) [6]

In this method, the variation of frequency $\left(\frac{df}{dt}\right)$ is monitored at the DG terminal. With a suitable threshold, this method is used for islanding detection. This method has a faster response time compared to the previously two discussed methods. However, this method has also a larger NDZ. The $\frac{df}{dt}$ with respect to the system frequency (f), active power mismatch (ΔP), system capacity (G), and system inertia (H) is given as

$$\frac{df}{dt} = ROCOF = \frac{\Delta P f}{2GH} \tag{3}$$

2.4 Rate of Change of Power (ROCOP) [19]

The rate of change of active- and reactive power is monitored at the DG terminals for islanding detection. With a suitable threshold, the ROCOP is used as a feature for islanding detection.

2.5 Voltage Unbalance and Total Harmonic Distortion (VU/THD) [7]

In this method, two parameters, voltage unbalance and total harmonic distortion of the current waveform at the relay terminal is monitored for islanding detection. The voltage unbalance (VU) is defined as

$$VU = \frac{V_2}{V_1} \times 100 \tag{4}$$

where V_1 and V_2 are the positive- and negative-sequence voltages at the DG terminal, respectively. Total harmonic distortion (THD) of the measured current at the DG terminal is computed as

$$THD_t = \frac{\sqrt{\sum_{h=2}^{H} I_h^2}}{I_1} \times 100 \tag{5}$$

where I_h and I_1 are the root mean square value of hth harmonic and fundamental components of the current signal, respectively. During islanding, the DG takes control of the local load as well as the other loads supplied by the grid. So, loading of the DG suddenly changes during islanding which causes fluctuations in voltage and introduces harmonics into the current signal. Thus, with suitable threshold setting, VU and THD are used together for islanding detection.

2.6 Rate of Change of Negative- and Positive-Sequence Current (ROCONSI/ROCOPSI) [14]

In this technique, the rate of change of negative- and positive-sequence components of current computed at the DG terminal are compared with preset thresholds for accomplishing the islanding detection. The computational steps of the method are as follows. First, the positive-, negative-, and zero-sequence currents (I_1, I_2 *and* I_0) are extracted from the measured three-phase currents (I_a, I_b, and I_c) as

$$\begin{bmatrix} I_0 \\ I_1 \\ I_2 \end{bmatrix} = \frac{1}{3} \times \begin{bmatrix} 1 & 1 & 1 \\ 1 & \alpha & \alpha^2 \\ 1 & \alpha^2 & \alpha \end{bmatrix} \times \begin{bmatrix} I_a \\ I_b \\ I_c \end{bmatrix} \tag{6}$$

where $\alpha = e^{j2\pi/3}$. Let during islanding, phase currents I_a, I_b, and I_c are changed to $I_a + \Delta I_a$, $I_b + \Delta I_b$, and $I_c + \Delta I_c$, respectively. The change in phase currents due to islanding will also change the sequence currents in (6). The sequence components of currents in (6) are modified during islanding as

$$\begin{bmatrix} I'_0 \\ I'_1 \\ I'_2 \end{bmatrix} = \frac{1}{3} \times \begin{bmatrix} 1 & 1 & 1 \\ 1 & \alpha & \alpha^2 \\ 1 & \alpha^2 & \alpha \end{bmatrix} \times \begin{bmatrix} I_a + \Delta I_a \\ I_b + \Delta I_b \\ I_c + \Delta I_c \end{bmatrix} \tag{7}$$

Now, the rate of change of positive- and negative-sequence currents $\left(\frac{dI_1}{dt} \, and \, \frac{dI_2}{dt}\right)$ are computed as

$$\frac{dI_1}{dt} = \frac{I'_1 - I_1}{\Delta t} = \frac{\Delta I_a + \alpha \Delta I_b + \alpha^2 \Delta I_c}{\Delta t} \tag{8}$$

$$\frac{dI_2}{dt} = \frac{I_2' - I_2}{\Delta t} = \frac{\Delta I_a + \alpha^2 \Delta I_b + \alpha \Delta I_c}{\Delta t} \tag{9}$$

With a suitable threshold, $\frac{dI_1}{dt}$ and $\frac{dI_2}{dt}$ are used together for islanding detection.

2.7 *Rate of Change of Phase Angle Difference (ROCOPAD) [9]*

In this technique, the *ROCOPAD* is monitored at the DG terminal for islanding detection. The method employs synchronous transformation based on phasor estimation technique for *ROCOPAD* computation. The computation steps are as follows. The power system signal $x(t)$ with a frequency of f can be expressed as

$$x(t) = \sum_{k=1}^{\infty} A_k \sin(2\pi k f t + \delta_k) \tag{10}$$

where A_k and δ_k are the amplitude and phase angle of kth-order waveform, respectively. Now, the three-phase to two-phase transformation is performed using $d - q$ transformation as

$$\begin{bmatrix} x_d \\ x_q \end{bmatrix} = \begin{bmatrix} \sin(\omega_0 t) & -\cos(\omega_0 t) \\ -\cos(\omega_0 t) & -\sin(\omega_0 t) \end{bmatrix} \times \begin{bmatrix} 1 & \frac{-1}{2} & \frac{-1}{2} \\ 0 & \frac{\sqrt{3}}{2} & \frac{-\sqrt{3}}{2} \end{bmatrix} \times \begin{bmatrix} x_a \\ x_b \\ x_c \end{bmatrix} \tag{11}$$

where $\omega_0 = 2\pi f_0$, and f_0 is the power system frequency (=50 Hz). At the mth instant, $t = mT_s$, where T_s is the sampling interval. So, (11) can be rewritten as

$$\begin{bmatrix} x_d(m) \\ x_q(m) \end{bmatrix} = \frac{3}{2} \begin{bmatrix} \sum_{k=1}^{\infty} A_k \cos[2\pi(kf - f_0)mT_s + \delta_k] \\ -\sum_{k=1}^{\infty} A_k \sin[2\pi(kf - f_0)mT_s + \delta_k] \end{bmatrix} \tag{12}$$

Fundamental quantities can be determined by putting $k = 1$,

$$x_{d1}(m) = 1.5A_1 \cos[2\pi(f - f_0)mT_s + \delta_1] \text{ and}$$

$$x_{q1}(m) = -1.5A_1 \sin[2\pi(f - f_0)mT_s + \delta_1] \tag{13}$$

The amplitude ($A1$), phase ($\delta 1$), and frequency (f) of the signal is computed using the $d - q$ quantities as

$$A_1 = \frac{2}{3}\sqrt{x_{d1}^2 + x_{q1}^2}, \delta_1 = \tan^{-1}\left[\frac{-x_{q1}(0)}{x_{d1}(0)}\right] \text{and}$$

$$f = \frac{[\delta_1(m) - \delta_1(m-p)]}{2\pi p T_s} + f_0 \tag{14}$$

Now, *ROCOPAD* can be calculated as

$$ROCOPAD = \frac{\Delta(\delta_V - \delta_i)}{\Delta t} \tag{15}$$

With a suitable threshold, *ROCOPAD* is utilized as a criterion for islanding detection.

3 Results and Discussions

For comparative assessment of the above-discussed passive islanding techniques, different possible islanding and non-islanding cases are simulated using PSCAD/EMTDC software on the modified IEEE standard 399-1997 system (Fig. 2) [20]. The DG1 connected at PCC1 is a synchronous generator, whereas the DG2 connected at PCC2 is inverter interfaced PV power plant. The feeders are represented with lumped R-L elements. The combinations of linear and nonlinear loads (L1–L5) are connected through the three radial feeders of the subsystem. The comparative results for three typical cases are provided below.

3.1 Performance of Islanding Detection Techniques for Large Power Mismatch

For the performance comparison of islanding causing large power mismatch, the main circuit breaker of the utility grid is disconnected at 0.5 s. This causes islanding to both DG1 and DG2. The comparative results of different passive techniques are shown in Fig. 3. From Fig. 3a, it is observed that the voltage at DG2 terminal is dropped below 0.85 p.u. in 25 ms. Thus, islanding detection at DG2 terminal can be accomplished using UV/OV detection technique. However, using UV/OV detection technique, islanding detection at DG1 terminal may not be performed as the change in voltage at the DG1 terminal is insignificant.

Figure 3b shows that the frequency of DG2 is dropped to 49.5 Hz in 25 ms. Thus, islanding detection at DG2 terminal can be accomplished using UF/OF detection technique. However, using UF/OF detection technique, islanding detection at DG1 terminal may not be performed as the change in frequency at the DG1 terminal is insignificant. Further, from Fig. 3c, it is observed that the ROCOF at DG2 terminal is around 0.5 and the time taken is around 20 ms after the opening of the main CB.

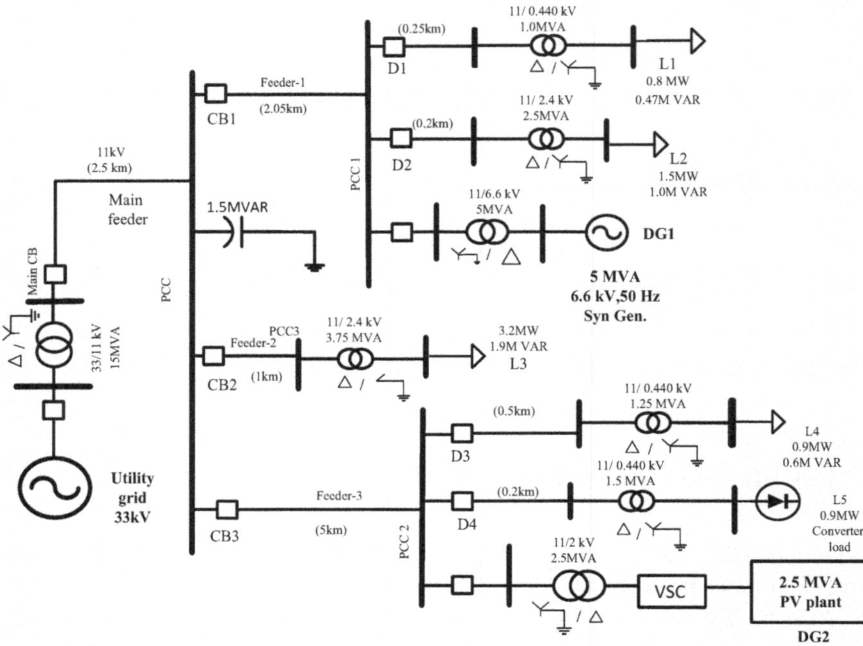

Fig. 2 Schematics of the test system studied

Thus, using ROCOF technique, islanding detection can be accomplished at the DG2 terminal. On the other hand, using ROCOF technique, islanding detection at DG1 terminal may not be accomplished as there is no such change in frequency is observed at the DG1 terminal.

The VU and THD of current at the DG terminals are shown in Fig. 3d and Fig. 3e, respectively. From the figures, it is observed that a variation of 4% in VU and 5% in THD of current is observed at the DG2 terminal within 20 ms after the initiation of the islanding. However, no such change in VU is observed at the DG1 terminal, but a change of 2.5% in THD of current is observed within 20 ms after the initiation of islanding. Thus, using VU and THD of the current technique, islanding can be detected at both DG1 and DG2 terminals.

The performance of ROCOP method is shown Fig. 3h and Fig. 3j, respectively. The rates of change of active- and reactive powers at both the DG terminals are shown in Fig. 3f and Fig. 3g, respectively. From the figures, it is observed that at the DG2 terminal a change of 0.5 MW/s and 0.5 MVAr/s are observed in the active- and reactive power, respectively. The change in the active power at the DG1 terminal is minimal. However, a change of 0.5 MVAr/s in reactive power is observed at the DG1 terminal within 50 ms after the islanding. Thus, using ROCOP-based islanding detection method, islanding detection at both the DG terminals can be accomplished.

The performances of ROCOPSI/ROCONSI are shown in Fig. 3h and Fig. 3i, respectively. The figures clearly show a variation of 1 pu/s in both ROCOPSI and

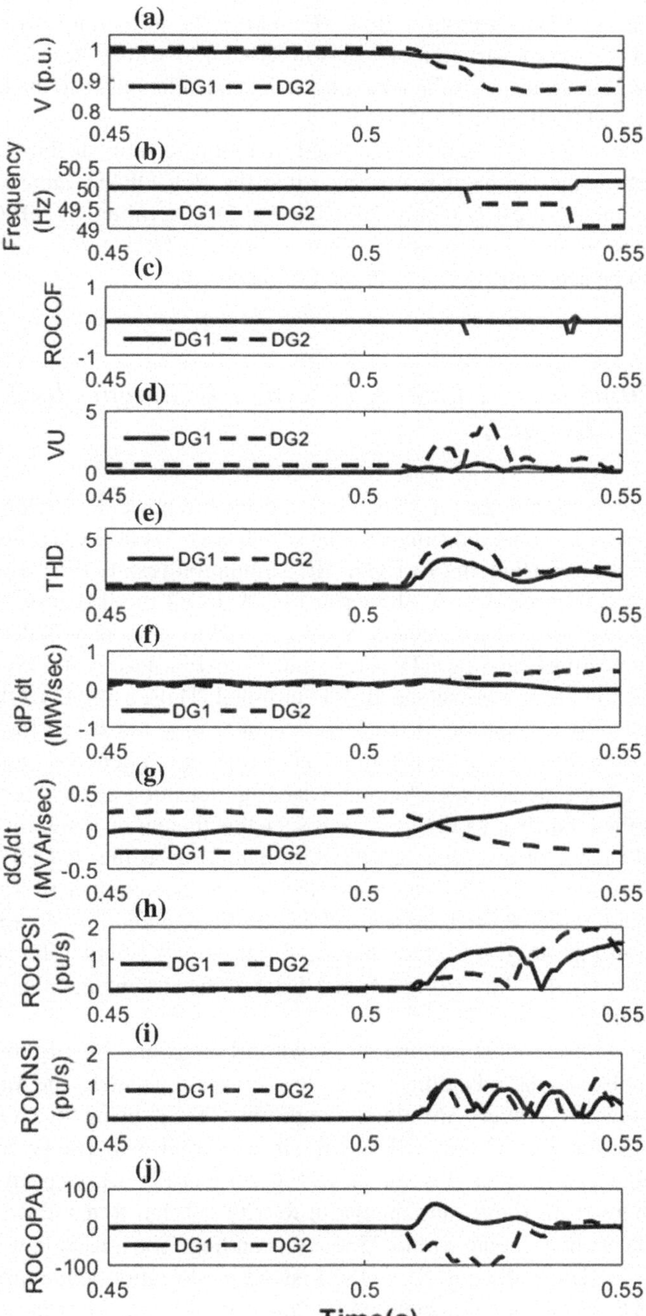

Fig. 3 Performance comparison for the islanding causing large power mismatch, **a** voltage, **b** frequency, **c** rate of change of frequency, **d** voltage unbalance, **e** total harmonic distortion, **f** rate of change of active power, **g** rate of change of reactive power, **h** rate of change of positive-sequence current, **i** rate of change of negative-sequence current, and **j** rate of change of phase angle difference

ROCONSI at the DG1 terminal within 20 ms after the initiation of the islanding. Similarly, DG2 shows a variation of 1 pu/s in ROCOPSI within 30 ms and 1 pu/s in ROCONSI within 20 ms after the islanding. Thus, islanding at both the DGs can be detected using ROCOPSI/ROCONSI technique.

The performance of the ROCOPAD based islanding detection technique is shown in Fig. 3j. From the figure, it is observed that the ROCOPAD value at the DG1 terminal is changed at 50°/s within 20 ms and at DG2 terminal at −100°/s within 30 ms after the initiation of islanding. Thus, using ROCOPAD technique islanding detection can be accomplished at both the DG terminals.

3.2 Performance of Islanding Detection Techniques for Low Power Mismatch

For the performance comparison of islanding causing low power mismatch, the circuit breaker CB3 which connects the main feeder with feeder-3 at the PCC is disconnected at 0.5 s. This causes islanding to DG2. The simulation results for various passive islanding detection methods are shown in Fig. 4. In all the detection techniques, there is no change in DG1 parameters. So, the available methods will definitely find difficulty in detecting islanding at DG1 terminal. From Fig. 4a, it is observed that there is an insignificant change in voltage at DG2 terminal. Thus, using UV/OV detection technique, islanding detection at the DG2 terminal may not be performed. From Fig. 4b, it is seen that there is an insignificant change in frequency at DG2 terminal. Thus, using UF/OF detection technique, islanding detection at DG2 terminal may not be performed. Further, from Fig. 4c, it is seen that there is an insignificant change in the rate of change of frequency at DG2 terminal. Thus, using ROCOF technique, islanding detection at the DG2 terminal may not be accomplished. The VU and THD of current at the DG terminals are shown in Fig. 4d and Fig. 4e, respectively. However, from Fig. 4e, f, it is clear that the variation of VU and THD at ther DG2 terminal is very small. Thus, using VU and THD of the current technique, islanding may not be detected.

The ROCOP at the DG2 terminal is shown in Fig. 4f and Fig. 4g, respectively. In this case also as the islanding causes low power mismatch, the variations of active- and reactive powers are almost negligible. Thus, using ROCOP method, islanding detection at DG2 terminal may not be accomplished. The performance of ROCOPSI/ROCONSI method is shown in Fig. 4h and Fig. 4i, respectively. From Fig. 4h, i, there is no significant change in ROCONSI, but ROCOPSI at the DG2 terminal changes by 0.5 pu/s within 25 ms of islanding. Thus, islanding at the DG2 terminal can be detected using ROCONSI islanding detection technique. The performance of the ROCOPAD technique is shown in Fig. 4j. From the figure, it is observed that there is a very small change in the ROCOPAD at the DG2 terminal. Thus, using ROCOPAD method, low power mismatch islanding conditions cannot

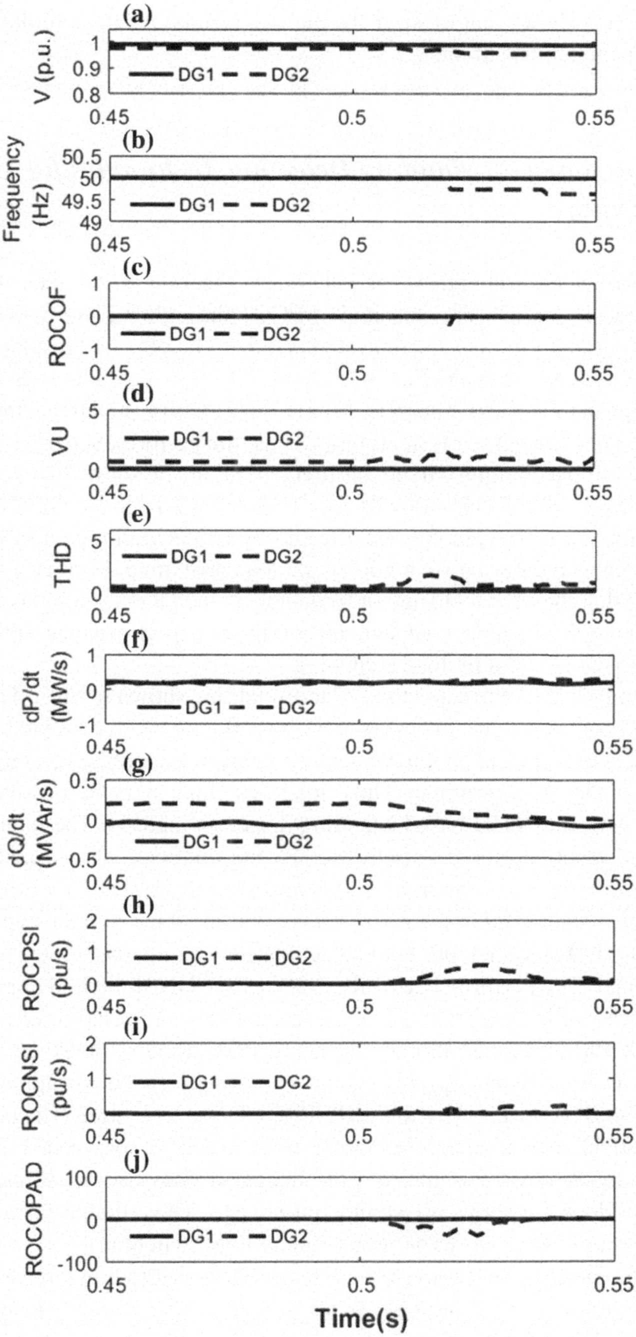

Fig. 4 Performance comparison for the islanding causing low power mismatch, **a** voltage, **b** frequency, **c** rate of change of frequency, **d** voltage unbalance, **e** total harmonic distortion, **f** rate of change of active power, **g** rate of change of reactive power, **h** rate of change of positive-sequence current, **i** rate of change of negative-sequence current, and **j** rate of change of phase angle difference

be detected. This shows that most of the passive techniques have limitation during low power mismatch islanding.

3.3 *Performance of Islanding Detection Techniques for Load Switching*

For the performance comparison of islanding detection techniques during load switching, load L_2, which is connected at PCC1 through 0.2 km distribution line and 11/2.4 kV transformer, is disconnected at 0.5 s. The obtained results for various passive techniques are shown in Fig. 5. In all the detection techniques, there is no significant change in the DG2 parameters. From Fig. 5a, it is observed that the change of voltage due to load switching is negligible. Thus, the performance of UV/OV-based islanding detection technique will not be affected. Figure 5b shows that the frequency of DG1 increases to 50.5 Hz within 25 ms. Thus, the performance of UF/OF-based islanding detection technique may be affected by load switching and DG1 may be wrongly disconnected during such non-island load switching. Further, from Fig. 5c, it is observed that the rate of change of frequency at the DG1 terminal is around 0.5, and time taken is 25 ms after load switching. Thus, the performance of the ROCOF technique may be affected by load switching.

The VU and THD of current at the DG terminals are shown in Fig. 5d and Fig. 5e, respectively. From the figure, it is observed that the due to load switching, a very small change is observed in VU, however, more than 6% increase in THD of current is observed at the DG1 terminal. Thus, load switching may be misinterpreted as islanding by VU- and THD-based islanding detection method. The rates of change of active- and reactive power at both the DG terminals are shown in Fig. 5f and Fig. 5g, respectively. From the figures, it is observed that at DG1 terminal a change of −0.8 MW/s is observed in the active power within 50 ms of load switching.

Thus, using ROCOP method, load switching may be misinterpreted as islanding at DG1 terminal. The performance of ROCOPSI/ROCONSI method during the load switching is shown in Fig. 5h and Fig. 5i, respectively. A variation of more than 2 pu/s in ROCOPSI and a variation of 2 pu/s in ROCONSI is observed at the DG1 terminal due to load switching. Thus, load switching may be wrongly detected as islanding at the DG1 terminal by ROCOPSI/ROCONSI technique. The performance of the ROCOPAD-based islanding detection technique is shown in Fig. 5j. From the figure, it is observed that the rate of change of phase angle difference at the DG1 terminal does not show any significant change. Thus, the performance of the ROCOPAD technique seems to be robust due to load switching.

The performances of the passive islanding detection methods demonstrated for the islanding and non-islanding cases above are summarized in Table 1. From the table, it is clear that all the methods are able to detect the islanding causing large power mismatch. However, all the methods except ROCONSI/ROCOPSI and ROCOPAD

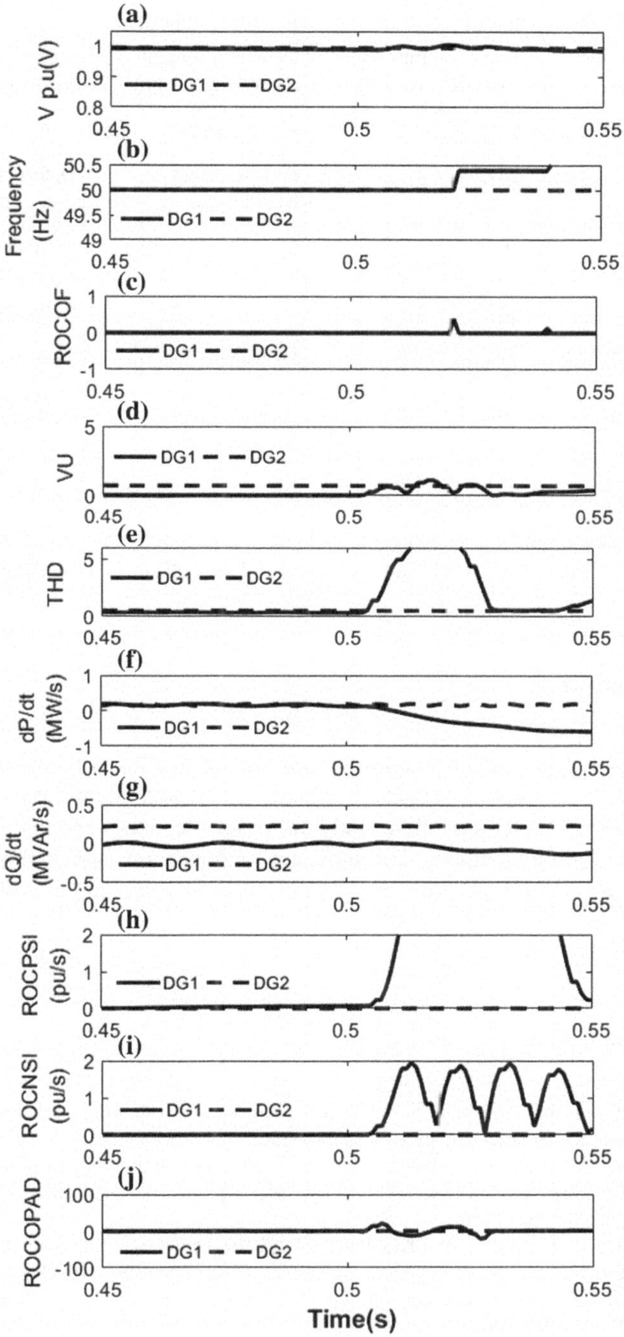

Fig. 5 Performance comparison during load switching, **a** voltage, **b** frequency, **c** rate of change of frequency, **d** voltage unbalance, **e** total harmonic distortion, **f** rate of change of active power, **g** rate of change of reactive power, **h** rate of change of positive-sequence current, **i** rate of change of negative-sequence current, and **j** rate of change of phase angle difference

Table 1 Performance comparison of passive islanding detection methods

<table>
<tr><th>Islanding detection techniques</th><th>Detection time (ms) (islanding causing large power mismatch)</th><th>Detection time (ms) (islanding causing small power mismatch)</th><th>Large load switching (non-islanding case)</th></tr>
<tr><td>UV/OV [5]</td><td>25</td><td rowspan="5">Fails to detect islanding</td><td rowspan="2">Detects as islanding (false operation)</td></tr>
<tr><td>UF/OF [6]</td><td>25</td></tr>
<tr><td>ROCOF [6]</td><td>20</td><td>Remains silent</td></tr>
<tr><td>ROCOP [20]</td><td>50</td><td rowspan="3">Detects as islanding (false operation)</td></tr>
<tr><td>VU and THD [7]</td><td>20</td></tr>
<tr><td>ROCONSI/ROCOPSI [15]</td><td>20</td><td>Detects in 25 ms</td></tr>
<tr><td>ROCOPAD [10]</td><td>30</td><td>Fails to detect islanding</td><td>Remains silent</td></tr>
</table>

fail to detect the islanding causing small power mismatch. Again, all the methods except ROCOF and ROCOPAD detect large load switching as islanding.

4 Conclusion

This paper presents a comparative study on some important passive islanding detection methods by simulating numerous islanding and non-islanding cases through PSCAD/EMTDC. The comparative study discusses the merits and demerits of each of the passive islanding methods. This indicates for further research in this area for the development of much more efficient islanding detection technique with negligibly small NDZ.

References

1. Li C, Cao C, Kuang Y, Zeng L, Fang B (2014) A review of islanding detection methods for microgrid. Renew Sustain Energy Rev 35:211–220
2. Xu W, Zhang G, Li C, Wang W, Wang G, Kliber J (2007) A power line signaling based technique for anti-islanding protection of distributed generators—part i: scheme and analysis. IEEE Trans Power Del 22(3):1758–1766
3. Chiang WJ, Jou HL, Wu JC, Wu KD, Feng YT (2010) Active islanding detection method for the grid-connected photovoltaic generation system. Electr Power Syst Res 80(4):372–379
4. IEEE Standard for Interconnecting Distributed Resources With Electric Power Systems, IEEE Standard P1547, June 2003
5. Zeineldin HH, Kirtley JL (2009) Performance of the OVP/UVP and OFP/UFP method with voltage and frequency dependent loads. IEEE Trans Power Del 24(2):772–778
6. Ye Z, Kolwalkar A, Zhang Y, Du P, Walling R (2004) Evaluation of anti-islanding schemes based on nondetection zone concept. IEEE Trans Power Electron 19(5):1171–1176

7. Jang SI, Kim KH (2004) An islanding detection method for distributed generations using voltage unbalance and total harmonic distortion of current. IEEE Trans Power Deliv 19(2):745–752
8. Danandeh A, Seyedi H, Babaei E (2012) Islanding detection using combined algorithm based on rate of change of reactive power and current THD techniques. In: Proceedings of Asia-Pacific power energy engineering conference, pp 1–4
9. Samui A, Samantaray SR (2011) Assessment of ROCPAD relay for islanding detection in distributed generation. IEEE Trans Smart Grid 2(2):391–398
10. Yin J, Diduch CP, Chang L (2008) Islanding detection using proportional power spectral density. IEEE Trans Power Del 23(2):776–784
11. Bejmert D, Sidhu TS (2014) Investigation into islanding detection with capacitor insertion-based method. IEEE Trans Power Del 29(6):2485–2492
12. Liu D, Aljankawey A, Diduch C, Chang L, Su J (2015) Passive islanding detection approach based on tracking the frequency-dependent impedance change. IEEE Trans Power Del 30(6):2570–2580
13. Liu N, Aljankawey AS, Diduch CP, Chang L, Su J, Mao M A new impedance-based approach for passive islanding detection scheme. In: Proceedings of 4th IEEE International Symposium on Power Electronics Distribution Generation Systems, July 2013, pp 1–7
14. Maheshwari RP, Bhalja BR, Sareen K (2016) Universal islanding detection technique based on rate of change of sequence components of currents for distributed generations. IET Renew Power Gener 10(2):228–237
15. Samui A, Samantary SR (2013) Wavelet singular entropy-based islanding detection in distributed generation. IEEE Trans. Power Del. 28(1):411–418
16. Ray PK, Kishor N, Mohanty SR (2012) Islanding and power quality disturbance detection in grid-connected hybrid power system using wavelet and S-transform. IEEE Trans Smart Grid 3(3):1082–1094
17. Samantaray SR, Babu C, Dash PK (2011) Probabilistic neural network based islanding detection in distributed generation'. Electric Power Comp Syst 39(3):191–203
18. El-Arroudi K, Joos G, Kamwa I, McGillis DT (2007) Intelligent based approach to islanding detection in distributed generation. IEEE Trans Power Del 22(2):828–835
19. Patil P, Kulkarni SU (2016) A novel method for islanding detection of distribution system with distributed generator. Int J Sci Res Pub 6(5)
20. Katiraei F, Iravani MR, Lehn PW, Motors SI (2005) Micro-grid autonomous operation during and subsequent to islanding process. Electronics 20(2):248–257

Image Segmentation by Fuzzy Edge Detection and Region Growing Technique

Amitab Khwairakpam, Ruhul Amin Hazarika and Debdatta Kandar

Abstract Image segmentation is one of the most important steps in computer vision and commonly used for applications like object detection, object recognition, etc. Some of the frequently used image segmentation techniques include region splitting and merging, k-means clustering, and region growing. Region growing can effectively segment the region of interest. However, due to lack of edge information, region growing process may terminate at the incorrect region boundary. To overcome the termination problem, edge detection and region growing techniques can be combined. In this paper, a new image segmentation technique is proposed. The proposed technique has two basic stages; the first stage detects edges and in the second stage, region growing technique is applied. We have used fuzzy logic to decide edge point in the image. The fuzzy membership function's parameters are determined by genetic algorithm. In the second stage, seeded region growing technique is applied at the edge detected image, where the seed point and the threshold value for comparing neighboring pixels are computed by the algorithm. The performance of the proposed technique is compared with some of the commonly used existing segmentation techniques based on segmentation accuracy and sensitivity, and it is found that the result of the proposed technique yields better result.

Keywords Image segmentation · Edge detection · Fuzzy logic · Genetic algorithm · Region growing

---

A. Khwairakpam (✉) · R. A. Hazarika
Department of Information Technology, North-Eastern Hill University, Shillong, India
e-mail: khamitab@gmail.com

D. Kandar
Department of Electrical and Electronic Engineering Technology, University of Johannesburg, Johannesburg, South Africa

© Springer Nature Singapore Pte Ltd. 2019
V. Nath and J. K. Mandal (eds.), *Proceedings of the Third International Conference on Microelectronics, Computing and Communication Systems*,
Lecture Notes in Electrical Engineering 556, https://doi.org/10.1007/978-981-13-7091-5_5

1 Introduction

In computer vision, image segmentation is the process of partitioning an image into multiple segments where each segment contains the pixels of the image which shares some common properties [1]. The main goal of image segmentation is to achieve dimensional reduction so that important features of an image can be easily processed. Moreover, it also helps to improve the pictorial information for human interpretation. There are various applications of image segmentation. In medical science, segmentation helps in analyzing the image and provides clues to the radiologist to detect the important structures [1]. Segmentation helps in analyzing of remote sensing image from satellites to identify and measure the region of interest (ROI).

Edges play an important role in image segmentation. An edge in an image is a set of pixels where there is a significant change in pixel intensity values [2]. Edge separates a region from another; it represents the boundary of an object in the image. Fuzzy edge detection [3] is one of the techniques used for identifying edge pixels of an image using fuzzy rules. Fuzzy logic is a computing approach based on "degrees of truth" rather than the usual "true" or "false" (1 or 0). Membership value determines the degree of belongingness. Sometimes because of the lack of sufficient information about a pixel, it is difficult to decide whether it is an edge pixel or not. Fuzzy logic technique improves the accuracy of detecting edge pixel by determining membership values for each pixel [4]. Again, the membership values are determined by using the membership function. To determine the parameters of the membership function, genetic algorithm (GA) can be used [5]. The genetic algorithm is a technique that imitates the process of natural selection.

Region growing [6] is one of the region-based image segmentation techniques where the pixels are compared with their neighborhood pixels and the similar pixels are merged together as one region. But, the major drawback of region growing algorithm is that sometimes it keeps growing beyond the boundary of the images and because of this, the accuracy could be adversely affected. For example, if one uses region growing technique only, the lack of edge information would terminate the region growing process at the wrong place. If the similarity criteria were too strict, many false edges would be generated [7]. In other words, the region growing process may not stop at the contour of the object. In order to correctly segment an image, combination of region growing and edge detection techniques can be used.

In this paper, a new image segmentation technique is proposed which is based on the fuzzy edge detection and the region growing approach. The edges are first identified by using fuzzy logic, which is followed by the region growing approach. The region growing process will be terminated at the edges. In the proposed technique, the seed point for region growing and the threshold value of the similarity criteria are chosen automatically.

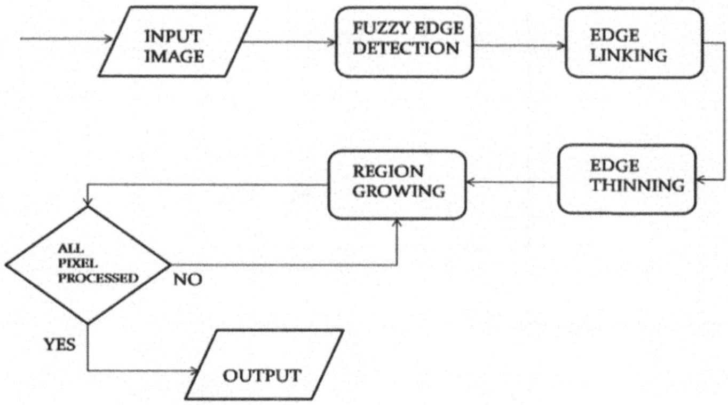

Fig. 1 Block diagram of the proposed technique

2 Proposed Image Segmentation Technique

The proposed image segmentation technique consists of four steps; fuzzy edge detection, edge linking, edge thinning, and region growing as shown in Fig. 1. The input image can be a color image or grayscale image. If the input is a color image, it is converted to gray scale for further processing.

2.1 Fuzzy Edge Detection

There are various edge detection techniques available in the literature [8, 9]. We have used a fuzzy-based technique [10] to detect the edge points. This method performs logical reasoning based on approximations rather than crisp values. Hence, the technique reduces the complexity of problems where fixed values cannot be predicted [11]. The followings are the steps for edge detection using fuzzy technique:

(1) *Obtain image gradient*: The first step is to find the gradient [12] values for each pixel in the horizontal (X) and vertical (Y) directions. An image gradient is a directional change in the intensity values. The gradient value of a pixel ("p") in X-direction is calculated by convolving "p" and its immediate neighbor with a gradient filter mask $[-1\ 1]$. Similarly, to calculate the gradient value of "p" in Y-direction, we have considered a gradient filter mask $[-1\ 1]^T$. The image formed by gradient value of all the pixels is the gradient image.

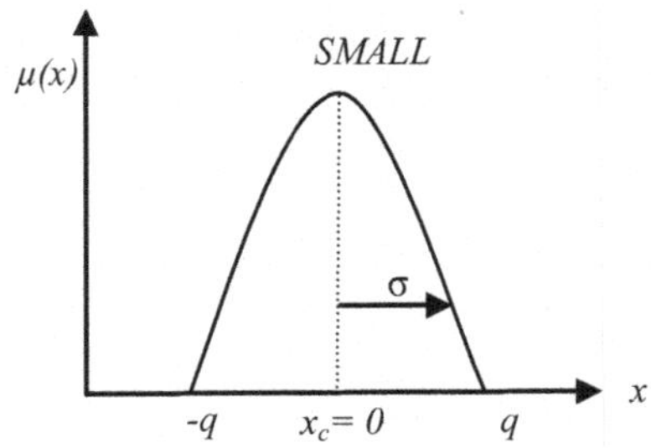

Fig. 2 Gaussian membership function

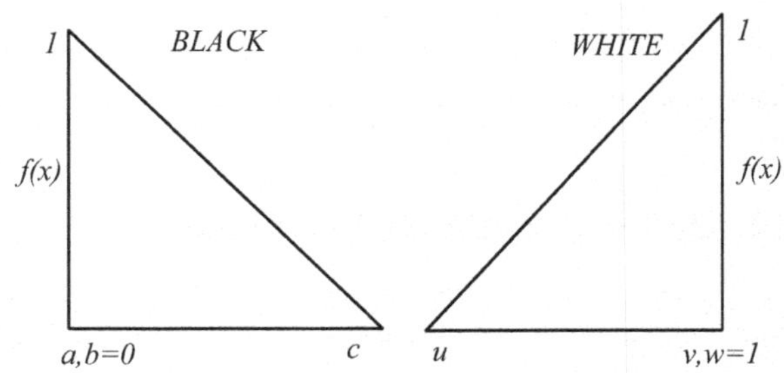

Fig. 3 Triangular membership function for BLACK and WHITE

(2) *Define Fuzzy rules*: Fuzzy rule allows dealing complex problem in a simpler way by using linguistic terms [13]. Following are the fuzzy rules used for identifying edge pixels:

Rule 1: "If Ix is SMALL and Iy is SMALL then Io is WHITE".
Rule 2: "If Ix is not SMALL or Iy is not SMALL then Io is BLACK".

In the above rules, Ix is the gradient value in X-direction, Iy is the gradient value in Y-direction, and Io is the output. The linguistic terms SMALL, WHITE, and BLACK are fuzzy sets defined by using membership functions. The fuzzy set SMALL is defined by a Gaussian membership function [14] and the fuzzy sets WHITE and BLACK are defined by two triangular membership functions [15]. The Gaussian membership function is defined in (1) and graphically shown in Fig. 2.

$$\mu(x) = e^{\frac{-(x-x_C)^2}{2\sigma^2}} \tag{1}$$

where "x_c" is the center point and "σ" is width.

If the gradient value (G) for a pixel is equal to x_c, then it belongs to the fuzzy set SMALL with a degree of 1, if $-q < G < q$, then it will partially belong to SMALL and if $-q > G > q$, then the degree of membership will be 0.

The triangular membership function [16] is defined in (2) and the corresponding graphical representation is shown in Fig. 3.

$$f(x) = \begin{cases} 0, x \leq a \\ \frac{x-a}{b-a}, a \leq x \leq b \\ \frac{c-x}{c-b}, b \leq x \leq c \\ 0, c \leq x \end{cases} \tag{2}$$

where a, b, and c define the foot of the triangle.

If the membership value of Io in BLACK is 0, then it belongs to the fuzzy set BLACK with a degree of 1. If $0 < Io < c$, then it will partially belong to BLACK and if $0 > Io > c$, then the degree of membership will be 0. If the membership value of Io in WHITE is 1, then it belongs to the fuzzy set WHITE with a degree of 1. If $u < Io < 1$, then it will partially belong to WHITE and if $u > Io > 1$, then the degree of membership will be 0.

(3) *Finding the fuzzy membership function parameters*: The performance of the fuzzy edge detection depends on the parameters used for defining the fuzzy membership function. These parameters can be accurately determined by using GA [5]. To produce a better approximation to the solution, GA applies the principle of survival of the fittest. The individuals whose fitness values are greater than others are chosen to produce the next generation [17].

The fuzzy edge detection uses four parameters c, u, σ_1 (for Ix), and σ_2 (for Iy) to define SMALL in (1), WHITE and BLACK in (2). These parameters are determined by the GA. In GA [18], individuals are expressed by binary strings of 0's and 1's. A population is a group of individuals; initially, individuals of the population are randomly encoded. In the reproduction phase, a fitness value is assigned to each individual determined by an objective function. The objective function checks the accuracy of the output (edge detected by fuzzy Rule 1 and Rule 2) by comparing with its corresponding ground truth image. The fittest individuals having high accuracy values survive and reproduce and the individuals having less accuracy value die. The process will terminate when the termination criterion is satisfied. The subsequent new generation is produced after performing the crossover and mutation operation on the individuals. In the crossover operations, the individuals are divided into two segments and the segments are swapped to form a new pair of individuals. By using mutation, the biological evolution is simulated. To mutate an individual, randomly choose the point to mutate and complement the bit in the encoded binary string.

2.2 *Edge Linking*

After detecting the edges, there may be broken edges. Next step is to link the broken edges. For edge linking, we have used morphological dilation [19] operation. Dilation increases the sizes of objects, filling in holes, and broken areas by comparing with the structure element.

Structure elements (SE) are small sets or sub-images used to interact with the original image. The SE can be of any size and of any shape. We have considered 2 × 2 square-shaped SE. Suppose A is a binary image in which "1" represents edges, and S is the SE, the dilation operation is discussed in the following example:

$$A = \begin{matrix} 0\,1\,0 \\ 1\,0\,0 \\ 0\,0\,0 \end{matrix} \text{ and } S = \begin{matrix} 1\,1 \\ 0\,0 \end{matrix}$$

Let A1 = {(0, 1), (1, 0)} be the position of pixel having value "1" in A, and S1 = {(0, 0), (0, 1)} in S. Next step is to add A1 and S1.

$$\begin{aligned} (0, 1) + (0, 0) &= (0, 1) \\ (0, 1) + (0, 1) &= (0, 2) \\ (1, 0) + (0, 0) &= (1, 0) \\ (1, 0) + (0, 1) &= (1, 1) \end{aligned}$$

The result is A2 having pixel value "1" in four positions

$$A2 = \begin{matrix} 0\,1\,0 \\ 1\,1\,0 \\ 1\,0\,0 \end{matrix}$$

If we compare the original image A with dilated image A2, it can be seen that it has some new edge points which may link the broken edges.

2.3 Edge Thinning

Next step is to make the edges in the image thin by removing extra edge points. For edge thinning, we have used morphological erosion [19] operation. Erosion generally decreases the sizes of objects and removes small unwanted image pixels. Suppose we have a binary image A3 in which "1" represents edge pixels and a structure element S2, the dilation operation is performed as follows:

$$A3 = \begin{matrix} 0\ 1\ 1 \\ 1\ 0\ 1 \\ 0\ 0\ 1 \end{matrix} \text{ and } S2 = \begin{matrix} 0\ 1 \\ 0\ 1 \end{matrix}$$

S2 moves through the whole image A3 by comparing the pixel values. S2 will be placed in the first position of A3 and will check if they are same with S2 or not. If they are same, the values will not change, otherwise all the values will become "0". In this example image, it is not matching for the first position so it becomes all "0".

$$A3 = \begin{matrix} 0\ 0\ 1 \\ 0\ 0\ 1 \\ 0\ 0\ 1 \end{matrix} \text{ and } S2 = \begin{matrix} 0\ 1 \\ 0\ 1 \end{matrix}$$

S2 will move on to the next position of A3 and will check the similarity and accordingly the pixel values will be changed. After processing the entire pixels in A3, the resulting image is A4.

$$A4 = \begin{matrix} 0\ 0\ 1 \\ 0\ 0\ 1 \\ 0\ 0\ 1 \end{matrix}$$

From the example, it can be seen that the resultant image has thinner edge.

2.4 Region Growing

The final step is to apply region growing technique on the edge thinned image presented in the following algorithm.

Algorithm 1 Region growing
Inputs: Edge thinned image I of size r × c, Largest connected component L in the image I
Output: Segmented image

```
1: Choose an initial seed point p(x, y).
        For i =1: r
        For j=1: c
                If I(xi, yj) ∈ L
                        p(x, y)=I(xi, yj)
                        Return p(x,y)
                End
        End
        End
2: Find the threshold value (T).
        Avg = (Sum of all the pixel intensity values in the
        image) ÷ (Total no. of pixels in the image)
        T = Avg ÷ 2

3: Growing the region by comparing with the neighboring pixels.
        neigb = [-1,0; 0,-1; 1,0; 0,1; -1,-1; 1,-1; 1,1; -1,1]
        count = 1
        reg_size = 1
        cl=number of pixels in L
        While (count<=cl)
        For j = 1:8
                xn =  x + neigb (j, 1)
                yn =  y + neigb (j, 2)
                dist = | I (yn, xn) - P (x, y)|
                If (dist<=T)
                        reg_size = reg_size+1
                End
                count = count+1
        End
        End
```

The proposed region growing technique determines a seed point (non-edge) within the largest region in the edge detected image. And the seed grows, the region growing process stops at the edge points as the difference between the seed and the edge point is large. The proposed technique segments the largest region in the image which is generally the background region. Subtracting the background from the image gives the segmented foreground, i.e., ROI.

3 Experimental Results

The proposed image segmentation technique is implemented in two phases; training and testing. Training phase determines the four membership parameters (c, u, $\sigma1$, and $\sigma2$) associated with the fuzzy sets SMALL, WHITE, and BLACK using GA.

For training, we have taken input images and its corresponding ground truth images from the Berkeley Segmentation Dataset (BSD) [20, 21]. The population size is initialized to 10; probability of mutation is set to 0.05. The optimum value of the parameters is found after 100 generations. In every generation, the algorithm checks the accuracy (3) of individuals to find the fitness value. The individuals which give the highest accuracy are selected to reproduce the next generation. In the testing phase, the value of parameters determined by the GA is used for defining the fuzzy membership functions (1) and (2). To validate the performance of the proposed segmentation technique, we have taken the standard ground truth images [20] from BSD and compared the result with several existing segmentation techniques such as region growing, region splitting–merging, and K-means algorithm.

In K-means clustering [22]-based image segmentation technique, initially the whole image is divided into K number of clusters. The center pixel (centroid) for each cluster is determined and the rest pixels of the image are compared with the centroids and assigned to the clusters having the most similar intensity value. The process is repeated by redetermining the centroids. For performance evaluation, we have taken K as 2. Seeded region growing [6, 23] is a region-based image segmentation technique where initially seed pixel is chosen and all the neighboring pixels of the seed point are compared. If a neighboring pixel is found to be similar to the seed pixel, then add the pixel in the same region. Newly added pixels will be the new seed points and the process will continue till all the pixels are being compared. For performance comparison, an interactive program was implemented where the users have to choose the seed point by clicking on the ROI.

Region splitting–merging technique [24] is an image segmentation technique where initially the whole image is considered as a single region (R). A predicate (P) is decided and P is applied in R. If R does not satisfy P then divide R into four equal parts. For each part of R, the same P will be again applied. If a part does not satisfy then that part will be subdivided into four subparts. The process stops when the predicate is satisfied. Then, all the regions satisfying P are merged together. For the comparative evaluation of the result, the P is defined as “The standard deviation of the region should be less than ‘10’ and the mean pixel intensity value of the region should be greater than ‘0’ and less than the average intensity of the image”.

The performance of the proposed technique and different segmentation techniques is compared by using accuracy and sensitivity [25].

Accuracy (3) gives the extent of true classification as compared to the total cases taken. It is defined as

$$\text{Accuracy} = (\text{TP} + \text{TN})/(\text{TP} + \text{TN} + \text{FP} + \text{FN}) \tag{3}$$

And sensitivity (4) is given by the extent of correctly segmented pixels to the total pixels which were actually deserved to be segmented. It is defined as

$$\text{Sensitivity} = (\text{TP})/(\text{TP} + \text{FN}) \tag{4}$$

Fig. 4 Original image (teddy.jpg)

Fig. 5 Ground truth image

where true positive pixels (TP) are the pixels which are common to ground truth and the segmented image. False positive pixels (FP) are the pixels which are the mistakenly segmented image (over segmented pixels). True negative pixels (TN) are the pixels which are correctly rejected (pixels which are not in ground truth as well as in the segmented image). False negative pixels (FN) are the pixels which are mistakenly rejected (pixels which are in the ground truth image but not in segmented image).

The performance of different segmentation techniques is shown in Table 1. We have used MATLAB for implementing the algorithms. As we can see from the performance comparison table, the accuracy and the sensitivity are highest for the proposed technique for all the above testing images. An original image (teddy.jpg)

Table 1 Performance comparison

Name of the image	Name of the algorithm	True positive value	False positive value	True negative value	False negative value	Accuracy	Sensitivity
Teddy.jpg	K-means	18770.00	648.00	88221.00	5393.00	0.95	0.78
	Region growing	21332.00	533.00	88336.00	2831.00	0.97	0.88
	Region splitting–merging	20378.00	7910.00	80959.00	3785.00	0.90	0.84
	Proposed method	23993.00	2622.00	86247.00	170.00	0.98	0.99
Fullmoon.jpg	K-means	3395.00	29.00	144920.00	6356.00	0.96	0.35
	Region growing	9589.00	85.00	144864.00	162.00	0.99	0.98
	Region splitting–merging	1198.00	592.00	144357.00	8553.00	0.94	0.12
	Proposed method	9751.00	1025.00	143924.00	0.00	0.99	1.00
Aeroplane.jpg	K-means	5481.00	3205.00	141736.00	3979.00	0.95	0.58
	Region growing	7390.00	245.00	144696.00	2070.00	0.99	0.78
	Region splitting–merging	3115.00	1619.00	143322.00	6345.00	0.95	0.33
	Proposed method	9278.00	1636.00	143305.00	182.00	0.99	0.98
Aeroplane2.jpg	K-means	23377.00	39268.00	91008.00	748.00	0.74	0.97
	Region growing	22325.00	1800.00	126254.00	4022.00	0.96	0.85
	Region splitting–merging	7027.00	6108.00	124168.00	17098.00	0.85	0.29
	Proposed method	23414.00	5128.00	125148.00	711.00	0.96	0.97

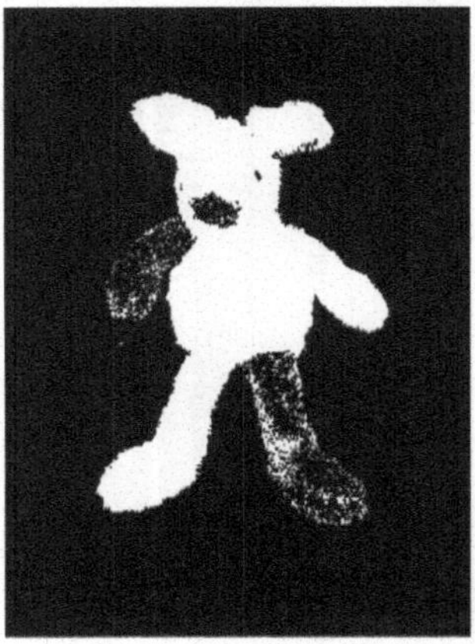

Fig. 6 Segmentation using K-means algorithm

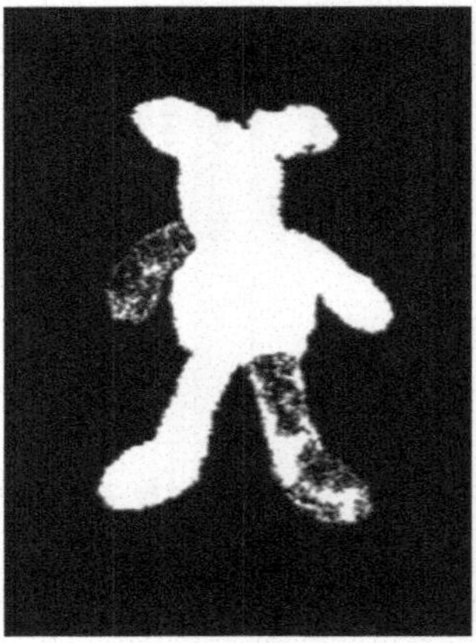

Fig. 7 Segmentation using region growing

and the corresponding ground truth images are presented in Figs. 4, 5. The visual results of the implemented algorithms are shown in Figs. 6, 7, 8, and 9.

From the visual result in Figs. 6, 7, 8, and 9, it is observed that the segmentation result of the proposed algorithm, Fig. 9, is very similar to the ground truth image Fig. 5.

4 Conclusion

An image segmentation technique based on fuzzy edge detection and region growing is proposed. The proposed technique is comparatively evaluated with K-means, region growing, and region splitting–merging based on segmentation accuracy and sensitivity. We have used the test images from BSD. It is observed that K-means algorithm is not able to differentiate between the ROI and the background as it is based on grouping similar pixel intensity value. Sometimes the region growing

Fig. 8 Segmentation using region splitting–merging algorithm

Fig. 9 Segmentation using proposed technique

technique keeps on growing beyond the region as the region growing process may terminate at the wrong place. The major drawback of region splitting and merging technique is to define appropriate predicate P which works for different images. The proposed technique improves the region growing technique by including fuzzy edge detection which defines the accurate terminating criteria of the region growing process. The comparative analysis table and the visual representation show that the proposed algorithm has more accuracy and sensitivity value, which implies to be a better segmentation algorithm among the four abovementioned algorithm.

References

1. Kaur D, Kaur Y (2014) Various image segmentation techniques: a review. Int J Comput Sci Mob Comput 3(5):809–814
2. Senthilkumaran N, Rajesh R (2009) Edge detection techniques for image segmentation–a survey of soft computing approaches. Int J Recent Trends Eng 1(2):250–254

3. Kaur R, Singh M, Singh B (2016) Comparative analysis of color edge detection techniques based on fuzzy logic. An Int J Eng Sci 17:119–126
4. Verma OP, Parihar AS (2017) An optimal fuzzy system for edge detection in color images using bacterial foraging algorithm. IEEE Tran Fuzzy Syst 25(1):114–127
5. Khwairakpam A, Kandar D, Paul B (2017) Noise reduction in synthetic aperture radar images using fuzzy logic and genetic algorithm. Microsyst Technol 1–10
6. Zhang X, Li X, Feng Y (2015) A medical image segmentation algorithm based on bi-directional region growing. Int J Light Electron Optics 126(20):2398–2404
7. Jeevakala S, Rangasami R (2018) A novel segmentation of cochlear nerve using region growing algorithm. Biomed Signal Process Control 39:117–129
8. Anitha U et al (2016) Comparison of standard edge detection techniques along with morphological processing and pseudo coloring in sonar image. In: International conference on emerging trends in engineering, technology and science (ICETETS). IEEE
9. Muthukrishnan R, Radha M (2011) Edge detection techniques for image segmentation. Int J Comput Sci Inf Technol 3(6):259–267
10. Alimohammadi M, Pourdeilami J, Pouyan AA (2013) Edge detection using fuzzy inference rules and first order derivation. In: 13th Iranian conference on fuzzy systems (IFSC). IEEE
11. Haq I, Anwar S, Shah K, Khan MT, Shah, SA (2015) Fuzzy logic based edge detection in smooth and noisy clinical images. PloS one 10(9)
12. Neggers J, Blaysat B, Hoefnagels JPM, Geers MGD (2016) On image gradients in digital image correlation. Int J Numer Meth Eng 105(4):243–260
13. Chaudhari S, Patil M, Bambhori J (2014) Study and review of fuzzy inference systems for decision making and control. Am Int J Res Sci Technol Eng Math 88–92
14. Viattchenin DA, Tati R, Damaratski A (2013) Designing Gaussian membership functions for fuzzy classifier generated by heuristic possibilistic clustering. J Inf Organ Sci 37(2):127–139
15. Barua A, Mudunuri LS, Kosheleva O (2013) Why trapezoidal and triangular membership functions work so well: towards a theoretical explanation. J Uncertain Syst 8:1–5
16. Ali OAM, Ali AY, Sumait BS (2015) Comparison between the effects of different types of membership functions on fuzzy logic controller performance. Int J Emerg Eng Res Technol 3(3):76–83
17. Lin WC, Wang JW (2018) Edge detection in medical images with quasi high-pass filter based on local statistics. Biomed Signal Process Control 39:294–302
18. Guo C, Yang X (2011) A programming of genetic algorithm in matlab7. 0. Mod Appl Sci 5(1):230–235
19. Soille P (2013) Morphological image analysis: principles and applications. Springer Science & Business Media
20. Li H, Cai J, Nguyen TNA, Zheng J (2013) A benchmark for semantic image segmentation. In: IEEE ICME
21. Martin D, Fowlkes C, Tal D, Malik J (2001) A database of human segmented natural images and its application to evaluating segmentation algorithms and measuring ecological statistics. In: IEEE international conference on computer vision, eighth 2:416–423
22. Celebi ME, Kingravi HA, Vela PA (2013) A comparative study of efficient initialization methods for the k-means clustering algorithm. Expert Syst Appl 40(1):200–210
23. Kandwal R, Kumar A, Bhargava S (2014) Existing image segmentation techniques. Int J Adv Res Comput Sci Softw Eng 4(4):153–156
24. Sonka M, Hlavac V, Boyle R (2014) Image processing, analysis, and machine vision. Cengage Learn
25. Irshad, Jaffery ZA (2015) Performance comparison of image segmentation techniques for infrared images. In: IEEE INDICON

Performance Analysis of Triband Elliptical Patch Microstrip Antenna for GPS and Radar Application

Mitchell S. Prajapati, Sunil Kumar Shesma and Abhishek Rawat

Abstract In this paper, we propose the design of triband microstrip patch antenna for 1.1870 GHz L-band, 2.1670 and 3.5580 GHz S-band application. This antenna consists of elliptical patch with major axis of 47 mm and axial ratio of 0.6 with desired gain and bandwidth. This small microstip patch antenna gives very good response for three bands with VSWR value less than 2. The elliptical patch antenna consists of single patch with multiple applications and increased bandwidth. This antenna can be used for GPS, satellite mobile phones, and some RADAR applications. The HFSS 15.0 is used for design and simulation of this triband antenna.

Keywords Elliptical patch · Antenna design · Triband antenna · HFSS

1 Introduction

For advanced wireless applications, microstrip patch antennas have become popular choice due to its several advantages as it is small in size, low in weight, easy to design, fabricate and install, and low in cost. In present advanced wireless communication systems, we need multiband antennas to communicate with multiple applications simultaneously [1]. For advanced wireless applications, bandwidth enhancement is the key challenge in microstrip antennas. Microstrip patch antenna is having a simple structure, and the substrate is sandwiched between ground plane and patch. Microstrip patch antennas radiate primarily because of the fringing fields between the ground plane and the patch edge. The efficiency of a microstrip antenna is affected

---

M. S. Prajapati (✉) · S. K. Shesma · A. Rawat
Department of Electrical Engineering, Institute of Infrastructure Technology Research and Management, Ahmedabad, India
e-mail: mishelprajapati@gmail.com

S. K. Shesma
e-mail: sunilshesma@gmail.com

A. Rawat
e-mail: arawat@iitram.ac.in

© Springer Nature Singapore Pte Ltd. 2019

V. Nath and J. K. Mandal (eds.), *Proceedings of the Third International Conference on Microelectronics, Computing and Communication Systems*, Lecture Notes in Electrical Engineering 556, https://doi.org/10.1007/978-981-13-7091-5_6

by patch size, geometry, substrate thickness, dielectric constant of substrate, feeding techniques, and location. For good antenna performance, a thick dielectric substrate having a low dielectric constant is desirable because this provides better efficiency, larger bandwidth, and better radiation [2]. But a thick dielectric substrate leads to a larger antenna size. In order to design a compact microstrip patch antenna, substrates with higher dielectric constants must be used which are less efficient and result in narrower bandwidth. Hence, according to the antenna application the parameters can be set to get desired output. Here, single patch and single feed are used to get three desired frequency band operations.

2 Elliptical-Shaped Microstrip Patch Antenna Design

In microstrip patch antenna, the patch can be chosen as circular, square, rectangular, triangular, elliptical, or hexagon in shape with some modifications. In presented design, elliptical patch is used to increase bandwidth and gain. The elliptical patch antennas have several advantages like providing larger flexibility in the design with single feed and give circular polarization [3]. Here, elliptical microstrip patch antenna with a as semi-major axis length and b as semi-minor axis length is designed. Circular polarization is easily achieved by elliptical patch. The formulas for calculation of resonance frequency for odd and even modes are available but in reverse case, from specified resonance frequency, formulas for calculation of major and minor axes are not available. Here, the elliptical patch with major axis of 47 mm and axial ratio of 0.6 is designed, and the substrate dimensions are 120 mm × 120 mm. The geometry of elliptical patch microstrip antenna is shown in Fig. 1.

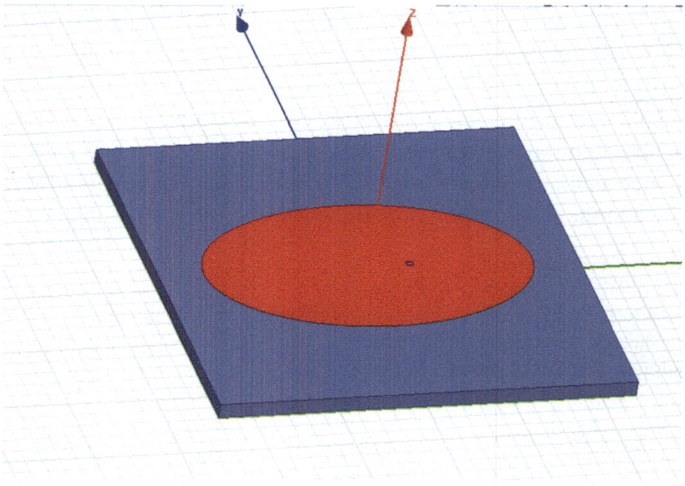

Fig. 1 Design of elliptical patch microstrip antenna using HFSS

Table 1 Dimensions and specification of proposed antenna

Sr. no.	Antenna parameters	Value
1	Operating frequency in GHz	1.1870, 2.1670, 3.5570
2	Substrate material	Rogers/RTduroid 5880 (tm)
3	Dielectric constant	2.2
4	Loss tangent	0.0009
5	Patch dimension	Major axis: 47 mm
6	Substrate dimension	120 mm × 120 mm

Here, the substrate material Roger/duroid 5880 is used, with dielectric constant of 2.2. The efficiency of a microstrip antenna depends upon patch size, shape, substrate thickness, dielectric constant of substrate, feed point type, and its location. Microstrip patch antennas can be feed by different feeding techniques. These methods can be classified into two categories: contacting and non-contacting. In the contacting method, the RF power is fed directly to the radiating patch using a connecting element such as a coaxial cable or microstrip line. In non-contacting scheme, electromagnetic field coupling is done to transfer power between the microstrip line and the radiating patch. There are most popular feed techniques used for the microstrip line, coaxial probe, aperture coupling, and proximity coupling. In this design, the coaxial feed is used. The dimensions for proposed antenna are given in following table (Table 1).

3 Results and Discussion

In this design, simulation result shows the excellent working on three resonance frequencies, 1.1870, 2.1670, and 3.5570 GHz. Important antenna characteristics like return losses, gain, VSWR, radiation pattern, and electric field distribution of patch and substrate are briefly expressed with their simulation results (Table 2).

3.1 Return Loss

Return loss is the difference between forward and reflected power, in dB, generally measured at the input to the coaxial cable connected to the antenna. For maximum power transfer, return loss must be as small as possible, so in dB it must be as large as negative number. Here, we got the value of return loss −41.6725 dB for 1.1870 GHz, −28.1794 dB for 2.1670 GHz, and −38.2886 dB for 3.5570 GHz, which shows the good performance of triband microstrip patch antenna (Fig. 2).

Table 2 Simulated result for proposed antenna

Sr. no.	Resonant frequency in GHz	Return loss in dB	VSWR	Gain in dB	Bandwidth in MHz	Application of band
1	1.1870	−41.6725	1.0166	2.570639	26.5	GPS, satellite mobile phones
2	2.1670	−28.1794	1.0812	3.540217	45.9	Radar
3	3.5570	−38.2886	1.0247	7.565323	23.64	Weather radar, surface ship radar, communication satellite

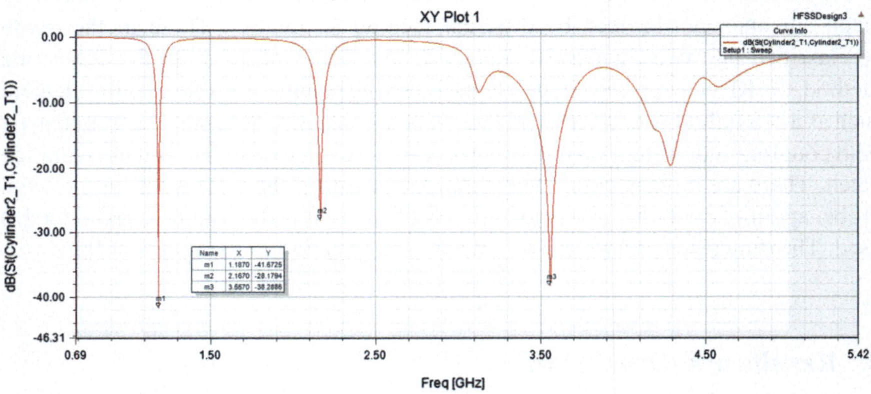

Fig. 2 Reflection coefficient versus frequency

3.2 *VSWR*

Voltage standing wave ratio shows the power reflected from the antenna. Standing waves represent power which is not accepted by the load and reflected back along the transmission line or feeder. Here, we got the value of VSWR like 1.0166, 1.0812, and 1.0247 dB which shows that less than 0.2% power is reflected back and also shows that a good matching of impedance is done here.

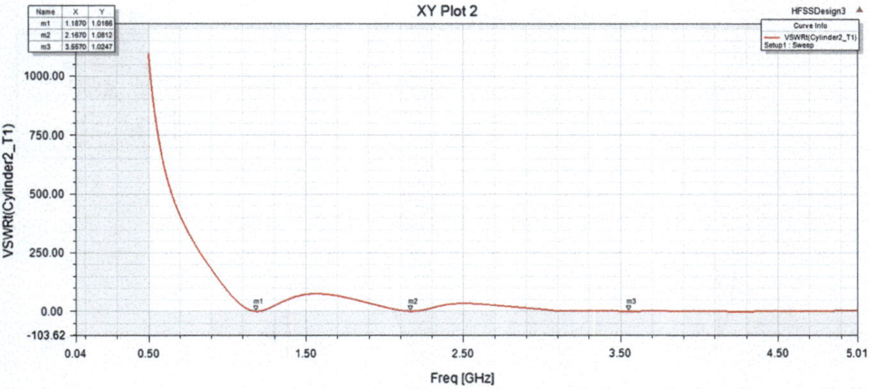

Fig. 3 VSWR plot

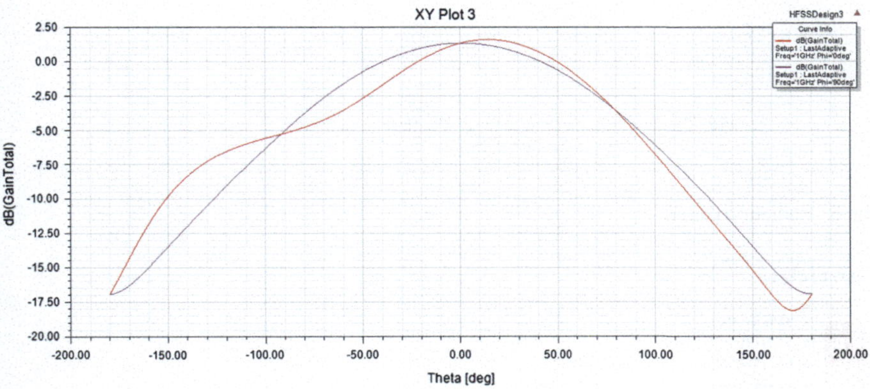

Fig. 4 Gain plot

3.3 Gain

Gain is the key parameter of an antenna. It is the combined measure of directivity and efficiency. For transmitting antenna, the gain describes how well the antenna converts input power into radio waves in a specified direction. Here, the values of gain are 2.570639, 3.540217, and 7.565323 dB for three desired frequencies (Figs. 3 and 4).

3.4 Radiation Pattern

The following Figs. 5 and 6 show the 2D and 3D radiation patterns for triband microstrip patch antenna.

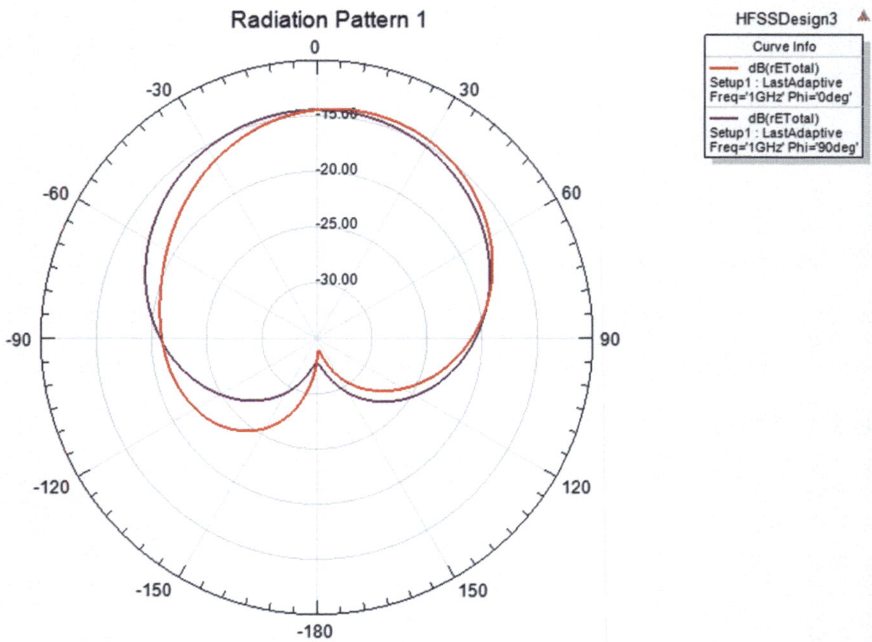

Fig. 5 Radiation pattern

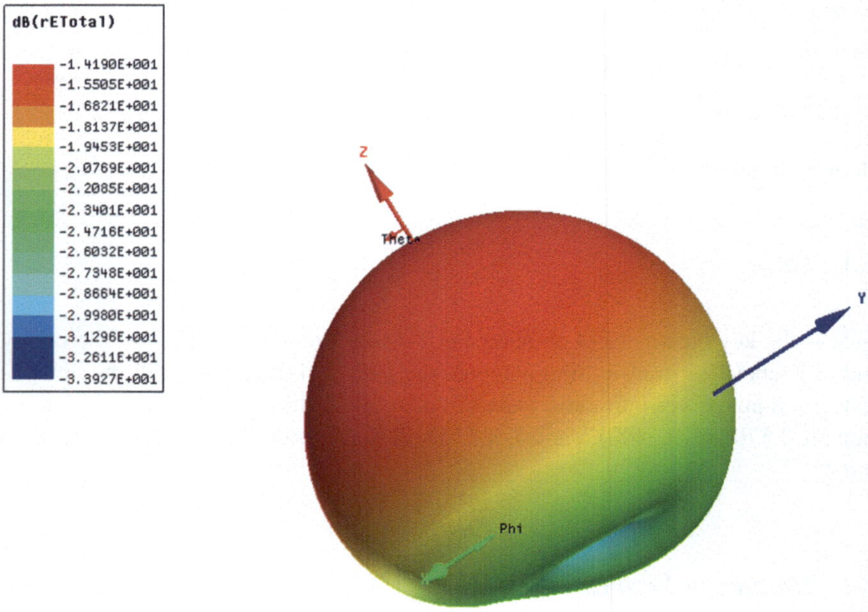

Fig. 6 3D radiation pattern

3.5 Electric Field Distribution

The following Figs. 7 and 8 represent electric field distribution across patch and ground. Here, we are getting that electric field distribution at patch is directly proportional to ground field.

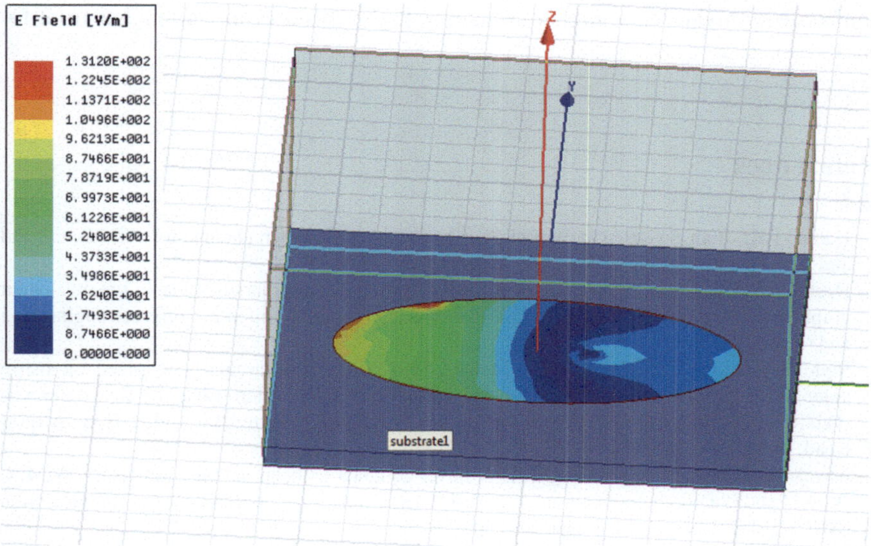

Fig. 7 Electrical field distribution across patch

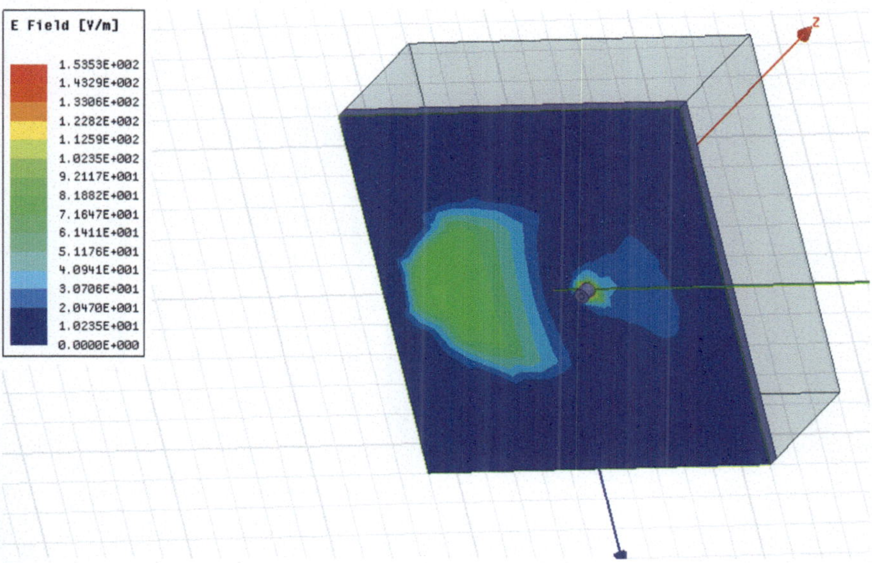

Fig. 8 Electrical field distribution across ground

4 Conclusion

In this design, the parameters of microstrip patch antenna are enhanced for triband applications using elliptical patch. This antenna can be used for multiple applications simultaneously and can also be modified to work as a reconfigurable antenna.

In the design of multiband frequencies, elliptical band is the most preferable antenna with higher gain, higher efficiency, higher bandwidth, and higher front-to-back ratio.

References

1. Hamad EKI, Badr S (2018) Design of multiband microstrip patch antenna for WiMax, C-band and X-band applications. Aswan Eng J (AswEJ)
2. Deshmukh AA, Zaveri P, Verma P, Agrawal S (2016) Circular polarized tunable square slot cut elliptical patch microstrip antenna. In: 2016 online international conference on green engineering and technologies (IC-GET)
3. Reddy BS, Kumar VS, Srinivasan VV, Mehta Y (2015) Dual band circularly polarized microstrip antenna for IRNSS reference receiver. In: IEEE international microwave and RF conference (IMaRC), Communication Systems Group, ISRO Satellite Centre, Bangalore, HAL Airport Road, India, PIN-560017
4. Nayak PB, Verma S, Kumar P (2014) A novel compact tri-band antenna design for WiMax, WLAN and bluetooth applications. In: 2014 twentieth national conference on communication (NCC). IEEE, Kanpur (Uttar Pradesh), pp 1–6

Performance Analysis of Block Matching Algorithms

Vivek Jain, Randheer, Rahul Priyadarshi and Ankush Thakur

Abstract To attain high compression ratio in videos, a scheme called as block matching algorithm is extensively used in innumerable coding principles. Numerous fast block matching algorithms have been recommended and industrialized. In block matching motion valuation, search patterns with altered shapes and center-biased features of motion vector dissemination have a huge impression on examining speed and eminence enactment. In this paper, we will examine three-step search, four-step search, and full search. Further simulation of all these three algorithms have been done using MATLAB tool and comparison of all three algorithms has been done which further reflects that the full search algorithm is performing fine associated with all three discussed algorithms in this paper with respect to PSNR value as well as computational intricacy.

Keywords Motion estimation · Block · Search pattern · Three-step search · Full search · Video coding

V. Jain
Electronics & Communication Engineering Department, Delhi Technological University, Delhi, India
e-mail: vivekjain1192@gmail.com

Randheer · R. Priyadarshi (✉) · A. Thakur
Electronics & Communication Engineering Department, National Institute of Technology, Hamirpur, HP, India
e-mail: rahul.glorious91@gmail.com

Randheer
e-mail: randheerbundela427@gmail.com

A. Thakur
e-mail: ankushlucky766@gmail.com

© Springer Nature Singapore Pte Ltd. 2019

V. Nath and J. K. Mandal (eds.), *Proceedings of the Third International Conference on Microelectronics, Computing and Communication Systems*, Lecture Notes in Electrical Engineering 556, https://doi.org/10.1007/978-981-13-7091-5_7

1 Introduction

Video coding is an imperative development in several multimedia uses. Moreover, to spatial redundancy, time-based redundancy shows a significant part in the communication of video frames. An operative and widespread process to lessen time-based redundancy is called Block Matching Algorithm (BMA), and motion valuation is the technique which is used to diminish the time-based redundancy. It practices the correlation between the consecutive frames to expect the content of frames. It has been extensively implemented in innumerable video coding standards [1]. As a result, fast and precise block-based search method is exceedingly required to guarantee much abridged handing out delay even though sustaining respectable remodeled image superiority. In motion estimation procedure, frame is separated into a number of nonoverlapping zones recognized as macro-blocks [1]. Every single macro-block has a typical size of 16 × 16. Generally, macro-block is reserved as square side which equals to 16 pixels and search parameter p equals to 7 pixels which are shown in Fig. 1. The difference among present frame and expected frame is computed in motion estimation [2]. Additionally to motion estimation, some supplementary info is required to specify any deviations in expectation procedure, called as motion compensation [3].

Motion estimate is also a progression to evaluate pixels of present frame with respect to reference frame. BMA is time-based redundancy exclusion method among two or more consecutive frames, which is an essential measure for motion-compensated video [4]. Frames are separated into systematic sized blocks called as macro-blocks. Block matching technique is to search for finest coordinated block from preceding frame, generally the first solitary frame, contained by a stationary size of search window [5, 6]. The finest match is mostly calculated by a cost function based on distortion measure like Mean Absolute Difference (MAD) [7, 8]. Because of high exhaustive calculation, FS algorithm that implements, examine all candidate blocks contained by search window comprehensively. A common mechanism of BMA is shown in Fig. 2.

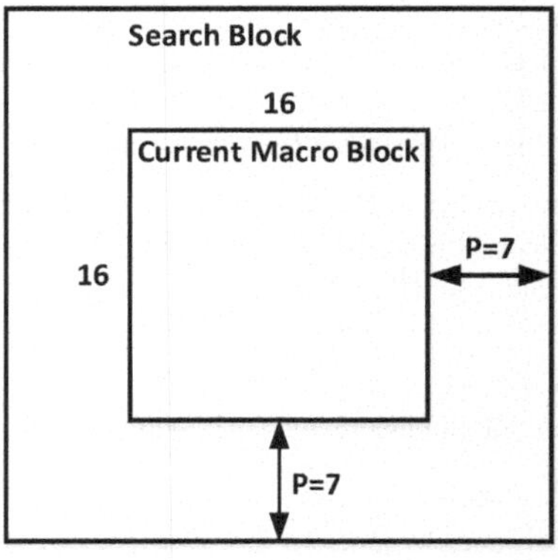

Fig. 1 Block matching macro-block

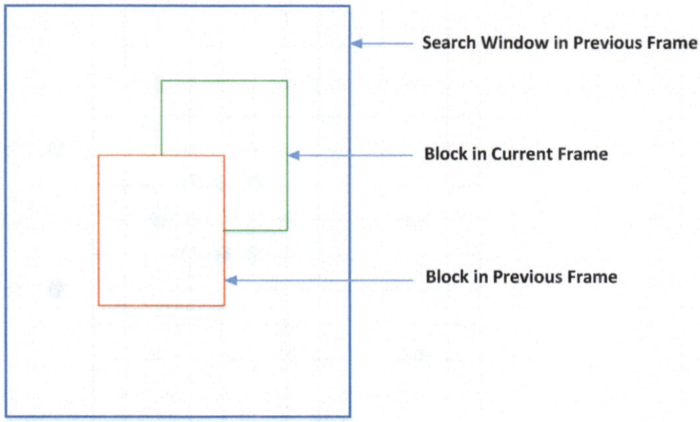

Fig. 2 Mechanism of BMA

Several fast motion estimation methods [9–11] have been projected to provide a faster estimate with parallel block falsification related to FS. The most recognized fast BMA are three-step search [12], four-step search [13], and the FS algorithm [14]. As distinguishing of center-biased motion vector dissemination that enthused many fast block matching algorithm in past few decades, over 85% of blocks may be considered as stationary.

This center-biased distinguishing can even be originated in fast motion arrangements. To exploit this occurrence, three-step search added eight center-adjacent blocks and presented an intermediate stop method to accomplish critical speedup for static blocks. Four-step searches also exploit midpoint-influenced belongings of motion vectors distribution by means of intermediate stop methods and lesser square search arrangement related to three-step search.

2 Block Matching Algorithms

2.1 Three-Step Search

This is the initial effort at fast BMAs [15]. It turns out to be widespread for the reason that of its straightforwardness, robust, and near-optimum performance.

It examines for finest motion vectors in a progression to reasonable search pattern [16]. It is a search method which generally uses three steps. While comparing to FS algorithm, it is having the huge saving factor of almost larger than a 100. Three-step search algorithm has fixed number of search step and it is 25. The main shortcoming of three-step search is that it practices an evenly allotted inspection point pattern in the first step that makes it incompetent to estimate of small motion. Initially, a fixed

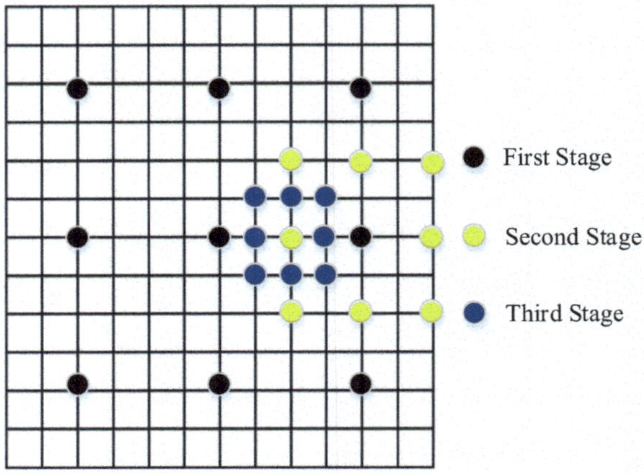

Fig. 3 Three-step algorithm search pattern

step size is selected, generally initially step size is considered as S = 4. Now, eight blocks from the center with step size S = 4 is considered and compared in the first step, and hence we find the best match.

Once best match is obtained then that point becomes center point and again step size is considered as S = 2 (now it became half of initial step size). Again eight blocks from the center with step size S = 2 is considered and compared in the second step and again we find the best match. Once best match is obtained then that point becomes center point and again step size is considered as S = 1 (now it became half of second step size). So, finally we stop at this point, and at this point it discovers position with smallest cost function and block at this position is finest match. Figure 3 shows three-step search algorithm to find the finest match.

Various advantages and disadvantages of three-step search algorithms are listed below:

Advantages

- Straightforwardness and optimum performance and
- Give best motion vectors.

Disadvantages

- Inefficient for the estimation of small motion.

2.2 *Four-Step Search*

Authors in [13, 17] initially proposed four-step search algorithm which is centered on real-domain image categorization's distinguishing of midpoint-biased motion and halfway stop techniques. This technique is comparable to three-step search but it has smaller step size compared to the three-step search. Because of smaller preliminary step size, this algorithm required four searching stages to touch periphery of a search

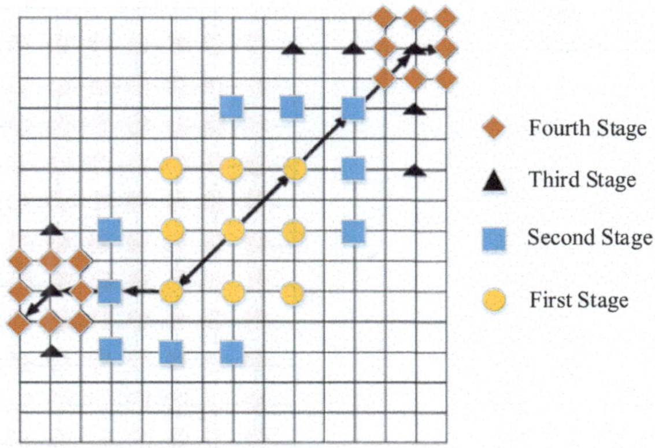

Fig. 4 Four-step algorithm search pattern

window with pixel size 7. Four-step search algorithm [18] generally uses a halfway stop method in its second and third step search stages.

Figure 4 displays the two different search paths for four-step search algorithm. It is clearly seen from the plot that left-side path required total of 25 points and in another hand right-side path required total of 27 points to reach the periphery of a search window. Four-step search algorithm generally uses four stages to reach boundary of window to find the best match.

Various advantages and disadvantages of three-step search algorithms are listed below:

Advantages

- More robust as compared with three-step search algorithm and
- PSNR points nearer to full search.

Disadvantages

- Processing time is longer compared to three-step search algorithm and is complex to implement.

2.3 *Full Search Algorithm*

FS algorithm is generally used in hardware operation of motion estimation [19] because of its effortlessness, reliability, and optimal consequence. The furthermost regularly used parameter to define best match for FS is MAD criterion [20].

This method is utmost computationally costly because it determines cost function at every single probable locality in search window. Therefore, it discovers best probable match and provides maximum PSNR among any BMA [21]. Figure 5 displays the search pattern for FS algorithm where one can easily find that the algorithm is searching all the candidate block in entire window to find the best match.

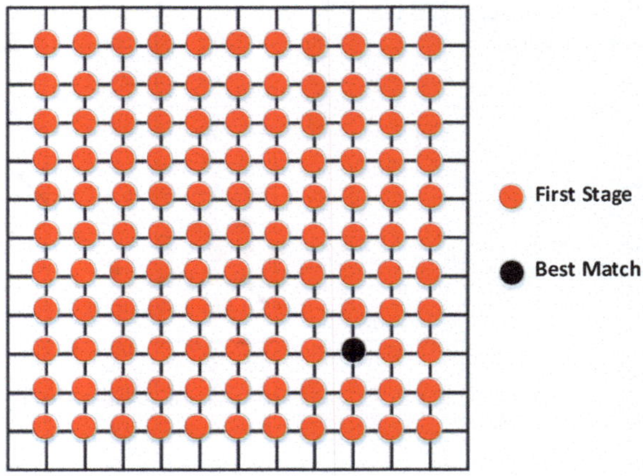

Fig. 5 Full search algorithm search pattern

Various advantages and disadvantages of FS algorithm are listed below:

Advantages

- More robust as compared with all existing BMA,
- Best search and more accurate result, and
- Best motion vector.

Disadvantages

- Computation complexity and slow processing and
- Costly and bit complex for hardware implementation.

3 Simulation Results and Analysis

Simulation has been done using MATLAB tool for all three algorithms and results have been compared and discussed to find which algorithms which are better in terms of PSNR value and computational intricacy. For simulation, there are 33 numbers of frames with size of block considered as 16 * 16. Error function is taken as MAD criterion. Simulation parameters are listed below in Table 1.

Table 1 Simulation parameters

Parameters	Values
Number of frames	33
Search parameters	7
Size of macro-block	16 * 16
Error function	MAD criterion

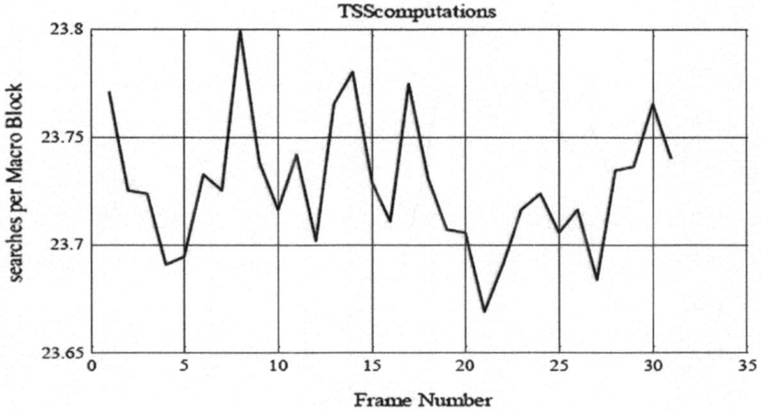

Fig. 6 Computation of three-step search

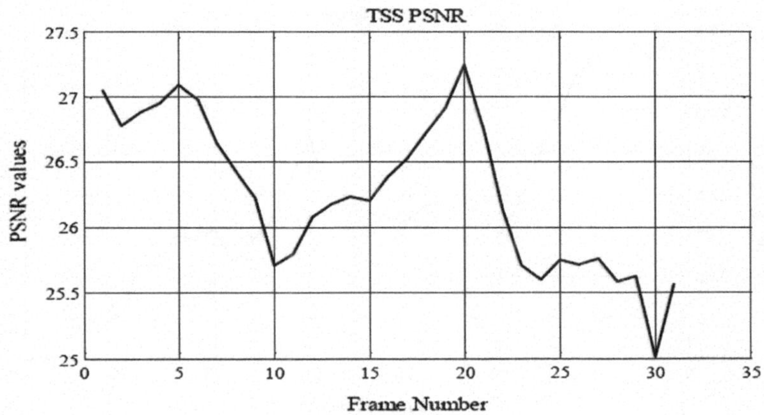

Fig. 7 PSNR of three-step search

Figures 6, 8, and 10 show the computational level of three-step search, four-step search, and FS algorithm, respectively, where it is clearly seen that it is more in case of FS algorithm compared to three-step search algorithm and four-step search algorithm but at the same time FS algorithm gives the best and most accurate results and it is one of the algorithm that is widely used in real-time hardware implementation. Figures 7, 9, and 11 show the PSNR plot for three-step search, four-step search, and FS algorithm, respectively, where it is clearly seen that PSNR value achieved in four-step search and FS algorithm is almost similar but higher than three-step search.

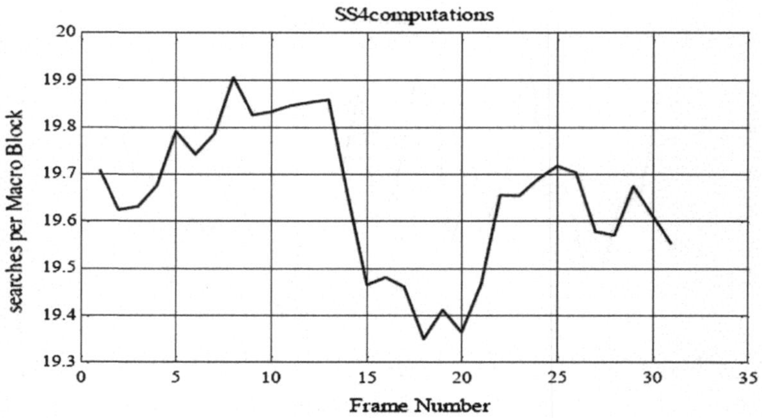

Fig. 8 Computation of four-step search

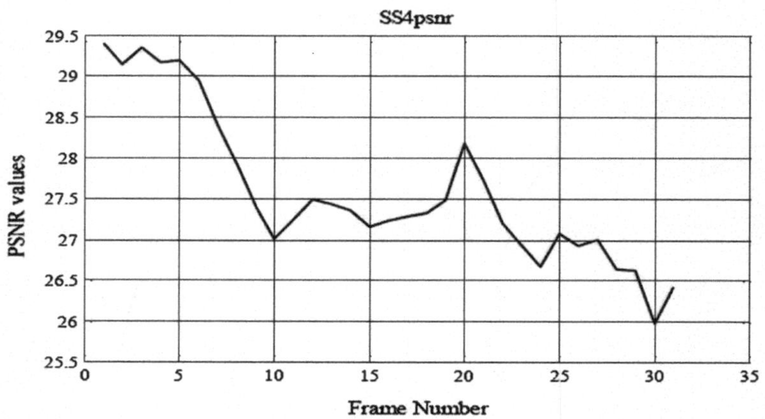

Fig. 9 PSNR of four-step search

4 Conclusion

In this paper, three BMA: three-step search, four-step search, and FS algorithm have been discussed to attain high compression ratio in video compression. Simulation has been done using MATLAB tool. Comparison of all three algorithms has been done in reference to computational intricacy and PSNR values which further reflects that FS algorithm is outperforming well compared to the all three discussed algorithms. FS algorithm finds best possible match because it searches the candidate in entire window size and makes system more accurate in terms of results but it makes system more complex and slower which can be overcome in future to find new kind of algorithm.

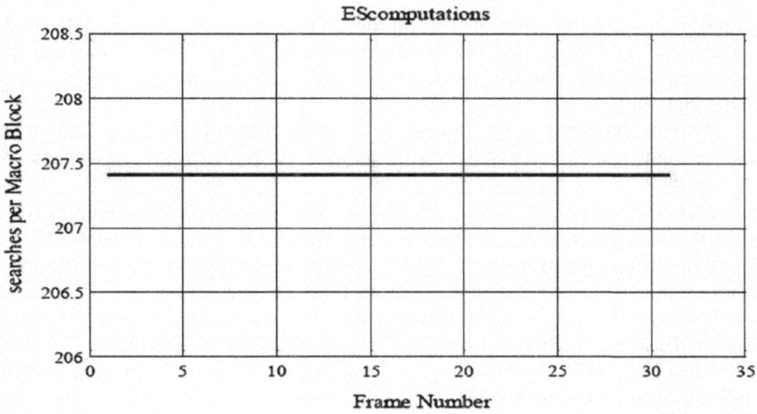

Fig. 10 Computation of full search

Fig. 11 PSNR of full search

References

1. Ohm JR, Sullivan GJ, Schwarz H, Tan TK, Wiegand T (2012) Comparison of the coding efficiency of video coding standards-including high efficiency video coding (HEVC). IEEE Trans Circuits Syst Video Technol 22(12):1669–1684
2. Barjatya A (2004) Block matching algorithms for motion estimation. IEEE Trans Evol Comput 8(3):225–229
3. Chan SH, Võ DT, Nguyen TQ (2010) Subpixel motion estimation without interpolation. In: ICASSP, IEEE international conference on acoustics, speech and signal processing—proceedings, pp 722–725
4. Sullivan GJ, Ohm JR, Han WJ, Wiegand T (2012) Overview of the high efficiency video coding (HEVC) standard. IEEE Trans Circuits Syst Video Technol 22(12):1649–1668
5. Côté G, Erol B, Gallant M, Kossentini F (1998) H.263+: Video coding at low bit rates. IEEE Trans Circuits Syst Video Technol 8(7):849–866

6. ITU (2005) ITU-T H.263 : video coding for low bit rate communication. Itu-T Sg 16 263
7. Vassiliadis S, Hakkennes EA, Wong JSSM, Pechanek GG (1998) The sum-absolute-difference motion estimation accelerator. In: Proceedings 24th EUROMICRO conference (Cat. No.98EX204), vol 2
8. Vanne J, Aho E, Hämäläinen TD, Kuusilinna K (2006) A high-performance sum of absolute difference implementation for motion estimation. IEEE Trans Circuits Syst Video Technol 16(7):876–883
9. Purnachand N, Alves LN, Navarro A (2012) Fast motion estimation algorithm for HEVC. In: 2012 IEEE second international conference consumer electronics. Berlin, pp 34–37
10. Wang D, Zhang L, Vincent A (2010) Motion-compensated frame rate up-conversion—part I: fast multi-frame motion estimation. IEEE Trans Broadcast 56(2):133–141
11. Luo J, Konofagou E (2010) A fast normalized cross-correlation calculation method for motion estimation. IEEE Trans Ultrason Ferroelectr Freq Control 57(6):1347–1357
12. Jing X, Chau LP (2004) An efficient three-step search algorithm for block motion estimation. IEEE Trans Multimed 6(3):435–438
13. Po LM, Ma WC (1996) A novel four-step search algorithm for fast block motion estimation. IEEE Trans Circuits Syst Video Technol 6(3):313–317
14. Kim JN, Choi TS (2000) A fast full-search motion-estimation algorithm using representative pixels and adaptive matching scan. IEEE Trans Circuits Syst Video Technol 10(7):1040–1048
15. Li R, Zeng B, Liou ML (1994) A new three-step search algorithm for block motion estimation. IEEE Trans Circuits Syst Video Technol 4(4):438–442
16. Chen TH (1998) A cost-effective three-step hierarchical search block-matching chip for motion estimation. IEEE J Solid-State Circuits 33(8):1253–1258
17. Duanmu CJ (2007) Fast scheme for the four-step search algorithm in video coding. Conf Proc IEEE Int Conf Syst Man Cybern 4:3181–3185
18. Wu A, So S (2003) VLSI implementation of genetic four-step search for block matching algorithm. IEEE Trans Consum Electron 49(4):1474–1481
19. Ahn TG, Moon YH, Kim JH (2004) Fast full-search motion estimation based on multilevel successive elimination algorithm. IEEE Trans Circuits Syst Video Technol 14(11):1265–1269
20. Kim JN, Byun SC, Kim YH, Ahn BH (2002) Fast full search motion estimation algorithm using early detection of impossible candidate vectors. IEEE Trans Signal Process 50(9):2355–2365
21. Priyadarshi R, Soni SK, Bhadu R, Nath V (2017) Performance analysis of diamond search algorithm over full search algorithm. Microsyst Technol, pp 1–9, Nov. 2017

Performance Evaluation Space-Time Interest Points Using Branching Particle Filters

Rahul Priyadarshi, Ankush Thakur and Anjila Deonath Singh

Abstract In digital video communication systems, it is vital that a video is to be compressed for the reason that of storing capacities along with bit-rate constrictions. Video processing is completed by means of sum of absolute differences and with the assistance of image processing block set. Branching particle filtering is basically one which is used for background extraction that is further detracted from motion frames for object detection. The distance traveled by vehicle is calculated using movement of centroid over frames. In this paper, we developed object tracing for actual time video which demonstrates motion-compensated video dealing out by means of key point descriptor. Simulation outcomes verify that proposed method can trace objects underneath several circumstances like noisy and low-contrast situation, and it may also trace object of interest with partial obstruction and byzantine contextual.

Keywords Motion vector · Object detection · Branching particle filter · Sum of absolute differences · Key point descriptor

1 Introduction

In everyday life, there has been an expanding notice for image tracing and movement acknowledgment frameworks; because of huge measure of uses those compounds can be utilized [1]. Average calculations are not pragmatic to utilize for picture following

R. Priyadarshi (✉) · A. Thakur
Electronics & Communication Engineering Department, National Institute of Technology, Hamirpur, HP, India
e-mail: rahul.glorious91@gmail.com

A. Thakur
e-mail: ankushlucky766@gmail.com

A. D. Singh
Electronics & Communication Engineering Department,
Shree L.R. Tiwari College of Engineering, Mira Road (E), Mumbai, India
e-mail: singh.anjila333@gmail.com

© Springer Nature Singapore Pte Ltd. 2019

V. Nath and J. K. Mandal (eds.), *Proceedings of the Third International Conference on Microelectronics, Computing and Communication Systems*,
Lecture Notes in Electrical Engineering 556, https://doi.org/10.1007/978-981-13-7091-5_8

because of the computational cost that emerges from the high number of degrees of opportunity of moving articles and from the vagueness of the pictures gotten from a solitary camera [2]. Imperatives in the arrangement of the moving items can be utilized to lessen its many-sided quality. The limitations can be concluded from exhibit, in view of various exercises. An image tracing scheme is created utilizing this sort of requirements and after that it is assessed. By image tracking, we allude to the capacity of a computer to recover the position and introduction of the object from an arrangement of pictures [3]. There have been a few distinctive ways to deal with enable PCs to infer naturally the kinematics posture and action from image arrangements. Video tracking is the way toward finding a moving item (or a few ones) in time utilizing a camera. A calculation examines the video casings and yields the area of moving focuses inside the video outline. The fundamental trouble in video following is to partner target areas in back-to-back video outlines, particularly when the items are moving quick in respect to the casing rate. Here, video tracking frameworks ordinarily utilize a movement model which portrays how the picture of the objective may change for various conceivable movements of the object to track [4].

The part of the tracking algorithm is to break down the video outlines to assess the movement parameters. These parameters describe the area of the objective [5]. Imperatives in the arrangement of the moving items can be utilized to lessen its multifaceted nature. The imperatives can be derived from exhibit, in light of various exercises. An image tracking framework is created utilizing this sort of limitations and afterward assessed. Image tracking and action response are getting expanding consideration among PC researchers because of the wide range of uses where they can be utilized, running from athletic execution investigation to video observation [6]. By image tracking, we allude to the capacity of a PC to recoup the position and orientation of the object from an arrangement of pictures. There have been a few distinctive ways to deal with enable PCs to infer consequently the kinematics posture and action from picture groupings. Video tracking [7] is the way toward finding a moving item in time utilizing a camera.

An algorithm examines the video edges and yields the area of moving focuses inside the video outline. The primary trouble in video following is to partner target areas in back-to-back video outlines, particularly when the articles are moving quick with respect to the casing rate. Here, video tracking frameworks typically utilize a movement model which depicts how the picture of the objective may change for various conceivable movements of the protest track [8]. The part of the tracking algorithm is to investigate the video edges so as to gage the movement parameters. These parameters portray the area of the target. Normally, the computational multifaceted nature for these calculations is low. The accompanying are some basic target representation and localization algorithms which are as follows:

- Blob tracking [9]: Segmentation of protest inside (for instance, blob location, square-based relationship, or optical stream).
- Kernel-based tracking [10]: An iterative limitation methodology in view of the expansion of a closeness measure [11].

- Contour tracking [12]: Detection of protest limits (e.g., dynamic forms or condensation calculation).
- Visual include coordinating: Registration.

Fleeting imperatives allude to the way that an object can just climb to a specific speed, along these lines given a specific arrangement, the protest can just achieve a subset of all the conceivable setups in whenever step.

There are sure requirements in tracking the object movement, for instance, joint edges limitations are the imperatives that arrangement with the confinements on the setups of the human body because of its tendency or to some other reality (i.e., movement performed, area of the application, and so forth). The arrangement of requirements that create the most exact tracking is the one that depicts better the activity performed. Movement acknowledgment frameworks depend on the idea of human tracking. These can be utilized in various applications, going from apply autonomy, PC liveliness, video observation, and so on to video ordering and even sports.

2 Related Work

Authors in [13] have proposed an automobile recognition structure utilizing dynamic Bayesian network in aerial observation. This technique has adaptability and great speculation capacities yet identification results can be settled by performing vehicle following on the vehicle location.

Authors in [14] have proposed the general approach of voting-centered movement valuation, moving article boundaries recognition, and substance-centered testing coding at transient and three-dimensional measures. These calculations confirmed that, under constrained system transfer speed, the transmitted picture quality can be step-by-step accomplished and the transmission data transfer capacity usage can be adequately diminished. In any case, the downside of this strategy is that there is no controlled detecting.

Authors in [15] have proposed another way to deal with video-based activity reconnaissance utilizing a fluffy crossbreed data induction component. There are three techniques for this proposed approach foundation refreshing, vehicle identification with square-based division, and vehicle following with mistake remuneration [16]. Authors in [17] have exhibited highlight extraction and highlight coordinating procedures in this paper. In the component extraction, four conditions for angle vectors are given [18, 19].

Authors in [20] have proposed an original model and calculation for summed up limit identification. Gb successfully consolidates various short- and central-level understanding layers of a picture in an upright way, and tenacities their requirements together, in shut frame, with a specific end goal to figure the correct limit quality and introduction. This strategy gives the effective estimation of limit quality and introduction in a solitary shut frame calculation yet there is no broad detailing of the issue that incorporates a wide range of limits.

Authors in [21] proposed a calculation in the region of video ordering, which requires the effective division of video into scenes. They utilize run of the mill scene location calculations [22]. Authors in [23] propose Mean-shift (MS)algorithm, measure just object's shading actual data, and did not comprise object's space data, so when focused on protest's shading circulation was like the foundation surface, the conventional calculations effectively brought about question following an incorrect outcome or lost.

Authors in [24] propose a piece-based strategy to deal with impediments. In this strategy, histograms are separated for every layout fix and after that these histograms are contrasted and those extricated from different locales in target picture. The format object is signified by various picture sections or fixes. Be that as it may, the layout is not refreshed and along these lines this approach is delicate to extensive appearance varieties.

Authors in [25] show an edge section created tracking process which is utilized to recognize moving items in picture grouping. In spite of the fact that this technique can track moving article or some portion of it viably under shifting light and incomplete impediment, it cannot manage full impediment.

Authors in [26] propose an online powerful and optimizing calculation utilizing a two-phase inadequate advancement approach. This following strategy chooses a meager and discriminative arrangement of elements to enhance productivity and power.

Authors in [27] propose a calculation in light of histograms of nearby inadequate portrayal for item following where the objective item is found by means of mode chasing of elective charts developed by reproduction blunders. Be that as it may, this technique is less viable in separating the frontal area patches from the foundation ones, and therefore of generative methodologies in view meager demonstration.

3 Proposed Algorithm

Branching particle method is recognized in nonlinear filtering scheme, un-normalized optimal filter $P_T(\varphi)$, that is a solution of Eq. (1), which is specified by Eq. (2).

$$dp_T(\varphi) = p_T(A\varphi)dT + p_T\left(h^*\varphi\right)dY_T \tag{1}$$

$$\tilde{E}\left[\varphi(X_T)exp\left(\int_0^T h^*(X_s)dY_x - \frac{1}{2}\int_0^T h^*(X_s)h(X_s)ds\right)|y_T\right] \tag{2}$$

where expectancy is considered relating to degree P that marks Y_T a Brownian motion. For the reason that one requests to estimate integration E[$|Y_T$] relating to measure P. Nevertheless, this expression delivers a recursive relation to develop a mathematical solution; we will create a series of branching particle schemes U_m by means of in that may be verified to tactic the solution p_t, which is

$$\lim_{n \to q} U_m(T) = p_T$$

Let $\{U_m(T), F_T; 0 \leq \mathrm{T} \leq 1\}$ be a series of branching particle schemes on (Ω, F, P). Preliminary condition is specified as

$U_m(T)$ which is observed degree of n particles of mass 1/m, that is, $U_m(T) = \sum_{i=1}^{m} X_i^m$, where $X_i^m \in E$, for every single i, m $\in N$ and $\delta X_i^m(x)$ is delta function pinpointed at X_i^m.

Development in interval [i/m, (i + 1)/m], i = 0, 1, …, m – 1

- At time i/m, the progression comprises the livelihood measure of $m_m(i/m)$ particles of mass $\frac{1}{m}$ $m_m(i/m)$ signifies number of elements alive at time T.
- Throughout the interval, particle travels self-reliantly with similar rule as in system Eq. (1). Let Z(s), s ∈ [i/m, (i + 1)/m], be the curve of a nonspecific particle, throughout this interval

$$dX_T = f(X_T)dT + \sigma(X_T)dW_T$$

- The mean number of offsprings for particle is given by

$$\mu_m^i = E\left(\varepsilon_m^i\right) = exp\left(\int h^*(Z(T))dY_T - \frac{1}{2}\int h^*h(Z(T))dT\right)$$

As a result, variance $v_m^i(V)$ is insignificant, where variance take place because of the off-rounding of $v_m^i(V)$ to calculate numeral value ϵ_m^i. The character * symbolizes complex conjugate. Determination of the number ϵ_m^i of offspring by

$$\varepsilon_m^i = \begin{cases} \left[\mu_m^i\right] & with\ probability\ \mu_m^i - \left[\mu_m^i\right] \\ \left[\mu_m^i\right] + 1 & with\ probability\ 1 - \mu_m^i + \left[\mu_m^i\right] \end{cases}$$

where [] represents the rounding operator.

4 Simulation and Results

Simulations have been done using MATLAB tool. One of the frames of video sequence used while doing simulation is shown in Fig. 1.

As shown in Fig. 2, final output is motion detected in input video for all three methods where one can see proposed model performing well compared to the existing algorithm.

Fig. 1 One of the frames of input original video

Also, performance evaluations of all three methods have been shown in Fig. 3 which is obviously better in proposed algorithm.

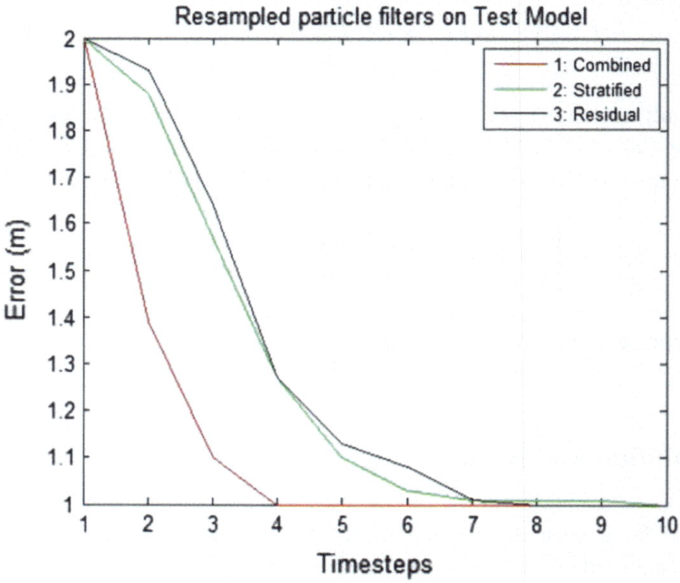

Fig. 2 Resampled particle filter on test model

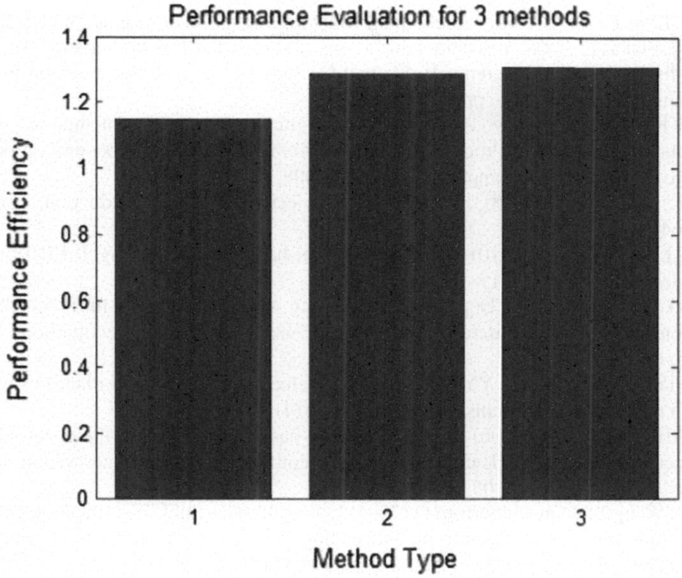

Fig. 3 Performance evaluation for three methods

5 Conclusion

We offered a new technique for tracing individually inflexible as well as deformable substances in video series. The projected tracing algorithm sectors object's area centered on motion, excerpts feature points, and envisages conforming feature points in the subsequent frame by means of optical flow, amends, and renovates erroneously prophesied feature points, and as a final point applies space-time interest point to handle the obstruction problematic. Space-time interest point, that is one of the foremost assistances of this paper, eradicates the off-line, preprocessing stage for engendering a priori preparation set. Future work focuses on tracing numerous objects at same time and on enhancing tracker accuracy for the duration of camera motion.

References

1. Comaniciu D, Ramesh V, Meer P (2003) Kernel-based object tracking. IEEE Trans Pattern Anal Mach Intell 25(5):564–577
2. Yilmaz A, Javed O, Shah M (2006) Object tracking. ACM Comput Surv 38(4):13–es
3. Yilmaz A, Javed O, Shah M (2006) Object tracking: a survey. ACM Comput Surv 38(4):13
4. Wu Y, Lim J, Yang MH (2013) Online object tracking: a benchmark. In: Proceedings of the IEEE computer society conference on computer vision and pattern recognition, pp 2411–2418
5. Wu Y, Lim J, Yang MH (2015) Object tracking benchmark. IEEE Trans Pattern Anal Mach Intell 37(9):1834–1848

6. Black J, Ellis T (2006) Multi camera image tracking. Image Vis Comput 24(11):1256–1267
7. Trucco E, Plakas K (2006) Video tracking: a concise survey. IEEE J Oceanic Eng 31(2):520–529
8. Nikolaidis N, Krinidis M, Loutas E, Stamou G, Pitas I (2009) Motion tracking in video. In: Essent Guide Video Process, pp 175–230
9. Wang F, Ren X, Liu Z (2008) A robust blob recognition and tracking method in vision-based multi-touch technique. In: Proceedings of the 2008 international symposium on parallel and distributed processing with applications, ISPA 2008, pp 971–974
10. Comaniciu D, Meer P (2003) Kernel-based object tracking 1 introduction. Comput Eng 25(5):564–577
11. Shen C, Kim J, Wang H (2010) Generalized kernel-based visual tracking. IEEE Trans Circuits Syst Video Technol 20(1):119–130
12. Zhou QY, Koltun V (2015) Depth camera tracking with contour cues. In: Proceedings of the IEEE computer society conference on computer vision and pattern recognition, 7-12-NaN-2015, pp 632–638
13. Cheng HY, Weng CC, Chen YY (2012) Vehicle detection in aerial surveillance using dynamic bayesian networks. IEEE Trans Image Process 21(4):2152–2159
14. Rivlin AE, Shimshoni I (2006) Robust fragments-based tracking using the integral histogram. In: Proceedings of the IEEE computer society conference on computer vision and pattern recognition, vol 1, pp 798–805
15. Avery R, Wang Y, Scott Rutherford G (2004) Length-based vehicle classification using images from uncalibrated video cameras. In: Proceedings the 7th international IEEE conference on intelligent transportation systems (IEEE Cat. No.04TH8749), pp 737–742
16. Lipton AJ, Fujiyoshi H, Patil RS (1998) Moving target classification and tracking from real-time video. In: Proceedings fourth IEEE work application computer vision WACV'98 (Cat. No.98EX201), vol 98, no 2, pp 8–14
17. Lian FL, Lin YC, Kuo CT, Jean JH (2013) Voting-based motion estimation for real-time video transmission in networked mobile camera systems. IEEE Trans Ind Inform 9(1):172–180
18. Wu BF, Kao CC, Juang JH, Huang YS (2013) A new approach to video-based traffic surveillance using fuzzy hybrid information inference mechanism. IEEE Trans Intell Transp Syst 14(1):485–491
19. Liu B, Yang L, Huang J, Meer P, Gong L, Kulikowski C (2010) Robust and fast collaborative tracking with two stage sparse optimization. In: Lecture notes in computer science (including subseries lecture notes in artificial intelligence and lecture notes in bioinformatics), vol 6314 LNCS, no. PART 4, pp 624–637
20. Liu B, Huang J, Yang L, Kulikowsk C (2011) Robust tracking using local sparse appearance model and K-selection. Cvpr 2011:1313–1320
21. Subudhi BN, Nanda PK, Ghosh A (2011) A change information based fast algorithm for video object detection and tracking. IEEE Trans Circuits Syst Video Technol 21(7):993–1004
22. Sakaino H (2013) Video-based tracking, learning, and recognition method for multiple moving objects. IEEE Trans Circuits Syst Video Technol 23(10):1661–1674
23. Hwang YT, Tsai BC, Pai YT, Sheu MH (2014) Feature points based video object tracking for dynamic scenes and its FPGA system prototyping. In: 2014 tenth international conference on intelligent information hiding and multimedia signal processing (IIH-MSP), pp 325–328
24. Wang CC, Thorpe C, Suppe A (2003) LADAR-based detection and tracking of moving objects from a ground vehicle at high speeds. IEEE IV2003 intelligent vehicles symposium proceedings Cat No 03TH8683, pp 416–421
25. Chavez-Roman H, Ponomaryov V (2014) Super resolution image generation using wavelet domain interpolation with edge extraction via a sparse representation. IEEE Geosci Remote Sens Lett 11(10):1777–1781
26. Gupte S, Masoud O, Martin RFK, Papanikolopoulos NP (2002) Detection and classification of vehicles. IEEE Trans Intell Transp Syst 3(1):37–47
27. Welch G, Bishop G (2006) An introduction to the kalman filter. Practical 7(1):1–16

Granularity-Based Assessment of Similarity Between Short Text Strings

Harpreet Kaur and Raman Maini

Abstract The capacity to discover the similarity between two textual bases, or inside one textual base, has much utilization including plagiarism detection and in the area of reused text (strings) in a database manageable to the removal of duplication. Past structure-metric methodologies have used either suffix trees or variance of longest common subsequence algorithms to recognize duplicate text. In this paper, different string distance metrics have been investigated: Levenshtein Distance (L. Dist.), Cosine Similarity (C.S.), and Hamming Distance (H. Dist) and also Hashes (ASCII-based hashing) on token sequences to detect matching of strings were used. Similarity index techniques vary on the basis of granularity: some techniques work on character level, word level, and some work on corpus-based granularity. The benefit of the approaches evaluated is to handle multiples patterns for similarity at a time. The work has been carried out on strings. From the simulation, it has been observed that ASCII-based hashing performs better than other techniques in terms of running time and accuracy. All techniques face one issue of increase in similarity searching time linearly with database size, whereas hashing handles this issue efficiently. ASCII-based hashing handles the issue of scalability very well.

Keywords Similarity index · Local similarity · Global similarity · Granularity

---

H. Kaur (✉) · R. Maini
Department of Computer Engineering, Punjabi University, Patiala, India
e-mail: harpreet.ce@pbi.ac.in

R. Maini
e-mail: researchraman@gmail.com

© Springer Nature Singapore Pte Ltd. 2019

V. Nath and J. K. Mandal (eds.), *Proceedings of the Third International Conference on Microelectronics, Computing and Communication Systems*,
Lecture Notes in Electrical Engineering 556, https://doi.org/10.1007/978-981-13-7091-5_9

1 Introduction

The issue of similarity in the text is important to numerous regions of the processing milieu. In an academic domain, comparability between student assignments may demonstrate written falsification or arrangement between the understudies. In the software industry, the textual similarity between documentation of two software products made by two distinct associations may show theft of strings and other licensed innovation. Also, similarity detection in strings is applicable to code similarity in various segments of a product. It may highlight the requirement for the source code to base refactored to expel any code duplication [1].

There are several approaches and tools to addresses the issue of similar string portions that might exist because of copy–paste or any other reason. In the textual comparison approach, lines of source code are analyzed literarily, and a similarity metric is generated on the basis of common lines in the documents [2]. In the metric approach, various features such as the quantity of identifiers, keywords, and language structure (syntax) components are extracted to create an arrangement of features vectors. Based on the distance among these vectors, the similarity of code portions is detected [3]. Tung et al. discussed the already proposed researches of similarity in strings. Author contrasted the effectiveness of algorithms used to compute the closeness between two archives and observed that the performance of fingerprinting and winnowing is superior to the cosine comparability [4]. Stein et al. have built up a fingerprint-based technique for a very effective nearly similar search and connected this strategy to distinguish appropriated sections in huge report accumulations. Different string matching techniques: Substring Matching, Keyword Matching, and Exact Finger Match and MD5 hashing have been discussed, but each technique has some limitations. MD5 hashing experiences two extreme problems: (1) it is computationally costly, (2) a little chunk size (3–10 words) must be distinguished to identify matches. But the proposed fuzzy fingerprinting conquers these constraints; it is a better suitable method to find duplicated and slightly modified sections in large documents. Substring selection, substring number, substring size, and substring encoding are the input parameters which serve as basis for experimentation [5]. Hussein et al. proposed a content-based strategy for document similarity investigation and visualization. The proposed technique depends on demonstrating the relationship between documents and their n-gram phrases, which are created from the normalized text. Latent Semantic Analysis (LSA) and Singular Value Decomposition (SVD) techniques are used to detect pairwise similarity. The investigation has been done on Arabic Language Documents. The proposed technique shows solid abilities in finding literal similarity in sentences [6]. Willassen et al. implemented hashing to generate fingerprints of source code. The hashing approach is much suitable because it is much easier for a computer to sort and analyze a list of hash codes than a list of files directly; this makes it simpler to find files with equal content among a large corpus [7].

To assess similarity in the token arrangement: string investigation and similarity location methods play a vital role. A few techniques address the issue of recognizing comparable strings. But the existing techniques for processing string comparability are centered on long textual records and are useful to detect plagiarism in documents. Textual similarity can be detected in various ways: Character-based similarity approach, term-based approach, corpus-based, and knowledge-based approach. Under each category, there are numerous methods to measure similarity among the two texts. Various algorithms of character-based similarity consist of N-gram, Smith–Waterman, Levenshtein Distance, Hamming distance, Hashing techniques, Cosine Similarity, etc. Different algorithms under term-based approach are Cosine Similarity, Block Distance, Dice's Coefficient, Jaccard Similarity, etc. [8]. Corpus-based methods measure the similarity on the basis of information gained from large corpora. It includes techniques such as Latent Semantic Analysis (LSA), Generalized Latent Semantic Analysis (GLSA), Explicit Semantic Analysis (ESA), etc. [9]. Semantic-Based measure relies on identifying the degree of similarity between words using information derived from semantic network such as WordNet [10]. WordNet gives a more powerful blend of conventional lexicographic data and modern computing. WordNet is an online lexical database intended for use under program control.

Generally, similarity techniques perform on long textual arrangements, and long text lines have a higher level of likeness: on the grounds that long messages contain co-occurring words. *But local identical regions cannot be detected by performing long textual similarity detection. Second, the co-occurrence of words in short messages is extremely uncommon, and if any change or modification occurs in any word in the document, then it is difficult to identify on long textual strings.* These two issues issue directs the research to compute similarity among short strings/messages. This paper focuses on detecting similarity between short texts (a sentence at a time) which uses character-based and term-based approaches and compares the strings lexically. Words are similar lexically if they would have the same character sequence. String-based measures operate on string sequences and character composition. A string metric is the measures of similarity or dissimilarity (distance) between two textual strings. In this paper, Character-based methods use Levenshtein Distance; Hamming Distance, and Fingerprinting technique generate a hash code for the whole string by adding ASCII values of every character in a string. Term-based approach uses Cosine Similarity measures.

Implementing ASCII-based hashing makes the similarity detection among two complete files easier because sometime comparing complete files is not the best way, even a slight change in text in one file, can produce different hash values for two files. Because of this, the similarity among two files will remain unnoticed. Therefore, line-based/word-based hash function is used for reporting similarity. And furthermore, this approach gives a chance to find more complex source lines that if observed to be equivalent in the two arrangements of documents, would show that the two sets have a common source.

The paper is organized as follows. Section 2 presents the algorithms and techniques used to measure the similarity. Simulation and Evaluation are described in Sect. 3. Results and discussions are shown in Sect. 4. Section 5 shows the proposed work. Conclusion and future work are discussed in Sect. 6.

2 Algorithm and Techniques to Detect String Similarity

The aim of current research is to detect similarity between code fragments taken as input, by converting the input string into tokens. Different Metrics: Levenshtein Distance, Cosine Similarity, Hamming Distance, and Fingerprint (Hash-Code) are used for similarity detection of tokenized source code lines.

2.1 Levenshtein Method

Token groupings can be seen as strings, and so distinguishing code similitude is a type of finding the Longest Common Subsequence (LCS) between two strings [11]. Numbers of LCS algorithms have been elaborated previously; Levenshtein Distance is such an algorithm which finds the number of differences (distance) among two strings. Levenshtein distance is named after the Russian scientist Vladimir Levenshtein, who proposed this algorithm [12]. If a portion of string is moved to another string, the algorithm detects a number of differences instead of a single change in the string [12]. Levenshtein Distance (LD) is a measure of the closeness between two strings. In the event that there are two strings: source string (s) and the objective string (t). *The L. Distance between s and t is the number of deletions, insertions, or substitutions required to change s into t.* Greater the L. Distance, more dissimilar the two strings are. Worst-case complexity of L. Distance is $O(n^2)$ which is prohibitive in most applications. LD is similar to finding the Longest Common Subsequence (LCS), it gives a lower bound on the LCS length [13]. It has also been rec as a commended as a reference to benchmark clone detection. Tiarks et al. considered a small reference set inspired by the Levenshtein Distance in [14]. An exhaustive set of all sets fulfilling a chosen Levenshtein distance threshold was proposed by Lavoie et al. [15]. *In Levenshtein distance, a distance function maps a pair of strings s1 and s2 to a real number "R", where a smaller value of "R" indicates greater similarity between s1 and s2.*

Definition 1 The string edit distance or Levenshtein distance is defined as the minimum aggregate cost of transforming x into y [16].

Definition 2 When strings of length len(s) and len(w) are considered, the total amount of work required to calculate their edit distance is equal to 4*len(s)*len(w) [17].

Definition 3 Minimal number of insertions, deletions, and replacements needed for transforming string a into string b.

Cost is associated with every operation performed on strings:

Operation	Cost
Copy character from string 1 to string 2	0
Deleting a character from string 1	1
Inserting a character in string 2	1
Substitute one character for another	1

Steps of Levenshtein Distance:

Step 1. *String s* *String t*
Step 2. *N = len(s)*
M = len(t)
Step 3. *If M = 0, return N, exit*
If N = 0, return M, exit
Step 4. *Construct Matrix Mat[M * N]*
Step 5. *Initialize the first row and column of a matrix to 0, …, N and 0, …, M respectively.*
Step 6. *Search for each character of s for i = 1 to N*
Search for each character of t for j = 1 to M
Step 7. *If s[i] = = t[j], then cost = 0,*
Else cost = 1.
Step 8. *Set cell d[i, j] of the matrix equal to the minimum of:*
a. The cell immediately above plus 1: d[i−1, j] + 1
b. The cell immediately to the left plus 1: d[i, j−1] + 1.
c. The cell diagonally above and to the left plus the cost: d[i−1, j−1] + cost.
Step 9. *After the iteration steps (3, 4, 5, 6) are complete, the distance is found in cell d[N, M]. Maintaining the Integrity of the Specifications.*

2.2 Cosine Similarity Measure

Tokenized sequences can also be converted into vectors and then the similarity is computed on the basis of the *cosine of the angle between two vectors* known as *Cosine Similarity* (C.S.). Cosine Similarity demonstrates the orientation between two strings, that how firmly two strings are oriented toward each other. Cosine closeness comes under the searching strategy "Sack of words". Sack of words represents the adoption of the vector space model to the area of plagiarism detection. Records are spoken to as one or numerous vectors, e.g., for various pairwise calculations.

Similarity calculation may then depend on the customary cosine closeness measure. As discussed by Salton et al., "term frequency–inverse document frequency" (tf–idf) is computed. Both the original text and user query are converted into vectors; afterward cosine angle is computed between original document *d* and query string *q in N-dimensional term space* [18] using Eqs. (1) and (2).

$$CosMatch\ : CM\,(s1,\ s2) = [0,\ 1] \tag{1}$$

Another value, Limit: L (s1, s2) = [0, 1] is defined such as:
If CM (s1, s2) > L (s1, s2), strings are similar
And if CM (s1, s2) < L (s1, s2), strings are dissimilar.

If t1 and t2 are two texts, and then to calculate C.S. among these two texts, First these texts are to be converted into vectors $\overrightarrow{t1}$ and $\overrightarrow{t2}$ and cosine similarity is calculated by Eq. (2), where $\overrightarrow{t1}$ and $\overrightarrow{t2}$ represents the term frequencies (tfs) in strings s1 and s2, respectively. ||t1|| and ||t2|| $\overrightarrow{t2}$ represent square roots of sum of squares of each term's frequency calculated in strings s1 and s2, respectively.

$$\cos(\cos\left(\overrightarrow{t1},\ \overrightarrow{t2}\right) = \left(\overrightarrow{t1}, *\overrightarrow{t2}\right)/\left\|\overrightarrow{t1}\right\|\left\|\overrightarrow{t2}\right\| \tag{2}$$

Closer the value of cos $\left(\underset{t1}{\rightarrow}, \underset{t2}{\rightarrow}\right)$ to 1, more similar the strings would be. Every word in this metric defines a dimension in Euclidean Space and the frequency of each word corresponds to the value in that dimension.

2.3 *Fingerprint Method (ASCII-Based Hashing)*

An important issue related to similarity detection is **scalability** (growing size of files) because clone detection can be performed on small as well as large projects. Both space and time complexity is covered in this scalability issue. Fingerprint generation using hashing algorithms are suitable to solve this issue. And also hashing reduces collisions, therefore minimizes false positives (where two different inputs map to the same key value) [19]. Similarity detection becomes easier when strings are represented using integers, and this is feasible using hashing methods. It means strings can be treated as an array of integers. But comparing strings (represented as numeric values) is not as easy as comparing simple integers. To compare two strings *A* and *B*, all of the elements of the strings should be traversed; therefore *length of string* is an important factor to be considered [20]. Extensive research on similarity search has been proposed in recent years. Among them, hash-based methods have achieved more consideration due to its ability to detect similarity search in high-dimensional space. *Hashing is the transformation of a string into a shorter fixed-length code called key, and that key represents an original string in the whole document. The key produced acts as a unique impression for the particular string. Hashing is used to* *index and retrieve strings in a database since it is faster to find the string (utilizing the shorter hashed key) than to discover it using the original value. Hashing can also be interpreted as a fingerprint of a substring.* Fingerprinting is currently the most widely applied approach to plagiarism detection and also for faster code clone detection. Hashing method used in this work *(termed as ASCII-based hashing)* can also be treated the same as *Locality-sensitive hashing* (LSH).

Let h(s1) and h(s2) be the hash codes(in terms of real number) generated for two sentences s1 and s2, respectively.

$$Sentences\, s1\, and\, s2\, are\, nearly\, similar\, if\; h(s1) \approx h(s2). \tag{3}$$

In present work, the Hash code for each string is calculated by adding ASCII value of every character in the string (**named it as ASCII-based hashing**). Only those strings are assumed similar, which have hash codes close to each other as described by Eqs. (3) and (4).

$$Hash\, Code = Sum\, of\; ASCII\; values\; of\; characters\; of\; a\, string \tag{4}$$

2.4 *Hamming Distance*

Another distance measure to detect similar string portions explored in this paper is Hamming Distance (H. dist). Hamming Distance between two strings of *equal length* is the number of positions at which the corresponding symbols are different. It is much easier to compute and also an efficient algorithm for near-neighbor queries for numerical vectors [21]. Processing steps for all techniques are described in (Fig. 1).

3 Simulation and Evaluation

This section summarizes the execution of methods on a desktop PC with an AMD A6-6310 APU with AMD Radeon 1.80 GHz processor and 4 GB of RAM. To measure lexical similarity among strings, four techniques said above have been analyzed and executed utilizing C++. Set of 100 strings (each having two subparts) are taken as input information. Table 1 shows test data set [9]. Figure 2 demonstrates the approach to assess all the four systems. Experimentation has been performed on simple text. The utilization of string similarity strategies has been actualized in this examination, also by measuring similarity among source code strings. Granularity is considered both at the character and word levels.

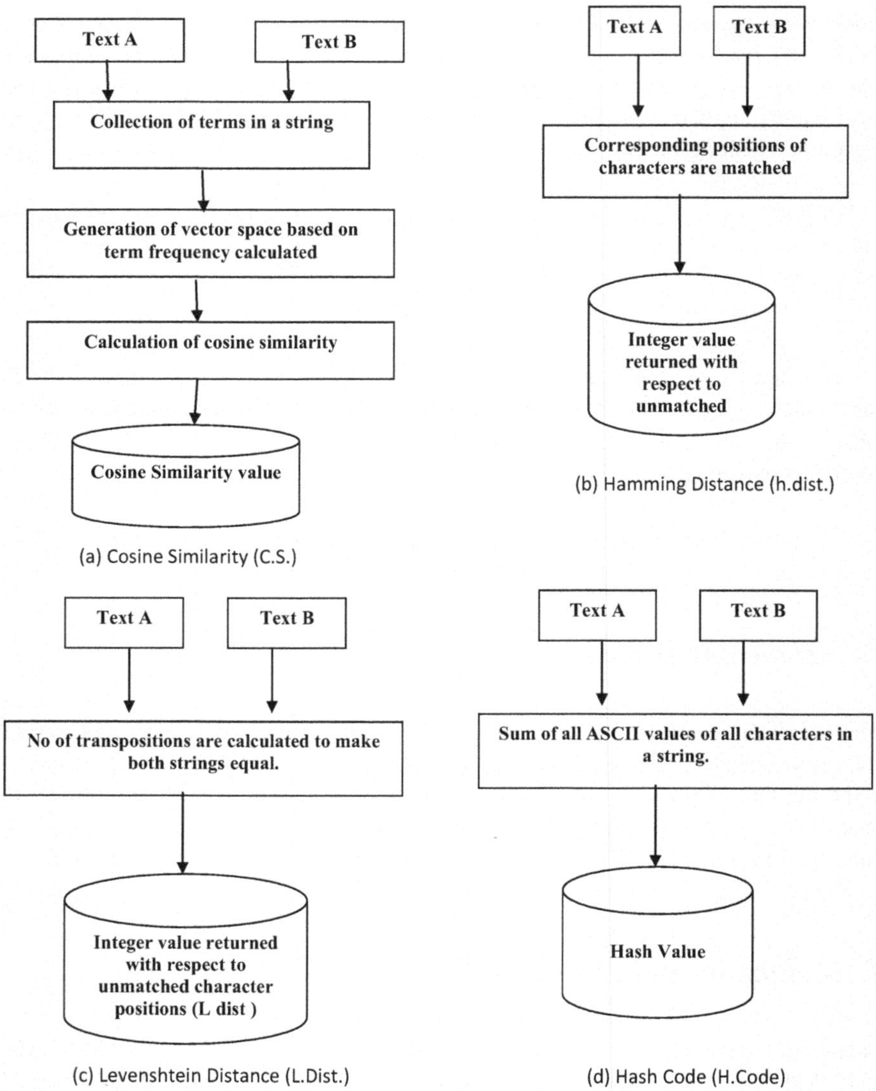

Fig. 1 Processing steps of all evaluated techniques

The values of L. dist, C.S., H. dist, and H. Code (ASCII-based hashing) are computed for each string. The output is reported as Similarity Index between strings as shown in Table 2. All techniques have also been compared on the basis of execution time as shown in Table 3.

Table 1 Test data set

S. no	Sentence pair	S. no	Sentence pair
St(1, 2)	I like that bachelor. I like that unmarried man.	St(3, 4)	I have a pen. Where do you live?
St(5, 6)	John is very nice. Is John very nice?	St(7, 8)	Red alcoholic drink. A bottle of wine.
St(9, 10)	It is a dog. That must be your dog.	St(11, 12)	Red alcoholic drink. Fresh orange juice.
St(13, 14)	It is a dog. It is a log.	St(15, 16)	Red alcoholic drink. An English dictionary.
St(17, 18)	It is a dog. It is a pig.	St(19, 20)	Dogs are animals. They are common pets.
St(21, 22)	I have a hammer. Take some nails.	St(23, 24)	Canis familiaris are animals. Dogs are common pets.
St(25, 26)	I have a pen. Where is ink?	St(27, 28)	Red alcoholic drink. Fresh apple juice.
St(29, 30)	A glass of cider. A full cup of apple juice.	St(31, 32)	I have a hammer. Take some apples.

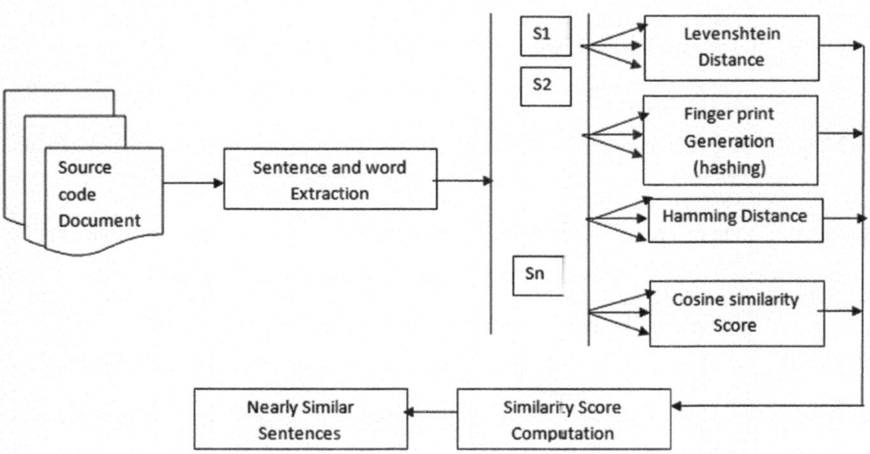

Fig. 2 Schematic diagram of methodology

Table 2 Similarity index for simple textual strings

S. no	Sentence pair	L. dist	H. dist	C.S.	H. Code
St(1, 2)	I like that bachelor. I like that unmarried man.	12	Strings are of unequal size	0.73	1887, 2338
St(3, 4)	I have a pen. Where do you live?	15	Strings are of unequal size	0	1041, 1690
St(5, 6)	John is very nice. Is John very nice?	7	7	0.8	1584, 1615
St(7, 8)	Red alcoholic drink. A bottle of wine.	16	Strings are of unequal size	0.44	1825, 1459
St(9, 10)	It is a dog. That must be your dog.	20	Strings are of unequal size	0.65	916, 1962
St(11, 12)	Red alcoholic drink. Fresh orange juice.	17	16	0.12	1825, 1459
St(13, 14)	It is a dog. It is a log.	1	1	0.92	916, 924
St(15, 16)	Red alcoholic drink. An English dictionary.	17	Strings are of unequal size	0.13	1825, 2063
St(17, 18)	It is a dog. It is a pig.	2	2	0.89	916, 922
St(19, 20)	Dogs are animals. They are common pets.	14	Strings are of unequal size	0.19	1546, 1943
St(21,22)	I have a hammer. Take some nails.	13	15	0.52	1352, 1424
St(23,24)	Canis familiarities are animals. Dogs are common pets.	25	Strings are of unequal size	0.36	2732, 1930
St(25,26)	I have a pen. Where is ink?	9	11	0.40	1009, 1113
St(27,28)	Red alcoholic drink. Fresh apple juice.	15	Strings are of unequal size	0.42	1825, 1658
St(29,30)	A glass of cider. A full cup of apple juice.	21	Strings are of unequal size	0.69	1463, 2323
St(31,32)	I have a hammer. Take some apples.	14	Strings are of unequal size	0.15	1352, 1598

4 Results and Discussions

First of all, the comparison of evaluated techniques has been performed on the basis of *nearly similar strings* reported by each technique. The hashing technique returned 8 similar strings from the given data set. Levenshtein Distance technique reported 7 similar techniques and Cosine Similarity measure algorithm also returned 7 nearly similar strings as shown in Fig. 3a–c for simple text. The detection rate of Hamming Distance (shown by Table 2) was very low because hamming distance technique

Table 3 Execution time of techniques

Granularity	S. no.	No of strings (max 50 words in each)	Execution time (s)	Type of input
Character and term based	H. code	3	1.69	Character and numeric strings Equal and unequal strings
		15	2.11	
		32	4.15	
		55	4.22	
		70	4.48	
	L. distance	3	1.87	Character and numeric strings Equal and unequal strings
		15	2.18	
		32	4.19	
		55	4.50	
		70	4.61	
	H. distance	3	1.92	Character and numeric strings both, but of equal size only
		15	2.32	
		32	4.88	
		55	4.66	Character and numeric strings Equal and unequal strings
		70	4.75	
	C.S.	3	1.89	
		15	2.21	
		32	4.16	
		55	4.33	
		70	4.36	

evaluates only those strings having equal length. Results have also been demonstrated for source code strings as shown in Fig. 3a–c. The mean value of L. distance computed for source code training set strings is 7 and the number of nearly similar strings reported below this threshold value are 7. A number of similar strings reported by cosine similarity above a threshold value of 0.93 are 7. The recognition rate of Levenshtein distance is 75% and of cosine similarity is 63%. Mean value of hash code differences is 245, therefore the number of similar strings reported below this value by *ASCII-based hashing* is 7. ASCII-hashing technique also returns the same number of similar strings as recognized by Cosine Similarity and Levenshtein Distance and its recognition rate is also same, i.e., 63%. Detection rate is described by Eq. (5).

$$\%\, of\ detection = \frac{number\ of\ matched\ strings\ above\ threshold}{total\ number\ of\ strings} \tag{5}$$

Another parameter chosen for comparison is: *Execution time.* The bar chart in Fig. 4 depicts that the hashing technique processes the database in less time as com-

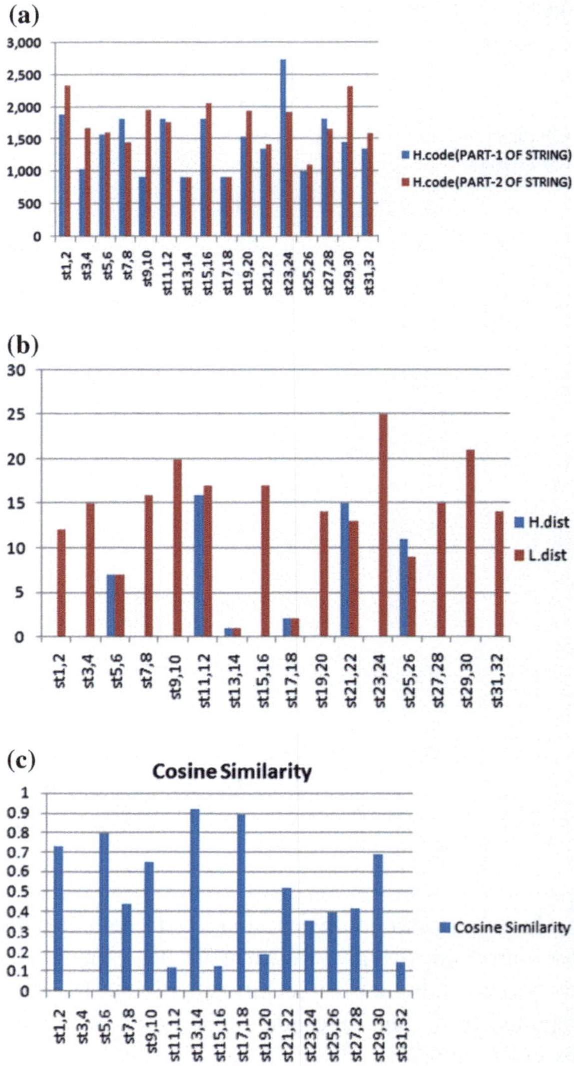

Fig. 3 **a** Hash code generation for simple text, **b** L. distance, and Hamming distance (H. dist.) **c** cosine similarity measure

pared to other three techniques with the growing size of the database. The Levenshtein technique takes more time for execution as compared to others, *because numerous steps of matrix calculation have to be performed*, such as: matching first, then transpositions, deletions, and insertions of characters to make both strings 100% similar. Execution time has also been demonstrated in Table 3.

The third parameter for evaluation is: *accuracy*. The nearly similar strings computed by each technique were also compared manually with the dataset taken as input. Hashing technique reported 65% accuracy, which is higher as compared to

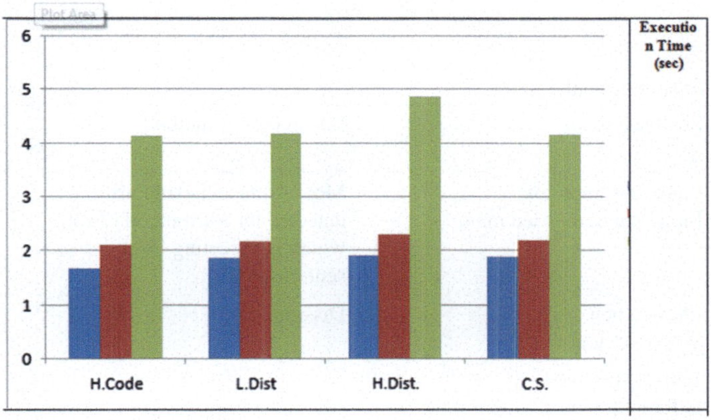

Fig. 4 Results of execution time

Fig. 5 Accuracy of techniques

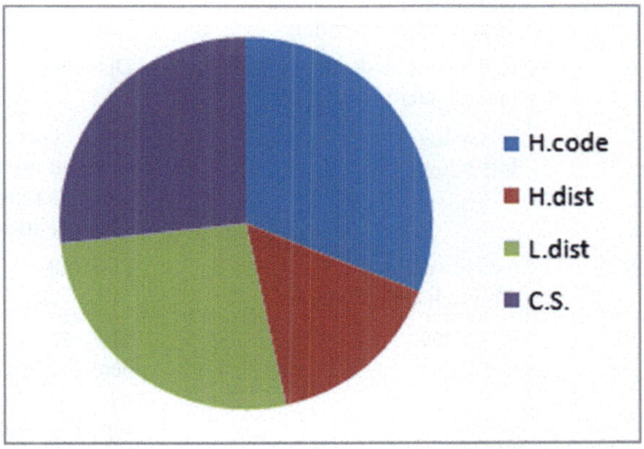

other evaluated techniques as depicted by the pie chart in Fig. 5. Hamming distance reported minimum accuracy.

ASCII-based hashing computes hash code of string in terms of each character which is closer to the desired matching string, thus it provides higher accuracy than Levenshtein Distance and Cosine Similarity (C.S.). Instead, C.S. operates on the basis of overall words matched in the string; therefore *it explores global similarity among strings, thus less accurate than hashing*.

Comparative analysis with literature has been shown in Table 4.

Table 4 Comparative analysis

S. no	Sentence pairs	Result by our method	Li et al. [9]
1	I like that bachelor. I like that unmarried man.	More syntactical similarity detected by our method (73%), which is reflecting from the sentence itself	0.561
2	I have a pen. Where do you live?	Dissimilar	0
3	John is very nice. Is John very nice?		
4	Red alcoholic drink. A bottle of wine.	Dissimilar	0.585
5	It is a dog. That must be your dog.		
6	Red alcoholic drink. Fresh orange juice.	Dissimilar	0.611
7	It is a dog. It is a log.	More syntactical similarity detected by our method (92%), which is reflecting from the sentence itself	
8	Red alcoholic drink. An English dictionary.	Dissimilar	0
9	It is a dog. It is a pig.	More syntactical similarity detected by our method (89%), which is reflecting from the sentence itself	0.790
10	Dogs are animals. They are common pets.		
11	I have a hammer. Take some nails.		
12	Canis familiaris are animals. Dogs are common pets.	Dissimilar	0.362
13	I have a pen. Where is ink?	Dissimilar	0.129
14	Red alcoholic drink. Fresh apple juice.	Dissimilar	0.420
15	A glass of cider. A full cup of apple juice.		
16	I have a hammer. Take some apples.	Dissimilar	0.121

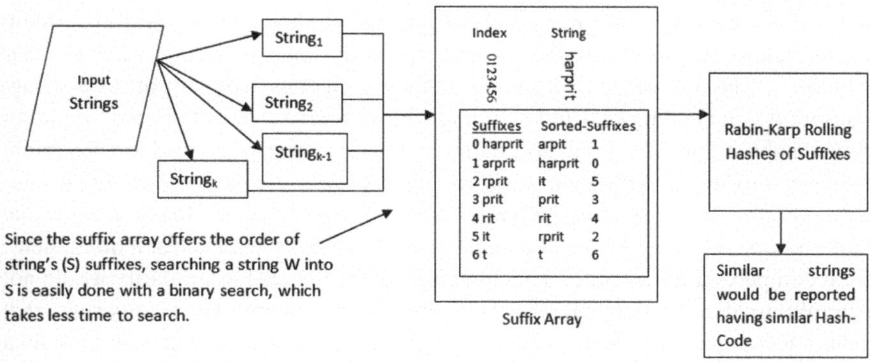

Fig. 6 Schematic diagram of proposed work for efficient similarity detection

5 Proposed Work

Several hashing methods exist in literature (MD5, Spectral Hashing, Rabin-Karp Rolling Hashing, etc.). In the proposed work, similarity among these strings will be identified using Rabin-Karp Rolling Hash. Rabin-Karp takes advantage of Learning to Hash (LTH) method. Detection of matches would be faster using RK-algo because it uses a hash of previously calculated text. Moreover, in implemented work, simple array has been used as a data structure for comparison. Using array may be time-consuming for comparison of long two text documents when hashes for every word would have to be stored for comparison. Therefore, an efficient data structure *"Suffix Array"* would be used for comparison. Suffix Array is the most widely used data structures for text indexing. **It is basically a data structure used to search a substring in a faster way**. It helps to search a pattern or substring from a given data source. It uses linear space for searching. **Suffix Array** will sort **sorts all suffixes of text and then the comparison would be made**. Suffix-Array even takes less time than Binary Search Trees (BSTs). Because in BSTs, most of the time is devoted in comparing subtrees which is not the case with Suffix Array. Proposed Methodology is shown below in Fig. 6.

6 Conclusions

Simulation has been performed for similarity detection in short textual strings using various string metrics: Levenshtein Distance, Cosine Similarity, Hamming Distance and ASCII-based hashing because these algorithms are widely used for string similarity detection. Character, Word, and Statement level granularity has been considered for evaluation. From exploratory results, all the techniques yield the same recognition rate, but it has been observed that hashing gives better results in terms of execution time and accuracy as compared to other evaluated methods. However, Levenshtein distance takes more time with the growing size of the database as compared to other techniques. The advantage of hashing is that it generates a fingerprint of a string, therefore it will become easy to fetch only similar source strings from the whole set of the database. Further, it will speed up the whole process of code clone detection. *Moreover, hashing detects local similarity as it operates on each character sequentially, therefore it will be more helpful to detect identical regions*. In contrast, cosine similarity operates on a whole document or on a complete sentence at a time; therefore it compares a statement or a document globally. So, cosine similarity would not be helpful to identify identical regions within two documents. The results show that hashing techniques handle growing size of the database in less execution time than other three techniques. Evaluation has been performed on the simple text as well as source code strings, in both types of case studies hashing comes out to be best. All techniques face one issue of increase in similarity searching time linearly with database size. Only ASCII-based hashing handles this issue efficiently.

References

1. Toomey W (2010) Code similarity detection in multiple large source trees using token hashes. In: PAN-09 3rd workshop on uncovering plagiarism, authorship and social software misuse and 1st international competition on plagiarism detection
2. Ducasse S, Rieger M, Demeyer S (1999) A language independent approach for detecting duplicated code. In: Proceedings of the IEEE international conference on software maintenance, p 109
3. Mayrand J, Leblanc C, Merlo E (1996) Experiment on the automatic detection of function clones in a software system using metrics. In: Proceedings of the IEEE international conference on software maintenance, pp 244–253
4. Tung KT, Hung ND, Hanh (2015) A Comparison of Algorithms used to measure the Similarity between two documents. Int J Adv Res Comput Eng Technol (IJARCET) 4(4)
5. Stein B, SM zu Eissen (2007) Fingerprint-based similarity search and its applications. 85–98
6. Hussein AS (2016) Visualizing document similarity using N-grams and latent semantic analysis. In: SAI computing conference 2016
7. Willassen SY (2009) Line based hash analysis of source code infringement. Dig Evid Electron Signat Law Rev 6:210–213
8. Gomaa WH, Fahmy AA (2013) A survey of text similarity approaches. Int J Comput Appl 68(13)
9. Li Y, McLean D, Bandar Z, O'Shea J, Crockett K (2006) Sentence similarity based on semantic nets and corpus statistics. IEEE Trans Knowl Data Eng 1138–1150
10. Miller GA (1995) WordNet: a lexical database for English. Commun ACM 39–41
11. Bergroth L, Hakonen H, Raita T (2000) A survey of longest common subsequence algorithms. In: International symposium on string processing and information retrieval, vol 39
12. Levenshtein VI (1966) Binary codes capable of correcting deletions, insertions, and reversals. Sov Phys Dokl 10(8):707–710
13. Lavoie T, Merlo E (2012) An accurate estimation of the Levenshtein distance using metric trees and, Manhattan distance. In: IWSC 2012, Zurich, Switzerland, 978-1-4673-1795-5/12/$31.00 © 2012. IEEE
14. Tiarks R, Koschke R, Falke R (2009) An assessment of type-3 clones as detected by state-of-the-art tools. In: Workshop on source code analysis and manipulation. IEEE Computer Society Press, pp 67–76

15. Lavoie T, Merlo E (2011) Automated type-3 clone oracle using Levenshtein metric, pp 25–32
16. Udagawa Y (2013) Source code retrieval using sequence-based similarity. Int J Data Min Knowl Manag Process 3(4)
17. Tung KT, Hung ND, Hanh LTM (2015) A comparison of algorithms used to measure the similarity between two documents. Int J Adv Res Comput Eng Technol (IJARCET) 4(4)
18. Salton G, Buckley C (1988) Term-weighting approaches in automatic text retrieval. Inf Process Manag 24(5):513–523
19. Sadowski C, Levin G (2007) SimHash: hash-based similarity detection
20. Rolling Hash (Rabin-Karp Algorithm) 6.006 Intro to Algorithms Recitation 06 February 18, 2011
21. Jiang L, Misherghi G, Su Z, Glondu S (2007) DECKARD: scalable and accurate tree-based detection of code clones

Smart Monitoring of Automatic Teller Machine—A Survey

Avinav Gupta and J. Priyadarshini

Abstract The idea of smart monitoring is born with the observation in civilian's real-life incidents happening around civilians. Financial transaction, ease of money exchange can be good example. So, security is very important factor. The crime rates have increased tremendously in Financial Organizations. This security system has a new concept that helps the banks and civilians in transaction at the ATM [1]. The drawback of present security system is that the password and PIN number might get leaked which leads to the loss of money in account. This paper focuses on the security measures to be followed in the ATM. With the help of behavior analysis using machine learning techniques, One-Time Password (OTP) to the user and automatic door locking system for the enhancement of security of both users and banks.

Keywords Machine learning techniques · Vibration sensors · Motion sensors · GSM technique · Arduino · OTP generation, etc

1 Introduction

In the real world, today people are concerned about their safety, for their valuable things. Old concepts and devices are getting modified as per requirement of people. Nowadays, everyone belong to the edge of digitized and smart world. People are getting smarter day by day with the help of new technology and new innovations. Main reason behind the upgradation of new technologies is nothing but to overcome the existing problems. Economic growth of world makes the life smarter and better

A. Gupta · J. Priyadarshini (✉)
School of Computing Science and Engineering, Vellore Institute of Technology, Chennai, India
e-mail: priyadarshini.j@vit.ac.in

A. Gupta
e-mail: avinavgupta05@gmail.com

© Springer Nature Singapore Pte Ltd. 2019

V. Nath and J. K. Mandal (eds.), *Proceedings of the Third International Conference on Microelectronics, Computing and Communication Systems*, Lecture Notes in Electrical Engineering 556, https://doi.org/10.1007/978-981-13-7091-5_10

as compared to previous lifestyle. A smart step toward economy is the introduction of Automated Teller Machine (ATM), for faster and easier money transfer. But a group of people do malpractices over this ATM system to put people, organization, or bank into a millions pounds of loses. Recent surveys says that—"90% of Crimes is of Robbery at ATM center". This is a serious threat to both bank and civilians.

Therefore, this security system proposes a solution to minimize the ATM robbery crimes by means of behavior analysis using machine learning, image processing [2], and Internet of Things (IOT) [3]. These techniques will help bank and civilians to achieve good results for the security system in ATM.

There are some restrictions for civilians in our security system. There is also setup of OTP generation [4] at the time of transaction. After swiping the ATM card, a One-Time Password (OTP) will be generated and sent to the registered mobile number of account holder. After the OTP being entered by the user in the display, if the user correctly enters the OTP, the transaction can proceed. There is also added security of automatic door locking system.

2 Literature Survey

Orah [5] has proposed a security system with three-factor authentication, first two authentication is smart card verification and PIN verification. Mainly, this security system integrates biometric authentication work as the third-level authentication step. The drawback in working of fingerprint sensor is the errors that it shows false non-match and false match. It means that false non-match occurs when two fingerprint samples from the same individual are having low similarity that the system cannot correctly match them while false match occurs when two fingerprint samples from different persons have high similarity, system matches them which is false.

Sivakumar et al. [6] gave a security system to monitoring ATM by using use of vibration sensors, arm-controller-based embedded system, and Keil tool is used to run stepper motor to leak the gas inside ATM to make the thief unconscious, and then the GSM technique sends message to nearby police station within time, and camera is used for video recording. The drawback of this work is that the system does not have any algorithm to detect more than one person in ATM room. This also cannot detect that person having any objectionable item with them and if gas leaks accidently may harm the civilians.

Aru and Gozie [7] have presented an ATM security system that used facial recognition technology into the identity verification process used in ATMs. The major drawback of this system can be seen in different situations like if the cardholder cannot go to the ATM due to emergency, no other family members or trusted person will be able to perform transaction, and also if there is any change in the face which the system cannot detect and do not authenticate the face, and then the owner will get into trouble as the system not using Artificial Intelligence (AI).

Sudipta et al. [8] introduced a security system for ATM theft that is taking place all over the world. In this system, they are using motion sensor to detect if there is more than one person into the room and then the transaction will be blocked, also using vibration sensor to detect if person doing any abnormal activity with the machine then buzzer blows and using GPS technology they are sending exact location of ATM to nearby police station and corresponding bank. The main limitation of this system is that it cannot detect if the person is carrying any baggage and by the time police reaches to ATM location the thief may escape. This problem can be solved by automatic door locking system.

Jantre and Kharade [9] have introduced an ATM security based on GSM technique. This system works as when ATM user insert the ATM in ATM machine, he can type the password and second, and then the message goes through cardholder via GSM and then the cardholder can send the security code or OTP using GSM through main system and start the transaction via cardholder. The main limitation of this system is that if person tries to tamper the cameras or ATM machine, no alarm system is there and also it cannot detect if the person is carrying any baggage and by the time police reaches to ATM location the thief may escape. This problem can be solved by behavior analysis and automatic door locking system.

Dutta et al. [10] gave a security system based on fingerprint recognition. This system proposes the idea of using fingerprints of customers as password included with traditional PIN number. After authorized verification, the customer will be able to proceed for transaction else after three successive wrong attempts, the ATM card will be blocked for 24 h and a message will be sent to the registered mobile number. The major drawback of this system can be seen in different situations like if the cardholder cannot go to the ATM due to emergency, no other family members or trusted person will be able to perform transaction also if person tries to tamper the cameras or ATM machine or doing wrong practice, no alarm system is there.

3 Proposed Security System

The proposed system mainly concentrates on the security of the ATM. The behavior analysis is based on certain criteria. The abnormalities considered are as follows:

- Person entering with any type of cap,
- Person carry any type of baggage,
- If there is only one machine in ATM room, then only one person is allowed (and kid with age up to 8 years)

  or if two ATM machine then only two persons are allowed, and

- Person trying to harm machine or camera.

The proposed system will work as follows:

First step is behavior analysis using machine learning techniques, image processing, and IOT to recognize the restrictions for the civilian as given below:

1. More than one person entry in ATM room if only one machine is there and only one kid is allowed (age up to 7 years)
2. When someone is wearing any type of cap.
3. When someone is carrying baggage or having any weapons.

Whenever the abovementioned situations occur, using vibration sensors and motion sensor, a low alarm will ring for 5 s which will be warning them that they are violating any of the above conditions. If still alarm rings after 5 s, then main alarm will ring inside and outside the ATM center that informs the security guard about the situation. After 15 s of continuous alarm ringing, using the GSM technique [11] a message will be sent to security agency and bank.

In this proposed system, there will be another authentication level which is OTP verification, and the OTP is generated after inserting smart card if OTP matches, then further transaction proceeds if not then only two chances will be given to enter correct OTP. If both chances failed then transaction will not be done, and then immediately through GSM technique, a message will be sent to the bank and account holder informing about the situation.

Also, in this suggested model, there will be an automatic door locking system within minimum time by using Arduino when someone tries to tamper the ATM machine or camera, so that thief cannot escape.

4 Comparison Table

Proposed system	Advantages	Disadvantages	Overcoming solution
Three-factor authentication for automated teller machine system	It helps to limit the risk related with money	In fingerprint sensor, the main error is of false non-match and false match	OTP generation at the time of transaction
Design and implementation of security-based ATM theft monitoring system	System has vibration technique and also automatic door locking system	System does not have any algorithm to detect more than one person in ATM room. This also cannot detect that person having any objectionable item with them	By behavior analysis using machine learning techniques

(continued)

(continued)

Proposed system	Advantages	Disadvantages	Overcoming solution
Facial verification technology for the use in ATM transactions	No one can withdraw money if card is stolen and got the smart card PIN number	This system can be seen in different situations like if the cardholder cannot go to the ATM due to emergency, no other family member or trusted person will be able to perform transaction	Instead of facial verification. Use OTP generation system for verification
ATM robbery prevention using advance security	Geological location will always be traced of an ATM machine Maintain the entry of only necessary person	This system cannot detect if the person is carrying any baggage and by the time police reaches to ATM location the thief may escape	By behavior analysis using machine learning techniques, image processing
GSM-based ATM security ATM banking	Instantaneous action by using GSM technology Essay to use GSM-related transaction technique Direct mobile communication system involves	This system is that if person tries to tamper the cameras or ATM machine, no alarm system is there	Alarm system for different situations
ATM transaction security using fingerprint recognition	As everyone fingerprints are different, so it is impossible to withdraw money by stealing the card and knowing the PIN number	This system can be seen in different situations like if the cardholder finger got injured then no other trusted person will be able to perform transaction	By behavior analysis using machine learning techniques and IOT

5 Work Flow Diagram

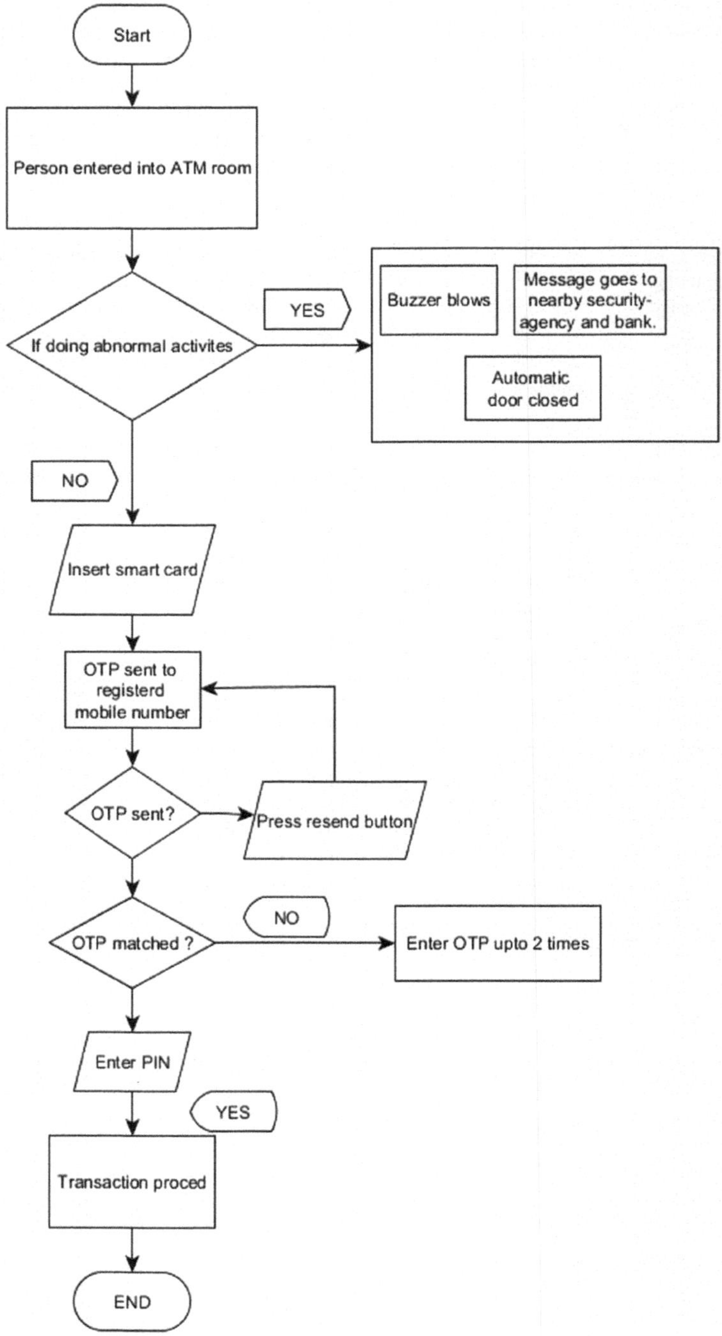

6 Conclusion

As everyone is seeing, the rate of ATM robbery is increasing year by year and thus security plays a very important role.

Our proposed system will provide security to civilians and banks. This system is very much effective to reduce the ATM theft. By doing behavior analysis using machine learning techniques, OTP generation and automatic door locking system will provide us a security system by which authority can prevent our society from ATM robberies.

References

1. https://en.wikipedia.org/wiki/Automated_teller_machine
2. https://en.wikipedia.org/wiki/Digital_image_processing
3. https://en.wikipedia.org/wiki/Internet_of_things
4. https://en.wikipedia.org/wiki/One-time_password
5. Orah JN (2014) Three factor authentication for automated teller machine system. Int J Comput Sci Inform Technol Secur (IJCSITS) 4(6). ISSN: 2249-9555
6. Sivakumar T, Ashok G, Venuprathap KS (2013) Design and implementation of security based ATM theft monitoring system. Int J Eng Invent 3(1). e-ISSN 2278-7461, p-ISSN 2319-6491
7. Aru OE, Gozie I (2013) Facial verification technology for the use in ATM transactions. Am J Eng Res (AJER) 2(5). e-ISSN: 2320-0847, p-ISSN: 2320-0936
8. Maiti S, Vaishnav M, Ingale L, Suryawanshi P (2016) ATM robbery prevention using advance security. Int Res J Eng Technol (IRJET) 03(02). e-ISSN: 2395-0056, p-ISSN: 2395-0072. www.irjet.net
9. Jantre SR, Kharade RA (20017) GSM based ATM security ATM banking. Int Adv Res J Sci Eng Technol (IARJSET) 4(2). e-ISSN: 2393-8021, p-ISSN: 2394-1588
10. Dutta M, Psyche KK, Yasmin S (2017) ATM transaction security using fingerprint recognition. Am J Eng Res (AJER) 6(8). e-ISSN: 2320-0847, p-ISSN: 2320-0936
11. https://www.elprocus.com/gsm-architecture-features-working/

Enhancing Readability of Dyslexic Children by Machine Learning Techniques—A Survey

Geeta Bhimrao Atkar and J. Priyadarshini

Abstract Dyslexia is type of disorder that occurs normally in children, and is also called as learning disability. Children who are having dyslexia, requires more time to read, write as compared to normal children. There are different kinds of disabilities which come under dyslexia like dysgraphia, and dyscalculia but this paper focuses on reading disability in children. Here, children between 8 and 10 years will be taught on most frequently used Hindi words. First, frequently used Hindi words will be created as a database and then the system is trained with those input words. Speech synthesis is applied to those words to get speech output from the trained system. Children will be given a set of words to read. If children are saying the wrong word three times then the system will give correct speech and train the children. This makes the process automatic and the system acts as a personal trainer to the dyslexic children. By doing this model, the children can learn in a more interactive manner and will remember the words as the system can show different pictures for the same word. The use of machine learning techniques makes this system more accurate and the system will be trained for speech recognition and phonetics of each alphabet and words in Hindi.

Keywords Acoustic · Dyslexia · Machine learning · Speech recognition · Hindi

G. B. Atkar (✉) · J. Priyadarshini
School of Computing Science and Engineering, Vellore Institute of Technology, Chennai, India
e-mail: geeta.atkar2016@vitstudent.ac.in

J. Priyadarshini
e-mail: Priyadarshini.J@vit.ac.in

© Springer Nature Singapore Pte Ltd. 2019

V. Nath and J. K. Mandal (eds.), *Proceedings of the Third International Conference on Microelectronics, Computing and Communication Systems*,
Lecture Notes in Electrical Engineering 556, https://doi.org/10.1007/978-981-13-7091-5_11

1 Introduction

1.1 Dyslexia

It occurs in people of all ages. There are number of disabilities which will come under dyslexia like writing problem, reading problem, memory management, processing speed, time management, etc. The main reason is lack of knowledge about phonological components. This disorder might be minimized by using some assistive techniques. In 1920, Samuel Torrey carton has another definition of this disorder that left side brain does the functions which should be done by right side brain [1].

There are different causes of dyslexia, according to the National Health Service [1]. It may be due to parent genes. That is a genetic disorder. The issue is in phonetics, which leads to lifelong problems. Most of the children are not able to differentiate between mirror image like "W", "M", and "b", "d". Some require time to literate new word which they are reading or learning first time. Nowadays most of the students are going to English medium school even if their mother tongue is like Hindi, Marathi, Kannada, and they are facing problem of learning English language. Such children are having dyslexia.

1.2 Readability

Reading is the most important thing which is necessary for the study of language. Reading may be affected because of less self-esteem or less self-confidence. In the early ages, children of age 8–10 can learn more quickly if they are taught with pictures only. If children do not have a picture in their mind they will find difficulty in reading. In this way, understandability and readability are totally dependent.

In general terms, using graphical schemes contains—Collection of demographic part, reading test, two questions which are design for understandability and survey which contain impression that how graphical schemes helps to read easily [2]. Using different font size also improves readability. The speed of reading is improved by using attractive fonts [3]. If we consider reading performance, some of the languages like "times new roman", "monotype cursive" enhance readability, whereas italic will minimize reading performance "Arial" font decrease readability. If we apply text to speech software on some of the dyslexic children, reading speed, understandability, vocabulary of children will improve. Also, it will improve the fluency of speaking a particular language.

2 Literature Survey

There are number of text to speech techniques are available in different languages like English, kannada, Malay, Hindi. This paper focuses on Hindi language reading disability. In Hindi language, there are normally 33 consonants and 13 vowels are present. Some of the Text to speech techniques are there as follows:

"Ramesh Kagalkar" and "Kaveri Kamble" gave results on Hindi language Text to speech system. In this system, they have taken total of 1540 Hindi words as input from hind to English dictionary. Hindi speeches are recorded by using PRATT software, which analyses speech in phonetics of any language and it is open source, freely available. Then mapping of symbols with Unicode and its decimal equivalent is done. In short, algorithm contains some steps like, If given input length is zero then no text is entered. Otherwise, calculate the length of input text and map it with Unicode and decimal. Positive point of this paper. Drawback is paper helpful for text to speech conversion of Hindi language only [4].

Darshna Badhe, P. M. Ghate worked on three steps are there for text to speech conversion [5]:

(1) **Preparing Hindi database**: Here, two files are used one is audio and other one is text file. In text file vowels and consonant are stored along with that their English translation is stored. In audio database sound of text is stored [6].
(2) **Processing of raw text**: After database creation, second step is text processing, in which one excel sheet is maintained. Excel sheet contains hexadecimal values of text and its transliteration. Then whatever input is given the hexadecimal values of input are compared with values in database. Text is divided into consonant and vowel pattern. After that for every consonant or vowel .wav file will be attached. Like gee.wav". Above strings are stored in text files for taking inputs.
(3) **Audio processing**: Read files with .wav extension. Then silence part is removed from signal and signal made ready to play [5].

Snehali K. Nandurkar, Zaki M. Shaikh, told that, while considering PDF data, PDF data is taken as an input in Hindi language. PDF may contain Hindi text written in English characters or Hindi text in Devnagri. Text is extracted by using PDF box by using files PDFBox.jar, fontbox.jar, org_apache_commons_loggong.jar. At last

(1) Segmentation
(2) Tokenization
(3) Speech generation

Disadvantage of this paper is it is having very lengthy process of taking input and suitable for Hindi language only [6].

Archana Balyan, Amita Dev, Ruchika Kumari, S. S. Agrawal explained some technique for text to speech conversion, following steps are included:

(1) Text analysis
(2) Database Creation
(3) Selection of Speech Unit

Database of <Previous phoneme> <phoneme> <following phoneme> <start time> <end time> <duration> <.wav file> is maintained and then speech unit selection is done by considering different cases. The drawback of this paper is Hindi transliteration of English text or vice versa is not done in this paper [7].

John Colac, Sangam Borkar told that whenever input will be taken as an image, following steps are included:

- Image acquisition
- Image Reading
- Convert into grayscale
- Image Filtering
- Count no of components of image
- Extract text and resize letter
- Write into text file
- Apply TTS

Input is taken in the form of image of letter or word you wanted to give as input. As input is not directly from keyboard, input image reading needs to be done. If the image is in color, it is converted into grayscale image. As grayscale images are having only one dimension and require less storage space as compared to color. It is easy to transform gray image rather than color. Image filtering is done. No of components in image are counted. Text is extracted from image and then letter is resized. Now we got output in text form. Text to speech techniques are applied to get speech output. Drawback of this paper is its lengthy process of getting speech output and useful in Konkani language [8].

3 Methodologies Needed

1. Speech Recognition:

Speech recognition means converting speech signals into words. It is also called as Automatic Speech Recognition algorithm. Such that, computer system can recognizes speech.

Figure 1 shows speech Recognition system's mathematical model. It contains four parts language model unit, front end unit, model unit, and search unit.

Let the specified word sequence: V,
Acoustic observation sequence X,
Probability: P(V, X).

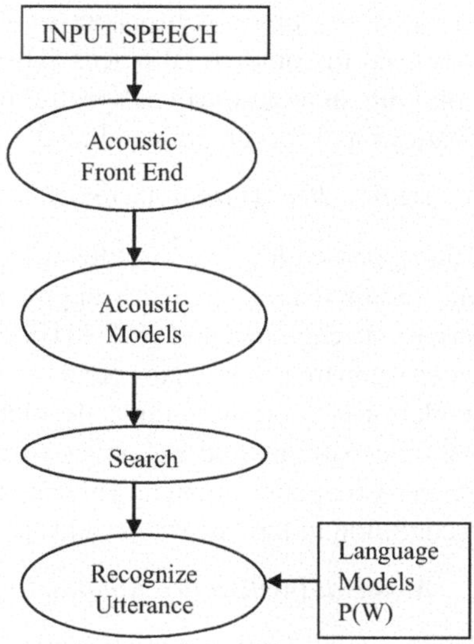

Fig. 1 Speech recognition system

Target is then to decode the word string, based on the acoustic observation sequence, in such a way that the decoded string has the maximum a posteriori probability [9].

$$P(V/B) = \arg\max P(V/B). \tag{1}$$

Applying Bay's rule, Eq. (1) will be

$$P(V/B) = P(A/V)P(W)/P(B) \tag{2}$$

P(B) is independent of V,
The decoding rule of Eq. (1) is

$$V = \arg\max P(B/V)P(V) \tag{3}$$

2. Approaches for Speech Recognition:

a. *Acoustic Approach*
b. *Pattern Recognition Approach*
c. *Artificial Intelligence Approach (AI)*

a. *Acoustic Approach*

Speech and feature detection are combined which converts the spectral measurement into feature will set. Properties of phonetic units are described by these features. After that, each segment region is divided into stable acoustic regions and then each region is labeled. More than one label is also attached to each phonetic region. This is called "lattice characterization of speech". Finally, string of words is obtained from labels.

In validation process, this approach is used at very lesser places [9, 10]. Mostly, it is used in commercial places. This is easy to grasp or learn by machines within minimum amount of time. Spectral measurements are transferred in phonetic units. Features and speech are combined. Labeling and segmentation is done.

b. *Pattern Recognition Approach:*

This approach has two vital features pattern comparison and pattern training. Vital and required feature of this approach is that it establishes consistent speech pattern representations and HMM (Hidden Markov Model) is used to represent speech pattern and then applied to sound (word, phase). Well formatted mathematical framework is used. Formal training algorithms are used to label. Here comparison between possible patterns and unknown speeches is made. Quality of unknown speeches depends on pattern match. This approach is very much popular method for speech recognition in the last six decades [9, 10].

c. *Artificial Intelligence Approach (AI)*

The Artificial Intelligence approach combines the above two approaches [9].

3. Machine Learning Techniques for Speech Recognition

A. *Deep Belief Network:*

It has a special type of neural network algorithm. It contains one visible layer at the bottom and number of hidden layers. Higher the layers, complex are neural network. The more higher the layers, the more complex are the neural network representations. Learning process is difficult in this approach. Here ordinary algorithm does not work properly. Here greedy and layerwise pre-training is used. After that supervised learning will be done by fine tuning, using the renowned gradient descent. There is an impact of pre-training phase on choice of values for supervised learning. Practically, it works better than the conventional random initialization of the weights, and causes to avoid local minima while using gradient descent in backpropagation DBNs are constructed by stacking many layers of restricted Boltzmann machines. An RBM contain two layers, out of that one layer is the hidden and the other is visible. Main goal here is to distribute given input among them. This leads to multiple layers of nonlinearity. This gives accurate input representation as well. Greedy bottom-up approach is applied for pertaining of greedy DBN. But top-down approach is used to hide information [11].

B. *Convolution Neural Networks (CNNs):*

Convolution Neural Networks is feedforward network, which makes hybrid of three architectural ideas to get surety of shared weights, some degree of shift and distortion invariance, Also along with temporal subsampling and spatial subsampling and local fields. Here, concept of shift window comes which is also called as small portion of data in local fields. CNN mainly contains convolution process and subsampling process. Trainable filters are input in convolution process. Trainable bias is used to produce convolution layer. In subsampling process adding of nearer four different pixels weights also trainable bias and passes it through sigmoid function so as to get small feature map. Here convolution layer is input data and transferred into smaller dimension feature maps by using subsampling [11]. In CNN spatial and temporal subsampling can shrink feature maps to previous level. New set of map is generated.

4 Three Steps to Recognize Speech

(1) *Enhancement of Signals*:

First step is to extract raw signals after that noise removal process is done which will remove different types of noise. It will be useful for improving recognition rate. To enhance speech frame adaptive probabilistic model is used on speech. Here speech window is set up and used moved on speech. This window is used for signal variation analysis. Signal averaging and thresholding techniques are used to improve signal. This improvement identifies low level and high-level signal noise. Noise suppression is done [12].

(2) *Extraction of features*:

There are two types of filters high-pass filter and low-pass filter. Out which is getting will be sampled down by factor two. These two filters are called as high- and low-frequency components. Sizes are N and N/2 along with rows. Along columns again these filters are downsampled by two. Sym6 function is used here to divide function into low- and high-frequency bands. Also, signal exploration is performed [12].

(3) Support Vector Machine (SVM):

Support vector machine is used to recognize speech and it is a classifier. Property of every classifier. It provides the ability to classify during working with multidimensional data. Therefore, it is called as a classifier. This platform models the data which is in a different format and coming from multiple types of sources. It used kernel-based approach. On linear boundaries of every decision are defined in SVM. Here data dependency is used to generate feature space. Linear classifier is a very simple form of SVM. These classifiers are applied on dataset which is balanced [12]. This will be explained along with vectors are represented with fixed dimension, structural representation, and generation of sequence. This method identifies data clarity. To controlled classification process kernel function is defined.

Table 1 Comparative analysis of machine learning techniques

Sr. no	Method	Application	Future use for easy reading of dyslexic child
1	Bag of features which uses frontal face for image classification [13]	Used for checking engage or disengage of child	Frontal face images would be helped to dyslexic read the word correctly
2	Intelligent tutoring systems [14]	Improves the learning process according to the specific needs of each learner	System could be used for easy reading
3	"Bijak Membaca" [15]	By using phonetic reading and multisensory approach reading quality is improved here	It includes simple letters to reduce complexity. Because by using complex letters complexity is improved
4	Off-task behaviors detection [16]	Log records are maintained to check children behavior containing (1) a model that only utilizes time features and (2) a model that uses time and performance features together	While reading- technique off-task behavior is detected then it will be easy to help children for faster reading
5	Adaptive and intelligent systems for collaborative learning support	Used for easy learning	It can be applied for reading purpose

5 Comparative Analysis of Previous Techniques

See Table 1.

6 Proposed Workflow Model

Here, in Fig. 3, prototype model is decided in which two or three letters Hindi words are taken as an input to train the system. As soon as the machine trained with the words. During the phase of training the mapping between Devnagri language and English language will be done. So that system will know about English transliteration of Hindi. Dyslexic child will be given a word to read. If child speak wrong word first time, he/she will be given another chance to read. If same thing is repeated three times then System will speak out the word along with some picture will get display, so that child will get to know that word. Here speech recognition technique of machine learning will be used to recognize the word spoken by child. In this way, the same session will be repeated for 20 min for child for different words.

```
static String[] DevText = { "अ", "आ", "इ", "ई", "ऑ", "ओ", "उ", "ऊ", "ऐ",
"ए", "ऐ", "व", "वा", "वि", "वी", "वें", "वे", "वों", "वो", "वु",
"वू", "रा", "रें", "रे", "रि", "री", "रों", "रो", "रु", "रू", "र्य",
"र्यो", "र्यि", "र्यी", "र्यें", "र्यु", "र्यू", "र्यों", "र्यों",
"त", "ता", "तें", "ते", "ति", "ती", "तों", "तो", "तु", "तू", "त्य",
"त्या", "त्यें", "त्ये", "त्यु", "त्यू", "त्यों", "त्यो", "त्यि",
"त्यी", "य", "या", "यें", "ये", "यि", "यी", "यु", "यू", "यों", "यो",
"प", "पा", "पें", "पे", "पि", "पी", "पु", "पू", "पों", "पो", "प्य",
"प्या", "प्यें", "प्ये", "प्यि", "प्यी", "प्यु", "प्यू", "प्यों",
"प्यो", "प्रा", "प्री", "प्रि", "प्रें", "प्रे", "प्रु", "प्रू",
"प्रों", "प्रो", "प्रु", "प्रू", "सा", "सि", "सी", "सु", "सू", "सों",
"सो", "सें", "से", "स्र", "स्रा", "स्रि", "स्री", "स्रु", "स्रू",
"स्रों", "स्रो", "स्रें", "स्रे", "स्य", "स्या", "स्यि", "स्यी",
"स्यें", "स्ये", "स्यु", "स्यू", "स्यों", "स्यो", "द", "दा", "दि",
"दी", "दों", "दो", "दु", "दू", "दें", "दे", "ध", "धा", "धि", "धी",
"धों", "धो", "धु", "धू", "धें", "धे", "फ", "फा", "फि", "फी", "फों",
"फो", "फु", "फू", "फें", "फे", "फ्र", "फ्रा", "फ्रि", "फ्री",
"फ्रों", "फ्रो", "फ्रु", "फ्रू", "फ्रें", "फ्रे", "ग", "गा", "गि",
"गी", "गों", "गो", "गु", "गू", "गें", "गे", "ग्य", "ग्या", "ग्यि",
"ग्यी", "ग्यो", "ग्यों", "ग्यु", "ग्यू", "ग्यें", "ग्ये", "ग्रा",
"ग्रि", "ग्री", "ग्रों", "ग्रो", "ग्रु", "ग्रू", "ग्रें", "ग्रे",
"घ", "घा", "घि", "घी", "घों", "घो", "घु", "घू", "घें", "घे", "ह",
```

Fig. 2 Devnagri (Hindi) script

Above Fig. 2 is the snapshot of Hindi vowels and some consonants. As we know vowels in English language are "a", "e", "i", "o", "u", where as "अ", "आ", "इ", "ई", "ऑ", "ओ", "उ", "ऊ", "ऐ" are vowels in Hindi language. Some of the letters are given in Fig. 3. In this way, all the Hindi consonants and vowels will be taken as an input and two, three letters Hindi words will be prepared from them as input to our system.

7 Conclusion and Future Scope

In this way, the paper gives technique which will be helpful for easy reading of dyslexic children having age in between 8 and 10 years. The child is allowed to read three or two letters word by which our system will be trained. If he/she is not saying correct word three times. System will give correct word. Here, whenever a child is speaking, Speech recognition by machine learning technique will be done. In the future, we can design algorithm which will be helpful for easy writing as well for the children who are having a problem in writing and easy leaning.

Fig. 3 Prototype model for easy readability

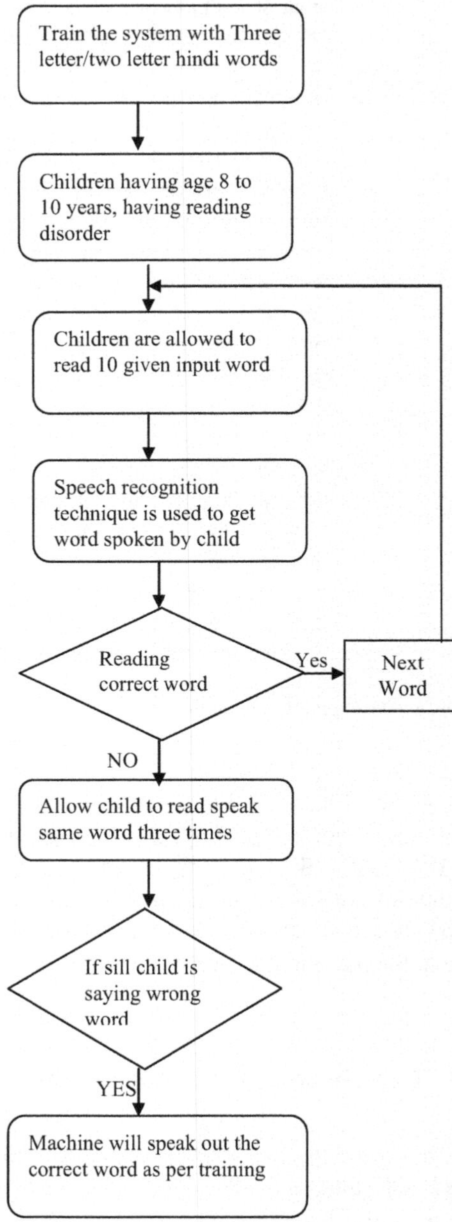

References

1. Davis RD, Braun EM (2010) The gift of dyslexia: why some of the brightest people can't read and how they can learn English, 3rd edn. Souvenir, London
2. Rello L, Saggion H, Baeza-Yates R, Graells E (2012) Graphical schemes may improve readability but not understandability for people with dyslexia. In: NAACL-HLT workshop on predicting and improving text readability for target population, pp 25–32, June 2012
3. Rello L, Baeza-Yates R (2013) Good fonts for dyslexia. ACM 987-1-4503-2405-2/13, October 2013
4. Kamble K, Kagalkar R (2014) A review: translation of text to speech conversion for Hindi language. Int J Sci Res (IJSR) 3(11)
5. Badhe D, Ghate PM (2015) Marathi text to speech synthesis—using Matlab®. IJCSN Int J Comput Sci Netw 4(4)
6. Nandurkar SK, Shaikh ZM (2015) Hindi PDF to speech. Int J Appl Innov Eng Manag (IJAIEM) 4(2)
7. Balyan A, Dev A, Kumari R, Agrawal SS (2015) Development and implementation of Hindi TTS. IEEE. 978-9-3805-4416-8/15
8. Colac J, Borkar S (2016) Design and implementation of Konkani text to speech generation system using OCR technique. Int J Sci Res Publ 6(9)
9. Anusuya MA, Katti SK (2009) Speech recognition by machine: a review. (IJCSIS) Int J Comput Sci Inform Secur 6(3)
10. Alhawit KM (2015) Advances in artificial intelligence using speech recognition. World Acad Sci Eng Technol Int J Comput Inform Eng 09(06)
11. Rubi CR (2015) A review: speech recognition with deep learning methods, India. Int J Comput Sci Mobile Comput 4(5): 1017–1024
12. Rubi CR et al (2015) A review on speech recognition with deep learning methods. Int J Comput Sci Mobile Comput 4(8): 301–307
13. Hamid SSA, Admodisastro N, Manshor N, Kamaruddin A, Abd Ghani AA (2018) Dyslexia adaptive learning model: student engagement prediction using machine learning approach. In: Ghazali R et al (eds) Recent advances on soft computing and data mining. Advances in intelligent systems and computing 700, Springer International Publishing AG
14. Slavuj V, Kovai B, Jugo I (2015) Intelligent tutoring systems for language learning. Opatija, Croatia, MIPRO 2015, 25–29 May 2015
15. Ahmad SZ, Ludin NNAAN, Ekhsan HM, Rosmani AF, Ismail MH (2012) Bejak Membaka—Applying phonetics reading techniques and multisensory approach with interactive multimedia for dyslexia children. In: IEEE colloquium on humanities, science and engineering, Dec 2012
16. Cetintas S, Si L, Xin YP, Hord C (2010) Automatic detection of off-task behaviors in intelligent tutoring systems with machine learning techniques. IEEE Trans Learn Technol 03

Multi-attached Network Topology with Different Routing Protocols and Stub Network Resolution in OSPF Routing

K. Pavani, Himanshu Mishra and Ramkumar Karsh

Abstract The multi-homing technology has always been an area for research interest in computer networking after its implementation to the Internet. Various multi-homing topologies are proposed for better understanding and approach to solving the issues raised by multi-homing technology. Relating to one such issue caused by stub networks, the performances of various routing protocols are analyzed and packet flow is studied in a multi-homed network. This article aims toward a multi-attached network topology as a type in multi-homing technology which is configured using different routing protocols for backup routes and reliable data flow. This paper also compares the static routing protocol with all other dynamic routing protocols and analyses the performance using the proposed topology.

Keywords Multi-attached networks · Routing · RIP · EIGRP · OSPF · Stub hosts · Loopback

1 Introduction

A routing protocol specifies how the networking devices communicate amongst each other, distributing useful data that helps them to specify routes between nodes in a computer network. Routing algorithms make the choice of routes. Each router initially has knowledge of only directly connected networks. A routing protocol shares this information in the whole network after sharing with the immediate neighbors and all routers are updated with the whole topology information [1].

K. Pavani (✉) · H. Mishra · R. Karsh
Department of Electronics and Communication Engineering, National Institute of Technology Silchar, Silchar, India
e-mail: kpavani.1811@gmail.com

H. Mishra
e-mail: him.nits@gmail.com

R. Karsh
e-mail: tnramkarsh@gmail.com

© Springer Nature Singapore Pte Ltd. 2019

V. Nath and J. K. Mandal (eds.), *Proceedings of the Third International Conference on Microelectronics, Computing and Communication Systems*,
Lecture Notes in Electrical Engineering 556, https://doi.org/10.1007/978-981-13-7091-5_12

Routing methods are default routing, static routing, and dynamic routing and discussing about routing protocols they are primarily categorized as interior gateway routing protocol and exterior gateway routing protocol. Broadly they are discussed as EIGRP, RIP, RIPv2, RIPng, IGRP, OSPF, IS-IS, EGP and BGP. All the routing protocols maintain routing tables. Routing table is built based upon three factors serially being: administrative distances of respective routing protocols, metric values of respective routing protocols and prefix length. The detailed explanation and functioning of these are discussed later on, in this document.

Now, the kind of network which only passes the data to or from a local host is called a stub network. Stub networks don't carry the data if they are not the source or destination. As most of the networks have single connection with the Internet and if this connection happens to fail, the stub networks are disconnected, and users suffer long network delays [2]. A stub network, i.e., with a /32 subnet mask generally follows the largest prefix rule while in routing table. The stub networks are classified into two categories based on data flow: data providers, that more specifically send data out of the sub-networks: and enterprises, that more specifically receive data from out of the stub network.

In this paper, it is observed that considering a multi-attached network topology, a loopback interface network once advertised on OSPF is declared as a stub host and thus the lowest AD (Administrative Distance) rule is violated while choosing the best route. Instead, the largest prefix rule is followed for packet forwarding, considering the same destination network id as different network ids. This technique in turn also increases the size of routing table. Larger routing tables require more memory and bandwidth for lookup for packet forwarding. Various software-based schemes were proposed by the researchers to speed up the lookup functionality. All the approaches required a minimum of four or six memory accesses. The need for higher throughput, latency and higher bandwidth of present memory architecture actually limited the count of affordable memory accesses [3]. Taking subnet masks other than /32 for loopback interfaces when working with other routing protocols leads to wastage of available IPv4 addresses. This issue is also considered and avoided in the proposed network model/topology. All the network topology simulation is done on Cisco Packet Tracer and verified on real-time Cisco routers. Hence, all the commands given further in the document are more specific to Cisco Packet Tracer environment.

2 Types of Network Models

Traditionally, most stub networks are connected to the Internet via only one connection or single route. In such cases, if that route happens to fail due to some technical or physical faults, the users suffer many delays. To avoid these situations, multiple connections or routes are established between the Internet and stub network. This methodology is referred to as multi-homing technology. Multi-homing is a technology in which a single network makes multiple connections with the Internet [2]. Depending on the number of ISPs the network models can be classified as:

2.1 Multi-attached Networks

The network that is connected to an Internet service provider (ISP) with several (multiple) connections is termed as a multi-attached network as shown in Fig. 1.

2.2 Multi-homed Networks

The network that is connected to multiple ISPs is termed as a multi-homed network as shown in Fig. 2 [4].

A multi-attached network can be connected to an ISP via various methods. One way is configuring all the routes with one routing protocol that satisfies the requirements of both ISP and users in which there will always be a backup route. Another way is to configure different routes with different routing protocols which will assist

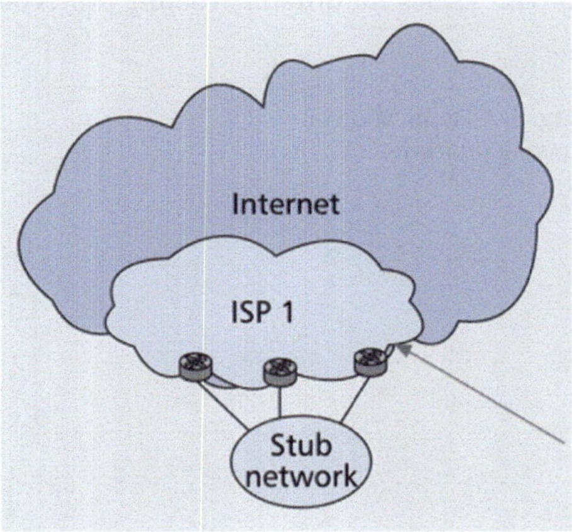

Fig. 1 Multi-attached network

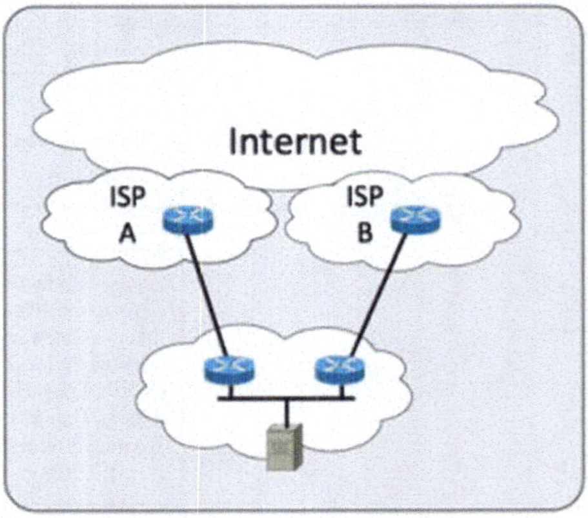

Fig. 2 Multi-homed network

us to use any route if the requirements happen to change anytime. By some minor changes in the topology, it is possible to use the required route without a requirement of installation of a whole new route.

3 Overview of Routing Protocols

Routing protocols are established rules for the data packets to flow from the source machine to the destination machine. Based on various deciding factors a route or the path to be selected by the packets is decided. The classification of routing protocols is depicted in the flowchart as shown in Fig. 3.

Beginning with the default routing method, it has AD value of 1. The AD is administrative distance. Administrative distance is defined as trustworthiness otherwise reliability of a routing protocol. Lower AD value implies higher preference of the route. While making routing decisions the route with the least AD value is always preferred.

AD values for different routing protocols are shown in Table 1.

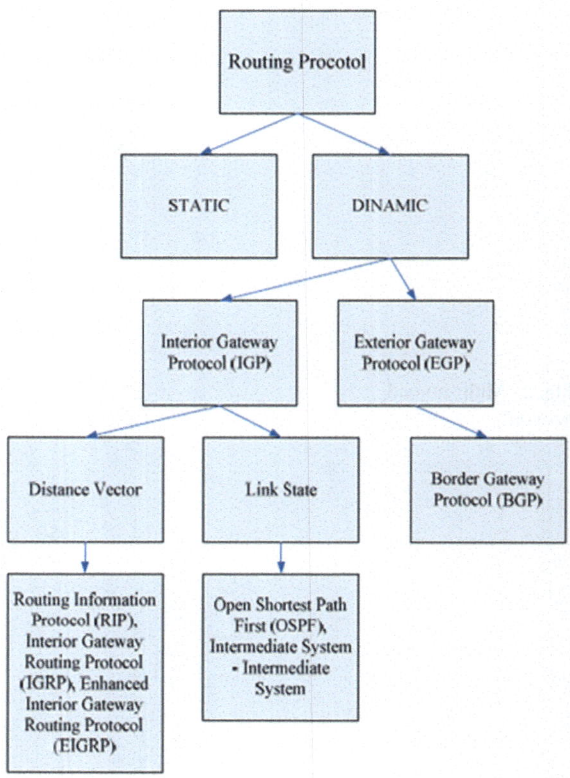

Fig. 3 Classification of routing protocols

Table 1 AD values for different routing protocols

Routing protocol	AD
Directly connected	0
Static route	1
Internal EIGRP	90
OSPF	110
RIP	120
External EIGRP	170
Unknown	225

Default routing is mainly preferred when the topology incorporates some unknown network and also when a network topology is being installed for the very first time. Default routing has no security as any unknown network can join the network topology.

Default route of configured as:

```
Router>en
Router# configure terminal
Router(conf)# ip route 0.0.0.0 0.0.0.0 <next hop address/exit interface>
```

Static routing is when all the remote networks are advertised manually on the routers. It is secured but affordable only in case of small network topologies. Its maximum hop count is 100 and has an AD value of 1 or 0 if exit configured on an exit interface. Suppose if two static routes are configured between the user and ISP and thus we want one route to serve as a backup when the other fails, we can configure the backup with higher AD value than the main route.

Dynamic Routing is when routing protocols are used for updating routing tables of the routers instead of manual entries. The routing protocols are further classified as IGP (Interior Gateway Protocol) and EGP (Exterior Gateway Protocol). IGP is when information exchange takes place within AS (Autonomous System) and BGP is when information is shared between external ASs where, AS is defined as the single administrative domain of an individual network or a group of networks.

3.1 Distance-Vector Routing Protocols

Distance-vector routing protocols utilize Bellman–Ford algorithm. Here no router has the complete information of the topology. Each routers calculates the distance value (DV) and shares that information with its neighbor router and this process goes on with every router. Utilizing this shared information, router fill and update their routing tables. This process continues up to when the routing tables of each router attain stable values [5].

(1) *Routing Information Protocol (RIP)*: It is the actual distance-vector routing protocol. Routing updates are sent by RIP via the configured interface that is active, every 30 s. It chooses the best path based on hop count and the maximum hop count is 15. So, in a complete RIP configured topology 16 routers can be connected. It works well with small network topologies but can't with large network toxpologies and variable bandwidth. RIP version 1 supports only class-full routing whereas RIPv2 supports both class-full and classless routing. RIPv2 does prefix routing which means sending subnet information along with IPs.

Comparison table of RIPv1 and RIPv2 features are shown in Table 2.
The routing table of RIP is checked by the command:

Router# show ip route

And details of the protocol can be checked using the command:

Router# show ip protocol

(2) *Enhance Interior Gateway Routing Protocol (EIGRP)*: It is a classless distance-vector routing protocol that defines sharing of routing information within a contiguous set of routers or an AS. It is also described as the advanced distance-vector routing protocol because it has properties of both link-state and distance-vector. It only sends the full updates while startup and later only shares the specific updates when a change in topology occurs. Its maximum hop is 255 with default being 100. Its features are as shown in Table 3.

It maintains three tables: neighbor table checked by the following:

Router# show ip eigrp neighbor

Table 2 RIPv1 and RIPv2 features

RIPv1	RIPv2
Distance-vector protocol	Distance-vector protocol
Maximum Hops is 15	Maximum hops is 15
Maximum Path is 16	Maximum path is 16
Broadcast IP is 255.255.255.255	Multicast and unicast IP is 224.0.0.9
Classful routing	Supports CIDR and VLSM by default it is classful but supports classless by #no auto summary
Hello timer = 30 s	Hello timer = 30 s
Invalid timer = 180 s	Invalid timer = 180 s
Hold-down timer = 180 s	Hold-down timer = 180 s
Flush after timer = 240 s	Flush after timer = 240 s
Split horizon is enabled	Split horizon is enabled
Sends version 1 packets and receives all packets	Sends version 2 packets and receives 2 packets

Table 3 EIGRP features

Support for IPv4 and IPv6 via protocol-dependent modules
Considered classless (same as RIPv2)
Support for VLSM/CIDR
Support for summaries and discontinuous networks
Efficient neighbors discovery
Data communication via reliable transport protocol (RTP)
Best path selection via diffusion update algorithm (DUAL)
Reduced bandwidth usage with bounded updates
Multicast IP is 224.0.0.10
Authentication type is MD5
Hop count is 255 per AS. By default it is 100

Topology table checked by:

Router# show ip eigrp topology

And route table checked by:

Router# show ip eigrp route.

Every router has the state information of the neighbor router. When any change in neighbor-ship occurs, it is updated in the neighbor table. Neighbor table thus then updates the topology table. Best loop-free path is then chosen by the DUAL. The route table records all the routes presently in use to make local routing decisions. Anything that the routing table consists of is considered as the next route. It has five types of packets namely, Hello packets, Update packets, Query packets, Reply packets and Acknowledgement packets. The range of AS numbers used is <1-65535>. It has five types of metrics $k1 = 1$, $K2 = 0$, $k3 = 1$, $k4 = 0$ and $k5 = 0$ where k1 is bandwidth, K2 is load, k3 is delay, k4 is reliability and k5 is MTU (Maximum Transmission Unit). The metric value is calculated by using the formula:

$$[(10^7/\text{k1}) + (\text{k3}/10)] * 25 \tag{4}$$

Router# show ip eigrp protocols

Prior to EIGRP routers can share routes among them, they should be neighbors, and also there exists other three conditions: Hello and receive of ACK, match of AS numbers and identical k values (metric).

Router# show ip eigrp traffic

Hello packets are continuously exchanged between routers every 5 s. When any change occurs, the prior is holding down and later declared dead and neighbor table is always maintained on the running router.

3.2 Link-State Routing Protocols

Link-state routing protocols are when the best path to destination is calculated using all the best paths calculated by every single router in the topology. Here, every router has topology information and calculate the best path to destination individually. This contrasts with distance-vector routing protocols, which function by information sharing by all the nodes with their neighbors. Link-state protocols share information of construction of connectivity maps between nodes [5].

(1) *Open Shortest Path First (OSPF)*: It functions on the Dijkstra algorithm in which a first formulated a shortest path tree. For the routing table to be updated with best possible paths.

The OSPF features are shown in Table 4.

OSPF should have an area 0, which is called backbone area and all other areas should be connected to it. A router which helps in connecting all other areas in the network to the backbone area and which itself also possess a connection with area 0 within an AS is known an area border router (ABR). OSPF also connects multiple autonomous systems and the router that bridges these ASs is known as an autonomous system boundary router (ASBR). Working of OSPF is like, it is first checked on which interfaces, OSPF is configured. It can be checked by:

Router# show ip ospf interface

And then neighbor-ship or adjacency is checked. Then LSA flooding takes place followed by SPF tree calculation. It can be checked by the command:

Router# show ip ospf

If an adjacency is found active Hello packets are sent every 10 s for point-to-point networks and every 30 s for point-to-multipoint networks. The address used for this is 224.0.0.5.

OSPF uses LSA flooding method to share routing information [6]. Topology database is maintained by flooding of LSA (Link-State Advertisement) packets within the area. All routers maintain the same topology for SPF calculation. The network type determines the multicast address associated with LSA flooding as shown in Table 5. Point-to-multipoint networks make the use of unicast IP address of next router.

Table 4 OSPF features

Supports creation of autonomous systems and areas
Routing update traffic is minimized
Highly versatile, scalable and flexible
Supports VLSM/CIDR
Hop count is unlimited
Open standard protocol with supports for multi-vendor deployment

Table 5 LSA updates addresses

Network type	Multicast address
Point-to-point	224.0.0.5
Broadcast	224.0.0.6
Point-to-multipoint	NA

After the LSA updates are shared in the network, receiver acknowledges receive of the update. Validation of the LSA update is equally mandatory for the receiver. OSPF fills LSDB (Link-State Database) with LSAs. There are 7 types of LSAs.

LSA 1 is referred to as Router LSA. This happens within the area when one router calls other within the area. One internal router informs an update to another internal router.
LSA 2 is Network LSA. In this, update is sent by DR to other routers within the same area similar to LSA 1.
LSA 3 is Summary LSA. Unnecessary flooding of LSA 1 is sopped by ABR.
LSA 4 is Summary ASBR LSA. LSA 5 is out by LSA 4.
LSA 5 is Autonomous System External LSA. Summarized network of other routing protocol is put on ASBR.
LSA 6 is referred to as MOSPF (Multicast Open Shortest Path First). This works while purchasing Class D and while multi-casting.
LSA 7 is Not-so-stubby area LSA.

In a given area, best path is calculated by every router using SPF (Shortest Path First) algorithm. OSPF uses cost as a metric. A cost is defined for all the outgoing interfaces in the SPF tree. The overall cost is defined as the sum of costs of all the interfaces in the path. Equation for the calculation of the cost at OSPF interface is

$$10^8/\text{bandwidth} \tag{5}$$

where, in Eq. 5, bandwidth represents user configured bandwidth for that particular interface.

OSPF maintains three tables: neighbor table, checked by the command:

Router# show ip ospf neighbor

Database table, checked by the command:

Router# show ip ospf database

And route table, checked by the command:

Router# show ip route.

While neighbor table shows the states of OSPF neighbors, database table shows advertised routers/linked routers, age and sequence number.

The other OSPF information including Router is checked by the command:

Router# show ip protocols

4 Experimental Setup

A multi-attached network topology is setup as shown in Fig. 4.

As shown in the figure above, a network router and an ISP router is configured with four paths between them. Each path is configured with a different routing protocol. The topmost paths are configured with RIP. The second from top is configured with OSPF with AS number 100 and area 0. The third from top is configured with EIGRP with AS number 100 and the bottom most is configured with static routing.

All the ip addresses used are class-full. A loopback interface L0 is declared on router Network with ip address 10.1.1.1/8 and a loopback interface L0 is declared on router ISP with ip address 20.1.1.1/8 [7].

The loopback of router ISP is pinged from router Network and the route is traced as shown in Fig. 5.

5 Result and Comparison

In the above topology, we have seen that instead of opting for the lowest AD path, i.e., the static route (AD = 1) path the packet flowed through the OSPF (AD = 110) path. The primary rationale behind this is that the OSPF routing protocol declares the loopback interfaces as stub hosts. The stub hosts have a subnet mask /32. As, /32 is the largest subnet mask available for the IPv4 addresses, the packets follow the largest prefix rule [specific] ignoring the lowest AD rule [trustworthiness] [8]. This

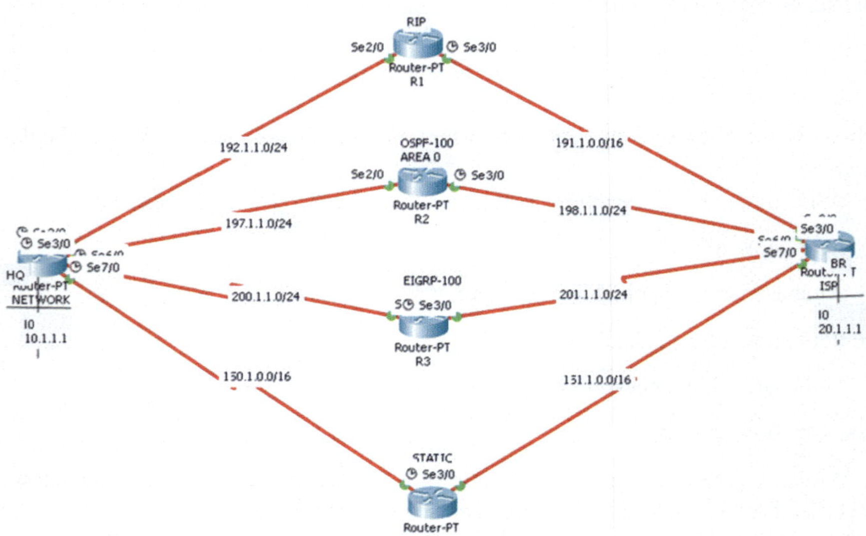

Fig. 4 Multi-attached network topology

```
network#traceroute 20.1.1.1
Type escape sequence to abort.
Tracing the route to 20.1.1.1

  1  197.1.1.1      65 msec   2 msec    4 msec
  2  198.1.1.2      1 msec    0 msec    2 msec
network#traceroute 20.1.1.1
Type escape sequence to abort.
Tracing the route to 20.1.1.1

  1  197.1.1.1      717 msec  1 msec    1 msec
  2  198.1.1.2      1 msec    2 msec    *
network#traceroute 20.1.1.1
Type escape sequence to abort.
Tracing the route to 20.1.1.1

  1  197.1.1.1      1 msec    1 msec    1 msec
  2  198.1.1.2      0 msec    1 msec    2 msec
network#traceroute 20.1.1.1
Type escape sequence to abort.
Tracing the route to 20.1.1.1

  1  197.1.1.1      12 msec   1 msec    2 msec
  2  198.1.1.2      1 msec    2884 msec 1 msec
network#
```

Fig. 5 Trace route from network to ISP

```
network#show ip ospf interface

Loopback0 is up, line protocol is up
  Internet address is 10.1.1.1/8, Area 0
  Process ID 100, Router ID 10.1.1.1, Network Type LOOPBACK, Cost: 1
  Loopback interface is treated as a stub Host
Serial3/0 is up, line protocol is up
  Internet address is 197.1.1.2/24, Area 0
  Process ID 100, Router ID 10.1.1.1, Network Type POINT-TO-POINT, Cost: 64
  Transmit Delay is 1 sec, State POINT-TO-POINT, Priority 0
  No designated router on this network
  No backup designated router on this network
  Timer intervals configured, Hello 10, Dead 40, Wait 40, Retransmit 5
    Hello due in 00:00:07
  Index 2/2, flood queue length 0
  Next 0x0(0)/0x0(0)
  Last flood scan length is 1, maximum is 1
  Last flood scan time is 0 msec, maximum is 0 msec
  Neighbor Count is 1 , Adjacent neighbor count is 1
    Adjacent with neighbor 198.1.1.1
  Suppress hello for 0 neighbor(s)
network#
```

Fig. 6 OSPF interface of router network

can be verified by checking the ospf interfaces on the router Network as shown in Fig. 6 and also by checking the routing table of router Network as shown in Fig. 7.

This issue is overcome by using a Cisco proprietary configuration. It is by going to the L0 interface of router Network and L0 interface of router ISP and giving the command:

Router (config-if) # ip ospf network point-to-point

```
network#sh ip rou
Codes: C - connected, S - static, I - IGRP, R - RIP, M - mobile, B - BGP
       D - EIGRP, EX - EIGRP external, O - OSPF, IA - OSPF inter area
       N1 - OSPF NSSA external type 1, N2 - OSPF NSSA external type 2
       E1 - OSPF external type 1, E2 - OSPF external type 2, E - EGP
       i - IS-IS, L1 - IS-IS level-1, L2 - IS-IS level-2, ia - IS-IS inter area
       * - candidate default, U - per-user static route, o - ODR
       P - periodic downloaded static route

Gateway of last resort is not set

C    10.0.0.0/8 is directly connected, Loopback0
     20.0.0.0/8 is variably subnetted, 2 subnets, 2 masks
S       20.0.0.0/8 is directly connected, Serial7/0
O       20.1.1.1/32 [110/129] via 197.1.1.1, 00:31:38, Serial3/0
C    150.1.0.0/16 is directly connected, Serial7/0
S    151.1.0.0/16 is directly connected, Serial7/0
R    191.1.0.0/16 [120/1] via 192.1.1.1, 00:00:15, Serial2/0
C    192.1.1.0/24 is directly connected, Serial2/0
C    197.1.1.0/24 is directly connected, Serial3/0
O    198.1.1.0/24 [110/128] via 197.1.1.1, 00:35:00, Serial3/0
C    200.1.1.0/24 is directly connected, Serial6/0
D    201.1.1.0/24 [90/21024000] via 200.1.1.1, 00:31:58, Serial6/0
```

Fig. 7 Routing table of router network

Table 6 Performance comparison of the proposed method with state-of-the-art methods

Parameter	[10]	[11]	[12]	[13]	Propose
Size of the Internet routing table	Large	Large	Large	Small	Small
Spoofing and network intruder	Yes	Yes	No	Yes	No
Routing based on AD value (Trustworthiness)	Yes	No	No	No	Yes

In this paper, this issue has been resolved by configuring the loopbacks as /32 before-hands and advertising it on all protocols as such. By doing so it is seen that the packets follow static route path first. This also reduces the size of routing table by displaying the 20.1.1.1 network only once in the static route [9].

Comparative study is shown in Table 6.

Further, the declaration of /32 subnet masks for loopbacks conserves the available IPv4 address space as for loopback only one IP is required.

The above topology is one that can be used to study all the routing protocols at once by comparing and analyzing their performance by using the command:

Router# show ip protocols

on any of the Network or ISP routers.

6 Conclusion

So, after having the proper knowledge of various routing protocols and their performances, here a multi-attached network topology is proposed which is configured using different routing protocol for each one of the paths. Using this topology the performance and best path detection mechanism is studied for the routing protocols. Stub networks are considered while tracing the route and OSPF properties are further looked into. Size of routing table and conservation of IPv4 addresses are examined and /32 subnet for all loopbacks is proposed for advertising on all routing protocols.

References

1. Liu H (2002) Routing table compaction in ternary CAM. IEEE Micro 22(1):58–64
2. Kumar R, Vats J, Kumar A (2011) A comparative study of routing protocols. 2: 1962–1964
3. CCNA by Todd Lammle, 5th edition
4. Liu X, Xiao L (2007) A survey of multi-homing technology in stub networks: current research and open issues. IEEE Netw 21(3)
5. Lehane A, Garcia F, Sventek J (2006) Agilent Technologies Inc, assignee. Identifying network routers and paths. United States patent application US 10/521,777, 16 Mar 2006
6. Cisco (2008) Route Selection in Cisco Routers, Document ID: 8651, 02 Jan 2008
7. Aggarwal R, Kompella K (2010) Advertising a router's local addresses in OSPF traffic engineering (TE) extensions
8. Shen J, Tan HW, Wang J, Wang JW, Lee SY (2015) A novel routing protocol providing good transmission reliability in underwater sensor networks. 16(1): 171–178
9. Hiromori A, Yamaguchi H, Yasumoto K, Higashino T, Taniguchi K (2003) Reducing the size of routing tables for large-scale network simulation. In: Proceedings of the seventeenth workshop on parallel and distributed simulation, 10 June 2013. IEEE Computer Society, p 115
10. Çakmak G, Aydin MN (2017) A country-specific analysis on internet interconnection ecosystems. In: 2017 9th international congress on ultra modern telecommunications and control systems and workshops (ICUMT). IEEE
11. Lone Q et al (2017) Using loops observed in traceroute to infer the ability to spoof. In: International conference on passive and active network measurement. Springer, Cham
12. Axnäs J et al (2017) Route determination in a multi-hop network using multiple routing metrics. U.S. Patent no. 9,647,930, 9 May 2017
13. Absar N, Wahab A, Sikder KU (2017) Performance measurement of open shortest path first (OSPF) protocol in IP networks. Int J Eng Res 6(2):110–115

A Parallel Bit-Plane Operation Based Chaotic Image Encryption Scheme

K. Abhimanyu Kumar Patro, Ajay Singh Raghuvanshi and Bibhudendra Acharya

Abstract Nowadays, the Internet is the medium through which peoples are sharing digital images. However, security is an issue in sharing of those digital images. So the main objective is to provide security in digital images. This paper proposes a bit-plane operation based encryption scheme which to provide security in digital images. In this encryption scheme, first, the bits in bit-planes are confused by using 4D hyper-chaotic map and then diffused by using PWLCM (Piecewise Linear Chaotic Map) system. Both the confusion and diffusion operations are performed in parallel and independent to get better security in images. The main advantage of this scheme is the use of separate chaotic sequences and separate key images to perform parallel and independent bit-level operation. This type of parallel and independent operation confuses pixels, reduces in attacking the cryptosystem by attackers, and thus increases the level of security. Moreover, the use of 4-D hyper-chaotic map increases the key space of the algorithm and thus more resists the algorithm against brute-force attack. Apart from that, a "Secure Hash Algorithm SHA-256" is used in this scheme to generate the secret keys which to provide resistivity against "Known-Plaintext Attack (KPA)" and "Chosen-Plaintext Attack (CPA)". The proposed method has better performance in terms of encryption efficiency, security, and resistivity for most of the common attacks.

Keywords Security · Image encryption · Bit-level operation · Chaotic system · Hyper-chaotic system · SHA-256

K. Abhimanyu Kumar Patro (✉) · A. S. Raghuvanshi · B. Acharya
Department of Electronics and Telecommunication Engineering, National Institute of Technology Raipur, Raipur 492010, Chhattisgarh, India
e-mail: abhimanyu.patro@gmail.com

A. S. Raghuvanshi
e-mail: asraghuvanshi.etc@nitrr.ac.in

B. Acharya
e-mail: bacharya.etc@nitrr.ac.in

© Springer Nature Singapore Pte Ltd. 2019

V. Nath and J. K. Mandal (eds.), *Proceedings of the Third International Conference on Microelectronics, Computing and Communication Systems*, Lecture Notes in Electrical Engineering 556, https://doi.org/10.1007/978-981-13-7091-5_13

1 Introduction

Due to the speedy advancements in the Internet, large numbers of images are stored and transmitted through them. But the major challenge is in their security. Encryption is one of the techniques which provide security in storage and transmission of images through the Internet. Various traditional techniques of encryption like "DES" [1], "3DES", "AES" [2], etc. are present but they do not perform better for encrypting images because of certain inherent properties of images like high redundancy, stronger correlation in adjacent pixels, massive data capacity, etc. [3–6].

In order to overcome these challenges, from the past few years, image encryption schemes are widely used chaos to perform encryption and decryption operation. Chaotic maps have certain prominent features like "sensitivity to initial conditions", "ergodicity", "pseudo-random property", "non-periodicity", etc. [7, 8]. These features of chaotic maps are very much essential to make a secure and stronger cryptosystem. Matthews [9] first proposed chaos-based encryption technique, since then, chaos-based encryption techniques are followed by many researchers. According to dimension, chaotic maps are classified into "one-dimensional chaotic map" and "high-dimensional chaotic map" [10]. Both have several pros and cons. "One-dimensional chaotic maps" has better resource utilization because of its simplicity in structure. It also has better speed and efficiency. But "one-dimensional chaotic system" provides smaller key space and weaker security [11, 12]. On the other hand, high chaotic performance, greater randomness, high key space, and key sensitivity are provided by "high-dimensional chaotic map" (mostly, hyper-chaotic map). But the complex structure of "high-dimensional chaotic map" requires large numbers of hardware resources [12]. A combination of "one-dimensional chaotic map" and "high-dimensional chaotic map" based cryptosystem is proposed in this paper for security and speed trade-offs.

"Pixel-level permutation operation" or "bit-level permutation operation" using chaotic maps has been used by different researchers in image encryption. In [13, 14], pixel-level permutation operations are used whereas in [15, 16], bit-level permutation operations are used. Only pixel positions are changed in "pixel-level permutation operation" whereas both pixel positions and values are changed in "bit-level permutation operation". Hence, this paper uses bit-level operations to take the advantages of changing both pixel positions and values simultaneously.

Most of the image encryption techniques are used a single key image to perform bit-diffusion operation which somehow reduces the level of security. But this paper uses 8 different key images to perform bit-diffusion operations which increase the level of security. Apart from that eight different hyper-chaotic sequences are used to perform bit-shuffling operations in eight bit-planes independently which more confuses pixels.

This paper contributes the following:

- A 4D hyper-chaotic map is used to permute bits in each of the bit-planes independently. This independent permutation operation more confuses pixels.

- The utilization of 4D hyper-chaotic map increases the key space so as strongly resists brute-force attack.
- A PWLCM system is utilized to generate eight key images to perform bit-diffusion operation independently so as increased the security level.
- The "Secure Hash Algorithm SHA-256" is utilized to resist the proposed algorithm against "Known-Plaintext Attack (KPA)" and "Chosen-Plaintext Attack (CPA)".

The organization of this paper is as follows. Section 2 introduces Chen's 4D hyper-chaotic map and the PWLCM system. The proposed encryption algorithm is discussed in Sect. 3. In Sect. 4, security analyses results and computer simulation outputs are discussed. Section 5 concludes the paper.

2 Preliminaries

2.1 Chen's 4D Hyper-Chaotic Map

It can be defined as [17, 18],

$$\begin{cases} \dot{p} = -ap + aq \\ \dot{q} = cq - pr + dp - s \\ \dot{r} = -br + pq \\ \dot{s} = k + p \end{cases} \tag{1}$$

where $a, c, d, b,$ and k are the control parameters and $p, q, r,$ and s are the initial values.

2.2 PWLCM System

Most encryption algorithms are widely used PWLCM system because of their good characteristic such as "less sensitivity toward external perturbation" than conventional Logistic map [19]. The expression for PWLCM system is

$$u_{n+1} = \begin{cases} \frac{u_n}{\mu} & \textit{if } 0 \le u_n < \mu \\ \frac{u_n - \mu}{0.5 - \mu} & \textit{if } \mu \le u_n < 0.5 \\ (1 - u_n) & \textit{if } 0.5 \le u_n < 1 \end{cases} \tag{2}$$

where the control parameter μ lies in the range (0, 0.5).

3 Proposed Methodology

Figure 1 illustrates the proposed encryption algorithm. The encryption steps are,

Select $M \times N$ sized a grayscale image A as an input.

Step-2: Perform binary bit-plane decomposition operation on this input grayscale image A to generate 8 bit-planes. Let the 8 bit-planes are denoted as

$$BPGI1, BPGI2, BPGI3, BPGI4, BPGI5, BPGI6, BPGI7, BPGI8 \quad (3)$$

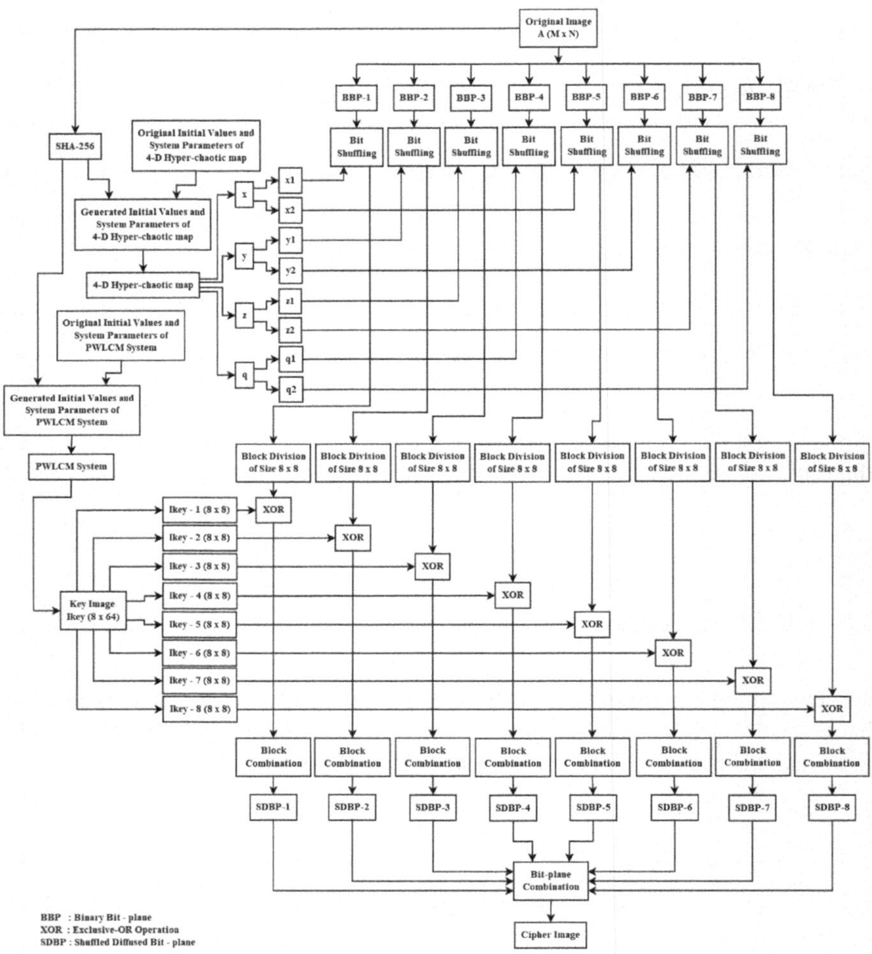

Fig. 1 Block diagram of encryption operation

Step-3: Apply hash algorithm (SHA-256) on A, to generate hash values of 256-bits. Convert hash values of 256-bits into 32-decimal values. The decimal values are

$$hash = h_1, h_2, h_3, \ldots, h_{31}, h_{32} \tag{4}$$

Step-4: Generate the secret keys of 4D hyper-chaotic map.

$$\begin{cases} a = a' + \left(\frac{mod(sum(h1:h3),256)}{2^9}\right) \times 0.1 \\ b = b' + \left(\frac{mod(sum(h4:h6),256)}{2^9}\right) \times 0.1 \\ c = c' + \left(\frac{mod(sum(h7:h9),256)}{2^9}\right) \times 0.1 \\ d = d' + \left(\frac{mod(sum(h10:h12),256)}{2^9}\right) \times 0.1 \\ k = k' + \left(\frac{mod(sum(h13:h15),256)}{2^9}\right) \times 0.1 \end{cases} \tag{5}$$

$$\begin{cases} p0 = p0' + \left(\frac{mod(sum(h16:h18),256)}{2^9}\right) \times 0.1 \\ q0 = q0' + \left(\frac{mod(sum(h19:h21),256)}{2^9}\right) \times 0.1 \\ r0 = r0' + \left(\frac{mod(sum(h22:h24),256)}{2^9}\right) \times 0.1 \\ s0 = s0' + \left(\frac{mod(sum(h25:h27),256)}{2^9}\right) \times 0.1 \end{cases} \tag{6}$$

Step-5: Iterate the 4D hyper-chaotic map for $M \times N \times 2$ times to generate four hyper-chaotic sequences p, q, r, and s.

Step-6: Equally divide each of the hyper-chaotic sequences p, q, r, and s into two parts having sizes $M \times N$. Let the divisional hyper-chaotic sequences are denoted as

$$p1\,\&\,p2\,(for\,p),\,q1\,\&\,q2\,(for\,q),\,r1\,\&\,r2\,(for\,r),\,\text{and}\,s1\,\&\,s2\,(for\,s) \tag{7}$$

Step-7: Generate the index values of each of the divisional hyper-chaotic sequences $p1, p2, q1, q2, r1, r2, s1,$ *and* $s2$. Let the index values are denoted as

$$\begin{gathered} p1index\,(for\,p1),\,p2index\,(for\,p2),\,q1index\,(for\,q1),\,q2index\,(for\,q2), \\ r1index\,(for\,r1),\,r2index\,(for\,r2),\,s1index\,(for\,s1),\,s2index\,(for\,s2). \end{gathered} \tag{8}$$

Step-8: Shuffle the bits of each of the bit-planes $BPGI1, BPGI2, BPGI3, BPGI4, BPGI5, BPGI6, BPGI7, BPGI8$ using the index values $p1index, q1index, r1index, s1index, p2index, q2index, r2index, s2index$, respectively. The shuffled bit-planes are

$$SBPGI1, SBPGI2, SBPGI3, SBPGI4,$$
$$SBPGI5, SBPGI6, SBPGI7, SBPGI8 \tag{9}$$

Step-9: Perform block division of size 8×8 on each of the eight shuffled bit-planes.
Step-10: Generate the keys of PWLCM system.

$$\begin{cases} mue = mue' + \left(\frac{mod(sum(h28:h30),256)}{2^9}\right) \times 0.1 \\ u0 = u0' + \left(\frac{mod(sum(h31:h32),256)}{2^9}\right) \times 0.1 \end{cases} \tag{10}$$

Step-11: Generate a key image *Ikey* of size 8×64 using the PWLCM system based keys.
Step-12: Divide the key image *Ikey* of size 8×64 into eight sub-key images of sizes 8×8. The eight sub-key images are denoted as $Ikey-1$, $Ikey-2$, $Ikey-3$, $Ikey-4$, $Ikey-5$, $Ikey-6$, $Ikey-7$, *and* $Ikey-8$. Each of the sub-key images is used to perform block-diffusion with each of the shuffled bit-planes independently. The sub-key image $Ikey-1$ is used to perform block-diffusion with the shuffled bit-plane *SBPGI*1, similarly, $Ikey-2$ is used for *SBPGI*2, and so on.
The diffusion processes in each of the shuffled bit-planes are as follows. First, the sub-key images are XORed with the first block of each of the shuffled bit-planes. Second, the outputs of the first XOR operation are XORed with the second block of each the shuffled bit-planes till all the blocks in each of the shuffled bit-planes are diffused.
Step-13: Combine all the blocks in each of the bit-planes to generate shuffled-diffused bit-planes. Let the shuffled-diffused bit-planes are denoted as

$$SDBPGI1, SDBPGI2, SDBPGI3, SDBPGI4,$$
$$SDBPGI5, SDBPGI6, SDBPGI7, SDBPGI8 \tag{11}$$

Step-14: Finally combine *SDBPGI*1, *SDBPGI*2, *SDBPGI*3, *SDBPGI*4, *SDBPGI*5, *SDBPGI*6, *SDBPGI*7, *SDBPGI*8, to generate a cipher image.

On reversing the above mentioned steps for encryption, we get the steps for the decryption process.

4 Computer Simulations and Security Analysis

The algorithm is verified by doing "simulation and security analysis" on two grayscale images. They are "Cameraman.tif" and "Lena.tiff" of size 256×256 and 512×512, respectively. The device used to carry out simulation processes is a personal computer having i7 processor (3.40 GHz), 64-bit operating system,

4 GB RAM and using MATLAB version R2012a. The key values used in this algorithm are $a' = 36.0, c' = 28.0, d' = -16.0, b' = 3.0, k' = 0.2, p0' = 0.3$, $q0' = -0.4, r0' = 1.2, s0' = 1$ for 4D hyper-chaotic map, $u0' = 0.2, mue' = 0.3$ for PWLCM system. The simulation results are depicted in Fig. 2 which shows a better encryption performance in the proposed method.

4.1 Key Space Analysis

The details of the keys for the proposed method are,

- Initial values $p0'$, $q0'$, $r0'$, $s0'$ and system parameters a', c', d', b', k' of 4D hyper-chaotic map.
- Initial value $u0'$, system parameter mue', and threshold value of PWLCM system.
- Hash value of 256-bits.

In this algorithm, the precision for all the keys is 10^{-15}, which is IEEE standard for floating point data representation [20] and a precision of 10^{-2} for the "Threshold value" of PWLCM system. Moreover, the "SHA-256 hash algorithm" uses a key space of 2^{128} for resisting the best attack. So, from the above information, the overall key space of our method is $10^{135} \times 10^{32} \times 2^{128} = 10^{167} \times 2^{128} = 1.6958 \times 2^{554} \times 2^{128} = 1.6958 \times 2^{682}$ which is very larger than 2^{128} [21]. Table 1 shows the key space comparison results using our scheme and the schemes in [22, 23]. From Table 1, it can be concluded that our method has best results of key space as comparison to

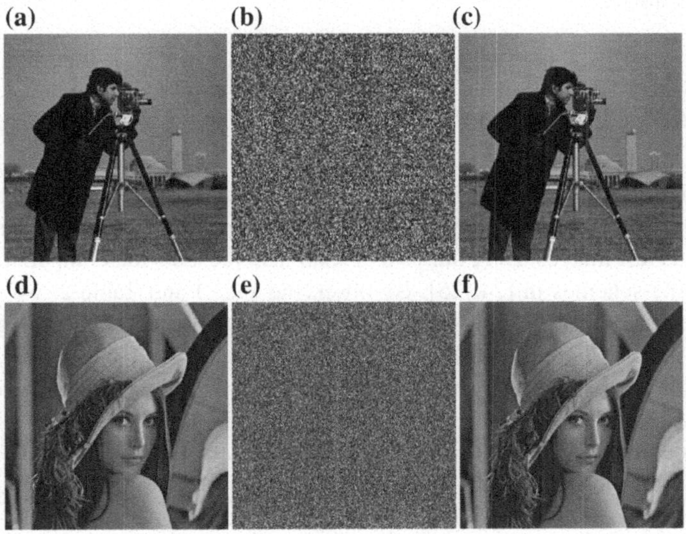

Fig. 2 Simulation output of "Cameraman" image: **a** Org., **b** Enc., **c** Dec.; Simulation output of "Lena" image: **d** Org., **e** Enc., **f** Dec.

Table 1 Comparison of key space results

Algorithms	Key space
Ours	1.6958×2^{682}
Ref. [22]	More than 2^{349}
Ref. [23]	2^{512}

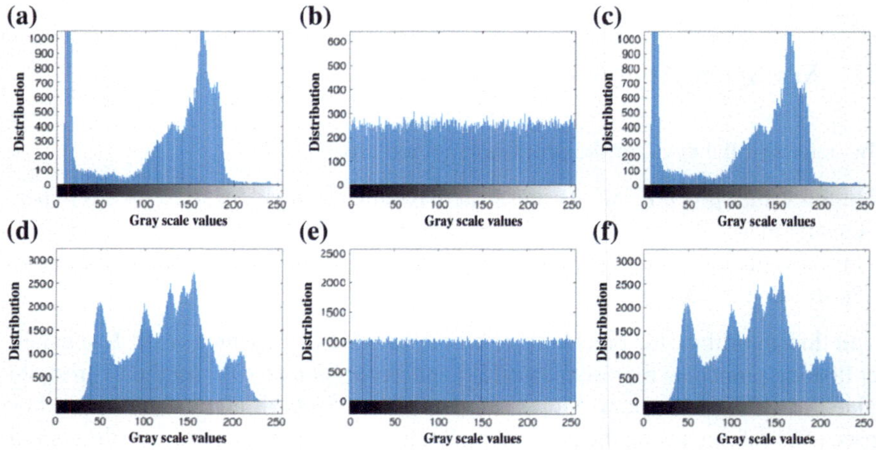

Fig. 3 Histogram plot of "Cameraman" image: **a** Org., **b** Enc., **c** Dec.; Histogram plot of "Lena" image: **d** Org., **e** Enc., **f** Dec.

the schemes in [22, 23]. This shows that our method has strong resistance against brute-force attack.

4.2 *Statistical Attack Analysis*

(1) *Histogram Analysis*: The histogram plot of "Cameraman" and "Lena" image using our scheme is as shown in Fig. 3. Table 2 presents the variance results of various test images using our scheme and also presents the comparison results with the schemes in [24, 25]. By observing Fig. 3 and Table 2, we can realize the greater uniformity in the histograms of ciphered (Enc.) images (gray scale values) than the schemes in [24, 25]. In Fig. 3, we can also observe large differences of histograms of ciphered and plain (Org.) images and greater similarities of histogram of decrypted (Dec.) and plain images. This shows that our method has strong resistance against statistical attacks and it also has integrity in transmission.

(2) *Correlation Analysis*: Table 3 presents the correlation results of various test images using our scheme and also presents the comparison results with the schemes in [26, 27]. In Table 3, we can realize the weak correlation (approxi-

Table 2 Comparison of variance results

Algorithms		Variance of test images	
		Org.	Enc.
Ours	"Lena"	6.3340e+05	994.5547
	"Cameraman"	1.1097e+05	299.7422
Ref. [24]	"Lena"	–	5554.8293
Ref. [25]	"Lena"	–	5335.8309

Table 3 Comparison of correlation results

Algorithms		Correlation of test images			
		Ours (Pairs of adjacent pixels = 10,000)		Ref. [27]	Ref. [26]
		"Cameraman"	"Lena"	"Lena"	"Lena"
Org.	Hz.	0.9319	0.9749	0.9761	0.9597
	Vt.	0.9609	0.9837	0.9626	0.9792
	Dg.	0.9077	0.9592	0.9448	0.9570
Enc.	Hz.	0.0049	0.0042	−0.0285	0.1257
	Vt.	0.0007	0.0026	0.0014	0.0581
	Dg.	0.0016	0.0006	0.0013	0.0504

mately to zero) of neighboring pixels in ciphered images and strong correlation (approximately to one) of neighboring pixels in plain images along "diagonal (Dg.)", "vertical (Vt.)", and "horizontal (Hz.)" directions. Table 3 also shows the better values of correlation of our scheme than the schemes in [26, 27].

Figure 4 depicts the correlation plot of plain and ciphered "Lena" image using our scheme. In Fig. 4, we can realize the weak and strong correlation of neighboring pixels along "diagonal", "vertical", and "horizontal" directions in ciphered and plain images, respectively. This shows that our method has strong resistance against statistical attacks.

4.3 *Differential Attack Analysis*

The differential attack performance measures are "NPCR (Numbers of Pixel Changing Rate)" and "UACI (Unified Average Changing Intensity)". Table 4 shows the average results of both the measures of various test images using our scheme and also presents the comparison results with the schemes in [28, 29]. The UACI and NPCR results of [29] shown in Table 4 are the average results of "Red", "Green", and "Blue" components. By observing Table 4, we can understand that our scheme provides larger UACI and NPCR values than the expected values and also better the values in [28, 29]. In this scheme, these two measures are calculated by averaging

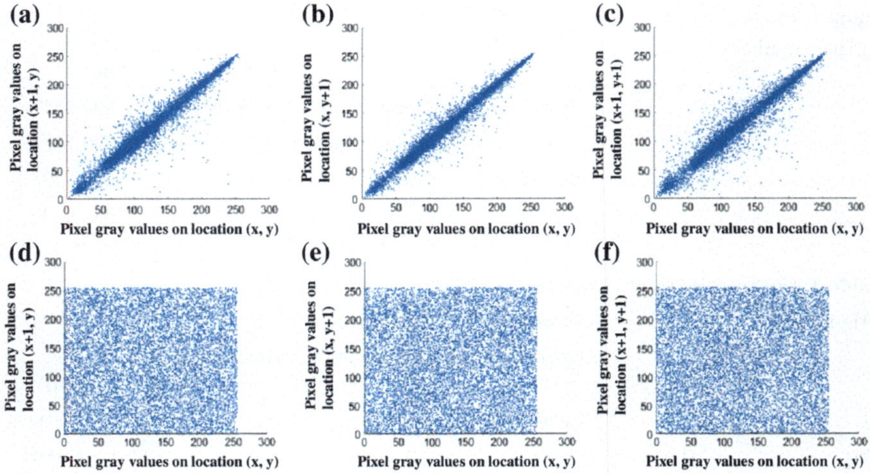

Fig. 4 Adjacent pixel correlation plot of plain "Lena" image: **a** Hz., **b** Vt., **c** Dg.; Ciphered "Lena" image: **d** Hz., **e** Vt., **f** Dg.

Table 4 Comparison of NPCR (Ideal 99.6094%) and UACI (Ideal 33.4635%) results

Algorithms		Average results in percentage	
		UACI	NPCR
Ours	"Cameraman"	33.4799	99.6098
	"Lena"	33.4969	99.6124
Ref. [28]	"Lena"	33.41	99.60
Ref. [29]	"Lena"	33.4342	99.6075

100 randomly pixel changing UACI and NPCR values, respectively. This shows that our method has strong resistance against differential attacks.

4.4 Entropy Attack Analysis

Table 5 shows the entropy results of various test images using our scheme and also presents the comparison results with the scheme in [29]. The ciphered image entropy result of [29] shown in Table 5 is the average entropy result of "Red", "Green", and "Blue" components. In Table 5, we can see that the information entropy results after encryption are very close to 8 using our scheme and also better than the scheme in [29]. This shows that our method has strong resistance against entropy attack.

Table 5 Comparison of entropy results

Algorithms		Entropy of test images	
		Org.	Enc.
Ours	"Lena"	7.4451	7.9993
	"Cameraman"	7.0097	7.9967
Ref. [29]	"Lena"	–	7.9894

4.5 KPA and CPA

The theme behind these attacks is that, if the cryptanalyst has access to both a part of plaintext/ciphertext and its corresponding ciphertext/plaintext, respectively, then he/she is capable of decrypting the image. In this algorithm, all the keys that are generated are dependent on the plaintext using the hash values. So, the keys are not fixed but changed for every input of plaintext. Even though now, if the cryptanalyst has some chosen plaintext/ciphertext, it is not possible to extract all other keys to decrypt the whole information. Because of its high dependability over plaintext which varies every time, make it more efficient toward KPA and CPA.

5 Conclusion

This paper proposed an image encryption using bit-plane "permutation and diffusion" operations. In this scheme, the bit-planes permuted using 4-D hyper-chaotic map and diffused using PWLCM system. The "permutation and diffusion" operations performed in parallel to get a better encryption output and also security. The security analyses show a higher protection of our scheme than the other referred schemes. This proves the suitability of our scheme to encrypt images.

Acknowledgements This work was supported by "Information Security Education and Awareness (ISEA) project phase-II" under MeitY, Govt. of India.

References

1. Coppersmith D (1994) The data encryption standard (DES) and its strengths against attacks. IBM J Res Dev 38(3):243–250
2. Pub NF. 197 (2001) Advanced encryption standard (AES), vol 197. Federal Information Processing Standards Publication, pp 441-0311
3. Gao H, Zhang Y, Liang S, Li D (2006) A new chaotic algorithm for image encryption, vol 29, pp 393–399
4. Samhita P, Prasad P, Abhimanyu Kumar Patro K, Acharya B (2016) A secure chaos-based image encryption and decryption using crossover and mutation operator. Int J Control Theory Appl 9(34):17–28

5. Gupta A, Thawait R, Patro KAK (2016) A novel image encryption based on bit-shuffled improved tent map. Int J Control Theory Appl 9(34):1–16
6. Shadangi V, Choudhary SK, Patro KAK (2017) Novel Arnold scrambling based CBC-AES image encryption. Int J Control Theory Appl 10(15):93–105
7. Guesmi R, Amine M, Farah B, Kachouri A, Samet M (2016) Hash key-based image encryption using crossover operator and chaos. Multimed Tools Appl 4753–4769
8. Guesmi R, Farah MAB, Kachouri A, Samet M (2016) A novel chaos-based image encryption using DNA sequence operation and Secure Hash Algorithm SHA-2. Nonlinear Dyn 83(3):1123–1136
9. Matthews R (2010) Cryptologia on the derivation of a ' chaotic' encryption algorithm, June 2012, pp. 37–41
10. Liu W, Sun K, Zhu C (2016) A fast image encryption algorithm based on chaotic map. Opt Lasers Eng 84:26–36
11. Özkaynak F, Bedri A (2016) Cryptanalysis of a new image encryption algorithm based on chaos. Opt—Int J Light Electron Opt 127:5190–5192
12. Wang X, Wang S, Zhang Y, Guo K (2017) A novel image encryption algorithm based on chaotic shuffling method. Inf Secur J A Global Perspect 1–10
13. Wang X, Guo K (2014) A new image alternate encryption algorithm based on chaotic map. Nonlinear Dyn 76:1943–1950
14. Zhang X, Zhao Z (2014) Chaos-based image encryption with total shuffling and bidirectional diffusion. Nonlinear Dyn 75:319–330
15. Zhu Z, Zhang W, Wong K, Yu H (2011) A chaos-based symmetric image encryption scheme using a bit-level permutation. Inf Sci (NY) 181:1171–1186
16. Fu C, Lin B, Miao Y et al (2011) A novel chaos-based bit-level permutation scheme for digital image encryption. Opt Commun 284:5415–5423
17. Gao T, Chen Z, Yuan Z, Chen G (2006) A hyperchaos generated from Chen's system. Int J Mod Phys C 17:471–478
18. Zhou N, Pan S, Cheng S, Zhou Z (2016) Image compression—encryption scheme based on hyper-chaotic system and 2D compressive sensing. Opt Laser Technol 82:121–133
19. Wang X-Y, Yang L (2012) Design of pseudo-random bit generator based on chaotic maps. Int J Mod Phys B 26(32):1250208
20. Floating-Point Working Group (1985) IEEE standard for binary floating-point arithmetic. ANSI. IEEE Std., pp 754–1985
21. Kulsoom A, Xiao D, Aqeel-ur-Rehman, Abbas SA (2016) An efficient and noise resistive selective image encryption scheme for gray images based on chaotic maps and DNA complementary rules. Multimed Tools Appl 75(1):1–23
22. El-latif AAA, Li L, Zhang T, Wang N, Song X, Niu X (2012) Digital image encryption scheme based on multiple chaotic systems. Sens Imaging 13:67–88
23. Zahmoul R, Ejbali R, Zaied M (2017) Image encryption based on new Beta chaotic maps. Opt Lasers Eng 96:39–49
24. Zhu Z, Zhang W, Wong K, Yu H (2011) A chaos-based symmetric image encryption scheme using a bit-level permutation. Inf Sci (NY) 181(6):1171–1186
25. Zhang Y-Q, Wang X-Y (2014) A symmetric image encryption algorithm based on mixed linear–nonlinear coupled map lattice. Inf Sci (NY) 273:329–351
26. Huang CK, Nien HH (2009) Multi chaotic systems based pixel shuffle for image encryption. Opt Commun 282(11):2123–2127
27. Chai X (2017) An image encryption algorithm based on bit level Brownian motion and new chaotic systems. Multimed Tools Appl 76(1):1159–1175
28. Brindha M, Ammasai Gounden N (2016) A chaos based image encryption and lossless compression algorithm using hash table and Chinese Remainder Theorem. Appl Soft Comput J 40:379–390
29. Wu X, Kurths J, Kan H (2017) A robust and lossless DNA encryption scheme for color images. Multimed Tools Appl

Hight Speed Low Area VLSI Architecture for LEA Encryption Algorithm

Zeesha Mishra, Gandu Ramu and B. Acharya

Abstract A novel hardware architecture for LEA encryption algorithm has been implemented in this paper. LEA algorithm is available in three key sizes of 128 bit, 192 bit and 256 bit. This paper focuses on the architecture of 128 bit key with 128-bit block size. The hardware implementation utilizes multiplexers to select the data, shift registers to store 32-bit data as hardware elements. The implementation results have been verified using Xilinx ISE. Virtex5 and Virtex4 FPGA family are used to test and validate the results. It is evident from the results that the proposed architecture utilizes 1% of the total slices available. Further, the processing speed for Virtex4 implementation is less in comparison to the processing speed of Virtex5. The total power consumption of virtex-5 and virtex-4 are 0.566 w and 0.348 w, respectively. The maximum operating frequency of the proposed architecture is compared with the existing literature and better performance is achieved.

Index Terms LEA algorithm · Virtex-5 · LUTs · Flipflop

1 Introduction

Performance of the computing devices needs to be upgraded in order to enhance the device performance in comparison to earlier devices. The popularization of these devices have not only replaced mobile phones but also provide additional services such as security, data storage facilities and network services. This results in an increase in the data needed for business purposes. Smart meter is one of the

Z. Mishra (✉) · G. Ramu · B. Acharya
Department of Electronics & Telecommunication Engineering, National Institute of Technology Raipur, Raipur 492010, Chhattisgarh, India
e-mail: zmishra.phd2016.etc@nitrr.ac.in

G. Ramu
e-mail: ramg1793@gmail.com

B. Acharya
e-mail: bacharya.etc@nitrr.ac.in

© Springer Nature Singapore Pte Ltd. 2019

V. Nath and J. K. Mandal (eds.), *Proceedings of the Third International Conference on Microelectronics, Computing and Communication Systems*,
Lecture Notes in Electrical Engineering 556, https://doi.org/10.1007/978-981-13-7091-5_14

important application which forms its basis on lightweight cryptography. The utilization of portable devices is growing at tremendous rate and the requirement of secure data is increasing on daily basis. To realize such applications, block ciphers with energy efficient techniques are used for better security. Advanced Encryption Standard (AES) [1] is the most popular cryptographic algorithm whose performance if good both in hardware as well as software implementations. Many Lightweight block ciphers are proposed such as PRESENT [2], LED [3], Piccolo [4], LiCi [5], QTL [6], and LEA [7]. In LEA algorithm, security is more against all the existing attacks. Furthermore, high-speed encryption on minimum chip area is possible with LEA in comparison with AES. LEA is based on (Addition, Rotation, and XOR) ARX model which uses round functions and simple key schedule. LEA technique is faster than AES. Some more ARX-based algorithms are also available in the literature, such as TEA [8] and XTEA [9]. Both are feistel-based ciphers. Their performance in terms of encryption speed is not too good because of 64-bit block size and 64 rounds more than LEA. Since HIGHT [10] is also a block cipher which is based on 8-bit ARX model, high-speed operations on 32-bit CPUs is not possible.

SPECK and SIMON [11] are two block cipher recently proposed by NSA. The basic building block of SPECK is ARX module and SIMON comprises of ANDs, Rotations, and XORs. Since the block size for LEA algorithm is 128 bit, hence all the algorithms with same block size can be easily compared. Processing speed of LEA algorithm is faster that SIMON algorithm when implemented on 32-bit and 64-bit processors. However, if SPECK is compared with LEA, then it is observed that the former algorithm performs better in 64-bit processer, but latter algorithm performs better on 32-bit processors.

Recently, PRINCE [12] block cipher was proposed with high processing speed, when implemented in both hardware and software provides better performance. In 2017, Carlos Andres Lara-Nino et al. [13] proposed hardware implementation of two new architectures and three state-of-art designs and elaborated comparison of all designs is done on the parameters which include area, performance, energy, and efficiency.

The overview of the paper can be outlined as Sect. 1 provides an overview of lightweight cryptographic algorithms proposed till date and its application. Section 2 describes the LEA algorithm including key scheduling and round function. The results are discussed in Sect. 3. The paper is finally concluded in last section.

2 LEA Algorithm for 128 Bit

The complete algorithm comprises of two parts: ***Key Scheduling*** and ***Round function***.

1. **Key Scheduling** This section generates intermediate 128 key sequences Rk_i of 32 bit each. The process followed for the generation of this sequence is mentioned

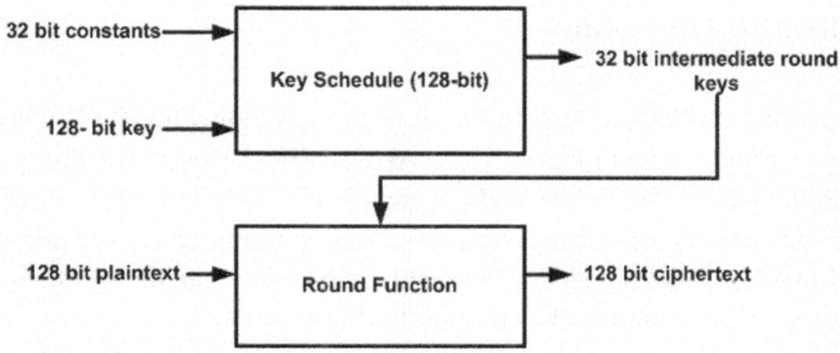

Fig. 1 LEA block diagram

below [7] *Constants.* Some constants are utilized for the round key generation (Fig. 1).

The key constants are

d[0] = 0x3efe9db
d[1] = 0x44626b02
d[2] = 0x79e27c8a
d[3] = 0x78df30ec
d[4] = 0x715ea49e
d[5] = 0xc785da0a
d[6] = 0xe04ef22a
d[7] = 0xe5c40957

Key Schedule with a 128-Bit Key

Let $K = (K[0], K[1], K[2], K[3])$ be a 128-bit key. We set $T[i] = K[i]$ for $0 \leq i < 4$. Round key $RKi = (RKi[0], RKi[1], \ldots, Rki[5])$ for $0 \leq i < 24$ are produced through the following relations:

$S[0] \leftarrow \mathrm{ROT1}\ (S[0] \boxplus \mathrm{ROL}i\ (d[i \bmod 4]))$,
$S[1] \leftarrow \mathrm{ROT3}\ (S[1] \boxplus \mathrm{ROL}i{+}1\ (d[i \bmod 4]))$,
$S[2] \leftarrow \mathrm{ROT6}\ (S[2] \boxplus \mathrm{ROL}i{+}2\ (d[i \bmod 4]))$,
$S[3] \leftarrow \mathrm{ROT}_{11}\ (S[3] \boxplus \mathrm{ROL}i{+}3\ (d[i \bmod 4]))$,
$ROKi \leftarrow (S[0], S[1], S[2], S[1], S[3], S[1])$.

Round Function

The cipher text C is represented in the following format: C = C(0), C(1), C(2), C(3). The initial plaintext P from step 1 is assigned to X. The value of i varies from 0 to 31.

$Xi{+}1[0] \leftarrow \mathrm{ROTL9}((Xi[0] \oplus ROKi[0]) \boxplus (Xi[1] \oplus ROKi[1]))$,
$Xi{+}1[1] \leftarrow \mathrm{ROTR5}((Xi[1] \oplus ROKi[2]) \boxplus (Xi[2] \oplus ROKi[3]))$,
$Xi{+}1[2] \leftarrow \mathrm{ROTR3}((Xi[2] \oplus ROKi[4]) \boxplus (Xi[3] \oplus ROKi[5]))$,
$Xi{+}1[3] \leftarrow Xi[0]$.

The ciphertext C is obtained from last set of iteration in key scheduling part.

3 Results and Discussion

We have performed hardware implementation of LEA algorithm for key size of 128 bit. Family of Field Programmable Gate Array (FPGA) used for this purpose is Virtex5 with device XC5VLX50T and package is FF1136. This section focusses on the performance parameters such as frequency of operation, and hardware utilization in terms of Look Up Tables (LUTs), slices and flip flop. Further, the results for Virtex5 are also compared with another FPGA family that is Virtex4.

Figure 2 represents the proposed architecture for 128-bit key size. Here, d0, d1, d2, and d3 represents 4 constant rotated values which act as input to multiplexer present the value of

d0 = 11000011111011111110100111011011,
d1 = 10001000110001001101011000000100,
d2 = 11100111100010011111001000101001,
d3 = 11000110111110011000011101100011.

These multiplexers along with the 32-bit shift registers, provide key constants for different rounds. At the extreme left, 4 multiplexers, with t0–t3 and Rk0–Rk3 as input are utilized. 128-bit key is divided into 4 32-bit parts t0, t1, t2, and t3. Input plaintext is provided in the architecture through the multiplexers at the extreme left. X0, X1, X2, and X3 represents 32-bit plaintext. The complete flow of data for each round can be easily understood through Fig. 2

Figure 3 displays the ciphertext as output when clk and plaintext are provided as input. Here, c0, c1, c2, and c3 represents the 32-bit ciphertext and x0, x1, x2, and x3 represent the 32-bit plaintext. Table 1 tabulates the hardware utilization of the architecture when implemented in Virtex5 and Virtex6 FPGA family. The parameters under consideration are number of LUTs, number of slices and number of flip flops

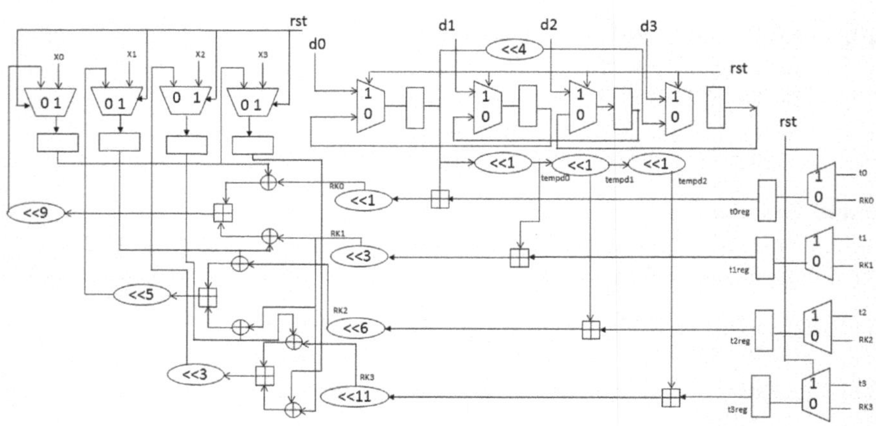

Fig. 2 Hardware architecture of LEA

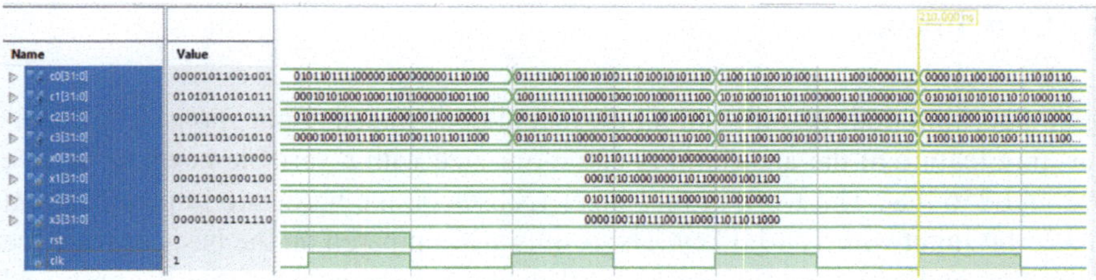

Fig. 3 Simulation result of LEA algorithm

Table 1 Hardware utilization comparison between virtex-5 proposed and virtex-5 previous

Parameters	Virtex-5 proposed work	Virtex-5 previous work [14]
Number of LUTs	360	713
Number of slice reg	382	388
Max. freq (MHz)	225.038	217.806

Table 2 Hardware utilization comparison between virtex5 and virtex4

Parameters	Virtex5	Virtex4
Number of slices	122/7200	332/18432
Number of LUTs	360/28800	453/36864
Number of flip flops	382/28800	382/36864
Static power	0.562	0.331
Dynamic power	0.004	0.017
Total power	0.566	0.348

used out of the available numbers. Only 1% hardware is utilized when the architecture is implemented in Virtex4 and Virtex5 FPGA families.

The paper focusses on operating frequency of the architecture. Higher is the frequency, less time would be required by the algorithm to encrypt data. 177.024 MHz of frequency is utilized in Virtex4 family whereas 225.038 MHz is used in Virtex5 FPGA. It can be inferred that the encryption process in Virtex5 is faster in comparison to Virtex4 family of FPGA. Further, the resource utilization of Virtex4 and Virtex5 FPGA is mentioned in III. In [14], the maximum frequency reported is 217.806 MHz. Our proposed architecture offer 3% increment in the operating frequency than the reported literature. It is evident from Table 2, that the number of LUTs and Slice Reg in previous work is 360 and 383 and proposed work LUTs and Slice Reg is 713 and 388. Power has been calculated by X-power analyzer from xilinx 14.6.

4 Conclusion

LEA algorithm finds its utilization in resource-constrained applications. The distinguishing feature of the algorithm is 128 block size with a variable key size of 128, 192 and 256 bits. The work in this paper focuses on the hardware implementation of 128-bit input with 128-bit key. The results are compared on the basis of resource utilization and operating frequency. Further, it can be concluded that the proposed architecture performs better in terms of speed and area occupancy.

References

1. Lee D, Kim D-C, Kwon D, Kim H (2001) Advanced encryption standard (AES). Federal Information Processing Standards Publication, vol 197, pp 1–51
2. Bogdanov A, Knudsen LR, Leander G, Paar C, Poschmann A, Robshaw MJB, Seurin Y, Vikkelsoe C (2007) Present: an ultra-lightweight block cipher. In: Paillier P, Verbauwhede I (eds). Springer, Berlin, pp 450–466
3. Guo J, Peyrin T, Poschmann A, Robshaw M (2011) The led block cipher. In: Preneel B, Takagi T (eds) Springer, Berlin, pp 326–341
4. Shibutani K, Isobe T, Hiwatari H, Mitsuda A, Akishita T, Shirai T (2011) Piccolo: an ultra-lightweight blockcipher. In: Preneel B, Takagi T (eds) Springer, Berlin, pp 342–357
5. Patil J, Bansod G, Kant KS (2017) LiCi: a new ultra-lightweight block cipher. In: 2017 international conference on emerging trends innovation in ICT (ICEI), pp 40–45
6. Li L, Liu B, Wang H (2016) QTL: a new ultra-lightweight block cipher. Microprocess Microsyst 45:45–55
7. Hong D, Lee J-K, Kim D-C, Kwon D, Ryu KH, Lee D-G (2014) Lea: A 128-bit block cipher for fast encryption on common processors. In: Kim Y, Lee H, Perrig A (eds) Information security applications. Springer International Publishing, Cham, pp 3–27
8. Wheeler DJ, Needham RM (1995) Tea, a tiny encryption algorithm. In: Preneel B (ed) Fast software encryption. Springer, Berlin, pp 363–366
9. Wheeler DJ, Needham RM (1998) Correction of xtea. Technical report, Computer Laboratory, Universoty of Cambridge
10. Hong D, Sung J, Hong S, Lim J, Lee S, Koo B-S, Lee C, Chang D, Lee J, Jeong K, Kim H, Kim J, Chee S (2006) Hight: a new block cipher suitable for low-resource device. In: Cryptographic hardware and embedded systems—CHES 2006. Springer, Berlin, pp 46–59
11. Beaulieu R, Treatman-Clark S, Shors D, Weeks B, Smith J, Wingers L (2015) The simon and speck lightweight block ciphers. In: 2015 52nd ACM/EDAC/IEEE design automation conference (DAC), June 2015, pp 1–6
12. Borghoff J, Canteaut A, Güneysu T, Kavun EB, Knezevic M, Knudsen LR, Leander G, Nikov V, Paar C, Rechberger C, Rombouts P, Thomsen SS, Yalçin T, Prince—A low-latency block cipher for pervasive computing applications. In: Wang X, Sako K (eds) Springer, Berlin, pp 208–225
13. Lara-Nino CA, Diaz-Perez A, Morales-Sandoval M (2017) Lightweight hardware architectures for the present cipher in FPGA. IEEE Trans Circuits Syst I Regul Pap 64(9):2544–2555
14. Lee D, Kim D-C, Kwon D, Kim H (2014) Efficient hardware implementation of the lightweight block encryption algorithm lea. Sensors 14:975–994

Quantum-Confined States of Electrons in Semiconductor Quantum Ring Under Crossed Electromagnetic Field

Arpan Deyasi and Swapan Bhattacharyya

Abstract Quantum-confined states of electrons inside quantum ring-like structure have been analytically investigated subjected to crossed electromagnetic field in order to study the effect of field quantization. Schrödinger equation without considering time-dependency factor is solved with boundary conditions based on the cylindrical geometry of the structure, and lowest two orders Bessel functions are considered for the evaluation of electron energy states. Result reveals that while perpendicular electric field lowers the energy subbands, parallel magnetic field makes the opposite effect; thus creating an interesting trade-off situation. The result is weighted against with that acquired independently in absence of electric field and magnetic field respectively for identical dimensions. The present work reflects this energy optimization by simultaneous tailoring of both the fields in order to obtain absorption and emission spectra.

Keywords Crossed field · Quantum ring · Eigenergy · Cylindrical geometry · Field quantization

1 Introduction

Study of nanostructure devices in recent years is curved in the direction of investigating properties of quantum boxes, dashes, dots or rings. This is probably due to the rapid progress in experimental methodologies, precisely in self-assembly techniques [1–3]. In nanostructures, smaller dimensions in any measured spatial direction

A. Deyasi (✉)
Department of Electronics and Communication Engineering, RCC Institute of Information Technology, Kolkata 700015, India
e-mail: deyasi_arpan@yahoo.co.in

S. Bhattacharyya
Department of Electronics and Communication Engineering, JIS College of Engineering, Kalyani, Nadia 741235, India
e-mail: swapanbhattacharyya@ieee.org

© Springer Nature Singapore Pte Ltd. 2019
V. Nath and J. K. Mandal (eds.), *Proceedings of the Third International Conference on Microelectronics, Computing and Communication Systems*, Lecture Notes in Electrical Engineering 556, https://doi.org/10.1007/978-981-13-7091-5_15

are closely equal to or maybe little less than the de-Bröglie wavelength [4, 5] and corresponding transport of carriers along those reduced dimensions are obstructed resulting quantization of energy states [6, 7]. The novel change in the fundamental property of the structure, in turn, generates meaningful alterations in both macro-world and micro-world properties of the ultimate device. Quantum ring is a new kind of device where solid material is surrounded of a small 'hole' placed exactly at the center, and this specific class of quantum geometry is investigated both experimentally as well as theoretically due to its novel exhibition of Aharonov–Bohm effect under certain external conditions [8]. Considerable attention has been received by semiconductor nanostructures with ring-geometry both in cases of computational and experimental analysis for their unique electrical as well as optical properties [9, 10]. Quantum ring is a special type of dot where all the natures of dot are reflected, which ultimately results in quantization of eigenstates. In order to find out those electronic properties, detailed works are carried out in the last decade [9, 11]. This special type of dot (ring) has already used in design of laser emitters [12], storage devices [13], fluorescent markers [14] on the basis of their interband optical transitions.

All the dimensions of quantum ring can be minutely controlled at the time of fabrication owing to the latest developments in CVD and MBE techniques along with photolithographic process. From the technology of quantum structure growth, it may be said that cylindrical ring can be grown quite easily than that of toroidal geometry. The quantum-confined states basically depend on the figure and volume of the nanostructure. These bound states can be altered by externally applying electric/ magnetic field or both by changing the configuration potential, which results in net change of their (energies) magnitude. Change of resonant states simultaneously change the tunneling probability of the carriers, and that controls the resonant tunneling property and RT-based devices constructed using quantum rings.

Following technical details contain a calculative model is provided to study the transition states of the quantum ring in presence of externally applied crossed electric and magnetic field. A few preliminary results of the study have been reported earlier [15, 16]. Analytical findings reveal that fields have opposing effect on the tuning of energy states for obtaining desired blueshift or redshift, which can be used in photonic [15] as well as biological [16] applications.

2 Mathematical Modeling

We consider a cylindrical quantum ring (as displayed in Fig. 1) where electric field is applied along the axis of the structure, and magnetic field is working along the plane where the ring is kept.

Using the conventional cylindrical system (ρ, φ, z), magnetic potential may be written in the form

$$A_z = B\rho \cdot \sin\theta \tag{1}$$

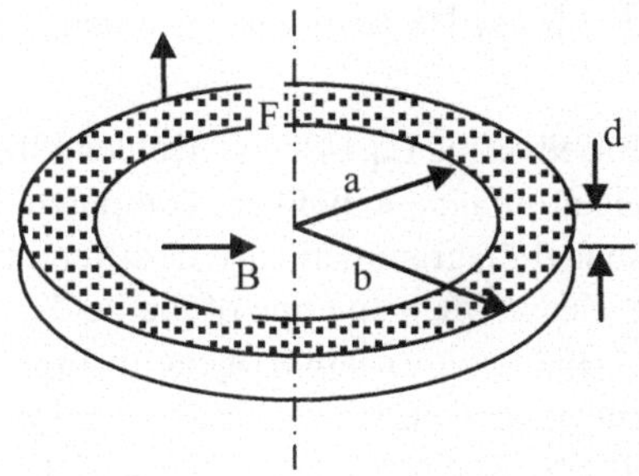

Fig. 1 The proposed ring with basic outline subjected to crossed EM field

In order to solve the ring structure, we first consider the time-independent Schrödinger equation with wave function Ψ as

$$-\frac{\hbar^2}{2m^*}\left[\frac{1}{\rho}\frac{\partial}{\partial\rho}\left(\rho\frac{\partial\Psi}{\partial\rho}\right)+\frac{1}{\rho^2}\frac{\partial^2\Psi}{\partial\theta^2}+\frac{\partial^2\Psi}{\partial z^2}\right]-qF\Psi + \frac{iqB\hbar}{m^*}\rho\sin\theta\cdot\frac{\partial\Psi}{\partial z}+\left(\frac{q^2B^2}{2m^*}\rho^2\sin^2\theta\right)\Psi = E\Psi \tag{2}$$

where ρ is the variable radius with boundary condition [a, b], other symbols are quite known and have norammly usable. F and B are externally applied electric and magnetic field, respectively.

Then perturbed Hamiltonian function may be written as Eq. (2) if we consider only the effect of magnetic field

$$\widehat{H}' = \frac{iqB\hbar}{m^*}\rho(\sin\theta)\frac{\partial}{\partial z}+\frac{q^2B^2}{2m^*}\rho^2\sin^2\theta \tag{3}$$

Using separation of variable methods, and for not very high electric field, final expression of eigenenergy is obtained as

$$E_{lmn} = \frac{\hbar^2}{2m^*}\lambda_{ml}^2+\frac{\hbar^2}{2m^*}\left(\frac{n\pi}{d}\right)^2 - \sqrt{\frac{\hbar^2}{2m^*}\left(\frac{n\pi}{d}\right)^2\cdot\left\{\frac{\hbar^2}{2m^*}\left(\frac{n\pi}{d}\right)^2-eFd\right\}} + \frac{1}{2}\frac{q^2B^2}{2m^*}\frac{\int_a^b\left[J_m(\lambda\rho)-\frac{J_m(\lambda a)Y_m(\lambda\rho)}{Y_m(\lambda a)}\right]^2\rho^2 d\rho}{\int_a^b\left[J_m(\lambda\rho)-\frac{J_m(\lambda a)Y_m(\lambda\rho)}{Y_m(\lambda a)}\right]^2 d\rho} \tag{4}$$

The minimum value of quantum-confined state E_{lmn} is represented with the set of notation $l = 1$, $m = 0$, $n = 1$.

3 Results and Discussions

Based on the Eq. (4), eigenvalues of the quantum structure for electrons under crossed electromagnetic field are computed. The three lowermost quantum states are represented diagrammatically in different combinations of the fields and evaluated with the case when any one of the fields, either electric or magnetic, is totally absent.

In Fig. 2, variation of energy eigenvalues with the inner radius 'a' of the quantum ring placed in crossed electric field and magnetic field are represented. It is interesting to note that application of parallel (in-plane) magnetic field shifts the bound states upward, while the electric field causes the lowering of energy states. For the values of F and B chosen, the shifting with B is more at higher b and less at lower b than those caused by electric field.

Similar variation with thickness (d) and width (b) of the quantum ring is shown in Fig. 3 and Fig. 4, respectively. Here, it is seen that the shifting of F is more at higher d and less at lower d than those caused by magnetic field. In fact, it is the choice of not only the fields, but also the dimension that determines the extent of shift caused by electric and magnetic fields.

The two lowest energy states are plotted versus electric field for different choices of magnetic field in Fig. 4. The combination of up-shifting and downshifting by the two fields can be utilized to swing the energy states over a range on both sides of the energy states in the absence of any external field. In the example shown, if the electric field is varied up to 100 kV/m and the magnetic field is varied up to 1T, then the lowest subband can experience a swing between −1 and +0.7 meV.

The variation of the two lowest energy states with magnetic field is shown for different choices of electric field in Fig. 5. The energy states increase with magnetic

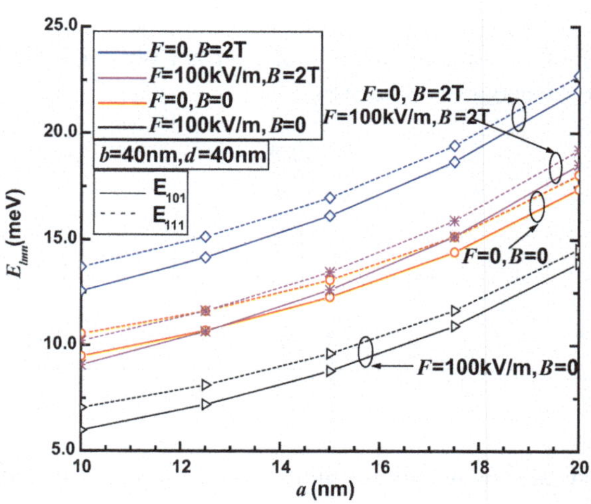

Fig. 2 Lowest energy states as a function inner radius (a) of the ring of constant b and d under different combinations of electric and magnetic fields

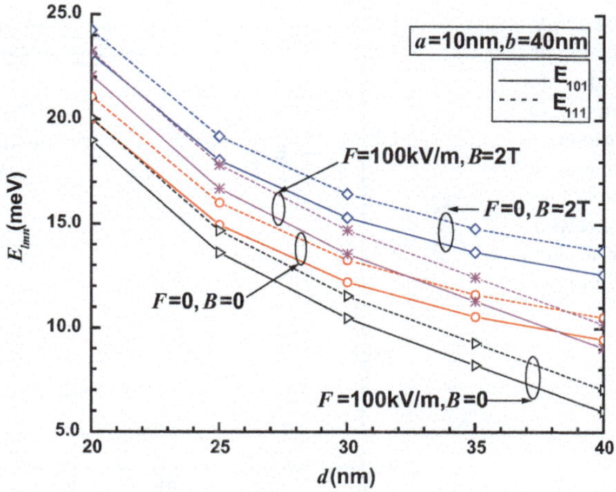

Fig. 3 Lowest two energy states (E_{101}, E_{111}) with thickness (d) of the ring for constant a and b under different combinations of electric and magnetic fields

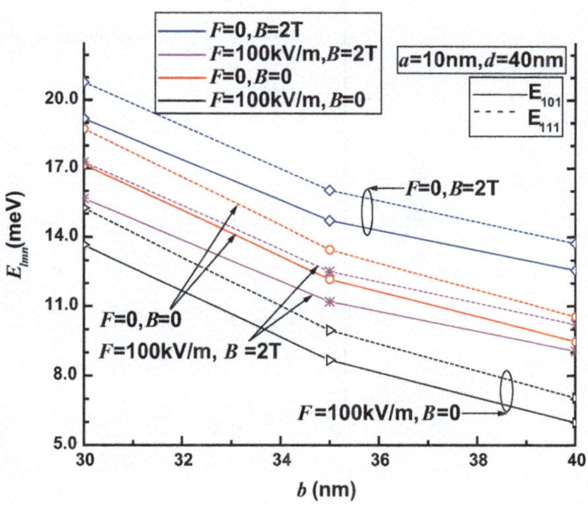

Fig. 4 Lowest two energy states (E_{101}, E_{111}) as a function of width (b) of the ring for constant a and b under different combinations of electric and magnetic fields

field, while decrease when perpendicular electric field is applied. Application of suitable electric field can lower a higher state even below the original position (in absence of field) of its lower state. As shown in the figure, the red solid line (E_{111} with field) comes below the black solid line (E_{101} with no field). A possible swing in E_{lmn} is also shown in the figure around B ~ 1T (Fig. 6).

A comparative study is carried out for eigenenergies when both the electric and magnetic fields are not present with the condition when all fields are totally absent.

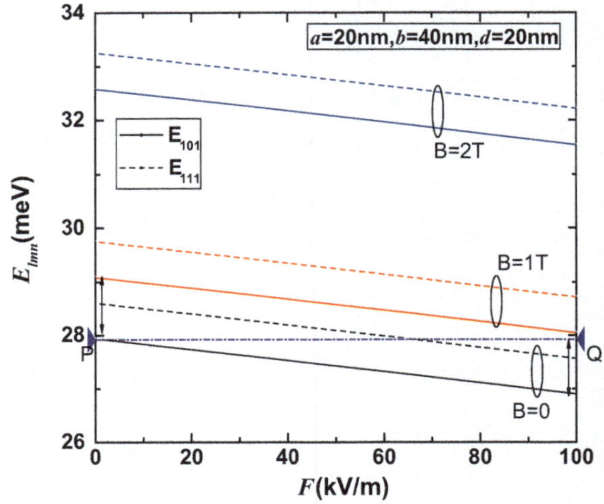

Fig. 5 Lowest two energy states E_{lmn} as a function of external electric field F of a cylindrical quantum ring in crossed field for various physical values of magnetic field B. An axis line PQ is drawn and double arrowed lines at two ends of F axis indicates possible swing in energy

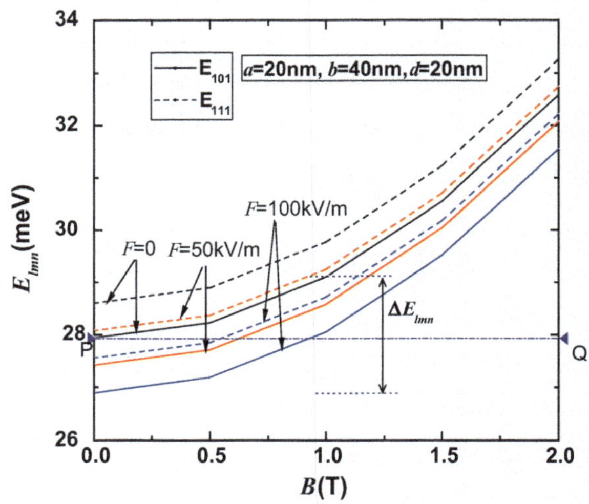

Fig. 6 Lowest two quantum-confined states E_{lmn} with external magnetic field B of the quantum ring in crossed field for various physical values of electric field F. An axis line PQ is drawn and possible swing ΔE_{lmn} around B ~ 1T is shown by double arrowed line

Data are represented in Table 1 where inner and outer radius of the ring is kept constant and only thickness of the device is varied within fabrication limit.

In Table 2, data are given keeping thickness constant and varying width (width is the different between outer and inner radii of the structure). In both the tables, it is observed that for lower dimension of the structure, quantum confinement is not possible by applying the magnetic field only, and hence eigenenergies can't be obtained.

Table 1 Eigenenergies in absence of field, in presence of electric field only, in presence of magnetic field only for varying thickness [d]

d (nm)	No. field			With El. field (F = 200 kV/m)			With Mag. field (B = 6 T)		
	Eigenenergy (a = 10 nm, b = 20 nm)								
	E_1 (meV)	E_2 (meV)	E_3 (meV)	E_1 (meV)	E_2 (meV)	E_3 (meV)	E_1 (meV)	E_2 (meV)	E_3 (meV)
10	99.45	101.51	107.55	98.66	100.73	106.78	–	–	–
11	91.77	93.88	100.07	90.88	93.00	99.21	–	–	–
12	85.80	87.95	94.26	84.81	86.97	93.29	–	–	–
13	81.07	83.26	89.66	79.98	82.17	88.59	–	–	–
14	77.28	79.49	85.96	76.08	78.30	84.79	–	–	–
15	74.19	76.41	82.95	72.88	75.12	81.68	–	–	–
16	71.63	73.88	80.46	70.23	72.48	79.09	32.62	98.62	208.63
17	69.51	71.76	78.38	67.99	70.26	76.90	30.11	88.57	186.02
18	67.71	69.98	76.63	66.08	68.36	75.05	28.00	80.15	167.07
19	66.19	68.47	75.15	64.44	66.73	73.44	26.22	73.03	151.04
20	64.88	67.17	73.87	63.01	65.31	72.05	24.70	66.94	137.35
21	63.75	66.05	72.77	61.75	64.06	70.82	23.39	61.70	125.56
22	62.77	65.07	71.82	60.64	62.95	69.73	22.25	57.17	115.35
23	61.91	64.22	70.98	59.63	61.95	68.75	21.26	53.21	106.44
24	61.16	63.47	70.24	58.72	61.04	67.86	20.39	49.73	98.62
25	60.49	62.81	69.59	57.87	60.20	67.04	19.63	46.66	91.72
30	58.09	60.42	67.25	53.69	56.05	62.96	16.87	35.65	66.94

Table 2 Eigenenergies in absence of field, in presence of electric field only, in presence of magnetic field only for varying width [w]

w (nm)	No. field			With el. field (F = 200 kV/m)			With mag. field (B = 6 T)		
	Eigenenergy (d = 30 nm)								
	E_1 (meV)	E_2 (meV)	E_3 (meV)	E_1 (meV)	E_2 (meV)	E_3 (meV)	E_1 (meV)	E_2 (meV)	E_3 (meV)
12.00	43.10	44.02	46.77	38.54	39.47	42.25	–	–	–
13.00	37.78	38.77	41.69	33.16	34.16	37.11	–	–	–
14.00	33.51	34.56	37.66	28.85	29.90	33.04	–	–	–
15.00	30.04	31.15	34.43	25.33	26.46	29.77	–	–	–
16.00	27.17	28.35	31.82	22.44	23.63	27.14	–	–	–
17.00	24.77	26.02	29.71	20.02	21.28	25.01	–	–	–
18.00	22.75	24.08	28.00	17.97	19.31	23.27	–	–	–
19.00	21.03	22.45	26.61	16.23	17.66	21.87	–	–	–
20.00	19.54	21.06	25.49	14.73	16.27	20.74	36.00	36.70	38.78
21.00	18.26	19.89	24.60	13.43	15.08	19.84	34.03	34.77	36.96
22.00	17.13	18.89	23.90	12.29	14.07	19.14	32.25	33.02	35.33
23.00	16.14	18.04	23.37	11.29	13.21	18.60	30.61	31.43	33.87

4 Conclusion

Quantum-confined eigenstates having dimensions within the practical range of fabrication of the proposed quantum ring have been worked out and graphically signified for the applicable range of crossed electric and magnetic fields with the consideration of n-GaAs as it is already considered for quantum ring manufacture. It has been observed that value of energy rapidly decreases while we increase diameter of that ring, and the rate becomes slow for larger diameter. According to this present analysis, the energy states increase with magnetic field but decreases when external field is perpendicularly applied. Moreover, the effect of electric field appears through the thickness of the ring (d), and the effect of magnetic field appears through inner radius (a) and outer radius (b). The control can be such with the electric field that a higher state may also come below the original position (in absence of field) of its lower state. At last, it may be stated that with constant electric field, if appropriate magnetic field is applied, then it gives energy swing, which is suitable for optical transitions.

References

1. Jeannin M, Cremel T, Häyrynen T, Gregersen N, Bellet-Amalric E, Nogues G, Kheng K (2017) Enhanced photon extraction from a nanowire quantum dot using a bottom-up photonic shell. Phys Rev Appl 8:054022
2. Patel M, Sahu S, Verma AK, Agnihotri P, Tiwari S (2016) Fabrication and characterization of CdSe quantum dot-sensitized solar cells by successive ionic layer adsorption and reaction (SILAR) process. In: 13th international conference on fiber optics and photonics, OSA technical digest (online), W3A.37
3. Zheng Z, Ji H, Yu P, Wang Z (2016) Recent progress towards quantum dot solar cells with enhanced optical absorption. Nanoscale Res Lett 11:266
4. Chukwuocha EO, Onyeaju MC, Harry TST (2012) Theoretical studies on the effect of confinement on quantum dots using the brus equation. World J Condens Matter Phys 2:96–100
5. Sultana S, Alam S (2015) Confinement effects and tunability of quantum dots within strong confinement regime. In: IEEE international WIE conference on electrical and computer engineering
6. Bhattacharyya S, Halder S, Deyasi A (2018) Effect of surface-to-volume ratio on eigenenergy in quantum ring. In: Industry innovative innovations in science, engineering and technology. Lecture notes in networks and systems, pp 71–78
7. Deyasi A, Bhattacharyya S (2013) Effect of band nonparabolicty on electron energies of a quantum disk in presence of electric field. INDICON 13:1–5
8. Kotimäki V, Räsänen E (2010) Aharonov-Bohm effect in many-electron quantum rings. Phys Rev B 81:245316
9. Park BH, Baek SD, Kim JY, Bae J, Han H (2002) Quantized electron-states in a semiconductor quantum ring of cylindrical symmetry in external electric and magnetic field. Opt Eng 41:1339–1345
10. Seo M, Chung Y (2018) Transport properties of a quantum dot and a quantum ring in series. J Korean Phys Soc 72(1):138–143
11. Bejan D, Stan C, Niculescu EC (2018) Optical properties of an elliptic quantum ring: eccentricity and electric field effects. Opt Mater 78:207–219

12. Wan Y, Jung D, Norman J, Shang C, MacFarlane I, Li Q, Kennedy MJ, Gossard AC, Lau KM, Bowers JE (2017) O-band electrically injected quantum dot micro-ring lasers on on-axis (001) GaP/Si and V-groove Si. Opt Express 25:26853–26860
13. Eslami L, Faizabadi E, Ahmadi S (2016) Quantum nano ring composed of quantum dots as a source of pure persistent spin or charge current. Phys Lett A 380(45):3854–3860
14. Toh KC, Stojković EA, van Stokkum IHM, Moffatbc K, Kennis JTM (2011) Fluorescence quantum yield and photochemistry of bacteriophytochrome constructs. 13: 11985–11997
15. Stanciu GA, Stanciu SG, Hristu R, Kwon O, Kim DK (2008) Investigation on photonic-corral-mode quantum ring lasers by laser scanning microscopy. In: IEEE 10th anniversary international conference on transparent optical networks
16. Jamieson T, Bakhshi R, Petrova D, Pocock R (2007) Biological applications of quantum dots. Biomaterials 28(31):4717–4732

Design of Ultra Low Power CMOS Sigma Delta ADC for Aerospace Applications

Aditya Kumar and Vijay Nath

Abstract In today's advancing world ultra-low power CMOS designs are the leading challenge in the electronic field industries and defense. The power dissipation plays a vital role in the terms of growth of battery-powered system, mobility, portability, reliability, cost, performance, environmental effects, and security. Generally, there are many techniques which are used in designing low power CMOS circuit and system. This paper emphasis on the various methodologies and power administration techniques for the design of ultra low power CMOS Sigma Delta ADC which will definitely meet the challenges in future days in aerospace application to design and fabricate high-speed low power circuits. ADC is regarded as the king of aerospace and defense applications.

Keywords CMOS · Sigma-delta ADC

1 Introduction

The basic block of the sigma-delta modulator [1] is shown in Fig. 1 given which include integrator, comparator, and a 1-bit DAC.

The main block of the sigma-delta modulator [2] is integrator and comparator which integrate and compare the input signal with reference to giving the modulated output. Sigma-delta modulation is a technique for concealing analog data signal into a digital data signal which is there in ADC [3]. In this type of modulation, the veracity of this modulation is improved very much by passing the digital output through the reference DAC and sigma attend analog data signal to the inputting signal which reduces the error delta modulation. Pulse-density modulation or pulse- frequency modulation is a sigma-delta modulation which is responsible for the conversion of analog voltage into a train of pluses depending on the use. As the name suggests

A. Kumar (✉) · V. Nath
Department of Electronics & Communication Engineering, Birla Institute of Technology Mesra, Ranchi 835215, Jharkhand, India
e-mail: aditya88952@gmail.com

© Springer Nature Singapore Pte Ltd. 2019

V. Nath and J. K. Mandal (eds.), *Proceedings of the Third International Conference on Microelectronics, Computing and Communication Systems*, Lecture Notes in Electrical Engineering 556, https://doi.org/10.1007/978-981-13-7091-5_16

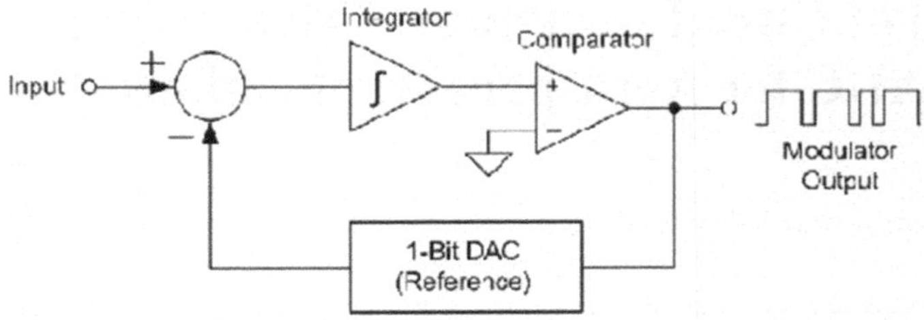

Fig. 1 Block diagram of sigma-delta modulator

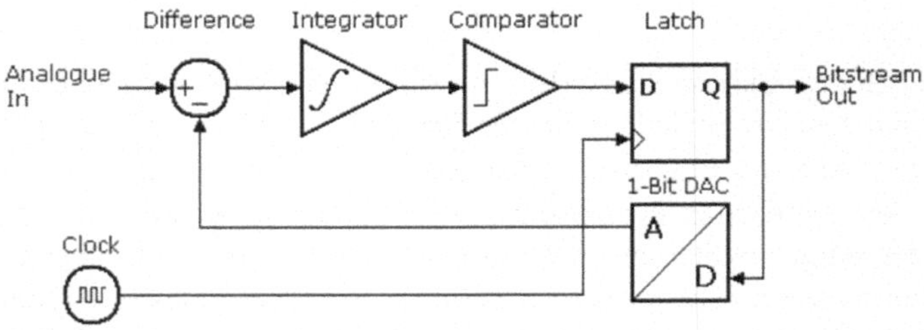

Fig. 2 ADC modulator circuit

comparator in the modulator compare the two voltages. When the voltage difference arises, the output of the comparator circuit is in one state, and when the condition changes the output switches to zero. These comparators are used as voltage sensors and detectors. ADC modulator circuit is shown in Fig. 2.

In ADC modulator there is an integrator which is based on operational amplifier, it can do mathematical operation (addition, subtraction) of integration w.r.t. time. The output voltage is directly proportional to the input voltage over a period of time. The analog computers, ADC and wave shaping circuits use the integrator circuits.

To remove the instability a resistor is connected in parallel to the capacitor [4] in the ideal integrator circuit. The output result is triangular wave when the input is a square wave in the integrator. While output of sine wave as an input is cosine wave. The ideal integrator differs from practical integrator circuit. Ideal integrator does not have the parameter like a finite open loop gain, an input offset voltage and input-biased current, while practical integrator has all these parameter. Due to this ideal integrator have many issues like when both input biased current and the output offset voltage will be responsible for the current to flow through the capacitor resulting in the drift of output voltage over a specific time period until the operational amplifier saturates. Likewise, if the signal is without DC component, there will be no drift in ideal circuit but it will occur in real circuit.

2 Methodology

This paper proposed a design of ultra-low power CMOS sigma-delta modulator which include comparator and most important integrator which is responsible for the ultra-low power CMOS sigma-delta ADC and ADC contain the important circuit of CMOS Op-amp [5–8]. For military application in aerospace field, we have to cope with a wide variation of temperature range (223.15–423.15 K). For this, an analog temperature sensor is designed with the help of ADC which will generate the output in digital form. The characteristic of integrator is same as the characteristic of a low pass filter for input signal and for noise quantization it acts as a high pass filter. In this paper cadence virtuoso for design of circuitry and mentor graphics caliber for testing of circuitry.

An op-amp is used in this research paper. A capacitor which has charging and discharging property is used to produce output. A resistance is used in parallel to stabilize the integrator. The output voltage is proportional to the input time integral voltage and applicable for sigma-delta ADC [9]. Actually, this circuit designed to work in harsh environment and consume low power [10]. This circuit required high linearity, therefore, gain enhancement cross-coupled architecture are utilized [11]. For direct utilizing in current mode, CCII controller for TITO system [12] is used. For improving the performance energy efficient 75.17 dB two-stage operational amplifier concept are utilized [13]. When this circuit is utilized so many others factors are also affected the ADC output, therefore, A CMOS temperature sensor and auto-zeroing circuit [14]. Complete circuits are work in low power environment therefore 40 nW CMOS temperature sensor [15] is utilized. Practical circuit of integrator shown in Fig. 3 and ideal integrator circuit is shown in Fig. 4. Integrator circuit internal block diagram shown in Figs. 5 and 6. Integrator circuit output responses are shown Figs. 7 and 8.

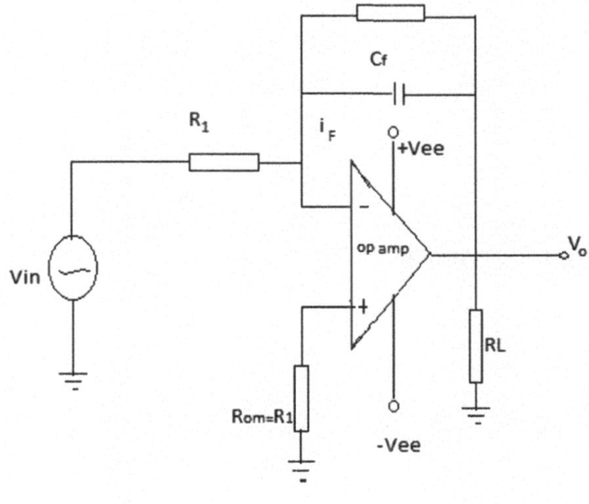

Fig. 3 Practical integrator circuit

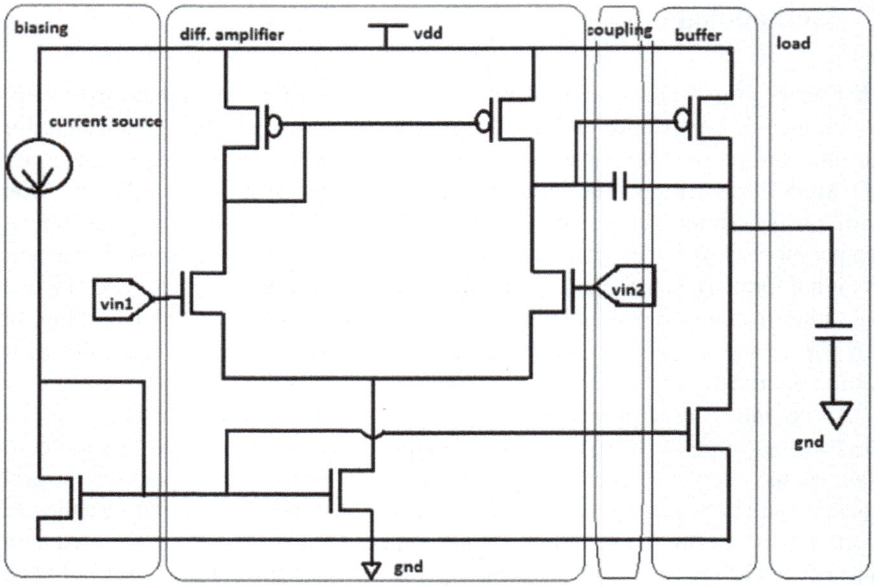

Fig. 4 Ideal integrator

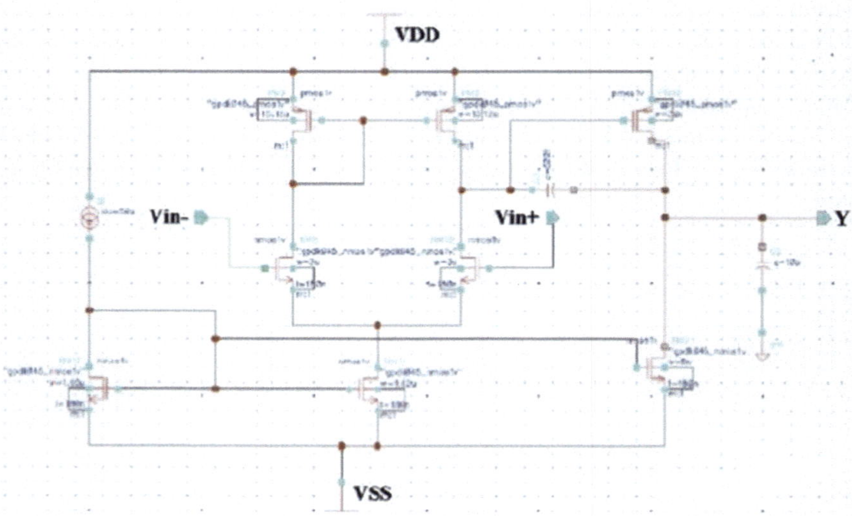

Fig. 5 OP-AMP circuit

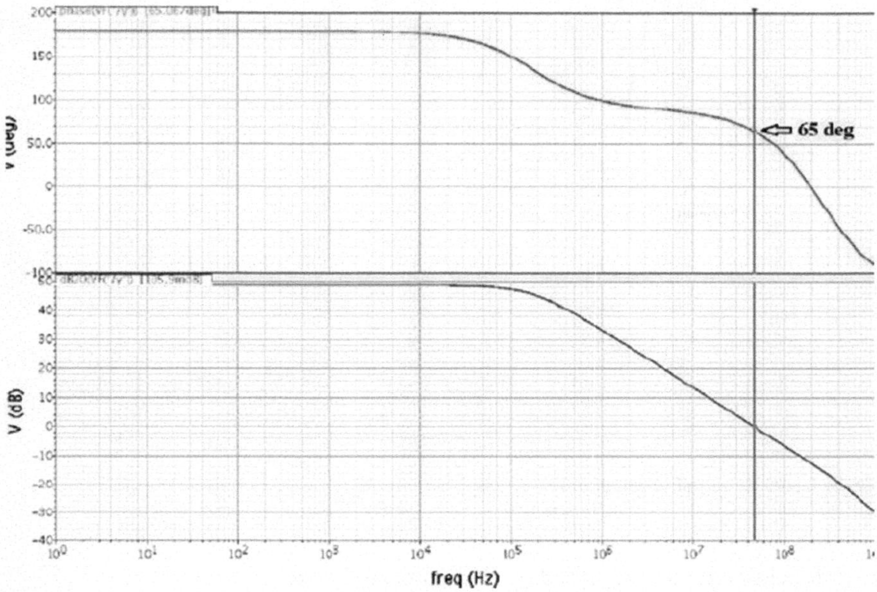

Fig. 6 Cadence stimulation of op-amp

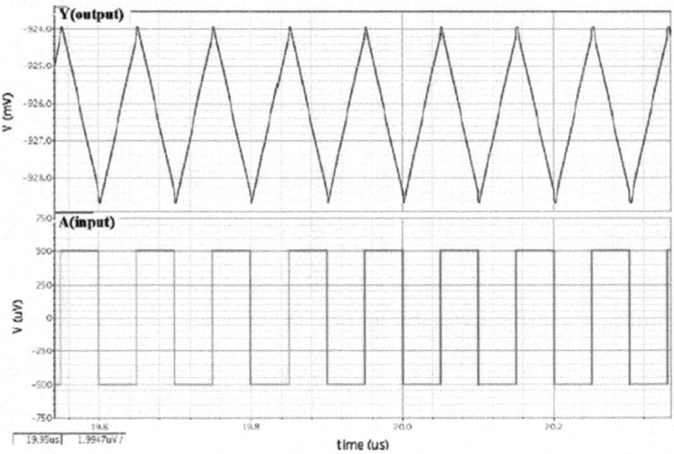

Fig. 7 Simulation result of Fig. 6

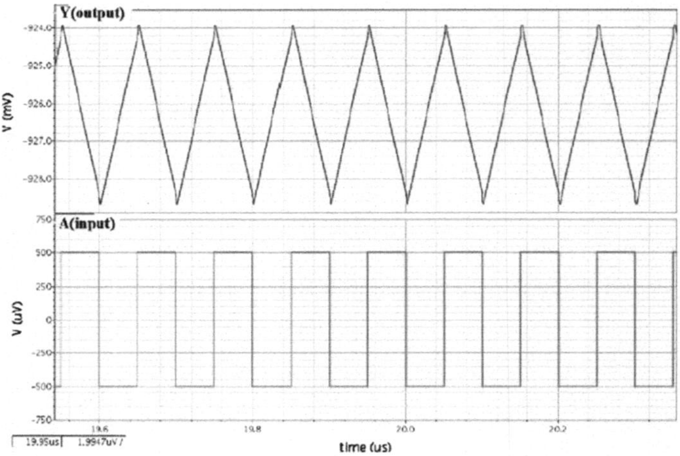

Fig. 8 Cadence output of integrator

If the ideal circuit is analyzed by KCL at node 2 then

$$I_1 = I_B + I_F \tag{1}$$

$$I_B = 0 \text{ (IDEAL OPAMP)} \tag{2}$$

$$I_1 = I_F \tag{3}$$

The relationship between current and voltage in a capacitor is given by

$$I_C = C\frac{dV}{dt} \tag{4}$$

By substituting the proper variable in (4).

$$\frac{V_{IN} - V_2}{R_1} = C_F\frac{d(V_2 - V_0)}{dt} \tag{5}$$

$$V_2 = V_1 = 0 \text{ (IDEAL OPAMP)} \tag{6}$$

$$\frac{V_{IN}}{R_1} = -C_F\frac{dV_0}{dt} \tag{7}$$

$$\int_0^t \frac{V_{IN}}{R_1}dt = -C_F\left(\frac{dV_0}{dt}\right)\cdot dt \tag{8}$$

$$V_0 = -\left(\frac{1}{R_1C_F}\right)\int_0^t V_{IN}dt \tag{9}$$

But the output voltage has some error due to input biased current. A resistance R_{on} is required which is parallel in connection with R_1, R_F and R_L.

The error in voltage is

$$V_E = \frac{R_F + R_1}{R_1} * V_{ios} \tag{10}$$

where V_{ios} is the input offset voltage.

There is a voltage drop at both the end of +ve and –ve terminal of op-amp due to the input biased current (I_{B1}). The capacitor acts as an open circuit in a DC steady state condition. To balance this, a large R_F is connected to the capacitor in parallel. Due to this DC gain is limited in the circuit to a finite value which results in the change of output drift to a finite value, rather very small DC error.

$$V_E = (R_F + R_1)/R_1 * (V_{IOS} + I_{B1}\{R_F \text{ parallel } R_1\}).$$

3 Results and Discussion

When in input a square pulse of train is given to the integrator, there will be charging and discharging of capacitor will take place in the input. As a result of this, a saw-tooth signal will appear whose frequency will depend on the RC time constant circuit. In the proposed scheme the output stability is undistorted and very good for 48.5 dB gain which is not very good in the previous work done because it was implemented on ideal integrator. Whereas in the proposed scheme practical integrator is used.

4 Conclusion

The above circuit was designed and has worked according to our expectation in **cadence virtuoso**. The output stability is very good for 48.5 dB gain for our designed op-amp. For our integrator also we get the desired output (triangular wave) on providing input (square wave).

Acknowledgements We are thankful to our Vice-chancellor Prof. M. K. Mishra for providing the infrastructure facility to carry out this research work. I am also thankful Birla Institute of Technology Mesra, Patna Campus faculty for guiding me on this paper. I am grateful to my parent and my friends to support me in this research work.

References

1. Singor F, Snelgrove WM (1995) Switched-capacitor bandpass delta-sigma A/D modulation at 10.7 MHz. IEEE J Solid-State Circuits 30:184–192
2. Ong A, Wooley B (1997) A two-path bandpass sigma-delta modulator for digital IF extraction at 20 MHz. IEEE J Solid-State Circuits 32:1920–1934
3. Analog integrated circuits and signal processing. 35(2–3):179–187
4. Shem B, Kozak M, Friedman G (2004) A high-speed CMOS op-amp design technique using negative miller capacitance. In: IEEE
5. Kundu Ramkrishna, Pandey Abhishek, Chakraborty Subhra, Nath Vijay (2017) A current mirror based two stage CMOS cascade op-amp for high frequency application. J Eng Sci Technol (JESTEC). 12(3):686–700
6. Chakraborty S, Pandey A, Nath V (2017) Ultra high gain CMOS op-amp design using self-cascading and positive feedback. Microsyst Technol 23(3):541–545. https://doi.org/10.1007/s00542-016-2971-7. Springer, Berlin. Online ISSN: 1432-1858
7. Prasad D, Pranav A, Nimbargi A, Singh A, Kumari Ray M, Mishra M, Kumar M, Nath V (2017) Design of 30 MHz CMOS operational amplifier. In: Singh R, Choudhury S (eds) Proceeding of international conference on intelligent communication, control and devices. Advances in intelligent systems and computing, vol 479, pp 519–525. https://doi.org/10.1007/978-981-10-1708-7_59. Springer, Singapore
8. Tyagi S, Saurav S, Pandey A, Priyadarshini P, Ray M, Pal BB, Nath V (2017) A 21 nW CMOS operational amplifier for biomedical application. In: Nath V (eds) Proceedings of the international conference on nano-electronics, circuits and communication systems. Lecture notes in electrical engineering, vol 403, pp 389–396. https://doi.org/10.1007/978-981-10-2999-8_33. Springer, Singapore
9. Prasad D, Nath V (2018) Design of CMOS integrator circuit for sigma delta ADC for aerospace application. In: Bhattacharyya S, Sen S, Dutta M, Biswas P, Chattopadhyay H (eds) Industry interactive innovations in science, engineering and technology. Lecture notes in networks and systems, vol 11, pp 377–383. https://doi.org/10.1007/978-981-10-3953-9_36. Springer, Singapore
10. Prasad D, Nath V (2018) An ultra-low power high-performance CMOS temperature sensor with an inaccuracy of −0.3/+0.1 °C for aerospace applications. Microsyst Technol 24(3):1553–1563. https://doi.org/10.1007/s00542-017-3564-9. Springer, Berlin. Online ISSN: 1432-1858
11. Chakraborty S, Pandey A, Prasad D, Vedam V, Nath V (2018) Linearity improvement of gain enhanced op-amp using cross-coupled architecture. Microsyst Technol 1–10. https://doi.org/10.1007/s00542-018-3885-3. Springer, Berlin. Online ISSN: 1432-1858
12. Verma VK, Ranjan RK, Gupta P, Priyadarshini B, Nath V (2018) A series expansion method aided design of CCII controller for TITO system. Microsyst Technol 1–10. https://doi.org/10.1007/s00542-018-3869-3. Springer, Berlin. Online ISSN: 1432-1858
13. Nidhi N, Prasad D, Nath V (2019) A high-performance energy-efficient 75.17 dB two-stage operational amplifier. In: Nath V, Mandal J (eds) Nanoelectronics, circuits and communication systems. Lecture notes in electrical engineering, vol 511, pp 469–474. https://doi.org/10.1007/978-981-13-0776-8_43. Springer, Singapore. Online ISBN: 978-981-13-0776-8
14. Pandey A, Nath V (2016) A CMOS temperature sensor and auto-zeroing circuit with inaccuracy of −1/+0.7 °C between −30 °C to 150 °C. Microsyst Technol (Springer), Impact factor 1.192, ISSN: 1432-1858, pp 1–9, May 2016, https://doi.org/10.1007/s00542-016-2968-2
15. Pandey A, Nath V (2015) Study and design of 40 nW CMOS temperature sensor for space applications. 13(3):813–819. http://dx.doi.org/10.12928/telkomnika.v13i3.1426. ISSN: 1693-6930

Grey Wolf Optimizer and Its Applications: A Survey

Madhusmita Panda and Bikramaditya Das

Abstract This paper describes Grey Wolf Optimizer (GWO) as a nature inspired metaheuristic algorithm. GWO is proposed in 2014. The leadership hierarchy and hunting behavior of the grey wolves is explained in GWO by developing a mathematical model. The mathematical model of GWO and the pseudocode are also discussed in this paper. Different applications of GWO have also been studied in this paper. This review will assist the future researcher to apply this algorithm in new fields of research.

Keywords Metaheuristic · Grey Wolf Optimizer (GWO) · Evolutionary algorithm · Social hierarchy

1 Introduction

A metaheuristic is a nature inspired algorithmic framework designed to provide a near-optimal solution to an optimization problem. These algorithms are preferred especially when the available information is not complete and computational capacities are limited [1]. Metaheuristic is different from heuristic as it is not application specific and may be applied for a variety of problems with a few assumptions. These algorithms are applied when the solution space is too large to be completely sampled [2]. These are nondeterministic in nature. Metaheuristic algorithms include simple local search procedures and also complicated learning phases. Many metaheuristic algorithms have been studied in literature and new variants are still emerging. Genetic Algorithm (GA), simulated annealing, Particle Swarm Technique of Optimization (PSO), tabu search, memetic algorithm, Ant Colony Algorithms for optimization (ACO) are some of the popularly studied algorithms in this field.

M. Panda (✉) · B. Das
Department of Electronics and Telecommunication, VSSUT, Burla, Odisha, India
e-mail: honey.smita@gmail.com

© Springer Nature Singapore Pte Ltd. 2019

V. Nath and J. K. Mandal (eds.), *Proceedings of the Third International Conference on Microelectronics, Computing and Communication Systems*, Lecture Notes in Electrical Engineering 556, https://doi.org/10.1007/978-981-13-7091-5_17

Metaheuristics are simple, flexible, and derivation free mechanisms. These algorithms are capable of avoiding local minima and can be categorized into the following three classes [3].

- First class of algorithms is inspired by the evolution process of nature and known as evolutionary algorithms. GA is the most popular algorithm in this class.
- Second class based on theories of physics. The most popular algorithms in this class are big-bang-big-crunch (BBBC), Blackhole algorithms.
- Swarm intelligence (SI) methods are classified as the third class. These are population-based metaheuristics that copy the natural behavior of swarms, flocks, herds, and schools of creature in nature. Examples of such algorithms include PSO, ACO, Cuckoo search (CS), Firefly Algorithm (FA), etc.

Grey Wolf Optimizer (GWO) is a new metaheuristic algorithm belongs to the third category. This algorithm is inspired by the hunting process found in Grey Wolves. This is unique as it follows the leadership hierarchy of the grey wolves. Grey wolves are well known for pack hunting and no other SI methods were proposed to follow this hierarchical hunting behavior [3]. This paper is an attempt to review the GWO algorithm and its proposed applications which help the researcher in future to apply it in other new and promising fields of application such as path planning and path following control of autonomous underwater vehicles [4–8].

The proceeding sections of this paper are arranged such that: Sect. 2 of this paper describes social behavior of the grey wolves. Section 3 provides mathematical model of the algorithm, Sect. 4 describes the pseudocode, and Sect. 5 discusses the various applications of GWO respectively.

2 Social Behavior of the Grey Wolves

Grey wolf belongs to Canidae family. They prefer to live in a group, generally called a pack. The group consists of 12–15 members on an average. They follow a strict social dominant hierarchy. The leader wolves are represented by Alpha (A_l). Alpha wolves constitute a male and a female wolf. They have the authority to decide the resting place, time of hunting, etc., for the group. They may not be the strongest ones but are the best ones to manage the group. The orders of the leader wolves are followed by the pack. The Betas (B_t) are next in rank to the alpha's and assist the leader in taking decision. The beta is an advisor to alpha and discipliner for the group. They are the best wolves next to be the alpha to take their position in the case of death or retirement of an alpha wolf. Omega (O_m) wolves are the lowest in rank. They have to follow the orders of all other dominant wolves and allowed to eat at last. If a wolf does not belong to any of the above categories, then it is called as delta (D_l). Delta wolves dominate omega and follow the orders of alphas and betas [3] (Fig. 1).

Grey wolves hunt in groups. The main phases of hunting include [9]:

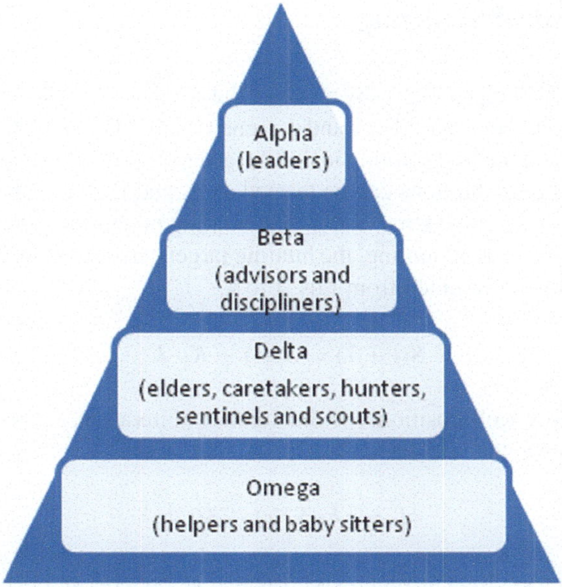

Fig. 1 Leadership hierarchy of grey wolves [3]

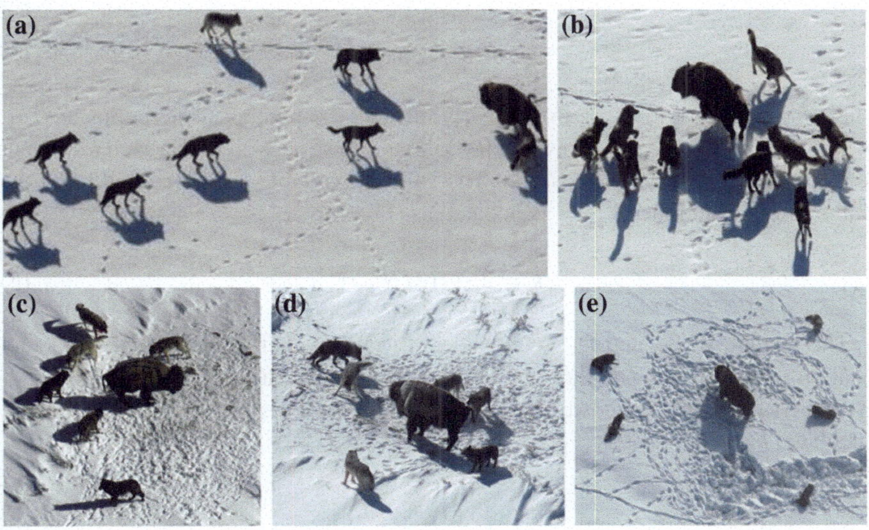

Fig. 2 Group hunting behavior of grey wolves [9]

Phase I: Approach the hunting target while continuously tracking and chasing.
Phase II: Start pursuing the target by encircling, and harassing it until it stops moving.
Phase III: Attack the target (Fig. 2).

3 Mathematical Modeling

Mirzali et al., in 2014 mathematically modeled these behaviors of social hierarchy and hunting observed in Grey Wolf Pack and designed the GWO [3]. As GWO suggested A_l wolves represent the best solution inside the group of possible solutions, while the second and third best solutions can be termed as B_t and D_l, respectively. The other possible solutions are considered as O_m. The hunting process is led by the leader A_l wolves. As the first step of hunting, the hunting target gets surrounded by A_l, B_t, and D_l. This process is represented from Eqs. 1 to 4.

$$S(t+1) = S_p(t) - K \cdot L \tag{1}$$

where S is the grey wolf position, t is the number of iteration, S_p is target position, and L is distance vector calculated as

$$L = \left| J \cdot S_p(t) - S(t) \right| \tag{2}$$

K and J are the coefficient vectors represented as Eqs. (3) and (4), respectively.

$$K = 2 \cdot k \cdot p_1 - k \tag{3}$$

$$J = 2 \cdot p_2 \tag{4}$$

where k is decreased from 2 to 0 linearly through the number of iterations and p_1, p_2 are two random vectors between [0, 1] that allow the wolves to reach any point between current and target position. The value of variable k for iteration t and the total number of iterations N_{iter} is updated as (Fig. 3)

$$k = 2 - t(2/N_{iter}) \tag{5}$$

The results A_l, B_t, and D_l are considered to have a better idea about the actual location of the target. The three best solutions at iteration t are represented by S_1, S_2, and S_3 represent the position of A_l, B_t, and D_l wolves, respectively. These are calculated as follows:

$$S_1 = |S_A - K_1 \cdot L_A| \tag{6}$$

$$S_2 = |S_B - K_2 \cdot L_B| \tag{7}$$

$$S_3 = |S_D - K_3 \cdot L_D| \tag{8}$$

The values of K_1, K_2, and K_3 are evaluated as per Eq. (3). The values of distance vectors L_A, L_B and L_D are given by Eqs. 9, 10, and 11, respectively, as follows:

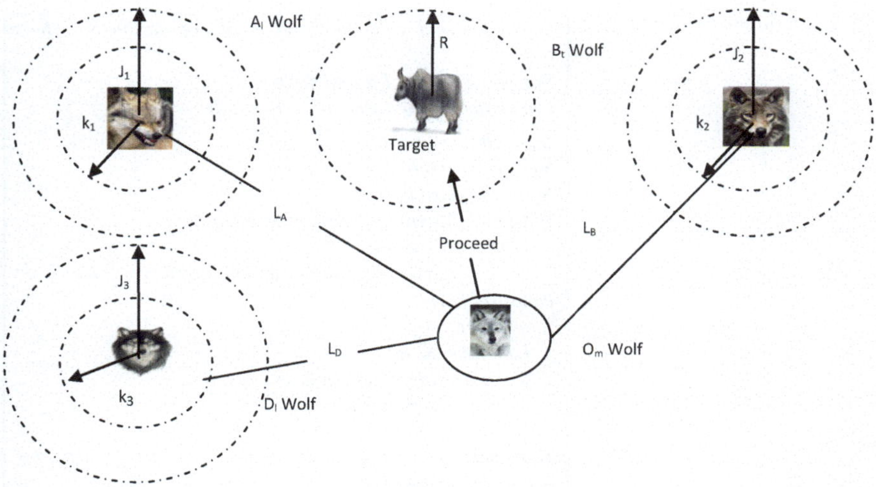

Fig. 3 Position updation using GWO algorithm [3]

$$L_A = |J_1 \cdot S_A - S| \tag{9}$$

$$L_B = |J_2 \cdot S_B - S| \tag{10}$$

$$L_D = |J_3 \cdot S_D - S| \tag{11}$$

The O_m wolves updated their positions according to the positions of A_l, B_t, and D_l wolves. The updated positions of wolves are given by Eq. (12) as

$$S(t+1) = (S_1 + S_2 + S_3)/3 \tag{12}$$

4 The Algorithm

The algorithm for the GWO [3] and the corresponding flow chart [10] are presented here (Figs. 4 and 5).

5 GWO Applications

GWO is a simple, flexible, and robust algorithm. It is very easy to implement. Thus, it can be applied in various field of application. There are many applications of GWO proposed in literature in various fields of science that can be summarized as follows:

Fig. 4 Algorithm for GWO [3]

Suggest initial value for grey wolf population S_i *where* $i=1,2\ldots\ldots m$.
Suggest k,K, and J initial value
Check the fitness value of each result
S_A *is the best result*
S_B *is the second best result*
S_D *is the third best result*
While *(t< Maximum number of iterations)*
  for *each result*
    Revise the value of the current result by equation (12)
  end for
  Update k ,K, and J
  Calculate the fitness of each result
  Update S_A, S_B & S_D
  $t=t+1$
end while
return the value of S_A

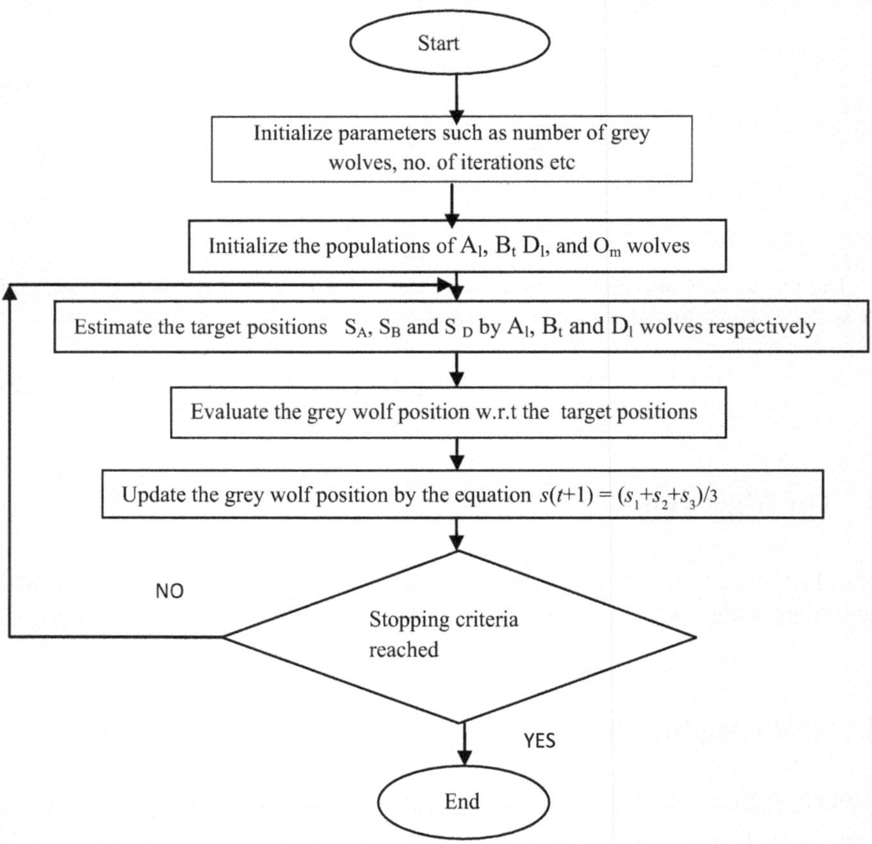

Fig. 5 Flow chart of GWO

5.1 Neural Network

Muangkote et al. [11] modified GWO algorithm to train neural networks based on Radial basis function (RBF). Mirjalili [12] employed GWO for Multilayer Perceptron (MLP) training. The obtained results have been proved to provide highly accurate results in a comparative study to other evolutionary algorithms such as GA, PSO, ACO, etc. He also employed "Evolutionary Population Dynamics (EPD)" to remove the poor solutions of GWO and rearrange the position of alpha, beta, or delta wolves to improve performance [13].

5.2 Power System

Song et al. [14] applied GWO in a bi-objective problem of power system. Economic dispatch (ED) and emission dispatch (EMD) objectives are merged to form one objective function by utilizing price penalty factor. This problem has been termed as "Combined Economic Emission Dispatched (CEED)" problem. El-Gaafary et al. [15] applied GWO to Multi-Input Multi-Output (MIMO) electric power systems. He optimized the location and sizing of installing "STATCOM" devices that improvised the system voltage stability.

The global peak (GP) of a photovoltaic (PV) array under partial shading conditions (PSCs) is determined in [16] using the "Maximum Power Point Tracking (MPPT)" technique based on GWO. Application of GWO to distribution and generation (DG) system has been proposed by Sultana et al. [17] that determines the best suitable site and size of DGs.

5.3 Surface Wave Dispersion Curve

Song et al. [18] suggested a "Surface Wave Dispersion Curve Inversion" scheme based on GWO. The proposed method has been verified on environments both with and without noise and proved to be efficient.

5.4 Scheduling and Routing Applications

Komaki and Kayvanfar [19] addressed the bi-stage assembly flow shop scheduling problem using GWO. The GWO-based approach performed better than other approaches and existing heuristic approach for the problem of two machine flow shop scheduling taking into consideration makespan and idle time of both machines [20]. Petrović and Miljković [21] applied GWO to the nondetermistic polynomial hard (NP-hard) problem of single robot scheduling. It has been found effective in minimizing the total transportation time of the mobile robot while it transferring raw material, goods, and parts in a manufacturing system. GWO also combined with "K-means clustering" algorithm termed as 'K-GWO' algorithm to address "capacitated vehicle routing problem (CVRP)" with capacity constraint [22].

5.5 Optimization Problems

Gholizadeh [23] proposed a grey wolf algorithm (GWA) termed as sequential GWA (SGWA) to achieve the optimization task for a sequence of optimization processes. The initial set of solutions for each process has been chosen from the neighboring region of the best results found in the previous optimization process. This procedure has been repeated until the stopping criterion satisfied. Mirjalili et al. [24] proposed a "Multi-Objective Grey Wolf Optimizer (MOGWO)" in 2016 to multiple objectives optimization problems. Kumar et al. [25] solved a few complex reliability optimization problems using GWO. A hybrid GWO-based learning strategy and simplex method using "elite opposition (EOGWO)" has also been proposed in literature to overcome the poor population diversity and slow convergence rate of standard GWO [26]. When used with high-dimensional data sets GWO algorithm has been found to fall into local optimum easily. Thus, an improved GWO algorithm has been proposed in [27] that combined the global-search ability of Cuckoo Search (CS) algorithm with GWO. The best three solutions of GWO have been updated by the proposed methodology and the searchability of the GWO has also been improved.

5.6 Feature Classification

Emary et al. [28] modified GWO termed as binary GWO (bGWO) method to obtain subsets of optimal features for feature classification problems. Two approaches have been used for finding optimum feature subset. Long and Xu [29] proposed a modified GWO (MGWO) method that improves the quality of solution and convergence rate.

5.7 Tuning of Controller Parameter

A "hybrid GWO and Pattern Search Technique (hGWO-PS)" has been suggested by Soni et al. [30] to improve the tuning of parameters of Proportional–Integral–Derivative Controller with two degree of freedom (2DOF-PID). It has been applied in "Automatic Generation Control (AGC)" for interconnected power system that controls the load frequency. GWO has also been applied for setting the parameters of PI controller based on Takagi-Sugeno fuzzy logic (T-S PI-FCs) [31].

5.8 Support Vector Machine (SVM) Classification

"Support Vector Machines (SVMs)" classification problems have also been addressed using GWO algorithm that improves the accuracy of parameter selection [32, 33]. The proposed GWO-SVM model has been applied for predicting solute solubility in supercritical carbon dioxide (SCCO2) [34] and an improved GWO (IGWO)-SVM has been used for the prediction of second major [35].

5.9 Wireless Sensor Networks (WSNs)

The GWO can also be applied to address rising problems of wireless sensor networks (WSNs). An energy efficient communication for WSNs has been proposed by Diwan and Khan [36] that employed fuzzy logic for cluster formation and GWO for cluster head (CH) election to achieve energy efficiency. GWO algorithm has also been incorporated to handle the node localization problem [37] by spotting the correct position of unknown nodes. The application of GWO has also been illustrated to enhance the performance of ZigBee networks in [38].

5.10 Path Planning

Zhang et al. [39] solved path planning problem in two-dimensional plane for the unmanned combat aerial vehicle (UCAV) using GWO. The proposed methodology considered the problems and limitations in the combat field and obtained an optimal or near-optimal path. Radmanesh and Kumar [40] proposed GWO to find the optimal trajectory for an unmanned aerial vehicle (UAV) in the presence of moving the obstacle. These moving obstacles were termed as intruder aircrafts (IAs) whose trajectory is not known. Yao and Wang [41] proposed a hybrid method for three-dimensional path planning of the UAV in complex terrain environment. He combined the improved interfered fluid dynamical system and GWO. The GWO has been found effective in optimizing reactive parameters of obstacles.

5.11 Biomedical Research

GWO can also be employed for modeling modular granular neural networks (MGNN). The proposed design performed optimal granulation of data for human recognition [42]. Parsian et al. [43] proposed a hybrid method that utilized GWO to optimize MLP network to detect the malignant melanoma images from the images. Kernel extreme learning machine (KELM) integrated with a modified GWO (IGWO) and termed as IGWO-KELM, and used in the field of medical diagnosis [44]. In the proposed approach GA has been used to provide different initial locations, and the recent locations of the population have been updated using GWO in the discrete searching space.

5.12 Knapsack Problem

Yassien et al. [45] investigated the knapsack problem (01KP) in networks and generated optimum solution by using GWO constraint of 5–12 members in one cluster. K-means clustering algorithm has been used for clustering. The total carrying capacity of knapsack has been maximized without exceeding a predefined capacity (Table 1).

6 Discussion

In this paper GWO, its mathematical model and the pseudocode have been discussed in details. Here, an attempt has been made to review the available literature to describe various applications of the GWO algorithm. As this is a review on Grey

Table 1 Summary of GWO applications

Year	Researchers	Applications	Remarks
2014	N. Muangkote et al.	q-Gaussian RBF-link nets neural network training [11]	The improved GWO algorithm obtained impressive results when used with multiclass classification problem
2014	H. M. Song et al.	CEED problems of power system [14]	Total generation cost has been by minimized due to less fuel cost and optimum power generation is possible because of emission at the same time
2015	S. Mirjalili	Training MLP [12]	Due to improved local optima avoidance very accurate results have been obtained
2015	S. Saremi et al.	GWO–EPD algorithm [13]	Shows good improvement in the performance by means of local minima avoidance and convergence rate

(continued)

Table 1 (continued)

Year	Researchers	Applications	Remarks
2015	X. Song et al.	Parameter estimation in surface waves [18]	Better results obtained as it avoids local optima and provides fast convergence rate
2015	G. M. Komaki et al.	Two-stage assembly flow shop scheduling problem with release time of jobs [19]	Outperforms other metaheuristic algorithms result
2015	A. A. El-Gaafary et al.	MIMO Systems [15]	Good quality solutions have been obtained with stable and fast convergence characteristics and are computational efficient
2015	L. Korayem et al.	Capacitated K-GWO algorithm for CVRP [22]	The total path cost or distance covered by the vehicles has been minimized in a constrained field
2015	S. Gholizadeh	Sequential GWA [23]	Multistage implementation of GWA that converges to better solution when employed for design optimization of double-layer grids
2016	Emary	Binary GWO [28]	Classification accuracy is maximum while minimum number of features are selected
2016	W. Long et al.	MGWO for global optimization problem [29]	Obtained precise solution precision with fast convergence speed
2016	S. Mirjalili et al.	MOGWO [24]	Provided very competitive results and outperforms Multi-objective Evolutionary Algorithm Based on Decomposition (MOEA/D) and Multi-Objective Particle Swarm Optimization (MOPSO) algorithms
2016	S. Mohanty et al.	MPPT design for a PV system [16]	The problems encountered in perturb and observe (P & O) and improved PSO (IPSO) techniques like lower tracking efficiency, steady-state oscillations, and transients have been overcome

(continued)

Table 1 (continued)

Year	Researchers	Applications	Remarks
2016	U. Sultana et al.	GWO applied to DG system [17]	Reactive power loss and voltage deviation has been minimized with optimum DG placement
2016	V. Soni et al.	hGWO-PS for Parameter optimization of 2DOF-PIDcontroller [30]	The thermal power system became robust to variations in system variables and operating load conditions from their specified values
2016	S. Eswaramoorthy et al.	GWO-SVM [33]	The optimal tuning of classifier parameters minimizes the gross errors
2016	P. Diwan et al.	Energy efficient communication for WSNs [36]	An energy efficient CH election scheme that outperforms conventional scheme in both stable and unstable period of network operation
2016	S. Zhang et al.	Path panning of UCAV [39]	The UCAV can find the safest two-dimensional route through the selected nodes along with threats areas avoidance and less fuel cost
2016	M. Radmanesh et al.	UAV path planning [40]	Utilized the formulation of dynamic Bayesian and distance based value function (DBVF) algorithm along with GWO to generate an effective path with collision avoidance for UAV
2016	P. Yao et al.	Three-dimensional UAV path planning [41]	The planned three-dimensional route is smooth, feasible and of good obstacle avoidance behavior
2017	A. Kumar et al.	System reliability optimization [25]	Better or competitive results have been obtained compared to well-known metaheuristic algorithms
2017	S. Zhang et al.	EOGWO [26]	The diversity of grey wolf population is increased and exploration ability is improved
2017	H. Xu et al.	GWO integrated with cuckoo search [27]	The search ability of GWO strengthened

(continued)

Table 1 (continued)

Year	Researchers	Applications	Remarks
2017	R. E. Precup et al.	GWO applied to T-S PI-FCs [31]	Improved fuzzy control of a family of nonlinear servo systems
2017	K. Jeet	Flow shop scheduling [20]	Applicable to real-time situations due with uncertainties involved
2017	M. Petrović et al.	Mobile robot scheduling [21]	Shown a significant improvement in the performance of the manufacturing system and provide optimal scheduling of mobile robot with static obstacles
2017	X. Q. Bian et al.	GWO-SVM [34]	The overall average absolute relative deviation of the proposed model is the lowest (3.20%) and less dependent on material type
2017	Y. Wei et al.	IGWO-SVM [35]	The developed approach has achieved better classification accuracy, the area under the receiver operating characteristic curve (AUC), sensitivity, and specificity over other advanced machine learning methods
2017	R. Rajakumar et al.	GWO-LPWSN [37]	Provided a better convergence rate and success rate in comparison to PSO and MBA algorithms
2017	N. R. Solomon Jebaraj et al.	GWA [38]	Optimizes the performance of ZigBee networks
2017	D. Sánchez et al.	MGNN for human recognition [42]	Better results when applied to human recognition
2017	A. Parsian, M. Ramezani, and N. Ghadimi	Hybrid GWO and neural network [43]	Minimizes the root mean square error
2017	Q. Li et al.	IGWO-KELM [44]	The optimal feature subset from medical data has been obtained by using the IGWO feature selection method
2017	E. Yassien et al.	GWO and K-means clustering [45]	Satisfactory performance and shows great compatibility between experimental and theoretical results

wolf optimization and its various applications there is no simulation has been done. From all the studied applications, the GWO is found to provide efficient and competitive results. This paper may help in discovering new aspects and field of application where GWO may be applied and also help in future research.

References

1. Bianchi L, Dorigo M, Gambardella LM, Gutjahr WJ (2009) A survey on metaheuristics for stochastic combinatorial optimization. Nat Comput 8(2):239–287
2. Blum C, Roli A (2003) Metaheuristics in combinatorial optimization: overview and conceptual comparison. ACM Comput Surv (CSUR) 35(3):268–308
3. Mirjalili S, Mirjalili SM, Lewis A (2014) Grey Wolf optimizer. Adv Eng Softw 69:46–61
4. Das B, Subudhi B, Pati BB (2016) Cooperative formation control of autonomous underwater vehicles: an overview. Int J Autom Comput 13(3):199–225
5. Das B, Subudhi B, Bhusan Pati B (2014) Adaptive sliding mode formation control of multiple underwater robots. Arch Control Sci 24(4):515–543
6. Das B, Subudhi B, Pati BB (2015) Employing nonlinear observer for formation control of AUVs under communication constraints. Int J Intell Unmanned Syst 3(2/3):122–155
7. Das B, Subudhi B, Pati BB (2016) Co-operative control of a team of autonomous underwater vehicles in an obstacle-rich environment. J Mar Eng Technol 15(3):135–151
8. Das B, Subudhi B, Pati BB (2016) Co-operative control coordination of a team of underwater vehicles with communication constraints. Trans Inst Meas Control 38(4):463–481
9. Muro C, Escobedo R, Spector L, Coppinger RP (2011) Wolf-pack (Canis lupus) hunting strategies emerge from simple rules in computational simulations. Behav Proc 88(3):192–197
10. Rezaei H, Bozorg-Haddad O, Chu X (2018) Grey Wolf optimization (GWO) algorithm. In: Bozorg-Haddad O (eds) Advanced optimization by nature-inspired algorithms. Studies in computational intelligence, vol 720. Springer, Singapore
11. Muangkote N, Sunat K, Chiewchanwattana S (2014) An improved Grey Wolf optimizer for training q-Gaussian radial basis functional-link nets. In: 2014 international computer science and engineering conference (ICSEC). IEEE, pp 209–214
12. Mirjalili S (2015) How effective is the Grey Wolf optimizer in training multi-layer perceptrons. Appl Intell 43(1):150–161
13. Saremi S, Mirjalili SZ, Mirjalili SM (2015) Evolutionary population dynamics and Grey Wolf optimizer. Neural Comput Appl 26(5):1257–1263
14. Song HM, Sulaiman MH, Mohamed MR (2014) An application of Grey Wolf optimizer for solving combined economic emission dispatch problems. Int Rev Model Simul (IREMOS) 7(5):838–844
15. El-Gaafary AA, Mohamed YS, Hemeida AM, Mohamed AAA (2015) Grey Wolf optimization for multi input multi output system. Univers J Commun Netw 3(1):1–6
16. Mohanty S, Subudhi B, Ray PK (2016) A new MPPT design using Grey Wolf optimization technique for photovoltaic system under partial shading conditions. IEEE Trans Sustain Energy 7(1):181–188
17. Sultana U, Khairuddin AB, Mokhtar AS, Zareen N, Sultana B (2016) Grey Wolf optimizer based placement and sizing of multiple distributed generation in the distribution system. Energy 111:525–536
18. Song X, Tang L, Zhao S, Zhang X, Li L, Huang J, Cai W (2015) Grey Wolf optimizer for parameter estimation in surface waves. Soil Dyn Earthq Eng 75:147–157
19. Komaki GM, Kayvanfar V (2015) Grey Wolf optimizer algorithm for the two-stage assembly flow shop scheduling problem with release time. J Comput Sci 8:109–120
20. Jeet K (2017) Fuzzy flow shop scheduling using Grey Wolf optimization algorithm. Indian J Sci Res 7(2):167–171

21. Petrović M, Miljković Z (2017) Grey Wolf optimization algorithm for single mobile robot scheduling. In: Proceedings of 4th international conference on electrical, electronics and computing engineering, IcETRAN 2017, Kladovo, Serbia, 05–08 June
22. Korayem L, Khorsid M, Kassem SS (2015) Using Grey Wolf algorithm to solve the capacitated vehicle routing problem. In: IOP conference series: materials science and engineering, vol 83, no 1. IOP Publishing, p 012014
23. Gholizadeh S (2015) Optimal design of double layer grids considering nonlinear behaviour by sequential grey wolf algorithm. J Optim Civ Eng 5(4):511–523
24. Mirjalili S, Saremi S, Mirjalili SM, Coelho LDS (2016) Multi-objective Grey Wolf optimizer: a novel algorithm for multi-criterion optimization. Expert Syst Appl 47:106–119
25. Kumar A, Pant S, Ram M (2017) System reliability optimization using Gray Wolf optimizer algorithm. Qual Reliab Eng Int 33(7):1327–1335
26. Zhang S, Luo Q, Zhou Y (2017) Hybrid Grey Wolf optimizer using elite opposition-based learning strategy and simplex method. Int J Comput Intell Appl 1750012
27. Xu H, Liu X, Su J (2017) An improved grey wolf optimizer algorithm integrated with Cuckoo search. In: 2017 9th IEEE international conference on intelligent data acquisition and advanced computing systems: technology and applications (IDAACS), vol. 1. IEEE, pp 490–493
28. Emary E, Zawbaa HM, Hassanien AE (2016) Binary Grey Wolf optimization approaches for feature selection. Neurocomputing 172:371–381
29. Long W, Xu S (2016) A novel Grey Wolf optimizer for global optimization problems. In: 2016 IEEE advanced information management, communicates, electronic and automation control conference (IMCEC). IEEE, pp 1266–1270
30. Soni V, Parmar G, Kumar M, Panda S (2016) Hybrid Grey Wolf optimization-pattern search (hGWO-PS) optimized 2DOF-PID controllers for load frequency control (LFC) in interconnected thermal power plants. ICTACT J Soft Comput 6(3)
31. Precup RE, David RC, Szedlak-Stinean AI, Petriu EM, Dragan F (2017) An easily understandable Grey Wolf optimizer and its application to fuzzy controller tuning. Algorithms 10(2):68
32. Elhariri E, El-Bendary N, Hassanien AE, Abraham A (2015) Grey Wolf optimization for one-against-one multi-class support vector machines. In 2015 7th international conference of soft computing and pattern recognition (SoCPaR). IEEE, pp 7–12
33. Eswaramoorthy S, Sivakumaran N, Sekaran S (2016) Grey Wolf optimization based parameter selection for support vector machines. COMPEL- Int J Comput Math Electr Electron Eng 35(5):1513–1523
34. Bian XQ, Zhang Q, Zhang L, Chen J (2017) A Grey Wolf optimizer-based support vector machine for the solubility of aromatic compounds in supercritical carbon dioxide. Chem Eng Res Des
35. Wei Y, Ni N, Liu D, Chen H, Wang M, Li Q, Ye H (2017) An improved Grey Wolf optimization strategy enhanced SVM and its application in predicting the second major. Math Probl Eng
36. Diwan P, Khan MR (2016) Energy efficient communication for WSNs using Grey-Wolf optimization algorithm. Int J Eng Comput Sci 5(12)
37. Rajakumar R, Amudhavel J, Dhavachelvan P, Vengattaraman T (2017) GWO-LPWSN: Grey Wolf optimization algorithm for node localization problem in wireless sensor networks. J Comput Netw Commun
38. Solomon Jebaraj NR, Keshavan HR (2017) Optimizing the ZigBee networks lifetime with a Grey Wolf algorithm. Asian J Inf Technol 16:472–478
39. Zhang S, Zhou Y, Li Z, Pan W (2016) Grey Wolf optimizer for unmanned combat aerial vehicle path planning. Adv Eng Softw 99:121–136
40. Radmanesh M, Kumar M (2016) Grey Wolf optimization based sense and avoid algorithm for UAV path planning in uncertain environment using a Bayesian framework. In: 2016 international conference on unmanned aircraft systems (ICUAS). IEEE, pp 68–76
41. Yao P, Wang HL (2016) Three-dimensional path planning for UAV based on improved interfered fluid dynamical system and Grey Wolf optimizer. J Opt Soc Am B: Opt Phys 30(3):615
42. Sánchez D, Melin P, Castillo O (2017) A Grey Wolf optimizer for modular granular neural networks for human recognition. Comput Intell Neurosci

43. Parsian A, Ramezani M, Ghadimi N (2017) A hybrid neural network-Gray Wolf optimization algorithm for melanoma detection. Biomed Res 28(8)
44. Li Q, Chen H, Huang H, Zhao X, Cai Z, Tong C, Tian X (2017) An enhanced grey wolf optimization based feature selection wrapped kernel extreme learning machine for medical diagnosis. Comput Math Methods Med
45. Yassien E, Masadeh R, Alzaqebah A, Shaheen (2017) A Grey Wolf optimization applied to the 0/1 knapsack problem. Int J Comput Appl (0975–8887) 169(5)

Development of Human Detection System for Security and Military Applications

Vidushi Goel, Harsh Raj, Kiran Muthigi, S. Sanjay Kumar, Deepak Prasad and Vijay Nath

Abstract In this research article workout on the problem of human detection, face detection, face recognition and tracking an individual. This article is capable of detecting a human and its face in a given video. Once a human is detected in video, it can be tracked that person assigning them a label. In this way, it can be tracked an individual in the video taken by a camera. Our whole work is based on the application of machine learning and image processing with the help of open CV. This research article focus was to use this technique for surveillance of restricted areas by the military and also use this technique for domestic security purposes.

1 Introduction

The observation or monitoring of the activity, behavior, and other information by a system which include several Closed Circuit Television (CCTV) cameras for observation and a set of algorithms to track a person is called surveillance system. Technology has evolved a lot in the last few decades; previously there were no security cameras neither in Banks nor Railway Stations and other places. There were only security guards which protect these areas. Once the security cameras came into existence, it became easy to monitor people passing within the range of CCTV camera by simply searching through the videos recorded. Inventions increase people's expectations, although security camera reduces human effort but one has to search for an individual through entire video which takes a considerable amount of time. So we thought what if searching task can be accomplished by machine; it would save both human effort and time. A combination of machine learning with image processing is used to make a machine learn to recognize a person and track that person in the

V. Goel (✉) · S. Sanjay Kumar · D. Prasad · V. Nath
Department of ECE, Birla Institute of Technology, Mesra, Ranchi 835215, Jharkhand, India
e-mail: vidushigoggle@gmail.com

H. Raj · K. Muthigi
Department of Computer Science, Birla Institute of Technology, Mesra, Ranchi 835215, Jharkhand, India

© Springer Nature Singapore Pte Ltd. 2019

V. Nath and J. K. Mandal (eds.), *Proceedings of the Third International Conference on Microelectronics, Computing and Communication Systems*, Lecture Notes in Electrical Engineering 556, https://doi.org/10.1007/978-981-13-7091-5_18

given footage. Our research is all about a system which has been designed to track human in the given video.

The input is given in the form of a video or a sequence of videos. The main objective of our project is to detect people and track them in a given video sequence. We have considered a few constraints in the interest of providing a successful output. First, our work is mostly restricted to homogeneous domain, i.e., the videos provided by the user must be shot with a homogeneous background, i.e., plain background. Irrespective of whether the human faces, the camera or not, he/she is successfully detected. There are certain major challenges that we faced while doing this project. First, detecting humans in all possible postures like standing, sitting, lying, etc., was a great problem. Second, detecting humans in all possible orientations such as when their front side is facing the camera, when their backside is facing the camera, when they are facing the cameras sideways, etc. Reducing the false positive rate to minimum possible value is still a greater challenge [1, 2]. Automated Toll Plaza Using Barcode-Laser Scanning Technology can be utilized in SVM [3]. Design of All-Terrain Rover Quadcopter for Military Engineering Services used to enhance the application of HOG [4]. Design of Robot Monitoring System for Aviation can be used to improve the activity of surveillance [5]. Electronic Toll Collection System Using Barcode Technology can be utilized to detect vehicles all around the city and country and reduce the theft and improve the billing environment [6]. Design of Smartphone Controlled Robot Using Bluetooth can be applicable to enhance the communication systems in moll and cities [7]. Study and Design of Smart Embedded System for Aviation System: A Review shows the ideas to improve surveillance function in any highly secure places for security purposes [8]. Study and Design of Smart Industry: A review helps in establishing the automated industry and improves the quality of products and increase the quantity of products [9]. Design of Smart Security Systems for Home Automation guides the SVM and HOG for perfect hospitality and high level security [10]. Auto-Train Track Fault Detection System describes the security systems which provide the guidance to improve the applicability of SVM and HOG as per the need of social current digital market [11].

2 Research Methodology

In this section, the different phases through which the system goes through before providing the output to the user are discussed. This provides an overview of how the system works and how the output is provided. This section gives both overview and detailed information about our project.

Input to our program is a video or a sequence of videos. The program tries to detect humans in the input videos. If a human is detected, the program draws a rectangle around them and tracks the human in the video. For human detection, the program takes a video or a sequence of videos as input. In the next step, it sets up a Histogram of Oriented Gradients (HOG) descriptor which is a feature descriptor used for detecting objects with the help of a Support Vector Machine (SVM). HOG along with SVM is used in our program with the help of open CV library to detect humans [12]. As the program then detects humans and a rectangle is drawn around the human.

2.1 *Support Vector Machine (SVM) Classifier*

In some cases, our program draws more than one rectangle for single person because of false positive cases, to minimize such false positive cases, in the next step our program uses non-maximum suppression to select a rectangle based on maximum overlapping criteria.

SVM is a set of supervised learning method used for classification, regression. SVM can efficiently perform both linear and nonlinear classification. When it is not possible to classify data linearly in the same dimensional space in which data point resides, SVM goes into higher dimensional space so that classification can be done. For nonlinear classification, regression and other task SVM construct a hyperplane or a set of hyperplanes in a high or in infinite-dimensional spaces. The result of SVM is shown in Fig. 1.

2.2 *Histogram Oriented Gradient (HOG)*

It is one of the features descriptor used in computer vision and image processing. Histogram of oriented gradients descriptor can be described by the distribution of intensity gradients or edge directions in an image. The image is divided into small connected region called as cells. A cell can contain several pixels and for each of the

Fig. 1 Result of support vector machine

Fig. 2 Result of histogram oriented gradient

pixel a histogram of gradient is made. The descriptor contains histogram of gradient of each and every pixel. For better accuracy the HOG is contrast-normalized, this is done by calculating intensity over a larger area several cells known as block, then this value is used to normalize all cells within that block. The normalized result gives better performance on variation in illumination and intensity [13, 14]. HOG descriptors have some advantages over other descriptors such as it is invariant to geometric and photometric transformations except object orientation. It is particularly suited for human detection in images. This is already discussed the program methodology and how the program works in the previous sections. In this section, it is provided the details about the results obtained by users while using this program against some of the test cases. Here, explained the output of each and every test case using the screenshots of the output provided by this new program. While making this model, user faced a lot of challenges and it has countered majority of these problems. There are some flaws in the output provided by the program but it has tried to minimize as much as possible. The program requires a lot of computations so the output of the program is also system dependent. Users are tried the program on one system which is available with us-Mac Book Air [15].

As mentioned before, designers have considered a constraint that the input video must be shot in a homogenous domain [16]. Here, design is tested and outputs result of histogram oriented gradient is shown in Fig. 2.

3 Results and Conclusions

The paper mainly focuses on tracking a person in a video. The program takes a video or a sequence of videos as input and outputs the names of the people who are seen in the videos. The assumption is that the videos are shot in a homogeneous environment. The program needs some processing time so for an input video of 30 frames per second the output comes out to be a video of 10–15 fps, i.e., the program is system dependent. The program tries to avoid the false positives as much as possible but for certain textures it still becomes confused. The program successfully detects the presence of a person in a video. The program uses the HOG Algorithm to detect humans found in a video and if the program succeeds in locating a human it tracks the human and a box around the person is outputted on the terminal thus helping the user to track any person given a sequence of videos. The focus will now be on reducing the amount of false positives and making the program robust for any kind of environment.

References

1. https://docs.opencv.org/2.4/doc/tutorials/ml/introduction_to_svm/introduction_to_svm.html
2. https://docs.opencv.org/3.4.1/d7/d8b/tutorial_py_face_detection.html
3. Raj U, Nidhi N, Nath V (2019) Automated toll plaza using barcode-laser scanning technology. In: Nath V, Mandal J (eds) Nanoelectronics, circuits and communication systems. Lecture notes in electrical engineering, vol 511. Springer, Singapore, pp 475–481. https://doi.org/10.1007/978-981-13-0776-8_44. Online ISBN: 978-981-13-0776-8
4. Chaudhary S, Prava A, Nidhi N, Nath V (2019) Design of all-terrain rover quadcopter for military engineering services. In: Nath V, Mandal J (eds) Nanoelectronics, circuits and communication systems. Lecture notes in electrical engineering, vol 511. Springer, Singapore, pp 507–513. https://doi.org/10.1007/978-981-13-0776-8_47. Online ISBN: 978-981-13-0776-8
5. Mohan R, Suraj AK, Agarawal S, Majumdar S, Nath V (2019) Design of robot monitoring system for aviation. In: Nath V, Mandal J (eds) Nanoelectronics, circuits and communication systems. Lecture notes in electrical engineering, vol 511. Springer, Singapore, pp 535–547. https://doi.org/10.1007/978-981-13-0776-8_50. Online ISBN: 978-981-13-0776-8
6. Hari Charan EVV, Pal I, Sinha A, Baro RKR, Nath V (2019) Electronic toll collection system using barcode technology. In: Nath V, Mandal J (eds) Nanoelectronics, circuits and communication systems. Lecture notes in electrical engineering, vol 511. Springer, Singapore, pp 549–556. https://doi.org/10.1007/978-981-13-0776-8_51. Online ISBN: 978-981-13-0776-8
7. Goel V, Riya, Kumari P, Shikha P, Tanushree, Prasad D, Nath V (2019) Design of smartphone controlled robot using bluetooth. In: Nath V, Mandal J (eds) Nanoelectronics, circuits and communication systems. Lecture notes in electrical engineering, vol 511. Springer, Singapore, pp 557–563. https://doi.org/10.1007/978-981-13-0776-8_52. Online ISBN: 978-981-13-0776-8
8. Raju D, Eleswarapu L, Saiv R, Nath V (2019) Study and design of smart embedded system for aviation system: a review. In: Nath V, Mandal J (eds) Nanoelectronics, circuits and communication systems. Lecture notes in electrical engineering, vol 511. Springer, Singapore, pp 573–590. https://doi.org/10.1007/978-981-13-0776-8_54. Online ISBN: 978-981-13-0776-8
9. Maurya DK, Kumar A, Kaunoujiya S, Prasad D, Nath V (2019) Study and design of smart industry: a review. In: Nath V, Mandal J (eds) Nanoelectronics, circuits and communication systems. Lecture notes in electrical engineering, vol 511. Springer, Singapore, pp 591–598 https://doi.org/10.1007/978-981-13-0776-8_55. Online ISBN: 978-981-13-0776-8
10. Sanjay Kumar S, Khalkho A, Agarawal S, Prakash S, Prasad D, Nath V (2019) Design of smart security systems for home automation. In: Nath V, Mandal J (eds) Nanoelectronics, circuits and communication systems. Lecture notes in electrical engineering, vol 511. Springer, Singapore, pp 599–604. https://doi.org/10.1007/978-981-13-0776-8_56. Online ISBN: 978-981-13-0776-8
11. Goel V, Kumar S, Muralidharan A, Markham N, Prasad D, Nath V (2019) Auto-train track fault detection system. In: Nath V, Mandal J (eds) Nanoelectronics, circuits and communication systems. Lecture notes in electrical engineering, vol 511. Springer, Singapore, pp 605–610. https://doi.org/10.1007/978-981-13-0776-8_57. Online ISBN: 978-981-13-0776-8

12. https://www.learnopencv.com/histogram-of-oriented-gradients/
13. Deep learning Coursera by Andrew NG
14. Deep learning A-Z™ hands-on artificial neural networks
15. https://www.coursera.org/learn/linear-algebra-machine-learning
16. Machine learning A-Z™ hands-on Python & R in data science

Bots a New Evolution of Robots: A Survey

Anjali Kumari, Partha Paul and Kumar Rajnish

Abstract Today is the age of virtual robots, i.e., bots. Bots are the software or app which runs automated job or task over the internet. A surprising change made by AI (Artificial Intelligence) in different fields is automation. Developing of bots is one of them. Different AI techniques are used to develop different bots. This paper will try to explore different bots based on AI algorithms and try to know why the specific algorithms used in bots and also sketch a view of different types of bots in growing demand for automation.

Keywords AI · Bots · Botnet · GA · Neural network · Reinforcement learning

1 Introduction

During the past few years, the use of bots is increased to automate the work. PennyCat, Hi Poncho and Alexa are the examples of bots. Either it is Facebook or Infosys, companies are now concerned to develop the different bots for their work. Facebook has developed bots which can interact with people in groups and chat. These bots are called chatbots. The Facebook messenger develops these chatbots to automate conversation. The Facebook user has first to register or contact the company admin to gain access. Infosys chatbots "Nia" [1] provide a framework to rapidly create and deploy conversational interfaces. It understands natural learning and responds accurately to the queries of the user.

The first question came to mind is *what is bots*? a bot is a collection of program that runs automatically over the internet. It is like a human which can talk. It is sometimes called web crawlers or spiders when it tracks webpages.

A. Kumari (✉) · P. Paul · K. Rajnish
Department of Computer Science & Engineering, BIT, Mesra, Ranchi, India
e-mail: anjalikumari2066@gmail.com

© Springer Nature Singapore Pte Ltd. 2019

V. Nath and J. K. Mandal (eds.), *Proceedings of the Third International Conference on Microelectronics, Computing and Communication Systems*,
Lecture Notes in Electrical Engineering 556, https://doi.org/10.1007/978-981-13-7091-5_19

2 Background of Bots

Different papers show different approaches in designing bots. One paper suggests that bots can be designed using "Free Will" to select a real number based on selection capability and frame expanding potential between machine and human [2]. It uses Free Will similar to the Epsilon-delta definition to avoid deadlock among agents.

2.1 Classification of Bots

Sometimes bots are classified on the basis of authentication: legal bots and illegal bots. Good or legal bots are used for a legal purpose like collecting information like web crawlers, telling weather forecast, alarm or act as an instant messaging tool like relay chat. Chatbots are bots which can talk to a human in textual form. On the other hand, bad or illegal bots are self-multiplied bots that are created for the purpose of infecting a host and get access right to a server. These types of bots are generally known as malicious bots. Malicious bots hack password, capture packets, and launch DoS attacks. They open back doors for the hackers. After malicious bots connect to the server(s), the server acts as a botnet. Signatures and characteristics of a botnet can be detected by leveraging bots spam server traffic properties and spam payload [3].

According to botsnerds.com [4], bots are of two types: good bots and bad bots (Table 1).

2.2 Chatbots

This bots are designed to chat with the human. This bot is human like AI which interacts with a text message. ELIZA was the first chatbots which were created by Weizenbaum [5] of the MIT Artificial Intelligence Lab. Other AI-based chatbots are Tay, Cleverbots and Niki.ai bot. Some of which can be used for ordering jobs via chat such as cab booking, recharge and bill payment.

Table 1 Good bots and bad bots

Good bots	Bad bots
Chatbots	Hackers
Crawlers	Spammers
Transactional bots	Impersonators
Entertainment bots	Scrapers
Informational bots	

2.3 Crawlers

Crawlers or spiders are the programs which track certain hyperlink and collect information as it goes. They can gather all sorts of data like broken links, HTML code validation, link and content on a page. Googlebots, DuckDuckBots, MSNbots, and Yahoo Slurp are some spider bots that gathers information about a web page for their search engine. This gathered page is sometimes called SERPs (Search Engine Results Page).

Googlebots use HREF or SRC tag of the webpage to track and find pages or images that are linked to any specified site. A file called robots.txt tells crawler what to follow by indexing or not. This file comes into play and controls the flow of the crawling traffic. The web crawler traffic can be examined by referring logs of User-agent field of HTTP.

2.4 Transactional Bots

The bots which work on the behalf of human to accomplish task or transaction by moving information from one platform to other are called transactional bots. Transactional bots are personal assistant bots. X.ai and Birdly are examples of transactional bots.

2.5 Entertainment Bots

Game bots and movies reviews are entertainment bots. Deep learning algorithm bots called, "Deep Drumpf", are used to transcripts of speeches and learn to speak like Donald Trump.

A game bot is a type of AI-based bots which plays game in place of a human. Alpha Go is one of the popular game.

2.6 Spammers or Spambot

These types of bots are specially designed to harvest e-mail address from newsgroup, website, and chatroom to send unsolicited email or spam.

3 AI Algorithm Used in Bots

3.1 Neural Network (NN)

A combination of a layered network of an Artificial Neuron (AN) is called Artificial Neural Network (ANN). There are three layers that are present in a simple NN: an input layer, hidden layers, and an output layer. It is good for probability distributions approximation [6]. A Neural Network Approach can respond to generate program that can well-trained end to end conversations of unstructured Twitter data.

In chatbots designing, first training data and target data have to fetch for training model then using an NN result of input data is obtained.

Job performed by NN in bots designing:

- To derive remarkable meaning from complicated or emprise data, Neural Network uses to extract or detect trends from it which cannot be detected easily by human or machine.
- A trained neural network can be used to categorized information and analysis it. So NN can be called an expert. This expert can be used to answer "what if" questions based on specific situation.
- Recurrent Neural Network (RNNs) of NN is used to train character-level language objects. Using a chunk of text and probability distribution of RNN derivation of the next character in the line using the previous one. The output is the generation of a new character at a time for chatbots.
- Advantages of NN:

  Adaptive learning: *the* ability to perform a task based on given information or previous experience.
  Self-Organization: during learning time, it creates own organization.
  Real-Time Operation: it is based on parallel computation.
  Fault Tolerance via Redundant Information Coding: in this ANN, when some network are destructed it retain its capability.

3.2 Genetic Algorithm and Genetic Programming Algorithm

Genetic algorithm is based on genetic evolution. Genetic programming algorithm is based on GA using a tree. GA is a similarity between genetic structure and characteristic of chromosomes within a population of individuals having the following points:

- Each individual has competition for resources and mates.
- Producing more offspring in each competition will be more successful.
- Good gene individuals reproduce throughout the population so that two good parents will produce offspring which is better than either of the parents.

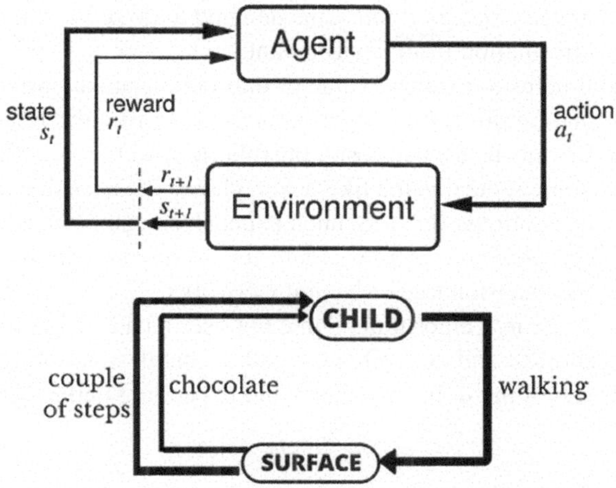

Fig. 1 Description of a reinforcement learning problem [7]

- This way successive generation will survive in the environment.

In genetics, fitness is defined as the rating of a species with evolution it transforms from existing to higher fitness value species.

3.3 Reinforcement Learning

It is a reward-oriented learning based on interaction with environment. Let us take an example to understand reinforcement learning. Suppose a child start to learn how to walk. His mother inspired him by saying you got chocolate when you walk. This way he get reward of chocolate based on the walking step (Fig. 1).

Simple formula to find a candidate optimal value function is

$$\mathbf{V}^*(\mathbf{s}) = \max_{\mathbf{a} \in \mathbf{A}} \{\mathbf{R}\,(\mathbf{s},\ \mathbf{a}) + \gamma \sum_{s' \in S} \mathbf{T}(\mathbf{s},\ \mathbf{a},\ \mathbf{s}')\,\mathbf{V}^*(\mathbf{s}')\},\ \mathbf{s} \in \mathbf{S}$$

where

A	is set of all possible actions,
S	is a set of environmental states,
R(s, a)	is reward function,
T(s, a, s′)	is the transition function

Job performed by Reinforcement learning in bots designing:

- Reinforcement learning allows bots or machine to learn from environment and based on feedback it attain optimization. This cannot be done in genetic algorithm in game design if the game demand to do so then this technique act after attaining information from environment.

- It helps to estimate value of particular state at particular time that is used to estimation time of propagation of bots at particular time.
- It is mostly used in designing robotic arms for controlling the game bots navigation. Some logical game like poker, chess, etc. uses this technique.
- Limitations: It is too much expensive to store data like its state and due to which complexity increases. Again for solving this approximation technique is a need-like decision tree or Neural Networks.

  A* search algorithm: Game bot uses a path searching algorithm that will find the shortest and cost effective path to the destination (goal). A* search algorithm is one of them. It is used to avoid expanding paths that are already expensive.

4 Comparative Analysis

For comparison of different algorithms, different papers are surveyed. From a paper titled "Comparison between genetic algorithm and self-organizing map to detect botsnet network traffic" [8], we find that how genetic algorithm gives output on particular dataset and it different from self-organizing map by winner-takes-all neuron concept.

From a paper titled, "Creating AI Characters for Fighting Games Using Genetic Programming" [9] how genetic programing is used in the implementation of bots. In this paper, it is explained that to generate a random population of valid group is triggered to place the changes in the character state. Using this idea test case of testing bot can be generated. Based on the different trigger different test case can be obtained based on the fittest test case. Then using selection method best test criteria be chosen.

From a paper titled, "A Deep Reinforcement Learning Chatbots" [11] and "Creating a multipurpose First Person Shooter Bots with Reinforcement Learning" [12], how reinforcement technique can be used to design is discussed.

From a paper titled "Turing computer Gaming agents using Q-learning" [13], how well bot can be tested or learn to fight and instead of rewarding how bot change is tested. Using Q-learning how bot learns or trained itself is another interesting concept. This paper gives an idea how a bot can learn for a particular job and get a reward in terms of another job and learning.

From a paper titled "Using Machine Learning to Detect Fake Identities: Bots vs Humans" [14], we get information on how to apply supervised learning in fake twitter account. During detection of account its data are mined and after applying supervised learning model result are evaluated.

Author of the paper titled "A Speech Intelligence conversation bots for interactive media information" [15] used corpus selection and then it provides test. If the result of the program is expected execution then the value of test is 1 if not it is 0.

4.1 *Comparative Analysis on Different Algorithm of AI Used in Bots*

See Table 2.

5 Future Scope of Our Work

Artificial Intelligence is used to automate the work and the best way to design bots. Either moving object or chatting like human all depend on the algorithm. So each algorithm has its own importance. Even though it makes revolution in bots development somewhat is lagged behind human brain working. Because all these bots do the specified task it cannot take its decision on its own. For example Chatbot is computer program that conducts a conversation via textual or auditory methods. One of the great work in bot development is the AI bots that were created by the Elon Musk-founded research lab OpenAI. Most notable work was the defeat of the world's best Go players by DeepMind's AlphaGo. They get information from players; take place in complex, ever-changing environments; and require the sort of strategic thinking

Table 2 Comparative Analysis on different algorithms of AI used in bots

Research paper based on bots	Algorithm/technique used	Year of publication/journals	Type	Result
Comparison between genetic algorithm and self-organizing map to detect botsnet network traffic [8]	Genetic algorithm (GA) and self-organizing map (SOM)	2017 [ICSET]	Botsnet detection	This paper uses sample dataset KDD99 to detect botsnet. GA is used to perform clustering which although takes time but gave an accurate result and identify maximum used IP address
Creating AI characters for fighting games using genetic programming [9]	Genetic programming algorithm	2016 [IEEE]	Game bots	This paper uses genetic programming in game to refines a character strategies into tournament selection randomly. This AI generated character was tested and rated based on result and game engaging or not

(continued)

Table 2 (continued)

Research paper based on bots	Algorithm/technique used	Year of publication/journals	Type	Result
s-Birds Avengers: a dynamic heuristic engine-based agent for the angry birds problem [10]	Heuristic techniques	2016 [IEEE]	Game bots	Heuristic techniques were used to analyze bots to decide where to hit the target(structure) with the birds
A deep reinforcement learning chatbots [11]	Reinforcement learning	2017 [IEEE]	Chatbots	"Milabots" is designed using Reinforcement Learning. It is capable of conversing with human. Reinforcement Learning is used to retrieve and generate model using source data and real-world user interaction. Lastly, it evaluated using A/B testing
Creating a multipurpose First Person Shooter Bots with Reinforcement Learning [12]	Reinforcement learning	2008 [IEEE]	Game Bots	It is based on Reinforcement Learning algorithm. It compares 3 different learning approaches to design a bots. Based on these approaches result is obtained using hierarchical or rule-based approach. Lastly, it gives risk additive behavior set
Turing computer gaming agents using Q-learning [13]	Q-learning	2011 [IEEE]	Game bots	Q-learning is used to evolve behavior of the game like learning to fight, planting the bomb in game bots
Using machine learning to detect fake identities: bots vs humans [14]	Machine learning	2017 [IEEE]	Bots to detect fake tweeter account	This paper uses machine learning algorithm like
A speech intelligence conversation bots for interactive media information [15]	speech recognition by Sphinx-4	2017 [IEEE]	Chatbots	Speech recognization is used in chatbots system with the modified MegaHal style

that cannot be easily simulated. OpenAI's Dota 2 is a great testing ground that uses the reinforcement learning to train the bot. As a result, the bots start out playing completely randomly, and over time, they learn to connect certain behaviors to rewards. This gives a win for artificial intelligence over human to some extent.

As bots are developing in almost every field either for good purpose or bad, so in near future testing of bots is required. This would eventually require developing new bots which test other bots, website or automate the testing. The idea is to develop bots which have different test cases and test script using AI algorithm which will automate the work. One site called diffbots.com [16] takes a step toward automation using bots in testing like API testing. This site has a feature to determine "page-type" and extract API which matches will be automatically extracted. This API testing is now free to test but other features are in the paid version. But as advancement in AI increases, the testing bots development will also increase.

6 Conclusion

After surveying different paper it is found that AI algorithm is one of the demanding technique in the bots creation and designing. Most of the bots are based on different branches of AI technique. If we get a closure look on the algorithms, algorithms can be combined and then compare its working to enhance bots functionalities. Genetic algorithm can be combined with neural network and neural to reinforcement learning, then we get a new version of bots. Some author did this according to their requirements.

References

1. Chatbots 101: building conversational interfaces, "purposeful chatbots for your enterprise". Forrester Research, December 2016
2. Engelbrecht AP (2007) Computational intelligence an introduction, 2nd edn., pp 68–73
3. Xie Y, Yu F, Achan K, Panigrahy R, Hulten G, Osipkov I (2008) Spamming botsnets: signatures and characteristics SIGCOMM '08, Seattle, Washington, USA, 17–22 August 2008
4. http://botsnerds.com/types-of-bots/
5. Weizenbaum J (1966) ELIZA—a computer program for the study of natural language communications between man and machine. Commun ACM 9(1):36–45
6. Jia J (2009) CSIEC: a computer assisted English learning chatbots based on textual knowledge and reasoning. Elsevier
7. https://www.analyticsvidhya.com/blog/2017/01/introduction-to-reinforcement-learning-implementation/
8. Prabhakar SY, Parganiha P, Viswanatham VM, Nirmala M (2017) Comparison between genetic algorithm and self organizing map to detect botsnet network traffic. In: IOP conference series: materials science and engineering, vol 263, p 042103
9. Martínez-Arellano G, Cant R, Woods D (2016) Creating AI characters for fighting games using genetic programming. IEEE Trans Comput Intell AI Games 9

10. Dasgupta S, Vaghela S, Modi V, Kanakia H (2016) s-Birds avengers: a dynamic heuristic engine-based agent for the angry birds problem. IEEE Trans Comput Intell AI Games
11. Serban IV, Sankar C, Germain M, Zhang S, Lin Z, Subramanian S, Kim T, Pieper M, Chandar S, Ke NR, Rajeshwar S, de Brebisson A, Sotelo JMR, Suhubdy D, Michalski V, Nguyen A, Pineau J, Bengio Y (2017) A deep reinforcement learning chatbots. IEEE
12. McPartland M, Gallagher M (2008) Creating a multi-purpose first person shooter bots with reinforcement learning. In: IEEE symposium on computational intelligence and games, 2008. CIG '08, 15–18 December 2008
13. Yorozu Y, Hirano M, Oka K, Tagawa Y (1987) Turing computer gaming agents using Q-learning. IEEE Transl J Magn Jpn 2:740–741. (Digests 9th Annual Conf. Magnetics Japan, p 301, 1982)
14. Van Der Walt E, Eloff J (2018) Using machine learning to detect fake identities: bots vs humans. IEEE Access 6:6540–6549
15. Yushendri J, Hanif AR, Siswadi AAP, Musa P, Kusuma TM, Wibowo EP (2018) A speech intelligence conversation bots for interactive media information. In: 2017 second international conference on informatics and computing (ICIC), 5 February 2018
16. https://www.diffbots.com/products/automatic/

Electronic Applications of Conducting Polymer Nanocomposites

Subhadra Panda and Bibhudendra Acharya

Abstract Recently, conducting polymers (CPs) have economic importance because of their functional elements, good environmental stability, flexibility, and electrical conductivity as well as because of their useful mechanical, optical, and electronic properties. In conducting polymer nanocomposites, polymer function as a matrix and different nanofillers are embedded in the matrix results in the formation of new material and can be used in various applications as per requirement. Some of the applications of conducting polymer nanocomposite include: conducting adhesives, electrostatic materials, electromagnetic shielding materials, transistors, supercapacitors, sensors, diodes, etc. This review includes some of the important applications of conducting polymers with nanofibers and nanotubes in sensors, field effect transistors, nanodiodes, field emission, energy storage, actuators, and super capacitors.

Keywords Conducting polymer · Electronic properties nanodevices · Nanofillers

1 Introduction

Nowadays, nanotechnology has become a current field of research because of its considerable potential for various applications in different fields. Attention towards conducting polymers increased due to their strong potential and as they are working as an alternative to their inorganic counterparts. Conducting polymers have some unique properties such as electrical properties, controllable chemical, and electrochemical properties, lighter weight, superior stability, resistance to corrosion, better workability, and satisfactory electrical conductivity and easy process ability. There are different types of conducting polymers like Poly Acetylene (PA), Poly Di Methyl

S. Panda · B. Acharya (✉)
Department of Electronics and Telecommunication, National Institute of Technology Raipur, Raipur 492010, Chhattisgarh, India
e-mail: bacharya.etc@nitrr.ac.in

S. Panda
e-mail: spanda109@gmail.com

© Springer Nature Singapore Pte Ltd. 2019

V. Nath and J. K. Mandal (eds.), *Proceedings of the Third International Conference on Microelectronics, Computing and Communication Systems*,
Lecture Notes in Electrical Engineering 556, https://doi.org/10.1007/978-981-13-7091-5_20

Siloxane (PDMS), Poly Pyrrole (PPy), Poly Phenylene Vinylene (PPV), Poly Aniline (PANI), Poly Ethylene Di Oxy Thiophene (PEDOT), Poly Thiophene (PTh) and Poly Furan (PF) derivatives, etc., have achieved distinctive importance in the field of nanoelectronics and nanotechnology [1–6].

Synthesis of CPs with other materials can result in new materials which have some different properties and the changed properties of the materials can be used for a particular application. Conducting polymers are the polymer matrix and the material added to it is nanofillers combined by some chemical and/or electrochemical techniques. CP nanocomposites contain different carbon nanofillers such as carbon nanofibers, graphene and carbon nanotubes, etc. [7–10]. Many synthetic strategies (both physical and chemical) have been adopted for the preparation of conducting polymer (CP) nanocomposites and nanowires.

2 Properties of Conducting Polymer Nanocomposites

Conducting polymers nanocomposites exhibit conduction properties and also exhibit some amazing properties such as optical properties, electronic, magnetic, mechanical and microwave-absorbing properties.

2.1 Electrical Conductivity Properties

Percentage of doping, conjugation length, arrangement of polymer chains and the purity of the samples are some important factors on which electrical conductivity of polymer depends. Electrical CPs are molecular in nature. As the polymers have conjugate bond, i.e., alternate single and double bond produces electronic movement create high conductivity. The higher conductivities depend upon the amount of fillers in the polymers. CPs have two ground states one of them is degenerate ground state such as charged solitons and trans-polyacetylene, another one is nondegenerate ground state such as cis-PA, PTh, PPy or PPV primarily polarons [11]. As conducting polymer provides desired properties due to their high potential for molecular engineering and cost-effective phenomenon, polymers are considered as wonderful materials for electrically conductive applications.

2.2 Optical Properties

Conducting polymers have vast application in nanophotonic devices because of their optical properties. The nanocomposite materials which are semiconductor having optical properties used for the fabrication of photochemical sensors, photodetectors, and photonic wire lasers [12]. The optical properties of CdS and PANI composite

Table 1 Conductivity and application of polymer nanocomposite samples

Samples	Properties	Conductivity(S/m)	Application
Polyphenylene	(i) Unusual electrical and optical properties (ii) High temperature resistant	1.36×10^{-4}	Electroluminescen-ce, Light emitting diode
Polyaniline	(i) Electrical properties	2.45×10^{1}	For EMI shielding of electronic circuit, corrosion inhibitor
Polypyrrole	(i) Electrical properties, chemically and thermally stable	1.25×10^{-1}	Gas sensor and anti-electrostatic coating, solid electrolytic capacitor
Polythiophene	(i) Low electrical conductive and high conductive even with less doping	4.76×10^{-1}	Emission and absorption can be tuned from UV to IR by changing substituent
PEDOT	(i) Dielectric properties	2.641×10^{1}	Actuators

nanocables have studied by Xi et al. [13–15] and he found that CDs nanowires had similar properties in the photoluminescence spectrum, but the signal intensities were enhanced. The enhancement of signal intensities occurred is because of the transformation of photogenerated carriers from the PANI layer into CdS nanowires. Turac et al. synthesized a new polythiophene derivative have been determined by electrochemical oxidative polymerization and the optical contrast, bandgap, and switching time.

Nowadays, conducting polymer is mostly used as its properties can be modified according to the application by adding various filler to the CPs. Conductivity of different polymer nanocomposite samples with its applications is shown in Table 1.

2.3 *Microwave-Absorbing Properties*

As the conducting polymers are having lower density and easy possibility nature so it has been explored as new microwave-absorbing materials. Polyaniline-Naphthalene Sulfonic Acid (NSA) and PANI-NSA = glucose micro nanotubes show an excellent electromagnetic loss prepared by a template-free method found by Wan et al. [16, 17]. It was reported by Liu et al. [18–20] that when fiber doped with polyaniline gives better electromagnetic wave absorbing property than that of particle doped with polyaniline. The above results show that conducting polymer nanocomposites can be used as electromagnetic shielding material due to having a high reflection, high absorption, lightweight, wide frequency and microwave absorbent properties.

2.4 Mechanical Properties

Recently, the carbon nanotube has drawn much attention due to their unique mechanical properties [21]. Cuenot et al. [22] studied the force-curve measurement or resonance-frequency measurement of polypyrrole nanotube and reported its elastic tensile modulus. From the result, it was found that the elastic modulus is inversely proportional to the thickness of the polypyrrole nanotube wall or its outer diameter. The adhesion strength between the polypyrrole film and a substrate should maintain to get better enhance mechanical properties [23]. Controlled thickness and strongly the adherent PPy film is possible by treating the surface with 10% nitric acid as it leads to important passivation of the iron electrode.

3 Application of Conducting Polymers

From the last two decades, conductive polymers have shown great advancement. An ideal condition for conduction of electricity in case of conducting polymers is a half-filled valence band should form from a continuous delocalized π-system. However, it is possible for a π-conjugated polymer to lower its energy more efficiently and making it a high energy gap semiconductor by alternating the single and double bond. Conducting polymer nanocomposites have remarkable and multifunctional properties for which they can be used in different fields such as nanoelectronic devices, storage device, chemical, or biological sensors, catalysis or electrocatalysis, energy, microwave absorption, and EMI shielding.

3.1 Optoelectronic Devices

Optoelectronic devices are electronic devices that involve light (any range of the electromagnetic radiation) and depending on whether light is used or produced in the operation, they can be light sensitive or light generating devices. Light sensitive devices operate as optical-to-electrical transducers or devices including solar cells, photodiodes, and phototransistors, while light generating devices are electrical-to-optical, such as light emitting diodes (LEDs), laser diodes and cathode ray tubes. Optoelectronic devices can be applied to many areas, including automatic control systems, telecommunications, military services, and medical equipment, among others. Due to the outstanding electronic, optical and mechanical properties, high structural stabilities, easy processing features, and economical synthesis costs, conducting polymers and moreover their derivatives, conductive polymer nanocomposites are widely used in various optoelectronic devices. It is fitting to enhance their characteristics, achieve some new functionalities, and improve their performances by designing, tailoring and modifying their structures compositions and morphologies [11–14, 24–26].

3.2 *Electrochemical Energy Storage Devices (EES)*

Among various EES devices, lithium-ion batteries and supercapacitors are most promising. Li-ion batteries are known for their high energy density while supercapacitors possess high power density. To meet the increasing demands of high-performance EES, conducting polymers are expected to be a promising option. Battery life has always been a barrier that limits new high-performance materials from replacing commercial Li-ion batteries. Wu et al. recently reported a stable Li-ion battery anode material [15] in which they incorporated silicon nanoparticles into the conducting polymer hydrogel network. Due to the high specific capacity, silicon is considered to be an ideal Li-ion battery anode material. However, its poor cycling stability has limited its applications. With the electrically conducting polyaniline hydrogel as the network, Wu et al. demonstrated over 90% capacity retention after 5000 charging and discharging cycles. The advantages of this unique hierarchical three-dimensional materials include continuous conductive path, providing porous space for silicon volume expansion.

Conducting polymers have been shown to enhance the performance of other electrode materials like CO_3O_4. Xia et al. constructed a conducting polymer (polyaniline) layer onto the CO_3O_4 to build organic/inorganic core–shell nanostructures [16] CO_3O_4 suffers from poor rate capabilities because of their low intrinsic conductivity. With the incorporation of conducting polymer, the coaxial electrode material exhibited a superior cycling stability and a higher storage capacity than bare CO_3O_4 nanowires arrays. Murugan et al. synthesized graphene nanosheets/polyaniline nanocomposites and used them as electrode materials for Li-ion batteries and supercapacitors (Fig. 1).

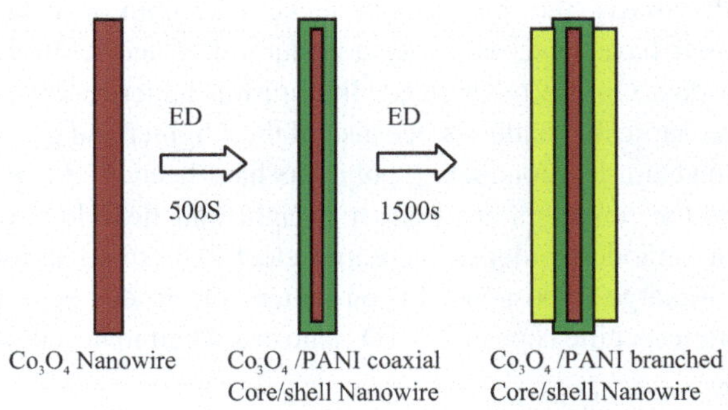

Fig. 1 Schematic of the synthesis of Co_3O_4/PANI coaxial and branched core/shell nanowire

3.3 Energy Storage

Some new materials are developed for advancement in energy conversion and storage, both of which are important in facing the challenge of global warming. Nanomaterials have unique properties as electrodes and electrolytes for various energy devices. The conducting polymer nanostructures are having a high electrical conductivity, short path length for ion transport, large surface area, superior electrochemical activity, and high specific capacitance value for which they are used in fuel cells, supercapacitors and lithium batteries, etc. So, conducting polymer nanostructures are considered as the next generation energy storage devices [27, 28].

Solar Cells
The solar cell is a device that converts sunlight into electric energy. Solar cell acts as a battery as it supplies DC power supply. When conducting polymers combined with metal oxide, they exhibit unique properties of light absorbance and hole transporting, result in the enhancement of the photovoltaic efficiencies [29–33]. As the metal oxide nanofibers have high electron mobility so they are used for the fabrication of solar cells. Nowadays, conducting polymers use metal oxide nanofibers as nanofillers for fabrication in solar cells as they have unique properties of metal oxides having the good light absorbing, good mechanical properties and hole transporting properties of conducting polymers, which enhances the photovoltaic efficiencies.

Fuel Cells
Fuel cell is a device that converts the chemical energy directly into electricity by electrochemical reactions. In the twenty-first century, fuel cells are gaining significant attention as improved and challenging technology for high energy conversion efficiency, environment friendliness, and fuel portability in the field of energy applications [34–36]. Most benefit thing about fuel cells is the direct energy conversion that enables the achievement of high efficiency. Nowadays, in the field of energy application, fuel cells have importance as it is used in electric vehicles. Recent focus on fuel cell is because of their high energy conversion efficiency.

Lithium-Ion Batteries
Previously, the rechargeable nickel-cadmium or nickel-metal hydride batteries are used have lower capacity and durability and it discharges up to 10% by self on a day. On the other hand, lithium-ion batteries are found as the most promising rechargeable batteries because of their lighter and greater capacity properties. In Li-ion batteries, conducting polymers have been used as potential cathodes for batteries as the polymers are light in weight and flexible. Conducting polymers are used in cathode results in increased electrical conductivity for which easy charge and discharge is possible. Li-ion battery electrodes have good cycle performance and high specific capacity as 1D-nanostructured materials are used as electrode [37–40].

Supercapacitors
Supercapacitors are one of the most challenging energy storage devices, also called as electrochemical capacitors. Supercapacitors have applications in various field such as uninterruptible power supplies, military applications, and electric vehicles etc.

Supercapacitors contain both the features of pseudocapacitor and electric double layer capacitor (EDLC). Conducting polymers are called as pseudocapacitive materials as they undergo a fast redox reaction to provide the capacitive response for which they exhibit superior specific energies to the carbon-based supercapacitors. Supercapacitors have higher specific power than lithium-ion batteries. Conducting polymers have high specific capacitance but having poor cyclic stability. This drawback has been overcome by the fabrication of conducting polymer nanocomposites [41–44].

3.4 Microwave Absorption and EMI Shielding

Electromagnetic frequency interference (EMI) is a serious problem and the reasons are a rapid increase in electronics, wireless systems and in space technology, etc. Electromagnetic radiation at high frequency has a tendency to hampers the electric device performance as well as various life forms including human beings. Therefore shielding materials such as metals, carbon materials, and conducting polymers have been used to prevent electromagnetic noise. The uses of conducting polymers as shielding materials have attracted the attention because of their good electrical conductivity which satisfies the reflection mechanism and processibility [45, 46]. It has been observed that for conducting polymers when combined with different nanofillers enhances EMI shielding performance. The nanofillers can be metal or carbon. Metal nanofillers used for shielding are Ag, Au, or Cu.

3.5 Actuators

An actuator is a specific type of transducer that converts electrical, chemical, or thermal energy into motion or mechanical energy. Different types of actuators are there like thermal, electrical, mechanical, etc. A Della Santa et al. describe performance analysis of conducting polymer actuator made of polypyrrole (Ppy). The performance of the actuator depends on its response time. The performance of actuator enhances by decreasing the response time. The response time is directly proportional to thickness of the film of conducting polymer.

Otero and his co-workers have used polypyrrole films in the design of a biomimetic dual sensing-actuator. Son et al. proposed PEDOT as transparent dielectric elastomer as a material for an actuator-driving and variable-focus lens system.

4 Conclusion

From last few years, it has seen that there is fast progress in discovery of new materials on hybrid of different conducting polymer with different fillers to form functional nanocomposites. In this review, application in electronics of different materials based on conducting polymers with different nanofiller form nanocomposites is described. It shows that in research and development field, nanofillers play a major role when synthesized with polymer because the newly formed nanocomposites have remarkable electronic properties, from excellent electrical conductivity to mechanical strength. The combination of these peculiar properties of nanofillers with various functional polymers produces novel materials with unusual electrical, magnetic, and optical properties. Some new applications are expected due to this versatility of these composite materials in the form of batteries, sensors, and microelectronics. In future, the prospective synthesis and fundamental studies of these novel materials are needed for better improvement of properties, which are still unknown for future unknown applications.

References

1. Tran HD, Li D, Kaner RD (2009) One-dimensional conducting polymer nanostructures: bulk synthesis and applications. Adv Mater 21:1487–1499
2. Li C, Bai H, Shi G (2009) Conducting polymer nanomaterials: electrosynthesis and applications. Chem Soc Rev 38:2397–2409
3. Wan M (2009) Some issues related to polyaniline micro-/nanostructures. Macromol Rapid Commun 30:963–975
4. Stejskal J, Sapurina I, Trchová M (2010) Polyaniline nanostructures and the role of aniline oligomers in their formation. Prog Polym Sci 35:1420–1481
5. MacDiarmid AG (2001) "Synthetic metals": a novel role for organic polymers. Curr Appl Phys 1:269–279
6. Heeger AJ (2001) Semiconducting and metallic polymers: the fourth generation of polymeric materials (Nobel lecture). Angew Chem Int Ed 40:2591–2611
7. Wang J, Dai J, Yarlagadda T (2005) Carbon nanotube– conducting-polymer composite nanowires. Langmuir 21:9–12
8. Lee BH, Lee JH, Kahng YH, Kim N, Kim YJ, Lee J, Lee T, Lee K (2014) Graphene-conducting polymer hybrid transparent electrodes for efficient organic optoelectronic devices. Adv Funct Mater 24:1847–1856
9. Gupta S, McDonald B, Carrizosa SB, Price C (2016) Microstructure, residual stress, and intermolecular force distribution maps of graphene/polymer hybrid composites: nanoscale morphology-promoted synergistic effects. Compos B Eng 92:175–192
10. Gupta S, Price C, Heintzman E (2016) Conducting polymer nanostructures and nanocomposites with carbon nanotubes: hierarchical assembly by molecular electrochemistry, growth aspects and property characterization. J Nanosci Nanotechnol 16:374–391
11. Yu G, Lu Y, Liu X, Wang W-J, Yang Q, Xing H, Ren Q, Li B-G, Zhu S (2014) Polyethylenimine-assisted extraction of α-tocopherol from tocopherol homologues and CO_2-triggered fast recovery of the extractant. Ind Eng Chem Res 53:16025–16032
12. Law M, Kind H, Messer B, Kim F, Yang P (2002) Photochemical sensing of NO_2 with SnO_2 nanoribbon nanosensors at room temperature. Angew Chem 114:2511–2514

13. Zhang Q, Yu G, Wang W-J, Yuan H, Li B-G, Zhu S (2012) Preparation of N_2/CO_2 triggered reversibly coagulatable and redispersible latexes by emulsion polymerization of styrene with a reactive switchable surfactant. Langmuir 28:5940–5946
14. Huang MH, Mao S, Feick H, Yan H, Wu Y, Kind H, Weber E, Russo R, Yang P (2001) Room-temperature ultraviolet nanowire nanolasers. Sci 292:1897–1899
15. Geetha S, Satheesh Kumar KK, Trivedi DC (2005) Polyaniline reinforced conducting E-glass fabric using 4-chloro-3-methyl phenol as secondary dopant for the control of electromagnetic radiations. Compos Sci Technol 65:973–980
16. Stafstrom S, Bredas JL, Epstein AJ, Woo HS, Tanner DB, Huang WS, MacDiarmid AG (1987) Polaron lattice in highly conducting polyaniline: theoretical and optical studies. Phys Rev Lett 59:1464
17. Saini P, Choudhary V, Singh BP, Mathur RB, Dhawan SK (2009) Polyaniline–MWCNT nanocomposites for microwave absorption and EMI shielding. Mater Chem Phys 113:919–926
18. Phang SW et al (2008) Microwave absorption behaviors of polyaniline nanocomposites containing TiO_2 nanoparticles. Curr Appl Phys 8:391–394
19. Phang SW, Tadokoro M, Watanabe J, Kuramoto N (2008) Synthesis, characterization and microwave absorption property of doped polyaniline nanocomposites containing TiO2 nanoparticles and carbon nanotubes. Synth Metals 158:251–258
20. Cuenot S, Demoustier-Champagne S, Nysten B (2000) Elastic modulus of polypyrrole nanotubes. Phys Rev Lett 85:1690
21. Muraoka M, Tobe R (2009) Mechanical characterization of nanowires based on optical diffraction images of the bent shape. J Nanosci Nanotechnol 9:4566–4574
22. Gordon MJ, Baron T, Dhalluin F, Gentile P, Ferret P (2009) Size effects in mechanical deformation and fracture of cantilevered silicon nanowires. Nano Lett 9:525–529
23. Olson DC, Piris J, Collins RT, Shaheen SE, Ginley DS (2006) Hybrid photovoltaic devices of polymer and ZnO nanofiber composites. Thin Solid Films 496:26–29
24. Lu Y, Yu G, Wang W-J, Ren Q, Li B-G, Zhu S (2015) Design and synthesis of thermoresponsive ionic liquid polymer in acetonitrile as a reusable extractant for separation of tocopherol homologues. Macromol 48:915–924
25. Martin CR (1994) Nanomaterials: a membrane-based synthetic approach. Sci 266:1961–1966
26. Cui Y, Lieber CM (2001) Functional nanoscale electronic devices assembled using silicon nanowire building blocks. Sci 291:851–853
27. Greene LE, Law M, Yuhas BD, Yang P (2007) ZnO–TiO2 core–shell nanorod/P3HT solar cells. J Phys Chem C 111:18451–18456
28. Shankar K, Mor GK, Paulose M, Varghese OK, Grimes CA (2008) Effect of device geometry on the performance of TiO2 nanotube array-organic semiconductor double heterojunction solar cells. J Non-Cryst Solids 354:2767–2771
29. Fan B, Mei X, Sun K, Ouyang J (2008) Conducting polymer/carbon nanotube composite as counter electrode of dye-sensitized solar cells. Appl Phys Lett 93:143103
30. Tepavcevic S, Darling SB, Dimitrijevic NM, Rajh T, Sibener SJ (2009) Improved hybrid solar cells via in situ UV polymerization. Small 5:1776–1783
31. Houarner-Rassin C, Blart E, Buvat P, Odobel F (2008) Solid-state dye-sensitized TiO2 solar cells based on a sensitizer covalently wired to a hole conducting polymer. Photochem Photobiol Sci 7:789–793
32. Khatri I, Adhikari S, Aryal HR, Soga T, Jimbo T, Umeno M (2009) Improving photovoltaic properties by incorporating both single walled carbon nanotubes and functionalized multiwalled carbon nanotubes. Appl Phys Lett 94:093509
33. Chen Z, Xu L, Li W, Waje M, Yan Y (2006) Polyaniline nanofibre supported platinum nanoelectrocatalysts for direct methanol fuel cells. Nanotechnol 17:5254
34. Ma Y, Jiang S, Jian G, Tao H, Yu L, Wang X, Wang X, Zhu J, Hu Z, Chen Y (2009) CNx nanofibers converted from polypyrrole nanowires as platinum support for methanol oxidation. Energy Environ Sci 2:224–229
35. Reddy ALM, Rajalakshmi N, Ramaprabhu S (2008) Cobalt-polypyrrole-multiwalled carbon nanotube catalysts for hydrogen and alcohol fuel cells. Carbon 46:2–11

36. Tarascon J–M, Armand M (2011) Issues and challenges facing rechargeable lithium batteries. In: Materials for sustainable energy: a collection of peer-reviewed research and review articles from nature publishing group. World Scientific, pp 171–179
37. Chan CK, Peng H, Liu G, McIlwrath K, Zhang XF, Huggins RA, Cui Y (2008) High-performance lithium battery anodes using silicon nanowires. Nat Nanotechnol 3:31
38. Malta M, Louarn G, Errien N, Torresi RM (2003) Nanofibers composite vanadium oxide/polyaniline: synthesis and characterization of an electroactive anisotropic structure. Electrochem Commun 5:1011–1015
39. Hughes M, Chen GZ, Shaffer MSP, Fray DJ, Windle AH (2002) Electrochemical capacitance of a nanoporous composite of carbon nanotubes and polypyrrole. Chem Mater 14:1610–1613
40. Ju Y-W, Choi G-R, Jung H-R, Lee W-J (2008) Electrochemical properties of electrospun PAN/MWCNT carbon nanofibers electrodes coated with polypyrrole. Electrochim Acta 53:5796–5803
41. Khomenko V, Frackowiak EB (2005) Determination of the specific capacitance of conducting polymer/nanotubes composite electrodes using different cell configurations. Electrochim Acta 50:2499–2506
42. Gupta V, Miura N (2006) Polyaniline/single-wall carbon nanotube (PANI/SWCNT) composites for high performance supercapacitors. Electrochim Acta 52:1721–1726
43. Chen L, Yuan C, Dou H, Gao B, Chen (2009) Synthesis and electrochemical capacitance of core–shell poly (3, 4-ethylenedioxythiophene)/poly (sodium 4-styrenesulfonate)-modified multiwalled carbon nanotube nanocomposites. Electrochim Acta 54:2335–2341
44. Liu R, Lee SB (2008) MnO2/poly (3, 4-ethylenedioxythiophene) coaxial nanowires by one-step coelectrodeposition for electrochemical energy storage. J Am Chem Soc 130:2942–2943
45. Wan M, Li J, Li S (2001) Microtubules of polyaniline as new microwave absorbent materials. 12:651–657
46. Liu C, Lee R, Bok S (2007) Electromaganic wave absorbing property of polyaniline/polystylene composites. Acta Metall Sin-Chin Ed- 43:409

Vehicle Detection in Aerial Images Using Selective Search with a Simple Deep Learning Based Combination Classifier

Tanuja Tewari, Kaustubh V. Sakhare and Vibha Vyas

Abstract Contrary to the growing zest for bringing in place a complex detection methodology, aiming at improvement in performance of existing methodologies for detecting vehicles in aerial images, this novel piece of work sets forward a much simpler approach with superior results. It was found that methods that showed exemplary performance on common benchmark datasets otherwise, their performance dropped remarkably on aerial images. To achieve performance at par or comparable with the state-of-the-art methods on common benchmark datasets, several adaptations have been suggested in literature to existing methods, for detecting small vehicle instances in aerial images. This ranges from adaptations to the object proposal methods to introduction of more complex deep learning-based classifiers such as fast RCNN and faster RCNN. However, these methods have their own limitations along with the growing increase in system complexity. In this work, a novice, simple and accurate method has been proposed for the detection of small vehicles from aerial images. The experiments have been performed on the publicly accessible and diverse Vehicle Detection in Aerial Imagery (VEDAI) database. This novice technique utilizes Selective Search algorithm as the object proposal method in combination with a deep learning-based framework for classification, which comprises of a simple Convolutional Neural Network (CNN) architecture proposed in combination with a simple Deep Neural Network (DNN) architecture. The DNN utilizes Histogram of Oriented Gradients (HoG) feature input to generate output features that combine with the CNN feature map for final classification. This method is much simpler and achieves a significant accuracy of 96% in vehicle detection, which is much superior to any of the methods tried for aerial images in literature so far.

Keywords Vehicle detection · VEDAI · Object proposal method · Deep learning-based classifier · Fast RCNN · Faster RCNN · Selective search algorithm · Simple CNN architecture · Simple DNN architecture · HoG

T. Tewari (✉) · K. V. Sakhare · V. Vyas
Department of Electronics and Telecommunication, College of Engineering Pune, Pune, India
e-mail: tanujat16.extc@coep.ac.in

© Springer Nature Singapore Pte Ltd. 2019

V. Nath and J. K. Mandal (eds.), *Proceedings of the Third International Conference on Microelectronics, Computing and Communication Systems*,
Lecture Notes in Electrical Engineering 556, https://doi.org/10.1007/978-981-13-7091-5_21

1 Introduction

In this era of automization and digitization, where there is an urge for processing huge amount of information in minimal time, aerial images have come as a savior [1, 2]. Owing to their ability to cover large areas in one go, they have been a point of growing interest among researchers for over a decade. They can access large areas altogether to determine the presence of object of interest within that range which has high utility for real-time applications such as traffic monitoring, environment pollution monitoring, inventory control, and surveillance purposes [1]. However, object detection accurately and instantly in aerial images for real time is a huge challenge. For example, for an aerial image of a traffic scenario, small instances of land vehicles such as cars, vans, and pickups will occupy a very few pixels as compared to the entire image size. Manual inspection will lead to missing of such small instances due to lack of concentration or fatigue which is obvious as human errors are inevitable in manual task. The top performing object detection methods on common standard datasets also fail to handle small instances in aerial images efficiently. This brings in place a high demand for an automatic object detection system which is robust, accurate and at the same time capable of handling small instances (Fig. 1).

Although huge work has been carried out in literature using different methods and techniques for vehicle detection in aerial images [3–14], there is always a scope

Fig. 1 Examples of aerial images in VEDAI with diverse backgrounds showing vehicles in urban and rural areas, forest, marshy, and agricultural lands

for improvement of detection accuracy, robustness and system complexity. All the methods proposed earlier were based on handcrafted features and a classifier combination which were totally dependent on human ingenuity for feature design [3–5]. These methods required manual analysis of real-world data to find an apt feature representation, thereby were not robust to meet the challenges of object detection in aerial images such as occlusion, viewpoint changes, shadow, illumination, background clutter etc. Hence, recent years see the transition to several deep learning-based frameworks such as CNNs [6, 7], which are capable of learning good features automatically from the training samples of most complex objects. However, Regional Convolutional Networks (RCNNs), the pioneers, are slow in the sense that they calculate convolutional features for each candidate region separately [8]. Fast RCNN and Faster RCNN methods, which use Object Proposal Methods to propose candidate regions to the classifier, evolved as the top performers for object detection on common benchmark datasets. However, on inspecting their potential on small occurrences, the performance was not that satisfactory [8–10, 15]. Hence, in this work, to increase the efficiency and robustness of vehicle detection system for handling small occurrences of vehicles in aerial images, a novice, efficient, simple and accurate method has been proposed, which utilizes the high-resolution proposals by Selective Search algorithm [16–18] and, a powerful complex feature representation and classification by a deep neural network framework, which is a combination of a simple CNN and DNN architecture [19–23]. The DNN architecture utilizes the HOG feature input [14], which is a powerful and classical edge information-based feature for detecting vehicles. The combination is powerful enough with its simple architecture to challenge the accuracy of highly complex deep learning-based detectors.

The organization of the paper is as follows. Section 2 comprises of the brief review of prior works and optimal object proposal method in literature. Section 3 discusses the database, performance evaluation parameters, HoG features used as input to the proposed DNN classifier, the proposed deep learning-based classifier combination (CNN + DNN) and the process flow for the novel detection methodology. Section 4 carries the experimentation results and comparison with prior methods whereas Sect. 5 derives the experiment-based conclusion and anticipates the future scope.

2 Related Work

2.1 Detection of Vehicles in Aerial Images

The advent of Unmanned Air Vehicles (UAVs) and drones have resulted in the upsurge of aerial photography which has accounted for numerous works on vehicle detection from aerial images for over a decade.

The wide-ranging survey by Cheng and Han [3], reviews traditional methods used for the detection of vehicles in aerial images and anticipates the much-needed transition to deep learning-based methods in this area. The initial methods in literature were generally based on mechanically derived features such as HoG, Texture, Bag-of-Words (BoW), Sparse Representation, Haar-like features, etc., and classifiers such as Support Vector Machines (SVMs), Adaboost, Artificial Neural Networks, and k-Nearest Neighbor, using a sliding window approach to generate candidate regions. The published work by Moranduzzo and Melgani [4], reviews the performance of various handcrafted features and classifier combination for detecting cars in UAV images. Razakarivony and Jurie [5] used handcrafted features and classifier combination such as (i) Haar wavelets and cascade of boosted classifiers (ii) HoG with SVM classifiers (iii) BoW model and (iv) Deformable Parts Model to counter the challenges in detecting vehicles in aerial images. However, these proposed methods might fail to give powerful feature representation for more complex instances.

Recent times perceive sliding window approach for generating candidate regions being replaced by object proposal methods which can generate regions with high objectness [16, 24]. This brought about a huge revolution in detecting objects in aerial images too, permitting the use of more complex classifiers like RCNNs [7, 20–23]. Thereafter, Fast RCNN [9] and Faster RCNN [10] emerged as the faster variants of RCNN. Instead of generating feature maps for all the proposals separately, they shared convolutional feature map for the entire image among the generated proposals. However, although they emerged to be the top performers on the common standard non-aerial datasets, their performance on detecting small- and medium-sized objects is found to be questionable [8, 15].

2.2 *Object Proposal Methods*

Among the state-of-the-art Object Proposal Methods based on handcrafted feature-based techniques, Selective Search [17, 18] has shown outstanding performance among all the other methods on common standard datasets as well as aerial datasets with some adaptations [8]. This method with its diverse grouping strategies[1] and hierarchical grouping is powerful enough to capture all the regions with high objectness.

$$s(r1, r2) = a_1 s_{colour}(r1, r2) + a_2 s_{texture}(r1, r2) + a_3 s_{size}(r1, r2) + a_4 s_{shape}(r1, r2) \quad (1)$$

$s(r1, r2)$ gives the final similarity measure used by Selective Search for grouping any two segmented regions $r1$ and $r2$ which is the weighted combination of color, texture, size, and shape based similarity measures.

For common benchmark dataset like PASCAL VOC, Selective Search gives a high recall of **99%** with Mean Average Best Overlap (MABO) as **0.879** [17]. On aerial images dataset such as VEDAI 1024, the performance of original Selective Search algorithm [17] drops. However, with the adaptation of parameters like initial segmentation size and minimum proposal width, recall value for Selective Search Algorithm

[1]Selective Search uses different similarity measures for hierarchical grouping of segmented regions.

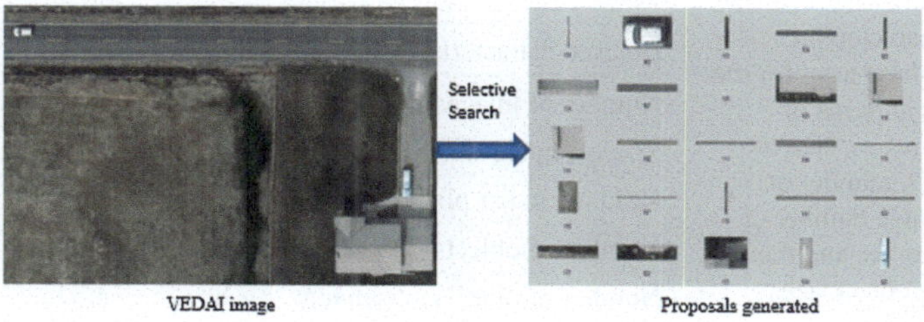

Fig. 2 Illustration of proposals generated by selective search from a VEDAI 1024 image

reaches close to 1 [8]. Hence adapted Selective Search algorithm is selected for generation of well captured and localized proposals to capture small instances in this work. The adaptations made to the algorithm to handle small instances in aerial images are discussed in detail in the subsequent section (Fig. 2).

3 Evaluation

3.1 Database for Aerial Images

All the experiments pertaining to vehicle detection in aerial images have been performed on VEDAI 1024 database. Contrary to prior databases, VEDAI, introduced for aerial images, is specially tailored to detect very small instances and to remove the drawbacks of earlier databases [5]. The dataset contains images for detection of small vehicles in an unrestricted environment, miscellaneous background conditions, images of vehicles affected by occlusions or masks and different orientations [8]. Collected over Utah in U.S., the images consist of varying backgrounds such as urban and rural; forests, marshy and agricultural lands. The images have a resolution of **1024 x 1024 pixels** and Ground Sampling Distance (GSD) of **12.5 cm** per pixel. Ground Truth annotations are available for all the classes in VEDAI. The classes selected for this experimentation are cars, vans, and pickups, since sufficient number of ground truths are available for these classes for training the classifier and performance evaluation. **70%** of the database is used for training, **20%** for validation, and the rest **10%** for testing. The available ground truths were aligned in a way for overlap calculation with proposal bounding boxes generated by the Selective Search algorithm (Table 1).

Table 1 Characteristics of VEDAI database used for the experiments in terms of number of images, size of images, GSD, number of objects, average number of objects per image and the approximate object sizes

Dataset characteristics	VEDAI
Number of images	1268
Size of image	1024 × 1024
GSD in cm per pixel	12.5
Number of objects	2950
Number of objects per image	2.88
Bounding box average width	33.40 ± 11.33
Bounding box average height	33.47 ± 11.68

3.2 Performance Evaluation

The performance of object detection depends on the performance of the Object Proposal Method as well as the classifier. How well the Selective Search algorithm captures the cars, vans, and pickups without missing them is determined by its Recall value and, how well the vehicles are localized in the proposals is determined by MABO. Adaptation of initial segmentation size in Selective Search as per the small size vehicles in VEDAI 1024 renders Recall value close to 1 and MABO close to 0.8 [8]. Intersection over Union (IoU) determines the measure of overlap between the ground truths and the proposals and is given by Eq. 2.

$$IoU = \frac{Ar_{proposal} \cap Ar_{Ground\ Truth}}{Ar_{proposal} \cup Ar_{Ground\ Truth}} \tag{2}$$

Here, $\boldsymbol{Ar_{proposal}}$ is the area of the proposal bounding box and $\boldsymbol{Ar_{Ground\ Truth}}$ is the area of the bounding box of Ground Truth annotation. Aligned ground truths which have an IoU greater than 0.5 are considered as covered in case of VEDAI according to PASCAL VOC criterion. [8, 25]. MABO gives the measure of localization of vehicle objects in the generated proposals, which is given by Eq. 3.

$$ABO = \frac{1}{|G|} \sum_{gt_i} \max IoU\left(gt_i, l_j\right);\ l_j \in L \tag{3}$$

It is calculated by finding the average of the best overlap between each Ground Truth annotation $\boldsymbol{gt_i} \in$ G where G is the set of ground truths and the corresponding object proposals $\boldsymbol{l_j} \in \boldsymbol{L}$ where $\boldsymbol{L}$ is the set of object proposals corresponding to $\boldsymbol{gt_i}$. Here |*G*| stands for the total number of Ground Truth annotations.

The classifier performance (combination of CNN and DNN using HoG features) is measured by the ***Average Precision*** metric, which gives the ratio or percentage of ground truth annotations, predicted as vehicles by the Selective Search and the deep learning-based classifier.

3.3 HoG Features

Features that can distinguish vehicles from most of the objects in the background, are most prominently the edges. Vehicles seem to contain more edges than the surrounding natural objects in the background [14]. Hence features based on edges can be exploited to train the DNN classifier to distinguish vehicles from most of the objects in diverse background scenarios. Many works have also utilized morphological features to enable classifier to distinguish vehicles from the background, but they failed in case of unconstrained background and worked well for a specific background [11–13]. HoG features have been selected as an input to DNN classifier owing to its ability to describe the edge directions and intensity gradient distribution in a localized area which works well to classify the vehicles and non-vehicles. This combination with CNN classifier has shown to improve the accuracy further by introducing feature engineering to assist the classifier (Fig. 3).

3.4 Deep Learning Framework for Classification

Machine learning based classifiers are dependent on manually designed features for classification which aim to reduce the complexity of the data so that the distinguishable patterns become more visible to the classifier [3, 4]. For classification of small instances (approximately **100 to 2000** pixels) in case of VEDAI 1024, merely human ingenuity does not suffice to derive distinguishable features. Hence, here deep

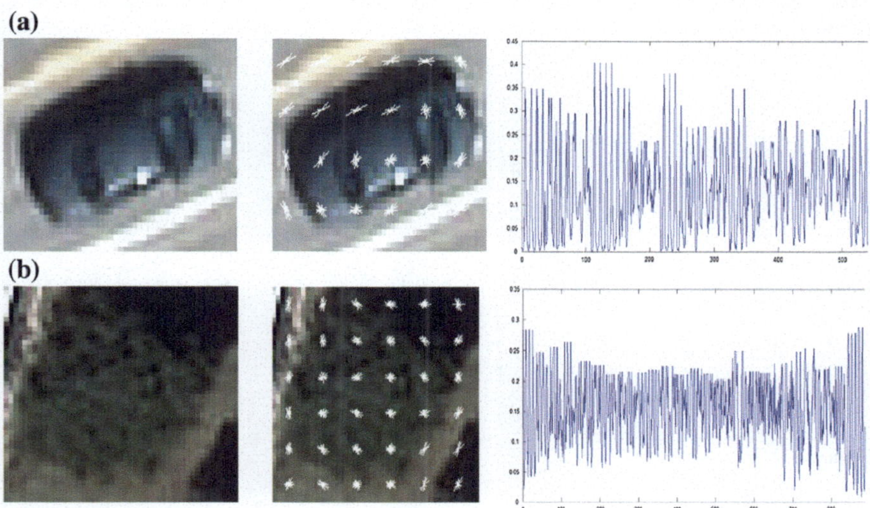

Fig. 3 **a** HoG feature visualization and plot of a vehicle image. **b** HoG feature visualization and plot of a non-vehicle image

learning-based classifiers come for rescue. Deep neural networks like CNN and DNN can extract high-level features to avoid underfitting of such complex sizes.

While CNNs are self-capable of extracting low-level features like edges and gradually construct more complex features, DNNs require low-level features to build up high-level features from them. Although the existing neural networks like RCNN [6, 7] and their new variants Fast RCNN [9] and Faster RCNN [10] have shown remarkable performance in classifying objects in common standard datasets, yet their performance is limited when it comes to small- and medium-sized objects owing to the low resolution of their feature map which leads to missed detections [8, 15]. Also, their complex and deep architecture require high-end machines and GPUs (Graphical Processing Units) for execution. In this work, the effort has been made to exploit the benefits of both feature engineering and deep learning by designing a simple but effective deep learning-based classifier combination that gives accuracy comparable to and above the existing methods. Also owing to its simple architecture, it does not require very high-end machines and GPU's for its execution. The combination

Table 2 Proposed CNN architecture summary for handling small vehicle sizes. Input size is 32 × 32 pixels

Layer type	Size of kernel	Pad, stride	Output shape
Conv2D Relu	3 × 3 × 32 –	0, 1 –	(No. of samples, 30, 30, 32) (No. of samples, 30, 30, 32)
Conv2D Relu MaxPool2D Dropout	3 × 3 × 32 – 2 × 2 –	0,1 – 0, None –	(No. of samples, 28, 28, 32) (No. of samples, 28, 28, 32) (No. of samples, 14, 14, 32) (No. of samples, 14, 14, 32)
Conv2D Relu	3 × 3 × 64 –	0, 1 –	(No. of samples, 12, 12, 64) (No. of samples, 12, 12, 64)
Conv2D Relu MaxPool2D Dropout	3 × 3 × 64 – 2 × 2 –	0,1 – 0, None –	(No. of samples, 10, 10, 64) (No. of samples, 10, 10, 64) (No. of samples, 5, 5, 64) (No. of samples, 5, 5, 64)
Flatten	–	–	(No. of samples, 1600)
FC Relu Dropout	–	–	(No. of samples, 256) (No. of samples, 256) (No. of samples, 256)
FC	–	–	(No. of samples, 2)

Table 3 Proposed DNN architecture summary for handling small vehicle sizes. Input is 324 HoG features corresponding to each CNN input of 32 × 32 pixels

Layer (type)	Output shape
FC	(No. of samples, 1024)
Relu	(No. of samples, 1024)
FC	(No. of samples, 256)
Relu	(No. of samples, 256)
FC	(No. of samples, 256)
Relu	(No. of samples, 256)
FC	(No. of samples, 2)

consists of a simple CNN architecture combined with a simple DNN architecture that uses the mechanically engineered HoG features as its low-level feature input (Tables 2 and 3).

3.5 Vehicle Detection

The classes in VEDAI 1024 considered for experimentation are cars, vans, and pick-ups due to their sufficient number of annotations available to train the CNN classifier and for performance evaluation. In Selective Search algorithm, which is a grouping-based object proposal method, the size of initial segmentation ***k*** which determines approximately the height or width of the vehicles of interest to be captured is adjusted to capture even the small-sized vehicles **(<200 pixels)**. The vehicles of interest in VEDAI 1024 ranges from **100 to 2000** pixels. The proposals generated even smaller than the size to be captured are eliminated by adjusting the value of ***minBoxWidth***. Also, a variable ***maxBoxWidth*** is introduced in the existing algorithm to set the upper limit. It does away with the proposals which are much larger than the vehicle sizes of interest in aerial images. The value of ***k, minBoxWidth*** and ***maxBoxWidth***, play a vital role in capturing well-localized vehicles in the proposals which is essential for classifier's performance in terms of *Average Precision*, since vehicles missed and with poor localization will definitely lead to missed detection at the classifier stage. For the training of the CNN classifier to classify the proposals generated from Selective Search, positive samples (vehicles) are obtained from the ground truth annotations available and positive proposals through Selective Search. The negative samples (non-vehicles) are obtained from the negative proposals generated by the Selective Search algorithm. The ratio of positive to negative samples is kept **1:3**, since the background of all the images in VEDAI is very diverse so to train the classifier on maximum diversity, negative samples are kept 3 times the positive ones. Out of the entire ground truth annotations available, eliminating the Ground Truths of the test images, the rest are used as positive samples for training the CNN and its validation. **70%** of the remaining Ground Truths are used as positive samples for the training dataset and 30% are used as positive samples for the validation dataset. The positive and negative samples to train and validate the classifier are resized to input size of **32 x 32 pixels** which is approximately equal to the sizes of small vehicles to be detected in the VEDAI dataset. The proposals generated by Selective Search for test images are also resized to **32 x 32 pixels** before feeding to the classifier for classification. Bounding boxes using the coordinates of the generated proposal are constructed if they are qualified as vehicles by the classifier.

To improve the validation accuracy of CNN classifier by reducing the data complexity, some feature engineering was done by introducing HoG features derived from the resized positive and negative samples of the training dataset along with their corresponding labels to train the proposed simple DNN architecture. After training, the validation accuracy of the DNN classifier is computed by feeding HOG features of the positive and negative samples in the corresponding validation dataset. Both the proposed CNN and DNN are trained until their best validation accuracy is obtained without overfitting. At this stage, the final softmax layer is removed and the final activations before the softmax layer from both the networks are combined. Using these activations, the final softmax layer is trained. The validation accuracy obtained from this combined model is much improved which gives an outstanding test accuracy of **96%** on the test dataset.

4 Results and Discussion

The train and validation accuracy of the proposed CNN and DNN model individually as well as that of the combined model is given in Table 4.

The combined CNN and DNN model is effective enough to give an Average Precision of **96%** which is above the accuracy obtained by the classifiers in aerial imagery vehicle detection so far. Examples of vehicles detected by the proposed method are shown in Fig. 4. Examples of missed detections and false positives are shown in Figs. 5 and 6. Table 5 shows a comparison of results of various detection models and the proposed model.

Table 4 Results of the experiments done for detecting vehicles in VEDAI database

Experiments	CNN classifier	DNN classifier	CNN + DNN combination classifier
Training accuracy	0.978	0.983	0.9971
Validation accuracy	0.956	0.9627	0.9709
Validation loss	0.1168	0.129	0.0936
Average precision (CNN + DNN classifier)			**0.96**

Fig. 4 Examples of small size vehicles detected in VEDAI images by the proposed architecture. The detected vehicles are bounded by red boxes

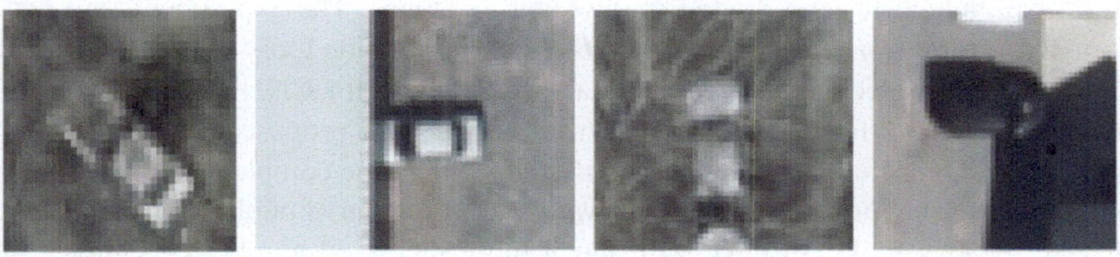

Fig. 5 Examples of missed detections caused due to partial occlusion or shadows

Fig. 6 Examples of false positives caused by objects of similar shapes such as houses, solar cells, trailers etc

Table 5 Comparison results for various detection models

Methods	Test accuracy (%)	System complexity	Database
SVM classifier with HoG + LBP [5]	76.8	Low	VEDAI 1024
Soft cascade structure of ICF + Adaboost classifier [26]	86.8	Low	DLR 3 K Munich
Faster RCNN with VGG-16 [8]	92.8	High. Expensive GPU machines required	VEDAI 1024
Proposed method CNN + DNN classifier with selective search	**96**	Low	VEDAI 1024

5 Future Scope and Conclusion

The scope of deep learning is ever progressing, but along with it, even classifier complexity is increasing by leaps and bounds. In the midst of this, this work proposes a Selective Search and deep learning classifier combination-based method for vehicle detection in aerial images, which gives a simplified architecture without compromising with accuracy, compared to prior methods. The proposed method can be generalized to any vehicle aerial images, by merely varying the value of initial segmentation size ***k***, minimum proposal size ***minBoxWidth*** and maximum proposal size ***maxBoxWidth*** in the Selective Search algorithm, depending upon the resolution of the input image and the class of vehicles to be detected. The architecture is simple enough to be executed on low-end machines. The experiments are performed on ***Intel Core i5 processor, 4 GB RAM***. Simulation tools are ***MATLAB 8.2*** for generating Selective Search proposals, ***Python 3.6*** for CNN model designing and ***Python Flask*** for demonstration. Further taking the proposed method to even low-end GPU machines, can subsequently reduce the time complexity for real-time applications. The future might look forward to the creation of publicly available and diverse databases like VEDAI, with much more increased number of samples of small instances or even consider data augmentation. This would further improve the performance of the proposed combination neural network which increases with the scale of data.

References

1. http://www.environmentalscience.org/principles-applications-aerial-photography
2. http://sciencing.com/difference-satellite-imagery-aerial-photography-8621214.html
3. Cheng G, Han J (2016) A survey on object detection in optical remote sensing images. ISPRS J Photogr Remote Sens 117:11–28
4. Moranduzzo T, Melgani F (2013) Comparison of different feature detectors and descriptors for car classification in UAV images. In: IEEE international geoscience and remote sensing symposium-IGARSS, pp 204–207
5. Razakarivony S, Jurie F (2016) Vehicle detection in aerial imagery: a small target detection benchmark. J Vis Commun Image Represent 34:187–203
6. Konoplich GV, Putin EO, Filchenkov AA (2016) Application of deep learning to the problem of vehicle detection in UAV images. In: XIX IEEE international conference on soft computing and measurements (SCM), pp 4–6
7. Chen X, Xiang S, Liu C-L, Pan C-H (2014) Vehicle detection in satellite images by hybrid deep convolutional neural networks. IEEE Geosci Remote Sens Lett 11(10):1797–1801
8. Sommer L, Schuchert T, Beyerer J (2017) Fast deep vehicle detection in aerial images. IEEE winter conference on applications of computer vision (WACV), pp 311–319
9. Girshick R (2015) Fast R-CNN. In: Proceedings of the IEEE international conference on computer vision, pp 1440–1448
10. Ren S, He K, Girshick R, Sun J (2015) Faster R-CNN: towards real-time object detection with region proposal networks. In: Advances in neural information processing systems, pp 91–99
11. Ajmal A, Hussain IM (2010) Vehicle detection using morphological image processing technique. In: IEEE international conference on multimedia computing and information technology

12. Pawar BD, Humbe VT (2015) Morphology based composite method for vehicle detection from high resolution aerial imagery. VNSGU J Sci Technol 4(1):50–56
13. Zheng H (2006) Automatic vehicle detection from high resolution satellite imagery using morphological neural networks. In: 10th WSEAS international conference on computers, Athens, Greece, pp 608–613
14. Bharathi TK, Yuvaraj S, Steffi DS, Perumal SK (2012) Vehicle detection in aerial surveillance using morphological shared-pixels neural (MSPN) networks. In: IEEE fourth international conference on advanced computing
15. Zhang L, Lin L, Liang X, He K (2016) Is faster R-CNN doing well for pedestrian detection? In: European conference on computer vision, pp 443–457. Springer
16. Hosang J, Benenson R, Dollár P, Schiele B (2016) What makes for effective detection proposals? IEEE Trans Pattern Anal Mach Intell 38(4):814–830
17. Uijlings JR, van de Sande KE, Gevers T, Smeulders AW (2013) Selective search for object recognition. Int J Comput Vis 104(2):154–171
18. van de Sande KE, Uijlings JR, Gevers T, Smeulders AW (2011) Segmentation as selective search for object recognition. IEEE international conference on computer vision
19. Lecun Y, Bengio Y, Hinton G (2015) Deep learning. Nature 521(7553):436–444
20. Krizhevsky A, Sutskever I, Hinton GE (2012) ImageNet classification with deep convolutional neural networks. In: Neural information processing systems, pp 1106–1114
21. Simonyan K, Zisserman A (2015) Very deep convolutional networks for large-scale image recognition. In: International conference on learning representations
22. Lecun Y, Bottou L, Bengio Y, Haffner P (1998) Gradient-based learning applied to document recognition. Proc IEEE 86(11):2278–2324
23. https://keras.io/
24. Alexe B, Deselaers T, Ferrari V (2010) What is an object? In: Proceedings of IEEE conference on computer vision and pattern recognition, pp 73–80
25. Everingham M, Van Gool L, Williams CK, Winn J, Jisserman A (2010) The Pascal visual object classes (VOC) challenge. Int J Comput Vis 88(2):303–338
26. Liu K, Mattyus G (2015) Fast multiclass vehicle detection in aerial images. Geosci Remote Sens Lett 1–5. IEEE

Modified Unit Cell Analysis Approach for EBG Structure Analysis for Gap Width Study Effect

Rajshri C. Mahajan, Vini Parashar and Vibha Vyas

Abstract Nowadays, the interesting properties such as inexpensive, low profile, and easily compatible with the microwave integrated circuits (MMIC) have made the microstrip antenna reasonably strong candidate for wireless communications regardless of main challenge of lower bandwidth. The recent studies show that using the electromagnetic band gap (EBG) structures along with the microstrip antennas can help to improve the performance. While using the EBG structures, it is vital to analyze the impacts of different factors such as gap width on the antenna parameters. In this paper, we have proposed the novel unit cell analysis method for the periodic EBG structure analysis. A novel unit cell analysis approach for periodic EBG structure analysis is presented which is a major step towards the study of the gap width effect on parameters of antenna like frequency band gap, reflection coefficient. This paper presents the complete EGB structure design and its analysis for the results validation. The outcome of this paper is that the modified unit cell analysis method effectively satisfies the EBG structure analysis as the validation results agree with each other strongly.

Keywords Gap width · EBG structure · Microstrip antenna · Unit cell analysis · Dispersion · Scattering

1 Introduction

The EBG structures act as the effective magnetic conductor to their resonance frequency; therefore, such EBG structures are used to mount an antenna close to them when the antenna currents are equivalent to the surface of EBG [1]. The concept of EBG structures is like the tank circuit with the capacitance is represented by the gap among the patches as well as the inductance represented by the distance among

---

R. C. Mahajan · V. Parashar (✉) · V. Vyas
Department of Electronics and Telecommunication, College of Engineering Pune, SPPU (University of Pune), Pune, India
e-mail: vinip16.extc@coep.ac.in

© Springer Nature Singapore Pte Ltd. 2019

V. Nath and J. K. Mandal (eds.), *Proceedings of the Third International Conference on Microelectronics, Computing and Communication Systems*,
Lecture Notes in Electrical Engineering 556, https://doi.org/10.1007/978-981-13-7091-5_22

the patches and ground plane [2, 3]. The EBG structure is the metalloid periodic structure printed on the substrate around the patch (radiating element). The EBG structure acts as a bandpass filter for suppressing the surface waves generated due to the metallic ground surface. This structure avoids the propagation of surface waves in a frequency band.

The dimensions of EBG cell and height of substrate decide the frequency band gap for EBG structure. A typical Sivenpiper's Mushroom EBG cell consists of a metallic patch of width 'w' which separated by gap width 'g' with another EBG cell. There is also a metallic wire connected between the center of EBG cell and ground surface of height 'h' which is also the height of the substrate. The width, gap width, and height of substrate form effective capacitance and inductance of a bandpass filter. Figure 1 shows the top view and Fig. 2 shows the side view of a typical EBG cell and effective capacitance and inductance of EBG cell [4].

Using Eqs. (1) and (2), the capacitance and inductance of EBG cell computed. The capacitance is given by:

$$C = \frac{w\varepsilon_0(1 + \varepsilon_r)}{\pi} \cosh^{-1}\left(\frac{w + g}{g}\right) \tag{1}$$

The inductance is given by:

$$L = \mu h \tag{2}$$

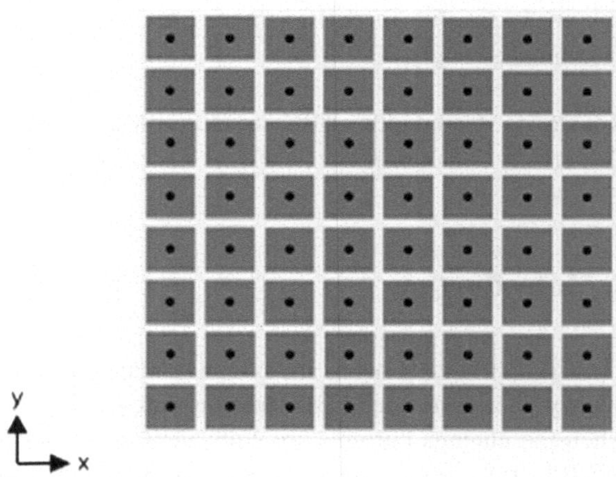

Fig. 1 Top view of mushroom-like electromagnetic band gap (EBG) structure

Fig. 2 Side view of mushroom-like electromagnetic band gap (EBG) structure

where w is unit cell patch width, ε_0 is free space permittivity, ε_r is relative permittivity of substrate, g is gap width, μ is permeability of substrate ($\mu = \mu_0\mu_r$) and h is the height of substrate. The LC model of EBG cell corresponding to physical parameters and L, C values are as shown in Fig. 3.

EBG structures are equivalent to a tank circuit. When C and L values are calculated through LC model of EBG cell then with the help of these values of impedance, frequency of resonance and bandwidth respectively of an EBG structure can also be calculated. The bandwidth of the EBG structure is defined as the band of frequencies where the reflection phase is between +90° and −90°. Using the Eqs. (3), (4) and (5), frequency of resonance, impedance, and bandwidth can be calculated, which are given by

$$\omega_0 = \frac{1}{\sqrt{LC}} \tag{3}$$

$$Z_s = \frac{j\omega L}{1 - \left(\frac{\omega}{\omega_0}\right)^2} \tag{4}$$

$$BW = \frac{1}{120\pi}\sqrt{\frac{L}{C}} \tag{5}$$

Using the EBG structure is complex analysis and hence requires rigorous experimentations. There are computerized mathematical tools like Finite Difference Time Domain (FDTD), Method of Moments (MOM) [5, 6]; Full-wave simulation based Finite Element Method (FEM) for analysis of EBG structure. High-Frequency Simulation Software (HFSS) is FEM based software which has developed a unit cell approach for such analysis as complete structure of EBG gives memory limitation problem. Using this approach, the electromagnetic behavior of periodic structure can be studied using properties of single unit cell structure. The unit cell is analyzed using two types of Ansoft HFSS simulation setups. The first is scattering simulation to determine the reflection phase characteristic of the unit cell and the second is an eigenmode simulation to determine the natural resonances and dispersion properties of the unit cell. The scattering simulation gives the plots of reflection coefficient magnitude and phase which ultimately decides the transmission bandwidth below −10 dB for magnitude plot. The reflection phase plot gives the range of frequency in which the reflected wave has +90° to −90° reflection phase. Both the plots agree for bandwidth for phase and magnitude [5]. The dispersion plot gives information about

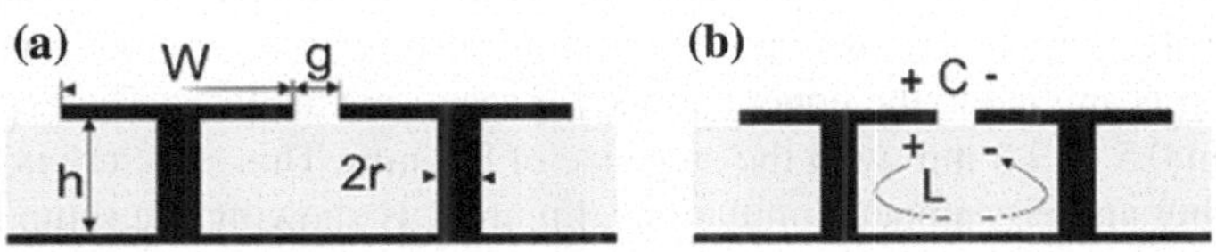

Fig. 3 LC model for the mushroom-like EBG structure: **a** EBG parameters and **b** LC model

the frequency band gap where surface waves are avoided. The frequency band gap of the unit cell should include the resonance frequency of antenna around which the EBG cells are placed [7].

From the literature study over the EBG structure analysis [8–14], it was noticed that unit cell method is used to perform the study of the parametric effect on unit cell patch width and height substrate. However, there is no single study till date that presented the effect of gap width on the antenna parameters. This motivated us to present the novel design set up of unit cell approach for studying the gap width and validate with simulations of complete EBG structure. In Sect. 2, the design of novel unit cell method for EBG structure analysis is presented. Section 3 presents the results achieved using the novel unit cell method. Finally, the conclusion based on results and future work discussed in Sect. 4.

2 Proposed Design

As the novel unit cell analysis method for the period EBG structure analysis is presented, in this paper, our main objective is to study the effect of the gap width on 6 mm x 6 mm square unit cell with the vias radius 0.5 mm located at the center of the unit cell and height of substrate is 1.6 mm. The dielectric material is Glass Epoxy FR_4 with loss tangent (tan $\delta = 0.02$) and relative permittivity $\varepsilon_r = 4.2$. For scattering mode analysis, the unit cell is placed with the airbox and Periodic Boundary Condition (PBC) is applied to the sidewalls. The top surface of the airbox is applied with Floquet port. For eigenmode analysis, the unit cell is placed with the airbox of height six times the height of the substrate. The air box is terminated with a Perfectly Matched Layer (PML) boundary. The airbox and PML boundary are enclosed with another airbox whose sidewalls are applied with PBC boundaries in x and y directions The phase shift transverse to the direction of propagation $(\beta d)_y = 0$ are kept constant and phase shift along the direction of propagation is varied from $0 \leq (\beta d)_x \leq \pi$. In both the setups, the PBC boundaries are kept just along the sidewalls of patch width w. The setups are predominantly designed for Frequency Selective Surfaces (FSS) in which the metallic unit cell is designed and the airbox is placed just around the metallic cell. The gap width effect is hardly studied in the earlier literature [15].

The novel change is incorporated in order to study the gap width effect in both such setups. The unit cell of patch width is considered and gap width ‘g’ is kept in all four directions of the unit cell. The air box is enclosed around the unit cell and the gap width. The PBC boundaries are applied to the sidewalls of the air box. Also, the PML is applied to the top surface of the airbox for eigenmode analysis and Floquet port is applied to the upper surface for scattering analysis. The gap width is varied from 0.5 to 3.5 mm with the step size of 0.5 mm. This structure is analyzed for scattering and eigenmode simulations. Figure 4 is showing the setup designed for new unit cell method.

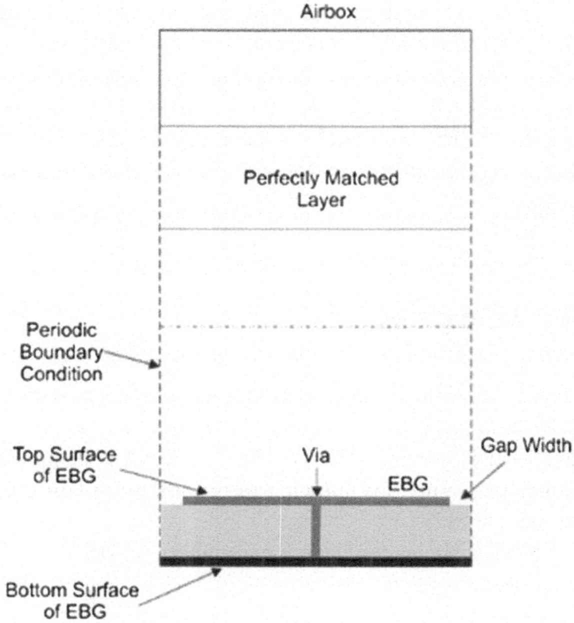

Fig. 4 Setup design of new unit cell approach Eigenmode analysis

3 Experimental Results

Using the proposed setup and design mentioned in the above section, the series of experiments is conducted to analyze the results of bandgap versus gap width among the EBG cells. Through the extensive numerical analysis and experiments, Table 1 shows the effect of gap width using the novel unit cell analysis approach presented in this paper.

Table 1 Effect of gap width using novel unit cell analysis approach

Gap width (mm)	Frequency band gap (Dispersion diagram results) (GHz)	Reflection coefficient S_{11} (Scattering diagram results) at 2.4 GHz
0.5	1.7–3.44	−29.5412
1	1.83–3.26	−29.1688
1.5	1.97–3.08	−29.0245
2	2.06–2.89	−28.6234
2.5	2.13–2.74	−28.0238
3	2.28–2.7	−27.5271
3.5	2.31–2.56	−27.0456

Table 1 shows that with changing the gap with dimension, the frequency band gap and reflection coefficient operating at 2.4 GHz is changing. The increasing gap width shows the impact of increasing bag gap frequency and reflection coefficient values. The impact is significant which is possible using the proposed unit cell analysis approach. The validation of the proposed unit cell analysis approach is performed based on the microstrip antenna (MSA). It is discussed in the below section of this paper.

4 Validations

The validation of the proposed unit cell analysis method is prepared by designing and simulating the complete EBG structure combined with the rectangular patch Microstrip antenna (MSA) [16]. The MSA is designed to operate at a frequency of 2.4 GHz. Figure 5 is showing the geometry of EBG placement along with the patch MSA.

Table 2 shows the design parameters of MSA.

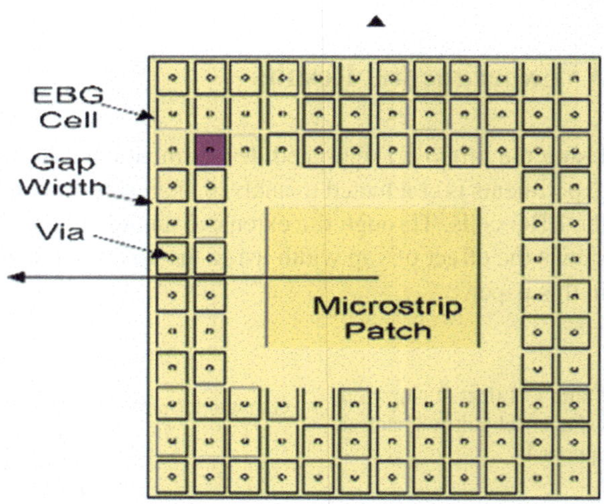

Fig. 5 Geometry of MSA with EBG structure (w = 6 mm, g = 1 mm)

Table 2 MSA design parameters

Parameter	Value
MSA patch width *W*	42 mm
MSA patch length *L*	28 mm
Ground plane dimension	80 mm × 80 mm
Height of substrate	1.6 mm
Feed location	7 mm
Feed type	Coaxial feed

Figure 5 shows the exemplary diagram for simulation in HFSS11 where gap width g is 1 mm. The similar simulations are carried out for gap width 0.5 mm to 3.5 mm with the step increment of 0.5 mm. The simulations are carried out to find the reflection coefficient and −10 dB transmission bandwidth for validation of results obtained for unit cell analysis approach. Table 3 shows the results obtained from these simulations. From Table 3, it is observed that S_{11} (reflection coefficient) decreases as gap width increases. These results are in good agreement with unit cell scattering results as shown in Table 1. The dispersion diagram results show the clear frequency band gap which corresponds to −10 dB transmission bandwidth of EBG structure with MSA.

Figure 6 shows the comparative analysis of return loss obtained due to Unit cell and MSA with EBG structure at a designed frequency of 2.4 GHz w.r.t gap width. Similarly, Fig. 7 shows the error in lower and upper frequencies for frequency band gap and impedance bandwidth are also plotted.

Table 3 Results for EBG structure with MSA

Gap width (mm)	Transmission bandwidth (−10 dB) (GHz)	Reflection coefficient S_{11} in dB at 2.4 GHz
0.5	2.1–3.1	−29.6512
1	2.22–2.9	−29.5588
1.5	2.25–2.87	−29.3245
2	2.27–2.76	−28.3434
2.5	2.30–2.65	−28.2038
3	2.37–2.55	−27.4371
3.5	2.39–2.49	−27.0256

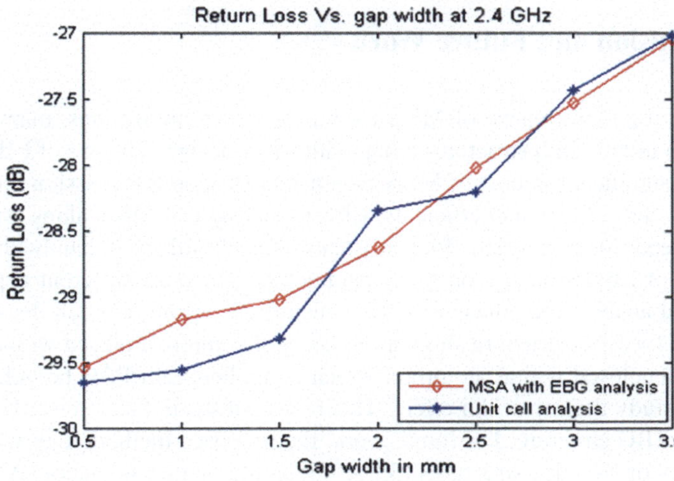

Fig. 6 Return loss versus gap width

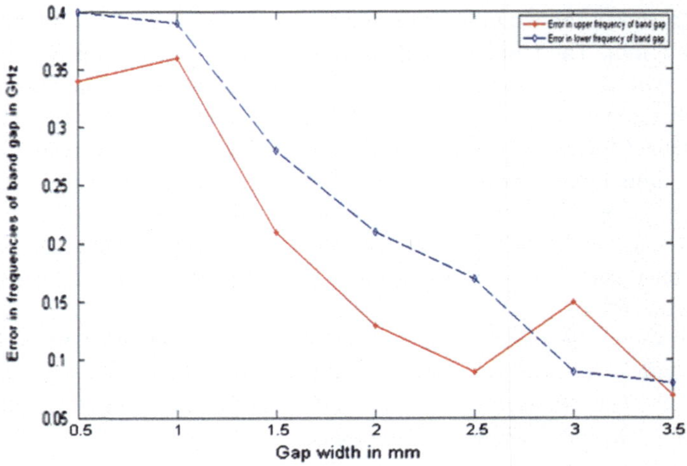

Fig. 7 Error in lower and upper frequencies of band gap

For gap width of 1 mm, the unit cell approach gives S_{11} as −29.1688 dB which is closely matching with −29.5588 dB for EBG structure with MSA. Also, the frequency band is from 1.83 to 3.26 GHz which is including the transmission bandwidth as 2.22–2.9 GHz as shown in Table 2. It is also observed that when gap width increases the reflection coefficient also increases for both unit cell-based approach and complete structure-based approach. The error between reflection coefficients obtained due to two approaches reduces as the gap width increases. Also, the error between upper and lower frequencies decreases as the gap width increases.

5 Conclusion and Future Work

To improve the performance of MSA for wireless communications, many studies reported the use of EBG structures along with MSA design. The use of EBG structures has a significant impact of MSA parameters such as transmission bandwidth frequencies and reflection coefficients. Effective design of MSA along with EBG structures needs to be analyzed efficiently to estimate the effects of bandwidth, shape parameters of EBG structure on MSA parameters. There are different approaches to study and analyze the effects of EBG structures. A novel unit cell design setup is proposed for EBG structure simulation. Using the unit cell placed with different gap widths enclosed within an airbox which is applied with PBC boundaries has resulted to study the gap width effect. This is the substantial step towards detailed analysis of EBG structure. For future work, the proposed methodology will aspire for the study of complex structures like Artificial Magnetic Conductor (AMC) and High Impedance Surface (HIS).

References

1. Yablonovitch E (1993) Photonic band-gap structure. J Opt Soc Am B Opt Phys 10:283–295
2. Winn JN, Meade RD, Joannopoulos JD (1995) Photonic crystals. Princeton Univ Press, Princeton, NJ
3. Zaghloul AI, Palreddy S (2013) Circuit analysis of electromagnetic band gap (EBG) structures. In: Proceedings of URSI-EMTS, pp 67–70
4. Sievenpiper D, Zhang L, Broas J, Alexopolous, Yablonovitch E (1999) High-impedance electromagnetic surfaces with a forbidden frequency band. IEEE Trans Microw Theory Tech 47(11)
5. Costa F, Monorchio A, Genovesi S (2013) A chipless RFID based on multiresonant high-impedance surfaces. IEEE Trans Microw Theory Tech 61(1)
6. Langley RJ, Lee H-J, Ford KL (2012) Independently tunable low-profile dual-band high-impedance surface antenna system for applications in UHF band. IEEE Trans Antennas Propag 60(9)
7. Vallecchi A, De Luis JR, Capolino F (2012) Low profile fully planar folded dipole antenna on a high impedance surface. IEEE Trans Antennas Propag 60(1)
8. Gregoryr MD, Bayraktar Z, Wang X, Werner DH (2012) A versatile design strategy for thin composite planar double-sided high-impedance surfaces. IEEE Trans Antennas Propag 60(6)
9. Weiss SJ, Kim IK, Wang H, Varadan VV (2012) Embedded wideband metaresonator antenna on a high-impedance ground plane for vehicular applications. IEEE Trans Veh Technol 61(4)
10. Deo P, Mehta A, Massey PJ, Nakano H, Mirshekar-Syahkal D (2010) Thickness reduction and performance enhancement of steerable square loop antenna using hybrid high impedance surface. IEEE Trans Antennas Propag 58(5)
11. Zhu C, Chen X, Li L, Liang CH, Su ZJ (2012) Dual-band high impedance surface with mushroom-type cells loaded by symmetric meandered slots. IEEE Trans Antennas Propag 60(10)
12. Mohaarram MA, Kishk AA (2017) Bandwidth study of the stacked mushroom EBG unit cell. IEEE Trans Antennas Propag 65(8)
13. Hao Y, Wen D, Munoz MO (2018) A compact and low-profile MIMO antenna using a miniature circular high-impedance surface for wearable applications. IEEE Trans Antennas Propag 66(1)
14. Garg R (2001) Microstrip antenna design handbook. Artech House, Inc. ISBN-13 978-0-521-88991-9
15. Yang F, Rahmat Y (2008) Electromagnetic band gap structures in antenna engineering. Cambridge University Press
16. Kumar G, Ray KP (2003) Broadband micro-strip antennas. Artech House, Inc. ISBN- 1-58053-244-6

Operational Transconductance Amplifier Structured Highly Linear Analog Multiplier

Amitkumar S. Khade and Vibha Vyas

Abstract Analog multiplier is a key element in modern communication systems. This paper presents the capability of cross-coupled operational transconductance amplifier (OTA) as an analog multiplier with performance analysis and design consideration. The proposed OTA structure is built and tested as a multiplier in Cadence Analog Design Environment (ADE) using standard 0.18 μm CMOS process. The simulation result shows that the proposed OTA structured multiplier has better linearity with comparable power consumption and noise performance.

Keywords Component · Formatting · Style · Styling

1 Introduction

The analog multipliers are used as subcircuits in electronics systems such as filters, neural networks and as mixer and modulators in communication systems [1]. Higher frequencies can be obtained by using frequency multipliers [2–4]. Varieties of multipliers structure have been proposed for different objectives [5–9] and some of them are driven by the early work of Gilbert [10–15]. The main motive behind these designs is to use electronic devices to process analog input signals, along with cancellation or error minimization caused by nonlinearity of devices. The CMOS multiplier implementation is still a challenging subject especially for low-voltage and low-power consumption [16–18].

In this paper first, few low-power multiplier structures are simulated in Cadence Analog Design Environment to obtain their characteristics such as input range, lin-

A. S. Khade (✉)
E&TC Department, Cummins College of Engineering for Women, Pune, Maharashtra, India
e-mail: amitkumar.khade@cumminscollege.in

V. Vyas
E&TC Department, College of Engineering Pune, Pune, Maharashtra, India
e-mail: vsv.extc@coep.ac.in

© Springer Nature Singapore Pte Ltd. 2019

V. Nath and J. K. Mandal (eds.), *Proceedings of the Third International Conference on Microelectronics, Computing and Communication Systems*, Lecture Notes in Electrical Engineering 556, https://doi.org/10.1007/978-981-13-7091-5_23

earity, input noise, and power consumption. The analysis on linearity, power consumption, and noise for proposed OTA multiplier is presented, which is verified by Spectre Simulator from Cadence tools using 0.18 μm technology.

2 CMOS Multiplier

Multiplication of two signals is one of the most important operations in analog signal processing. A multiplier is an active network, the output of which is proportional to the product of two input signals. A general representation of multiplier is

$$Z = K\,(XY) \tag{1}$$

where X and Y are the two input signals with K being multiplier constant with suitable dimensions. All signals, i.e., X, Y, and Z are measured with respect to ground. The signals can be either voltages or currents.

Though the digital technology dominates in modern electronics, analog circuits are preferred due to low-power and low-cost fabrication. As the advanced CMOS technology offers low-power consumption, MOS transistors are obvious choice for the circuit implementation, while differential circuits are widely used for nonlinearity cancellation [14]. For CMOS analog multiplier design, the MOS transistor output current is modeled as (2) and (3).

$$\begin{aligned} I_d &= K\,[V_{GS} - V_T - V_{DS}/2]V_{DS} \\ &\quad for\ V_{GS} > V_T,\ V_{DS} < V_{GS} - V_T \end{aligned} \tag{2}$$

$$\begin{aligned} I_d &= (K/2)[V_{GS} - V_T]^2 \\ &\quad for\ V_{GS} > V_T,\ V_{DS} > V_{GS} - V_T \end{aligned} \tag{3}$$

For MOS in its linear and saturation region, respectively, $K = \mu_0 Cox(W/L)$ and V_T are the conventional for the transconductance parameter and the threshold voltage of the MOS transistor, respectively.

3 Multiplier Structure I

Figure 1 shows multiplier structure I, which is the most recommended multiplier structure, provided in [14].

A fully differential structure improves the linearity as nonlinearity cancellation is achieved. In this fully differential structure, lower four NMOS transistor, i.e., N1-N4 biased in linear region where as upper four NMOS transistor, i.e., M1-M5 are biased

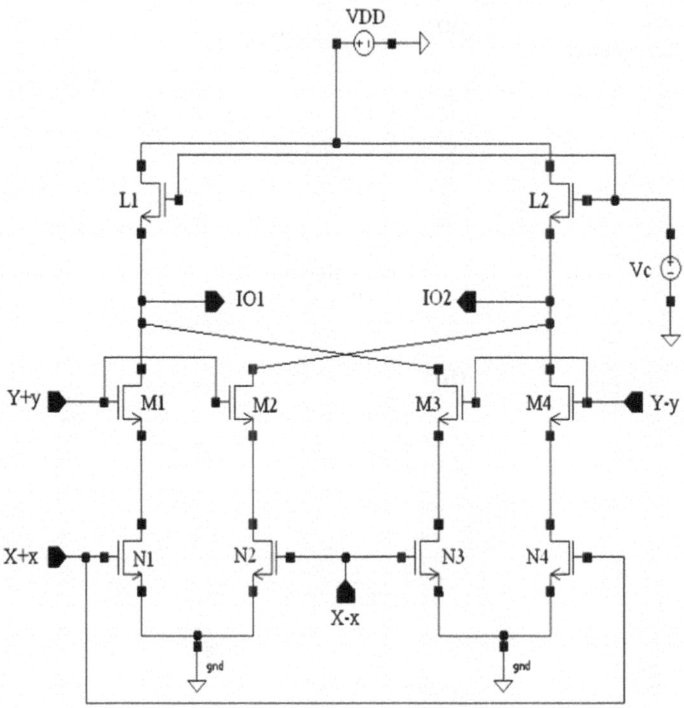

Fig. 1 Multiplier structure I

in saturation region. For the circuit shown in Fig. 1, the differential output current expressed by Eq. (4) shows the multiplication of two signals.

$$I_{out} = I_{01} - I_{02} = 4\,K\,(x \cdot y) \tag{4}$$

3.1 *Input Range*

The input range of a multiplier circuit is obtained from its bias conditions. Constraint set (5) gives the required bias conditions for the structure I.

$$\begin{aligned} &V_T < X \pm x \\ &Y \pm y - V_T < X \pm x - V_T \\ &V_T < Y \pm y \\ &Y \pm y - V_T < V_d \end{aligned} \tag{5}$$

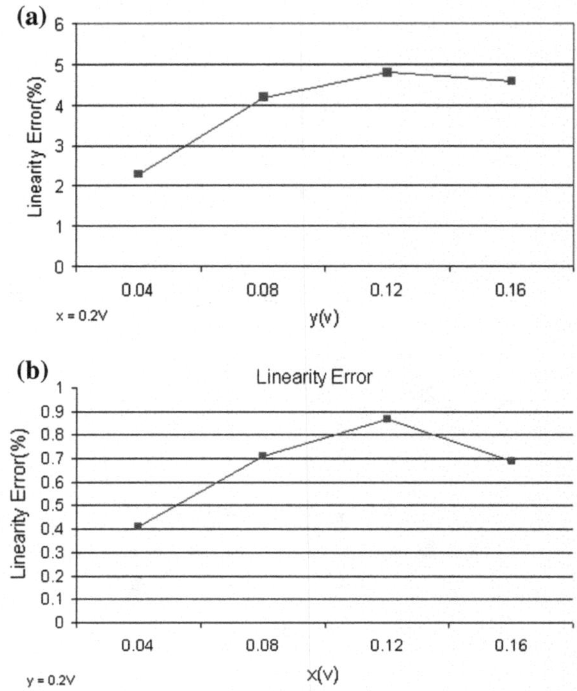

Fig. 2 **a** Linearity error versus y **b** linearity error versus x

3.2 Linearity Analysis

The linearity error with respect to y and x inputs is shown in Fig. 2a, b with. Typical biasing voltages used are: $V_{DD} = 1.8$ V, $X = 1.5$ V, $Y = 0.9$ V, $V_{THP} = 0.45$ V, and $V_{THN} = 0.5$ V when the transistors have same size of $W/L = 0.41$ μm/0.18 μm.

The linearity error increased as the value of both x and y inputs increases. In particular, when x = 0.12 and y = 0.2 V, the linearity error observed is 4.8% and when x = 0.2 V and y = 0.12 V, the linearity error observed is 0.89%.

The Total Harmonic Distortion (THD) is observed when a constant DC voltage is applied to x (or y) while a 100 kHz, 0.2 V p-p sinusoidal wave is applied to y (or x) is shown in Fig. 3a, b. Figure 3 proves that the multiplier has good linearity with respect to x input.

3.3 Power Consumption

For given power supply, the power consumed by multiplier can be estimated by checking the total supply voltage current, which is expressed as

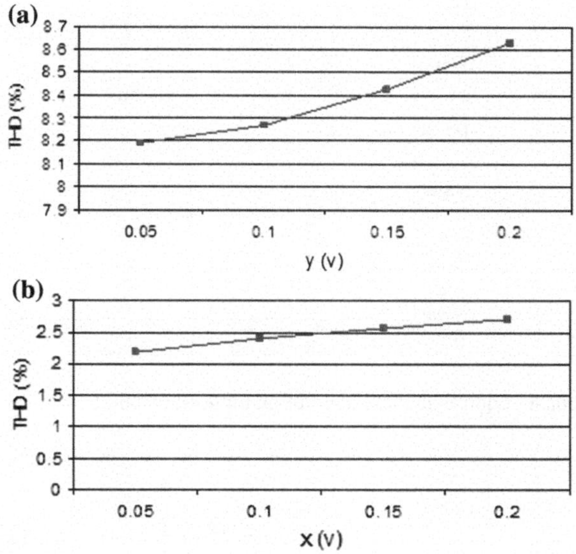

Fig. 3 **a** THD of signal y **b** THD of signal x

$$I_{total} = I_{O1} + I_{O2} = I_{dM1} + I_{dM2} + I_{dM3} + I_{dM4}$$
$$I_{total} = (2K_M)\left[(Y - V_T)^2 + y^2\right] \quad (7)$$

Relation (7) shows that the total power consumption is independent of signal x and DC bias X for transistors N1-N4.

The total power consumption of first multiplier structure is 34 μA at $V_{dd} = 1.8$ V when both x and y being a 100 kHz, 0.2 V p-p sinusoidal wave.

3.4 Noise

A thermal noise current power density of a MOS transistor biased in saturation region is modeled as shown in (8).

$$i^2_{nsat} = \frac{8}{3} kT g_m \, df \quad (8)$$

and (9) for transistor operated in linear regions respectively [19].

$$i^2_{nlin} = 4kT g_{ds} \, df \quad (9)$$

The total output noise current is expressed as

$$i^2_{no} = 4(i^2_{nsat} + i^2_{nlin})$$

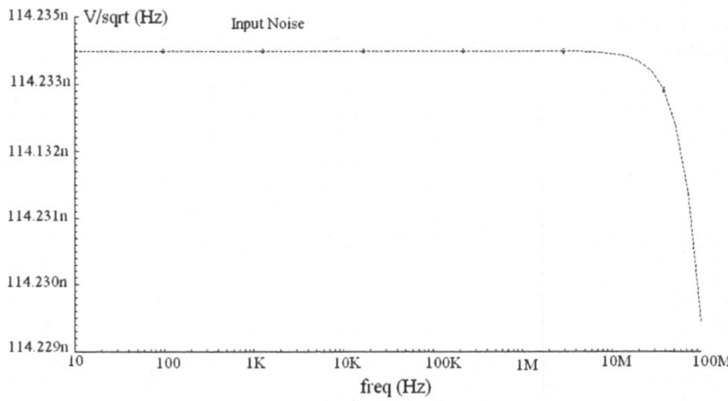

Fig. 4 Equivalent input noise of multiplier structure I

$$i_{no}^2 = 16kT\left(g_{ds} + \frac{2}{3}g_m\right)df$$

This suggest that output noise can be minimized by carefully designing the g_m and g_{ds}, which vary with X and Y. The noise performance of structure I was simulated. Figure 5 shows the equivalent input noise within range 10–100 MHz, which is 114.234 n V/sqrt (Hz) (Fig. 4).

4 Multiplier Structure II

The multiplier structure, shown in Fig. 5, shows a low-power multiplier reported in [18].

A low-power CMOS analog multiplier structure, consist of four PMOS transistors (P1-P4) biased in saturation region and eight NMOS transistors (N1-N4 and M1-M4) biased in linear region. Considering channel length modulation effect for MOS transistor, the drain current in saturation region is given by relation (10).

$$I_D = K/2\,(V_{GS} - V_T)^2(1 + \lambda V_{DS}) \tag{10}$$

where K = uCox(W/L) is the Transconductance parameter, V_T the threshold voltage of device and λ is the channel length modulation effect for device.

Biasing all the transistors in Fig. 6 in proper region, the differential voltage can be written as (11), which shows multiplication of two signals x and y.

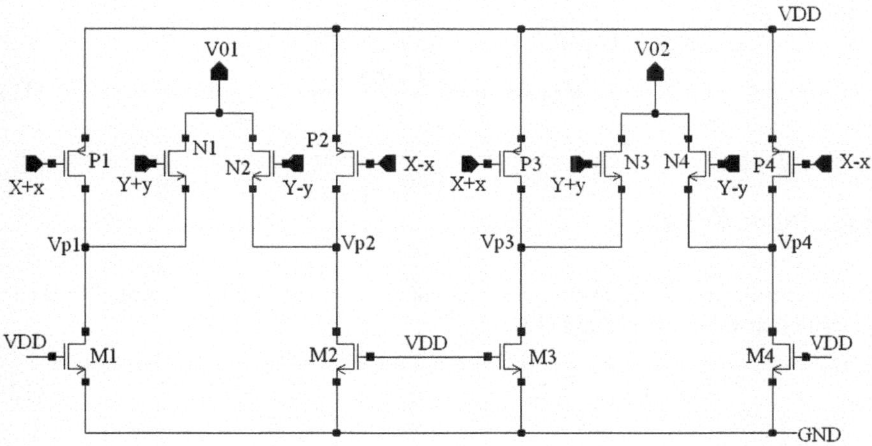

Fig. 5 Multiplier structure II

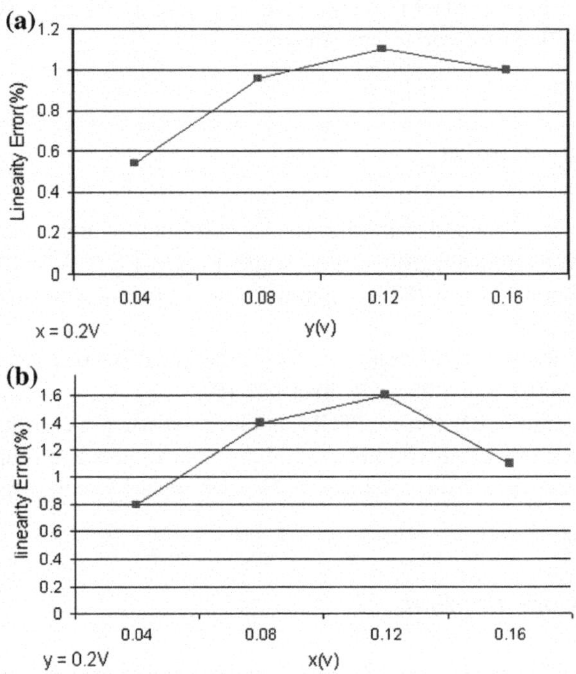

Fig. 6 **a** Linearity error versus y **b** linearity error versus x

$$\mathrm{V}_{O1} - \mathrm{V}_{O2} \propto \frac{K_N K_M}{K_P}(xy) \tag{11}$$

4.1 Input Range

The required bias conditions to obtain the input range, for multiplier structure II can be written as relation (12) and (13).

$$\mathrm{V_P} - |V_{THP}| \leq \mathrm{X} \pm \mathrm{x} \leq V_{DD} - |V_{THP}| \; for\; P1\text{-}P4 \tag{12}$$

$$\mathrm{Y} \pm \mathrm{y} \geq V_O + V_{THN} \; for\; N1\text{-}N4 \tag{13}$$

The biasing voltage of M1-M4 is chosen to be V_{DD} so as to keep V_P comparatively small, allowing P1-P4 for wider input range.

4.2 Linearity Analysis

The linearity error of second multiplier structure for both y and x inputs are shown in Fig. 6a, b. Typical biasing voltages used are: $V_{\mathrm{DD}} = 1.8$ V, $X = 0.5$ V, $Y = 1$ V, $V_{\mathrm{THP}} = 0.45$ V, and $V_{\mathrm{THN}} = 0.5$ V when the transistors have the same size of W/L = 0.41 μm/0.18 μm.

In particular, when x = 0.12 and y = 0.2 V, the linearity error is 1.62%. And for y = 0.12 and x = 0.2 V, the linearity error is 1.1%.

The Total Harmonic Distortion (THD) is observed when a constant DC voltage is applied to x (or y) while a 100 kHz, 0.2 V p-p sinusoidal wave is applied to y (or x) is shown in Fig. 7a, b. The structure shows good linearity with respect to x input.

4.3 Power Consumption

The total supply current for multiplier structure II is given by relation (14).

$$\begin{aligned} &\mathrm{I}_{\text{total}} = \mathrm{I}_{P1} + \mathrm{I}_{P2} + \mathrm{I}_{P3} + \mathrm{I}_{P4}, \\ &\mathrm{I}_{\text{total}} \propto 2\mathrm{K}_P\left[(\mathrm{V}_{DD} - \mathrm{X} - \mathrm{V}_T)^2 + \mathrm{x}^2\right] \end{aligned} \tag{14}$$

This indicates that the power consumption is independent with respect to signal y and DC bias Y for transistor N1-N4.

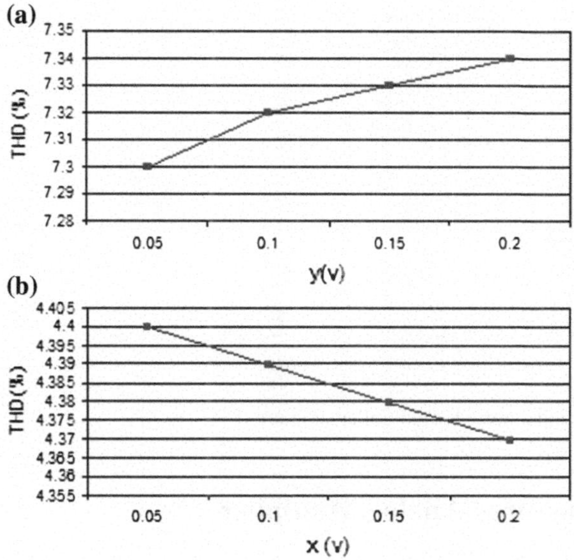

Fig. 7 **a** THD of output signal with y **b** THD of output signal with x

The total supply current when both x and y being a 100 kHz, 0.2 V p-p sinusoidal wave is 28 μA, which is less as compared to that of multiplier structure I.

4.4 *Noise*

The total output noise for multiplier structure II is given by relation set (15).

$$i_{no}^2 = 4(i_{nsat}^2 + 8i_{nlin}^2)$$
$$i_{no}^2 = 16K_N T g_{ds}\, df + 16K_M T g_{ds}\, df + \frac{32}{3} K_P T g_{ds}\, df$$

The output noise is strong function of g_{ds}, which can be controlled by current flowing through the PMOS and NMOS transistor.

Figure 8 shows the equivalent input-referred noise integrated within 10 Hz to 100 MHz. The equivalent input noise of the second multiplier structure is 867 n V/sqrt (Hz).

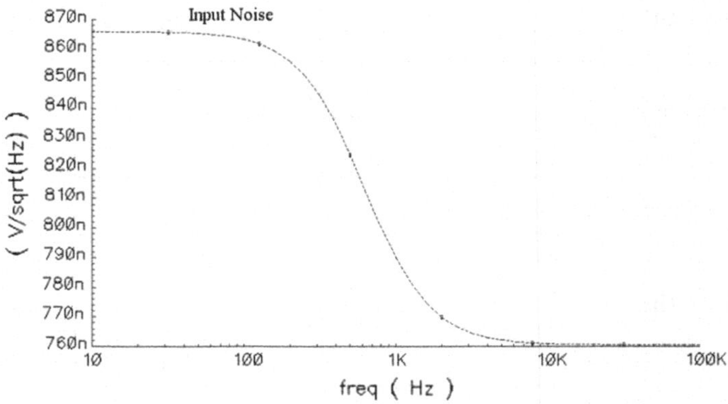

Fig. 8 Equivalent input noise for multiplier structure II

5 OTA Structured Analog Multiplier

The simplified schematic diagram of proposed OTA multiplier, based on two cross-coupled differential MOS pair, biased by current 2I is shown in Fig. 9. Transistor M1-M4 having same (W/L) ratio, Y + y is one input and X + x is second input of Cross-Couple OTA multiplier, and bias current 2I can be obtained sing programmable current source. OTA is a Voltage Controlled Current Source (VCCS) device and hence the output of OTA multiplier structure shown in Fig. 1 is

$$I_{out} = I_1 - I_2.$$

Fig. 9 Proposed OTA multiplier structure

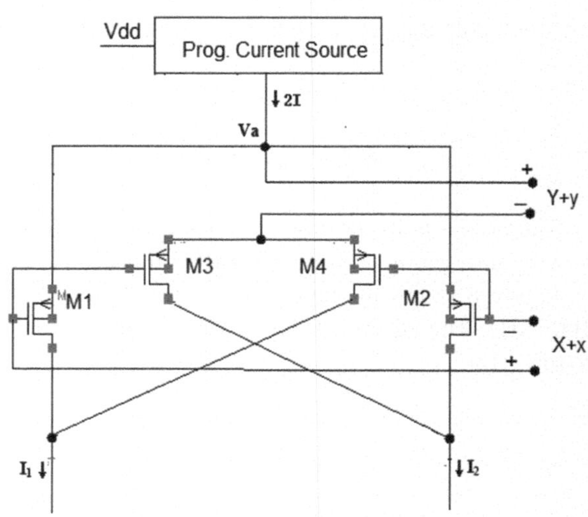

Also $I_1 = I_{DM1} + I_{DM4}$ and $I_2 = I_{DM2} + I_{DM3}$. Therefore I_1 and I_2 can be written as relation (4) and (5).

$$\begin{aligned} I_1 &= K(V_P - V_{TP})^2 + K(V_Q - (Y + y) - V_{TP})^2 \\ &= K(V_{PT})^2 + K(V_{QT} - (Y + y))^2 \end{aligned} \tag{15}$$

$$\begin{aligned} I_2 &= K(V_Q - V_{TP})^2 + K(V_P - (Y + y) - V_{TP})^2 \\ &= K(V_{QT})^2 + K(V_{PT} - (Y + y))^2 \end{aligned} \tag{16}$$

where V_P and V_Q are the gate to source voltages of M1 and M2 respectively, K = 0.5 uCox(W/L).

The differential output current is given as $I_{out} = I_1 - I_2$.

$$I_{out} = 2K(Y + y)(X + x) \tag{17}$$

The output Eq. (17) proves that the proposed OTA structure is able to work as a four-quadrant multiplier.

5.1 *Input Range*

The input range of a multiplier circuit is obtained from its bias conditions. Constraint set in (18) gives the required bias conditions for the OTA multiplier

$$V_T < X + x - (Y + y),\ Vd > X + x - V_T \tag{18}$$

where Vd the drain voltage of MOSFET M1-M4.

5.2 *DC Transfer Characteristics*

The DC transfer characteristics of I_{out} versus x and y are shown in Fig. 10a, b. Typical biasing voltage values to be used are: $V_{DD} = 1.8$ V, $X = 1.1$ V, $Y = 0.9$ V, $V_{THP} = 0.45$ V, and $V_{THN} = 0.5$ V when the transistors have the same size of $W/L = 0.41\ \mu m/0.18\ \mu m$. A DC transfer characteristic shows that multiplier has four-quadrant response.

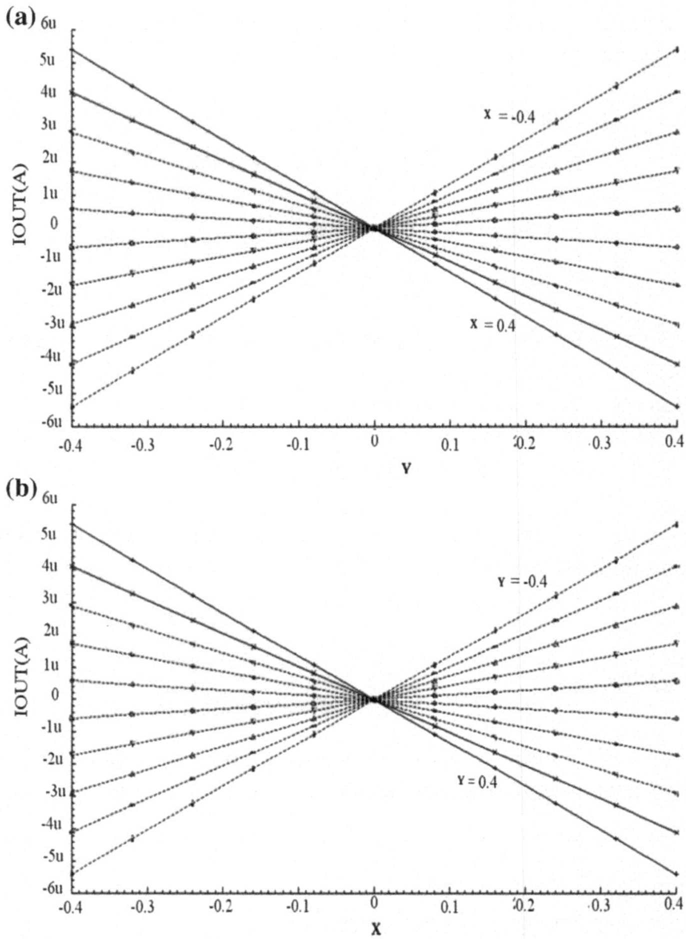

Fig. 10 **a** DC transfer characteristics I_{out} versus input y **b** DC transfer characteristics I_{out} versus input x

5.3 Linearity Analysis

The linearity error of proposed OTA multiplier with respect to inputs x and y is shown in Fig. 11a, b. In particular, when $y = 0.2$ and $x = 0.15$, the linearity error observed is 1.5%, which is very less as compared to multiplier structures I and II.

The Total Harmonic Distortion (THD) is estimated when a constant DC voltage is applied to x (or y) and a 100 kHz, 0.2 V p-p sinusoidal wave is applied to y (or x) is shown in Fig. 12a, b. The structure of the proposed OTA multiplier has good linearity with respect to y input. The value of THD increase as the input x increased.

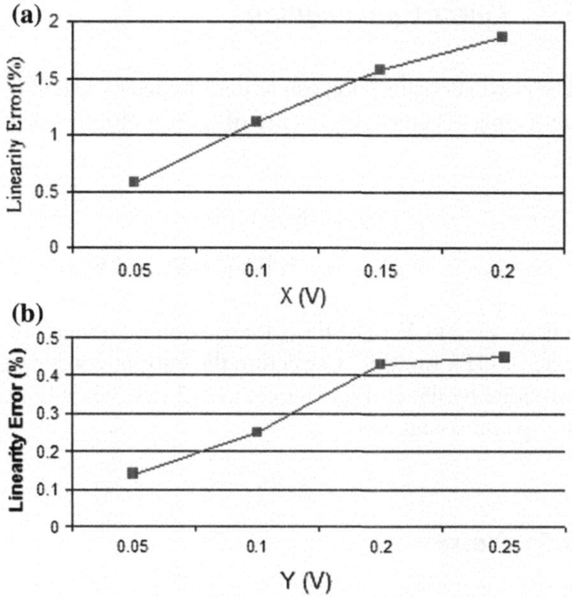

Fig. 11 **a** Linearity error versus x **b** linearity error versus x

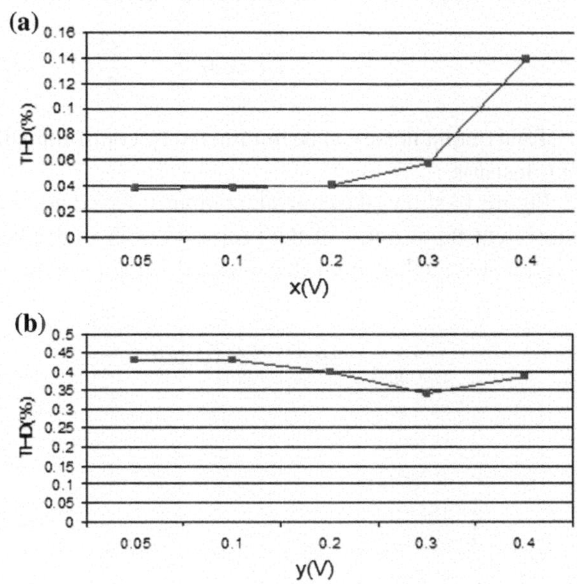

Fig. 12 **a** THD of output signal with x **b** THD of output signal with y

5.4 Power Consumption

The power consumption can be determined by calculating the total current when supply voltage is constant. In case of OTA multiplier structure, the power consumption = 2I.

$$
\begin{aligned}
I_{total} &= I_1 + I_2 \\
I_{total} &= 2K\left[(V_P + V_Q - 2V_{TP}) - 0.5(Y + y)\right]^2
\end{aligned}
\tag{19}
$$

where V_P and V_Q are the gate to source voltages of M1 and M2 respectively. In case of OTA multiplier structure the current consumption is almost the bias current provided by the current source, i.e. 42 μA, when both x and y being a 100 kHz, 0.2 V p-p sinusoidal wave.

5.5 Noise

The total output noise current for OTA multiplier is expressed as

$$
i_{no}^2 = 4i_{nsat}^2 = \frac{32}{3}kTg_m\,df
$$

Total output noise can be minimized by controlling the effective transconductance of transistor.

Figure 13 shows the equivalent input-referred noise from 10 Hz to 100 MHz. The equivalent input noise of the proposed OTA multiplier structure is 11.37 μV/sqrt (Hz), that is slightly high as compared to the previous two multiplier structures.

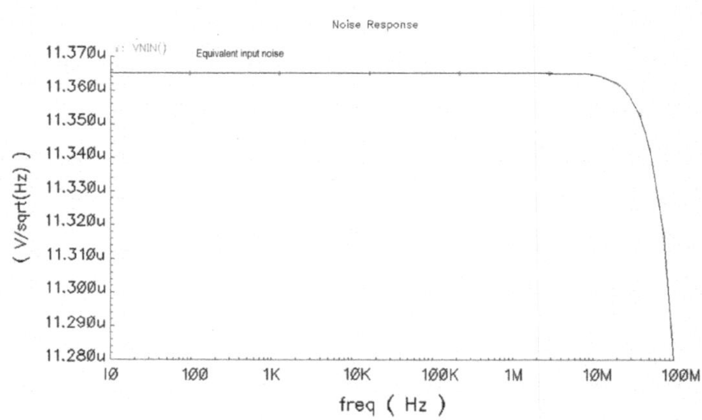

Fig. 13 Equivalent input noise for OTA multiplier structure

Table 1 Comparison of three multiplier structures ($V_{dd} = 1.8$ V)

Performance metric	1st structure	2nd structure	OTA multiplier
LE of x (%)	4.8	1.62	0.48
LE of y (%)	0.87	1.1	0.18
THD of x (%)	2.7	4.3	0.04
THD of y (%)	8.6	7.3	0.388
Input noise $\left(\frac{V}{\sqrt{Hz}}\right)$	114.1 n	867 n	11.37 μ
Consumption (I) (μA)	34	28	42

6 Comparison of Three Structures

The comparative study of three multipliers with respect to different performance metrics is given in Table 1 and it shows that the proposed OTA multiplier structure has good performance in term of linearity with comparable power consumption and input noise, as compared to multiplier structures I and II.

7 Conclusion

A highly linear OTA multiplier has been proposed and simulated in Cadence Analog Design Environment. The simulation proves that the proposed OTA structure is able to work as a four-quadrant multiplier. Several important performance metrics, i.e., input range, linearity, power consumption, noise, etc., have been analyzed, for proposed OTA multiplier structures. The experimental results have shown that proposed OTA multiplier consumes comparable power with the existing structures, while maintaining good linearity. The total supply current and noise can be further reduced by some design considerations.

Acknowledgements The support of the Management and the Director Dr. M. B. Khambete of MKSSS's Cummins College of Engineering for Women Karve Nagar, Pune is gratefully acknowledged.

References

1. Popa C (2014) Improved accuracy current-mode multiplier circuits with applications in analog signal processing. IEEE Trans Very Large Scale Integr (VLSI) Syst 22(2):443–447
2. Katayama K et al (2016) A 300 GHz CMOS transmitter with 32-QAM 17.5 Gb/s/ch capability over six channels. IEEE J Solid State Circuits 51(12):3037–3048
3. Hossain M et al (2016) A G-band high power frequency doubler in transferred substrate InP HBT technology. IEEE Microw Guided Wave Lett 26(1):49–51

4. Liu G et al (2015) Frequency doublers with 10.2/5.2 dBm peak power at 100/202 GHz in 45 nm SOI CMOS. In: Proceedings of IEEE radio frequency integrated circuits symposium (RFIC), pp 271–274, May 2015
5. Al-Absi MA, As-Sabban IA (2015) A new highly accurate CMOS current-mode four-quadrant multiplier. Arab J Sci Eng 40:551–558
6. Beyraghi N, Khoei A (2015) CMOS design of a low power and high precision four-quadrant analog multiplier. Int J Electron Commun (AEÜ) 69:400–407
7. Popa C (2009) Multiplier circuit with improved linearity using FGMOS transistors. In: Proceedings of international symposium ELMAR 2009, pp 159–162
8. Hidyat R, Dejhan K, Moungnoul P, Miyanaga Y (2008) OTA-based high frequency CMOS multiplier and squaring circuit. In: Proceedings of international symposium on intelligent signal processing and communication systems, pp 1–4
9. Naderi A, Khoei A, Hadidi K (2007) High speed, low power four quadrant CMOS current-mode multiplier. In: Proceedings of IEEE international conference on electronics circuits and systems, pp 1308–1311, Dec 2007
10. Maundy B, Aronhime P (2002) Useful multipliers for low-voltage applications. In: Proceedings of IEEE international symposium on circuits and systems, vol 1, pp 26–29, May 2002
11. Liu S, Chang C (1995) CMOS subthreshold four quadrant multiplier based on unbalanced source coupled pairs. Int J Electron 78:327–332
12. Mehrvarz HR, Kwok CY (1995) A large-input-dynamic-range multi input floating gate MOS four-quadrant analog multiplier. In: Proceedings of IEEE international solid-state circuits conference, pp 60–61, Feb 1995
13. Wang Z (1993) A four-transistor four-quadrant analog multiplier using MOS transistors operating in the saturation region. IEEE Trans Instrum Meas 42(1):75–77
14. Han G, Sanchez-Sinencio E (1998) CMOS transconductance multipliers: a tutorial. IEEE Trans Circuits Syst II Analog Digit Signal Process 45(12):1550–1563
15. Liu S, Hwang Y (1994) CMOC four-quadrant multiplier using bias feedback techniques. IEEE J Solid State Circuits 29:750–752
16. Sawigun C, Mahattanakul J (2008) A 1.5 V, wide-input range, high bandwidth, CMOS four-quadrant analog multiplier. In: Proceedings of IEEE international symposium on circuits and systems, pp 2318–2321, May 2008
17. Sawigun C, Demosthenous A, Pal D (2007) A low-voltage, low-power, high-linearity CMOS four-quadrant analog multiplier. In: Proceedings of 18th European conference on circuit theory and design, pp 751–754, Aug 2007
18. Chen C, Li Z (2006) A low-power CMOS analog multiplier. IEEE Trans Circuits Syst
19. Razavi B (2001) Design of analog CMOS integrated circuits. McGraw-Hill, New York

Design and Development of Low-Cost Patch Antenna Using Air Gap for Wireless Applications

Neela Chattoraj, Neha Rajak, Basudeo Mahato and Rahul Kumar

Abstract Rectangular patch antenna is designed in this paper. In this antenna, air is acting as a substrate between the patch and ground. The patch and the ground is made up of a copper sheet of thickness 0.2 mm. The dimension of the microstrip patch and ground is 148.5 mm × 147 mm and 155 mm × 159 mm, respectively. The air gap between the patch and substrate is 5 mm. This antenna resonates at 908 MHz. The antenna is designed for Machine-to-Machine (M2M) RFID application. The simulation and optimization are done in CST software and different parameters like reflection coefficient, radiation pattern and VSWR and Gain are analyzed.

Keywords Microstrip patch antenna · Air substrate · M2M

1 Introduction

Microstrip antennas are composed of metallic patch on a grounded substrate. The patch can be of any shape like rectangular, circular, square, triangular, etc. The main advantage of these antennas is low profile, lightweight, and conformal. These types of antenna are used mainly in aerospace, mobile applications and in low power transmitter and receiver systems [1]. In order to have an efficient antenna, the substrate should be thick and the dielectric constant should be low as it decreases surface waves [2]. There are no surface waves in the air [3]. The main factors which are related to the losses of the microstrip patch antenna are material of the conductor,

N. Chattoraj · N. Rajak (✉) · B. Mahato · R. Kumar
Department of ECE, Birla Institute of Technology, Mesra, Ranchi, India
e-mail: rajakneha@gmail.com

N. Chattoraj
e-mail: nchattoraj@bitmesra.ac.in

B. Mahato
e-mail: basu18may@gmail.com

R. Kumar
e-mail: rahulaero@gmail.com

© Springer Nature Singapore Pte Ltd. 2019

V. Nath and J. K. Mandal (eds.), *Proceedings of the Third International Conference on Microelectronics, Computing and Communication Systems*,
Lecture Notes in Electrical Engineering 556, https://doi.org/10.1007/978-981-13-7091-5_24

dielectric of the substrate and the surface waves [4]. The easiest way to suppress the surface waves is the use of air as substrate material or low range dielectric constant material [5]. The dielectric constant of air is 1 which is low and the thickness of the substrate, i.e., air height between the patch and the ground can be adjusted according to the requirement. The height between the patch and ground affects the bandwidth, resonant frequency and gain of the antenna [6]. The air gap has an effect on the impedance of the antenna. The resonant frequency of the antenna is linked with the impedance so there is a shift in the frequency [7]. There is also an improvement in the radiation pattern of the microstrip patch antenna when the air is taken as substrate [8, 9]. Air-based microstrip antenna have been made for various applications such as X band, S band communication, Wi-Fi, WLAN, etc. [9, 10]. There are various ways of increasing the gain of the antenna like layering of the dielectric material of the antenna, the use of metasurfaces and metamaterials, use of air cavity, etc. [11–14].

M2M is a part of the Internet of things (IOT) which allows the devices to communicate data with each other [15]. The M2M RFID has applications in Automotive, Transportation and Security industries as well as public safety, etc.

In this paper, the microstrip patch antenna has been designed and fabricated. The patch and ground of the antenna are made up of a copper sheet of thickness 0.2 mm with air as the substrate between them. This antenna is designed for M2M RFID application which has a frequency band from 902 to 928 MHz. The structure of the antenna is described in Sect. 2. The simulation and its discussion is done in Sect. 3. The fabricated design and its measured result are summarized in Sect. 4. At last conclusion is given in Sect. 5.

2 Antenna Design

The microstrip patch antenna consists of patch, substrate, ground and SMA connector port to excite the antenna. The dimension of the patch is 148.5 mm × 147 mm and ground is 155 mm × 159 mm. The gap between the patch and ground plane is 5 mm, i.e., the height of the air substrate. The patch and ground is connected by pin of the coaxial probe. The dielectric constant of the air substrate is 1. Patch and ground plane is made of a copper sheet of thickness 0.2 mm. The probe is drilled into the ground plane reaching up to the radiating patch.

The dimension of the patch is based on these formulae

$$L = v_o/2f_r\sqrt{\varepsilon_r} \tag{1}$$

$$w = v_o/2f_r\sqrt{\frac{2}{\varepsilon_r + 1}} \tag{2}$$

where length is represented by L, width by w, resonant frequency by f_r, speed of light by v_o, and relative permittivity of substrate by ε_r. The ground plane dimension is based on these formulae

$$L_g = 6h + L \quad (3)$$

$$W_g = 6h + w \quad (4)$$

where L_g and W_g represent length and width of the ground plane, respectively. The height of the substrate is given by h.

The normal electric field component of the copper sheet of the antenna is more intense in the air gap region than in the copper sheet leading to energy storage and less radiation from the back side of the antenna. There will be a negligible surface wave as no dielectric medium is present (Fig. 1).

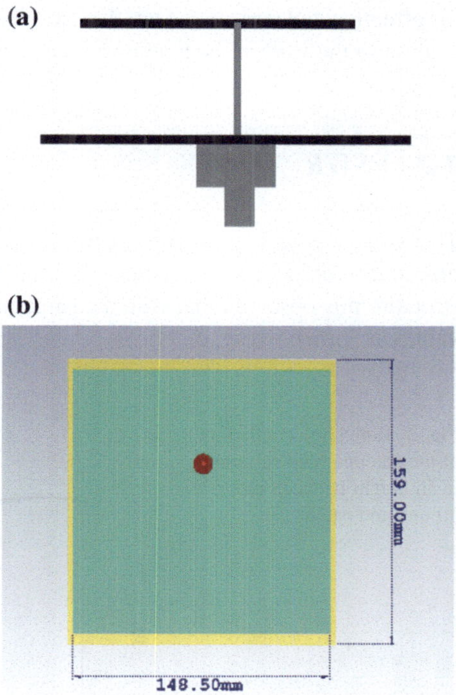

Fig. 1 Patch design **a** side view **b** top view

3 Simulated Result and Discussion

Simulation of the antenna is done in CST (Computer Simulation Technology) which is based on Time Domain Simulation.

3.1 Return Loss

The antenna operates at 908 MHz with a reflection coefficient of −25 dB and bandwidth of 30 MHz (900–930 MHz). Since the bandwidth of the antenna is narrow, the transmission rate of data will be high which is required for the M2M RFID application transmission. The air height between the patch and ground plays an important role in securing the resonant frequency. With the change in the height of the air gap, there is variation in impedance of the antenna which results in the shift in the resonant frequency as shown in Fig. 2. The change in the height of the air gap changes the effective dielectric constant, thereby changing the overall dimension of the patch leading to the shift in the resonant frequency.

3.2 VSWR

The Voltage Standing wave ratio which is the ratio of the maximum voltage and minimum voltage in the antenna. It should be between 1 and 2. In the proposed structure it is observed that with a change in the air gap, there is a change in VSWR values as shown in Fig. 3.

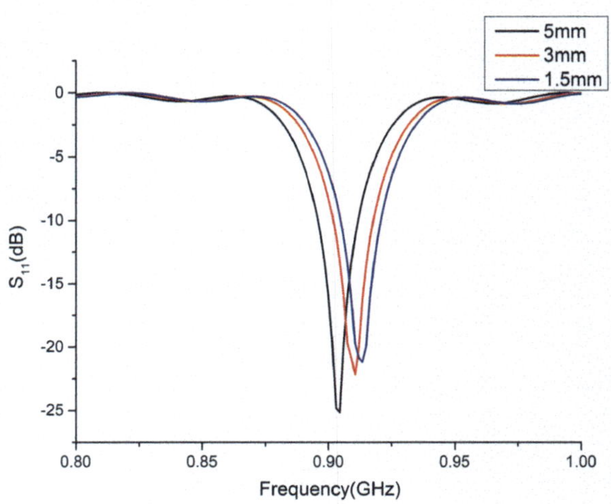

Fig. 2 Reflection coefficient of the antenna with variation in air height between the ground and patch

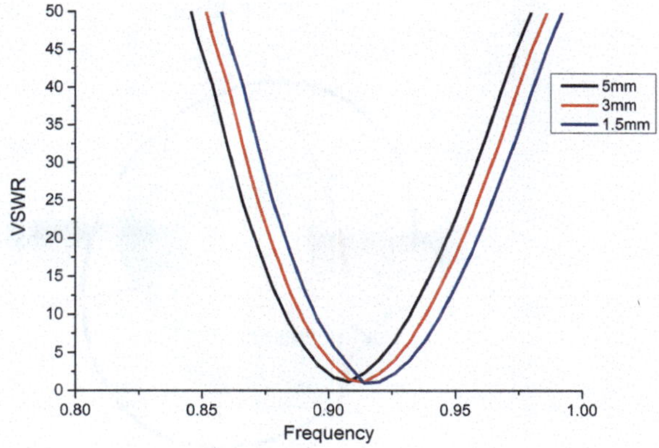

Fig. 3 VSWR with the variation in the air height between the patch and ground

3.3 *Radiation Pattern*

The radiation pattern is simulated on 908 MHz frequency which is shown in Fig. 4. H-plane is Omnidirectional and E-plane is bidirectional in nature. Both the patch and the ground plane are radiating. The back lobe which is due to the ground plane is less than the front lobe of the patch. The air gap between the ground and patch stores the energy within itself leading to a reduction in the surface wave and back radiation

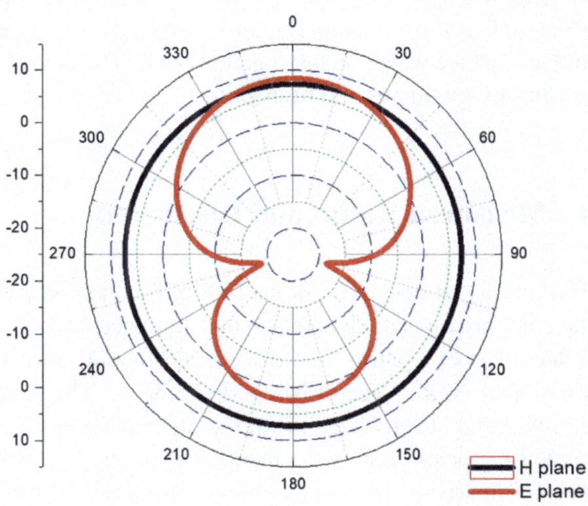

Fig. 4 Radiation pattern

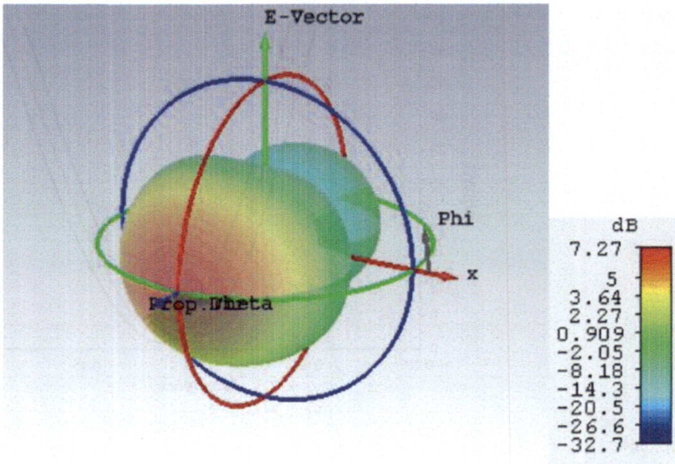

Fig. 5 Gain in far-field region

3.4 Gain

The gain of the antenna at the resonating frequency 908 MHz is 7.27 dB which is shown in Fig. 5 and radiation efficiency is 82%. Normally, the gain of the microstrip patch antenna is less than 4 dB but in our case due to the presence of air substrate the gain has increased. The reduction in the losses leads to an increase in gain and efficiency of the antenna. The air is used as a substrate so there is no dielectric loss and no surface waves as loss tangent is nil. The energy is stored between the air gap leading to low losses and high gain.

4 Measured Result and Discussion

The fabricated design is shown in Fig. 6 which is made of copper sheet. In Fig. 7, the comparison is done between the simulated and the measured result is given. The reflection coefficient is −25 dB at 908 MHz in simulation and −28 dB at 908 MHz when measured in Vector Network Analyzer. The bandwidth is 30 MHz which is a narrow band. There is a shift in the resonant frequency due to an external environmental factor present in the measuring room. The VSWR is 1.50 when measured. The Comparisons of simulated and measured VSWR is shown in Fig. 8. The gain measurement of the antenna is done using gain transfer method. The gain is 7.2 dBi which is a good result in the patch antenna as shown in Fig. 9.

The comparison table is given below in which the proposed antenna is compared with other antennas.

Fig. 6 Proposed design a antenna with VNA b top view of antenna

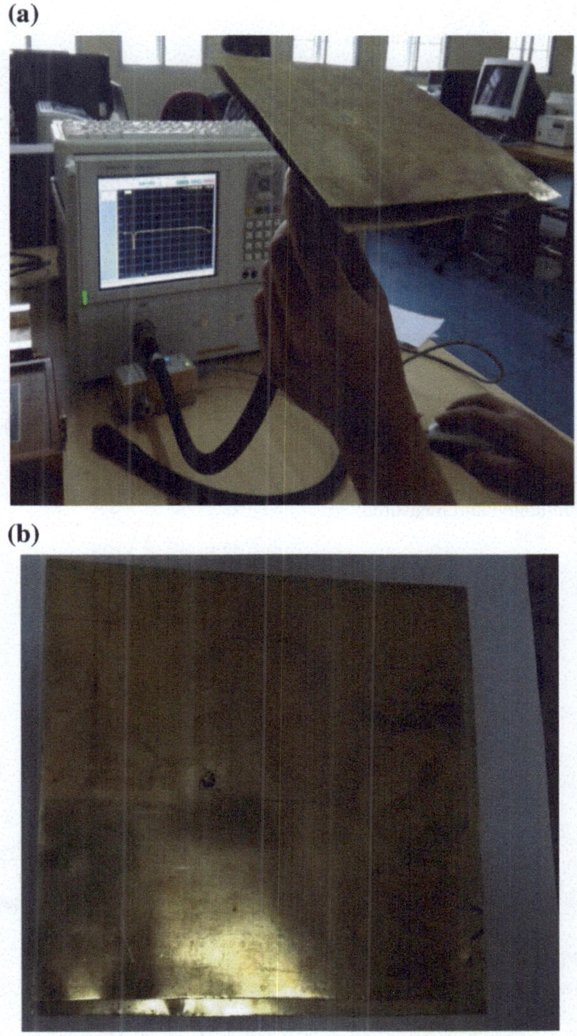

It can be seen in Table 1 that the gain of this antenna is better as compared to other antennas.

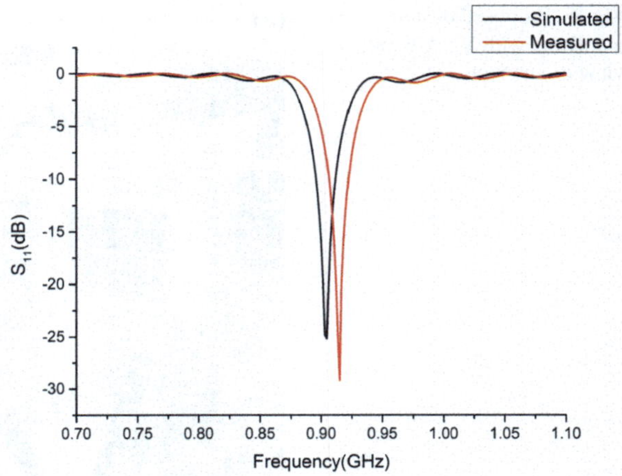

Fig. 7 Comparison of simulated and measured S_{11}

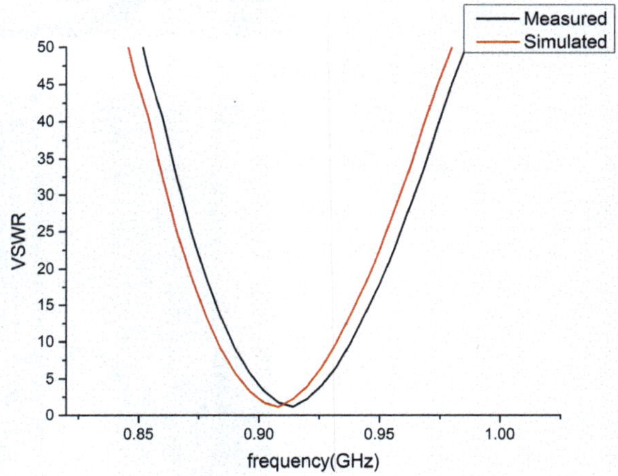

Fig. 8 Comparison of simulated and measured VSWR

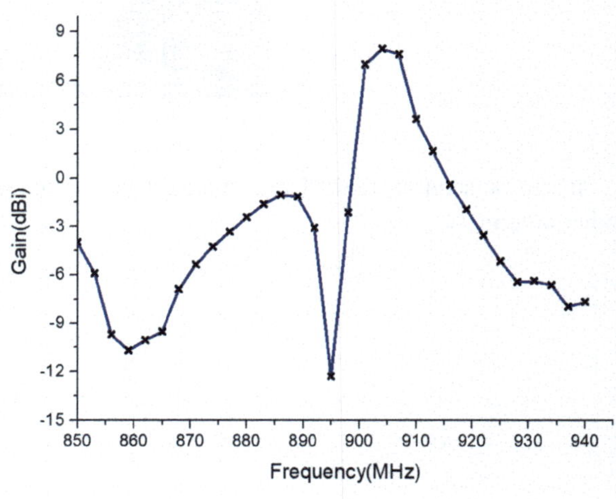

Fig. 9 Gain versus frequency plot

Table 1 Comparison between different antennas and proposed antenna

Refs. no.	Substrate	Dimension (mm^2)	Resonant frequency (GHz)	Gain (dBi)
[4]	Rogers 4003	125 × 125	2.4	6.7
[16]	FR4	25 × 30	5.12, 6.08	5
[17]	Taconic	60 × 60	3.6	6.8
[6]	Air	**320 × 280**	0.882	9.2
[18]	RT Duroid 5880	52 × 52	1.82	4.8
proposed	Air	**148.5 × 147**	0.908	**7.28**

5 Conclusion

The patch antenna using air as a substrate has been simulated in CST software. The fabricated design is cost efficient, economical, and lightweight. The parameters such as reflection coefficient, VSWR, radiation pattern, and gain have been examined. The parametric study of the variation of the air gap height has been studied. The proposed antenna works for M2M RFID band with a gain of 7.27 dB. The simulated and measured result is in close proximity to each other. This design is very simple which can be easily realized and utilized effectively.

Acknowledgements The authors are grateful to the Faculty of Department of Electronics and Communication Engineering, BIT Mesra, Ranchi for their support and encouragement.

References

1. Garg R, Bhartia P, Bahl I, Ittipiboon A (2001) Microstrip antenna design handbook. Artech House, Norwood, MA
2. Ali MT, Jaafar H, Subahir S, Yusof A (2012) Gain enhancement of air substrates at 5.8 GHz for microstrip antenna array. In: Asia-Pacific symposium on electromagnetic compatibility (APEMC), pp 477–480
3. Waterhouse RB (2003) Microstrip patch antennas-a designer's guide. Kluwer Academic Publishers, Boston, MA. Chapter 4
4. Yeap SB, Chen ZN (2010) Microstrip patch antennas with enhanced gain by partial substrate removal. IEEE Trans Antennas Propag 58(9):2811–2816
5. Panther A, Petosa A, Stubbs MG, Kautio K (2005) A wideband array of stacked patch antennas using embedded air cavities in LTCC. IEEE Microw Wirel Compon Lett 15(12):916–918
6. Ayoub AFA (2003) Analysis of rectangular microstrip antenna with air substrate. J Electromagn Waves Appl 17(12):1755–1766
7. Jamlos MF, Rahman TA, Kamarudin MR, Ali MT, Md Tan MN, Saad P (2010) The gain effects of air gap quadratic aperture coupled microstrip antenna array, In: Progress in electromagnetics research symposium proceedings, Cambridge, USA, 5–8 July 2010

8. Lee KF, Dahele JS (1983) Mode characteristics of annular-ring and circular-disc microstrip antenna with and without airgaps. In: IEEE antennas propagation society international symposium digest, pp 55–58
9. Baek YH, Truong LH, Park S-W, Lee S-J, Chae Y-S, Rhee E-H, Park H-C, Rhee J-K (2009) 94-GHz log-periodic antenna on GaAs substrate using air-bridge structure. IEEE Antennas Wirel Propag Lett 8:909–911
10. Chandan C, Ghosh A, Ghosh SK, Chattopadhyay S (2009) Radiation characteristics of rectangular patch antenna using air substrates. In: 2009 international conference on emerging trends in electronic and photonic devices & systems (ELECTRO-2009)
11. Liu IY-T, Su C-W, Wong K-L, Chen H-T (2004) An air substrate narrow patch microstrip antenna with radiation performance for 2.4 GHz WLAN access point. Microw Opt Technol Lett 43:189–192
12. Yeap SB, Chen ZN, Qing X (2011) Gain-enhanced 60-GHz LTCC antenna array with open air cavities. IEEE Trans Antennas Propag 59(9):3470–3473
13. Rajak N, Chattoraj N (2017) A bandwidth enhanced meta surface antenna for wireless applications. Microw Opt Technol Lett 59(10):2575–2580
14. Rajak N, Chattoraj N (2017) Design and analysis of a bandwidth enhanced antenna based on metasurface for wireless application. In: 2016 IEEE Uttar Pradesh section international conference on electrical, computers and electronics engineering (UPCON 2016), pp 367–371
15. Elloumi O, Song J, Ghamri-Doudane Y, Leung VCM (2015) IoT/M2M from research to standards: the next steps (part I) [Guest Editorial]. IEEE Commun Mag 53(9):8–9
16. Wi S-H, Lee Y-S, Yook J-G (2007) Wideband microstrip patch antenna with U-shaped parasitic elements. IEEE Trans Antennas Propag 55:1196–1199
17. Guha D, Biswas M, Antar YMM (2005) Microstrip patch antenna with defected ground structure for cross polarization suppression. IEEE Antennas Wirel Propag. Lett 4:455–458
18. Sung YJ, Kim M, Kim YS (2003) Harmonics reduction with defected ground structure for a microstrip patch antenna. IEEE Antennas Wirel Propag Lett 2:111–113

Evolutionary Decision Tree-Based Intrusion Detection System

Chandrashekhar Azad, Ashok Kumar Mehta and Vijay Kumar Jha

Abstract In this paper, an Evolutionary Decision Tree based Intrusion Detection System (IDS) has been proposed which is based on the concept of decision tree (DT) and genetic algorithm (GA). The goal of the projected IDS is to protect the computer network from the various types of cyberattacks. The main hurdles in today's decision tree-based IDS are the problem of small disjunct, preprocessing of the network logs, learning of the desired system for anomalous or signature detection, outlier handling on the training set, etc. In this research paper, the concept of interquartile range is used to preprocess the data to handle the outlier and extreme values in the training set, and then the decision tree is used to generate the DT. Further, the decision rules, which come in the category of small disjunct, who have a vital role in the accuracy are optimized using metaheuristic GA. Furthermore, the proposed system is compared with the existing IDS. The performance of the system is significant in terms of accuracy and classification error is compared to the existing systems.

Keywords Anomaly detection · Data mining · Decision tree · Genetic algorithm · IDS · Misuse detection

1 Introduction

Development in the field of science and technology has made the world technology dependent. The main reason behind the technological dependency is the innovations in the domain of smart devices such as mobile devices, electronic gadgets, IoT, etc.

C. Azad · A. K. Mehta (✉)
Department of Computer Applications, NIT Jamshedpur, Jamshedpur 831014, India
e-mail: akmehta.ca@nitjsr.ac.in

C. Azad
e-mail: csazad.ca@nitjsr.ac.in

V. K. Jha
Department of CSE, Birla Institute of Technology Mesra, Ranchi 835215, India
e-mail: vkjha@bitmesra.ac.in

© Springer Nature Singapore Pte Ltd. 2019

V. Nath and J. K. Mandal (eds.), *Proceedings of the Third International Conference on Microelectronics, Computing and Communication Systems*, Lecture Notes in Electrical Engineering 556, https://doi.org/10.1007/978-981-13-7091-5_25

Today, WWW is used by most of the organizations either corporate or social for sharing of data/services to the intended users. Data/services on the web server are today's most valuable assets in the digital world. Since the integrity of the organizations either corporate or social mainly depends on the integrity and reliability of the hosted data/services on the web servers. Today, the internet is the essential need for the human being like food, accommodation, and clothes. Without WWW, the organizations are helpless in the smooth running of the day-to-day activity of the business. Individual and corporate users use web applications and services not only for the purpose to generate a return but also often use to share valuable information with the stakeholders. Therefore, information security becomes the critical factor in computer networks. Intrusion detection and prevention is a crucial aspect of ensuring the security of the computer network. It ensures the integrity of resources in the computer networks, both in relation to security and privacy. Therefore, there is a need for a robust system to detect security threats and maintain the integrity and availability of the data and services in the web servers. WWW generate a huge volume of computer network data day-to-day, and its manual analysis is difficult. Therefore, the field of data mining is gaining importance as per literature [4, 18], data mining to classifications [1, 5], clustering's [16, 21] and association rule mining [19]. Finding the anomalous network traffic in the computer network is the most challenging task for the WWW consortium community. In such a scenario, data mining-based intrusion detection system is the best suitable options to overcome the problem of attack on the cyberinfrastructure. In this work, we have proposed an intrusion detection system using data mining and optimization techniques.

1.1 Security Assurance

The function of the security system is to bound access to network resources. A perfectly protected system ensures the integrity, availability, reliability of data and services from unauthorized person. Unfortunately, till date, there is not a single system that provides 100 percent security to the digital assets. Today, we need more than just a protection system. Today, we want not only mechanisms that safeguard in contrast to security violations, but also they can detect the security violations, and respond to the security violations. It is an important point to understand that security mechanism do not exist just to limit access to a system, but also to protect the data or on the web. It also requires that the users must be aware of security policies to follow then that support protection of the whole computing system as needed.

Keeping in mind security key points and security process cycle in this article an evolutionary Intrusion Detection System has been proposed which is based on Decision Tree and Genetic Algorithm. The rest of the paper is organized as follows:

Sect. 2 describes the related work in the domain of intrusion detection, Sect. 3 describes the related methods and materials, Sect. 4 describes the proposed work in terms of architecture, pseudo code, Sect. 5 describes the results and related discussion, and finally in Sect. 6 conclusion and future scope of the work is described.

2 Related Work

An IDS is a system that monitors the network activity in the WWW. This may alarm the system administrator if any online vulnerable network packet has been sniffed by the system which possibly violates the integrity, reliability, availability, confidentiality of the data and services hosted on the WWW. Data mining is the field of decision science used to extract the unseen information, facts, knowledge from the datamarts. The data extracted with the help of data mining may be financially valuable as well as previously hidden data may contribute to achieving the overall goal of the system, organization, or individual. First, Anderson [3] addressed about IDS in his paper Computer Security Threat Monitoring and Surveillance, after that till date many of the innovations are done in the field of the IDS by the researchers but still the data and services is not 100% secured in the web [6–9]. Nowadays, the domain of IDS is attracting the researchers. It is because today IDS plays the role of digital guardian in the WWW. Sindhu et al. [24] have proposed a decision tree-based IDS. The proposed system was assessed using detection and error percentage. In [23] authors have proposed rule-based IDS, which are based on the C4.5 decision tree algorithm. The effectiveness is evaluated using KDD CUP data. In 2010 Mulay et al. [20] have proposed a Decision Tree-SVM-based IDS. In [21] authors have proposed an anomaly detection system using k-means and C 4.5 decision tree algorithm. Jiang et al. [14] has proposed an incremental decision tree-based IDS, which is based on rough set theory. Koshal and Bag [17] have proposed a C 4.5 and SVM rule-based IDS. The decision tree C 4.5 is proposed by Quinlan [22], and this is a widely used classifier. The C 4.5-based classifier gives improved result in terms of accuracy. Many of the scholars use decision tree C 4.5, REP Tree, J48, etc., build IDS. It is widely used in data mining because decision tree is, easy to implement and parametric classifier. Rule extracted through decision tree or rule-based system are categorized into two clusters small disjunct and large disjunct. Rule set which covers a small number of training data, i.e., small disjunct and the large disjunct which covers a large number of training data [9, 10, 12, 13]. C 4.5 is biased toward the large disjunct in contrast to the small disjunct [9–11]. The IDS developed using the decision tree rule-based system; performance may be compromised due to the availability of small disjunct [2]. The small disjunct in the decision tree-based IDS is essential as much as a large disjunct, therefore the small disjunct cannot be ignored. If the small disjunct is ignored, the system may be error-prone and the availability of small disjunct may affect the accuracy of the system and may also increase the false alarm rate.

3 Materials and Methods

3.1 Decision Tree

The decision tree is used for the data classification task. The decision tree is composed of a root node, branches, and leaf nodes. It has one root node, 0, or more branches. The leaf node in the decision tree is labeled with target class and non-leaf nodes are labeled with outcomes of attributes. The number of branches in a non-leaf node is influenced by decision tree type and attribute type. The top node in the decision tree is called root node and which is at the start of the tree.

3.2 Attribute Selection Measures

3.2.1 Information Gain

$$Info(D) = -\sum_{i=1}^{n} P_i log_2 p_i \tag{1}$$

$$info_A(D) = -\sum_{j=1}^{v} \frac{|D_j|}{|D|} info(D_j) \tag{2}$$

$$Gain(A) = Info(D) - info_A(D) \tag{3}$$

3.2.2 Gain Ratio

$$SplitInfo_A(D) = -\sum_{j=1}^{v} \frac{|D_j|}{|D|} log_2\left(\frac{|D_j|}{|D|}\right) \tag{4}$$

$$GainRatio = \frac{Gain(A)}{SplitInfo_A(D)} \tag{5}$$

3.3 Genetic Algorithm

The Genetic Algorithm (GA) is a metaheuristic technique inspired by Darwin's theory "survival of the fittest." GA is used for the optimization purpose through the natural evaluation process. GA takes DNAs as an input and produces offspring from given DNAs through genetic processes. The produced offspring's fitness is calculated, and the best individuals who have better fitness value are kept for the next generation [7]. Here GA is used to handle the problem of small disjunct in the decision rules. The small disjunct can be handled using GA as follows: (i) Instances of each leaf nodes individually for optimization, i.e., GA Small and (ii) All instances is covered by all the leaf for optimization, i.e., GA Large. Here in this work, the focus is on the second situation.

3.3.1 Individual Encoding

A chromosome in the GA is the base for operations, and each consists of n genes, and each gene represents one attribute. The first gene corresponds to the first attribute, the second gene corresponds to the second attribute, and so on. Each gene is divided into the three subparts attribute, operator, and the value.

3.3.2 Selection

In the genetic algorithm, the selection operator plays a vital role in the further evolution of the individuals. If the worse fitness value of an individual is chosen, then it may not provide the optimal solution. Roulette-wheel selection, ranking-based, tournament selection, etc., are the different selection methods.

3.3.3 Crossover

The crossover operator is used to produce new individuals from the parents. This is the reproduction scheme in the genetic algorithm.

3.3.4 Mutation

The mutation operator is used to maintain genetic diversity among the individuals of the different generations.

3.3.5 Fitness Function

The fitness function estimates the quality of the rule and also controls the optimization process in the succeeding iterations. The Genetic Algorithm fitness function is an association of two well-known performance measures, i.e., sensitivity and specificity. The fitness function used is shown below:

$$\begin{aligned} Fitness\,function &= Sensitivity * Specificity \\ &= \frac{TP}{TP + FN} * \frac{TN}{FP + TN} \end{aligned}$$

		True Class	
		+	-
Predicted Class	+	TP	FP
	-	FN	TN

Fig. 1 Confusion matrix

The zeal behind the function is to strengthen the GA to define the best dominance rule having high specificity-sensitivity. The parameters True Positive (TP), False Negative (FN), False Positive (FP), and True Negative (TN) are the entries in the matrix shown in Fig. 1. It shows the ratio of predicted and actual. Columns of the confusion matrix mean the actual outcomes, and the row shows the predicted outcomes.

4 Proposed Work

4.1 System Design of the IDS

See Fig. 2.

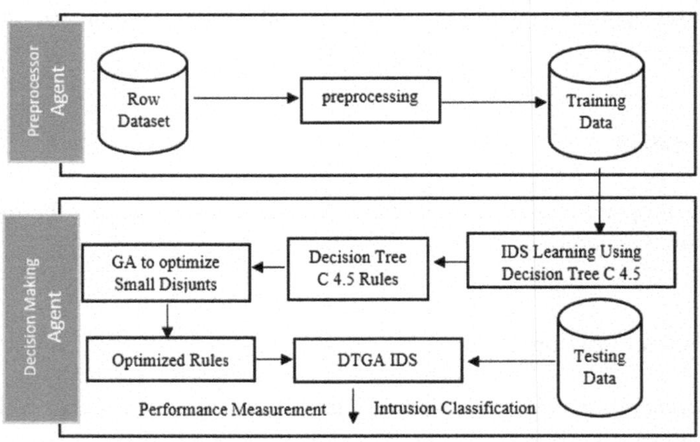

Fig. 2 Architecture of the proposed system

4.2 Algorithm for Proposed IDS

Step 1. Load Training data set.
Step 2. Create node N, instances in Training Set fit into alike class C then return node N by labeling as a leaf node by majority class C.
Step 3. For each feature calculate the gain ratio.
Step 4. Let feature A has the highest gain ratio, create a node with best splitting node A.
Step 5. Remove the splitting attribute from the attribute list.
Step 6. Repeat steps 2–5 for each outcome of split on A.
Step 7. Find the decision rules from the decision tree.
Step 8. If the rule falls into the category of the small disjunct rule, then encode the rule for optimization.
Step 9. Select parents.
Step 10. Perform crossover and mutation operation.
Step 11. Fitness evaluation and keep the best offspring's.
Step 12. If maximum iteration or stopping criteria fulfilled then go to step 13 else go to 10.
Step 13. Return optimized rules and C 4.5 decision rules.
Step 14. Stop.

5 Experimental Result and Discussion

The Decision Tree (DT) and the Genetic Algorithm (GA) are used to develop a classification system. By further training of the model, unlabeled data instance is labeled using the trained system.

5.1 Dataset

To test the effectiveness of the system, we have used KDD Cup dataset from the UCI repository.

5.2 Performance Parameters

The performance measures stipulate the abilities of the projected model. To test the competence of the system Classification Accuracy (CA) and the Classification Error (CE) are used and are conveyed in percentage.

$$CA = \frac{TP + TN}{TP + FP + TN + FN} * 100$$

$$CE = \frac{FP + FN}{TP + FP + TN + FN} * 100$$

5.3 Results

See Tables 1, 2 and 3.

Table 1 DT fivefold result

Fold	Sample dataset 1		Sample dataset 2		Sample dataset 3	
	CA	CE	CA	CE	CA	CE
1	96.93	03.07	94.46	05.54	97.80	02.20
2	96.87	03.13	94.54	05.46	97.29	02.71
3	95.72	04.28	94.82	05.18	97.80	02.20
4	96.93	03.07	93.88	06.12	98.66	01.34
5	95.72	04.28	94.46	05.54	96.87	03.13

Table 2 Result of the evolutionary IDS

D. size	Fold	Sample dataset 1		Sample dataset 2		Sample dataset 3	
		CA	CE	CA	CE	CA	CE
5	1	98.84	01.16	96.87	03.13	98.62	01.38
	2	98.29	01.71	97.80	02.20	97.65	02.35
	3	98.59	01.41	98.09	01.91	99.04	00.96
	4	98.84	01.16	97.80	02.20	98.77	01.23
	5	99.29	00.71	97.80	02.20	99.94	00.06
10	1	98.59	01.41	97.29	02.71	99.99	00.01
	2	98.84	01.16	98.84	01.16	98.32	01.68
	3	97.55	02.45	97.80	02.20	98.32	01.68
	4	97.29	02.71	98.09	01.91	97.55	02.45
	5	98.84	01.16	97.77	02.23	98.84	01.16
15	1	96.87	03.13	98.09	01.91	98.77	01.23
	2	97.77	02.23	97.80	02.20	98.09	01.91
	3	98.84	01.16	98.77	01.23	97.80	02.20
	4	98.59	01.41	97.29	02.71	98.77	01.23

(continued)

Table 2 (continued)

D. size	Fold	Sample dataset 1		Sample dataset 2		Sample dataset 3	
		CA	CE	CA	CE	CA	CE
	5	98.59	01.41	97.29	02.71	98.77	01.23
20	1	97.29	02.71	97.80	02.20	99.04	00.96
	2	98.07	01.93	97.80	02.20	99.04	00.96
	3	96.68	03.32	96.87	03.13	97.55	02.45
	4	98.84	01.16	98.09	01.91	97.55	02.45
	5	99.07	00.93	99.73	00.27	98.77	01.23

Table 3 Comparison with the existing models

Models	CA	CE (%)
DT (sample dataset 1)	096.93	003.07
DT (sample dataset 2)	094.82	005.18
DT (sample dataset 3)	098.66	001.34
[a]Proposed (D. size = **5**)	099.94	000.06
[a]Proposed (D. size = 10)	099.99	000.01
[a]Proposed (D. size = 15)	098.84	001.16
[a]Proposed (D. size = 20)	099.73	000.27
Naive Bayes[b]	092.27	007.73
Rep tree[b]	089.11	010.89
Random tree[b]	088.98	011.02
Random forest[b]	089.21	010.79
DTLW IDS[b]	098.38	001.62

[a]Best result at the threshold, [b]Sindhu et al. [24]

5.4 Discussion

In this segment, a detailed discussion of the evolutionary decision tree-based IDS have been presented. Figure 2 presents the architecture of the process flow of the proposed evolutionary decision tree-based IDS. The proposed IDS resolves the problem of the small disjunct in decision tree and rule-based intrusion detection system. The proposed IDS is tested simulated in the environment of Windows 8 operating system, i5 processor, 4 GB RAM, JDK 1.8, and Keel toot. The KDD Cup dataset is used for training and testing the proposed system. The KDD Cup Dataset is taken from the UCI repository. The system is evaluated using the k-fold cross-validation technique, and for training and testing, three sets of training and testing sets of subsample are taken from the KDD Cup dataset. In k-fold cross-validation, the whole data set is randomly partitioned into k identical-sized subsamples. Out of k subsamples, a subsample is used as a testing set, and the remaining k-1 subsamples are used for training purpose. The cross-validation procedure is repeated k-folds. We can say that the each of k subsamples is used exactly once as a testing set. After k validation, the outcomes from each are averaged to produce a cumulative outcome. The benefit of using this method over recurrent random subsampling is that all observations are used training and as well as testing. Here, fivefold cross-validation has been used for testing the effectiveness of the system, and the outcomes are recorded in the tables.

The data set is preprocessed before training the model. For preprocessing, outliers and extreme values are removed from the dataset. Outliers are extreme instances different from other instances in the dataset; they may specify an unevenness in a measurement, investigational mistakes or uniqueness in the data patterns. In other words, it is an observation that diverges from overall instances. Identification outlier is an important task in preprocessing because of the following reasons: It may represent bad data, e.g., the pattern may be encoded incorrectly, or it may affect the performance of the simulation. If an instance is observed as an outlier, then the outlier value must be removed from the dataset (or corrected if probable). To handle the outlier and extreme values, we use the concept of interquartile range from the Weka. In the experimental work training set is used to train the system then the GA is used to improve the rules of the trained system. For the genetic operations, 1% mutation probability and 80% crossover probability is used. Decision rule of decision tree is evaluated using the following small disjunct size (D. Size) 5, 10, 15, and 20.

The maximum GA is iterated 100, 150, and 200 and 250 number of times or the stopping criteria has been matched. The result established in the result provides the best result at 200 generations. CA and CE are used as a performance measure. The proposed decision tree and genetic algorithm-based IDS is simulated on all three samples of the KDD CUP data set, and the results are shown in Tables 1 and 2. The outcomes of the system are equated with the different system like Naïve Bayes, Random Tree, C4.5 tree, Random Forest, Rep Tree, etc., and is presented in Table 3. It is worth to mention that our evolutionary decision tree-based IDS delivers better accuracy and error rate as compared to the decision tree-based system. The high accurateness and the low error rate indicate the efficiency of the system. Table 1 shows the classification accuracy and classification error of the decision tree classification on all three samples. The decision tree is validated using the fivefold cross-validation. The decision tree classifier delivers the best performance 98.66% accuracy, and 01.34% error on sample 3 and its worst performance is 93.88% accuracy and 06.12% classification error on sample 2.

The best performance observed at threshold level 5 is 99.94% classification accuracy, 0.04% classification error and the worst performance observed at the same threshold is 96.87% accuracy, 03.13% classification error. The best performance observed at threshold level 10 is 99.99% classification accuracy, 0.01% classification error and the worst performance observed at the same threshold is 97.29% accuracy, 02.71% classification error. The best performance observed at threshold level 15 is 98.77% classification accuracy, 01.23% classification error and the worst performance observed at the same threshold is 96.87% accuracy, 03.13% classification error. The best performance observed at threshold level 20 is 99.73% classification accuracy, 00.27% classification error and the worst performance observed at the same threshold is 96.68% accuracy, 03.32% classification error. The third sample provides an improved result at threshold label 10 with accuracy 99.99% and error 0.01% and the second best result is at threshold level 5 is 99.94% classification accuracy, 0.04% classification error.

Table 3 shows the Comparison of Evolutionary Decision Tree-based IDS with Existing Models with C 4.5 decision tree, Decision Tree-based Lightweight (DTLW) IDS, Rep tree, Random tree, Naive Bayes, and Random tree. It is worth to mention that the proposed evolutionary decision tree-based IDS deliver better results in contrast to the existing systems as shown in Table 3. The proposed systems best performance is 99.99% classification accuracy, and the 0.01% classification error is at threshold level 10. The worst performance of the system is 96.87% classification accuracy, and the 03.13% classification error is at threshold level 5 and 20.

6 Conclusion

In this article, an IDS has been proposed which is based on the concept of Decision Tree and Genetic Algorithm. The proposed IDS overcomes one of the significant drawback associated with the decision tree-based IDS. The problem with the decision tree-based IDS is the problem of small disjunct. Due to the availability of small disjunct in the IDS, the effectiveness of the system may be compromised. If a small accuracy is comprised due to the availability of such disjunct in such a scenario, the organization may lose the integrity in the digital word. In this effort first, the first training samples are prepreprocessed to remove the outlier records and extreme values. After preprocessing, the decision tree classification algorithm is used to generate a model, then the rule those who come in the category of small disjunct are further optimized by using metaheuristic genetic algorithm. It is worth to mention that the proposed IDS gives a better outcome in contrast to the approaches such as the C 4.5 Decision Tree, Naïve Bayes, etc. Also, it is observed that the nature-inspired algorithm plays a significant role in better decision making. Other nature-inspired algorithms with decision tree-based system may be simulated in the near future.

References

1. Altwaijry H (2013) Bayesian-based intrusion detection system. IAENG transactions on engineering technologies. Springer, Netherlands, pp 29–44
2. Alcala-Fdez J, Sanchez L, Garcia S, del Jesus MJ, Ventura S, Garrell JM, Otero J, Romero C, Bacardit J, Rivas VM, Fernandez JC (2009) KEEL: a software tool to assess evolutionary algorithms for data mining problems. Soft Comput 13(3):307–318
3. Anderson JP Computer security threat monitoring and surveillance. Technical report, James P. Anderson Company, Fort Washington, Pennsylvania, 15 Apr 1980
4. Azad C, Jha VK (2013) Data mining in intrusion detection: a comparative study of methods, types and data sets. Int J Inf Technol Comput Sci (IJITCS) 5(8):75

5. Azad C, Jha VK (2014) Data mining based hybrid intrusion detection system. Indian J Sci Technol 7(6):781–789
6. Azad C, Jha VK (2016) Fuzzy min-max neural network and particle swarm optimization based intrusion detection system. Microsyst Technol 1–2
7. Azad C, Jha VK (2015) Genetic algorithm to solve the problem of small disjunct in the decision tree based intrusion detection system. Int J Comput Netw Inf Secur (IJCNIS) 7(8):56
8. Azad C, Jha VK (2016) A novel fuzzy min-max neural network and genetic algorithm-based intrusion detection system. In: Proceedings of the second international conference on computer and communication technologies. Springer, India, pp 429–439
9. Carvalho DR, Freitas AA (2000) A genetic algorithm-based solution for the problem of small disjuncts. In: Principles of data mining and knowledge discovery 13 Sept 2000. Springer, Berlin, pp 345–352
10. Carvalho DR, Freitas AA (2000, July) A hybrid decision tree/genetic algorithm for coping with the problem of small disjuncts in data mining. In: GECCO, pp 1061–1068
11. Carvalho DR, Freitas AA (2002) A genetic-algorithm for discovering small-disjunct rules in data mining. Appl Soft Comput 2(2):75–88
12. Carvalho DR, Freitas AA (2004) A hybrid decision tree/genetic algorithm method for data mining. Inf Sci 163(1):13–35
13. Holte RC, Acker L, Porter BW (1989) Concept learning and the problem of small disjuncts. In: IJCAI 20 Aug 1989, vol 89, pp 813–818
14. Jiang F, Sui Y, Cao C (2013) An incremental decision tree algorithm based on rough sets and its application in intrusion detection. Artif Intell Rev 40(4):517–530
15. KDD Cup (1999) Data Set. http://kdd.ics.uci.edu/databases/kddcup99/kddcup99.html
16. Khan L, Awad M, Thuraisingham B (2007) A new intrusion detection system using support vector machines and hierarchical clustering. VLDB J Int J Very Large Data Bases 16(4):507–521
17. Koshal J, Bag M (2012) Cascading of C4. 5 decision tree and support vector machine for rule-based intrusion detection system. Int J Comput Netw Inf Secur 4(8):8
18. Liao SH, Chu PH, Hsiao PY (2012) Data mining techniques and applications–a decade review from 2000 to 2011. Exp Syst Appl 39(12):11303–11311
19. Mabu S, Chen C, Lu N, Shimada K, Hirasawa K (2011) An intrusion-detection model based on fuzzy class-association-rule mining using genetic network programming. IEEE Trans Syst Man Cybern Part C Appl Rev 41(1):130–139
20. Mulay SA, Devale PR, Garje GV (2010) Intrusion detection system using support vector machine and decision tree. Int J Comput Appl 3(3):40–43
21. Muniyandi AP, Rajeswari R, Rajaram R (2012) Network anomaly detection by cascading kMeans clustering and C4. 5 decision tree algorithm. Procedia Eng 30:174–182
22. Quinlan JR C4. 5: programs for machine learning. Elsevier; 28 June 2014
23. Rajeswari LP, Arputharaj K (2008, Oct 31) An active rule approach for network intrusion detection with enhanced C4. 5 algorithm. Int J Commun Netw Syst Sci 1(04):314
24. Sindhu SS, Geetha S, Kannan A (2012) Decision tree based light weight intrusion detection using a wrapper approach. Exp Syst Appl 39(1):129–141

Analysis and Proposal of a Flash Subranging ADC Architecture

Farhana Begum, Sandeep Mishra, Md. Najrul Islam and Anup Dandapat

Abstract The paper describes the architectures of various subranging flash analog-to-digital converters (ADCs) along with the proposal of a novel subranging algorithm. A comparative study of the state-of-the-art designs are made with respect to the figure of merit (FoM) and excess delay (EXD) parameter. Simulations are carried out in generic process design kit (GPDK) 45-nm technology in SPECTRE environment. The proposed subranging ADC shows an overall improvement of power by 94% per conversion cycle as compared to flash ADC. The EXD of proposed subranging ADC shows an improvement by 74% compared to binary flash ADC but however is slower than flash ADC by 34%.

Keywords Binary search based flash ADC · Dynamic architecture and frequency scaling (DAFS) · Flash ADC · Subranging ADC · FoM · EXD

1 Introduction

The usage of battery operated devices like cell phones, laptops and many portable medical instruments has made power consumption a prominent issue. The devices that are commonly used in a chip process all data within the chip. But to interact with the outer world an analog-to-digital converter (ADC) is required. This stands as a fundamental interfacing device to communicate with the real world. For carrying out

---

F. Begum (✉) · S. Mishra · Md. Najrul Islam · A. Dandapat
Department of Electronics and Communication Engineering, National Institute of Technology Meghalaya, Shillong, India
e-mail: farhanajubi@gmail.com

S. Mishra
e-mail: ssandeep.mmishra@gmail.com

Md. Najrul Islam
e-mail: najrulislam095@gmail.com

A. Dandapat
e-mail: anup.dandapat@gmail.com

© Springer Nature Singapore Pte Ltd. 2019

V. Nath and J. K. Mandal (eds.), *Proceedings of the Third International Conference on Microelectronics, Computing and Communication Systems*,
Lecture Notes in Electrical Engineering 556, https://doi.org/10.1007/978-981-13-7091-5_26

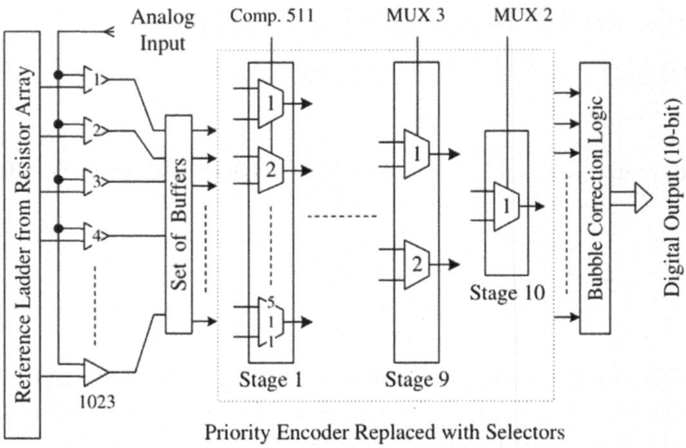

Fig. 1 Conventional (10-bit) flash ADC

high sampling operations, various architectural and algorithmic developments have been carried out in ADC as in [1–3].

Flash and time-interleaved [1] ADCs are used to provide high speed with a sampling rate above 1 GHz. Flash ADCs as shown in Fig. 1 consumes much high power compared to the conventional architecture and is suitable for medium resolution (6–8). So, time-interleaved [2] ADCs are used which consume lesser power however with an increase in resolution. High sampling operations are done basically using the flash, which operates at a sampling rate above 1 GHz but these flash ADCs dissipate the highest power among all the ADCs. Though successive approximation register (SAR).

ADCs dissipate the least power these have lower sampling rate. Subranging flash ADCs as in [4] are used whose operating frequency is more than SAR-ADC but lesser than flash ADC. However, the power consumed by these is more than SAR-ADC. Several subranging algorithms are used to modify the conventional flash ADC architecture. Subranging concept is based on the fact that all voltage levels are not used for most significant bits (MSB) determination and only certain levels are used [5, 6]. The pipelined ADC also can operate at high speeds however it imposes stringent measures with high operating speed.

The several timing mismatch can be reduced through the usage of fast Fourier transform (FFT) as in [7] which leads to power consumption with the usage of overhead. The design comprises a subranging architecture where initial levels are termed as coarse levels and subdivisions are fine levels. At first, the coarse levels are chosen and then the fine levels with the help of multiplexers. Various subranging algorithms like interpolation and folding architecture as shown in [8] are used. It integrated the distributed track-and-hold circuit. Multiplexers (MUXs) select fine levels which also increase the settling time of the multiplexers. With this architecture, a sampling rate of 160 MS/s was achieved with a power dissipation of 84 mW.

Many subranging architectures and pre-charging techniques are used for reducing the reference settling time of ADC.

2 Types of Subranging Algorithms

Many algorithms for subranging architectures are defined which are broadly categorized as the binary search and frequency scaling algorithm. In binary search algorithm, input voltage level is searched using divide and search mechanism. Another is the dynamic architecture and frequency scaling (DAFS) based algorithm [9] which constitutes of flash and binary search ADC. Proposed herein is subranging 5-bit ADC design. Improvement in the binary searching is made to achieve a performance, which is in between traditional flash and binary search.

2.1 *Flash ADC Algorithm*

The flash ADC comprises parallel architecture which makes the conversion of analog signals to digital signals. The flash conversion design comprises a number of comparators connected in parallel followed by a priority encoder as shown in Fig. 1. It constitutes 2^N − 1 of comparators for N resolution [3, 10, 11]. The parallel operation of the same improves the speed of operation. However, the increase in the resolution causes an exponential increase in power and area. As such, subranging schemes came into the picture. The figure of merit (FoM) and excess delay (EXD) also varies with it as shown in the simulation results of Sect. 3.

2.2 *Binary Search Algorithm*

Binary conversion scheme [7] involves the connection of all the voltages in a divide and search way. For a 5-bit ADC, the middle of the voltage level is taken as divided by two. 1/32 levels are used. At first, the MSB bit is determined using the middle bit of the comparator. 15/32 voltage comparator determines the first MSB.

Next, a multiplexer is decided which will help to choose if the first level of multiplexers is chosen or next level. Then again, the same process for the next bits, the multiplexer with 14/32 levels is needed. This means at one cycle of comparison, not all the stages are there and hence the energy required is less. The conversion delay includes the comparators reset time and will be very high. The FoM for this conversion is low compared to flash conversion. The delay is however more compared to the conventional flash as it includes a step by step conversion.

2.3 DAFS Algorithm

The dynamic architecture frequency scaling (DAFS) technique is based on two architectures, flash ADC and binary search ADC [12]. The previously illustrated are found to have a clear power and speed trade-off for resolutions under 6 bit. The illustration of the same two algorithms has been done in the previous subsections. The flash ADC's conversion delay is almost same as that of comparators plus the reset time of the comparator. This section draws a relation with dynamic frequency scaling.

The high-frequency scaling is possible only though flash conversion. DAFS [13, 14] is proposed to utilize the benefits of binary search and flash converter. An OR cell is used, which decides the converter (flash or binary converter). By unifying the conversion which is to be taken, DAFS performs its search. For example, since the binary search conversion is made, the counter uses the rising edge of the MUX output. The mapping codes are digital values which configure the comparator outputs. When the ADC is operating, mapping switches are switched every cycle to cancel the varying offset. By shorting the reference input with the reference voltage binary search can be conducted.

3 Proposed Subranging ADC and Analysis

3.1 Proposal of 5-Bit Subranging ADC

Based on the subranging concept, which entails that all reference levels are not necessary for the determination of MSB. For a 5-bit subranging ADC, three MSB are determined in the first level which is the coarse level. Next, two LSB is determined from the finer levels. The finer levels are determined using multiplexers. The analysis is done in the subsequent section in CADENCE using generic process design kit (GPDK) 45-nm CMOS technology at a temperature of 27 °C with a power supply of 1 V. In the design of subranging architecture, initial levels are termed as coarse levels and subdivisions of the coarse levels are termed as fine levels. A two-step time-interleaved ADC is introduced by [3, 9] to reduce the number of comparators. Two fine ADCs of the same size are connected after the MSB generation with a time gap of 1 phase required for settling. It reduces power dissipation to a great extent but additional delay limits the resolution limit. Multiple MSBs are determined in the first step to speed up the process.

3.2 Results and Discussions

F represents flash ADC, *BS* represents binary search ADC and *SF* represents subranging flash ADC in Fig. 2. Subranging technique is using the selected voltage

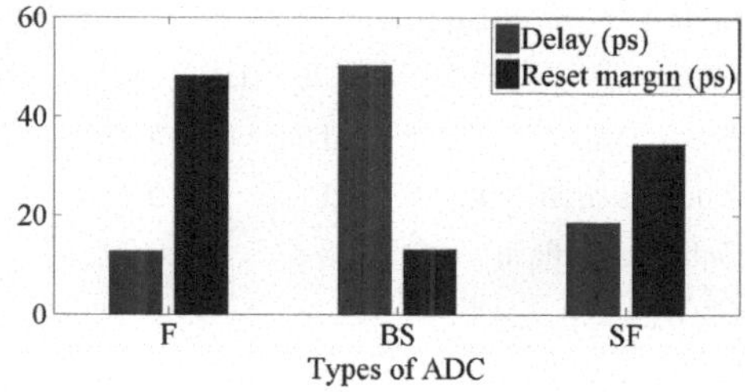

Fig. 2 Excess delay (EXD) representation

levels generated after the coarse comparisons. Power of more than 50% is wasted in sample and hold circuit. Thus, several efforts have been made to reduce the height (number of comparators per stage) of fine comparators eventually to reduce power consumption. These design considerations certainly improve subranging architecture performance but organizing fine comparators in terms of both height and width still remains a challenging task.

Excess delay (EXD) is a prime metric calculated for both the cycle delay and reset margin. As shown in Fig. 2 it reduces the EXD generated in the binary search subranging architecture. FoM is very high for the flash type of conversions. The excess delay is suppressed due to faster discharge rate of comparators which is primarily due to searching of the finest voltage. Though the binary search algorithm is based on dividing the voltages and searching at each step, it is slow due to multiple number of conversions. The power required for conversion is not proportional to the power required for parallel conversions. The searching process increases the delay in binary search based ADC. At each conversion step, one comparator is activated in each step. The flash ADC poses as a major drawback by using an array of comparators however increasing the number of comparisons though with a lesser time but increased number of conversion energy per step [8].

The subranging ADC conversion consists of S/H circuit, coarse conversion and fine conversion. The operation step is shown for a 5-bit ADC. At first the sampling switch is closed and then the applied signal is sampled to capacitance. At the next cycle of operation applied input signal is sampled to capacitance and the switch opens and starts at the first cycle. 3 MSB bits are determined from this cycle. The subsequent cycle determines the next bits. The next two mode fine bits and in this way the process goes on. As the timing diagram shows, the subranging ADC conversion consists of S/H circuit, coarse conversion and fine conversion steps. The operation is shown for a 5-bit ADC. The sampling switch is closed and the applied input signal is sampled to input and the switch opens at the start of the cycle P[N] as shown in Fig. 2. At the next cycle of operation, P[N], the 3-MSB bits are determined and then in the next cycle the fine conversion the next 2-LSB bits are determined. Then the cycle P[N+2] determines the next two fine mode of bits and in this way, the process goes on. Whereas in the next binary bits of conversion for binary search at one cycle, only the first MSB bit is activated and in the next cycle, the next two sets of comparator are activated. In this way, the process goes on. As the output is greatly dependent on the number of voltages levels. If the conversion at the cycle P[N+1] is

Table 1 FoM variation with cycles of conversion for 5-bit ADCs

CLK	Cycle 1 (fJ)	Cycle 2 (fJ)	Cycle 3 (fJ)	Cycle 4 (fJ)	Cycle 5 (fJ)
Flash ADC	5.71	9.37	22.49	33.65	39.72
Binary search ADC	0.21	0.73	1.67	2.01	2.37
Subranging flash ADC	0.18	0.29	1.51	1.99	2.12

prolonged, the flash operation continues for the cycle. The amount of EXD produced cancelled by flash operation decreases until the next cycle of operation. The number of subranging stages is proportional to the resolution of ADC [3]. The searches per second are also an important parameter of design. Thereafter, the EXD produced by flash operation decreases compared to the rest. The searches depend on the number of flash conversions. The ratio is given for 5-bit flash ADC

$$\textbf{Searches (Flash ADC)} = \text{Number of Flash conversions/Number of searches } (32/32 = 1) \quad (1)$$

$$\textbf{Searches (Binary Search ADC)} = \text{Number of Flash conversions/}\log(M/2) + 1(32/4 + 1 = 9) \quad (2)$$

$$\textbf{Searches (Subranging Flash ADC)} = \text{Log}\, M - 1(4 - 1 = 3) \quad (3)$$

where M represents the number of subranging levels.

Equations (1)–(3) shows the stage requirements (coarse + fine) of compared flash architectures. Design analysis of 5-bit flash ADCs is discussed and the performance is summarized in Table 1. Figure 2 shows the EXD plots of various subranging designs. The usage of higher number of fine levels using binary search scheme degrades the conversion speed which may dither the conversion at higher sampling frequency. Subranging flash, on the other hand, has two stages and can perform with lower additional delay of 5.8 ps. Faster flash ADC dissipates higher cycle energy (FoM of 39.72 fJ in just five cycles) that makes it less efficient at higher resolution converters. The proposed subranging flash provides the best energy-delay trade-off at medium to high resolutions.

The number of flash converters increases and the number of binary search operation. The delay produced is much more for binary operation, however, it gets reduced to cycle of 1 for flash ADC. Second, the power ADC consumption is high. So, DAFS-based technique is used. The conventional ADC must satisfy the results. A hybrid of ADCs are used which have low power MSB decision and also have the fine reference transition. The settling can be completed with the first cycle whereby each channel performs the ADC operation. As explained previously, the EXD is monitored in real time. However, it can cause challenging to install long timing thresholds. Since

Table 2 Comparative analysis of ADCs available in the literature

References	[10]	[11]	[12]	[13]	[14]	Proposed
Technology (nm)	55	65	130	130	130	45
Speed (MS/s)	1000	1000	1000	600	200	2500
Supply voltage (V)	1.2	1.1	2.2	1.2	1.2	1
Power (mW)	16	9.9	9.1	38	38	3.4
Figure of merit (FoM) (fJ)	197	278	–	552	552	2.12
SNDR (dB)	43.5	32.8	–	53	53	31.7

the output is greatly determined on the number of voltage levels, flash operation continues for another cycle. Flash ADC converts this in the second phase.

The converted waveforms of different flash architectures are analysed and results are depicted in Fig. 2. EXD presents per conversion over various operating cycles. The timing algorithms are described using the expressions below:

(i) **Flash conversion delay** = *Reset time of the comparator* + *Comparator conversion time.*
(ii) **Binary search delay** = *Delay at each step of conversion.*
(iii) **Dynamic architecture and frequency scaling timing:** *Delay at each step+binary units of conversion.*
(iv) ***Subranging flash ADC*** = *Multiple MSB conversion at each step.*

Table 2 draws a comparative analysis of ADCs available in literature. The proposed 5-bit subranging ADC has better conversion rate compared to the other ADCs. The FoM is the lowest at around 2.12 fJ/conversion and SNDR is 31.7.

4 Conclusion

The paper draws a comparative analysis between different flash ADC algorithms, namely conventional flash ADC, binary search ADC, DAFS and subranging flash ADC algorithm. The power of flash ADC is more with increase in the number of conversion cycles as seen from simulation results. The simulation is done with 5-bit ADC which comprises flash ADC, DAFS and binary search ADC. Flash ADC dissipates 39.72 fJ followed by binary search ADC which uses 2.37 fJ and subranging flash ADC uses 2.12 fJ. The power consumed by subranging flash ADC is 94% lesser than flash ADC and 12% lesser than binary flash along with an improvement of speed.

EXD is the sum of reset margin and delay and is a measure of speed efficiency of the architecture. The excess delay is inversely proportional to delay. The reset margin is highest for flash ADC. That means flash ADC has the least delay due to its parallel processing architecture. The subranging flash ADC has an average delay of 18.2 ps. Whereas, the delay of binary search is 50.2 ps on an average as seen to be 12.6 ps in flash ADC. Thus, the subranging ADC has a better performance over flash and binary search ADC.

Acknowledgements This research was supported in part by the Ministry of Electronics and Information Technology (MeitY) under Project SMDPC2SD 9(1)/2014-MDD, Government of India.

References

1. Dong VT et al (2017) Effect of offset mismatch in time-interleaved ADC circuits on OFDM-BER performance 64(8)
2. Martens E et al (2018) A 69-dB SNDR 300-MS/s two-time interleaved pipelined SAR ADC in 16-nm CMOS FinFET with capacitive reference stabilization 53(4)
3. Brandt BP, Lutsky J A 75-mW, 10-b, 20-MSPS CMOS subranging ADC with 9.5 effective bits at nyquist. In: IEEE J. Solid-State Circuits, vol 34, pp 1788–1795
4. Taft RC, Tursi MR (2001) A 100-MS/s 8-b CMOS subranging ADC with sustained parametric performance from 3.8 V down to 2.2 V. IEEE J Solid-State Circuits 36:331–338
5. Huber DJ, Chandler RJ, Abidi AA (2007) A 10b 160MS/s 84 mW 1 V subranging ADC in 90 nm CMOS. In: ISSCC digest of technical papers, vol 454–615
6. Chang HY, Yang CY (2016) A reference voltage interpolation-based calibration method for flash ADCs. IEEE Trans Very Large Scale Integr (VLSI) Syst 24:1728–1738
7. Vander Plas G, Verbruggen B (2008) A 150 MS/s 133 W 7 bit ADC in 90 nm digital CMOS. IEEE J Solid-State Circuits 43(12):2631–2640
8. Yoshioka K, Saito R, Danjo T, Tsukamoto S, Ishikuro H (2015) Dynamic architecture and frequency scaling in 0.8–1.2 GS/s 7 b Subranging ADC 50(4)
9. Figueiredo PM (2006) A 90 nm CMOS 1.2 V 6b IGS/s two-step subranging ADC. In: ISSCC digest of technical papers, pp 2320–2329
10. Chung YH, Tsorrng J (2015) A 16-mW 8-Bit 1-GS/s digital-subranging ADC in 55-nm CMOS. IEEE Trans Very Large Scale Integr Syst 23(3):557–566
11. Danjo T, Yoshioka M, Isogai M, Hoshino M (2014) A 6-bit, 1-GS/s, 9.9-mW, interpolated subranging ADC in 65-nm CMOS. IEEE J Solid State Circuits 49(3):673–683
12. Zahrai SA, Zlochisti M, Dortz NL, Onabajo M (2017) A low-power high-speed hybrid ADC with merged sample-and-hold and DAC functions for efficient subranging time-interleaved operation. IEEE Trans Very Large Scale Integr Syst 25(11):3193–3205
13. Venca A, Ghittori N, Bosi A, Nani C (2016) A 0.076 mm2 12 b 26.5 mW 600 MS/s 4-way interleaved subranging SAR-ADC with on-chip buffer in 28 nm CMOS. IEEE J Solid State Circuits 51(12):2951–2962
14. Chu M, Kim B, Lee BG (2015) A 10-bit 200-MS/s zero-crossing-based pipeline ADC in 0.13-μm CMOS technology. IEEE Trans Very Large Integr Syst 23(11):2671–2675

Competency-Based Talent Management—An Effective Management Tool

Thaya Madhavi and Rajesh Mehrotra

Abstract Management Executives and Human resource management as a whole have predominantly focused its attention on talent management, which involves acquiring, hiring and retaining well trained and talented employees in particular. To reap benefits for the well-being of the organizational success, the managements need to engage or appoint highly qualified employees. To instill confidence in the workforce and to prove their talents, they are to be aligned with company strategy, and to prove their leadership criteria across all its functional areas, they have recognized the precise field of competencies run on analytical, technical, educational, and experience-based to consolidate its growth on the business front. The essence of this paper is summing up of talent strategy management embedded therein with, competency profiling having its effects on critical areas of operation.

Keywords Competencies · Effective management · Talent management · Organizational success · Human resource

1 Introduction

Normally, the business executives tend to implement high-quality business talents over their counterparts to face bitter competition in the international market and consolidate their position exploring new opportunities. The true success is derived when they adopt long term trends, while they anticipate and capture a new type of opportunities in the market. To speed up their talents in the business arena, they combine all aspects of talent mindsets that involves all processes and systems contributing for the overall development of the organization [1].

T. Madhavi (✉) · R. Mehrotra
School of Business and Management, Jaipur National University, Jaipur 302017, Rajasthan, India
e-mail: madhavimbabit@gmail.com

R. Mehrotra
e-mail: rajeshmehrotra7@rediffmail.com

© Springer Nature Singapore Pte Ltd. 2019
V. Nath and J. K. Mandal (eds.), *Proceedings of the Third International Conference on Microelectronics, Computing and Communication Systems*, Lecture Notes in Electrical Engineering 556, https://doi.org/10.1007/978-981-13-7091-5_27

The strategic talent management promotes the following need-based opportunities.

- To be always proactive and reactive. To bring out changes as and when required.
- To point out exceptional skills in the employees to minimize training costs.
- Identification of highly qualified employees at the recruitment level, use their services at strategic points of operation to promote industrial competencies.

Competency-based talent management can develop both performance and productivity by recognizing key personalities of top performers and how those behavior differ from average employees.

1.1 Talent Management as a Tool for Effective Management

Talent management is considered as a tool to establish human resources and business planning processes to reach their business goals by integrating talent mindsets in the organization. Thus, in itself, improves the performance and potentiality of the people at the organization level. The performance concerns with the focus based on past and present, while potential represents the future. The potential exists among people, but it should be identified and developed [1]. The talent management is a combined effect of difference facility of fields, such as Ethos (embedded values and behavior), focus (knowing their job descriptions), positioning (acquiring high-level posts at management level in the organization, but not human resource initiatives), structure (this involves creating tools, methods, and techniques with accountability on their work), and system (to have long-term holistic approach in bringing out periodical change into organizational set up, so as to meet the challenges from its competitors in the market). Advanced human resource talent management starts with the conclusion in mind. The organization requires talented human resources. For this, three things have to be considered how the human resource is increasing value to the business at present, in future and how it is closing the gap in dealing with risk, opportunity, and performance [2].

1.2 Talent Management as a System

In talent management, the efforts put in by workforce are dissipated to all fields of activities, which fit in the talent management system. The integrated talent management have more facilitates and reach up to higher-ups in the organization. It would often attract the attention of board level management and senior teams in the organization. The talent planning runs side by side with business planning, to create people and strategy integration. To achieve this integration, five elements are needed which are need-based, data collection, planning, activities, and results. By introducing this type of system, the talent management can be regarded as a strategic differentiator, rather than a standard tool of human resource processes, in case right conditions, context, time scales, and offerings exist in the first place with the integration of system and alignment, the talent management efforts are fit in for the purpose [1].

2 Talent Management Process

Many human resource groups work on talent management strategy to map out their goals and ties with them for the management strategic plans and goals. It is obligatory on the part of the management to set up goals and plans for implementing organizational activities through human resource team, otherwise, talent management activities have no role to play.

The important steps included in the talent management processes are shown in Fig. 1.

Step 1. Identification of organizational goals and priorities, inviting new directions and initiatives for the organization.

Step 2. To earmark the organizational trendsetters and challenges.
It includes challenges both inner and outer such as highly competitive job market, change of rules and regulations, introducing updated know-how, etc.

Step 3. To perform a gap analysis.
To set human resource goals for the preceding period in support of the organization or else encounter the risk of not addressing the gap.

Step 4. To identify human resource priorities and targets.

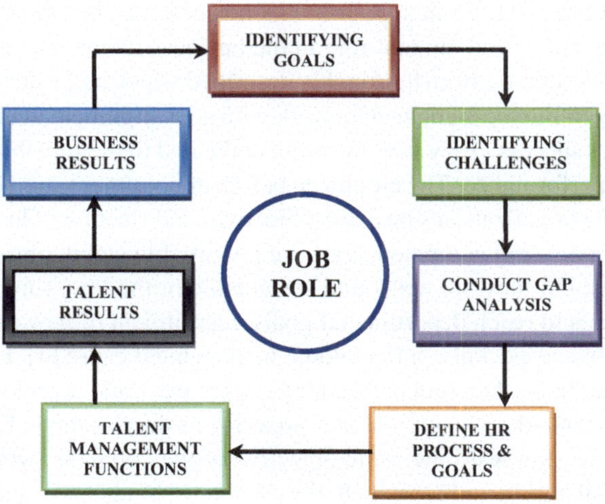

Fig. 1 Talent management process

While doing so, in support of the organization we have to and formulate a business strategy for the coming year to fulfill the goals. We have to link them by exploring ways and means for its implementation effectiveness to derive desired results.

Step 5. Talent management processes/functions to be included in the inventory: To prepare an inventory of talent management processes, bringing out any changes necessary in the existing processes to support the goals set in and addresses them.

Step 6. Identify results and communicate its success or contributions, if any.

It is equally important to measure and evaluate the successes both in its implementation and effectiveness, to have the desired effect on the organization performance, summing up the difference between successes and failures in a specific job implementation, the right and talented person will be put in a right place and such a person will be allowed to build in morals and work efficiency, skill, and dedication, doing his best to the organization. A top form of talent management indicates that it is involved in a continuous connection with other disconnected processes that shows it is not working alone but engaged in all people management practices [3].

3 Factors Which Influence Talent Management

3.1 The Individual Goals are Kept in Line with Corporate Strategy

The potential talent management coincides with the already framed company's strategic plan for the organization in support of its business needs. However, alignment of company goals plays a dominant role in the business sector. The job roles of the individual employees have been thoroughly described which add value for the organization. When an employee is engaged in work with goal alignment, it ultimately leads to the company's success. They become more dedicated to the job obligations with a display of high performance. The employee before appointment should be explained in clear terms, the organization's business objectives. Fabrication of human resources systems to make the core competencies more potential in stabilizing organizational values and goals, but they are not found sustainable in the long run. Implementing core competencies to reach departmental goals, recruitment of new employees have become productive, especially in the middle management cadre [4]. Further sharing of responsibilities helps departmental heads to invent ways and means for viable business cooperation and identification of new areas for its development. Everyone in the organization, if work toward the same objective of the company, would strengthen the organizational agility and for which, the company should focus on the following points.

- Efforts on the part of employees, for reaching the company's distinctive goals.
- They must be made more responsible and accountable to the company's specific goals.
- Strengthening accountability, by taking appropriate measures, to articulate goals clearly set before employees. Various skills are required among employees for diverse activities.

Frequent turnover of the manpower is the order of the day, in organizations nowadays. The candidates are content with short projected living due to lack of fit, which has to be looked into by recruiting and retaining the right persons, is a real challenge for human resource. This needs continuous education of its employees by imparting training at frequent intervals to update their technology and knowledge [5].

3.2 *Internal Talent Pools Among Employees to be Tapped*

Strategically planned organizations plan their activities with the skill development of their workforce with the right competencies. They have clear insight into the working model of its employees and place at strategic points to meet the urgent needs in the market place. At times, the management might find it difficult to assess the potential candidate due to the shortage of time and change in management or leadership or prevailing industry conditions. Furthermore, the prolonged leadership is found to be too risky. It often tends to raise questions about the availability of the internal talent pool.

Successful talent management always looks for available experts in the company where they could be considered as trustworthy and a reliable source of talent who in turn become pioneers in the planning process. Abled and talented pools can be drawn among the workforce which ensures high quality of performance at all levels of the organization. In order to maintain expertise in the workforce, the trained and experienced employees must be retained who can play a leadership role in the institution they work.

Human resource development has conducted research in this field on how the organizations use talent strategy by selection of best people for placement in vital positions, rather selecting explicit candidates for clear positions (Fig. 2) [6].

3.3 *Change in Employees Requirements and Expectations*

Nowadays, the organizations are greatly relying upon talent management strategies and practices. Employees in the present times are beset with certain qualities while working in various fields. They are the following:

Fig. 2 Skilled talent pool

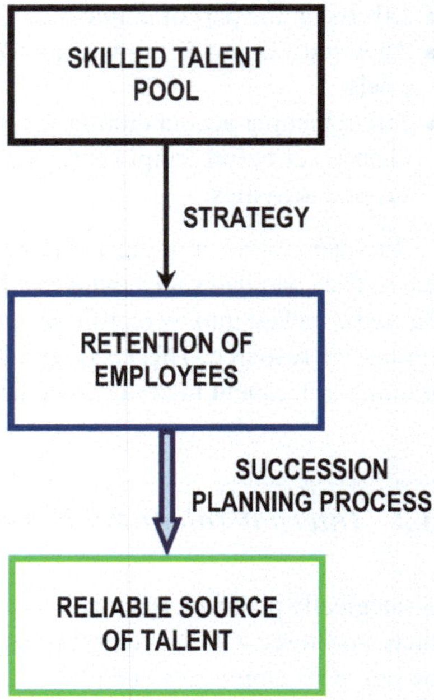

- Display often challenging qualities and play a positive role in meaningful work.
- They are found dedicated to their profession rather than an organization which is a sign of progress.
- Having a low profile over traditional structure and authority
- They are more concerned with balance work time period
- Prepare to come forward to take up the challenging tasks of their career and development.

With such evincing views of the employees with a myriad of challenges, it would be difficult to capture the heart and minds of today's workforce as well.

3.4 Disengagement of Information Centers and Enhancement of Coordinated Efforts

To corner success, the business leaders must able to overcome the organizational silos, to prevent leakage of company secrets from any source. To ensure quick delivery of goods to the customers they must be more flexible, knowledgeable, and experienced or even proactive in their business dealings with the people in contact with them. Some new innovations need the involvement of the right person at the right time, to tap the business. A collaborative atmosphere is created to make available the tools with transparency. These collaborative efforts need the involvement of employees and management with special emphasis on experience, interest and special skills and language communication ability. This would facilitate employees to approach various departments and offices for the acquisition of knowledge based experience and managers on their part work on information on talent management decisions to place the business profile of the organization on top.

3.5 To Encourage Pay for Performance Culture

Better perks for better performance is the order of the day and nowadays, it is adopted in many countries. Employees are sufficiently motivated to speed up their knowledge base, to fetch better emoluments from the management. This would facilitate the workforce to prove their mettle in any specific field, of work that ultimately leads to the all-round performance of the institution.

Human resource scholars have examined the concept of management, as they are found interlinked and the knowledge so acquired would advance talent management in Toto. The synthesis of the two concepts of talent management and knowledge management subsequently has been considered by human resource scholars, who have tended to examine how knowledge can be helpful to advance talent management. The recent research shows that synthesis of the two concepts can be done from a knowledge management perspective which covers identifying key employees, creation of knowledge, competencies development, retention, and sharing of knowledge that can be developed through application of talent management principles [7].

Creativity as a specific element of talent along with the environment influences the development of talent, which requires a balanced view of innate and learned talent. Talent management plans in many organizations do not involve low skilled employees who are working for a long time and experts in their fields. Such employees should be considered for their experience and give them sufficient time to develop who help the organization to run efficiently [8–15].

Recent research suggests there is a need for consideration of involving corporate social responsibility as a significant component in decision-making while planning talent management. Many organizations recognize the corporate social strategy to recruit and retain highly talented employees. It is a false notion that employees choose a company only for money but, select a socially reputed company and give importance for the quality of relationship in an intra-level organization such as confidence, belief, and outgrowth of mannerism should be given importance. Employees are motivated for various reasons in socially responsible companies as CSR activities differ in different countries [9, 16–21].

4 Conclusion

Business establishments who outperform their competitors realize the importance of strategic management in building character in employees while discharging their business obligations. However, business executives make use of analysis and diagnostic tools to reach beyond normal feelings while estimating the overall performance of the workforce. Imparting training courses frequently to the employees or workforce would bring out the benefit of strategic talent management processes, which lead to sources of the organization. The talent in the field of competency is well recognized and placed on priority in the business front. It is also considered as a strategic differentiator and its management; predominantly take more of a strategic role.

The management of talent has become a part and parcel of business culture, the human resource executives managing those initiatives, shot into predominance at a short notice. Further, the establishment of talent management and its advocation is not that easy task, but in course of time, it brings out laurels to both human resource and the organization as well. Keeping the talent management best practices in place, we can build a world-class workforce, which is aligned, motivated, and inspired to deliver exemplary results to the organization and achieve strategic goals set for the purpose. Future research could examine the link between corporate social responsibility and talent management, which would be successful in emerging markets of developing countries to retain employees.

References

1. Ashton C, Morton L (2005) Managing talent for competitive advantage: taking a systemic approach to talent management. Strat HR Rev 4(5):28–31. https://doi.org/10.1108/14754390580000819
2. Ulrich D, Younger J, Brockbank W, Ulrich M (2012) HR talent and the new HR competencies. Strat HR Rev 11(4):217–222. https://doi.org/10.1108/14754391211234940
3. Reilly P (2008) Identifying the right course for talent management. Public Pers Manag 37(4):381–388. https://doi.org/10.1177/009102600803700401
4. Rutledge L, LeMire S, Hawks M, Mowdood A (2016) Competency-based talent management: three perspectives in an academic library. J Libr Adm 56(3):235–250. https://doi.org/10.1080/01930826.2015.1105051
5. Sharma R, Bhatnagar J (2009) Talent management-competency development. Ind Commer Train J 41(3):118–132. https://doi.org/10.1108/00197850910950907
6. Tarique I, Schuler RS (2010) Global talent management: literature review, integrative framework, and suggestions for further research. J World Bus 45:122–133. https://doi.org/10.1016/j.jwb.2009.09.019
7. Whelan E, Carcary M (2011) Integrating talent and knowledge management: where are the benefits? J Knowl Manag 15(4):675–687. https://doi.org/10.1108/13673271111152018
8. Tansley C (2011) What do we mean by the term 'talent' in talent management? Ind Commer Train 43(5):266–274. https://doi.org/10.1108/00197851111145853
9. Vaiman V, Scullion H, Collings D (2012) Talent management decision making. Manag Decis 50(5):925–941. https://doi.org/10.1108/00251741211227663

10. Luo Y, Tung RL (2018) A general theory of springboard MNEs. J Int Bus Stud 49(2):129–152. https://doi.org/10.1057/s41267-017-0114-8
11. Henningsson S, Yetton PW, Wynne PJ (2018) A review of information system integration in mergers and acquisitions. J Inf Technol 1–49. https://doi.org/10.1057/s41265-017-0051-9
12. Madhavi T (2007) Indian warehousing of agricultural commodities–preservation of stocks and implementation WDRA 2007. In: INROADS (Int J Jaipur Natl Univ), vol. 06, no. 01, January–June 2017. ISSN: 2277-4912
13. Caligiuri P (2014) Many moving parts: Factors influencing the effectiveness of HRM practices designed to improve knowledge transfer within MNCs. J Int Bus Stud 45(1):63–72. https://doi.org/10.1057/jibs.2013.52
14. Achrol RS, Kotler P (2012) Frontiers of the marketing paradigm in the third millennium. J Acad Mark Sci 40(1):35–52. https://doi.org/10.1007/s11747-011-0255-4
15. Teece DJ (2014) A dynamic capabilities-based entrepreneurial theory of the multinational enterprise. J Int Bus Stud 45(1):8–37. https://doi.org/10.1057/jibs.2013.54
16. Zhao M, Park SH, Zhou N (2014) MNC strategy and social adaptation in emerging markets. J Int Bus Stud 45(7):842–861, Sept 2014. https://doi.org/10.1057/jibs.2014.8
17. Luftman J, Lyytinen K, Zvi TB (2017) Enhancing the measurement of information technology (IT) business alignment and its influence on company performance. J Inf Technol 32(1):26–46. https://doi.org/10.1057/jit.2015.23
18. Nurmi N, Hinds PJ (2016) Job complexity and learning opportunities: a silver lining in the design of global virtual work. J Int Bus Stud 47(6):631–654. https://doi.org/10.1057/jibs.2016.11
19. Madhavi T (2016) Prerequisite of green HRM for organizations long time sustainability. Int J Res Manage Sci Technol 4(1):54–57. ISSN: 2321-3264
20. Kumar V, Loonam J, Allen JP, Sawyer S (2016) Exploring enterprise social systems & organisational change: implementation in a digital age. J Inf Technol 31(2):97–100. https://doi.org/10.1057/jit.2016.13
21. Stern I, James SD (2016) Whom are you promoting? Positive voluntary public disclosures and executive turnover. Strat Manag J 37(7):1413–1430. https://doi.org/10.1002/smj.2393

Design of Light-Sensing Automatic Headlamps and Taillamps for Automobiles

Ankitanshu Swaroop, Abhishek Kumar, Sailaab Tirkey, Kumari Neelam, Deepak Prasad and Vijay Nath

Abstract This paper demonstrates a model of Light-Sensing Automatic Headlamps and Taillamps for Automobiles. Light-dependent resistor and other basic components are used for the basic functioning of the model. For avoiding the human eye effects, a prototype model has been developed for dimming down the bright headlight of the vehicles when opposite side vehicles are coming. Dimmer is a simple electronic circuit that consists of light sensors (LDR) and control circuit. It senses the switches headlight and function according to the condition applied. This circuit switches convert light intensity from high beam to low beam automatically and also minimized the glare effect by sensing the light intensity value and also eliminate the manual switches which are in control of diver. Drivers are not able to always handle this system while this designed circuit will do perfectly whenever is required and also help the driver to drive the vehicles on the road in fast speed without any barrier. The proposed circuit is designed and tested and it is working and sensing up to a distance of 147 m.

Keywords Light-dependent resistor · Voltage divider circuit · Transistor as a switch

1 Introduction

The requirement of implementation of Automatic Headlamps and Taillamps can be explained by the increasing number of vehicles on the road. Elaborating, rapid advancement in the automotive industry and in transportation systems has resulted

A. Swaroop · A. Kumar · S. Tirkey · K. Neelam · D. Prasad (✉) · V. Nath
VLSI Design Group, Department of ECE, Birla Institute of Technology, Mesra, Ranchi, India
e-mail: prasaddeepak007@gmail.com

K. Neelam
e-mail: kumarineelam375@rediff.com

V. Nath
e-mail: profvnath@gmail.com

© Springer Nature Singapore Pte Ltd. 2019

V. Nath and J. K. Mandal (eds.), *Proceedings of the Third International Conference on Microelectronics, Computing and Communication Systems*, Lecture Notes in Electrical Engineering 556, https://doi.org/10.1007/978-981-13-7091-5_28

in quicker means of travel. Safety is an important parameter in today's every vehicle According to survey done by the "Institute for Traffic Accidents Research and Data Analysis", Japan about 70% the road accidents occur at night time or foggy condition driver/motorists face huge problems due to high light beam which falls directly into their eyes. Their effect is that those people become temporary blindness, glare, and fading effect of the image for a while and sometimes causing accident leading to loss of many lives. Hence, there is a need to design and construct a prototype of this device that automatically dims the headlights for oncoming vehicles using light-dependent resistor-sensing technique to help solve this problem. The aim of this project is that the driver can judge the vehicles coming from the front side and stop collision. The logic behind developing new headlight system is simply to turn the headlight according to the rotation of the steering. The traditional vehicles has the manually operated headlights, means if the person is driving a vehicle and if any another vehicle comes in front of his vehicle, then he had to manually switch on the headlights. That may create the disturbance to that person while driving, and increases the possibility of the accidents. Also if the driver has got any curve road, then while passing through that curve road the traditional headlights can lighten only that region of the road which is in front of it and it cannot provide the equal brightness on the complete road. This may cause accidents. So to avoid all these problems, here, we are presenting new models that automatically increase the brightness of the headlights when another vehicle is coming in front of it. This is done with the help of the LDR sensor [1, 2].

Balaji Wattamwar in April 2016 investigated "Automatic Headlight Beam Shifter for Controlling High Beam to Low Beam of Vehicles". This system operated by the synergy of a microcontroller and a sensor, which make the system more scalable and advance to automatically shifting the upper and lower beam of the vehicle coming from the opposite direction and when it will be needed [3]. Hence, an automatic headlight beam shifter can play a crucial role in shifting the headlights from driving beam to meeting beam and vice versa. They are using some sensor to enhance the range of application.

O. Akinsanmi in August 2015 demonstrated "Design and Development of an Automatic Automobile Headlight Switching System". This designed system is automatically switching high beam to low beam as soon as vehicles are approaching from the opposite direction and switches it back when vehicles are passed to each other [4]. This model consists of 12 V battery, LDR, potential divider, NPN transistor, and SPDT. Here, a 12 V battery is used as a power supply for the car itself, Light-dependent register (LDR) work as sensor, and a potential divider network which serves as a comparator to NPN transistor joined with SPDT relay which does the switching. This model is successfully designed and tested for the range of 147 meters. Automobile industries are introducing this automatic headlight control in the production of new vehicles and minimize the road accidents in the night.

Okrah S. Kin April 2016 demonstrated "Design and Implementation of Automatic Headlight Dimmer for Vehicles using Light Dependent Resistor (LDR) Sensor". This model has been designed to dim the headlight of oncoming vehicles to avoid human eye effects [5]. This automatically switched the high beam into low beam, therefore reducing the glare effect by sensing the light intensity value of the approaching vehicle and also eliminated the requirement of manual switching by the driver which was not done at all times. MATLAB and KEIL software was employed in designing this model to program the microcontroller. The system device was able to automatically switch the headlight to low beam when it sensed a vehicle approaching from the opposite side using LDR sensor. It was observed that the maximum spread angle of the headlight was 135°.

2 Methodology

The following logical methods are applied to design and construct this model for optimum performance. The steps are as follows:

i. Response delay and high illumination effect on human eye.
ii. Headlamps luminous intensity.
iii. Choosing the correct parts/components for the effective switching of beams.
iv. The sensitivity of components minimize the switching time.
v. Addition of power supply block.
vi. Protection of the proposed module from electrical and environmental hazards.
vii. Backup system in case of module failure.

A. ***Light-Dependent Resistor (LDR)***

Light-dependent resistor is generally known as photoresistor. It is a light intensity-controlled variable resistor. Its resistance decreases when incident light intensity is increased; therefore, it has the property of photoconductivity. They are light-intensity-sensitive devices typically used to specify the presence or absence of light, or to measure the intensity of light. When the environment is dark, their resistance is very high, at times even up to 1 MΩ, but when the LDR sensor is exposed to light, the resistance drops significantly, to few ohms, depending on the intensity of light. LDR's sensitivity changes with the wavelength of the light falling on it and are nonlinear devices. The ohmic value of light-dependent resistor (LDR) changes with the intensity of light incident on it. This resistor is used in series with the biasing circuit of a switching transistor. The design of the circuit is such that the transistor is switched on and off depending on the threshold value of the intensity of light, the variable parameter being the LDR resistance. LDR is the best product of high resistive semiconductor materials. If the light is falling on the surface of high resistive semiconductor material with very high frequency, then photons are absorbed by materials and the resultant generated enough energy to jump electrons from valence band to conduction band, and also create low resistance path and build-up high conductive region.

The light-sensitive part of the LDR is a wavy track of cadmium sulfide. Cadmium sulfide cells depend on the materials ability to vary its resistance according to the amount of light striking cell on the surface. LDR is employed in the circuit to convert the intensity of the high beam headlight of the approaching vehicle into electrical

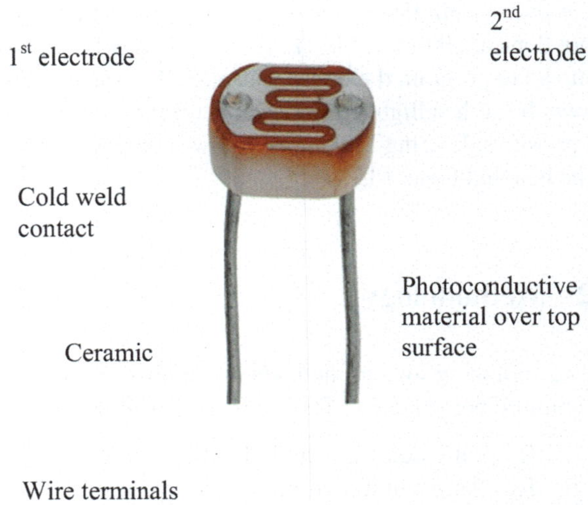

Fig. 1 Light-dependent resistor (LDR)

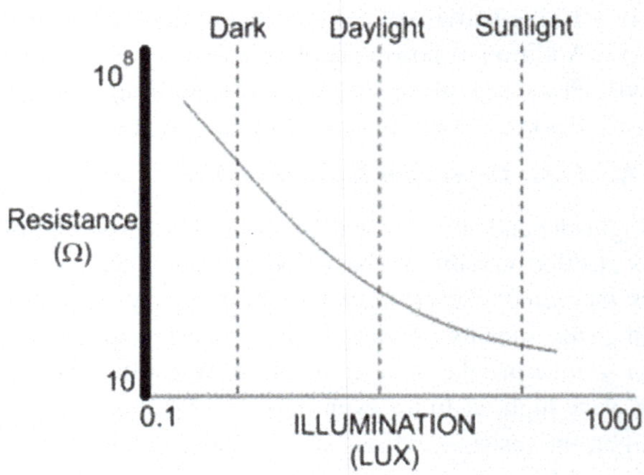

Fig. 2 A graph of resistance against Lux

signal. LDRs have several advantages such as cheaper, readily available in different sizes and shapes in market, practical LDRs is also available in a variety of sizes and package styles, the most popular size having a face diameter of roughly 10 mm and finally, they need very small power and voltage for their operations. Figure 1 shows the construction of LDR [6] while Fig. 2 shows the graph of resistance versus lux.

B. ***NPN Transistors (BC547) Transistor***

It is an NPN BJT, where a small amount of current is supplied at its base controls a larger amount of current at collector and emitter terminals, it is used for various applications. It can be used as the active component for switches and amplifiers along with resistors, capacitors, coils, etc. BC547 is mainly used for switching purposes. Biasing is done in the transistor to keep it fully on when a small signal is applied at the base. When no base signal is applied, it becomes completely off. In the circuit, transistor plays the role of switching the sensor.

C. ***Light-Emitting Diode (LED)***

It is a semiconductor device which acts as a source of light. It is a P–N junction diode which radiates light when active. When an appropriate amount of current is supplied to the leads, the electrons recombine with electron holes, which results in the release of energy in the form of photons. LEDs create light by electroluminescence in a semiconductor material. Electroluminescence is defined as the condition of a material which emits light when the electric current flows through it.

D. ***9 V Battery and 2 Resistors (50 KΩ, 1 KΩ)***

A battery is an electrical device which consists of several cells, which acts as a converter of chemical energy to electrical energy and also acts as a source to supply power. A resistor is an electrical device that offers electrical resistance to a circuit element. Resistors used in electronic circuits, reduce the flow of current in the circuits, and it is also used to modify signal levels, divide voltages, bias active elements, and terminate transmission lines.

3 Working

In this research article, NPN (BC547) transistor is used as a switch. In the circuit, the 1 KΩ resistor's one end is connected to the positive terminal (+) of the 9 V battery and the other end of the anode to the LED. Cathode of the LED is connected with Collector (C) of BC 547 transistor. Emitter (E) of BC547 transistor is connected with the negative terminal of the battery. Base (B) of the BC547 transistor is connected to the junction point of the 50 KΩ resistor and LDR. The complete working set up is shown in Fig. 3.

Considering the specification of BC547 transistor we observe that V_{be} max is 0.7 V. Therefore, if the base–emitter voltage < 0.7 V, there will be no current flow through the collector–emitter of the BC547 transistor and the transistor will remain in OFF state. When V_{be} > 0.7 V, there will be current flow from collector–emitter, hence, the transistor will be switched on. Now by connecting a load to the collector, it will be turned on. Now, after applying a voltage divider, the output of the voltage divider is connected to the base of the BC547. The voltage divider is the amalgamation of the resistance and LDR. Voltage divider crcuit is a very useful circuit which is applied in various applications. The combination of R1 and R2 will affect Vout. Vout = Vin*(R2/R1 + R2), so if we want to convert 5–2.5 V, R1 should be equal to R2. Here, take R1 = R2 = 1 KΩ, and it gives 2.5 V [7–9]. The resultant output of the voltage divider is usually the combination of R1 and LDR. It is connected to the base of the BC547 and a LED is connected to the collector. The 1 kΩ resistance is used to bind the current through the LED to avoid it from getting damaged. At the time when proper light is made to strike on the LDR, it offers negligible resistance. Now,

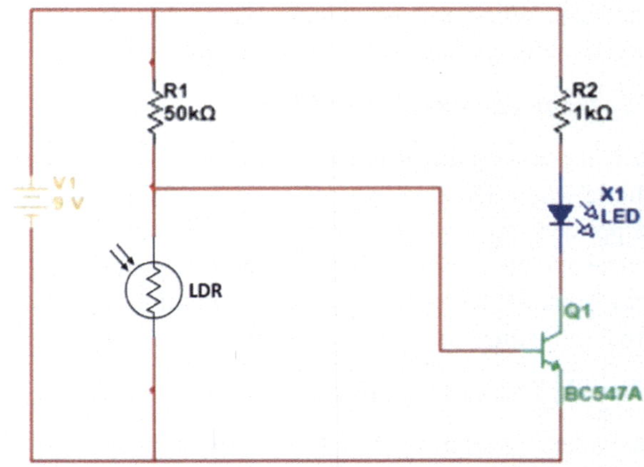

Fig. 3 Working setup of LDR

let us suppose, R1 = 220 KΩ and $R_{LDR} = 4.6$ K-Ω. If, the input voltage, $V_{in} = 9$ V, then the voltage on the base of BC547 = 0.184 V which is < 0.7 V so the BC547 will be in off state and therefore, the LED will not be glowing. Now, in case of absence of proper light, the resistance presented by LDR is quite large (sometimes even in order of Mega ohms). Now let us consider a case, when there is very less light available, then the resistance of the LDR will be high, let it be, $R_{LDR} = 150$ KΩ, so, the base voltage of transistor = 3.648 which is > 0.7 V. In this case, the current will flow from collector–emitter of the transistor and the LED will be glowing [10]. This circuit has enormous use in daily life such as detection of vehicles, tour and travels, schools, colleges/universities, offices, medical stores, hospitals, etc.

4 Advantages

I. Provide safe driving during bad weather conditions.
II. Economical to be installed.
III. Maintenance cost is low.
IV. Protect human eyes from high beam.

5 Conclusion

Although, the circuit for Light-Sensing Automatic Headlamps and Taillamps looks simple and easy to understand, and its application can be hugely beneficial for road transport and will surely bring down the rate of unfortunate incidences that occur every time on the road. The concept of Light-Sensing Automatic Headlamps and Taillamps can be used anywhere in a process industry with little modifications according to the requirements. Therefore, its uses are not limited. For example, this concept can implement for automatic switching on and off of street lamps, saving a lot of energy without any human input. This concept can also be applied in the field of automation and for long-time application as it has the flexibility to modify itself according to the required objective.

Acknowledgements We wish to thank our respected teachers, HOD ECE Prof. S. Pal, and Prof. M.K. Mishra VC BIT Mesra for their constant support and providing infrastructure facility to carry out this research work. We also thank ECE staff for their excellent contribution and immense support throughout this research work till the completion. We would also like to thank our prestigious university to give us the opportunity to learn something new explore the mighty world of science and technology.

References

1. Akinsanmi O (2015) Design and development of an automatic automobile headlight switching system. Int J Eng Appl Sci (IJEAS) 2(8). ISSN: 2394-3661
2. Muralikrishnan R (2014) Automatic headlight dimmer: a prototype for vehicles. Int J Res Eng Technol 03(03):85–90
3. Wattamwar B (2016) Automatic headlight beam shifter for controlling high beam to low beam of vehicles. Int J Innov Res comput Commun Eng 4(4), ISSN: 2320-9801
4. Prasad D, Nath V (2018) An ultra-low power high-performance CMOS temperature sensor with an inaccuracy of −0.3 °C/0.1 °C for aerospace applications. Microsyst Technol 24(3):1553–1563
5. Akinsanmi O (2016) Design and development of an automatic automobile headlight switching system. Int J Emerg Technol Innov Eng 2(4). ISSN: 2394–659
6. Okrah. SK (2016) Design and implementation of automatic headlight dimmer for vehicles using light dependent resistor (LDR) Sensor. Int J Emerg Technol Innov Eng 2(4). ISSN: 2394–6598
7. Prasad D, Nath V (2017) An overview of temperature sensors. In: Nath V, Mandal J (eds) Proceeding of the second international conference on microelectronics, computing & communication systems (MCCS 2017). Lecture notes in electrical engineering, vol 476. Springer, Singapore
8. Patil S Design and implementation of automatic street light control using sensor. Int J Eng Res
9. Light-dependent resistor. Int. J Res Adv Technol (IJRAT) (E-ISSN: 2321-9637) Special Issue National Conference 9th April 2017
10. How an LDR (Light Dependent Resistor) works. www.kitronik.co.uk

Change Over Time in Grey Levels of Multispectral Landsat $5^{TM}/8^{OLI}$ Satellite Images

Amit Kumar Shakya, Ayushman Ramola, Akhilesh Kandwal and Rishi Prakash

Abstract Change detection (CD) from the Earth's surface has developed into an important issue throughout the world. CD methods have found applicability in socio-economic assessments, environmental monitoring, exploring resources, etc., and till today, various numbers of CD methodologies utilising remotely sensed data have been developed. Researchers have categorised different CD techniques on the basis of distinct viewpoints. A common one is to associate different techniques in two categories pre-classification CD and post-classification CD. This paper examines the efficiency of post-classification, texture-based CD method. Changing conditions are investigated for the three cases of civic urbanisation, depleted lake's water and surging rivers and lake. Changes in the image parameters are first analysed with first-order statistics, i.e. (non-zero 'mean' and 'standard deviation'), and then changes in the visual characteristics are quantitatively measured through the grey level by second-order statistical parameters. Contrast, correlation, energy and homogeneity are the grey level co-occurrence matrix (GLCM) features which predict change in the 'visual' appearance of the texture. Texture gets converted from one form to another form represented by pre- and post-grey level image, respectively. Finally, we have established a novel changing 'pattern' of the texture from the aforesaid three cases.

Keywords Change detection · Post-classification · Grey level co-occurrence matrix · Contrast · Correlation · Energy · Homogeneity · Visual appearance · Changing pattern · Texture

Nearly two thousand years ago Greek brainy Heraclitus came forward with an exploration that ***"it is impossible to step in the same river twice"****. More than 1600 ages ago, a Chinese sage, Gehong experienced that* ***"the sea changes into mulberry fields, while the mulberry fields change into seas."*** *We have no idea that several years ago with no modern automation and machinery, exactly how Gehong recognize those landscape changes neither we know exactly if Heraclitus claimed changes in context*

A. K. Shakya (✉) · A. Ramola · A. Kandwal · R. Prakash
Department of Electronics & Communication Engineering, GEU, Dehradun, UK, India
e-mail: xlamitshakya.gate2014@ieee.org

© Springer Nature Singapore Pte Ltd. 2019

V. Nath and J. K. Mandal (eds.), *Proceedings of the Third International Conference on Microelectronics, Computing and Communication Systems*,
Lecture Notes in Electrical Engineering 556, https://doi.org/10.1007/978-981-13-7091-5_29

of flood and drought or changes in its morphology, but one thing is assuredly true, that human race was always apprehensive of these natural transformations in their living environment since thousands of years [1].

1 Introduction

Land cover (LC)/land use (LU) change estimation are important facets of scientific exploration and socio-budgetary assessments. Knowledge about the changes taking place in LC/LU is crucial to amuse appetite for up to date information about the Earth surface [1]. CD and their evaluation are gaining importance in the environmental exploration and monetary use, e.g. keeping an eye on environmental and ecological changes, urban planning, pre-disaster management, etc. for estimating changes that have occurred due to various factors, remote sensing (RS) has been developed as an optimal tool, as it offers information regarding huge areas at a shortened interval of time. A decisive and advantageous application of remotely sensed data is that it has made achievable to review the changes at slashed with limited time and superior accuracy. A few natural origin factors and human-induced factors which have compelling impacts on the dynamic plight of the Earth's surface are along these lines [2]. Natural disasters are generated by natural paradox which brings abuse to human society, damage and loss of valuable properties such as buildings, communication set-up's, forest cover, etc.; illustration of natural disasters includes earthquakes, volcanic eruption, hurricanes, lightning storms, windstorm, tsunami, forest fire, etc. [3]. Human-made disasters are generated by direct human activities answerable for bringing forth change conditions for the ecosystem; examples of human-made disasters incorporate industrial chemical accidents, major armed conflicts, nuclear accidents, landmines explosions, etc. [4]. Human cajoled calamities are natural disasters which are accelerated and influenced by indirect human activities, e.g. landslides, desertification, soil erosion, flooding, etc. [4]. Changeable conditions of the planet surfaces are not only due to disasters, but many other factors which support the evolution of human beings also produce significant changes in the physical pattern of the LC like the advancement of residential settlements, development of industrial areas for increasing employment, and deforestation for bolstering agricultural lands [5].

1.1 Types of Change Detection Methods

Previously, before the development of RS data and computers, tracing paper and topographic sheets were employed to detect the LU/LC changes. Nevertheless, this method demands lots of human activity and time in traditionally examining colossal areas [6]. Mapping of LU/LC has turned into achievable and simpler by RS technique [2]. The dominance of using RS data is to obtain information about inaccessible regions. Ecologists and environmentalists have got assistance by RS and geographical information system (GIS) automation techniques. RS data are also used for the allocating purpose of LU/LC. Thus, 'RS methods can be engaged to arrange types of land use in a realistic, prudent and repetitive fashion, over large areas' [1]. Thus, CD methods are used to classify the type and pattern of change. Selecting a CD method is also dependent upon the type of change we wish to monitor. CD methods are typically classified into two disparate categories pre-classification CD and post-classification CD [4]. In the pre-classification revision, information is reaped by pixel-based CD technique whereas post-classification methodology uses both pixel-based and object-based approach [7, 8]. Pre-classification CD approaches incorporate image differencing [8, 9] image rationing [10], image regression [11], vegetation index differencing [12], change vector analysis [13], principal component analysis [14], etc. Post-classification CD approaches cater us information in the planned manner about assorted changes that have raised on the Earth's surface in a terse duration of time [13]. Post-classification approaches provide us information in a detailed manner. Post-classification CD approaches include composite or multi-date classification [15], machine learning [16, 17] GIS-based [18, 19] and texture analysis based CD which is a trivial method used for the texture classification of remotely sensed images. An approach for texture classification is GLCM which is able to examine spatial as well as the spectral distribution of pixels. GLCM features appraised the visual characteristics of the textured surface [20].

Today, with the emergences of changeable high-resolution satellite images over the last decades have motivated the development of object-based change detection (OBCD) methods [21, 22]. OBCD is a two-step process

- The first step is the segmentation of the image into discrete objects.
- Second is the classification of the objects. The inference of OBCD method is that the image is composed of unruffled similar regions or patches, larger in size than the individual pixels.

A threat for the object-based approach is the segmentation process, as it deteriorates errors like under segmentation and over-segmentation such fault results in the consequent loss of information of objects that do not serve as real-world object [23]. Object-based post-classification CD techniques are classified as direct object change detection (DOCD) [24, 25], classified object change detection (COCD) [26, 27], multi-temporal and multi-date object change detection [28, 29].

1.2 Role of Change Detection in Texture Classification Through Remote Sensing

An appropriate and authentic CD of the Earth's surface administers us the authority for improved understanding and communication between human and natural paradox in order for better management and use of assets [7]. CD from the Earth's surface is also essential because it has numbers of practical applications like damage assessment, environmental and disaster monitoring, urban planning and land management [8]. Tardie and Congalton [30] defined CD as an approach used in RS to regulate the changes as a precise object of study medially between two time periods. CD found handy application for urban and environmental monitoring as the CD process provides significant analysis of the spatial distribution in the field of study [30]. RS dossier is an appropriate source of CD by virtue of its high temporal frequency and accessible for the computation purpose. RS data in the digital format possess the competence of selecting wide spatial and spectral resolution [21]. RS convinces to be a powerful and effective tool for the study of CD due to numerous backing factors such as economical, consuming limited time and resulting in better accuracy [1]. The difference is introduced in the area of interest due to factors like atmospheric conditions, illumination effects, etc. These transformations can be isolated out from the area of interest as these developments occur due to the spectral behaviour of RS data. This is the preeminent principle for using RS data for CD [8]. Hiroyuki Miura et al. [31] investigated the crisis of calamitous Haiti earthquakes buttoned up with texture characteristics and high-resolution satellite images. They matured an automatic damage detection approach based on the texture appearance of the pre-and post-earthquake images. Tomowski et al. [32] conferred a texture analysis based approach for calamity application. They figure out the texture of the pre-disaster and post-disaster images on the ground of four parameters contrast, correlation, energy and homogeneity and then they compared the capability of their approach with accepted CD methods like image differencing, image ratio, PCA. Soh et al. [33] conferred analysis of mapping sea ice pattern in SAR images. They used GLCM to appraise texture parameters. They implied that parameters like quantization, displacement and orientation are finest for SAR sea ice texture analysis. Baraldi and Parmiggiani [34] investigated six different statistical texture features (TF). They concluded that out of six different TF energy and entropy are contemplated best for defining texture. They also spawn a new statistical parameter known as 'reccursivity' which shows interaction with parameter contrast. Lendaris and Stanley [35] explained that remotely sensed images adopt texture to allocate LU league like (agriculture, forest, water bodies, urban regions, etc.). The texture is used to access important information of various LU by observing the modifying texture of the surface. J. S. Rawat used RS techniques to monitor the changes in LU/LC in Hawalbagh block of District Almora, Uttarakhand, India [36]. K. Rokni developed a new technique for detecting change in the water surface through pixel and image classification technique [37], O. R. Abd El-Kawya detected changes in LU/LC across the western Nile river delta of Egypt [38], K. S. Willis reviewed the changes that got developed in the ecology across the protected areas of United States [39]. On the contrary CD, methods are accessible, including the traditional pixel-based and contemporary object-based CD method.

1.3 Background of NASA Landsat Program

The Landsat program was started by National Aeronautics and Space Administration (NASA) in the era of the atom bomb and moon mission [40]. It was the Apollo Mission which inspiration resulted in the development of the Landsat Programs. During the Apollo Mission, the photographs of the Earth's surface were taken for the first time. In the year 1965, United States Geological Survey (USGS) William Pecora came up with the idea to develop the satellite program based on RS for the mother Earth [40].

The main objective of his proposal was to gather vital information about the Earth resources [41]. These motives give birth to the Landsat programs; the first Landsat satellite was launched on 23 July 1972 and was decommissioned on 6 January 1978. Since then, NASA has successfully launched 8 Landsat satellites.

Landsat 6 launched on 5 October 1993, was unable to reach its predefined orbit. This was also a reason which resulted in the extensive use of Landsat 5 beyond their design life [42]. NASA is expecting to launch their Landsat 9 in 2020. It will carry Operational Land Imager (OLI-2) and Thermal Infrared Sensor (TIRS-2) it is designed for the time period of 5 years. It's snapping capacity will be 700 scenes per days [43] (Fig. 1).

Landsat 5 was a low Earth orbital satellite. It was launched on 1 March 1984. The main purpose of this satellite was to collect information about the Earth surface, through images. It has sent about 2.5 million images back to Earth station- from the outer space. It has worked for 29 years and got retire from the active service on 5 June 2013 [44]. It uses thematic mapper (TM) sensor, which is an advanced

Fig. 1 Landsat 5 thematic mapper/Landsat 8 operational land imager spectrum band configuration

(a) (b)

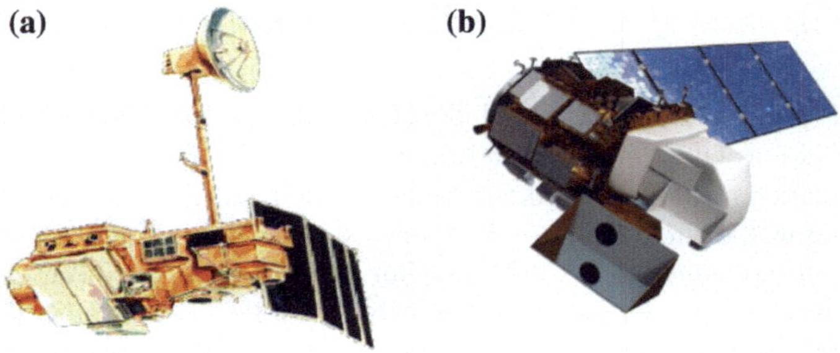

Fig. 2 **a** Landsat 5 and **b** Landsat 8 (Image Courtesy United States Geological Survey (USGS))

multispectral scanning, Earth resource sensor which was designed to capture high-resolution images. TM data is sensed in seven spectral bands. A TM sensor has an instantaneous field of view (IFOV) of 30 × 30 m in bands 1–5 and 7 while band 6 has an IFOV of 120 × 120 m on the ground [45] (Fig. 2).

Landsat 8 is an Earth observation satellite. It was launched on 11 February 2013. The satellite ensures the continuous acquisition and availability of the Earth imagery. Landsat 8 snaps more than 700 Earth shots daily. The most important task performed by Landsat 8 is the control, command and management of the data sent from satellite to the Earth command centre [46]. Landsat 8 uses OLI sensor, it contains nine spectral bands with the spatial resolution of 30 m for Band 1–7 and 9. The ultra-blue band 1 is useful for coastal and aerosol studies. Band 9 is used for cloud detection. Thermal bands 10 and 11 provide more accurate surface temperature and collected at 100 m. The approximate swath size is 170 km north-south by 183 km east-west (106 mi by 114 mi) [47].

Landsat satellites have played an important role in the detecting change in the LU/LC, coastline changes, forest cover, agricultural cover, vegetation cover, etc. Roy et al. [48] described the mission objectives of Landsat 8 with data acquisition technique, sensor performance, etc. He further investigates area near Dulles Airport, Baltimore Maryland, Ventura California for detecting changes in the LC. Jeffrey G. Masek [49] used time series Landsat 1 MSS and Landsat 7 ETM+ for the period of 25 years to check the stability of the boreal forest located in the Richmond Gulf, Quebec and Great Slave Lake in Northern Canada. Cohen and Goward [50] described the role of Landsat 1–7 satellites in the ecological applications. He further studied the changes developed in the Boreal forest situated in north Canada. Armston et al. [51] used Landsat 5 TM and Landsat 7 Enhanced Thematic Mapper (ETM+) imagery to predict the condition of the woody vegetation LC in the state of Queensland, Australia. Ekercin [52] studied the changes developed across Aegean Sea coasts in Turkey from the Landsat satellite sensors MSS, TM, ETM+ imagery. Huang et al. [53] used Landsat time series imagery to reconstruct the previous forest cover, with the objective to detect loss of forest cover in the period of 10 years. S. P. Healey [54] used Landsat data to estimate the partial agricultural harvest in the forest area of western Washington, USA, by application of two regression-based methods. Masek

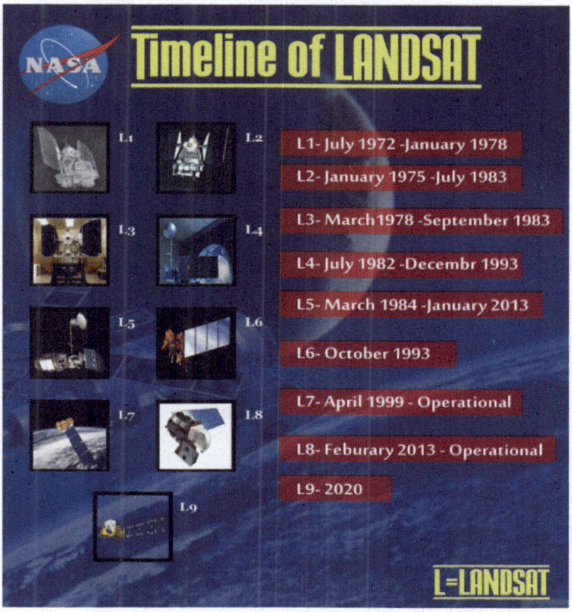

Fig. 3 Pictorial history of Landsat space program

et al. [55] used 'Landsat Ecosystem Disturbance Adaptive Processing System' to map the forest disturbance across North America during 1990–2000. Rokni et al. [56] used multi-temporal Landsat imagery to extract water features and determine changes occurred in Lake Urmia located in Iran. Fan et al. [57] monitored the changes that got developed in the LU/LC in the one of the most populated areas of Guangzhou located in China. Watts et al. [58] used the object-oriented approach to classify agricultural practices in Central Montana by Landsat 5-TM and Landsat 7-ETM+. Cleve et al. [59] investigated the changes caused due to human interference in the wildlife areas by monitoring the LU/LC [60].

In Fig. 3, history of Landsat program is presented which shows the contribution of the Landsat program from the late 70s in Earth monitoring and ecological management.

Table 1 shows the features comparison between parameters of the Landsat 5 and Landsat 8 satellites.

2 Statistical Methods for Texture Classification

Statistical methods refer to the classification based on the numbers of pixels deciding the local texture feature (TF) of an image. Statistical methods are broadly branched into three classes, first-order (one pixel), second-order (two pixels) and higher order [61]. However, third-order and higher order statistics are also presented but they

Table 1 Features comparison of Landsat 5 and Landsat 8

S. No.	Features	Landsat 5	Landsat 8
1.	Shoot date	1 March 1984	11 February 2013
2.	Current status	Retire	Active
3.	Sensors	Thematic mapper (TM), Multispectral scanner (MSS)	Operational land imager (OLI), Thermal infrared sensor (TIRS)
4.	Altitude	705 km	705 km
5.	Orientation	98.2°	98.2248°
6.	Orbital position	Polar, sun-synchronous	Geocentric, sun-synchronous
7.	Revolution period	99 min	98.8 min
8.	Repeat coverage	16 days	16 days

are not suitable from the point of human visual interpretation [62]. The discrepancy between these classes is, the first-order statistics estimate properties such as (mean and standard deviation) of an individual pixel, not contemplating the spatial interaction among image pixels, while the second-order statistics evaluate image properties by considering the spatial relationship bounded by neighbouring pixels of an image [33]. Texture scrutiny by histogram approach is done to determine the distribution of intensity values in complete parts of the image. Trivial histogram features include mean, variance, dispersion, mean square value or average energy, entropy, skewness and kurtosis [33]. Second-order statistics provides detail information about changing texture and are also referred to as 'textons' by [63]. A single GLCM is not at all acceptable to define the TF of an image. Therefore, multiple GLCM's are generated for the single input image. For the creation of a multiple GLCM arrays of offsets is specified. These offsets decide the relationship of pixels with waffling distance and orientation [34]. Haralick [64] was the patron of GLCM, as he developed GLCM-based set of '14' TF for an image.

2.1 Image Classification for the Supervised and Unsupervised Classification

The objective of satellite image classification is, to categorise pixels into several land cover classes or 'themes'. The satellite data is used to generate maps of LC present in form of classes. Multispectral data is used for the classification purpose and the spectral pattern in the data for each pixel is used as the numerical basis for categorization. The intent of the image classification is to identify and labelled unique grey level (or colour) to the objects, LC which actually represents on the ground [65].

Image classification is considered as a very interesting part of image analysis. 'Pretty pictures' or images showing variations of LC in terms of colour for CD, but all of this information is quite useless unless we don't know what these colours interpret [66], so there is a requirement of the image classification. There are two methods for image classifications.

- Supervised Classification
- Unsupervised Classification

In supervised classification LC classes are predefined; reference data is already available and are used as training samples [67]. These samples are also known as 'training sites' [68]. These sites generate the signature to train classifiers. These classifiers include maximum likelihood [69], the minimum distance [70], artificial neural network [71], and decision tree classifiers [72]. In unsupervised classification, large numbers of known pixels are examined and are divided on the basis of natural grouping. Unsupervised classification does not require any pre analysed dataset. The basic principle of this classification is value with in a given cover type should be close together, i.e. have similar grey levels whereas comparatively close should have different grey levels. These classifiers include ISODATA [73], k-mean clustering algorithm [74]. We will be using unsupervised classification GLCM-based approach in our experiment.

2.2 Grey Level Image Differencing Through Change Histogram

Grey level image differencing is defined as a process refers to the subtraction of multi-temporal images. Change histogram in Fig. 4 represents the frequency of the image pixels that get 'changed', 'no change' and 'do not change'. They are represented by red, black-dotted line and blue colour, respectively.

Mathematically it is expressed as follows:

$$I_d(t_1, t_2) = I_1(t_1) - I_2(t_2) \tag{1}$$

where t_1, t_2 represents the time intervals between the images. $I_d(t_1, t_2)$ represents the absolute difference images in our case, we have acquired difference by the image differencing of the grey levels of the pre- and post-event image. $I_1(t_1)$ is a pre-event image acquired at a time t_1 and $I_2(t_2)$ is post-event image acquired at the time interval t_2. Some important features of image differencing are as follows.

- Image differencing is a process of subtraction of imagery between two time intervals.

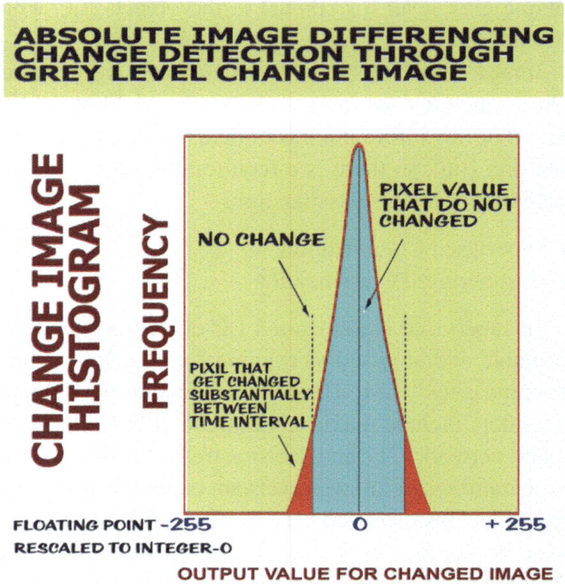

Fig. 4 Block diagram representation of absolute image differencing

- If the two images are having almost identical radiometric characteristics, the subtraction will results in positive and negative values in areas of radiance change and zero values in areas of no change.
- The result of image differencing will be a new image highlighting the changes occurred.
- The resultant image is called grey level difference image in this case.

2.3 Background of Grey Level Co-occurrence Matrix

GLCM is also known as 'Grey tone spatial dependency matrix' [75], it contains some unique properties which are listed as follows:

- It is square in dimension.
- It contains equal numbers of rows and column as the quantization level of the image [76].
- It is having symmetrical geometry along diagonal.
- It is used to evaluate TF of the image [75].

GLCM is having specific advantages from the point of view of textual evaluation.

- GLCM calculates high order distribution of grey values of pixels located at a predefined distance and at various orientations.
- GLCM, pixel of interest (POI), i.e. $p(x_i, y_i)$ represents the distribution of occurrence of the pair of grey values separated by specific distance vectors.

- Normalised values of the GLCM can be obtained by the total number of occurrence providing the probability of occurrence.
- TF are statistically computed with GLCM [77].

Here an input image test image of dimension (5 × 5) is created shown in Fig. 5a. GLCM of dimension (8 × 8) is created from the input image. Here grey level '1' followed by grey level '1' in the input image have assigned grey tone '0' in the GLCM image. Grey level '1' followed by grey level '4' have assigned grey tone '1' similarly grey level '6' followed by grey level '5' have assigned grey tone '2' and so on. Thus, following the same pattern across the input image, a GLCM from the input image is created.

The crucial aspect of counting GLCM of an image is through displacement 'd' and orientation 'θ'. A single GLCM do not provide sufficient information to describe the texture, therefore multiple GLCM having same distance and different orientation (0°, 45°, 90°, 135°), degrees (D) is computed for a single input image. Let POI is noted by $p(x_1, y_1)$ where (x_1, y_1) represents the pixel position of POI (Fig. 6).

We have formed $(R_r \times C_c)$ extended GLCM of the input image. It is represented by $GLCM_{Sat}(d, \theta)$ which represents that the new GLCM is created along orientation 'θ' and distance '$d = 2$'. In this newly formed GLCM, elements represent the co-occurrence probability of the image value (i, j) at the POI. Hence, the notations for four different directions at four different angular positions in general are expressed below:

$$GLCM_{Sat}(d, \theta) = \text{Prob}\{g(x_1) = i, g(y_1) = j\}$$

(a)

	1	2	3	4	5
1	1	3	2	8	8
2	6	5	8	2	2
3	7	1	4	6	5
4	3	7	8	3	1
5	8	4	8	1	2

(b)

	1	2	3	4	5	6	7	8
1	0	1	1	1	0	0	0	0
2	0	1	0	0	0	0	0	1
3	1	1	0	0	0	0	1	0
4	0	0	0	0	0	1	0	1
5	0	0	0	0	0	0	0	1
6	0	0	0	0	2	0	0	0
7	1	0	0	0	0	0	0	0
8	0	1	1	1	0	0	0	1

Fig. 5 **a** Input image of dimension (5 × 5) and **b** created (8 × 8) GLCM for input (5 × 5) image

$$\text{Distance d}\,(x_1, y_1) = d \text{ and } Angle^{Degree} = 0. \tag{2}$$

$$GLCM_{Sat}(d,\theta) = \text{Prob}\{g(x_2) = i, g(y_2) = j\}$$
$$\text{Distance d}\,(x_2, y_2) = d \text{ and } Angle^{Degree} = -d. \tag{3}$$

$$GLCM_{Sat}(d,\theta) = \text{Prob}\{g(x_3) = i, g(y_3) = j\}$$
$$\text{Distance d}\,(x_3, y_3) = 0 \text{ and } Angle^{Degree} = -d. \tag{4}$$

$$GLCM_{Sat}(d,\theta) = \text{Prob}\{g(x_1) = i, g(y_1) = j\}$$
$$\text{Distance d}\,(x_1, y_1) = -d \text{ and } Angle^{Degree} = -d. \tag{5}$$

One limitation of GLCM is that it correlates the relationship between pixels, but do not work efficiently for the structural textures [78].

Beside GLCM, there are various other texture-based features methods like Moran's I [79], Geary's C [80], triangular prism surface area (TPSA) [81], lacunarity [82], discrete wavelet transform DWT [83], granulometric [84] and local binary pattern LBP [85].

2.4 Statistics Derived from Grey Level Co-occurrence Matrix

1.

$$\text{Contrast: } fr_1 = \sum_{0}^{K_g-1} n^2 \left\{ \sum_{r=1}^{K_g} \sum_{c=1}^{K_g} p(r,c) \right\} \tag{6}$$

Here

$|r - c| = \text{n}$ Difference in grey level pair.
Kg Number of distinct grey levels in the quantized image.

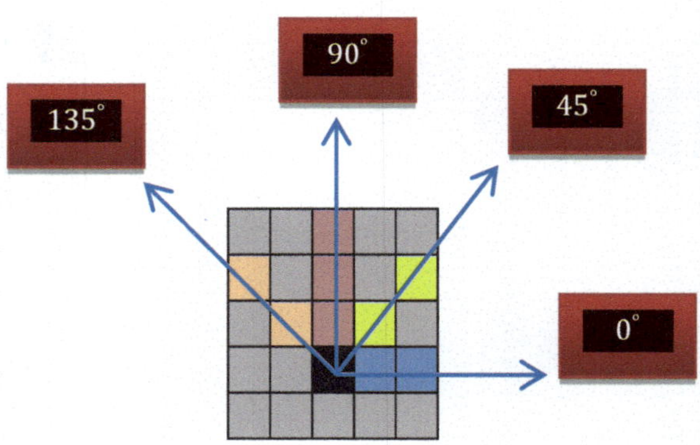

Fig. 6 Array of offsets at (0°, 45°, 90°, 135°)

Theoretical description: It is defined as the measure of the intensity between the pixel of interest (POI) and its neighbours over the entire image. When (r, c) are equal, then $|r - c| = 0$ which represents that pixel values are entirely identical to their neighbouring pixels. If '*r*' differs '*c*' by 1 there is small contrast, as a difference between '*r*' and '*c*' increases, contrast also increases.

2.

$$\text{Correlation: } fr_2 = \sum_r \sum_c \frac{(r, c)p(r, c) - \mu_x \mu_y}{\sigma_x \sigma_y} \tag{7}$$

Here $\mu_x, \sigma_x and\ \mu_y, \sigma_y$ are mean and standard deviation of rows and columns, respectively.

Theoretical description: It is defined as the measure of intensity, i.e. how correlated POI is with its neighbour over the entire image. Correlation lies in the range $(-1 \leq correlation \leq 1)$.

3.

$$\text{Energy: } fr_3 = \sum_r \sum_c \{p(r, c)\}^2 \tag{8}$$

Theoretical description: It is defined as the measure of the uniformity in the range $(0 \leq energy \leq 1)$. Energy is '1' for a constant image. It is also accepted as angular second moment (ASM).

4.

$$\text{Homogeneity: } fr_4 = \sum_r \sum_c \frac{1}{1 + (r - c)^2} p(r, c) \tag{9}$$

Theoretical description: It is a measure of the spatial closeness of the distribution of elements in GLCM along its diagonal, it is bound in the range $(0 \leq homogeneity \leq 1)$. It is also noted as inverse difference moment (IDM).

3 Methodology for the Change Detection

In our approach, we are using first-order image statistics to obtain information whether these statistics are showing variation or not for pre- and post-event images. We have concluded that if 'no change' have occurred than mean and standard deviation will have 'zero' value otherwise changes have occurred. But changes in these statistics are also influenced by many other factors like difference in illumination. So we proceed towards second-order statistics for obtaining detailed information about textural changes. Now we have enforced post-classification textural CD method in study areas to obtain detailed change information through GLCM. It is defined as 'second-order histogram of grey levels of pixel pair, which are detached by a fixed spatial relationship'. The information obtained using GLCM descriptors are used to 'create' a 'pattern' in the changing texture from pre-event to post-event.

Texture descriptor through GLCM is a measure of the properties of a surface. We have classified texture in three categories, i.e. smooth, coarse and regular. In this study, we are regulating direction independent GLCM by calculating the average of the four GLCM calculated in four different directions. We are using four features contrast, correlation, energy and homogeneity derived from GLCM for the quantitative estimation of the TF. Finally, the absolute image differencing technique is applied to obtain the change map. The grey levels of the pre- and post-images are subtracted to obtain this change map. This will visually confirm that the changes have occurred in the grey levels. In this experiment, we are taking nine cases divided in three different categories of urbanisation, lake dryness and flooding. These cases also contain three sub-cases like for urbanisation we are monitoring urbanisation pattern of Las Vegas, Dallas and Zhu San Jiao urbanisation.

The event lake dryness includes the study of Lake Poopo, Lake Eyre and Lake Mead. Finally, the flooding case includes studies of the flood caused by River Tigris, Lake Eyre and River Mekong.

We have quantified a number of changes that have occurred in the TF of pre- and post-event and on the basis of that 'we have established a pattern of changing texture', i.e. we have defined the status of the texture of pre-image, and after the event, we will redefine the new status of the texture. Finally, the pattern of the changing texture will define 'to-from' change.

Now a brief summary of the various areas of investigation can be obtained in Table 2 which represents test image 1, test image 2, areas under investigation, image dimensions and pre-event and post-event image snapping satellites.

Notations used: U# = Urbanisation, F# = Flooding Areas, Con# = Contrast, Corr# = Correlation, E# = Energy, H# = Homogeneity.

Now we have calculated the first-order statistics, i.e. mean and standard deviation for our Landsat images and obtained a conclusion that they have attained 'non-zero' value from this we have concluded that radiometric changes will appear in our absolute difference image.

Table 3 shows the variation in the first-order statistical parameter. These variations are expressed with the overlay plots shown in Fig. 7a, b.

Overlay plots in Fig. 7 suggest that the first-order statistical parameter mean have attained higher value for the post-event image, while the statistical parameter standard deviation show has the higher value for pre-event image. These variations in the first-order statistical parameters inform us about changes that got developed in LU/LC.

Therefore for the detailed evaluation of changes we are proceeding towards GLCM method for evaluation of the TF of the LC.

Table 2 Investigation sectors portraits with image locations, dimension, pixels and satellites

S. No.	Event	Study areas	Images dimension	Satellite (pre-image/post-image)
1.	Test image 1	Urbanisation Shanghai, China	4500 × 4500	Landsat 5/8
2	Test image 2	Urbanisation across Lake Mead, U.S.A	9783 × 6522	Landsat 5
3.	U#	Las Vegas, U.S.A	2149 × 2150	Landsat 5
4.		Dallas, U.S.A	4000 × 3000	Landsat 5
5.		Zhu San Jiao, China	5198 × 3465	Landsat 5/8
6.	Lake dryness	Lake Poopo, Bolivia	6000 × 9000	Landsat 8
7.		Lake Eyre, Australia	3600 × 4800	Landsat 5
8.		Lake Mead, U.S.A	3000 × 2400	Landsat 5
9.	F#	River Tigris, Iraq	4338 × 2892	Landsat 8
10.		Lake Eyre, Australia	2400 × 1600	Landsat 5
11.		River Mekong, Cambodia	646 × 1200	Terra Satellite

For this, we have taken test image 1 of Shanghai, China. The latitude and longitude of Shanghai is 31.2304° N and 121.4737° E. The pre- and post-images were taken from Landsat 5 and 8, respectively (Fig. 8).

Now Fig. 9a, b represent the greyscale images for the pre- and post-urbanisation image of Shanghai, Fig. 9c shows the change map obtained by the absolute image differencing which indicates variation in the patterns of the colours that have taken place during this time interval. ‘Blue’ symbolises that no change has occurred in that region whereas ‘green’ represents changes that have occurred in the grey levels, which finally symbolises the change in the texture of the LC.

Table 3 Mean and standard deviation values for study areas

S. No.	Study areas	Pre-event image		Post-event image	
		Mean	Standard deviation	Mean	Standard deviation
1.	Las Vegas	128.27	9.29	151.63	24.98
2.	Dallas	95.06	9.51	103.82	9.586
3.	Zhu San Jiao	98.06	9.58	109.92	10.78
4.	Lake Poopo	116.24	10.69	131.96	10.82
5.	Lake Eyre	139.92	9.98	156.03	10.01
6.	Lake Mead	112.24	12.68	132.48	13.05
7.	River Tigris	151.40	10.02	136.22	11.01
8.	Lake Eyre	160.24	10.15	176.56	10.92
9.	River Mekong	170.62	12.64	184.84	12.98

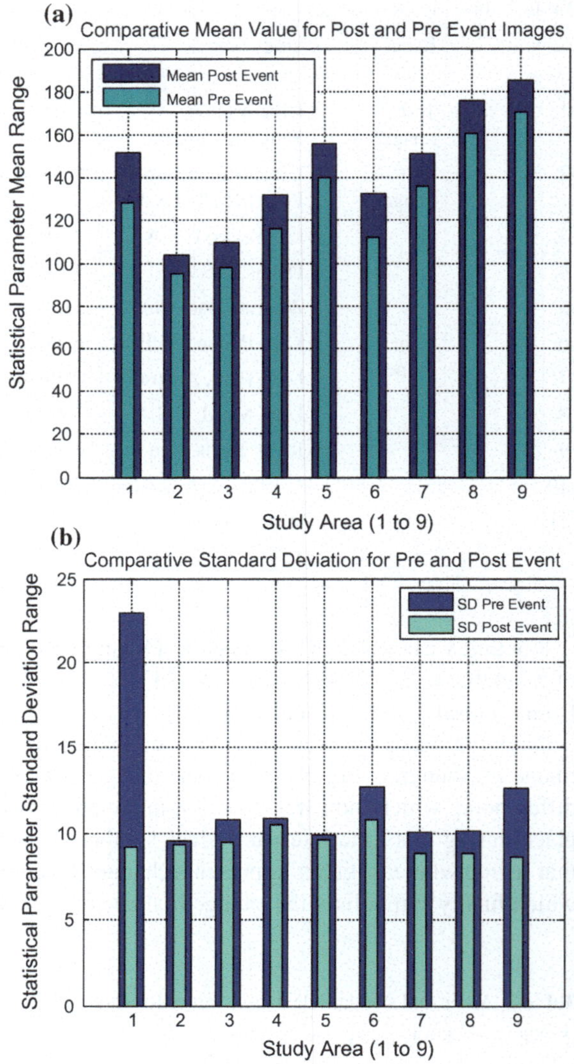

Fig. 7 **a** Comparative bar plot of the statistical parameter 'mean'. **b** Comparative bar plot of the statistical parameter 'standard deviation'

Now Table 4 represents the feature variation of I-order statistics of the test image 1. Mean and standard deviation is calculated for pre-, post- and difference urbanisation image. The difference image obtained show changes in the first-order, if no changes have developed in the LU/LC than the mean and standard deviation will obtain zero '0' value.

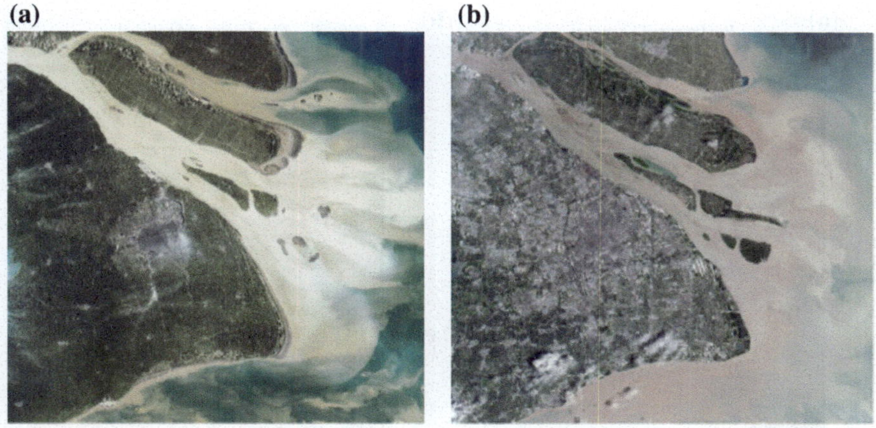

Fig. 8 **a** Pre-image obtained on 23 April 1984 and **b** post-image obtained on 20 July 2016

3.1 Measure of Texture 'Smoothness' and 'Coarseness' Through Grey Level Co-Occurrence Matrix

The texture of a surface can be very smooth, smooth, coarse, regular, etc. [86]. These types of changes in the texture of LC occur due to various reasons. Events like deforestation, urbanisation, floods, droughts, population overgrowth, etc., have made a significant impact on the changing texture of the LC [87], so here we are classifying texture in three different patterns coarse, regular and smooth. In our approach of texture classification, three particular subsections of the identical dimension of (100 × 100) are taken from the test image 2. These subsections contain properties of smoothness, regularity and coarseness, which can be optically identified. Visual interpretation from the 'yellow', 'green' and 'red' sub-sections can be understood as follows.

- Subsection 'yellow' shown in Fig. 10b symbolise a coarse surface as it contains too many settlement areas.
- Subsection 'green' shown in Fig. 10c symbolise regular surface (intermediate between coarse and smooth).
- Subsection 'red' shown in Fig. 10d symbolise a smooth surface compared to the yellow and green subsection.

The three subsections show smoothness, coarseness and regularity which is concluded on the basis of texture visual parameter as texture visual properties. When we compare pre- and post-event images, the behaviour of the TFs provides us information regarding texture and their variations provide us information about change in texture.

Table 5 demonstrates specific conclusions as section 'yellow' is a most coarse surface, and at the same time, it is also least smooth surface. Likewise, section 'red' is a most smooth surface, and meanwhile, it is also a least coarse surface. Section

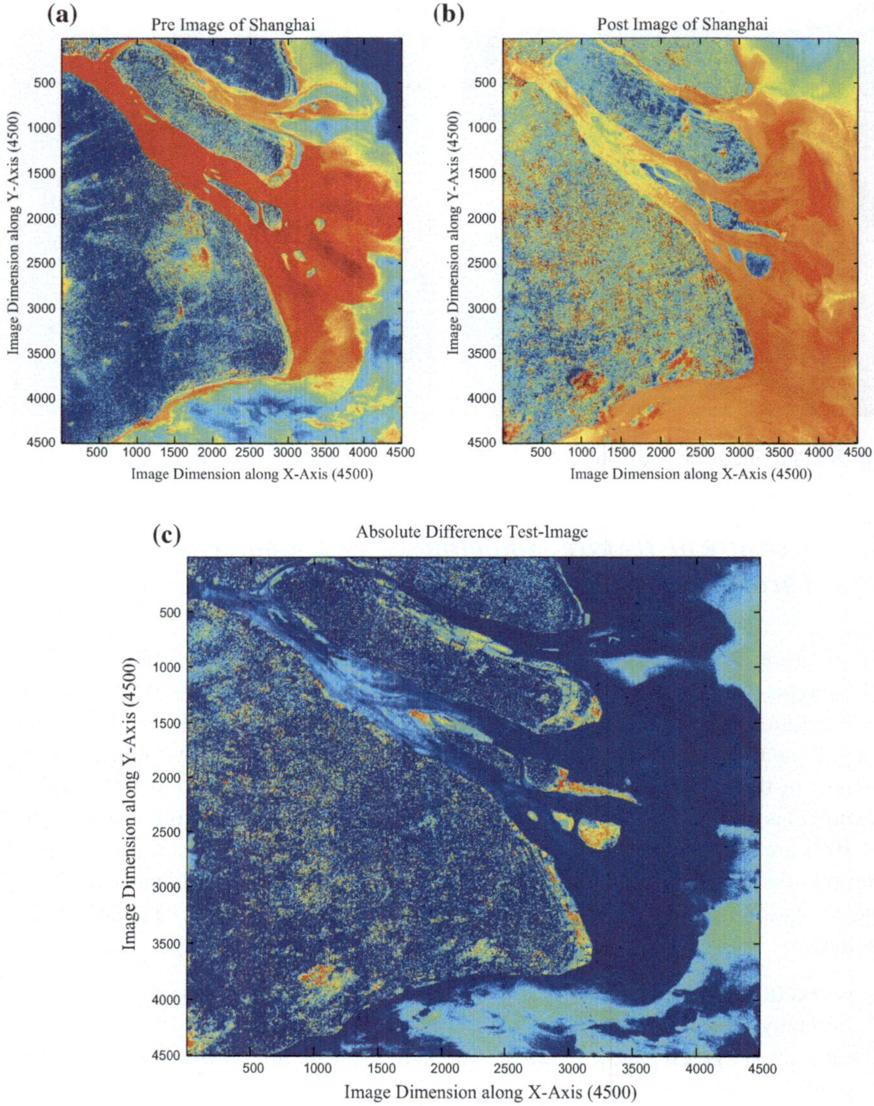

Fig. 9 **a** Greyscale pre-image of Shanghai, **b** greyscale post-image of Shanghai, and **c** change map obtained through absolute image differencing

Table 4 I-order features of urbanisation, Shanghai

S. No.	I-order statistics	Pre-image	Post-image	Difference image	No change
1	Mean	128.277	151.637	24.9885	'0'
2	Standard deviation	9.2968	9.1688	7.4173	'0'

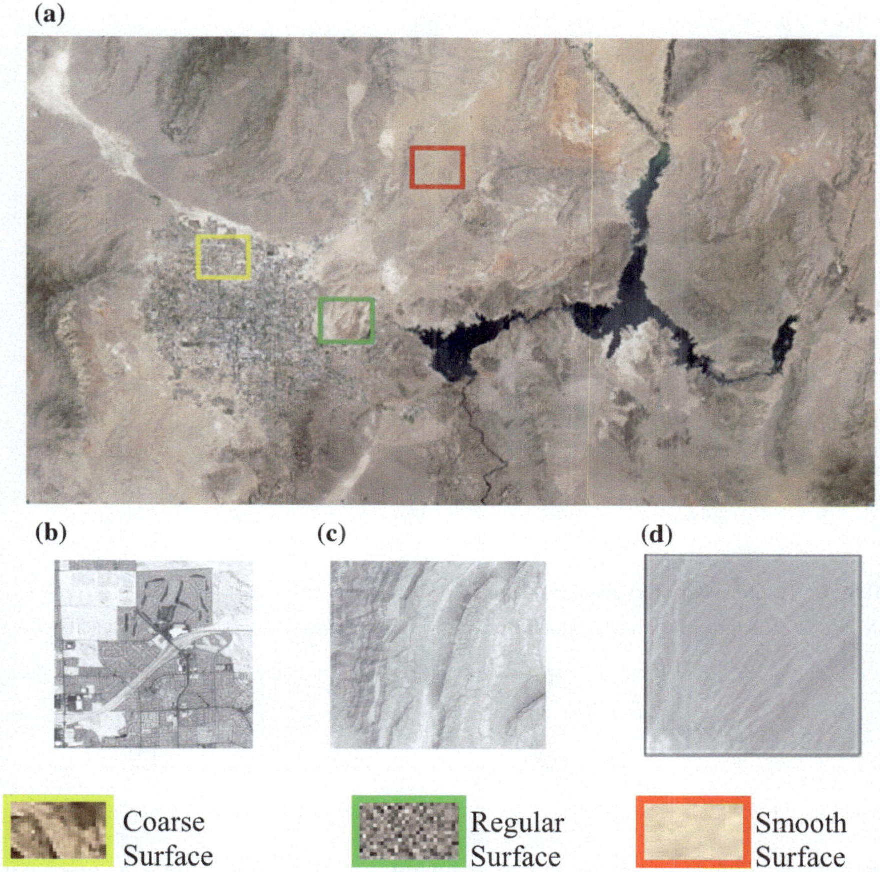

Fig. 10 **a** Test image 2, **b** coarse surface, **c** regular surface, and **d** smooth surface

Table 5 Measure of smoothness and coarseness

S. No.	Subsection	Coarse	Smooth
1	Yellow	Most	Least
2	Green	Less than yellow or more than red	Less than red or more than yellow
3	Red	Least	Most

'green' is an intermediate surface between yellow and red subsections, and it is called a regular surface, it is less coarse than yellow and more coarse then red and at the same time, it is also less smooth than red and more smooth than yellow. Table 5 simply conclude the above information in a systematic manner.

Tables 6 and 7 represents average values of GLCM features contrast, correlation, energy and homogeneity for both subsections 'yellow' and 'red', respectively.

Table 6 GLCM features for coarse 'yellow' surface

TF	Coarse (yellow) surface				
	0 D	45 D	90 D	135 D	Average
Con#	0.5961	0.8265	0.6163	0.8364	0.7188
Corr#	0.7762	0.6899	0.7691	0.6861	0.7298
E#	0.1541	0.1390	0.1497	0.1391	0.1454
H#	0.8055	0.7666	0.8004	0.7659	0.7846

Table 7 GLCM features for smooth 'red' surface

TF	Smooth (red) surface				
	0 D	45 D	90 D	135 D	Average
Con#	0.6180	0.8208	0.7593	0.8565	0.7636
Corr#	0.9038	0.8715	0.8814	0.8661	0.8807
E#	0.1842	0.1482	0.1397	0.1688	0.1602
H#	0.8156	0.7982	0.8112	0.7964	0.8056

Table 8 Texture descriptors changing pattern

S. No.	Texture behaviour	Contrast	Energy	Homogeneity
1.	Smooth-coarse			
2.	Coarse-smooth			

Among these contrast, energy and homogeneity are compared as only they represent texture visual descriptors [88].

GLCM descriptors are showing opposite nature, i.e. for the coarse surface value of contrast is lower than smooth surface while energy and homogeneity show the higher value for smooth surfaces. This symbolises texture, visual descriptor specific behaviour. The changeover of texture from 'smooth' to 'coarse' results in the increment of contrast and decrement of energy and homogeneity. Furthermore, alteration of texture from 'coarse' to 'smooth' results in the decrement of contrast and increment in energy and homogeneity (Table 8).

We have used Matlab 2013 (b) with CPU enabled with 4 GB RAM and 1 TB hard disk for the experimental purpose and Adobe Photoshop for creating the poster in Figs. 1, 3 and 4.

4 Case Study, Case A: Urbanisation

CD through texture classification for the cases of urbanisation comprises three distinct geographical locations.

- Las Vegas situated in the state of Nevada, U.S.A.
- Dallas located in the state of Texas, U.S.A.
- Zhu San Jiao based along the Pearl River Delta, China.

Las Vegas and Dallas lie in the continent of North America and Zhu San Jiao in the continent of Asia. These distinct images are captured from different Landsat sensors, which possess different latitude and longitude positions. The geographical details of these areas are given below.

4.1 Las Vegas, Nevada, U.S.A

The first case investigation is examining the increase in urbanisation of Las Vegas, Nevada, USA. The orientation of Las Vegas is 36° 10′ 30″ north and 115° 08′ 11″ west. The pre- and post-urbanisation images are shown in Fig. 11a and Fig. 11b was acquired on 29 May 1989, and 26 February 2009, respectively. These images were captured by the TM sensor on Landsat 5.

In these images the western portion of the Las Vegas metropolitan area has put on display, the spread of the city in the desert landscape can be seen. Undeveloped land appears along the left edge of the pre urbanisation image. Poster image shows a grassy area which appears less intense in colour. Las Vegas experienced rapid urban growth between 1989 and 2009. Building a city in the desert raised the challenge for meeting residents' water needs [89] (Fig. 12).

The computed GLCM textural features give us Global information for each direction, as these image matrixes are obtained from the grey level images. This will

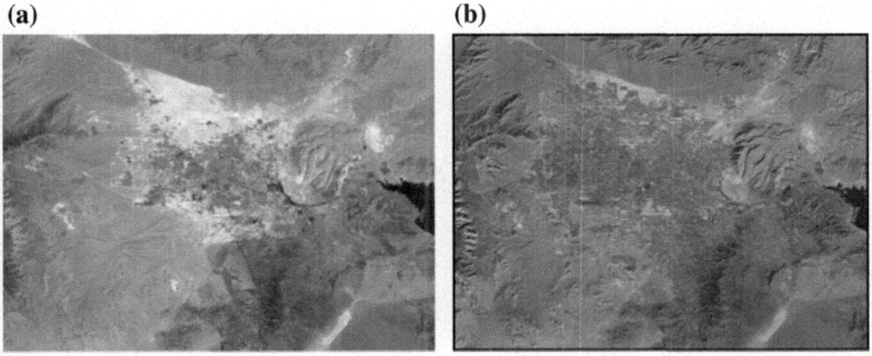

Fig. 11 **a** Pre-urbanisation image of Las Vegas and **b** post-urbanisation image of Las Vegas

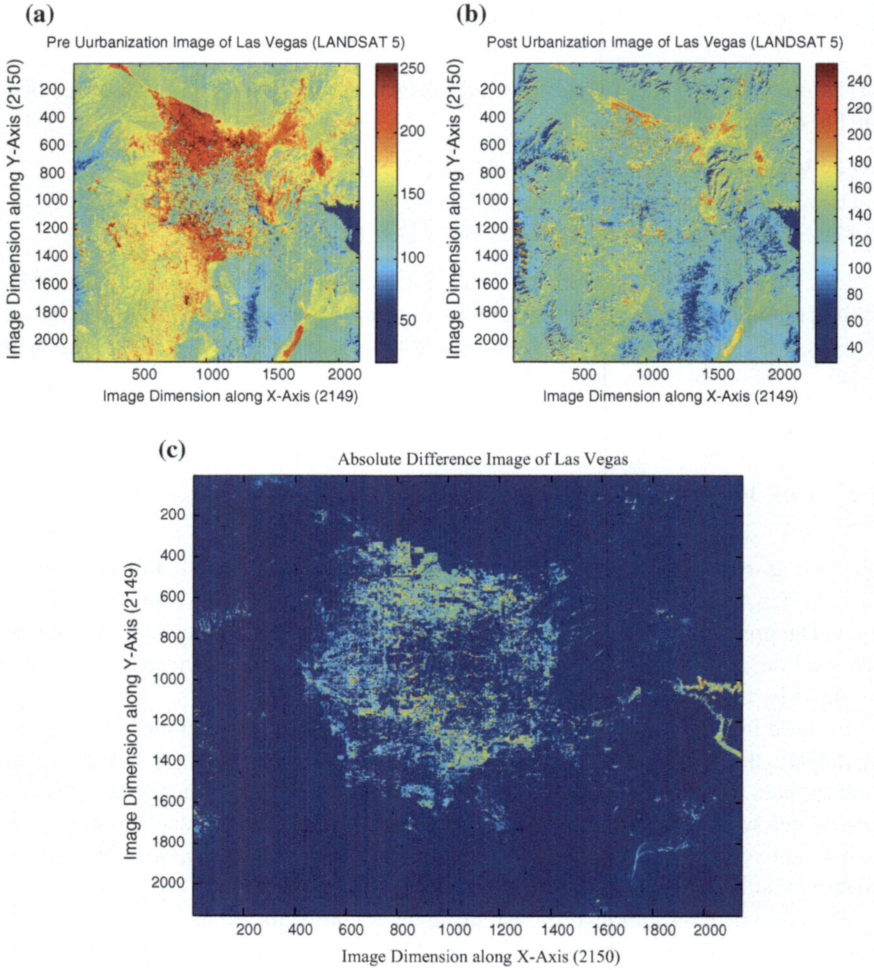

Fig. 12 **a** Greyscale pre-image of Las Vegas, **b** greyscale post-image of Las Vegas, and **c** grey level difference image of Las Vegas

follow the range as (a) Contrast > 0 (greater than zero) (b) Range of correlation [−1, +1] (c) Range of energy [0, 1] and (d) Range of homogeneity [0, 1]. This pattern of the change in TF will be followed in every upcoming case (Table 9).

4.2 Dallas, Texas, U.S.A

The second case study is to boost the investigation of the increase in urbanisation of Dallas, Texas, USA. The orientation of Dallas is 32° 46′ 33″ north and 96° 47′ 48″

Table 9 GLCM for pre- and post-urbanisation image of Las Vegas

TF	0 D	45 D	90 D	135 D	Average
Pre-image GLCM parameters					
Con#	0.3649	0.4652	0.3690	0.4790	0.4195
Corr#	0.8029	0.7488	0.8009	0.7413	0.7734
E#	0.1984	0.1859	0.1990	0.1812	0.1911
H#	0.8612	0.8396	0.8619	0.8338	0.8491
Post-image GLCM parameters					
Con#	0.4064	0.5139	0.4859	0.6333	0.5098
Corr#	0.7552	0.6905	0.7076	0.6186	0.6929
E#	0.1865	0.1725	0.1785	0.1633	0.1752
H#	0.8419	0.8173	0.8278	0.7990	0.8215

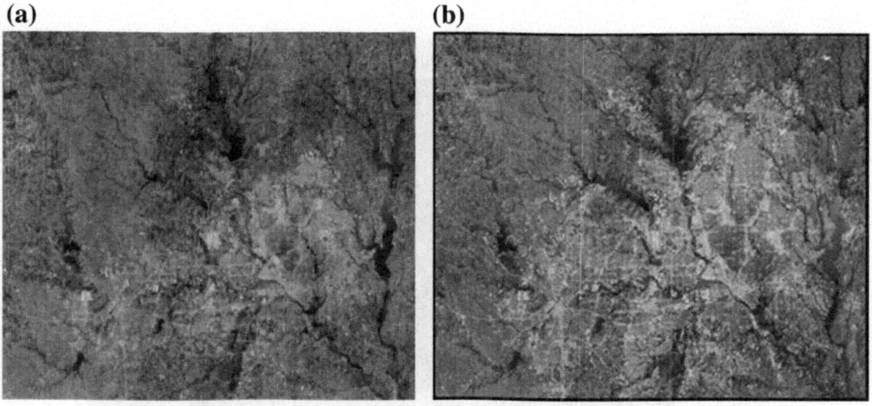

Fig. 13 **a** Pre-urbanisation image of Dallas and **b** post-urbanisation image of Dallas

west. The pre- and post-urbanisation images are shown in Fig. 13a and b which are acquired on 31 August 1984, and 4 August 2009. These images are captured by the TM sensor on Landsat 5 (Fig. 14).

These images show the expansion of the north-eastern margin region of the Dallas metropolitan area which advances explosively. These images highlight the development of three boom bergs Irving, Carrollton and Plano along the north-eastern margin of the Dallas metropolitan area. During the cycle of 15 years, these suburbs grew explosively filling the patchwork of the agricultural lands to the north and the west of Dallas [90] (Table 10).

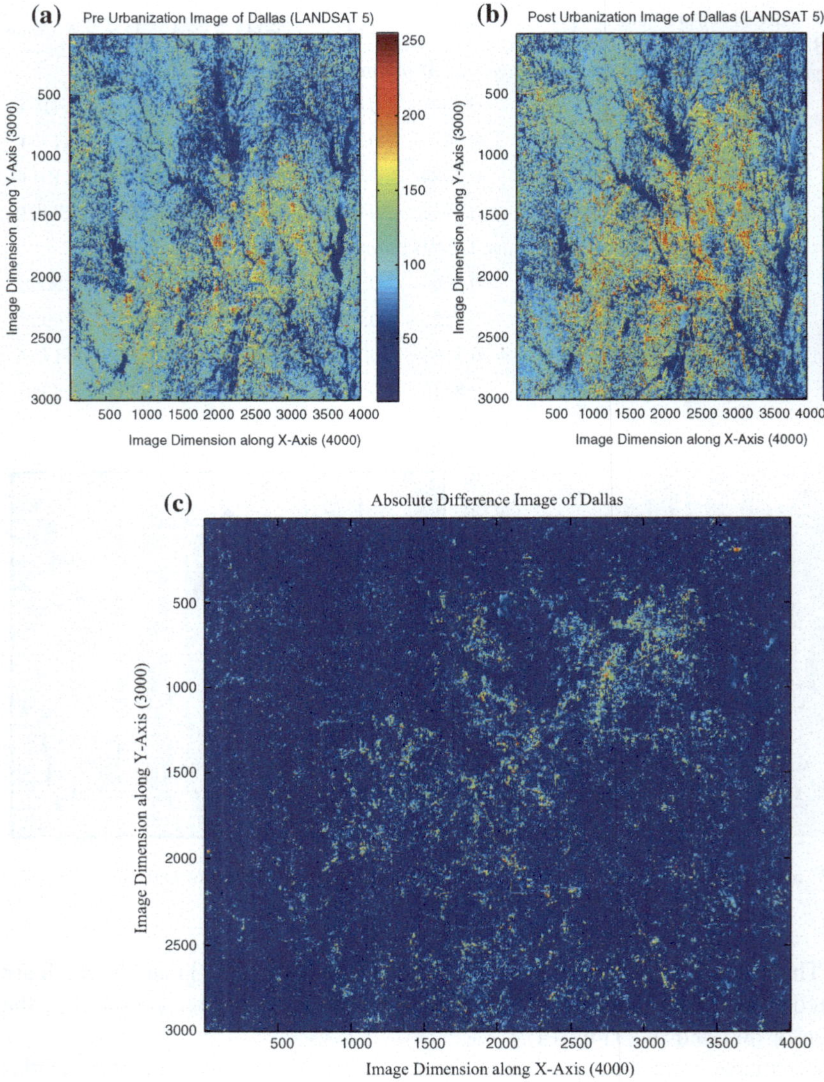

Fig. 14 **a** Greyscale pre-image of Dallas, **b** greyscale post-image of Dallas, and **c** grey level difference image of Dallas

Table 10 GLCM for pre- and post-urbanisation image of Dallas

TF	0 D	45 D	90 D	135 D	Average
Pre-image GLCM parameters					
Con#	0.6573	0.8705	0.6533	0.8669	0.7620
Corr#	0.7471	0.6652	0.7487	0.6665	0.7068
E#	0.1148	0.1013	0.1161	0.1016	0.1084
H#	0.7869	0.7505	0.7899	0.7515	0.7697
Post-image GLCM parameters					
Con#	0.8912	1.1990	0.9005	1.1900	1.0451
Corr#	0.7405	0.6510	0.7379	0.6536	0.6957
E#	0.0879	0.0770	0.0886	0.0773	0.0827
H#	0.7525	0.7128	0.7539	0.7138	0.7138

4.3 China Pearl River Delta (Zhu San Jiao)

The third and final case study is to inspect the increase in urbanisation along the Pearl River Delta (Zhu San Jiao) China. The region of Zhu San Jiao is one of the biggest hubs of electronics goods, resulting in the slow but steady increase of the urban population. The orientation of Zhu San Jiao is 22° 32′ north and 113° 44′ west. The pre- and post-urbanisation images shown in Fig. 15a and Fig. 15b are snapped on 24 November 1988 and 16 November 2014, respectively. The pre-image was captured by the TM sensor on Landsat 5 and post-image by the OLI on Landsat 8.

These images demonstrate rural areas, mainly forest and grassland which appears green and the urban area appears grey and white, around 1988 an interlacing network

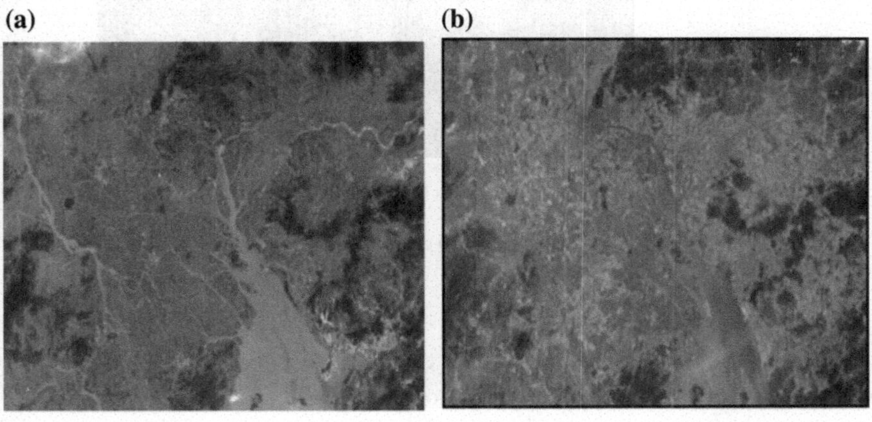

Fig. 15 **a** Pre-urbanisation image of Zhu San Jiao and **b** post-urbanisation image of Zhu San Jiao

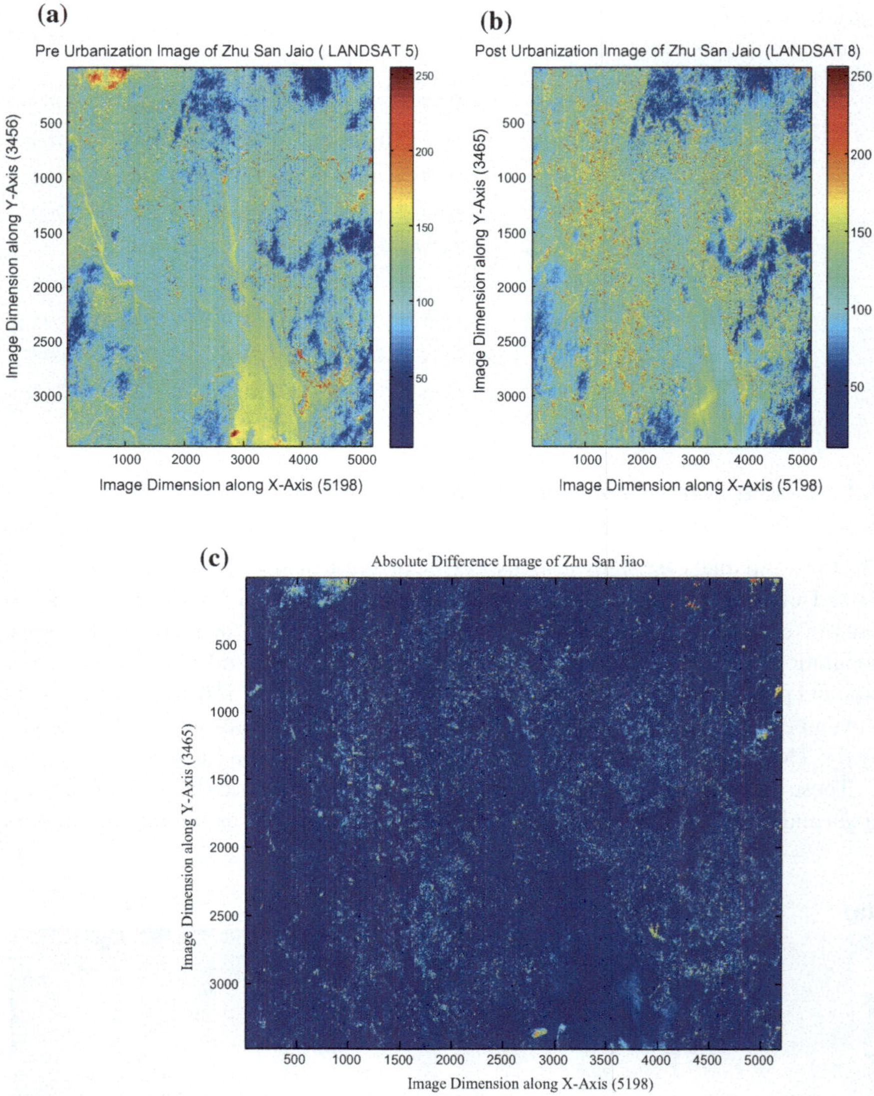

Fig. 16 **a** Greyscale pre-image of Zhu San Jiao, **b** greyscale post-image of Zhu San Jiao, and **c** grey level difference image of Zhu San Jiao

of stream flowed through fertile alluvial soil full of rice paddies and fish ponds. Around three decades later, this city has grown so rapidly that they have merged into an interconnected megalopolis. The Pearl River Delta has a very different pattern of growth compared to other fast-growing cities of China [91] (Fig. 16 and Table 11).

Table 11 GLCM value for pre- and post-image of Zhu San Jiao

TF	0 D	45 D	90 D	135 D	Average
Pre-image GLCM parameters					
Con#	0.5096	0.5600	0.5182	0.5737	0.5403
Corr#	0.6958	0.6656	0.6906	0.6574	0.6773
E#	0.1570	0.1509	0.1563	0.1494	0.1534
H#	0.8145	0.8025	0.8127	0.7987	0.8071
Post-image GLCM parameters					
Con#	0.6675	0.7471	0.6789	0.7533	0.7117
Corr#	0.6726	0.6335	0.6670	0.6305	0.6509
E#	0.1318	0.1243	0.1306	0.1241	0.1277
H#	0.7892	0.7717	0.7862	0.7705	0.7794

5 Case Study, Case B: Lake Dryness

CD through texture classification for the case of lake dryness incorporates three contrasting lakes streaming in two distinct continents. The lakes included in this case are as

- Lake Poopo, Bolivia, U.S.A.
- Lake Eyre, Queensland, Australia.
- Lake Mead, Texas, USA.

Lake Poopo and Lake Mead lie in the continent of North America, whereas Lake Eyre lies on the continent of Australia. Landsat which grabbed these images along with latitude and longitude locations are clarified below.

5.1 Lake Poopo, Bolivia, U.S.A

The first case study is to survey the decline rate of Lake Poopo streaming in the continent of South America, USA. The orientation of Lake Poopo is 18° 33′ north and 67° 05′ west. The pre- and post-lake dry images are shown in Fig. 17a and b are obtained on 12 April 2013, and 15 January 2016, respectively. These images were captured by OLI sensor on Landsat 8.

These images exhibit Lake Poopo of Bolivia, the second largest lake which has essentially dried up. This has happened due to drought and the diversion of the lake water for mining and agriculture. In a typical year rainfall during the wet season recharge the lake directly and via increased inflow from the Desaguadero River, but more than a month in 2015–2016 wet season droughts persist [92] (Fig. 18 and Table 12).

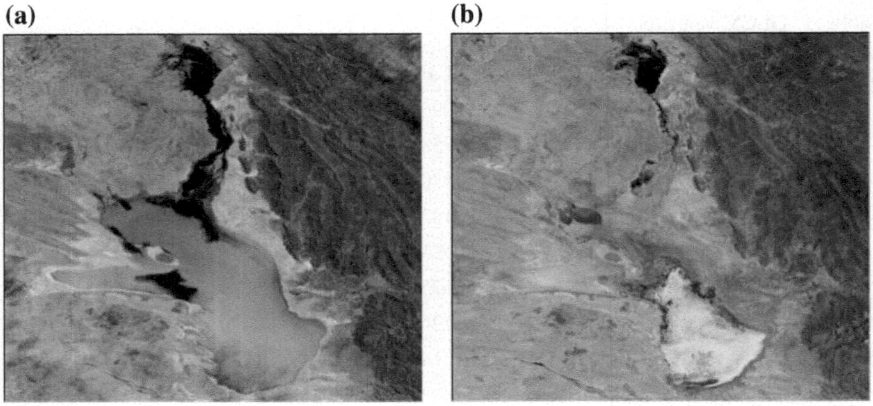

Fig. 17 **a** Pre-lake dry image of Lake Poopo and **b** post-lake dry image of Lake Poopo

5.2 Lake Eyre, Queensland, Australia

The second case study is to analyse the downturn rate of water level in Lake Eyre situated in the state of Queensland, Australia. The orientation of Lake Eyre is 137° 22′ east. The pre- and post-lake dry images shown in Fig. 19a and Fig. 19b were obtained on 22 May 2009, and 10 June 2009, respectively. The images were captured by TM on Landsat 5.

These images display that Lake Eyre contains water, which is tinted green and blue by sediments and algae, but this water does not hold for a long time and get dried up in a quick period of time. The white patches of the land in the post-image indicate that the lake has essentially dried up [93] (Fig. 20 and Table 13).

5.3 Lake Mead, Texas, U.S.A

The third case study is to inspect the drop in the water level of Lake Mead of Texas, USA. The orientation of Lake Mead is 36.25° north and 114.39° west. The pre- and post-images are shown in Fig. 21a and b are obtained on 2 August 1985 and 11 August 2010. The pre- and post-images were snapped by the TM sensor on Landsat 5.

In these images, Lake Mead has reached its lowest level since 1956. The largest reservoir in the United States was shrinking from persistent drought and increasing human water supply demand. Lake Mead provides power and water for human activities in Nevada, Arizona, Southern California and Northern Mexico. Lake Mead dries up because of the rapid population growth of the American south-west and 12 years of the persistent drought [94] (Fig. 22 and Table 14).

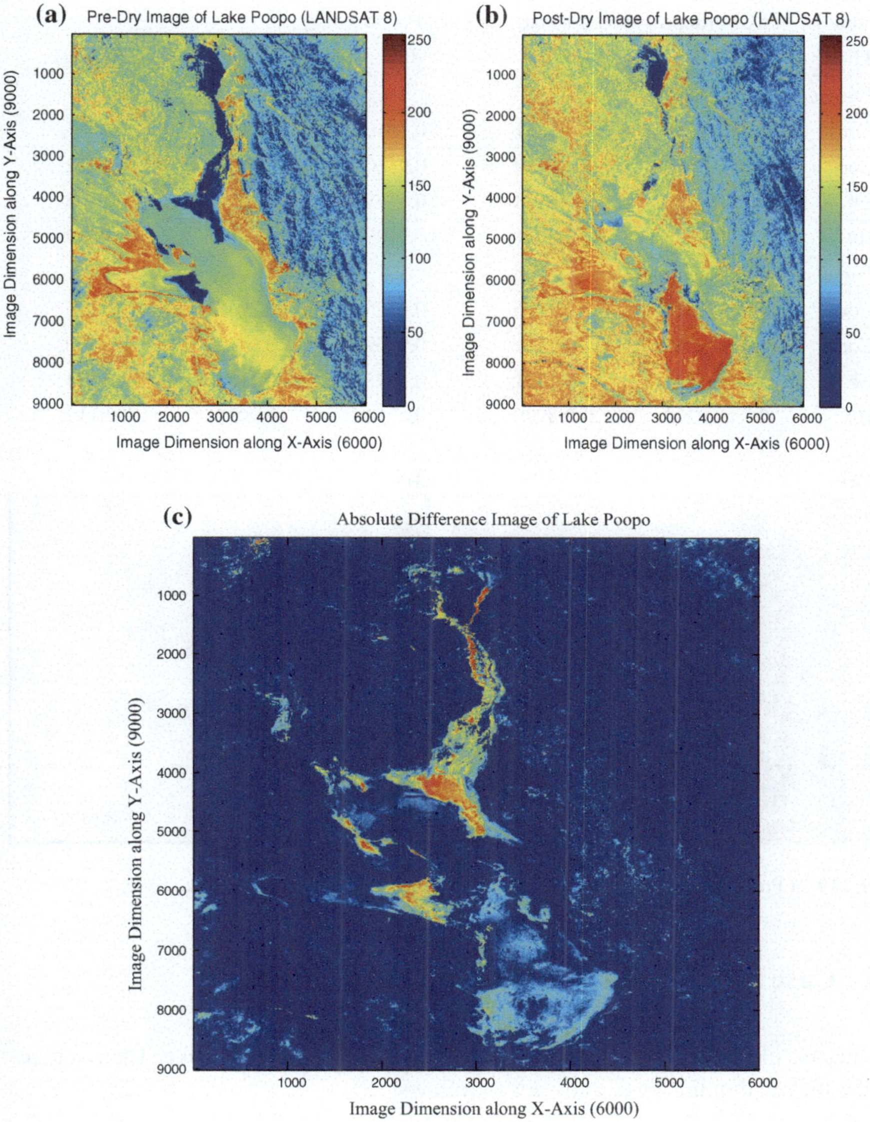

Fig. 18 **a** Greyscale pre-image of Lake Poopo, **b** greyscale post-image of Lake Poopo, and **c** grey level difference image of Lake Poopo

Table 12 GLCM for pre- and post-image of Lake Poopo

TF	0 D	45 D	90 D	135 D	Average
Pre-image GLCM parameters					
Con#	0.1710	0.2351	0.1640	0.2159	0.1965
Corr#	0.9541	0.9369	0.9560	0.9421	0.9472
E#	0.1566	0.1428	0.1578	0.1451	0.1505
H#	0.9176	0.8921	0.9199	0.8983	0.9069
Post-image GLCM parameters					
Con#	0.1628	0.2116	0.1525	0.2078	0.1836
Corr#	0.9539	0.9401	0.9569	0.9412	0.9480
E#	0.1652	0.1521	0.1675	0.1531	0.1594
H#	0.9207	0.8998	0.9252	0.9012	0.9117

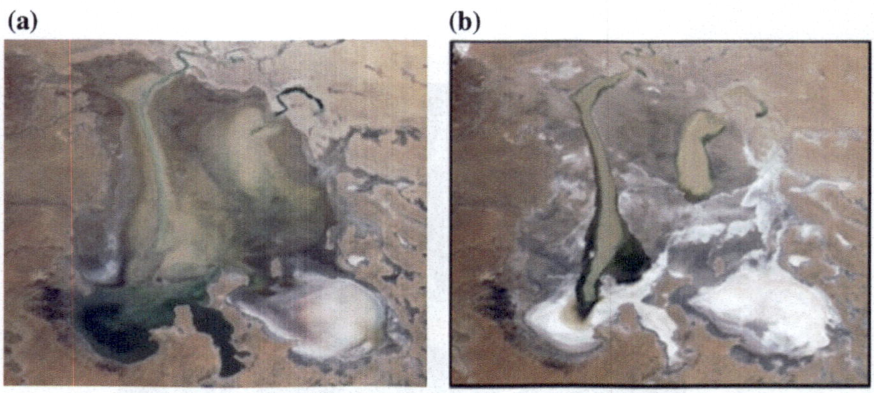

Fig. 19 **a** Pre-lake dry image of Lake Eyre and **b** post-lake dry image of Lake Eyre

6 Case Study, Case C: Flooding

The case of flooding includes floods caused by two rivers and a lake. The rivers and lake included in this case study are as follows:

- River Tigris, Wasit, Iraq.
- Lake Eyre, Queensland, Australia.
- River Mekong, Cambodia.

These rivers and lake lie in the continent of Asia and Australia, respectively. These images are captured using different satellite and sensors, the longitude and latitudes positions in these areas are given below.

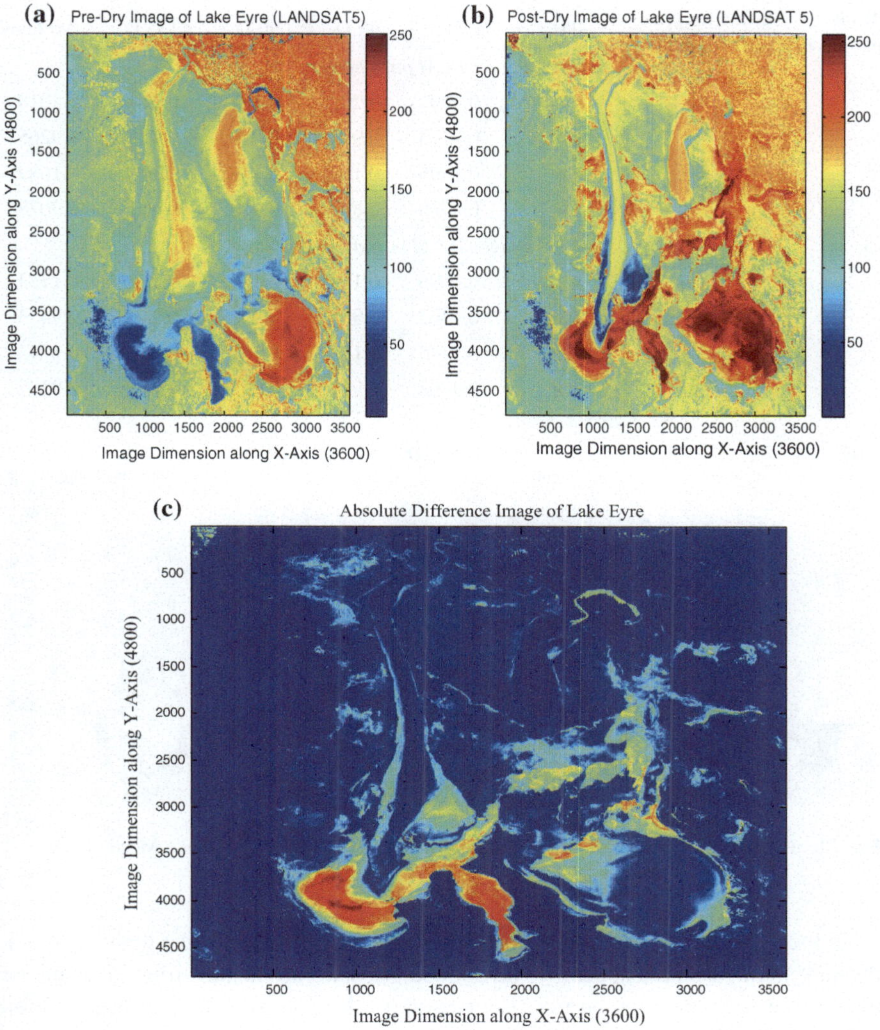

Fig. 20 **a** Greyscale pre-image of Lake Eyre, **b** greyscale post-image of Lake Eyre, and **c** grey level difference image of Lake Eyre

6.1 *River Tigris, Wasit, Iraq*

The first case study is to study the effects of flooding caused by River Tigris in the town of Wasit (also known as Kut) in Iraq from October 2015 to November 2015. The orientation of River Tigris is 32° 30′ 20″ north and 45° 49′ 29″ east. The pre- and post-river dry images are shown in Fig. 23a and Fig. 23b was obtained on 13 October 2015 and 14 November 2015, respectively. The pre- and post-images were captured by OLI sensor on Landsat 8.

Table 13 GLCM for pre- and post-image of Lake Eyre

TF	0 D	45 D	90 D	135 D	Average
Pre-image GLCM parameters					
Con#	0.1997	0.2285	0.1642	0.2149	0.2018
Corr#	0.9273	0.9168	0.9402	0.9217	0.9265
E#	0.1695	0.1636	0.1781	0.1647	0.1689
H#	0.9080	0.8984	0.9227	0.9023	0.9078
Post-image GLCM parameters					
Con#	0.1696	0.1945	0.1365	0.1893	0.1724
Corr#	0.9462	0.9383	0.9567	0.9400	0.9453
E#	0.1745	0.1685	0.1848	0.1694	0.1743
H#	0.9208	0.9116	0.9351	0.9133	0.9202

(a) **(b)**

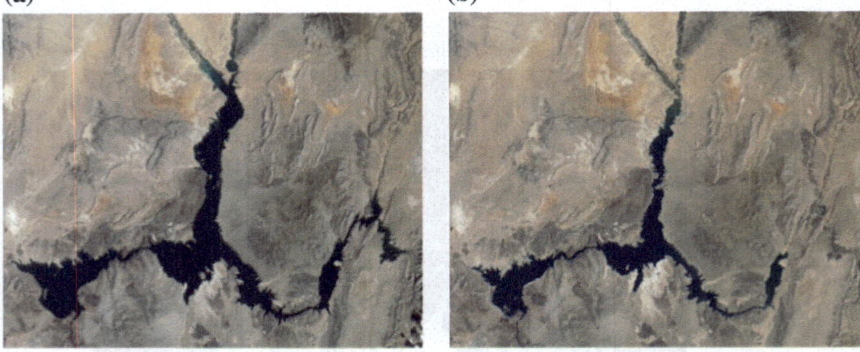

Fig. 21 **a** Pre-lake dry image of Lake Mead and **b** post-lake dry image of Lake Mead

In these images flood water rushes from the Tigris River into the town of Wasit, through some farmland appears to be flooded. Most homes in and around Kut appear to be protected by a flood barrier. The reason for flooding was the torrential rainfall in Iraq and many parts of the Middle East. The resulting floods prompted authorities to declare the State of Emergency, power outages, overflowing sewage and flooded streets were widespread in Bagdad during the time of flooding [95] (Fig. 24 and Table 15).

6.2 *Lake Eyre, Queensland, Australia*

The second case study is to inspect the aftermath of flooding caused by Lake Eyre in the state of Queensland, Australia from February 2009 to May 2009. The orientation of Lake Eyre is 137° 22′ east. The pre- and post-flooding images were obtained on

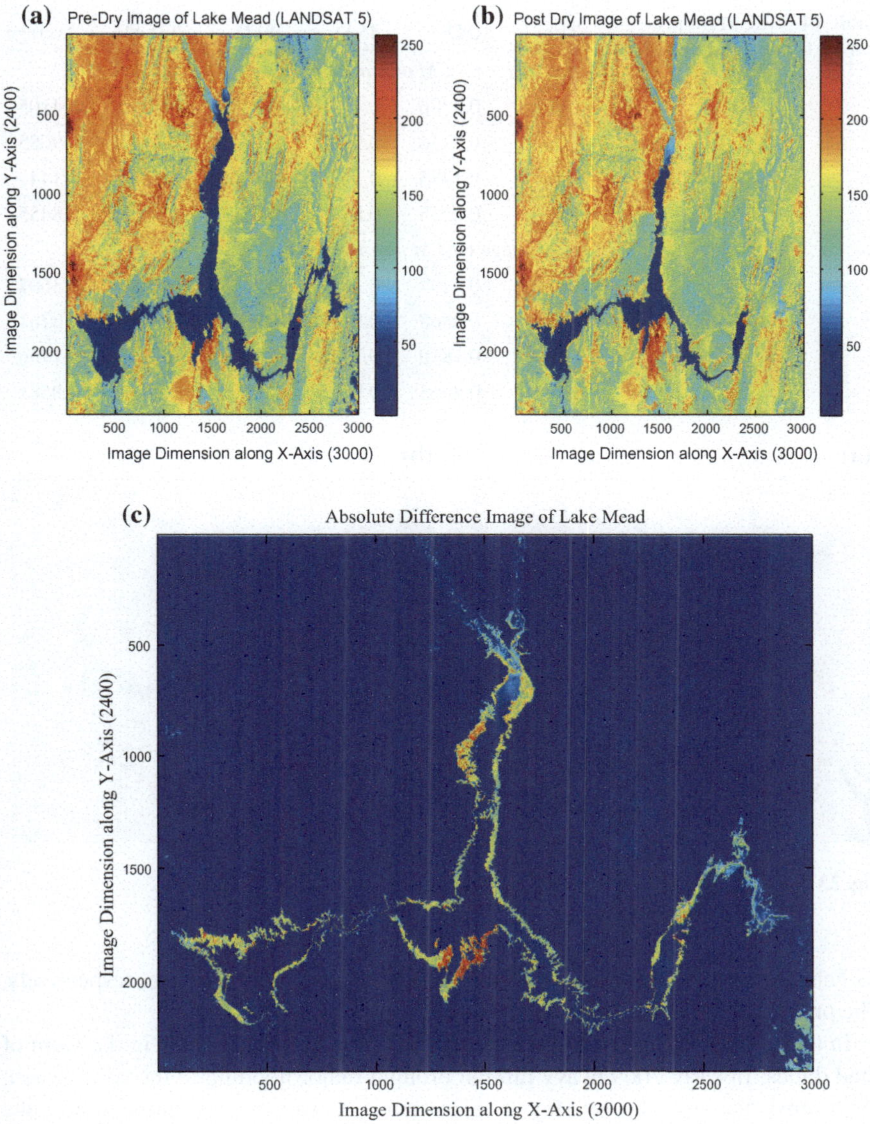

Fig. 22 **a** Greyscale pre-image of Lake Mead, **b** greyscale post-image of Lake Mead, and **c** grey level difference image of Lake Mead

Table 14 GLCM for pre- and post-image of Lake Mead

TF	0 D	45 D	90 D	135 D	Average
Pre-image GLCM parameters					
Con#	0.3786	0.4120	0.3505	0.5011	0.4105
Corr#	0.9156	0.9082	0.9219	0.8883	0.9085
E#	0.1485	0.1437	0.1506	0.1341	0.1442
H#	0.8535	0.8448	0.8586	0.8253	0.8455
Post-image GLCM parameters					
Con#	0.3297	0.3688	0.3158	0.4286	0.3607
Corr#	0.8868	0.8734	0.8915	0.8528	0.8761
E#	0.1810	0.1746	0.1824	0.1645	0.1756
H#	0.8665	0.8566	0.8694	0.8402	0.8581

(a) (b)

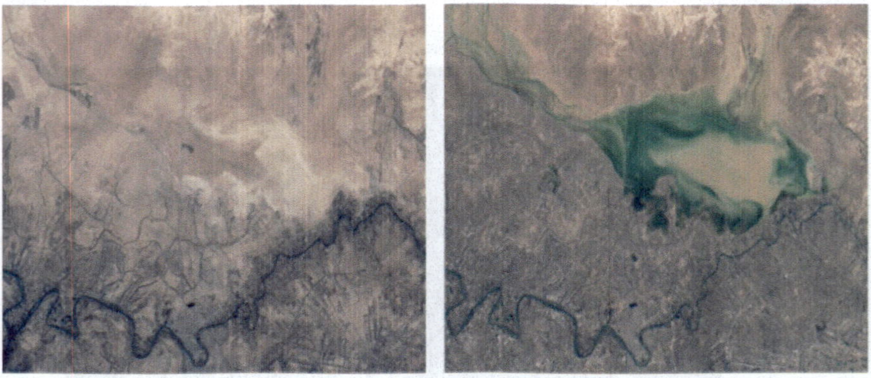

Fig. 23 **a** Pre-flood image of River Tigris and **b** post-flood image of River Tigris

18 February 2009, and 22 May 2009, shown in Fig. 25a and Fig. 25b, respectively. The pre- and post-images are captured by TM on Landsat 5.

In these images, Central Australia Simpson desert usually comes in the form of sand dunes. In early 2009 heavy rainfall brought major flooding to the river system of Queensland, Australia, by May water has made its way and has started to fill Lake Eyre, as the water flow through Lake Eyre and surrounding river channels, lake and lagoons, in the desert got transformed and the lake becomes the breeding ground for thousands of birds such as Pelicans [96] (Fig. 26 and Table 16).

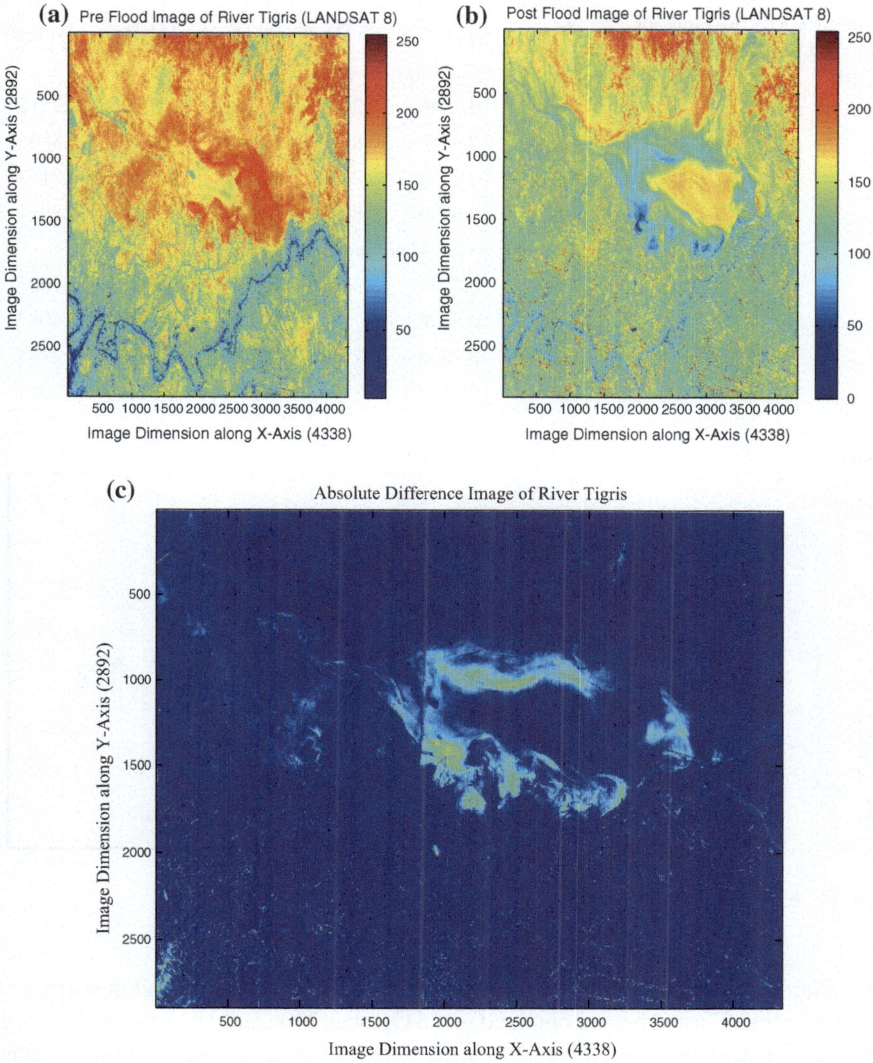

Fig. 24 **a** Greyscale pre-image of River Tigris, **b** greyscale post-image of River Tigris, and **c** grey level difference image of River Tigris

6.3 River, Mekong, Cambodia

The third case study is to study the outcome of flooding caused by River, Mekong and its surrounding area in Cambodia from October 2006 to October 2011. The

Table 15 GLCM value for pre- and post-image of River Tigris

TF	0 D	45 D	90 D	135 D	Average
Pre-image GLCM parameters					
Con#	0.2246	0.2801	0.2280	0.2721	0.2512
Corr#	0.9013	0.8769	0.8998	0.8804	0.8896
E#	0.2230	0.2118	0.2228	0.2133	0.2177
H#	0.8996	0.8817	0.8989	0.8844	0.8911
Post-image GLCM parameters					
Con#	0.2733	0.3343	0.2817	0.3232	0.3031
Corr#	0.8276	0.7891	0.8222	0.7960	0.8087
E#	0.2208	0.2068	0.2212	0.2090	0.2144
H#	0.8973	0.8795	0.8973	0.8824	0.8891

(a) (b)

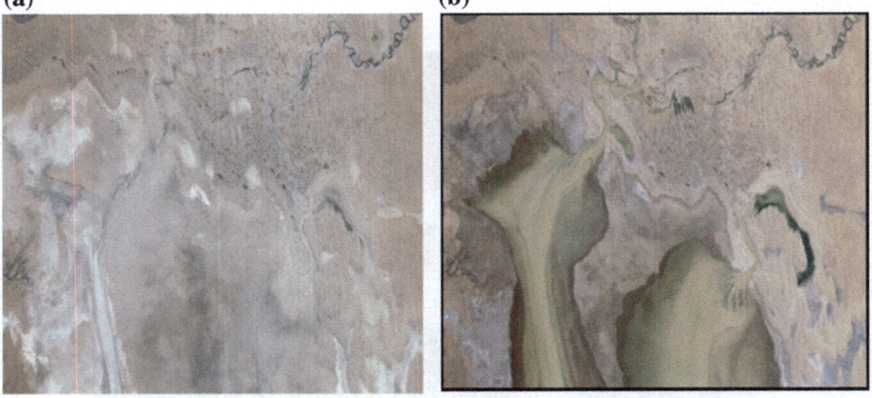

Fig. 25 **a** Pre-flood image of Lake Eyre and **b** post-flood image of Lake Eyre

orientation of River, Mekong is 13° north and 105° east. The pre- and post-images shown in Fig. 27a and b were obtained on 26 October 2006 and 18 October 2011; the pre- and post-images were captured by the MODIS (Moderate Resolution Imaging Spectroradiometer) of the NASA Terra Satellite.

In these images, seasonal rainfall results in the severe flooding in the area of Mekong River and Tonle Sap areas. The flooding causes heavy damage resulting at least 207 people dead and around 34,000 evacuated. Kampong Thom was the most severely impacted province in Cambodia during these floodings [97] (Fig. 28 and Table 17).

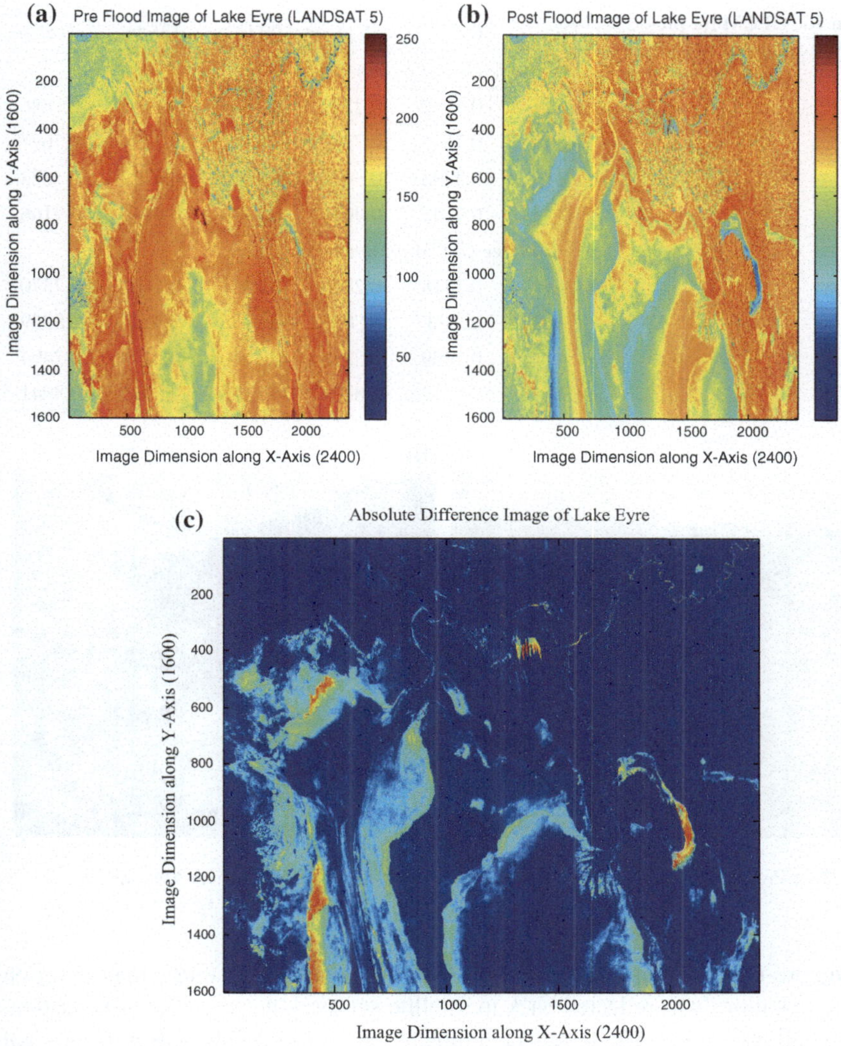

Fig. 26 **a** Greyscale pre-image of Lake Eyre, **b** greyscale post-image of Lake Eyre, and **c** grey level difference image of Lake Eyre

7 Summary for the Event Urbanisation

Now from the bar plot shown in Fig. 29, we observed that TF contrast get increased for the post-urbanisation image surface, whereas energy and homogeneity have greater value for the pre-urbanisation image, therefore we have concluded that initially, the textural behaviour of the surface was 'smooth' and later on with the rapid development of settlements the texture of the surface get converted to 'coarse'. Initially,

Table 16 GLCM for pre- and post-image of Lake Eyre

TF	0 D	45 D	90 D	135 D	Average
Pre-image GLCM parameters					
Con#	0.1897	0.2231	0.1464	0.1953	0.1886
Corr#	0.7666	0.7252	0.8197	0.7594	0.7677
E#	0.3793	0.3675	0.4033	0.3792	0.3823
H#	0.9147	0.9039	0.9339	0.9140	0.9166
Post-image GLCM parameters					
Con#	0.2418	0.2828	0.1834	0.2403	0.2370
Corr#	0..8337	0.8055	0.8738	0.8347	0.8369
E#	0.2469	0.2359	0.2676	0.2475	0.2494
H#	0.8912	0.8774	0.9155	0.8925	0.8941

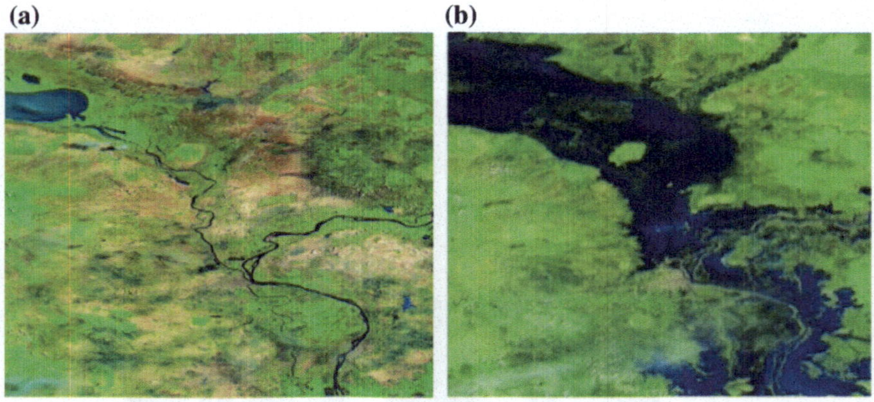

Fig. 27 **a** Pre-flood image of River Mekong and **b** post-flood image of River Mekong

during pre-urbanisation, there were no or very few settlements thus a major portion of the radiations was reflected back to satellite sensor. Later on, after urbanisation, the radiations are not able to reflect completely back to satellite sensor as they got scattered here and there after colliding with the urban settlements.

8 Summary for the Event Lake Dry

Now from the bar plot shown in Fig. 30, we observed that TF contrast is declining and energy and homogeneity are increasing. This variation symbolises the conversion of surface texture from 'coarse' to 'smooth'. This type of pattern is observed because initially with water content in the lake the satellite signals were absorbed by the lake water as water is having non-reflective property and later on when the water in the

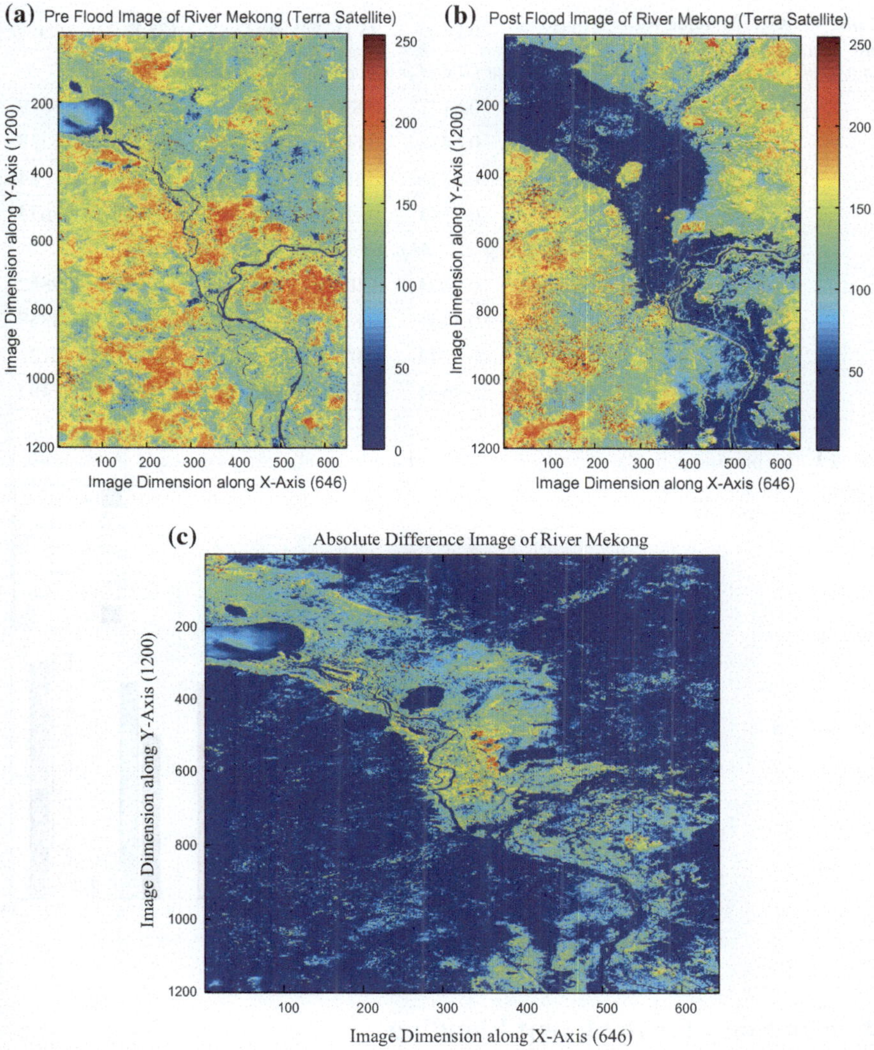

Fig. 28 **a** Greyscale pre-image of River Mekong, **b** greyscale post-image of River Mekong, and **c** grey level difference image of River Mekong

lake get dried out satellite sensors were able to retrieve most of their signals. This sought of the pattern is completely opposite to that of urbanisation.

Table 17 GLCM value for pre- and post-image of River Mekong

TF	0 D	45 D	90 D	135 D	Average
Pre-image GLCM parameters					
Con#	0.8137	0.8982	0.7596	0.8562	0.8319
Corr#	0.6418	0.6047	0.6656	0.6232	0.6338
E#	0.1316	0.1252	0.1316	0.1255	0.1284
H#	0.7744	0.7591	0.7754	0.7606	0.7673
Post-image GLCM parameters					
Con#	0.9094	1.0190	0.9183	1.0943	0.9852
Corr#	0.8465	0.8280	0.8450	0.8153	0.8337
E#	0.0744	0.0704	0.0742	0.0700	0.0722
H#	0.7584	0.7424	0.7576	0.7394	0.7494

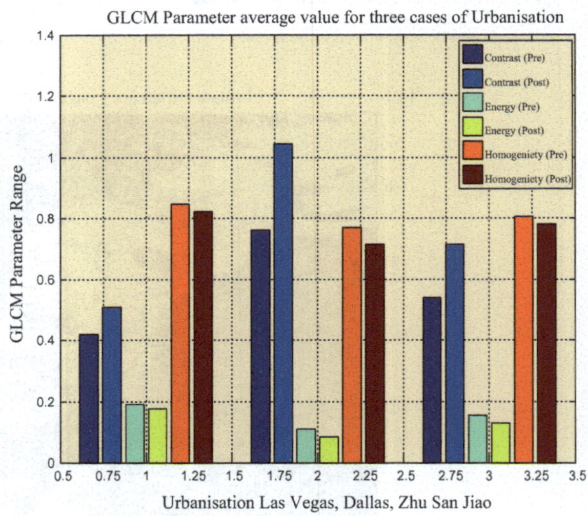

Fig. 29 Bar plot indicating a summary of texture features three cases of urbanisation

9 Summary for the Event Flooding

Now from the bar plot shown in Fig. 31, we observed that TF contrast is increasing whereas energy and homogeneity are showing decreasing nature for pre- and post-events. This sought of change in pattern suggest the conversion of texture from initially 'smooth' to later 'coarse'. This sought of pattern is opposite to that of event lake dry.

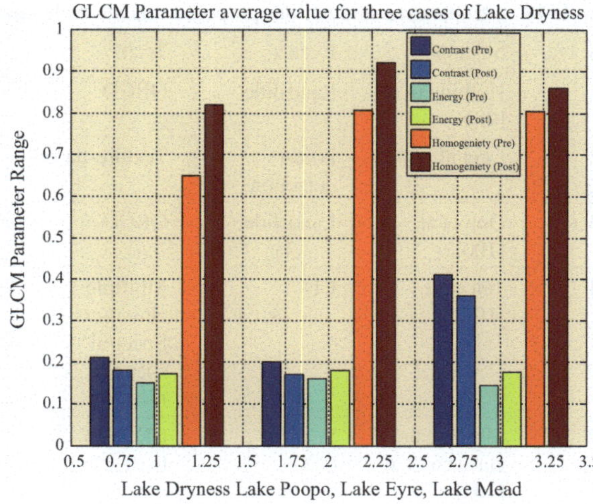

Fig. 30 Bar plot indicating a summary of texture features for three cases of lake dryness

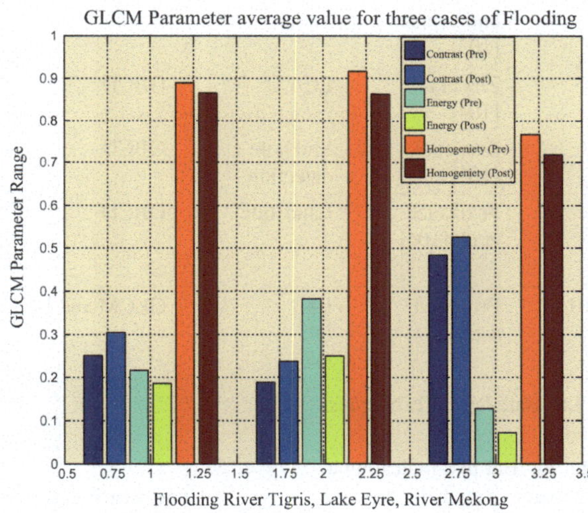

Fig. 31 Bar plot indicating a summary of texture features for three cases of flooding

10 A Comparative Study of the Current Research

In this section, we have provided a comparative overview of our proposed scheme and compared this with other researchers in terms of numbers of events, techniques, regions investigated and satellite data used. The table presents a systematic overview of this comparison (Table 18).

Now from the bar plot diagram shown in Fig. 32 we have concluded that our proposed study is better than other studies as in all the four cases, i.e. data, regions, techniques and events it is having the largest dataset under investigation.

Table 18 Comparative study of our proposed scheme with various other researchers

S. No.	Scientist	Events	Tech*	Region	Data
1	Holbing et al. [98]	Landslide	OBCD	Northern Taiwan	SPOT 5 DEM
2	Li et al. [99]	Forest detection	Texture-based	Tokyo Kuala Lumpur	ZY-3 WV-2
3	Dou et al. [100]	Landslide	OBCD	Conghua	SPOT 5 QB
4	Tan et al. [101]	CD	Multiple classifiers and Spectral spatial features	Random	ZY-3
5	Montereale-Gavazzi et al. [102]	Seafloor CD	CD from backscatter	Belgian-North Sea	Random
6	Shirowzhan and Trinder [103]	U#	SVM and ID	South Wales	Lidar
7	Zha et al. [104]	LC/LU	OBCD	Guangzhou China	SPOT HRV
8	Dou et al. [105]	Sinkhole detection	OBCD	Guangzhou China	Random
9	Fezizadeh et al. [106]	Landslide	OBCD	Urmia Lake	IRS—IDP6 SPOT 5 ALOS
10	Proposed scheme	U# F# D#	GLCM and ID	9 study area	L 5 L 8 MODIS

L Landsat, *Tech** Techniques

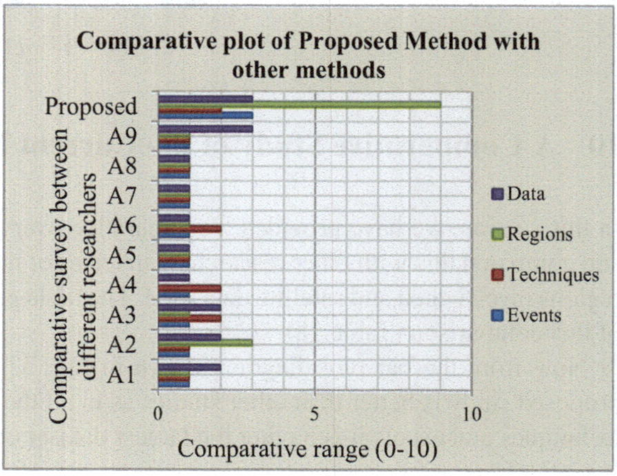

Fig. 32 Bar plot indicating a comparative study between proposed and other studies

11 Conclusion

The main motive for the CD from the pre- and post-event images is to identify, quantify and create a pattern of the changes for the quantification of the changes a post-classification, texture-based GLCM, CD techniques have been used. The assumption to use texture analysis for the CD is that texture is the leading property of the RS images which is affected by changes in the spectral behaviour of the image. Here we have shown CD through texture analysis by GLCM features, which is self-effective technique to provide detailed information about change variation. The three cases of urbanisation, lake dryness and flooding show some interesting results. The following statements can be drawn as a conclusion to this research work.

- In the analysis of the RS images, the texture is a prominent feature.
- Contrast value increases as the degree of coarseness increases, similarly on decreasing the contrast degree of smoothness of surface increases.
- Energy and homogeneity are strongly correlated to each other and inversely correlated with contrast.
- Image differencing proves to be a good pre-classification CD technique. It has gained popularity over other pre-classification because it is reliable, easy to implement and a useful tool for creating change maps.
- GLCM is a marvellous technique for post-classification CD technique because it works on the spectral and spatial relationship of pixels and satellite images get affected by these both properties.
- Event urbanisation has concluded that texture will always be considered 'smooth' before urbanisation and 'coarse' after urbanisation.
- Event Lake dryness and Event Flooding cases show opposite textural behaviour.

Future Scope

The recommended CD technique can be made more realistic by increasing the number of events. New case studies, like desertification, earthquake damage, volcanic eruption, etc., can be enumerated to increase the area of CD. Texture behaviour in the three cases of urbanisation, lake dryness and flooding suggests transformation. The query also arises why texture changes from 'smooth' to 'coarse' and vice versa, which can also be an another area of investigation. Finally, a graphical user interface (GUI) can also be created which can automatically identify the events and classify them accordingly.

Acknowledgements We would like to convey our sincere thanks to United States Geological Survey (U.S.G.S), NASA Earth Observatory and NASA Landsat Science for their valuable Landsat satellite images, data and guidance used in this research work. Authors also wish to express their deep gratitude to honourable reviewers for their useful and constructive suggestions in making this research more useful.

References

1. Yuan D, Elvidge CD, Lunetta RS (1999) Survey of multispectral methods for land cover change analysis. Taylor Fr, Remote Sens Chang Detect 21–22
2. Hussain M, Chen D, Cheng A et al (2013) Change detection from remotely sensed image: from pixel based to object based approaches. ISPRS J Photogramm Remote Sens 80:91–106
3. Alcantara-Ayala I (2002) Geomorphology, natural hazards, vulnerability and prevention of natural disasters in developing countries. Geomorphol 47(47):107–124
4. Van Western C (2000) Remote sensing for natural disaster mangement. International archives of photogrammetry and remote sensing, vol XXXIII, no Part B7, pp 1609–1617
5. Heimlich RE, Barnard CH (1992) Agricultural adaptation to urbanization: farm types in northeast metropolitan areas. North East J Agric Resour Econ 19(2):50–60
6. Majumder B (2010) Landuse and landcover change detection study at Sukinda valley using remote sensing and GIS. Mining Engineering, National Institute of Technology, Rourkela
7. Chen X, Chen J, Shi Y, Yamaguchi Y (2012) An automated approach for updating land cover maps based on integrated change detection and classification methods. ISPRS J Photogramm Remote Sens 71:86–95
8. Lu D, Mausel P, Brondízio E, Moran E (2004) Change detection techniques. Int J Remote Sens 25:2365–2401
9. Singh A (1989) Review article digital change detection techniques using remotely-sensed data. Int J Remote Sens 10:989–1003
10. Lu D, Mausel P, Batistella M (2005) Land-cover binary change detection methods for use in the moist tropical region of the Amazon: a comparative study. Int J Remote Sens 26:101–114
11. Rignot EJM, Van zyl JJ (1993) Change detection techniques for ERS-1 SAR data. IEEE Trans Geosci Remote Sens 31:896–906
12. Chen Z, Elvidge CD, Groeneveld DP (1999) Vegetation change detection using high spectral resolution vegetation indices. Environ Monit Methods Appl, pp 181–190. Taylor & Francis
13. Im J, Jensen JR (2005) A change detection model based on neighbourhood correlation image analysis and decision tree classification. Remote Sens Environ 99:326–340
14. Lillesand TM, Kiefer RW, Chipman JW (2008) Remote sensing and image interpretation, 6th edn. Wiley, Hoboken, New Jersy
15. Mas JF (1999) Monitoring land-cover changes: a comparison of change detection techniques. Int J Remote Sens 20:139–152
16. Huang C et al (2008) Use of a dark object concept and support vector machines to automate forest cover change analysis. Remote Sens Environ 112:970–985
17. Smith A (2010) Image segmentation scale parameter optimization and land cover classification using the random forest algorithm. J Spat Sci 55:69–79
18. Li D (2010) Remotely sensed images and GIS data fusion for automatic change detection. Int J Image Data Fusion 1:99–108
19. Bouziani M, Goïta K, He DC (2010) Automatic change detection of buildings in urban environment from very high spatial resolution images using existing geo database and prior knowledge. ISPRS J Photogramm Remote Sens 65:143–153
20. Caridade CMR, Marçal ARS, Mendonça T (2008) The use of texture for image classification of black & white air photographs. Int J Remote Sens 29:593–607
21. Chen G, Hay GJ, Carvalho LMT, Wulder MA (2012) Object-based change detection. Int J Remote Sens 33:4434–4457
22. Dronova I, Gong P, Wang L (2011) Object-based analysis and change detection of major wetland cover types and their classification uncertainty during the low water period at Poyang Lake, China. Remote Sens Environ 115:3220–3236
23. Villarreal NR (2016) Pixel based and object based classification methods for surveying wetland vegetation with a unmanned aerial system. Thesis, Texas State University, Texas
24. Tomowski D, Ehlers M, Klonus S (2011) Colour and texture based change detection for urban disaster analysis. In: Urban remote sensing event (JURSE) joint, pp 329–332

25. Zhou W, Troy A, Grove M. Object-based land-cover classification and change analysis in the baltimore metropolitan area using multitemporal high resolution remote sensing data. Sens 8:1613–1636
26. Hansen MC, Loveland TR (2012) A review of large area monitoring of land cover change using Landsat data. Remote Sens Environ 122:66–74
27. Xian G, Homer C (2009) Updating the 2001 national land cover database impervious surface products to 2006 using Landsat imagery change detection methods. Remote Sens Environ 114:1133–1147
28. Stow D, Hamada Y, Coulter L, Anguelova Z (2008) Monitoring shrubland habitat changes through object-based change identification with airborne multispectral imagery. Remote Sens Environ 112:1051–1061
29. Duveiller G, Defourny P, Desclee B, Mayaux P (2008) Deforestation in Central Africa: estimates at regional, national and landscape levels by advanced processing of systematically-distributed Landsat extracts. Remote Sens Environ 112:1969–1981
30. Tardie PS, Congalton RG (2002) A change-detection analysis using remotely sensed data to assess the progression of development in Essex County Massachusetts from 1990 to 2001. In: ACSM/ASPRS annual conference, 30 June 2005
31. Stauffer ML, McKinney RL (1978) Landsat image differencing as an automated land cover change detection technique. Computer Sciences Corporation, National Aeronautics and Space Administration, Goddard Space Flight Center, Greenbelt, Maryland
32. Tomowski D, Klonus S, Ehlers M (2010) Change visualization through a texture based analysis approach for disaster applicatins. In: ISPRS TC VII Symposium-100 years ISPRS, vol XXXVII, part 7A, Vienna, Austria, 5–7 July 2010
33. Soh LK, Tsatsoulis C (1999) Texture analysis of SAR sea ice imagery using gray level co-occurence matrices. Trans Geosci Remote Sens 37(2)
34. Barldi A, Parmiggiani F (1995) An investigation of the textural characteristics associated with Gray level co occurence matrix statistical parameters. Trans Geosci Remote Sens 33(2):293–304
35. Lendaris GG, Stanley GL (1978) Diffraction-pattern sampling for automatic pattern recognition. Proceeding IEEE 58(2):198–216
36. Rawat JS, Kumar M (2015) Monitoring land use/cover change using remote sensing and GIS techniques: a case study of Hawalbagh block, district Almora, Uttarakhand, India. Egypt J Remote Sens Space Sci 18:77–84
37. Roknia K, Ahmada A, Solaimanib K, Hazinica S (2014) A new approach for surface water change detection: Integration of pixel level image fusion and image classification techniques l. Int J Appl Earth Obs Geoinf 34:226–234
38. Abd El-Kawy OR, Rod JK, Ismail HA, Suliman AS (2011) Land use and land cover change detection in the western Nile delta of Egypt using remote sensing data. Appl Geogr 31:483–494
39. Willis KS (2015) Remote sensing change detection for ecological monitoring in United States protected areas. Biol Cons 182:233–242
40. Irons JR (2017, June) NASA Landsat science. https://landsat.gsfc.nasa.gov/about/history/
41. Irons JR, Taylor MP, Rocchio L (2017, May) NASA Landsat science. landsat.gsfc.nasa.gov
42. Wikipedia (2017, May) Wikipedia, The Free Encyclopedia. https://en.wikipedia.org/wiki/Landsat_6
43. NASA Official: JR Iron (2017, July) Landsat science. https://landsat.gsfc.nasa.gov/landsat-9/
44. Riebeek H (2013, June) NASA Landsat Science. landsat.gsfc.nasa.gov/historic-landsat-5-mission-ends
45. USGC (2017, June) USGC Landsat missions. https://landsat.usgs.gov/landsat-5-history
46. Wikipedia: The Free Encyclopedia (2017, May) Wikipedia. en.wikipedia.org/wiki/Landsat_8#cite_note-4
47. USGS (2015, January) USGS. https://lta.cr.usgs.gov
48. Roy DP, Wulder MA, Loveland TR, Woodcock CE (2014) Landsat-8: science and product vision for terrestrial global change research. Remote Sens Environ 145:154–172

49. Masek JG (2001) Stability of boreal forest stands during recent climate change: evidence from Landsat satellite imagery. J Biogeogr 28:967–976
50. Cohen WB, Goward SN (2004) Landsat's role in ecological applications of remote sensing. Biosci 54(6):535–545
51. Armston JD, Denham RJ, Danaher TJ, Scarth PF, Moffiet TN (2009) Prediction and validation of foliage projective cover from Landsat-5 TM and Landsat-7 ETM + imagery. J Appl Remote Sens 3(033540)
52. Ekercin S (2007) Coastline change assessment at the Aegean sea coasts in Turkey using multitemporal Landsat imagery. J Coastal Res 23(3):691–698
53. Huang C, Goward SN, Masek JG, Gao F (2009) Development of time series stacks of Landsat images for reconstructing forest disturbance history. Int J Digit Earth 2(3):195–218
54. Application of two regression-based methods to estimate the effects of partial harvest on forest structure using Landsat data. Remote Sens Environ 101:115–126 (2006)
55. Masek JG, Huang C, Wolfe R, Cohen W (2008) North American forest disturbance mapped from a decadal Landsat record. Remote Sens Environ 112:2914–2926
56. Rokni K, Ahmad A, Selamat A, Hazini S (2014) Water feature extraction and change detection using multitemporal landsat imagery. Remote Sens 6:4173–4189
57. Fan F, Weng Q, Wang Y (2007) Land use and land cover change in Guangzhou, China, from 1998 to 2003, based on Landsat TM /ETM + imagery. Sens 7:1323–1342
58. Watts JD, Lawrence RL, Miller PR, Montagne C (2009) Monitoring of cropland practices for carbon sequestration purposes in north central Montana by Landsat remote sensing. Remote Sens Environ 11:1843–1852
59. Cleve C, Kelly M, Kearns FR, Moritz M (2008) Classification of the wildland–urban interface: a comparison of pixel and object based classifications using high-resolution aerial photography. Comput Environ Urban Syst 32:317–326
60. Huang X, Zhang L, Lu Q (2014) A multi-index learning approach for classification of high-resolution remotely sensed images over urban areas. ISPRS J Photogramm Remote Sens 90:36–48
61. Van de Wouwer G, Scheunders P, Van Dyck D (1999) Statistical texture characterization from discrete wavelet representations. Trans Image Process 8(4):592–598
62. Julesz B (1981) Textons, the elements of texture perception, and their interactions. Nat 290:91–97
63. Sebe N, Lew MS (2003) Robust computer vision: theory and applications, vol. 26. Springer Science + Business Media, Dordrechi
64. Haralick RM (1979) Statistical and structural approaches to texture. IEEE 67(5)
65. Lillesand T, Kiefer RW, Chipman J (2015) Remote sensing and image interpretation, 7th edn. Wiley, United States of America
66. PCI, Using PCI software, 1997
67. Lu D, Weng Q (2007) A survey of image classification methods and techniques for improving classification performance. Int J Remote Sens 28(5):823–870
68. Eastman JR (2003) Guide to GIS and image processing, vol 1. Clark Labs, Clark University, Worcester, MA, USA, 14, pp 239–247
69. Otukei JR, Blaschke T (2009) Land cover change assessment using decision trees, support vector machines and maximum likelihood classification algorithms. Int J Appl Earth Obs Geoinf 12(5):527–531
70. Dai X, Khorram S (1999) A feature-based image registration algorithm using improved chain-code representation combined with invariant moments. Trans Geo-Sci Remote Sens 37(5):2351–2362
71. Hong Y, Hsu K-l, Sorooshian S, Gao X (2004) Precipitation estimation from remotely sensed imagery using an artificial neural network cloud classification system. J Appl Methodol 43:1834–1852
72. De Fries RS, Hansen M, Townshend JRG, Sohlberg R (1998) Global land cover classi®cations at 8 km spatial resolution: the use of training data derived from Landsat imagery in decision tree classifiers. Int J Remote Sens 19(16):3141–3168

73. Melesse AM, Jordan JD (2002) A comparison of fuzzy vs. augmented-ISODATA classification algorithms for cloud-shadow discrimination from Landsat images. Photogramm Eng & Remote Sens 68(9):905–911
74. Kanungo T et al (2001) An efficient k-means clustering algorithm analysis and implementation. Trans Pattern Anal Mach Intell 24(7):881–892
75. Hall-Beyer M (2008, August) The GLCM tutorial home page. http://www.fp.ucalgary.ca/mhallbey/
76. Clausi DA (2001) An analysis of co-occurrence texture statistics as a function of grey level quantization. Can J Remote Sens 28(1):45–62
77. Gadkari D (2000) Image quality analysis using GLCM. Thesis. College of Arts and Sciences, University of Central Florida, Orlando, Florida
78. Haralick RM, Shapiro LS (1991) Computer and robot vision, vol. I. Addison-Wesley
79. Iqbal J, Thomasson JA, Jenkins JN, Owens PR, Whisler FD (2005) Spatial variability analysis of soil physical properties of alluvial soils. Soil Water Manag Conserv 69
80. Myint SW (2002) Fractal approaches in texture analysis and classification of remotely sensed data: comparisons with spatial autocorrelation techniques and simple descriptive statistics. Int J Remote Sens 24(9)
81. Read JM (2003) Spatial analyses of logging impacts in Amazonia using remotely sensed data. Photogramm Eng Remote Sens 69(3):275–282
82. Myint SW, Lam N (2005) A study of lacunarity-based texture analysis approaches to improve urban image classification. Comput, Environ Urban Syst 29:501–523
83. Unser M (1995) Texture classification and segmentation using wavelet frames. Trans Image Process 4(11)
84. Kemmouche A, Mering C, Sansal B, Dewolf Y (2004) Macro-texture mapping from satellite images by morphological granulometries: application to vegetation density mapping in arid and semi-arid areas. Int J Remote Sens 25(23):5319–5335
85. Musci M, Feitosa RQ, Costa GAOP, Velloso MLF (2013) Assessment of binary coding techniques for texture characterization in remote sensing imagery. Geosci Remote Sens Lett 10(6)
86. Wang JR, O Neill PE, Jackson TJ, Engman ET (1982) A multi-frequency measurement of thermal microwave emission from soils: the effects of Soil texture and surface roughness. Goddard Space and Flight Centre, USDA Beltsville Agricultural Research Center, Maryland
87. Lambina EF, Turnerb BL, Geista HJ, Agbolac SB, Angelsend A (2001) The causes of land-use and land-cover change: moving beyond the myths. Glob Environ Chang 11:261–269
88. Julesz (1962) Visual pattern discription. IRE Trans Info Theory 8:84–92
89. Allen J, Simmon R, Scott M (2009, March) NASA earth observatory. https://earthobservatory.nasa.gov/IOTD//view.php?id=37228
90. Scott M, Lindsey R (2010, February) NASA earth observatory. https://earthobservatory.nasa.gov/IOTD/view.php?id=42551&src=ve
91. Voiland A (2015, September) NASA earth observatory. https://earthobservatory.nasa.gov/IOTD/view.php?id=86603
92. Hansen K (2016, January) NASA earth observatory. https://earthobservatory.nasa.gov/NaturalHazards/view.php?id=87363
93. Riebeek H (2009, June) NASA earth observatory. https://earthobservatory.nasa.gov/IOTD/view.php?id=38994&src=ve
94. Carlowicz M (2014, August) NASA earth observatory. https://earthobservatory.nasa.gov/IOTD/view.php?id=84105&src=ve
95. Voiland A (2015, November) NASA earth observatory. https://earthobservatory.nasa.gov/IOTD/view.php?id=87011
96. Riebeek H (2009, May) NASA earth observatory. https://earthobservatory.nasa.gov/IOTD/view.php?id=38717&src=ve
97. Riebeek H (2011, October) NASA earth observatory. https://earthobservatory.nasa.gov/IOTD/view.php?id=76212&eocn=image&eoci=related_image
98. Hölbling D, Friedl B, Eisank C (2015) An object-based approach for semi-automated landslide change detection and attribution of changes to landslide classes in northern Taiwan. Springer: Earth Sci Inf 8:327–335

99. Li Q, Huang X, Wen D, Liu H (2017) Integrating multiple textural features for remote sensing image change detection. Photogramm Eng & Remote Sens 83(2):23–35
100. Dou J et al (2015) Automatic case-based reasoning approach for landslide detection: integration of object-oriented image analysis and a Genetic Algorithm. MDPI: Remote Sens 7:4318–4342
101. Tan K, Jin X, Plaza A, Wang X, Xiao L (2016) Automatic change detection in high-resolution remote sensing images by using a multiple classifier system and spectral–spatial features. IEEE: J Sel Top Appl Obs Remote Sens 9(8):3439–3451
102. Montereale-Gavazzi G et al (2017) Seafloor change detection using multibeam echosounder backscatter: case study on the Belgian part of the North Sea. Mar Geophys Res 39:229–247
103. Shirowzhana S, Trinder J (2017) Building classification from Lidar data for spatio-temporal assessment of 3D urban developments. Elsevier: Procedia Eng 180:1453–1461
104. Zhu J, Yanjun S, Guo Q, Harmon TC (2017) Unsupervised object-based differencing for land-cover change detection. Photogramm Eng & Remote Sens 83(3):225–236
105. Dou J et al (2015) Automatic detection of sinkhole collapses at finer resolutions using a multi-component remote sensing approach. Springer: Nat Hazards 78(2):1021–1044
106. Feizizadeha B, Blaschkeb T, Tiedeb D, MHR Moghaddama (2017) Evaluating fuzzy operators of an object-based image analysis for detecting landslides and their changes. Elsevier: Geomorphol 293(2):240–254

User Interface Design for MSP430 IoT Hardware

K. Ram Tejaswini, J. Prathyusha, G. Rahul Kumar and N. Abid Ali Khan

Abstract Low-power microcontroller is one of the key challenges in designing Internet-of-Things (IoT) nodes. This paper proposes TI's MSP430 as the MCU and discusses the architectural features. Developing a user interface to interact with IoT using the 2 × 16 LCD, UART, ADC and Interrupts is highlighted in the second part of the paper. The choice of operating the LCD either in 8-bit mode or in 4-bit mode along with the detailed device driver implementation using CCS cloud IDE with the results was projected.

Keywords Microcontroller · Embedded system · IoT · Device drivers · Debugging

1 Introduction

Selection of microcontroller with low power consumption is one of the key differentiating factors for any sensory and consumer market segments. The configurations of LCD drivers, UART, interrupts, ADC which are usually done by c51 can be replaced by MSP430 to decrease the power consumption dramatically [1]. Thus, the world's first low-power MCU is used to its maximum potential to increase the battery life.

Experimental setup: The specific chip we are using is MSP430G2553 of MSP430 family given by TEXAS INSTRUMENTS in a Launchpad as in Fig. 1 [2]. The software we use is the cloud environment provided for, i.e., CCSV6, CCS CLOUD IDE. The software used for UART is PUTTY.

K. Ram Tejaswini (✉) · J. Prathyusha · G. Rahul Kumar · N. Abid Ali Khan
Department of Electronics & Communication Engineering, Vasavi College of Engineering (Autonomous), Hyderabad 500031, India
e-mail: ramtejaswinik123@gmail.com

N. Abid Ali Khan
e-mail: abid.net@gmail.com

© Springer Nature Singapore Pte Ltd. 2019

V. Nath and J. K. Mandal (eds.), *Proceedings of the Third International Conference on Microelectronics, Computing and Communication Systems*, Lecture Notes in Electrical Engineering 556, https://doi.org/10.1007/978-981-13-7091-5_30

Fig. 1 Proposed TI's MSP430G2553 MCU launchpad

2 MSP430 as IoT Reference Hardware

MSP430 is a mixed signal processor in which both analog and digital lie on a single semiconductor die. In contrast, with other 8-bit MCUs where trading on an equal footing of data and addresses is not possible, MSP430 has a 16-bit architecture throughout the address, the data, registers, and the timers in the CPU. The CPU has plenty of registers which can be used equally for data or address. It has orthogonal addressing.

The MSP430 Value Line controllers are cheaper than the higher end AT-Mega controllers used in the Arduino platform. You do not need an external crystal, because the MSP430 controllers can run at the full 16 MHz from their internal clock source. With the MSP430 Launchpad, TI offers a development board that includes everything you need for about a quarter of the cost of an Arduino board.

As projected in Fig. 2, this MCU consumes less power per instruction than others due to the lowest switching time; preference is given to it over other development boards.

3 LCD Driver Design and Implementation

LCD can be configured using 8-bit as well as 4-bit mode. The 4-bit mode is preferred as it uses less number of pins while using the other pins for other purposes.

Processor	16-bit RISC	32-bit ARM Cortex M3	16-bit H8/300H
Vendor	TI	ST	Renesas
Manufacturer part #	MSP430F2617TPMR	STM32F103CBT6	DF38099FP10V
Max speed	16 MHz	72 MHz	10 MHz
Flash memory	92 kbytes	128 kbytes	128 kbytes
RAM	8 kbytes	20 kbytes	4 kbytes
Speed & voltage	2.2 V @ 1 MHz	2.4 V @ 8 MHz	2.7 V @ 4 MHz
Active mode current	365 μA	8000 μA	3000 μA
Sleep mode current	0.5 μA	2.8 μA	1.5 μA
Wake-up time	1 μs	1.8 μs	Not specified
Package	64-LQFP	48-LQFP	100-LQFP
Dimensions (mm^2)	10 by 10 mm	7 by 7 mm	14 by 14 mm
IO Pins	48	37	75
On-chip ADC	12 bits, 8 channels	12 bits, 10 channels	10 bit, 8 channels
On-chip LCD driver	No	No	Yes
Compiler Vendor	IAR	Keil or IAR	Renesas
Compiler Cost	$2500	Free (Keil)	Free

Fig. 2 Features of MSP430 launchpad

*3.1 8-Bit 16 * 2 LCD*

First, we program to control the MCU from entering the five power modes available by setting watchdog timer as below (Fig. 3):

```
WDTCTL=WDTPW+WDT HOLD;
```

As the MSP430G2553 has only 20 pins, we decided to give P1.0 to P1.7 to data lines and P2.0, P2.1, P2.2 as the control pins. Initially, we are initializing the port1 and port2 by using the following instructions:

```
P1DIR|= 0xFF
P2DIR|= 0xFF
P2OUT&= ~ (EN)
_delay_cycles(int);
```

For delay, we used the functions which are predefined in MSP430G2553.h. We can call the function as indicated in the above source listing.

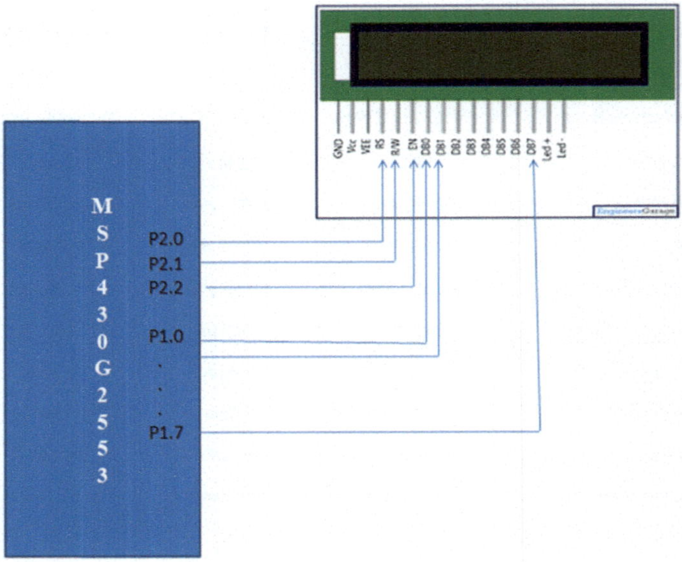

Fig. 3 Eight-bit 16 * 2 LCD connections

The LCD commands are similar to that of C51 MCU and are easy to program [3].

3.2 *4-Bit 16 * 2 LCD*

Now as the data to be sent is 8 bit and the data lines we have are 4 as shown in Fig. 4; hence, we are dividing the 8-bit data into two groups of 4 bits each called Nibbles and sending the higher nibble first through the data lines, then the second one and finally perform the action after combining both.

```
P1OUT = (P1OUT & 0xF0)|((cmd>>4) & 0x0F);
        // (send higher nibble)
P1OUT = (P1OUT & 0xF0)|(cmd & 0x0F)
        // (send lower nibble)
```

Now as we need only four data lines and three control lines, we used only the port1.

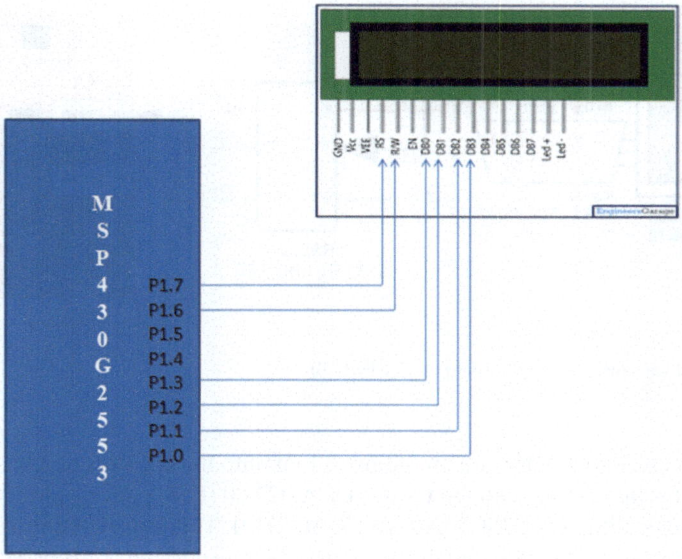

Fig. 4 Four-bit LCD mode connections

P1.0 to P1.3: Data lines
P1.4: RS
P1.6: EN

Here as we are always writing the command line R/ (~W) is permanently grounded, and hence, we need not write any code for the third control signal. Here also, the LCD commands are the same as that of C51 in 4-bit mode.

4 Built-In I/O Configuration and Integration

The MSP430G2553 Launchpad has some built-in I/O configurations such as UART, ADC, and Interrupts which can be configured using GPIO pins as per given instructions in MSP430G2553.H header file. The integration of these I/O devices is shown in the following.

4.1 UART Configuration and Integration

The UART pins P1.1 and P1.2 can be used for serial communication using the USB cable configured for the com port required and required baud rate [4]. After setting

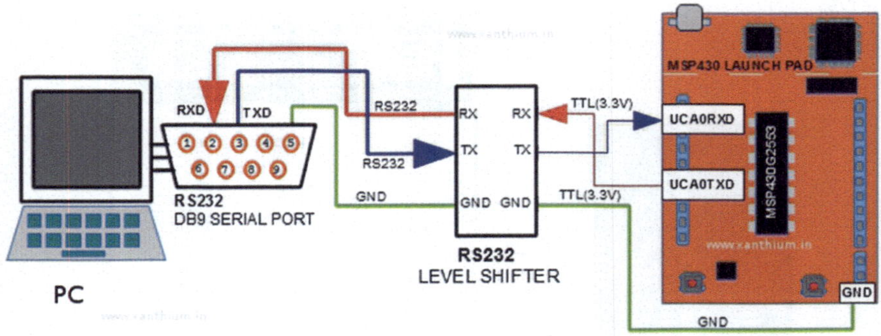

Fig. 5 Hardware interfacing of UART with MSP430

the system clock to 1 MHz, we are going to look into the registers and bits that are important for interacting with the UART. UCAxCTL0 (USCI_Ax Control register0) here we are dealing with USC1_A0, so UCA0CTL0. This register controls the settings for the parity selection, the direction of data transmission (LSB or MSB first), the character length, the number of stop bits, and the modes of serial transmission [5]. We choose the com port 6 and set the baud rate as 9600 while creating the project so that the hardware connected to that com port is derived (Fig. 5).

Recently, as USB ports are replacing DB9 serial ports in the computers, we are using a USB to serial converter board to connect PC to a Launchpad [6]. USB to serial converter boards provide a virtual serial port which you can read and write like a standard serial port. When you read or write to the virtual serial port, the data you have sent passes through the standard USB port to the converter chip (e.g., FT232) which translates to the serial data stream.

We use terminal program PUTTY for listening in on the bytes sent by a Launchpad. Here as getting the entire hardware is difficult we used the loop-back method and got the receiver connected to a transmitter.

```
UCA0STAT |= UCLISTEN; // loop-back mode enabled
UCA0TXBUF = 'A'; //Transmit a byte
```

We transmit a byte of data by using the above source listing. The data is transmitted and verified using a loop-back method (Fig. 6).

4.2 ADC Configuration and Integration

To understand the functionality of ADC we tried to measure the temperature of the room using ADC. At first, we figured that the temperature is always being measured by sensors internally and is stored in ADC to prevent the MCU from damaging. But

Fig. 6 Loop-back mode in MSP430G2553

that value stored is a long value and is not directly in degree centigrade or Fahrenheit, so we need to calibrate it. The temperature measured is stored in ADC memory.

```
temp = ADC10MEM; // temperature stored in memory
ADC10CTL1 = INCH_10 + ADC10DIV_3;
    // (Temp Sensor ADC10CLK/4)
ADC10CTL0 |= ENC + ADC10SC;
    // (Sampling and conversion start)
oF = ((A10/1024)*1500mV)-923mV)*1/1.97mV
    = A10*761/1024 – 468
    //calibration for degree Fahrenheit
oC = ((A10/1024)*1500mV)-986mV)*1/3.55mV
    = A10*423/1024 – 278
  // calibration for degree centigrade
```

The reference voltage for measuring is 1.5 V. The temperature sensor produces voltage concerning temperature. Thus finally, observations are made, and the calibrated results are verified and noted.

4.3 Interrupts Configuration and Implementation

An interrupt is anything that obstructs the flow of the program by giving an additional important task that is to be performed before the one that is already under process by CPU. The interrupt needs to be as small as possible and based on the nature of interrupt there are two types of interrupts called maskable (software) and non-maskable (hardware) interrupts. To write the interrupt service routine, we define it by using a directive # pragma and keyword vector. The non-maskable interrupt was meant to interpret the changes in the pins that are programmed to have interrupts. We can give maskable or software interrupts by using timers A0 and A1. Thus, we wrote a program for interrupts, one a software interrupt where we blink an led connected to a pin concerning the main clock, and the other being a hardware interrupt where we have a push button connected to a pin which triggers the action of letting an led connected to the pin on and off. Both actions happen simultaneously.

```
#pragma vector=TIMER0_A0_VECTOR
__interrupt void Timer_A (void)
{
  P1OUT ^= BIT0; // Toggle P1.0
}
//This is software Interrupt.

// Port 1 interrupt service routine
#pragma vector=PORT1_VECTOR
__interrupt void PORT1_ISR (void)
{
  P1OUT ^= BIT6; // Toggle P1.6
  P1IFG &= ~BIT3; // P1.3 IFG cleared
}
//This is a hardware interrupt.
```

We used pins in Port1. P1.3 is programmed to be a non-maskable interrupt and is triggered by high to low pulse.

```
P1IES |= BIT3; // P1.3 Hi/lo edge
```

Then we finally combined this with 4-bit LCD interface in a way that all of them seem to work simultaneously. The combination is possible because we are not giving any input to control signals hence even when we use the data lines of LCD for other purposes here P1.3 is a data line for LCD but is also a trigger input for led at pin P1.6.

5 Results

The results of the project are as follows.

5.1 LCD Driver Configuration and Implementation

See Figs. 7 and 8.

5.2 Built-In I/O Configuration and Integration

See Fig. 9.

The typical values measured by ADC are:

Temp	IntdegC	IntdegF
731	23	75
732	24	75

Fig. 7 Eight-bit 16 * 2 LCD

Fig. 8 Four-bit mode for LCD

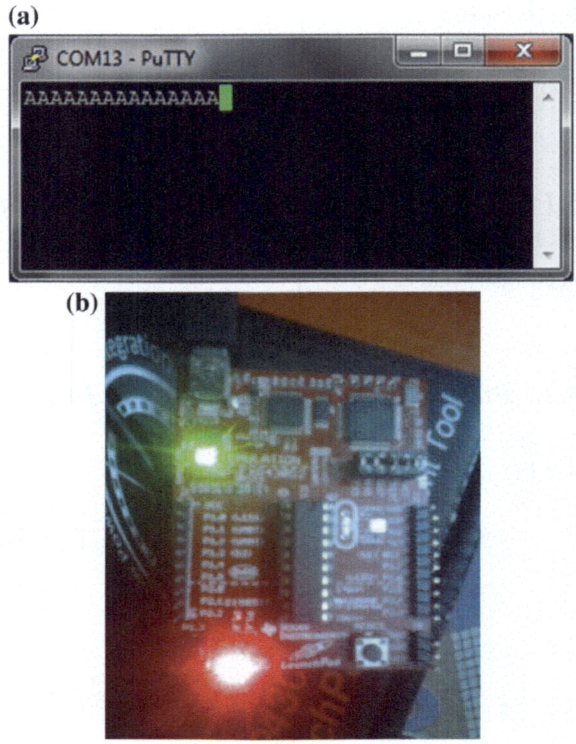

Fig. 9 **a** UART serial communication and **b** software interrupt (LED blinking)

6 Work in Progress and Future Scope

The development of designing input interface with a single key a single push button as an interrupt to get user ID and password as in case of old mobile phones is in progress. The driver needs to be implemented in such a way that the three characters on one button with the proportionate pressing and release time; the intended character needs to be detected.

As the MSP40 is titled the world's first low-power MCU, it can replace any existing C51-based designs. Hence, it is used in all the possible cases so that the power consumption can be reduced dramatically, and the battery lasts long.

References

1. Nagy C (2013) Embedded systems design using the TIMSP430 series. Embedded Technology Series, Elsevier (2003)
2. Rowley Associates Limited (2004) Cross Studio. Cross works for MSP430—user guide
3. IAR Systems (2004) MSP430 IAR embedded workbench TM IDE user guide for texas instruments. MSP430 Microcontroller Family
4. Texas Instruments (2008) User's guide, SLAU144E. MSP430x2xx Family
5. Texas Instruments (2007) User's guide, SLAU056G. MSP430x4xx Family
6. Texas Instruments (2008) User's guide, SLAU208. MSP430x5xx Family

Real-Time Energy Management Scheme for Dual Converter-Based Hybrid Solar/Battery/Ultra-Capacitor Vehicular System

Pankesh Bhargav and Shraddha Kaushik

Abstract Hybrid Electric Vehicle (EVs) adopting batteries, ultra-capacitor, and PV cells using a maximum power point track (MPPT) solar system is a unique power-sharing strategy technique which provides a promising solution, reduces power loss, and increases efficiency for upcoming automotive industry using hybrid electric vehicle. Batteries and solar photovoltaic's cells are integrated with DC–DC boost converter and ultra-capacitors with DC–DC bidirectional converter to stabilize the voltage. This paper deals with the modeling of an efficient DC–DC converter which can be used in applications involving high values of power while simultaneously reducing the size of the batteries used. Several DC–DC converters are described and compared with the proposed converter to verify its efficiency, reliability, and stiffness. In order to deal with fast load dynamics in vehicles, an online energy management scheme is proposed which aims at optimizing the circuit's internal flow of power and satisfies the power demand of the driver. Also, physical explanation about when to produce, consume, and store electric power during regenerative braking is also implemented and simulated in this paper using MATLAB/Simulink. Further, a DC motor is considered as a load to demonstrate the performance of this strategy.

Keywords DC–DC boost converter · DC–DC bidirectional converter · Modified interleaved boost converter · Energy management scheme · Ultra-capacitor · Photovoltaic (PV)

1 Introduction

Environmental impacts, escalating prices of petroleum-based fuels, emission restrictions, and the depletion of natural resources provide compelling impetus toward the development of alternatives of internal combustion engines in which our trans-

P. Bhargav (✉) · S. Kaushik
Department of Electrical Engineering, Bhilai Institute of Technology, Durg, India
e-mail: pankeshbhargav@gmail.com

S. Kaushik
e-mail: shraddha.kaushik01@gmail.com

© Springer Nature Singapore Pte Ltd. 2019

V. Nath and J. K. Mandal (eds.), *Proceedings of the Third International Conference on Microelectronics, Computing and Communication Systems*,
Lecture Notes in Electrical Engineering 556, https://doi.org/10.1007/978-981-13-7091-5_31

portation is highly dependent [1]. Also, the importance provided to "Sustainable development" which can be defined as "meeting the needs of the present without compromising the ability of future generations to meet their own needs" increases the need of such alternatives and represents one of the greatest challenges of today.

2 Proposed System for HEV Configuration

Hybrid electric vehicles, contrary to gasoline-based ones, are powered by more than one power source, and thus the name "hybrid". In the proposed method, one source, as the primary, is the photovoltaic (PV) while the tandem of battery and ultra-capacitors is used as the other or secondary source of power. This alliance aids the HEV with a sense of freedom enabling it to deliver power even at peak as well as during transient power demands [2]. The basic design feeds the permanent magnet DC (PMDC) from the conjunction of these two sources.

This paper shows a parallel structure in which three converters are connected in parallel and operated by three different sources. There are many advantages of parallel structure, i.e., reliable in nature simple methodology. In this paper, boost converter with MPPT controller is connected to the photovoltaic. To maintain a fixed voltage at the load side, the boost converter is connected to the battery and the bidirectional converter, which works in boost mode while delivering power and in buck mode while absorbing it, is connected to the ultra-capacitor [3]. For a certain load power demand ($\mathrm{Power}_{\mathrm{LOAD}}$), power can be delivered partially by photovoltaic ($\mathrm{Power}_{\mathrm{PV}}$), battery ($\mathrm{Power}_{\mathrm{BAT}}$), and ultra-capacitor ($\mathrm{Power}_{\mathrm{UC}}$), respectively, and the load power should satisfy the following condition ideally:

$$\mathrm{Power}_{\mathrm{LOAD}} = \mathrm{Power}_{\mathrm{BAT}} + \mathrm{Power}_{\mathrm{UC}} + \mathrm{Power}_{\mathrm{PV}}$$

Or more precisely, η_{BAT}, η_{PV}, and η_{UC} are the efficiencies of battery, photovoltaic (PV), and UC, respectively, and non-propulsion load $\mathrm{Power}_{\mathrm{NP}}$

$$\mathrm{Power}_{\mathrm{LOAD}} = \eta_{\mathrm{BAT}} \times \mathrm{Power}_{\mathrm{BAT}} + \eta_{\mathrm{UC}} \times \mathrm{Power}_{\mathrm{UC}} + \eta_{\mathrm{PV}} \times \mathrm{Power}_{\mathrm{PV}} + \mathrm{Power}_{\mathrm{NP}}$$

2.1 EV System Battery, Ultra-Capacitor, Photovoltaic, and PMDC Motor

This section comprises various elemental models and their dynamics used during simulation and in practical implementation. Figure 1 shows the parallel structure of hybrid battery/photovoltaic /UC power system [4].

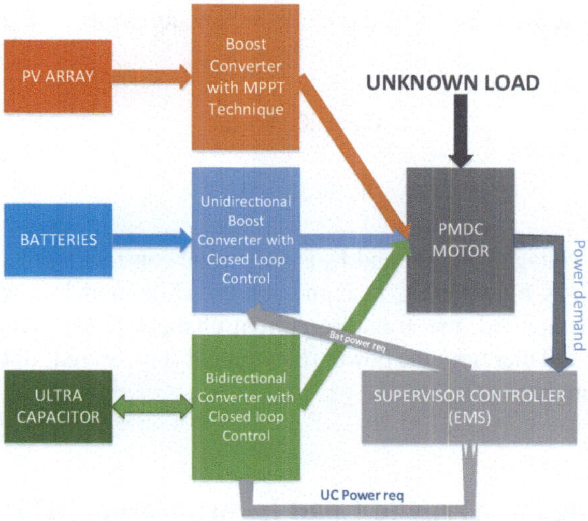

Fig. 1 Parallel structure of hybrid battery/photovoltaic/UC power system

(1) *Batteries*

The battery used is a lithium–ion one used to feed the DC motor. Its simplistic model makes it easier to imagine and connect to an equivalent series resistance (ESR). A ready-made lithium–ion battery model available in MATLAB is made use of as an ideal voltage source, related to its discharge characteristics, connected to an ESR in series [5].

(2) *Ultra-Capacitors*

The ultra-capacitor model presented consists of an RC ladder network in which the V_{UC} is dependent on the distributed capacitance value [6–8]. A series combination of capacitor with an ESR or EPR can be used to replace the ladder network in order to analyze the characteristics [9–11]. In MATLAB, a generic model of ultra-capacitor is supposed to be comprised of a bundle of capacitors joined together in series and parallel in order to meet the voltage and current demands [12, 13].

(3) *Photovoltaic*

The PV arrays are having small voltage ratings, while on the other hand the load which is a DC motor has larger voltage ratings. A DC–DC boost converter is used to interface the two, which steps up the small voltage to a considerable amount, thus incorporating MPPT or maximum power point tracker, which is an electronic DC-to-DC converter that optimizes the balance between the PV panels and the battery bank or the utility grid, which tracks the maximum power point and operates at a particular voltage and current [14].

(4) *PMDC Motor*

The PMDC motors are the eccentric choice for motor in applications regarding electric vehicles which is similar to a DC motor excited by a permanent magnet in the field [15]. The back emf and the electromagnetic torque of the motor can be expressed as

$$E_b = K_e \times \omega_m$$

$$T_e = K_t \times I_a$$

where K_e is the voltage constant and K_t is the torque constant and $K_e = K_t$.

The SIMULINK block of DC machine is used while modeling with MATLAB. In case a load is applied, the armature current changes in proportion to the load torque (Te α Ia) and therefore the power requirement (Power_req) of the system also changes accordingly [1].

3 Interfacing PV Array to Load Incorporating MPPT

3.1 Maximum Power Point Tracking

Maximum Power Point Tracking (MPPT) is an electronic system that operates the PV modules in such a way that the modules produce the maximum power possible at all times. Instead of moving the modules physically in order to track the maximum sunlight, MPPT varies the electrical operating point of the modules so as to deliver maximum power available [16].

Figure 2a shows the current v/s voltage characteristics of solar array. The PV characteristics are obtained as a product of voltage and current characteristics as shown in Fig. 2b. The power output is maximum at the point indicated as MPP [17].

3.2 Perturb & Observe (P&O) MPPT Algorithm

In this technique, a perturbation is periodically given to the module voltage and the corresponding output power is then compared to the previous perturbation cycle [18]. According to this technique, a slight perturbation is introduced in the system which causes the power of the PV module to vary. If the result is an increase in the power, then the perturbation to the system is continued in the same direction; otherwise, it is reversed. Once the peak power is reached, the power starts to decrease from the very next instant and there the direction of perturbation is reversed (Fig. 3).

When a stable condition is reached, the algorithm oscillates around the peak power point. In order to maintain the power variation, small amounts of perturbations are given to the system and the technique is advanced in such a way as to set the reference voltage of the module in correspondence to the peak voltage of the module.

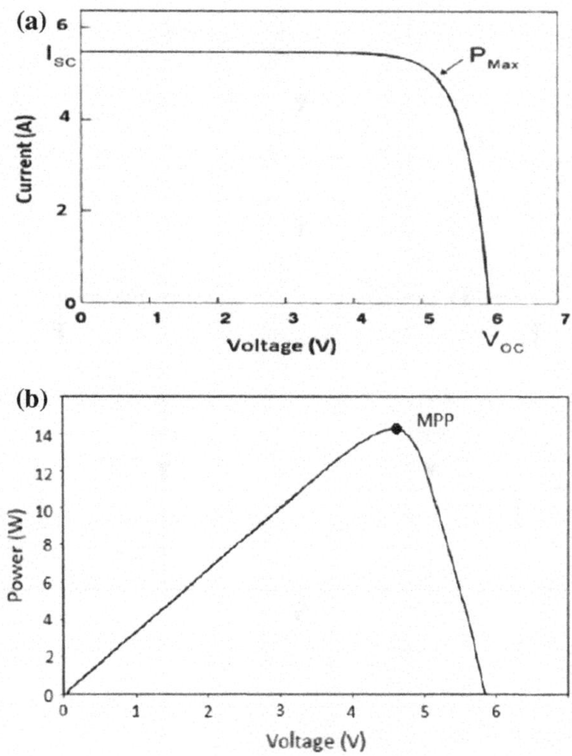

Fig. 2 **a** Current–Voltage characteristics of solar array, **b** power voltage characteristics of solar array

4 Frequency Decoupling Energy Management Scheme

Considering the same, frequency decoupling technique or bandwidth allocation technique is proposed which separates the power based on the frequency in which the source can work, thereby respecting dynamics of each source [4, 19–21]. During peak day hours, solar PV cell [7] will be in a standby mode to supply power demands (Fig. 4) [22, 23].

Figure 4 shows the power split between two sources by a frequency decoupling technique having low-frequency battery power and high-frequency UC power [24–26]. The required low-pass filter with transfer function $H(j\omega)$ is expressed as

$$H(j\omega) = \frac{1}{\sqrt{1 + \varepsilon \times \left(\frac{\omega}{\omega p}\right)^{2\times n}}}$$

where n represents the filter order, ω equals to $2\pi f_c$, and ε is the maximum band-pass gain.

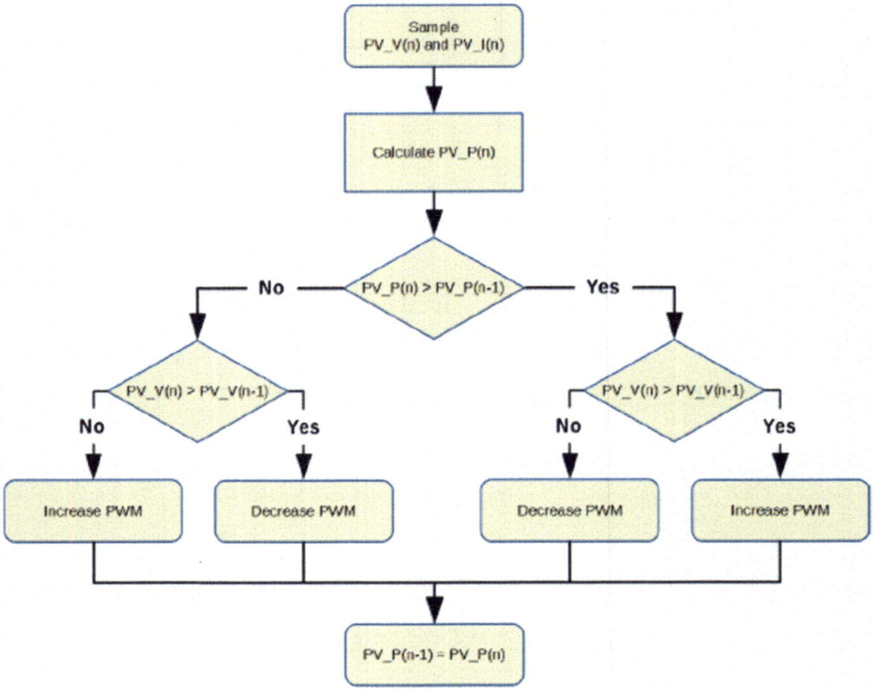

Fig. 3 P&O algorithm flowchart

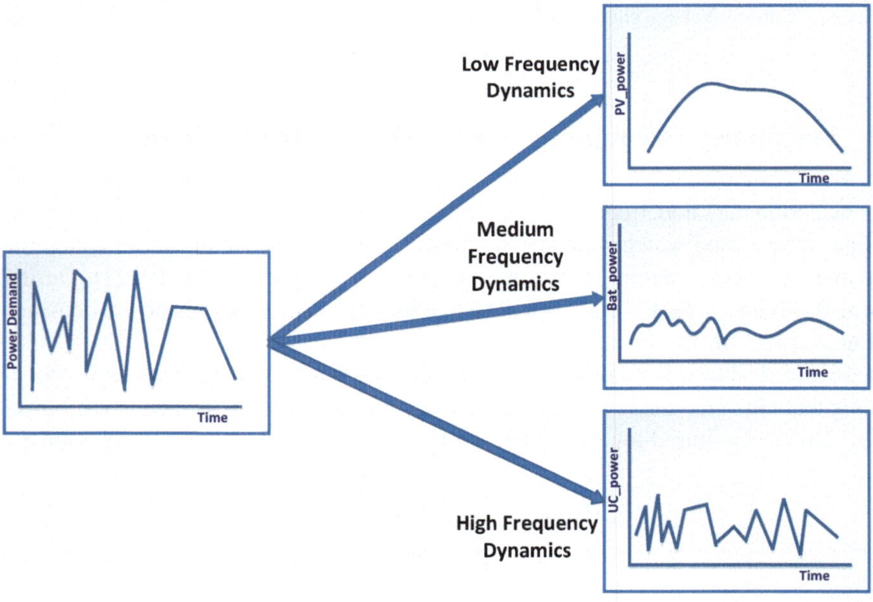

Fig. 4 Bandwidth allocation technique

5 Designing of DC–DC Boost Converter and Bidirectional Converter

5.1 Conventional Boost Converter

A boost converter (step-up converter) is a power converter which produces a DC output voltage greater than its input voltage. It is also called a step-up converter since it "steps up" the source voltage [27, 28] (Fig. 5).

5.2 Bidirectional Converters

A bidirectional DC–DC converter intended for use in electric vehicles allows an apt control of motor in both motoring and regenerative braking mode of operations and can significantly contribute to an increase in overall drive system efficiency [2].

Figure 6 shows a DC motor being fed by a bidirectional DC–DC converter designed to provide high capricious current which is done by a closed-loop current controller as well for local control of the converter (Fig. 6). The inductor current

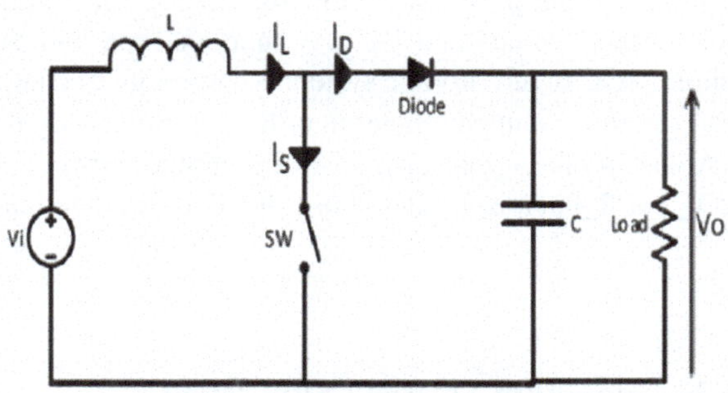

Fig. 5 Boost converter schematic

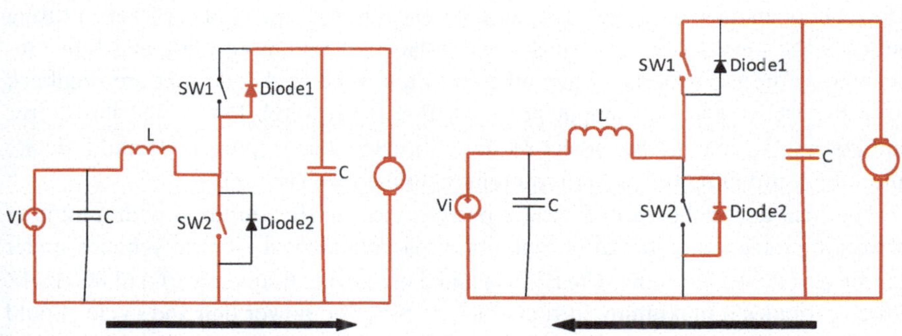

Fig. 6 Bidirectional DC–DC converter with battery and DC motor

is compared with its reference value, the error is fed to a PI controller, and a corresponding PWM signal serves as the trigger pulses for the switches.

6 Simulation Results and Discussion

The Simulink model of the system consists of four main subsystems and models of four components, namely, lithium–ion battery, permanent DC motor, solar panel, and ultra-capacitor. As discussed before, the battery voltage operating at 36 V is boosted to a voltage of 220 V using a boost converter, and the ultra-capacitor operating at 16 V is connected to the load using a bidirectional converter [29]. The load connected at the output is a permanent magnet DC motor whose parameters are varied by the load torque which is considered to be random. The voltage (V_dc) and current (I_dc) at the DC bus are measured and fed to the energy sharing scheme subsystem which provides power to be fed by each source, i.e., power required, Power_UC Power_solar, and Power_bat as output. These power requests are fed to the result verification section, and Power_UC is fed as the input to the buck/boost bidirectional converter subsystem which also have PWM signal, i.e., a sawtooth wave as the input. Power_UC is the power request which is to be tracked by the UC converter to feed the load, and the remaining power is fed by the battery.

To verify the results, input power should be matched to the output power of the system for which battery's solar' and UC's voltage, current, and State of Charge (SOC) are measured. SOC of any storage system is a measure of charge contained at the instant which is an important parameter in order to prevent deep discharge of the storage system. At last, all the signals, i.e., Power_required, Power_UC, Power_bat, V_uc*I_uc, and V_bat*I_bat are compared to validate the authenticity and stability of the system.

6.1 Energy Management Strategy Subsystem

The energy management strategy as discussed is to control the power splitting between the three loads. The subsystem consists of two input ports and four output ports. The input ports of the subsystem are V_dc and I_dc that are feedback from the DC motor, and output ports are Power_required, Power_bat Power_pv, and Power_UC that are the power requests for each load. Inputs V_dc and I_dc are multiplied providing the power requirement to the load (Fig. 7).

The load curve shown in Fig. 8 is power cycle of 10 s duration with a number of surging edges intended to resemble the load demand of electric vehicles under certain practical conditions. The EMS is tuned to a corner frequency (f_c) of 800 mHZ which provides a maximum current slope of 5.6. The power demand cycle should consist of all four modes of operation of the hybridization, i.e., constant power, gradually increase in power, abrupt increase in power, and regenerative braking. The load profile shown in Fig. 8 is generated by providing the mechanical input to the motor in the form of load torque which is shown in Fig. 8. The armature current of PMDC motor as discussed earlier is directly proportional to the load torque and changes the power requirement of the motor in the same proportion as that of load torque assuming the voltage at the motor terminal constant. The power demand cycle shown in Fig. 9 consists of two projections, three depressions, three instants when load power is constant, and regenerative braking period.

The starting of the DC motor also requires very high current and if singly fed by the battery will cause serious damage. Though the designing of the starter for the motor has not been considered to avoid complexity in the proposed model, it can

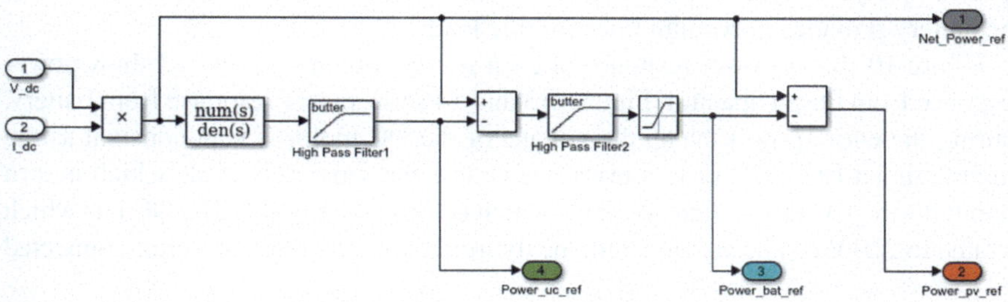

Fig. 7 Energy management strategy subsystem

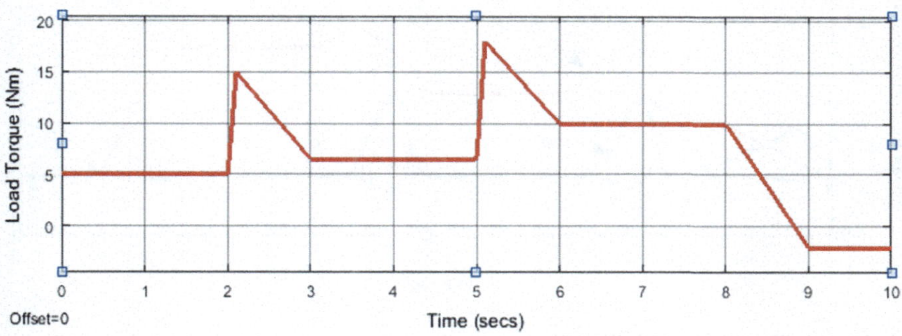

Fig. 8 Load torque input to motor

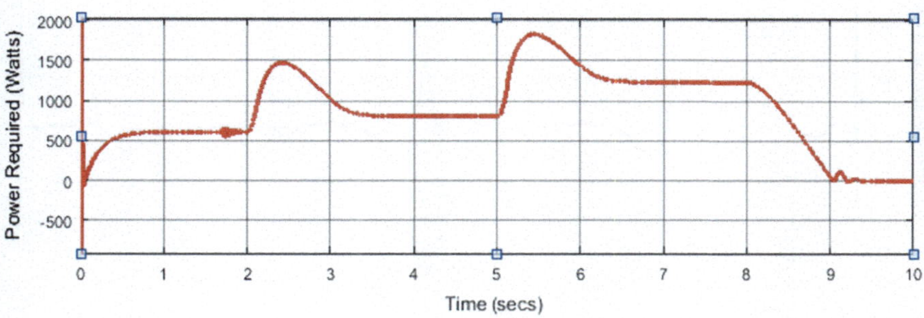

Fig. 9 Power demand cycle

be achieved in a way that the ultra-capacitor accompanies the battery to provide the large starting current.

The DC motor, in the proposed model, is considered to be in steady state, i.e., operating at rated speed, while constant load torque is applied to the motor. The initial load torque applied to the motor is 5 Nm for a duration of 1 s during which the battery delivers considerable amount of power and the ultra-capacitor remains dormant.

Next, an impulsive load torque is applied which reaches its peak and decays back to its initial value in the next 1 s time lapse. During this period, the EMS filters out the high-frequency component of the power signal which is considerably large and has to be supplied by the ultra-capacitor, while the battery power gradually increases to accompany the UC after a second, and only when the load power becomes still the battery provides maximum power to the load.

Figure 10 shows power response of each source. Figures 11 and 12 show power response from ultra-capacitor. Figures 13 and 14 show power response from battery. During the entire power cycle, the voltage of the DC motor is kept constant as the unidirectional DC–DC boost converter is kept at the same duty cycle which in turn maintains the DC bus voltage to 220 V which is shown in Fig. 11. The DC bus which is rated to 220 V can be verified with the figure shown. The filter capacitor connected

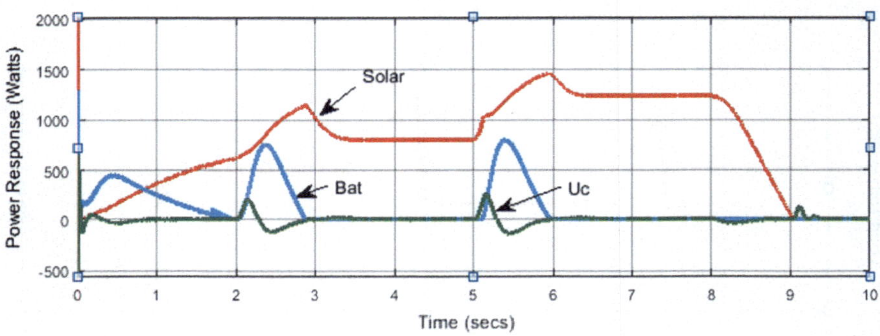

Fig. 10 Power response of each source

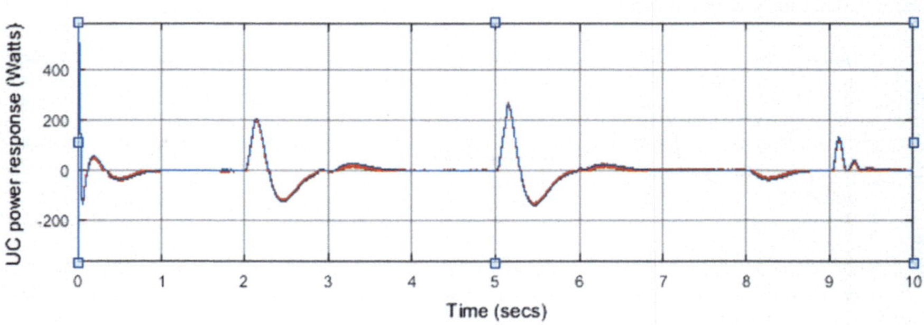

Fig. 11 Power response of ultra-capacitor

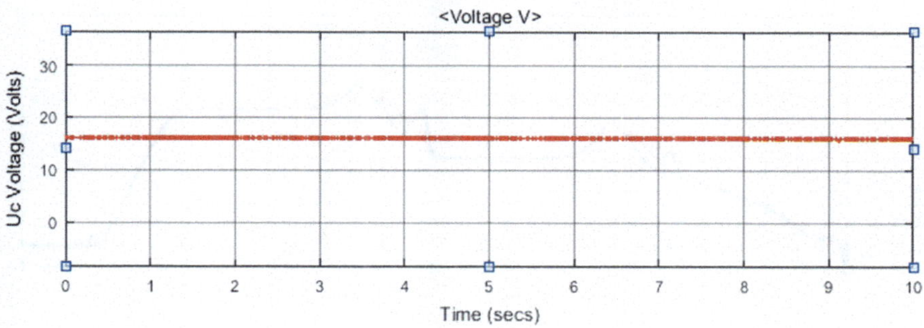

Fig. 12 Voltage of ultra-capacitor

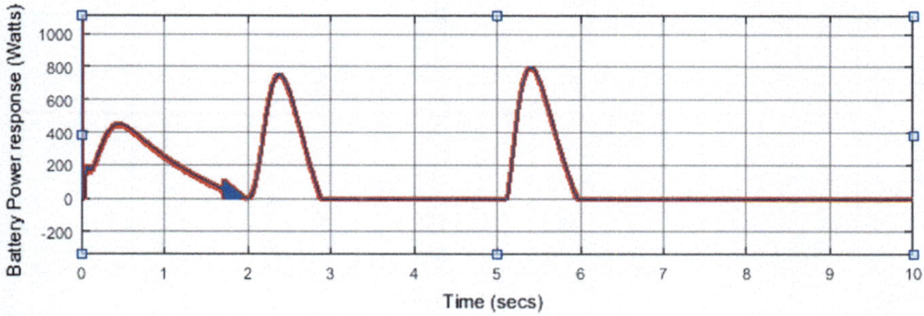

Fig. 13 Power response of battery

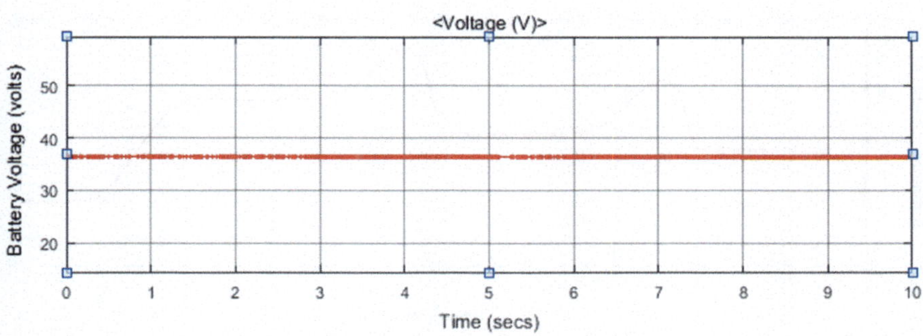

Fig. 14 Battery voltage

at the DC bus filters out the voltage ripple to produce a smooth DC at the terminal. As discussed earlier, the load torque modifies the armature current shown in Figs. 15, 16, and 17 of the PMDC motor which in turn changes the load power and has the same waveform as that of the load profile and load torque.

The photovoltaic block also consists of the same two measurement ports of voltage and current. Photovoltaic current shown in Fig. 18a, b is a similar waveform as that of the $\text{Power}_{\text{\_PV}}$ with amplified value because of the boost converter with MPPT controller.

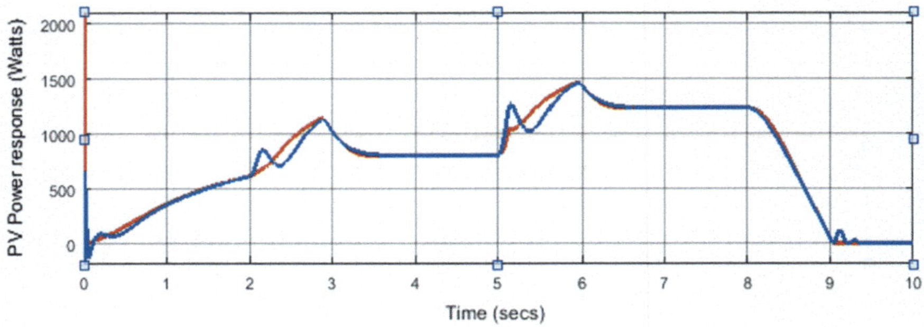

Fig. 15 Power response of photovoltaic

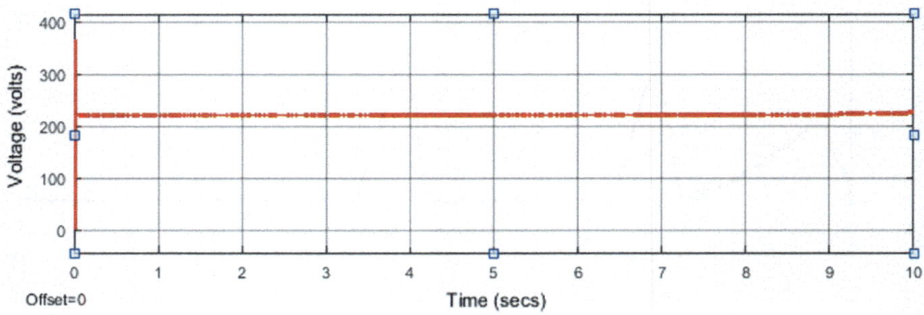

Fig. 16 Voltage profile of the DC bus

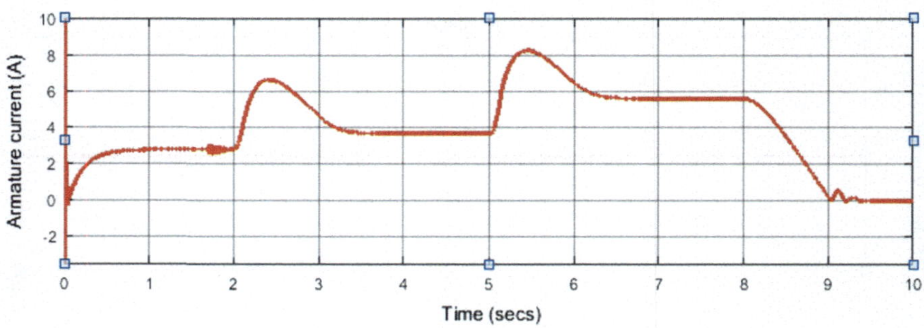

Fig. 17 Armature current of PMDC motor

The speed characteristic of the DC motor is an important part of the electric vehicle as speed is one of the controlling parameters. As we know that for constant power speed is inversely proportional to the load torque and thus in Fig. 19 speed drops can be seen where there is an increase in load torque, during the period of regenerative braking, speed started increasing to a constant value where the load torque is constant (Fig. 19).

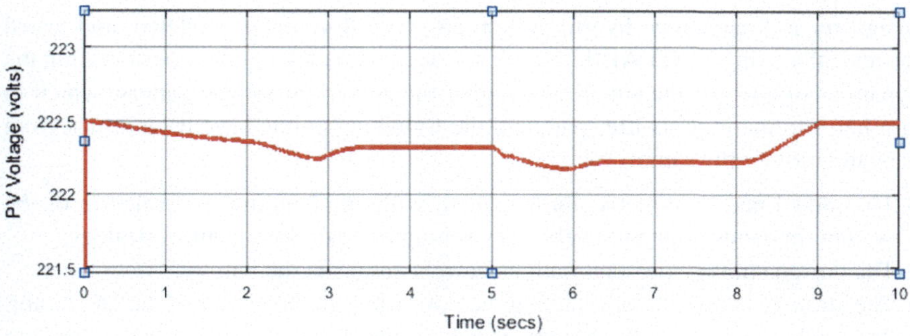

(a) PV voltage

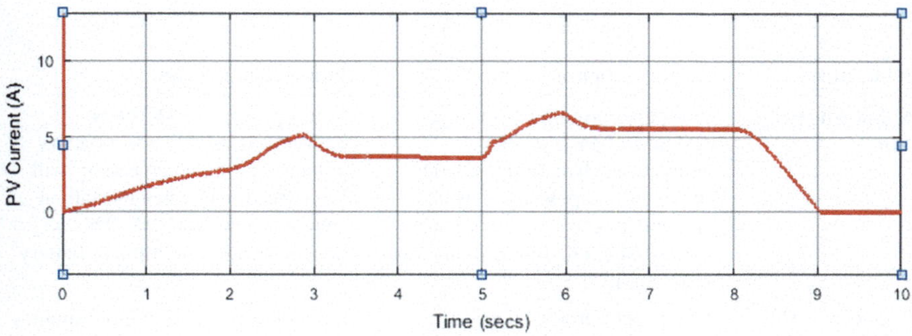

(b) PV Current

Fig. 18 **a** PV voltage, **b** PV current, characteristic waveforms of photovoltaic

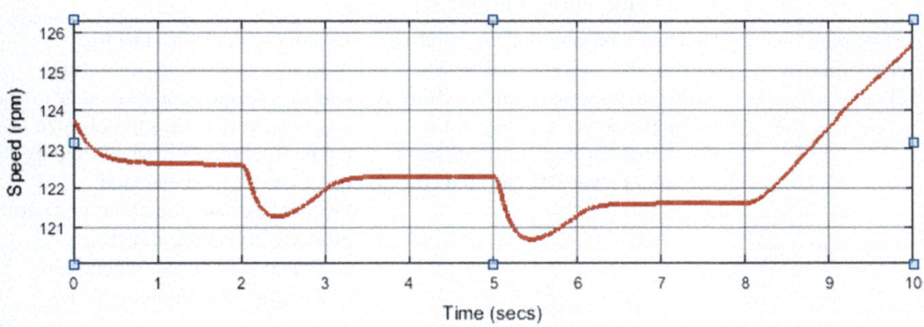

Fig. 19 Speed characteristics of DC motor

7 Conclusion

In this paper, an energy management strategy is developed to split the power in accordance with the abilities and limitations of each source and the EMS is incorporated in a multi-input multi-converter configuration of hybridization with two sources. The design for a three-source hybrid system has been developed, modeled, and tested successfully using MATLAB/Simulink software while taking into consideration the dynamics of each of the sources including the constraint for the battery, which in turn helps to increase its life span, and the following points have been drawn from the validation of the same:

- The model can be directly implemented, without changing the structure of the system, in battery-operated vehicles working on open-loop control strategy.
- The design strategy prevents sudden current surges in the battery.
- The strategy is capable of reducing the power loss in the starter of the DC motor.
- A complete cut-off of the battery is achievable during the regenerative braking while the excess power is fed to the ultra-capacitor.

Publications	Paper description	Proposed topologies
Al-Sheikh et al. [30]	This paper presents modeling, design, and analysis of a bidirectional half-bridge DC/DC converter composed of a battery unit and an ultra-capacitor pack with parallel DC-linked multi-input converter [30]	In this paper, DC–DC boost converter is used to boost battery voltage, due to which battery will be charged only when its state of charge is less than 20%. This is done to increase lifetime of battery
Chen et al. [31]	This paper introduced an ultra-capacitor boosted hybrid fuel cell vehicle drive train in which fuel cell provides basic power and ultra-capacitor provides peaking power to meet load demand [31]	In this paper, we have three sources with PV panel in standby mode. Battery provides base power and ultra-capacitor provides peak load power demand
Neenu and Muthukumaran [32]	This paper proposed a comparative study to the conventional HESS, and the new design is able to fully utilize the power stability of the ultra-capacitor without requiring a matching power DC/DC converter [32]	In order to deal with fast load dynamics in vehicles, an online energy management scheme is proposed which aims to optimize the internal power flow and satisfy the driver's power demand. Physical explanation about when to produce, consume, and store electric power during regenerative braking are also implemented
Uqur et al. [19]	A battery-based hybrid vehicular system prototype is proposed in this paper; hence, a battery and n UC are used. The objective of this arrangement is using batteries for the base power demand and UC for peak power demand and thus implementing load sharing strategy on a four-wheeled car with advanced battery technology [19]	In this paper, an energy management strategy has been developed. The objective of this arrangement is to use batteries for base power demand, UC for peak power demand, and PV for medium range frequencies. The energy management strategy proposed having different corner frequencies verified to be a stable, reliable, and robust technique

Appendix: Comparison Table with Standard Journals

Sr. no.	Components	
1	**Ultra-capacitor**	**Value**
	Capacitance	500
	Rated voltage	16
	ESR	2.1e–3
2	**Battery**	**Value**
	Voltage (open circuit)	36
	Capacity (rated)	50
	ESR	0.045 Ω
3	**PV panel**	**Value**
	Module type	Suntech STP270S-24_Vb
	V_{oc}, I_{sc}, V_{mp}, I_{mp}	44.45 V, 8.2 A, 35 V, 7.7 A
	ESR	0.32025 Ω
4	**Inductors and capacitors**	**Value**
	Inductors Luc/Lbat/Lpv	1mH/1mH/1mH
	Capacitor Cdc	100 μF
5	DC bus voltage	200 V

References

1. Singh A, Pattnaik S (2016) Design of a efficient power sharing strategy for a battery-ultra-capacitor hybrid energy storage system. In: IEEE international conference on power electronics, intelligent control and energy systems (ICPEICES-2016)
2. Lohner A, Evers W (2004) Intelligent power management of a super capacitor based hybrid power train for light-rail vehicles and city busses. In: Presented at IEEE Power electronics specialists conference, PESC 04
3. West MJ, Bingham CM, Schofield N (2003) Predictive control for energy management in all/more electric vehicles with multiple energy storage units. In: Presented at IEEE international electric machines and drives conference, IEMDC'03
4. Zhifeng B, Sun Y, Lin Y, Chen G, Cao B (2010) Research on ultra-capacitor-battery hybrid power system. IEEE
5. Oruganti SK et al (2016) Wireless power and data transfer system for internet of things over metal walls and metal shielded environments. In: 2016 URSI: APRASC 2016. https://doi.org/10.1109/ursiap-rasc.2016.7601321
6. Aziba T, Larouci C, Chaibet A, Boukhnifer M (2014) Online energy management strategy of a hybrid fuel cell/battery/ultra-capacitor vehicular power system. IEEJ Trans Electr Electron Eng

7. Endo M, Takeda T, Kim YJ, Koshiba K, Ishii K (2001) High power electric double layer capacitor (EDLCes) from operating principle to pore size control in advanced activated carbons. Carbon Sci 1:117–128
8. Rufer A, Barrade P, Hotellier D, Derron P (2002) Supercapacitive energy storage: power electronic solutions and applications. In: Presented at Associazione Nazionale Azionamenti Elettrici, 13o Seminario Interattivo, Azionamenti elettrici: Evoluzione Tecnologica e Problematiche Emergenti, Bressanone, Italy. http://leiwww.epfl.ch
9. Kotz R, Carlen M (2000) Principles and applications of electrochemical capacitors. Electrochim Acta Pergamon Elsevier Sci 45:2483–2498
10. Miller JM, Smith R (2003) Ultra-capacitor assisted electric drives for transportation. In: Presented at IEEE international electric machines and drives conference
11. Bullard GL, Sierra-Alcazar HB, Lee HL, Morris JL (1989) Operating principles of the ultracapacitor. IEEE Trans Magn 25:102–106
12. Burke AF (1996) Prospects for ultra-capacitors in electric and hybrid vehicles. In: Presented at 11th annual battery conference on applications and advances
13. Surewaard E, Tiller M, Lizen D, Linzen D (2003) A comparison of different methods for battery and supercapacitor modeling. In: Presented at SAE future transportation technology conference–SAE technical paper series 2003-01-2290
14. Advanced Energy Industries, Inc. (2012) Advanced energy, DC loading of PV powered inverters, Inc. 2012 All rights reserved. Printed in U.S.A. 55-600100-75C 0M 3/12. http://solarenergy.advanced-energy.com/upload/File/Application%20Notes/DCLoadingOfPVPinverters_55-600100-75C_AppNote.pdf>Advanced
15. Wu W et al (2012) DC motor parameter identification using speed step responses. Model Simul Eng 2012:5. Article ID 189757. https://doi.org/10.1155/2012/189757
16. Kute UT, Ratnaparkhi PS et al (2013) Literature survey on maximum power point tracking (mppt) technique for photovoltaic (pv) system. Int J Adv Res Eng Appl Sci 2(12)
17. Mollah AH, Panda GK et al (2015) Single phase grid-connected inverter for photovoltaic system with maximum power point tracking. Int J Adv Res Electr Electron Instrum Eng 4(2):2015
18. Maximum power point tracking. https://en.wikipedia.org/wiki/Maximum_power_point_tracking
19. Ugur E, Ousmez S, Vural B, Uzunoglu M (2012) Implementation of a reliable load sharing strategy between battery and ultra-capacitor on a prototype electric vehicle. IEEE
20. Luk PCK, Economou JT, White BA et al (2007) Power and energy management of multiple energy storage systems in electric vehicles. http://studylib.net/doc/18841257/power-and-energy-management-of-multiple
21. Rosario L et al (2007) Applying management methodology to electric vehicles with multiple energy storage systems. In: 2007 international conference on machine learning and cybernetics, vol 7. https://doi.org/10.1109/icmlc.2007.4370888
22. Salman M, Chang MF, Chen JY (2005) Predictive energy management strategies for hybrid vehicles. In: Presented at IEEE vehicle power and propulsion, VPPC
23. Lv YM, Yuan HW, Liu YY, Wang QS (2010) Fuzzy logic based energy management strategy of battery, ultra-capacitor composite power supply for HEV. In: 2010 first international conference on pervasive computing, signal processing and application
24. Chen JS, Salman M (2005) Learning energy management strategy for hybrid electric vehicles. In: Presented at IEEE vehicle power and propulsion, VPPC
25. Moreno J, Ortuzar ME, Dixon JW (2006) Energy-management system for a hybrid electric vehicle, using ultra-capacitors and neural networks. IEEE Trans Industr Electron 53:614–623
26. Gielniak MJ, Shen JZ (2004) Power management strategy based on game theory for fuel cell hybrid electric vehicles. In: Presented at IEEE vehicular technology conference, VTC Fall
27. Newlin DJS et al (2013) A performance comparison of interleaved boost converter and conventional boost converter for renewable energy application. In: Proceedings of 2013 international conference on green high performance computing
28. Sasikumar S, Krishnamoorthi K Design of power managing converter for fuel-cell/battery of hybrid electric vehicle. Int J Recent Trends Eng Res (IJRTER)

29. Pany P et al (2011) Bidirectional DC-DC converter fed drive for electric vehicle system. Int J Eng Sci Technol 3(3):101–110
30. Al-Sheikh H, Bennouna O, Hoblos G (2014) Modeling, design and fault analysis of bidirectional DC-DC converter for hybrid electric vehicles. In: 2014 IEEE 23rd international symposium on industrial electronics (ISIE), 1–4 June 2014, Istanbul, Turkey
31. Chen B, Gao Y, Ehsani M (2009) Design and control of a ultra-capacitor boosted hybrid fuel cell vehicle. In: Vehicle power and propulsion conference, VPPC'09. IEEE
32. Neenu M, Muthukumaran S (2012) A battery with ultra-capacitor hybrid energy storage system in electric vehicles. Adv Eng Sci Manage (ICAESM) IEEE Trans

Design and Functional Verification of Reversible Logic Based FFT Using OHRNS

Jaswanth Vuggirala, Ghanshyam N. Patil and H. V. Jayashree

Abstract This work shows importance of reversible logic circuits using One-Hot Residue Number System (OHRNS) for arithmetic circuit designing and signal processing (digital) for less delay and less power dissipation applications. OHRNS technique highlight is delay of one active (high) transistor which is equal to the implementation delay which shows better results in comparison to reversible logic design circuits and conventional methods. Another benefit of using OHR is lucidity of implementation when compared to reversible logic circuits and conventional methods. The advantage of using reversible logic design circuits is having minimal power dissipation. Design of arithmetic blocks adder and subtractor using one-hot encoding technique is focused mainly. Frequently these subsystems are used in the design of filters. In this work, the above subsystems implemented using OHR are utilized in designing of FFT computational blocks. The essential blocks required in designing a fast Fourier transform computation block using conventional, OHRNS, reversible logic circuits methods, are implemented in Tanner EDA tool with standard CMOS technology (250 nm), and comparative results are provided. Experimental results show the FFT block implementation using OHRNS technique which dissipates less power, consumes less transistors, and is faster and feasible option in design of efficient circuits when compared to similar implementation techniques such as reversible logic circuits and conventional techniques.

Keywords One-hot residue number system (OHRNS) · Fast fourier transform (FFT) · Reversible logic design (RLD)

Jaswanth Vuggirala (✉) · G. N. Patil · H. V. Jayashree
PES University, Bengaluru, India
e-mail: jaswanth341@gmail.com

G. N. Patil
e-mail: ghanshyam369@gmail.com

H. V. Jayashree
e-mail: jayashreehv@pes.edu

© Springer Nature Singapore Pte Ltd. 2019

V. Nath and J. K. Mandal (eds.), *Proceedings of the Third International Conference on Microelectronics, Computing and Communication Systems*,
Lecture Notes in Electrical Engineering 556, https://doi.org/10.1007/978-981-13-7091-5_32

1 Introduction

Our focus in the design of VLSI circuits is to minimize dissipation of power, minimize the size, and increase the speed of implementation inherently. Arithmetic operation milestones like high speed and less area can be achieved by adapting RNS [1–5]. RNS has more potential in ad hoc networks, FFT, signal processing (digital), and in many other applications.

Computational arithmetic circuits demand is increasing, finding ways to reduce the design size and increasing the processing speed becoming crucial. By utilizing this encoding method (one-hot) technique, subtraction and addition operations are faster due to less delay and reduction size of hardware design and less power dissipation using RNS moduli.

Recently, reversible logic is receiving greater attention because of its ability in reduction of the power dissipation. For every reversible logic gate, the number of inputs should be the same with outputs. Reversible logic gate inputs can be determined from its outputs. In general, conventional computation is irreversible. Landauer [6] pointed out that in conventional computation, there is loss of information during computation. For logic computations, the loss of every bit information in irreversible leads to generation of heat energy *KTln2* joules, where T is absolute temperature and K is Boltzmann's constant for every computation.

2 Arithmetic Component Design Using OHRNS and Reversible Logic

To design FFT, we need two components, Radix and Twiddle factors. In order to design Radix and Twiddle factors, we need to implement both arithmetic components such as adder and subtractor. Following sections explain the designs of adder and subtractor using OHRNS and reversible techniques.

2.1 One-Hot Residue Number System (OHRNS)

Legal combinations of group of bit values referring to one-hot encoding are all low (0) except a single high (1) bit in digital circuits. Set of natural numbers conferred in RNS are $A = \{a_1, a_2, a_3, \ldots, a_L\}$, where $a_i = A \bmod n_i$, $i = 1, 2, \ldots, L$. $[0 : N - 1]$ is the range of natural numbers which can be conferred in RNS, where $N = n_1, n_2, n_3, \ldots, n_L$ [7]. Arithmetic operations can be done easily in RNS.

RNS system rate can be improved by enhanced method called OHRNS. In OHRNS, n_i number of lines are introduced by each number in modulo-n_i, which are represented from 0 *to* n_{i-1}. According to it, particular/corresponding line number is activated.

Table 1 Decimal representation, binary representation and one hot residue moduli-n_i representation

Steps	Input-1	Input-2	Active stage	Result of step	Final output
Step-1	01000	$b_0 = 0$	A_0	00000	–
Step-2	01000	$b_1 = 0$	A_1	01000	–
			OR_0	01000	
Step-3	01000	$b_2 = 0$	A_2	00000	–
			OR_1	00100	
Step-4	01000	$b_3 = 0$	A_3	00000	–
			OR_2	00010	
Step-5 output	01000	$b_4 = 0$	A_4	00000	10000 (4)
			OR_3	00001	

Computation rate of arithmetic operation is improved by technique called one-hot encoding. For modulo-n, n signals are used in this technique, where at clock of each cycle there is only one high (active) signal and rest of signals are low. Decimal representation, binary representation, and one-hot residue modulo-n_i representation are shown in Table 1.

(1) *Design of OHRNS Modulo-n Adder*: Barrel shifters are used for OHRNS adders as their basic computational elements. The modulo-n circuit shown in Fig. 1 has two inputs of n-bits in one-hot encoding format each, data input as input-1 and

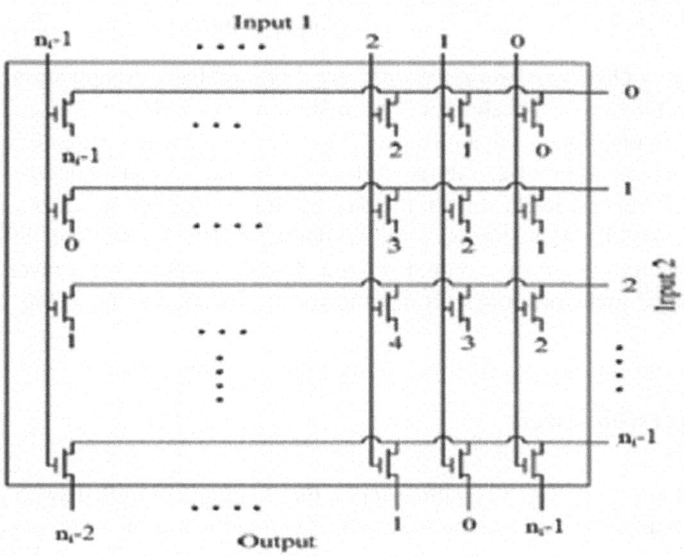

Fig. 1 Structure of OHRNS modulo-n adder [10]

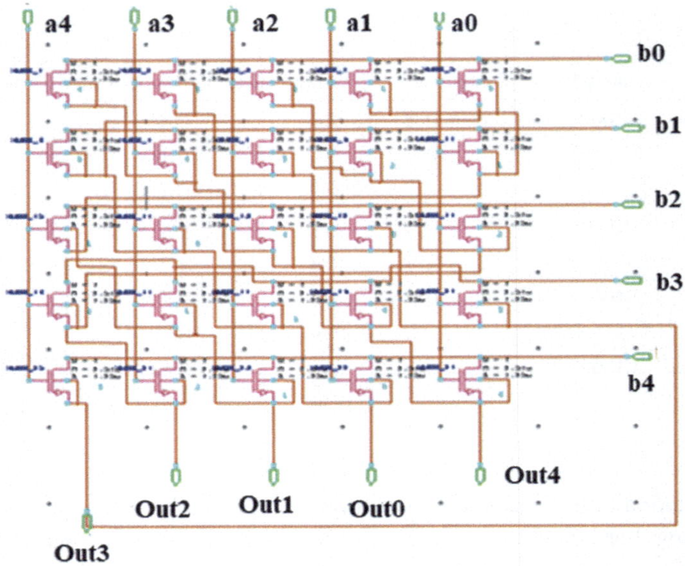

Fig. 2 Structure of OHRNS modulo-5 adder

shift control input as input-2. Data input got generated by all possible rotations by barrel shifter and selects output based on the shift input.

An example of modulo-n is shown using modulo-5 adder in Fig. 2. The data input and shift input are 5 bits each and the data input is shifted left based on the shift input, i.e., if shift input value = x then data input is shifted x number of times.

(2) *Design of OHRNS Modulo-n Subtractor*: For their basic computation, OHRNS subtractor uses barrel shifters. The modulo-n circuit shown in Fig. 3 has two inputs in one-hot encoded format of n-bits each, data input as input-1, and shift control input as input-2. Subtractor design is the same as adder design with only two differences. Right shift of the data input has occurred in adder circuit, but left shift of data input has occurred in subtractor. The second main difference is shift input order given as input to circuit. The shift input order is reversed when compared to the order of shift input in adder circuit as in Fig. 3 (Fig. 4).

2.2 Reversible Logic

3 * 3 Fredkin gate in Fig. 5 is a three-input, three-output controlled swap gate. The C input is mapped directly to the C output. If C input is low, no swap is performed else the other two inputs are swapped. Input of Fredkin gate is $X_v = (A, B, C)$ and output of Fredkin gate is $Y_v = (P = A, Q = {\sim}AB \oplus AC, R = {\sim}AC \oplus AB)$. Fredkin

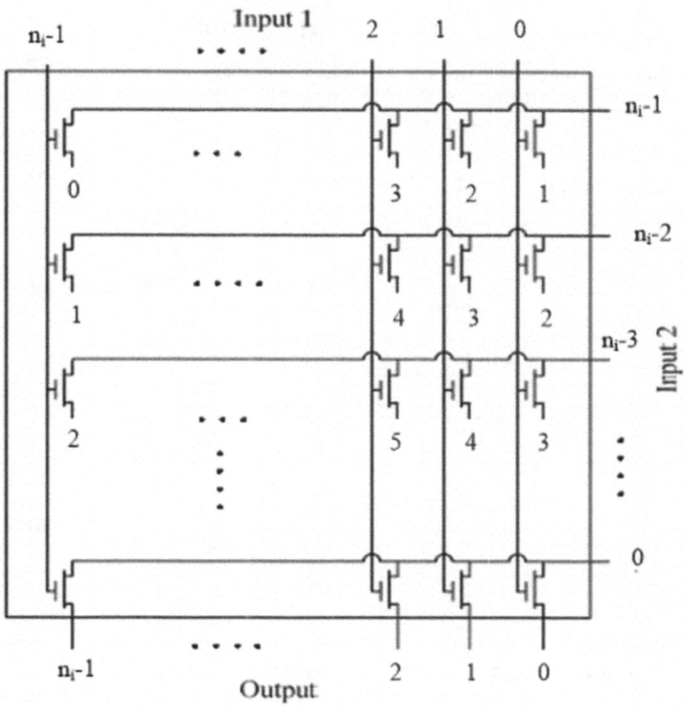

Fig. 3 Structure of OHRNS modulo-n subtractor

gate acts as OR gate (Q = A + B) when C is given as logic (1). Fredkin gate acts as AND gate (R = AB) when C is given as logic (0).

Fredkin gate implementation is shown in Fig. 6 using four transistors. From input A, P output is taken directly.

OR gate implementation using Fredkin gate is shown in Fig. 7. Input vector X_v = (A, B, C) of Fredkin gate, output vector Y_v = (P = garbage output, Q = A + B, R = garbage output). Fredkin gate acts as OR gate for inputs A, B by giving logic 1 as input to C.

Q = ~AB ⊕ AC (Initial)
Q = ~AB ⊕ A //(given C = 1)
Q = ~A (~AB) + A (A + ~B)
Q = ~AB + A + A ~B
Q = A + ~AB (1 + ~B)
Q = A + ~AB
Q = (A + B) (A + ~A) (DISTRIBUTION LAW)
Q = A + B //(A + ~A = 1).

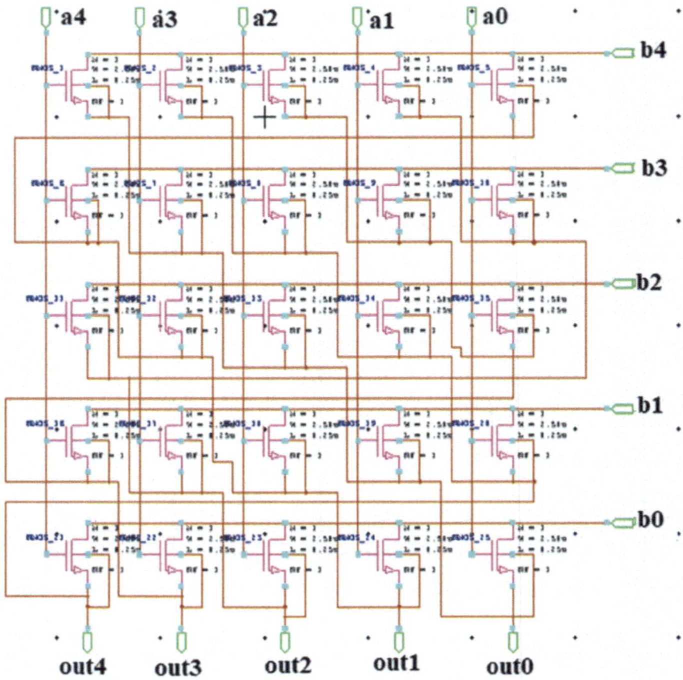

Fig. 4 Structure of OHRNS modulo-5 subtractor

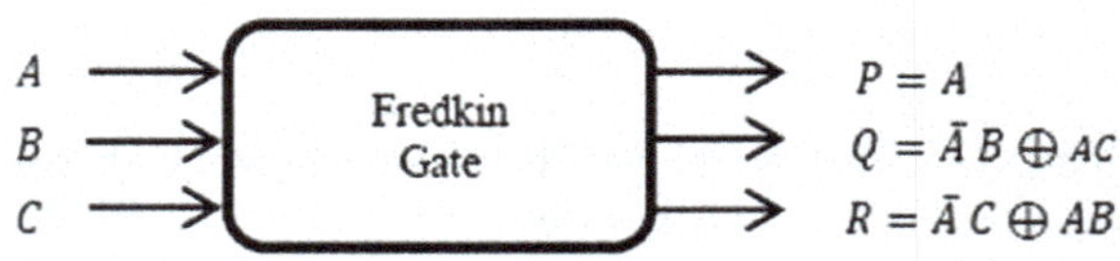

Fig. 5 3 * 3 Fredkin gate [11]

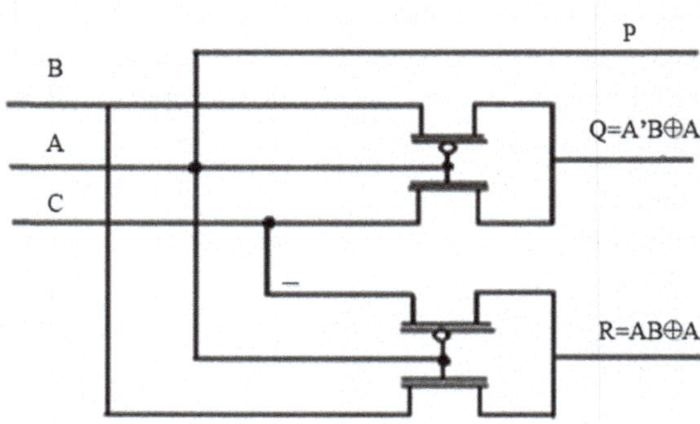

Fig. 6 Transistor level implementation of 3 * 3 Fredkin gate [12]

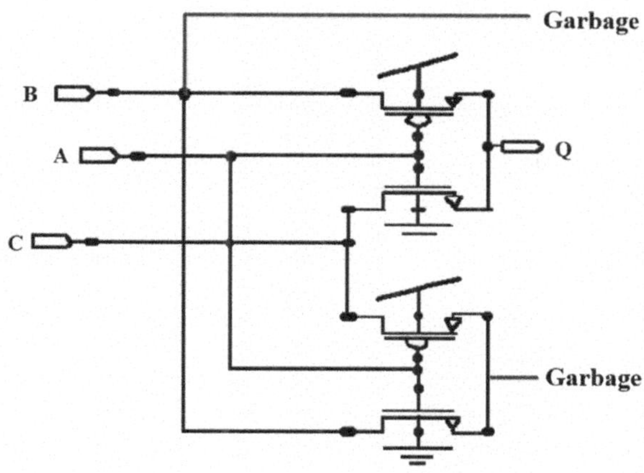

Fig. 7 Transistor level implementation of OR gate using Fredkin gate

Also, AND gate is implemented using Fredkin gate. Input X_v = (A, B, C) and output Y_v = (P = garbage output, Q = garbage output, R = AB). Giving input to C as 0 to make Fredkin gate as AND gate for the rest of other inputs A, B.

(3) *Design of Modulo-n One-Hot Adder using Reversible Logic Design*: Technique called barrel shifter is used in implementation of Modulo-n Reversible Logic Design adder circuit. Figure 8 shows the modulo-n adder circuit. The adder takes two n-bit inputs in one-hot encoding format, first is data input and second is shift input. Shifting left side of the data input addition can be done, based on the value of the shift input, i.e., if shift input value is x then data input is shifted x number of times.

Fredkin gate can be used as AND gate and OR gate for designing reversible logic design.

Algorithm:

```
Behavioral model for OHRNS addition using Reversible
Logic Design.
    Data Input - a, Control Input - b and Output - OUT.
        For(j = 0; j <= n − 1; j + +)
        {
            For(i = 0, i <= n − 1; i + +)
            {
                b_j.a_i = A_j [i]
            }
        }
```

$$A_1[n-1:0] + (A_0[n-1:0] >>> 1) = OR_0[n-1:0].$$
$$A_2[n-1:0] + (OR_0[n-1:0] >>> 1) = OR_1[n-1:0].$$
$$A_3[n-1:0] + (OR_1[n-1:0] >>> 1) = OR_2[n-1:0].$$
$$\ldots$$
$$A_{n-1}[n-1:0] + (OR_{n-3}[n-1:0] >>> 1) = OUT[n-1:0].$$

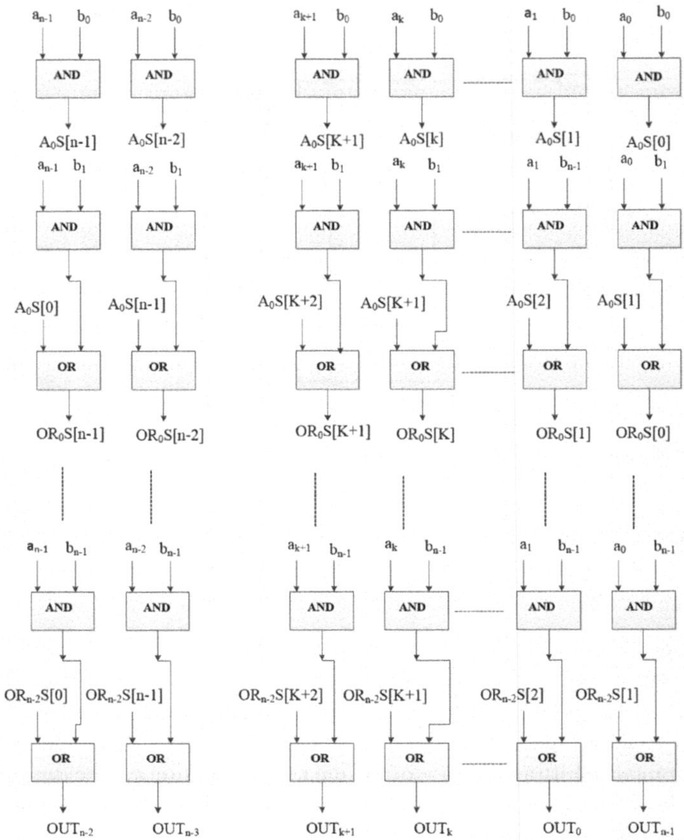

Fig. 8 Circuit diagram of modulo-n one-hot adder using reversible logic design

The illustration of modulo-n adder is shown by modulo-5 adder. The data input and shift input are 5 bits each and the methodology is shown below.

Methodology:
Data inputs (a_0 to a_4), control inputs (b_0 to b_4), and adder output- OUT[4 : 0].

Step 1: Data inputs (a_0 to a_4) and shift input b_0 are fed to AND gates resulting in

$$A_0S[4:0].$$
$$A_0[i] = b_0 \cdot a_i \text{ for } 0 \le i \le 4.$$

Step 2: Data inputs (a_0 to a_4) and shift input b_1 are fed to AND gates resulting in $A_1[4:0]$. $A_1[4:0]$ is first input and $A_0[4:0]$ is right shifted once and given as second input to OR gate producing $OR_0[4:0]$.

$$A_1[i] = b_1 \cdot a_i \text{ for } 0 \le i \le 4$$
$$A_1[4:0] + (A_0[4:0] >>> 1) = OR_0[4:0].$$

Step 3: Data inputs (a_0 to a_4) and shift input b_2 are fed to AND gates resulting in $A_2[4:0]$. $A_2[4:0]$ is first input and $OR_0[4:0]$ is right shifted once and given as second input to OR gate producing $OR_1[4:0]$.

$$A_2[i] = b_2 \cdot a_i \text{ for } 0 \le i \le 4$$
$$A_2[4:0] + (OR_0[4:0] >>> 1) = OR_1[4:0].$$

Step 4: Data inputs (a_0 to a_4) and shift input b_3 are fed to AND gates resulting in $A_3[4:0]$. $A_3[4:0]$ is first input and $OR_1[4:0]$ is right shifted once and given as second input to OR gate producing $OR_2[4:0]$.

$$A_3[i] = b_3 \cdot a_i \text{ for } 0 \le i \le 4$$
$$A_3[4:0] + (OR_1[4:0] >>> 1) = OR_2[4:0].$$

Step 5: Data inputs (a_0 to a_4) and shift input b_4 are fed to AND gates resulting in $A_4[4:0]$. $A_4[4:0]$ is first input and $OR_2[4:0]$ is right shifted once and given as second input to OR gate as producing outputs at port $OUT[4:0]$.

$$A_4[i] = b_1 \cdot a_i \text{ for } 0 \le i \le 4$$
$$A_4[4:0] + (OR_2[4:0] >>> 1) = OUT[4:0].$$

(4) *Design of Modulo-n One-Hot Subtractor using Reversible Logic Design*: The data inputs and control inputs information given in Modulo-n Reversible Logic design adder section are applicable for the design proposed here. Subtractor design shown in Fig. 9 is the same as adder design with only two differences. Right shift of the data input has occurred in adder circuit, but left shift of data input has occurred in subtractor. The second main difference is shift input order given as input to circuit. The shift input order is reversed when compared to the order of shift input in adder circuit.

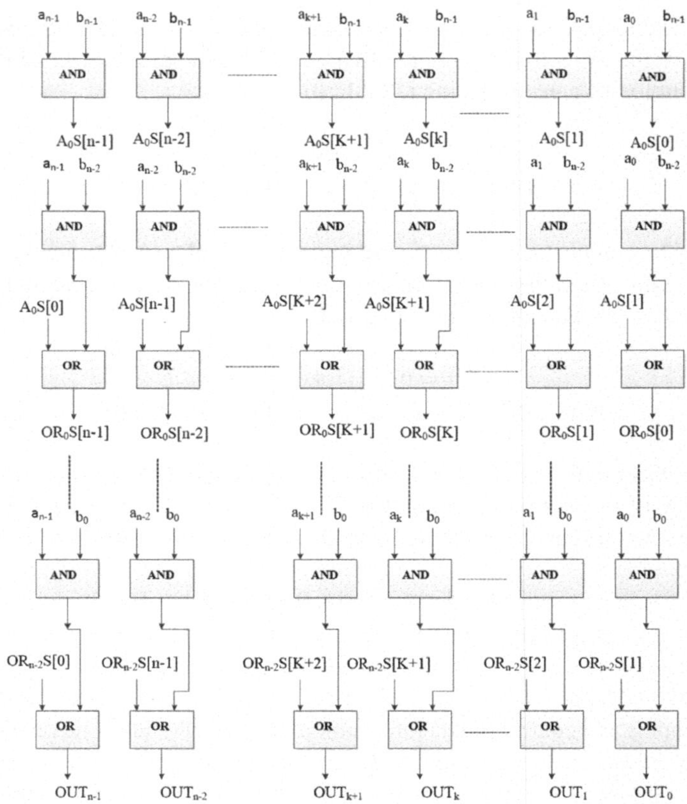

Fig. 9 Circuit diagram of modulo-n one-hot subtractor using reversible logic design

Algorithm:

Behavioral model for OHRNS subtraction using RLD
Data Input - a, Control Input - b and Output - OUT.

```
For(j = n-1; j <= 0; j - -)
{
    For(i = 0, i <= n – 1; i + +)
    {
        b_j.a_i = A_j [i]
    }
}
```

$A_{n-2}[n-1:0] + (A_{n-1}[n-1:0] >>> 1) = OR_0[n-1:0].$

$A_{n-3}[n-1:0] + (OR_1[n-1:0] >>> 1) = OR_1[n-1:0].$

$A_{n-4}[n-1:0] + (OR_2[n-1:0] >>> 1) = OR_2[n-1:0].$

...

$A_0[n-1:0] + (OR_{n-3}[n-1:0] >>> 1) = OUT[n-1:0].$

Methodology:
Data inputs (a_0 to a_4), control inputs (b_0 to b_4), and subtractor output-OUT[4 : 0].

Step 1: Data inputs (a_0 to a_4) and shift input b_4 are fed to AND gates resulting in $A_4S[4:0]$.

$$A_4[i] = b_4 \cdot a_i \text{ for } 0 \leq i \leq 4$$

Step 2: Data inputs (a_0 to a_4) and shift input b_1 are fed to AND gates resulting in $A_3[4:0]$. $A_3[4:0]$ is first input and $A_4[4:0]$ is right shifted once and given as second input to OR gate producing $OR_0[4:0]$.

$$A_3[i] = b_3 \cdot a_i \text{ for } 0 \leq i \leq 4$$
$$A_3[4:0] + (A_4[4:0] >>> 1) = OR_0[4:0]$$

Step 3: Data inputs (a_0 to a_4) and shift input b_2 are fed to AND gates resulting in $A_2[4:0]$. $A_2[4:0]$ is first input and $OR_0[4:0]$ is right shifted once and given as second input to OR gate producing $OR_1[4:0]$.

$$A_2[i] = b_2 \cdot a_i \text{ for } 0 \leq i \leq 4$$
$$A_2[4:0] + (OR_0[4:0] >>> 1) = OR_1[4:0].$$

Step 4: Data inputs (a_0 to a_4) and shift input b_3 are fed to AND gates resulting in $A_1[4:0]$. $A_1[4:0]$ is first input and $OR_1[4:0]$ is right shifted once and given as second input to OR gate producing $OR_2[4:0]$.

$$A_1[i] = b_1 \cdot a_i \text{ for } 0 \leq i \leq 4$$
$$A_1[4:0] + (OR_1[4:0] >>> 1) = OR_2[4:0].$$

Step 5: Data inputs (a_0 to a_4) and shift input b_4 are fed to AND gates resulting in $A_0[4:0]$. $A_0[4:0]$ is first input and $OR_2[4:0]$ is right shifted once and given as second input to OR gate as producing outputs at port OUT[4 : 0].

$$A_0[i] = b_0 \cdot a_i \text{ for } 0 \leq i \leq 4$$
$$A_0[4:0] + (OR_2[4:0] >>> 1) = OUT[4:0].$$

3 Proposed FFT Computation Block Using One-Hot Encoding Technique

Computation of four-point FFT is shown in Fig. 13 which mainly consists of computational block of butterfly block of Radix-2 and computational block of Twiddle factor implemented using reversible logic and OHRNS technique for modulo-5.

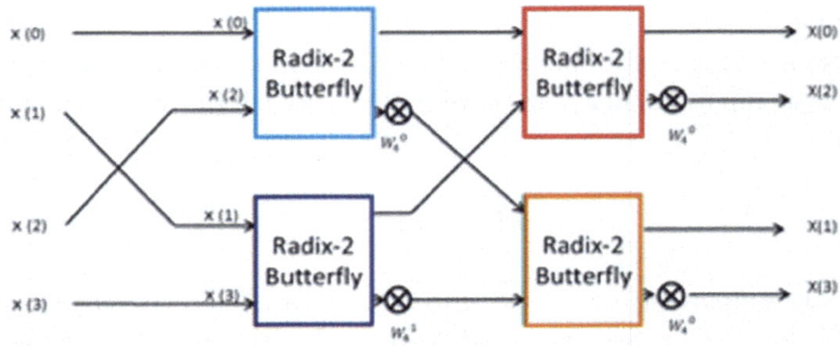

Fig. 10 Structure of four point FFT computation block [13]

Here, computation blocks can be visualized in two forms as shown in Fig. 10, Twiddle factor block (MULT) or Radix-2 block.

3.1 Design of Mod-5 Computational Block of Radix-2 Butterfly

In Fig. 11, implementation of computation of mod-5 butterfly Radix-2 using OHRNS and reversible logic is shown. This design has two inputs and two outputs in complex number. Therefore, as shown in Fig. 14, the real number and an imaginary number are provided as separate inputs to arithmetic block.

In AddSub block, real part in IN1-RE is added with IN2-RE to give output in OUT1-RE as a real part result, the same input IN1-RE is subtracted with IN2-RE in block AddSub2 producing the result in OUT2-RE. Two inputs IN1-IMG and IN2-IMG are used for Radix-2 addition in AddSub1 block resulting in OUT1-IMG and same inputs are used for subtraction in AddSub3 block producing the output result in OUT2-IMG.

3.2 Design of Mod-5 Computational Block of Twiddle Factor

Modulo-5 implementation of Twiddle factor computation block using OHRNS and reversible logic is shown in Fig. 12b. The Radix-2 computation output is a complex number and is divided into two inputs, In1-RE representing real part and In2-IMG representing imaginary part of complex number.

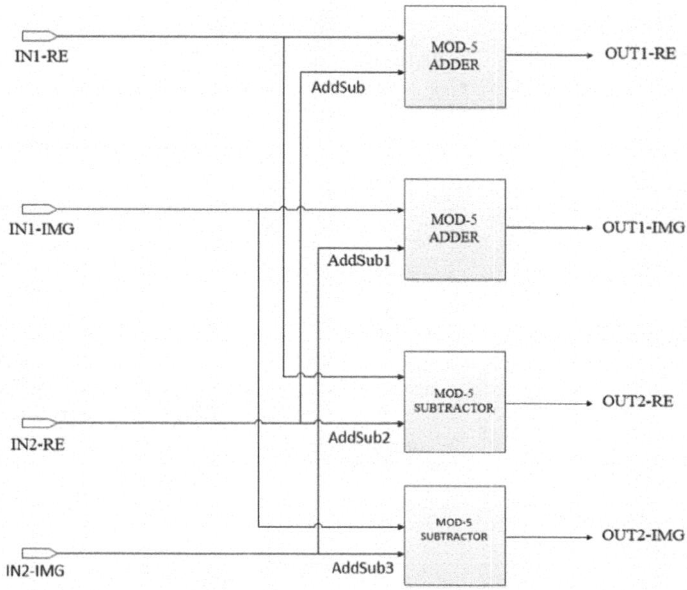

Fig. 11 Structure of computation of modulo-5 Radix-2 block [13]

Table 2 Parameters of twiddle factor

W	W_4^0	W_4^1
W-RE	1	0
W-IMG	0	−1

In the general block diagram of Fig. 12a, the inputs for W-RE and W-IMG are selected from Table 2 For example, consider W_4^0 the W-RE value will be 1 and W-IMG will be 0. In the proposed two-point FFT, we have come up with a design for Twiddle factor without using multipliers to give value for W_1^4 $(-1j)$ only. Since the output value of W_4^0 is 1 as used in general block diagram. Twiddle factor for two-point FFT using only adder and subtractor is proposed as shown in Fig. 12b.

4 Performance Metric Analysis

In this, the comparative analysis of OHRNS-based modulo-5 data path subsystems, such as modulo-5 adder, modulo-5 subtractor with the subsystems designed using conventional and reversible logic approach list is given in Table 3, Table 4, and Table 5, respectively. Comparison is based on the parameters such as number of transistors, propagation delay, and power dissipation.

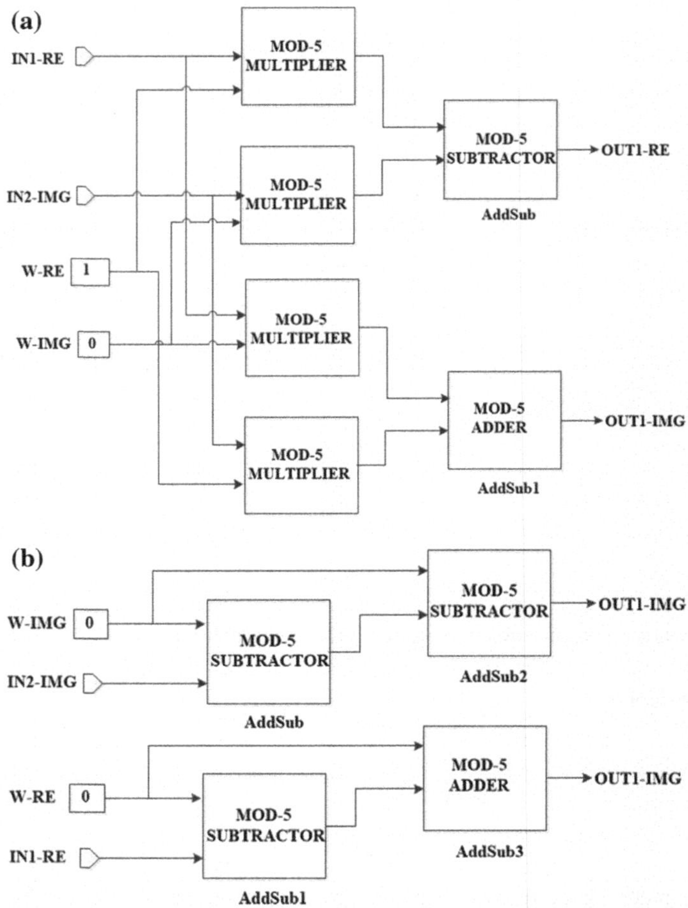

Fig. 12 **a** General twiddle factor with multiplier [13]. **b** Proposed multiplier-less twiddle factor computation block for W_4^1

Table 3 Performance metric analysis of OHRNS modulo-5 adder, reversible logic modulo-5 adder and conventional modulo-5 adder

Adder type	# Transistors	Propagation delay (ns)	Average power (mW)
OHRNS modulo-5	25	0.092	0.101
One-hot reversible logic design	180	0.108	0.122
Conventional modulo-5	342	10.51	0.619

Table 4 Performance metric analysis of OHRNS modulo-5 subtractor, reversible logic modulo-5 subtractor and conventional modulo-5 subtractor

Subtractor type	# Transistors	Propagation delay (ns)	Average power (mW)
OHRNS modulo-5	25	0.097	0.011
One-hot reversible logic design	180	0.381	0.195
Conventional modulo-5	352	10.84	0.645

Table 5 Performance metric analysis of one-hot modulo-5 FFT computation block and conventional modulo-5 FFT computation

FFT type	# Transistors	Propagation delay (ns)	Average power (mW)
OHRNS modulo-5	500	0.271	0.203
One-hot reversible logic design	3600	0.426	0.286
Conventional modulo-5 FFT	6246	6.611	0.316
Conventional FFT	2106	10.155	3.841

4.1 Performance Analysis of Modulo-5 Adder Based on OHRNS, Reversible Logic Design, and Conventional Technique

From Table 3, it is observed that OHRNS-based modulo-5 adder design consumes less number of transistors, delay, and power compared to reversible logic design and conventional.

4.2 Performance Analysis of Modulo-5 Subtractor Based on OHRNS, Reversible Logic Design, and Conventional Techniques

From Table 4, it is observed that OHRNS-based modulo-5 subtractor design consumes less number of transistors, delay, and power compared to reversible logic design and conventional.

Table 6 Comparison of modulo-5 FFT block results

FFT type	Implementation type	Propagation delay
OHRNS modulo-5 FFT block (Proposed)	ASIC design (250 nm)	0.271 (ns)
One hot reversible logic design modulo-5 FFT block (Proposed)	ASIC design (250 nm)	0.426 (ns)
Conventional modulo-5 FFT block (Proposed)	ASIC design (250 nm)	6.611 (ns)
Conventional FFT block (Proposed)	250 nm ASIC design	10.155 (ns)
Conventional Radix-2 FFT [14]	FPGA design	1.640 (ns)
Conventional Radix-4/2 FFT [15]	FPGA design	23.94 (μs)

4.3 Performance Analysis of Modulo-5 FFT Based on OHRNS, Reversible Logic Design, Conventional, and General Conventional Techniques

From Table 5, we can observe that OHRNS modulo-5 FFT circuit performance parameters results are promising compared to reversible logic design, conventional FFT, and conventional design of modulo-5 FFT.

In Table 6, conventional Radix-2 FFT [8, 9] is compared with proposed designs, that is, OHRNS mod-5 FFT, reversible logic design mod-5 FFT, and conventional mod-5 FFT. From the comparison, it is clear that the proposed design modulo-5 OHRNS FFT outperforms the Radix-2 conventional FFT [8, 9]. Thus, proving with the less processing delay (latency) and less number of transistors, the computation blocks are very much efficient in hardware using one-hot encoding.

5 Simulation Results

Operation of modulo-5 adder with different combinations of inputs is listed in Table 7. In Fig. 13, mod-5 OHRNS adder simulation waveform is shown with inputs listed in Table 7. Due to space constraints, only adder results of OHRNS are shown.

Table 7 OHRNS modulo-5 adder test cases result

Input-1	Input-2	Output adder
01000 (3)	10000 (4)	00100 (2)
01000 (3)	01000 (3)	00010 (1)
00010 (1)	00100 (2)	01000 (3)
00100 (2)	00100 (2)	10000 (4)

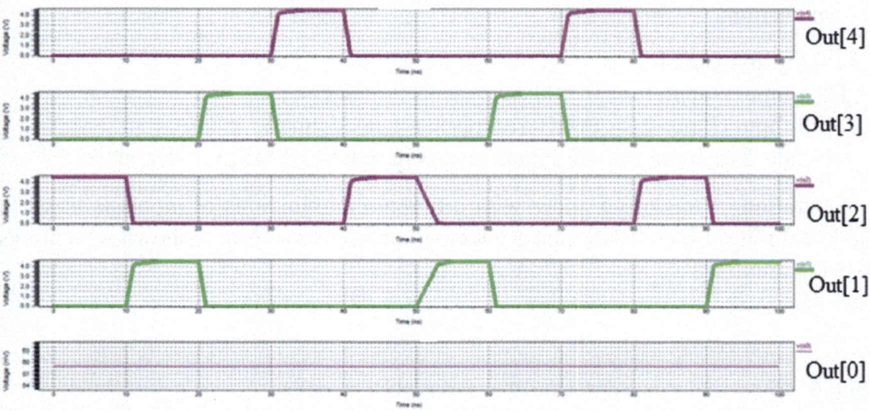

Fig. 13 Waveform of OHRNS modulo-5 adder

6 Conclusion

Novel design using OHRNS for computation of FFT is proposed in this work. OHRNS technique which is the proposed design outperforms in all comparison analysis in propagation delay, requires less transistors, and low power as compared to results of conventional techniques. These are crucial in any designing of circuits and also in DSP applications (image processing). It is also observed that reversible logic based implementation is close to OHRNS in terms of power but gives degraded result for speed. Proposed design gives better result in speed when compared with existing work in [8] and [9].

Acknowledgements We would like to thank the department of ECE, PES University for the tool support.

Appendix

Reversible Logic Design Modulo-5 One-Hot Adder

The illustration of modulo-5 addition using RLD with an example is shown in Table 8. The processing of data input vector with respect to shift input happens due to barrel shifting action.

Taking an example,
Data input (Input-1) = 01000 (represents 3 in decimal).
Shift input (Input-2) = 00010 (represents 1 in decimal).
Modulo-5 addition of both inputs results in output 10000 (4 in decimal).

Operation of modulo-5 adder with different combinations of inputs is listed in Table 9. In Fig. 14, OHRNS mod-5 adder simulation waveform is shown with inputs listed in Table 9.

Table 8 Modulo-5 addition operation using reversible logic design

Steps	Input-1	Input-2	Active stage	Result of step	Final output
Step-1	01000	$b_0 = 0$	A_0	00000	–
Step-2	01000	$b_1 = 0$	A_1	01000	–
			OR_0	01000	
Step-3	01000	$b_2 = 0$	A_2	00000	–
			OR_1	00100	
Step-4	01000	$b_3 = 0$	A_3	00000	–
			OR_2	00010	
Step-5 output	01000	$b_4 = 0$	A_4	00000	10000 (4)
			OR_3	00001	

Table 9 Reversible logic design modulo-5 adder test cases

Input-1	Input-2	Output adder
10000 (4)	00001 (0)	10000 (4)
01000 (3)	10000 (4)	00100 (2)
10000 (4)	00100 (2)	00010 (1)
10000 (4)	10000 (4)	01000 (3)

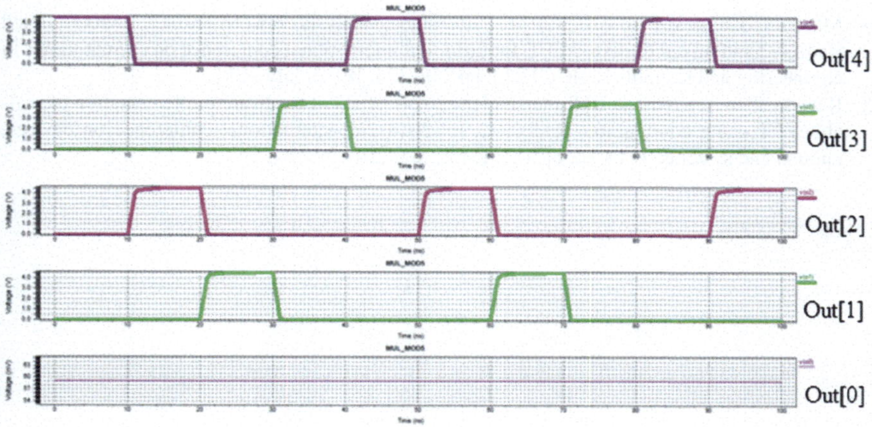

Fig. 14 Waveform of modulo-5 adder reversible logic design

References

1. Bajard J, Imbert L (2004) A full RNS implementation of RSA. IEEE Trans Comput 53(6):769–774
2. Ramírez J, García A, Meyer-Baese U, Lloris A (2002) Fast RNS FPL-based communications receiver design and implementation. In: International conference on field programmable logic and applications, pp 472–481. Springer, Berlin, Heidelberg, Sept 2002
3. Abdallah M, Skavantzos A (2005) On multi moduli residue number systems with moduli of forms r^a, $r^b - 1$, $r^c + 1$. IEEE Trans Circuits Syst I Regular Paper 52(7):1253–1266
4. Timarchi S, Navi K (2009) Arithmetic circuits of redundant SUT-RNS. IEEE Trans Instrum Meas 58(9):2959–2968
5. Parhami B (2001) RNS representations with redundant residues. In: Proceedings of the 35th Asilomar conference on signals, systems, and computers, Pacific Grove, CA, vol 2, pp 1651–1655. IEEE, Nov 2001
6. Landauer R (1961) Irreversibility and heat generation in the computing process. IBM J Res Dev 5(3):183–191
7. Labafniya M, Eshghi M (2010) An efficient adder/subtracter circuit for one-hot residue number system. In: International conference on electronic devices, systems and applications (ICEDSA), pp 121–124, Apr 2010
8. Taleshmekaeil DK, Safari A, Kong Y (2012) Using one hot residue number system (OHRNS) for digital image processing. In: The 16th CSI international symposium on artificial intelligence and signal processing (AISP 2012), pp 064–067. IEEE, May 2012
9. Chren William A (1998) One-hot residue coding for low delay-power product CMOS design. IEEE Trans Circuits Syst II Analog Digit Signal Process 45(3):303–313
10. Jassbi SJ, Hosseinzadeh M, Navi K (2007) A novel multiple valued logic OHRNS modulo r^n adder circuit. Proc World Acad Sci Eng Technol 1(4):245–249
11. Raju IBK, Rajesh Kumar P, Bhaskara Rao P (2014) Residue arithmetic's using reversible logic gates. In: 2014 2nd international conference on devices, circuits and systems (ICDCS), pp 1–6. IEEE, Mar 2014
12. Hashmi I, Babu HMH (2010) An efficient design of a reversible barrel shifter. In: 23rd international conference on VLSI design, 2010, VLSID'10, pp 93–98. IEEE, Jan 2010
13. The 15th LSI contest conducted in 2012 at Okinawa. http://www.lsicontest.com/2012/shiyou-4-1-1e.html

14. Mangaiyarkarasi V, Paul CKC (2014) Performance analysis between Radix2, Radix4, Mixed Radix4-2 and mixed Radix8-2 FFT. In: 2014 2nd international conference on current trends in engineering and technology (ICCTET), pp 430–434. IEEE, July 2014
15. Ranganathan S, Krishnan R, Sriharsha HS (2014) Efficient hardware implementation of scalable FFT using configurable Radix-4/2. In: 2014 2nd international conference on devices, circuits and systems (ICDCS), pp 1–5. IEEE, Mar 2014

Advancements and Challenges in Tunnel Field Effect Transistor

Nitika Sharma, Nidhi Garg and Gurpreet Kaur

Abstract TFET abbreviated for tunnel field effect transistor, is a p-i-n diode which functions as a transistor when operated in the reverse bias condition, output current of which depends upon quantum tunneling of the charge carriers across a barrier, also called band-to-band tunneling that occurs between the source and the channel which is responsible for the switching mechanism. TFETs suffer from lower ON-state currents, considerable amount of ambipolar conduction currents and poor RF performance. The purpose of this paper is to study the selected TFET models which list out the significant improvements to provide better switching capabilities. Section 1 gives an introduction to the device discussing band-to-band tunneling (BTBT) and subthreshold swing. Section 2 highlights the study of previous related work on various TFET models such as DG-TFET, DP-TFET, DMCG-TFET, HG-dielectric TFET, overlapping gate on drain TFET, PAC-TFET, DMCG-CPTFET, SP-TFET, and Multi-Fin TFET which provide improvements in I_{ON}/I_{OFF} ratio, ambipolar current suppression, and improved RF performance of the device. The models discussed make TFET a better candidate in terms of switching performance and potential model to substitute MOSFETs in low-power and high-speed switching circuits.

Keywords Tunnel field effect transistor (*TFET*) · Band-to-band tunneling (*BTBT*) · Subthreshold swing · Double gate (*DG*) · High-κ dielectric

1 Introduction

With consummation of the Moore's prediction, microelectronics are turned into nanoelectronics [1]. The size of MOSFET has been abridged to nanometer regimen. The main intent of scaling down of technology is to realize high chip density, reduced power consumption, improved ON-state currents, and enhanced analog performance.

N. Sharma (✉) · N. Garg · G. Kaur
Department of Electronics and Communication, University Institute of Engineering and Technology (UIET), Panjab University (PU), Chandigarh, India
e-mail: nitika.sharma.2792@gmail.com

© Springer Nature Singapore Pte Ltd. 2019

V. Nath and J. K. Mandal (eds.), *Proceedings of the Third International Conference on Microelectronics, Computing and Communication Systems*,
Lecture Notes in Electrical Engineering 556, https://doi.org/10.1007/978-981-13-7091-5_33

While scaling of MOSFET several problems such as short channel effects (SCEs), subthreshold leakage current, hot carrier effect (HCE), and drain-induced barrier lowering (DIBL), etc. emerged that rankles the device performance.

Downscaling of dimensions and operating voltages lead to the outset to novel devices. One such promising research device is tunnel FET, holding potential to transpire as an energy-efficient switch [2].

It is one of the emanating devices that may replace MOS technology which provides subthreshold swing below 60 mV/decade and low OFF-state currents. Band-to-band tunneling (BTBT) mechanism is the key phenomenon for the conduction of current in this quintessential reverse-biased gated p-i-n structure unlike MOSFETs where conduction is due to transition of charge [1, 2]. Before studying various models, it is essential to understand the phenomenon of band-to-band tunneling and understanding of subthreshold swing in TFET.

1.1 Band-to-Band Tunneling (BTBT)

For BTBT phenomenon, the transmission of carriers through tunneling can mathematically be achieved by Wentzel, Kramer's, and Brillouin (WKB) approximation given by

$$T_t \approx \exp\left[-\frac{4}{3}\frac{\sqrt{2m^*}}{qF\hbar}.(E_g)^{3/2}\right] \tag{1}$$

Equation 1 is the common method to describe band-to-band tunneling transmission.

Figure 1 shows a conventional TFET p-i-n structure showing band-to-band tunneling (BTBT).

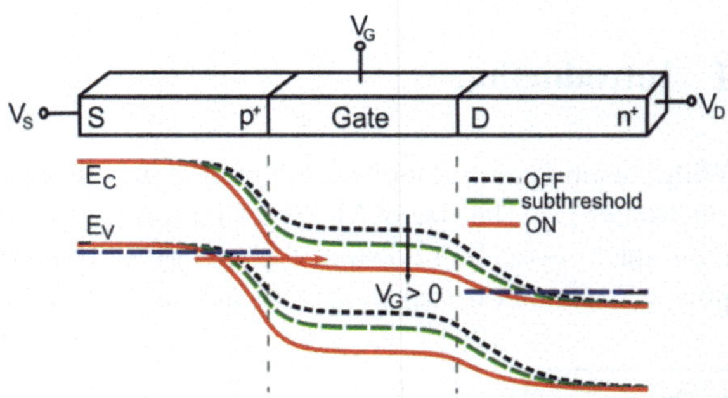

Fig. 1 Conventional TFET structure showing band-to-band tunneling phenomenon [17]

1.2 Subthreshold Swing (SS)

The subthreshold swing of a device is defined as the increase in drain current of the order of one decade brought by change in gate voltage (V_g), expressed in Eq. 2,

$$S = \frac{dV_g}{d(\log I_d)}[\mathrm{mV/dec}] \tag{2}$$

Subthreshold Swing of TFET
A TFET does not have diffusion currents taking place in weak inversion like in MOSFETs, since the current transition is based on the width of tunneling barrier and not on an inversion channel formation.

The expression for the BTBT current [2] is given below for reverse-biased p-n junction:

$$I = aV_{eff}E_e^{\frac{-b}{E}} \tag{4}$$

where

$$a = Aq^3\frac{\sqrt{\frac{2m^*}{E_g}}}{\hbar^2\pi^2} \tag{5}$$

where the device cross-sectional area is denoted by A, $\hbar$ denotes the Plank's constant, and m^* denotes the effective mass of the charge carrier.

$$b = 4\sqrt{m^*}\frac{E_g^{3/2}}{3q\hbar} \tag{6}$$

Correspondingly, the subthreshold swing in a TFET increases with increase in gate-source voltage and is much steeper at lower gate voltages.

In terms of the gate voltage, the subthreshold slope for a TFET can be expressed as

$$S_{\mathrm{TFET}} = \frac{V_{gs}^2}{5.75(V_{gs} + Const)}[\mathrm{mV/dec}] \tag{7}$$

where the term constant is determined by material parameters and dimensions taken for the device. Figure 2 shows the compared $I_d{-}V_g$ characteristics of TFET and MOSFET.

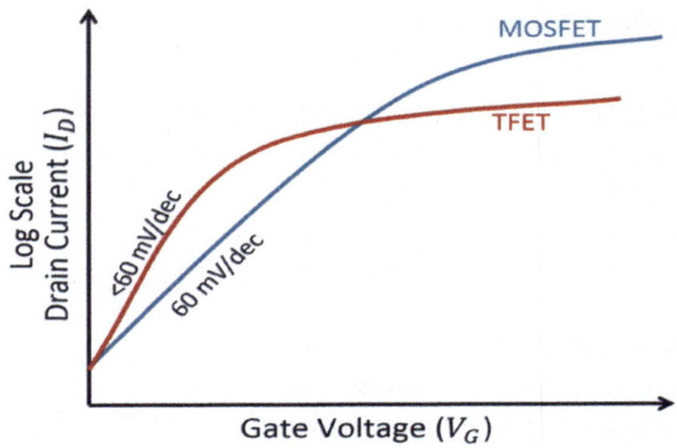

Fig. 2 Compared $I_D - V_G$ characteristics of TFET and MOSFET [18]

Despite all these advantages, TFETs suffer from ambipolar behavior, lower ON-state currents, and poor RF performance.

The study highlights the abovementioned areas and discusses some of the TFET models which proposed modifications in the TFET structure to bring out significant improvements in ON-state current, ambipolar current suppression, and RF performance. It is important to note that ON-state current has been discussed as I_{ON}/I_{OFF} ratio.

2 Previous Related Work

In earlier research work, the structures like single gated tunnel FET were studied. Many design variations were proposed in single gated TFET (SG-TFET) to improve ON-state currents and subthreshold swing. Bhuwalka et al. proposed a vertical tunnel FET structure, [3] also proposed incorporation of a thin layer of SiGe referred as delta layer at the boundary of the p+ region and channel region that would reduce the width of the tunneling barrier and would further improve the subthreshold swing and ON-state current [4].

Thereafter, double gate tunnel FETs were studied. Boucart et al. proposed double gate TFET by adding dielectrics with high-k values at gate oxide [4]. Figure 3a shows a conventional p-type TFET structure. There were several other models proposed, but the ON currents obtained were very low for a device to replace conventional MOSFETs.

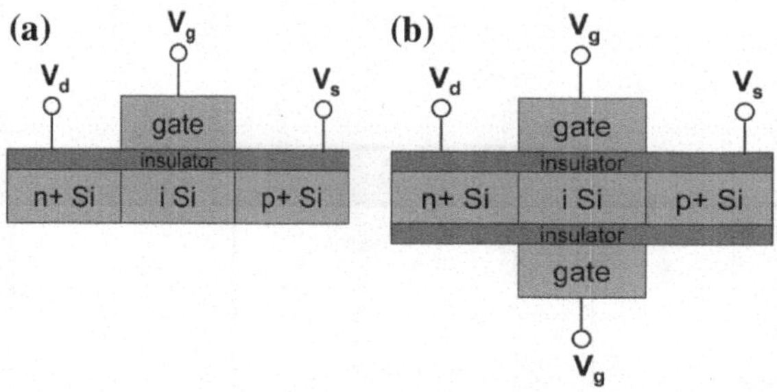

Fig. 3 **a** Single gated p-type TFET, **b** double gated p-type TFET [19]

2.1 I_{ON}/I_{OFF} Ratio Improvement

a. *Double Gate TFETs and Gate Dielectric Materials*

Figure 3b shows a double gated p-type TFET. This device used a high-k dielectric at gate, which supported small threshold voltage shifts on variation of temperature as compared to those using SiO_2 as gate dielectric.

b. *Gate Stack Architecture*

It was observed that by using gate dielectrics having high-k values, leakage current increased and the device suffered from deplorable performance lowering parameters such as high density of interface trap states, higher levels of bulk fixed charge, and low silicon interface carrier mobility. In order to remove these issues, Narang et al. proposed a gate stack architecture comprising a thin passivation SiO_2 layer between the channel and high-k gate dielectric that resulted in reduced effective oxide thickness and higher ON-state currents [5, 6].

c. *Dielectric Pocket TFET*

To increase the efficiency of the device, Gupta et al. proposed dielectric pocket (DP) TFET. This device gave enhanced performance in terms of increased ON-state current, increased I_{ON}/I_{OFF} ratio, reduced threshold voltage, reduced subthreshold swing, increased trans-conductance g_m, and reduced gate capacitance [7, 8]. Figure 4 shows the structure of dielectric pocket TFET.

Comparison of I_{ON}/I_{OFF} of different TFET device architectures values has been given in Table 1. It can be inferred from the table that gate stack architecture and dielectric pocket architecture give high values for I_{ON}/I_{OFF} ratio as compared to single gate TFET and double gate TFET.

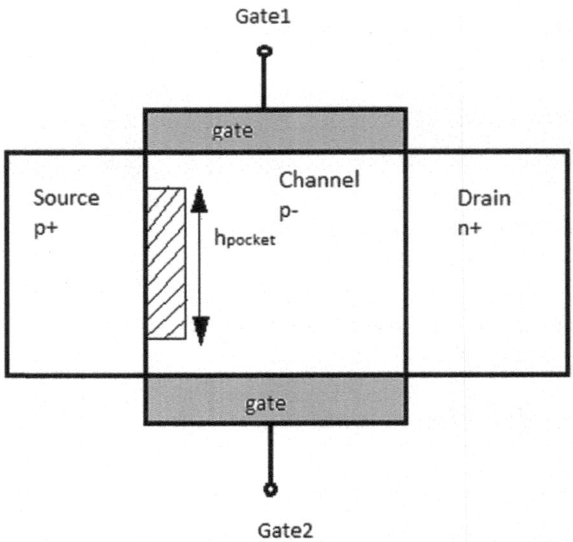

Fig. 4 Dielectric pocket (DP)-TFET

Table 1 Comparing I_{ON}/I_{OFF} and ambipolar conduction of different device architectures

Device architecture	Device parameters	
	I_{ON}/I_{OFF}	I_{amb} (A/μm)
SG-TFET	1×10^8	1×10^{-8}
DG-TFET	2×10^8	1×10^{-6}
GS-DG-TFET	1×10^{11}	–
DP-TFET	1×10^{12}	–
DG (gate overlap)-TFET	1×10^9	1×10^{-9}
PAC-TFET	1×10^9	1×10^{-9}
DMG-HD-TFET	6.46×10^8	1×10^{-12}
DMCG-TFET	1×10^8	1×10^{-13}
DMCG-CP-TFET	1×10^8	1.2×10^{-18}

2.2 *Temperature Dependency*

The comprehensive study revealed that when the current conduction is dominated by band-to-band tunneling such as TFET, the device shows weak dependence on temperature, whereas the dependence on temperature increases when the device is in its OFF state.

Narang et al. through their work highlighted the fact that the increasing temperature projects a constructive effect on the switching characteristics of TFET, due to reduction in the propagation delay time. As compared to the conventional p-i-n structure of TFET, the p-n-p-n TFET structure was more resilient to delay variations

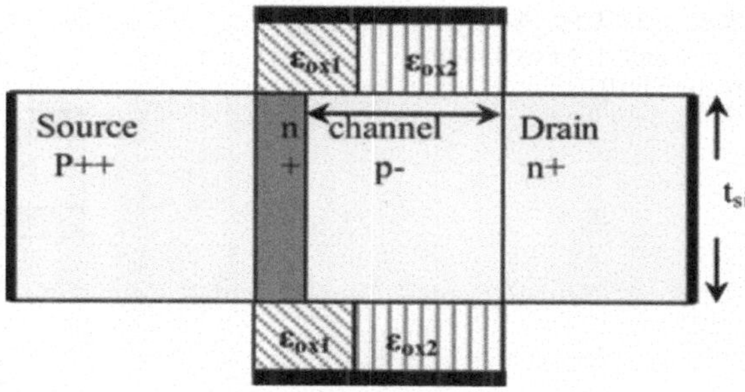

Fig. 5 Hetrodielectric double gate p-n-p-n TFET [9]

due to change in temperature and exhibited better circuit performance as observed by the author. The improved circuit performance was proposed further by insertion of heterogate (HG) dielectric [9] shown in Fig. 5.

2.3 Reduction in Ambipolar Conduction in TFETs

a. *Overlapping Gate on Drain*

TFET also possesses distinctive attribute of ambipolarity that is conduction of current, for both high positive as well as high negative gate voltages. The conduction at high negative voltages occurs due to tunneling of charge carriers from conduction band to valence band of drain to channel, respectively. Ambipolar conduction increases the leakage and thus the device TFET becomes less efficient in complementary circuit applications thereby restraining its efficacy in designing digital circuits. Abdi et al. came out with architecture by extending the gate partly covering the drain that would suppress ambipolar conduction in tunnel field effect transistor (TFET) [10]. Figure 6b explains the proposed model of overlapped gate where L_{OV} indicates the overlapped gate length.

b. *Polarity and Ambipolarity (PAC)-TFET*

In this work, to control the polarity and the ambipolar current conduction at once Narang et al. brought about a new-fangled architecture, i.e., PAC-TFET based on double gate geometry [11].

The author proposed an architecture which could configure the device polarity by change in biasing condition of the two independently driven gate electrodes along with suppression of ambipolar current conduction. The architecture depicted in Fig. 7 shows symmetric and independently driven PAC-TFET, where double gate structure is split into four gates. Gate1 and Gate3 being controlled by V_g1 and Gate2 and Gate4 by V_g2 biasing.

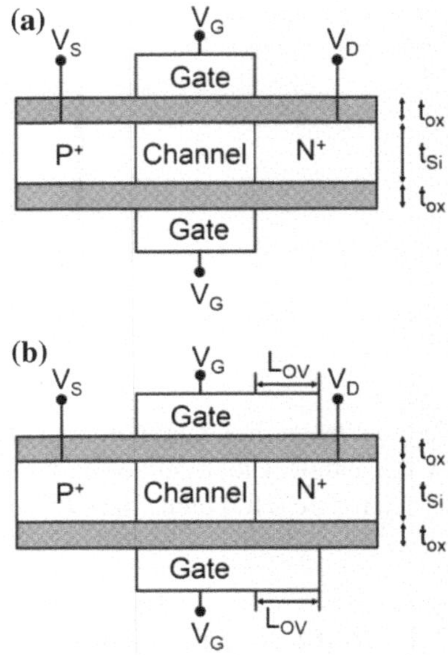

Fig. 6 (a) n-type DG-TFET (b) overlapped gate on drain structure [10]

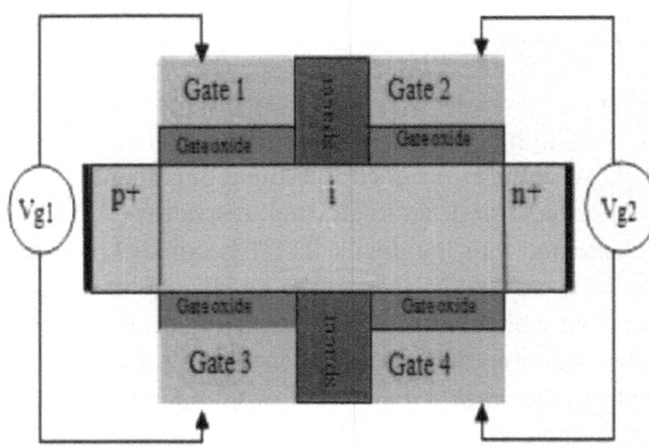

Fig. 7 Symmetric and independently driven PAC-TFET [11]

c. *DMCG-TFET*

Another generalized model that described the electrostatic behavior of device was proposed by Upasana et al. by taking three different double gate TFET architectures, i.e., dual material gate (DMG) TFET, heterodielectric (H-D) TFET, and the combination of both ideas dual material gate heterodielectric (DMG H-D). Through device simulation based on TCAD, the transient performance and capacitive behavior of the above three device architectures were analyzed. Both the modeling and simulation results revealed that DMG H-D TFET shown in Fig. 8 gives better results than the other two architectures [12, 13].

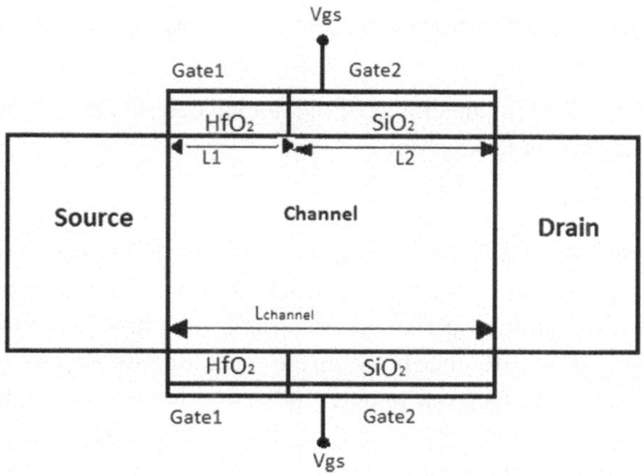

Fig. 8 DMG-HD-TFET

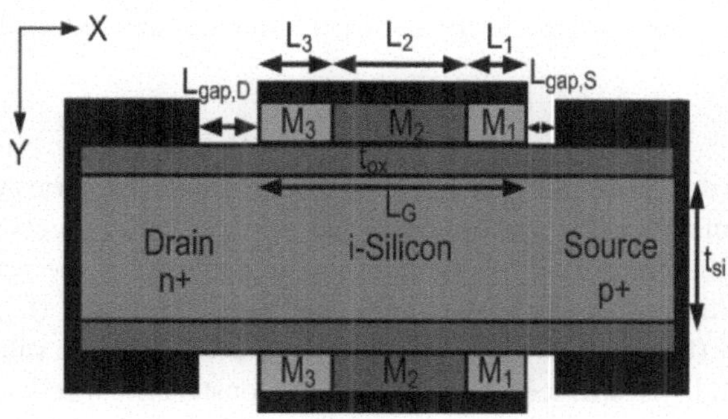

Fig. 9 Cross-sectional view of DMCG-CPTFET [13]

d. *DMCG-CPTFET*

For the suppression of ambipolarity and enhancement of analog or radio frequency (RF) performance and linearity performance of TFET, Nigam et al. proposed a dual material control gate charge plasma based TFET (DMCG-CPTFET) architecture [13]. In DMCG-CPTFET shown in Fig. 9, the gate was segmented into three parts, i.e., tunneling gate (M1), control gate (M2), and auxiliary gate (M3). The charge plasma concept based architecture proposes the lowered fabrication complexity. The tunneling gate and auxiliary gate were chosen with lower work function that helped in suppression of ambipolar conduction and increased ON-state currents.

Table 1 discusses values for ambipolar currents in different device architectures. It has been observed that DMCG-CPTFET architecture reduces the ambipolar conduction to the order of 10^{-18}.

2.4 RF Performance Improvement

For the RF performance enhancement of the device certain changes made in device structure are discussed in this section.

a. *Source Pocket TFET*

For further improvement in ON-state current and better RF figure of merit, different possible combinations of length for dual work function of double gate-source pocket TFET (Fig. 10) with symmetric and asymmetric doping profiles of the drain region were proposed [14]. The difference between the work function of the separate sections of a gate material created steep rise in potential at the channel region. In this concern, the work function of M2 gate section at the source side was kept lower as compared to the work function of the gate section at the drain end, and this created more bending of the energy band at the interface of source and channel as well as in source region, which increased the tunneling probability. Its consequences reflected in the increment of ON-state current, electrostatic performance, and reduction in threshold voltage [14, 15].

b. *Multi-fin TFETs*

In this work, the design of multi-fin TFET was studied and RF metrics were extracted using TCAD simulations. It was observed that the drive current gets multiplied for the increase in the number of fins without the degradation of the leakage current. RF metrics, intrinsic gain (A_0), maximum oscillation frequency (f_{max}), unity gain cutoff frequency (f_t), and Y-parameters were extracted for various values of inter fin separation (IFS) [16]. It was seen that A_0 remained constant irrespective of IFS values. Due to the electrostatic coupling between the adjacent fins, f_t and f_{max} offered higher values for the lesser IFS values. The Y-parameters were extracted at an operating frequency of 10 GHz and which inferred that both real and imaginary components of Y-parameters offer higher value for larger IFS values. Figure 11 shows the horizontal top view of multi-fin TFET.

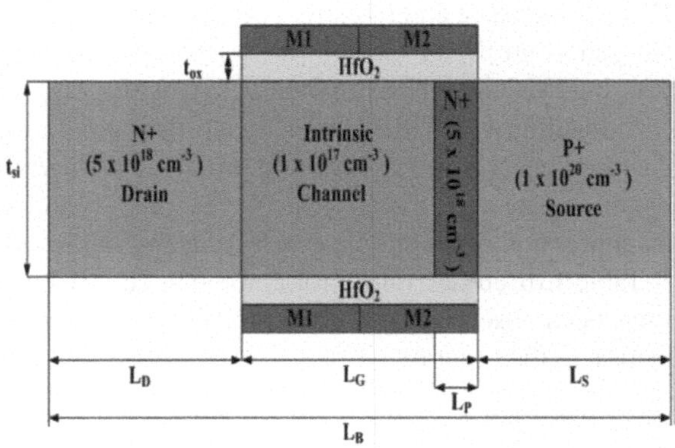

Fig. 10 Double screen gate-source pocket TFET (DSG SP TFET) [14]

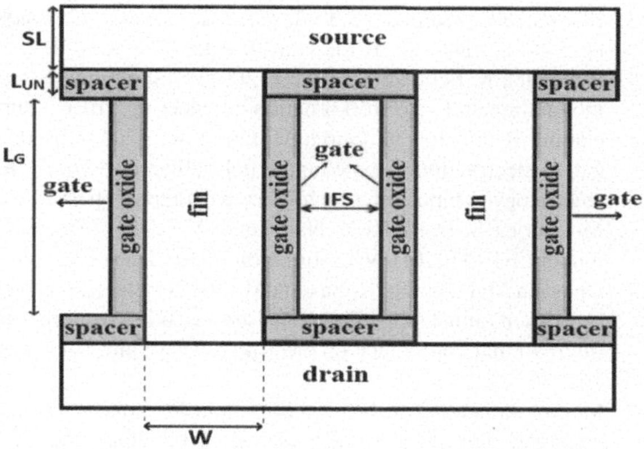

Fig. 11 Multi-fin TFET [16]

3 Proposed Work

Some of the previously proposed TFET models are discussed in this paper. All these models provide an insight to improve I_{ON}/I_{OFF} ratio, suppress ambipolar currents, and to improve RF performance which could be incorporated to develop a TFET model that could be incorporated in a high-speed and ultralow-power switching circuits.

4 Conclusion

Tunnel FET (TFET) is one of the flourishing devices to replace MOS technology as it provides low OFF-state current and subthreshold swing (SS) < 60 mV/decade. TFETs suffer from ambipolar behavior, lower ON-state current, and poor RF performance. For improving its switching capabilities and to remove its shortcomings some of TFET models have been discussed such as DG-TFET, DP-TFET, DMCG-TFET, HG-dielectric TFET, overlapping gate on drain TFET, PAC-TFET, DMCG-CPTFET, SP-TFET, and multi-fin TFET. The models are discussed with intent to build TFET a better candidate in terms of switching performance and potential model to replace MOSFETs in applications requiring high-speed switching without costing on low power.

References

1. ITRS Reports—International Technology Roadmap for Semiconductors. http://www.itrs2.net/itrs-reports.html. Accessed 28 Mar 2018
2. Mamidala JK, Vishnoi R, Pandey P (2016) Tunnel field effect transistors (TFET): modeling and simulation. ISBN 978-1-119-24629-9

3. Bhuwalka K, Sedlmaier S, Ludsteck AK, Tolksdorf C, Schulze J, Eisele I (2004) Vertical tunnel field-effect transistor. IEEE Trans Electron Devices 51(2)
4. Bhuwalka K, Schulze J, Eisele I (2005) Scaling the vertical tunnel FET with tunnel bandgap modulation and gate workfunction engineering. IEEE Trans Electron Devices 52(5):909–917
5. Narang R, Saxena M, Gupta M, Gupta RS (2011) Modeling and simulation of multi layer gate dielectric double gate tunnel field-effect transistor (DG-TFET). In: 2011 IEEE students' technology symposium (TechSym), Kharagpur, pp 281–285
6. Mookerjea S, Datta RKS, Narayanan V (2009) Effective capacitance and drive current for tunnel FET (TFET) CV/I estimation. IEEE Trans Electron Devices 56(9)
7. Upasana MG, Narang R, Saxena M (2015) Influence of dielectric pocket on electrical characteristics of tunnel field effect transistor: a study to optimize the device efficiency. In: 2015 IEEE International conference on electron devices and solid-state circuits (EDSSC), Singapore, pp 762–765
8. Mishra A, Narang R, Saxena M, Gupta M (2016) Impact of interfacial fixed charges on the electrical characteristics of pocket-doped double-gate tunnel FET. IEEE Trans Device Mater Reliab 16(2):117–122
9. Narang R, Saxena M, Gupta RS, Gupta M (2013) Impact of temperature variations on the device and circuit performance of tunnel FET: a simulation study. IEEE Trans Nanotechnol 12(6)
10. Abdi DB, Jagadesh Kumar M (2014) Controlling ambipolar current in tunneling FETs using overlapping gate-on-drain. IEEE J Electron Devices Soc 2(6):187–190
11. Narang R, Saxena M, Gupta M (2015) Polarity and ambipolarity controllable (PAC) tunnel field effect transistor. In: 2015 IEEE International conference on electron devices and solid-state circuits (EDSSC), Singapore, pp 333–336
12. Upasana MG, Narang R, Saxena M, Gupta M (2015) Modeling and TCAD assessment for gate material and gate dielectric engineered TFET architectures: circuit-level investigation for digital applications. IEEE Trans Electron Devices 62(10):3348–3356
13. Nigam K, Pandey S, Kondekar PN, Sharma D, Kumar Parte P (2017) A barrier controlled charge plasma-based TFET with gate engineering for ambipolar suppression and RF/linearity performance improvement. IEEE Trans Electron Dev 64(6):2751–2757
14. Raad BR, Sharma D, Kondekar P, Nigam K, Baronia S (2017) DC and analog/RF performance optimisation of source pocket dual work function TFET. Int J Electron 104(12):1992–2006
15. Yadav D, Verma A, Sharma D, Tirkey S, Raad B (2017) Comparative investigation of novel hetero gate dielectric and drain engineered charge plasma TFET for improved DC and RF performance. Superlattices Microstruct 111:123–133
16. Thoti N, Lakshmi B (2017) RF performance enhancement in multi-fin TFETs by scaling inter fin separation. Mater Sci Semicond Process 71:304–309
17. Sylvia SS (2013) The tunnel FET. http://somaiasylvia.com/wp-content/uploads/2014/06/bd-operation.png. Accessed 27 Mar 2018
18. Wikimedia Commons contributors (2016) File: TFET transfer characteristics.png. https://commons.wikimedia.org/w/index.php?title=File:TFET_transfer_characteristics.png&oldid=227012829. Accessed 28 Mar 2018
19. Boucart K, Ionescu AM (2007) Double-gate tunnel FET with high-κ gate dielectric. IEEE Trans Electron Devices 54(7)

A Brief Survey on Exploratory Search Systems

Ratna Kumari and Sanjay Kumar Jain

Abstract The definition of exploratory search is complex and multifarious. It is a process of seeking information to help the users who are unfamiliar to the domain of their search goal or have no idea how to achieve their goal. There exists several search systems which support exploratory search. In this rapidly developing field, there is a need to look at the achievements and tendencies in order to give a better angle to the future research works. This paper provides an overview of exploratory search systems and discusses the state of the art. We analyze these available systems based on features such as algorithms, result ranking, explanations, etc.

Keywords Classification · Information retrieval · Exploratory search system

1 Introduction

Searching on the web has become a day-to-day activity for all from children to researchers. On the Internet, searching is not just about finding the information as per your requirement or acquiring some knowledge in particular area, sometimes it also about discovering about some new terms, learning some new technology, etc. In these developing fields of technologies, many innovative search methodologies have been invented in traditional systems to reduce the unmanageable information. In traditional information retrieval system user possesses some queries, investigate the results reverted by the search engine, and alter the query to proceed further. In traditional searching process, when user poses a query, the search system assumes that they have decent knowledge of the schema, meaning, and contents of the database and they are aware of this specific query which will give the result as per their requirements. Nevertheless, the traditional system is not appropriate for many real-world problems, for example, it may not help if scientific researchers may want to explore new research topics, students may want to learn new subjects, travelers may

R. Kumari (✉) · S. K. Jain
Department of Computer Engineering, NIT Kurukshetra, Kurukshetra, India
e-mail: ratnakumari679@gmail.com

© Springer Nature Singapore Pte Ltd. 2019

V. Nath and J. K. Mandal (eds.), *Proceedings of the Third International Conference on Microelectronics, Computing and Communication Systems*,
Lecture Notes in Electrical Engineering 556, https://doi.org/10.1007/978-981-13-7091-5_34

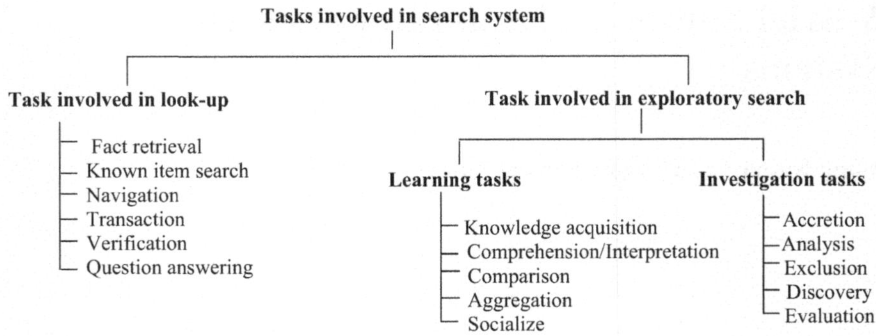

Fig. 1 List of search tasks categorized under lookup and exploratory search categories [2]

want to plan a budget-friendly trip, and so on. The main problem of searchers is to how to precise their query to proceed the search process. There is not any hard and fast rule to translate the problem into proper queries. Because people are not familiar with the domain of their search goals, they do not know the way to achieve their goal or goals are not clear to themselves. Here comes exploratory search which helps users with ill-phrased query.

Searching process is mainly categorized into two categories: lookup and exploratory [1]. Lookup search is a basic searching process which is well understood and presumed to have a specific goal. Lookup search is mainly for the users who have some idea about the area of the subject they are looking for. On the other hand, exploratory search is just opposite of the lookup search. When the user does not have any idea about the topic they are looking for, then they go for exploratory search. Exploratory search is executed at any time a user wants to discover a domain to grow his knowledge, to acquire knowledge about new topics, or to explore some new topics. So, exploratory search is mainly about learning and investigate [1].

Data exploration is efficiently extracting knowledge from data even if we do not know exactly what we are looking for [2]. Data exploration gave birth to the exploratory search. The definition of exploratory search is ambiguous. There is not any definite meaning of exploratory search. Based on state of the art of the area, it has given some definition, exploratory search is a specialization of information exploration which represents the activities carried out by the searchers who are unfamiliar with the domain of their goal, unsure about the ways to achieve their goal [1]. In exploratory search, users generally submit a tentative query to reach to relevant document in the collection, then explore the background to better recognize how to exploit it, selectively, seeking, and passively obtaining hints about their next step [2].

Marchionini [1] explains three types of search tasks lookup, learn, and investigate and argues that exploratory search includes learn and investigate task. Figure 1 exhibits different tasks included in lookup and exploratory searches.

Lookup search process is basic search process in which users have the idea about the goal and also predictable result. A classic example would be a user wanting to book a room in a hotel for their vacation; in this case, the user has the idea about their goal and specific result. On the other side, in exploratory search, users do not have any idea about the topic they want to explore. So we can say that exploratory search is open ended, the user starts with an unclear information and based on the search result they proceed further.

For example, if a user wants to know about the "machine learning", the user does not have any idea about this topic and he proceeds with single keyword and explore the topic in their search session and based on the search result they proceed further to learn more about this topic. Hence, the main goal of the exploratory search is learning through investigation.

There exists plenty of search systems which support exploratory search. The objective of this paper is to discuss the existing system to help the user in their searching process. Here, in this paper, we are going to discuss various exploratory search systems and their features so that user gets benefited while choosing the search system as per their criteria. Due to increasing rate of data, it becomes very complex to handle the data. In this paper, we will discuss various result classification methods which help to reduce the browsing burden of the user.

The structure of this paper is as follows: In next section, we discuss various result classification methods to reduce the density of result page. In Sect. 3, we have discussed various existing exploratory search methods. Last, we have conclusion and future scope of exploratory search systems.

2 Information Structuring for Exploratory Search System

Web search engine gives high preference to the higher ranked result but sometimes lower ranked page also satisfies users' requirement. Exploratory search gives preferences to the both higher and lower ranking pages. So that lower ranked pages also come under consideration and thus browsing burden to the user may increase. To give relief to the user by decreasing browsing, burden exploratory search has introduced several methods to classify search results. With many results divided into several groups, it is easy to identify the key information.

Classification is a general process of categorization, which includes an arrangement of entities based on analysis of necessary and sufficient condition that defines each class. Here, we have discussed various classification methods to reduce the resulting burden to the users. The exploratory search system mainly uses four main methods to reduce the burden of the results space, that is, hierarchical classification, faceted classification, dynamic clustering, and social classification.

2.1 Hierarchical Classification

Hierarchical classification is a process of organizing the system in non-overlapped classes. In hierarchical classification, problems are arranged in the hierarchy. With the help of hierarchical structure, users get to feel the predetermined ordering of reality. It uses tree data structure to divide the problem into several smaller problems. By dividing the problem into several smaller problems, users get to know more about the system and go ahead with the results of their specific requirement. Yahoo! directory and open directory are two most active users of the hierarchical classification for their web resources. Hierarchical classification is also used to enhance the internal search of the website. These features of hierarchical classification are helpful to make exploratory search system more effective and efficient. It helps to enhance the search result organization and users are able to create a mental model to the whole result space rapidly. But as we all know everything in this world has good as well as bad side. The bad side of hierarchical classification is that it needs high amount of money, time, and space to build non-overlapping classes. Thus, not many systems are able to use this information structure method to reduce the result space. Even if it is affordable, it is very difficult to properly reduce the result spaces in few classes with clear edges. So, this method of classification is not recommended by many systems.

2.2 Faceted Classification

Faceted classification is a process of categorizing facet into many facets and several categories come under each facet. Facets means characterizing the item into several categories and each category contains their related items. The faceted classification plays a very important role in the exploratory search system, and it is mostly used classification system in the exploratory system. In exploratory search, the purpose of faceted categorization is to represent a different attribute of search information by posing small list of hierarchies that relates to ideas contained within the depository [3]. Faceted classification is rapidly used by many online sites, mostly in online shopping sites. The best example is our day-to-day shopping sites, when we go to shop desktop in any shopping site we see that items are categorized based on their feature like desktop size, brand, processor, etc. Based on their feature there are many facets available to choose the user as per their requirement. Similarly, it is used in exploratory search systems; results are categorized in many facets and user chooses the facet as per their requirements.

2.3 Dynamic Clustering

Clustering refers to finding a structure in a collection of unlabeled data. The basic idea of clustering is organizing information item into a group by using some algorithms so that the group's members are similar in some way. Therefore, it is a collection of information which is similar to them and is "dissimilar" to the information items belonging to others. Clustering is famous for grouping search results. A clustering algorithm is applied to each query to categorize the search result based on their content. The best thing about this process is everything is done automatically and it makes it famous from all. Dynamic clustering removes the complexity and cost of constructing and retaining fixed. Clustering provides the user a best way to understand the result under particular topics, and it also resolves the problem of polysemy [4]. It may be possible of existing several meanings of the same word or phrase. Clustering differentiates the results according to their meaning, which enables users to make selective browsing. Furthermore, clustering also collects data from different pages, so that the user gets benefitted and gets the whole result pages in more organized way.

2.4 Social Classification

Social classification is also known as folksonomy. It is made up of people-contributed free tags. Tagging is a process in which user assigns meaning to the online objects including text, bookmark, and videos. Tagging process makes user, resource consumer as well as the provider. Tag given by the user helps other users in near future. Folksonomies comes from the bottom-up social tagging process, in which tags are applied by user to online items [5]. It is a method of distributed classification systems that helps to describe the item without any hierarchy or exclusive category. Today, the social tagging system is relying on tag cloud, and tags are usually displayed in alphabetical order with their relationship. There are advantages as well as disadvantages of folksonomies. It gives chances to everyone to express their personal view using tags and greets distinct or even conflicting ones to coexist. On the other hand, it also has disadvantages of "vocabulary problem" because of distributed tagging by everyone.

3 Exploratory Search Systems

Exploratory search is a process of seeking information where the search goal is not defined, and the domain of search area is unknown. In 2005, there was the first major event organized for exploratory search interface [6]. Exploratory search system uses variety of functionalities for exploration purposes. The system uses set of algorithm and ranking model to fulfill the users' requirements. In this section, we will discuss exploratory search systems, how information retrieved from the search system, and their features.

3.1 PULP

PULP [7] is an exploratory search system which is used for searching scientific literature. The system is also applicable to other types of literature. The pulp system uses reinforcement learning to help the user with an ill-phrased query to search required information. Pulp system uses topic models to give an overview of the dataset to the users. Topics are shown through the interactive diagram, and the search starts with the user selecting the topic as per their interest. The year-wise topic is arranged with a vertical list of research topics. Each topic is labeled with five most prominent key word. A user can select one or more topic at the same time. The keyword from the topic selected by the user will act as a query for that search session.

3.2 Expedition

An expedition [8] is a time-aware exploratory search system designed for scholars. It uses ad hoc and diversity-based retrieval method to help the user to retrieve the complex information. Expedition method uses various filters that help the user to preview and compare the large textual data. In expedition system, when user enters a key word query the system displays a drop-down list with a number of retrieval models such as textual relevance, temporal relevance, topical diversity, temporal diversity, and historical diversity. Furthermore, a list of silent entities is also available which act as a filter.

3.3 Expose

Expose [9] is an exploratory search system which is used to explore the past news for seminal events. They have developed time-aware exploratory search system to explore the events that happened in the past. They have used four approaches to get the accurate information as per the user requirements. The Expose system converts each of the methods as a search mode to make exploration with different temporal intents possible [5]. Text-Only, Publication-Date, Temporal-Expression, and Two-Stage Cascade are the four approaches that system goes through. In "Text-only" approach, system compares the query details with the textual terms present in the whole document. The system uses a ranking algorithm in Publication-Date approach to rank the document by linking their publication date with event date, if the publication date is near to the event date then user gets satisfied with the results. This method is most effective for occasions with short implications. Expose system also gives preferences to the news article that is not contemporary to an event, "Temporal-expression" method positions an article higher if many sequential expression in its content refers to the time period when the event occurred. And the last, "Two-stage cascade" approach targets an automatically classifying extra time points that are important to the event. It also uses feedback to get more accurate information as per user requirements.

3.4 Linked Jazz

Linked Jazz [10] is an exploratory search system used to find the relationship of jazz community. The system uses semantic-based linked data technology to improve the discovery of digital heritage. Another aim of the system is to find out the association between the musicians and disclose their community links. Thus, in linked jazz search system, it becomes easy to find out the relation concerning the professionals' lives and personal information of jazz musician from the details given by the musicians in their interview or from the jazz archives.

A built-in database RDF is used to store the information of jazz community. An RDF dataset describes these meaningful relations of jazz and community. The visualization tool used by the linked jazz provides number of modes such as fixed, free, similar, and dynamic modes. These modes are used to visualize and explore the jazz community network. The "free mode" groups them all together based on their number of connections and then "fixed mode" attaches the individual with the most connections which make the presentation clearer. The "similar mode" clutches them together based on their number of mutual connection and the last "dynamic mode" gives the functionalities of adding manually with their shared connections.

3.5 Discovery Hub

Discovery hub [11] is a linked data exploratory search system. It uses semantic-sensitive traversal algorithm which is based on live graph sampling technique. Discovery hub is multifeatured search system, it supports faceted browsing and multiple result explanation features. It gives suggestion to the users to proceed in many information directions of their interest. It helps users to get more information by redirecting to the third-party platform, and users can easily extend their search process to get more relevant information. In discovery hub, user types the query of their interest in the search box and when a query is entered in the search box, the system displays a drop-down list which contains semantically related data of the inputted query. The user chooses some relevant topics of their interest, and when the topic is selected it goes to the stack right to the search box.

3.6 Aeemo

Aeemo [12] is view-based exploratory search system. It is based on encyclopedic knowledge pattern (EKP) which is used to describe the entities of certain class. When a user inserts the keyword-based query into Aeemo search interface, the system explores DBpedia first to fix user query, if it is not fulfilled then it goes to other sources such as Wikipedia Google news to fulfill the requirement of the user. EKP helps the system to choose which information should be presented to the user so that the user gets specific information as per their requirement. Aeemo offers a function that displays additional information that is not included due to EKP filtration. Aeemo also displays explanation by using cross-references in Wikipedia pages between the topic of interest and results.

3.7 Semantic Wonder Cloud (SWOC)

Semantic web cloud [13] is a tool to explore the DBpedia. It helps the user in exploratory search. The SWOC has two main subsystems which act as front end and back end of the system. The "SWOC explore" is graphical user interface which acts as a front end of the system. The SWOC explore helps the user in exploratory search on DBpedia. The back end of the system is "DBpedia ranker", which is used to arrange the document in systematical order. The main task of "DBpedia rankers" is to compute the similarity between DBpedia nodes, and ignore the similar data to decrease the burden of the user to searching data.

3.8 Seevl

Seevl [14] is a latest semantic web technology, which uses linked data to discover musical information. It is a linked data based application which is used to discover the music as per their requirement. Seevl uses this technology to connect all the music so that users easily discover the music. In seevl search system, various web resources are collected and linked up with, massive graph of musical things, such as "Artist", "Labels", "Bands", etc. Some other functionalities like instance search and recommendation are also provided.

A "Triplestore" or "RDF store is" a built-in database which is used for storage and information retrieval based on users' queries. The "Elastic compute cloud" is used to get benefitted with their features like load balancing and elastic cache. The seevl helps in exploration with stored data with their recommendation of each artist. They have also given the detailed information of artist and the reason of their recommendation. The main part of the system is their dedicated "recommendation" which helps the user to explore the area in effective way.

3.9 Lookup Explore Discover (LED)

Lookup explorer discover [15] is a web-based tool that helps in query refinement and exploratory search. The LED search interface is concentrated on three main areas. The first one is "keyword search" where the user enters the keyword and obtains a complete list of related keyword and the second area consists of "tag cloud" which automatically generates tags that are semantically related to the keyword entered by the user. The "tag cloud" helps the user to discover new knowledge that is not possible with the traditional search systems. Finally, the third area of interface acts as a "metasearch" engine. It displays the result coming from the popular search engine, social network, and so on.

Table 1 shows the overview of the exploratory search systems that have been explored in previous sections. Every system has a specific purpose. The main purposes of the systems have been provided in the table. The main problem of exploratory search is how to precise their query. If user will be provided some query as a suggestions that will be very helpful for the user, which helps users to formulate the query. As we have given in the table, "Expedition", "PULP", and "LED" are the systems which provide query suggestion to the user. There are some other features like" ranking", which uses ranking algorithm to rank the search results, and "Offline-processing", "query model" are also discussed in the table.

4 Conclusion and Future Scope

User expectations toward search system are beyond the current capabilities of the existing systems. There is need for enhanced electronics and dynamics mechanism to help the user to reach their search goal more effectively and efficiently, and also there is a requirement to fulfilling long-term personal goals of the user. This paper describes many existing search system which helps users' in exploratory search. We have explained some well-known classification process which helps to reduce the browsing burden of the user. Many exploratory search systems have specific searching purposes which we have discussed in this paper so that user gets to know which system is suitable as per their requirement. As we have explained in the paper, exploratory search is advanced search approach to fulfill user requirements. There are many advanced novel interaction paradigms which have been introduced, and the exploratory search systems will apply these technologies to support human–machine communication during the search process. The users will not have any restriction of using a desktop computer and a mouse pointer to operate information display. By analyzing the state of the art of exploratory search, we can conclude that future of exploratory search is bright. Although researchers are still expanding exploratory search to make novel ideas a reality, the existing research on the searching system has been focusing on individual users, and they are ignoring the social support to

Table 1 Overview of exploratory search systems

Systems	Purpose	Query suggestion	Query model	Algorithm	Principal layout	Ranking	Offline proc.	API
Expedition [8]	Scholar	Yes	Lookup	Intent modeling	List	Yes	No	Yes, RESTful
Expose [9]	News	No	Lookup	Retrieval methods	List	Yes	No	Yes, RESTful
Pulp [7]	Scientific literature	**Yes**	Keyword	Reinforcement learning	List	Yes	No	No
Discovery Hub [11]	General	No	Lookup	Semantic spreading activation	List	Yes	No	Yes, JSON-LD
Linked jazz [10]	Jazz musicians	No	Manual selection from a list	Mapping and Curator Tool and the Transcript Analyzer	No	Graph	Yes	Yes, JSON, RDF
Aeemo [12]	General	No	Lookup	EKP filter	Graph	No	Yes	Yes, RESTful
Seevl [14]	Music	No	Lookup	LDSD, DBrec algorithm	List	Yes	Yes	No
SWOC [13]	IT Domain	No	Lookup	DBpedia ranker	Graph	Yes	Yes	No
LED [15]	ICT	Yes	Keyword search	DBpedia ranker	Tags cloud	Yes	Yes	Yes, RESTful

explore the information. We may achieve a countless information from the finding on both direct and indirect social triangulations. The interest in social interaction will diversify future research on exploratory search.

References

1. Marchionini G (2006) Exploratory search: from finding to understanding. ACM, pp 41–46
2. White RW, Roth RA (2010) Exploratory search: beyond the query-response paradigm. Synthesis lectures on information concepts, retrieval, and services, pp 1–98
3. Hearst MA (2006) Clustering versus faceted categories for information exploration. Commun ACM 59–61
4. Croft WB, Leouski AV (1996) An evaluation of techniques for clustering search results. Computer Science Department Faculty Publication Series, 36
5. Hammond T, Hannay T, Lund B, Scott J (2005) Social bookmarking tools. D-Lib Mag 11(4)
6. White RW, Kules B, Bederson B (2005) Exploratory search interfaces: categorization, clustering and beyond. ACM SIGIR Forum 39(2):52–56
7. Medlar A, Ilves K, Wang P, Buntine W (2016) PULP: a system for exploratory search of scientific literature. In: Proceedings of the 39th international ACM SIGIR conference on research and development in information retrieval. ACM
8. Singh J, Nejdl W, Anand A (2016) Expedition: a time-aware exploratory search system designed for scholars. In: Proceedings of the 39th international ACM SIGIR conference on research and development in information retrieval. ACM
9. Mishra A, Berberich K (2015) EXPOSÉ: exploring past news for seminal events. In: Proceedings of the 17th international conference on World Wide Web. ACM, pp 18–22
10. Pattuelli MC, Miller M, Lange L, Fitzell S, Li-Madeo C (2015) Crafting linked open data for cultural heritage: mapping and curation tools for the linked jazz project. Code4Lib J
11. Marie N, Gandon F, Ribière M, Rodio F (2013) Discovery hub: on-the-fly linked data exploratory search. In: Proceedings of the 9th international conference on semantic systems. ACM
12. Musetti A, Nuzzolese AG, Draicchio F, Presutti V, Blomqvist E, Gangemi A, Ciancarini P (2013) Aemoo: exploratory search based on knowledge patterns over the semantic web. In: Semantic Web Challenge
13. Mirizzi R, Ragone A, Noia TD, Sciascio ED (2010) Semantic wonder cloud: exploratory search in DBpedia. In: Daniel F, Facca FM (eds) Current trends in web engineering, pp 138–149
14. Passant A (2012) Seevl: mining music connections to bring context, search and discovery to the music you like
15. Mirizzi R, Ragone A, Di Noia T, Di Sciascio E (2010) Lookup, explore, discover: how DBpedia can improve your web search

Job Allocation on Cloud: A Comparative Study

M. Manasa and J. Priyadarshini

Abstract The rapid growth in the field of cloud computing leads to migration of people to cloud, which makes them feel more tensile and adaptive in the environment. Cloud makes the users give less importance to maintain their hardware and other resources because all these works are done by a service provider. This would restrict the users from spending more money on capital expenditure and so the people are ready to invest more in the cloud. The virtual machine (VM) is a buzz which replaced a traditional physical machine through the method called virtualization. Virtualization is the main objective for establishing cloud services. The core idea of virtualization is to create an instance or virtual machine according to user demands, and the number of servers needed is proportional to the amount need for the resource pool. This paper presents a heuristic algorithm—gravitational search algorithm (GSA)—which formulate an optimal solution for allocating task on the cloud. To analysis the efficiency of the proposed algorithm, a comparative study has been done with other heuristic algorithms like the ant colony and particle swarm optimization which is also used for job allocation.

Keywords Gravitational search algorithm · Job allocation · Particle swarm optimization · Task scheduling · Virtual machine

M. Manasa (✉) · J. Priyadarshini
SCSE, VIT, Chennai, India
e-mail: manasam.2017@vitstudent.ac.in

J. Priyadarshini
e-mail: priyadarshini.j@vit.ac.in

© Springer Nature Singapore Pte Ltd. 2019

V. Nath and J. K. Mandal (eds.), *Proceedings of the Third International Conference on Microelectronics, Computing and Communication Systems*, Lecture Notes in Electrical Engineering 556, https://doi.org/10.1007/978-981-13-7091-5_35

1 Introduction

The cloud computing is all about providing a wide range of computational platforms, storage capacity and network services. Around 100 servers form a farm in a data center. Numerous service providers are there in the market to provide many services for people based on their needs. One service provider competes with others to provide unique and best services for their customers. There are many challenging aspects to be concentrated by the service contributor which includes node failure, load balancing, and job allocation. This paper will concentrate on task scheduling where the incoming user request will be allocated to servers. The first and foremost condition of a customer would be less makespan time of a job. If the cloud contributor fails to satisfy a customer's requirement, then they will lose their reputation among their customers who use their services. For achieving this strategy each and every service providers are working on it. Till now, many optimization algorithms were proposed for performing job allocation on the cloud which includes particle swarm optimization, ant colony optimization, and artificial bee algorithms. Even though these algorithms are providing efficient allocation of a job, they consume more memory as algorithms, which leads to increased makespan time. In this paper, the gravitational search algorithm is been used to achieve an optimized task scheduling mechanism. This gravitational search algorithm is inspired by physical phenomena where Newton's law of gravity had been applied to an algorithm.

2 Literature Survey

The cloud provider should be trustworthy as most of the data of the organizations are stored on the cloud. All the data should be protected irrespective of their types. For each hosted page, trust values are calculated using PSO algorithm where the fitness function is calculated. And then based on its threshold value trustworthiness of website is provided [1].

The prior content gives information about cloud service provider and scheduling algorithm that had been used for allocating a particular job using round-robin algorithm which is known for implementing scheduling operations in CPU. But it became a failure approach as it can process only dynamic requests [2].

In order to overcome the previous drawback, another approach was proposed where dynamic round-robin algorithm should handle the requests successfully, irrespective of the nature of requests [3]. This paper concentrates only on scheduling which allocates jobs by processing the request. Load balancing was not concerned when incoming requests rate is high. To overcome this drawback, a new approach is proposed where slide modification has been done in the algorithm to support load balancing These are the drawbacks of using round-robin algorithm [4].

The allocation is done by using ant colony optimization algorithm for scheduling the task at real-time. The optimization level of this algorithm is based on the fitness function calculation [5]. Cloud computing derived from grid computing, ubiquitous computing, and mobile computing. These three computing are mixed up together and provide service for the people [6]. The gravitational search algorithm is used to form a cluster and allocate tasks to the grid.

In artificial bee algorithm-based task scheduling, the proposed method gives an idea about why this algorithm has been chosen and its feasible solution. Since it is a real-time algorithm, it could be supported for a dynamic and heterogeneous job. This implementation shows a reduction in makespan time even though the number of computer nodes had been increased. The cons of this paper were that when task allocation performed through the network, there was no reduction in makespan time [7].

The modified version of the classical scheduling process is proposed on this priority based scheduling where each and every job resources are scheduled based on priority given to them. Difficulty faced in this algorithm is that the complexity of a program increases when the number of resources and job level is increased [8].

Existing task scheduling algorithm's aim is to make a cloud platform in a large-scaled server cluster using virtual machines. The advantage of using a greedy algorithm is to schedule tasks to improve the overall quality of the cloud computing services when compared to the previous implementation of using a genetic algorithm. This greedy particle swarm algorithm produces a fast convergence, strong local and global ability search, balancing the system load, improving the efficiency of task scheduling and resource utilization [9].

3 Proposed System Architecture

As shown in Fig. 1, the proposed model consists of three modules: client side, cloud broker, and data centers. In the client side, the user sends the request for the resource to be allocated through the user interface and that has been received by cloud broker and then he has to trace and allocate particular virtual machine. The allocation details of virtual machine have been updated every fraction of seconds, and utilization of resources by the user has been monitored.

The major portion of implementation in this architecture is deploying the scheduling algorithm for the purpose of processing the request traffic from different location of this world.

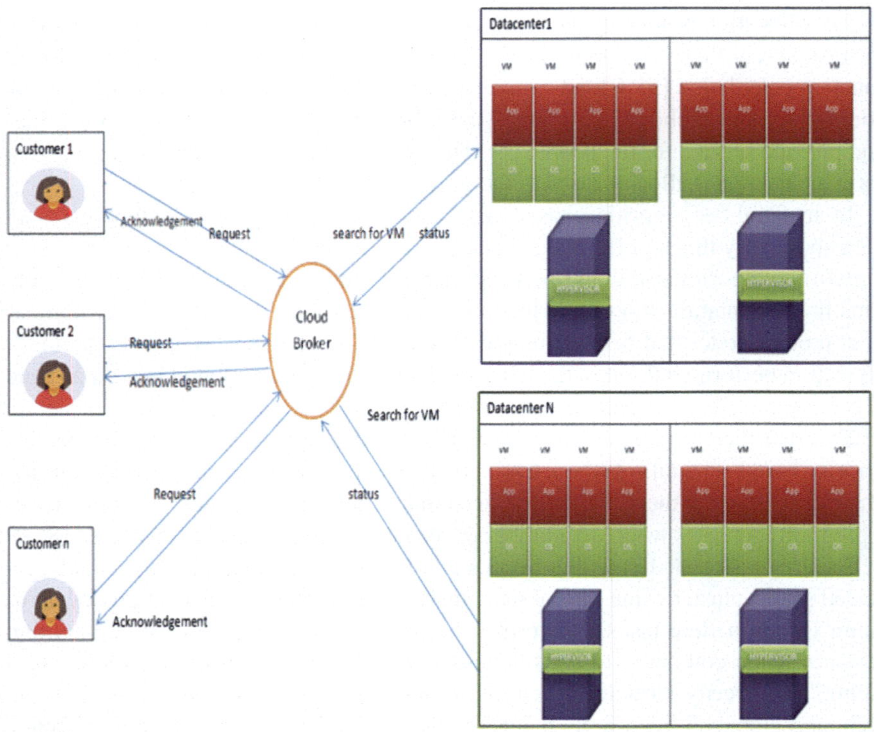

Fig. 1 Architecture of the proposed system

4 Comparison of Heuristic Algorithm

Table 1 shows the efficiency of the proposed algorithm—gravitational search algorithm—compared with other heuristic algorithms—particle swarm and ant colony optimization. The algorithm which had proposed in the table is an evolutionary algorithm, which increases the efficiency rate of the parameter for each iteration [10]. When gravitational search algorithm is compared with the particle swarm optimization algorithm, they achieved an accuracy rate of 85% and also compared with other heuristic algorithms. Since GSA is a memoryless algorithm which makes GSA greater than other algorithms [11] (Table 2).

5 Algorithm

Gravitational Search Algorithm:
The gravitational search algorithm was first proposed by Rashedi [10]. GSA is a heuristic search algorithm which works on the influence of Newton's second law of

Table 1 Summary of literature survey

SI no	Title of the paper	Main idea of papers	Drawbacks
1	Predicting of job failure in compute cloud based on (OELM), IEEE 2017, Sharif Chuanchang, Junliang Chen	Incremental fast learning algorithm can deal with a sequential arrival data and it also achieved rate of accuracy 93%	It does not support for distributed system
2	Towards a trust Prediction Framework for cloud services based on PSO-neural network, IEEE 2017, chengying Mao, Rogrulin, Changfu xu, Qiang he	Mapping of a public datasets in a neural network is faster	Performance of neural network is easily affected because of initial value
3	Dynamic round robin for load balancing in cloud, IJCSMC,2014, Ajay gulati, Ranjeev.k. Chopra	Optimal result than compare to round robin	Restricted to finite number of load
4	Scheduling virtual machines for load balancing in cloud using round robin algorithm, IJEIR, 2014, Bhavana bidarker, Shakheela attikeri	Remember the previous allocation details	Limited to finite number of virtual machine
5	Extended round robin load balancing cloud computing, IJEC, 2014, Priyanka Gautam, Rajiv Bansal	Presents algorithm for dynamic virtual machine	Finite number of virtual machine
6	Virtual machine-based task scheduling algorithm in a cloud compute, Zhifeng Zhang, Kun Chen, Xiaojun Zhai, and Shuange Zhou, IEEE, 2016	It based on historical experience of a directed virtual machine to a server using particle swarm optimization algorithm	The data transmission rate and bandwidth are not considered
7	A survey on VM scheduling in cloud computing, Li Liu, Zhe Qiu, IEEE, 2016 [13]	Virtual machine scheduling is classified into two mapping a task with VM and deploying a VM in a physical machine	Internal VM scheduling is not concentrated
8	"Trends in gravitational search algorithm", P.B. deMoura Oliveira, Josenalde Oliveira, José Boaventura Cunha, IEEE, 2016	Company a GSA with another heuristic algorithm PSO	PSO performance in convergence is less compare to GSA

(continued)

Table 1 (continued)

SI no	Title of the paper	Main idea of papers	Drawbacks
9	Antcolony optimization for VM to savings of energy and efficient use of different resources Xiao-Fang Liu, Zhi-Hui Zhan, Jeremiah D. Deng, Yun Li, Tianlong Gu, Jun Zhang, IEEE, February 2018	Reducing a energy usage by the virtual machine that is deployed in a data centers using a historical information of a virtual machine	It can prove efficient solution but not comparatively

Table 2 comparison table of Heuristic algorithm

Content	Gravitational search algorithm	Particle swarm optimization algorithm	Ant colony optimization algorithm
Optimisation value obtained by the search	Movement of the agent in a different direction	Movement of agent travel in the same direction	Movement of agent travel in the same direction
Research ideas	Inspired by physical phenomena	Social behaviour of birds	Focus on food
Memory	Memoryless about the current status	Use memory for updated search for calculating the velocity	Use memory for the path in a graph
Task	Dynamic	Dynamic	Dynamic
Parameters	The initial position of node, mass, velocity	Personal best, global best	Path cove rage on the graph
Type of jobs	Heterogeneous kind of job at a similar point of time	Heterogeneous kind of job at a similar point in time	The job is traced along the path

motion. The Gravitational Search algorithm's one of the efficient features is it supports for both a dynamic and heterogeneous type task scheduling. This algorithm also has another feature, that is, a faster convergence rate, which increases the processing speed along with less makespan time [12].

6 Performance Analysis

The feasibility and performance of the algorithms are analyzed using a cloud analyst and Net Beans IDE. The computer architecture of X86, Linux operating system, Xen hypervisor and 1 GB of RAM has been set up for simulating the scheduling for the two different scenarios: request for data center within the same region and request for data center within the different regions, both the scenario's task has been scheduled using round-robin algorithm. The following figure shows the implementation result of a virtual machine allocation which shows a number of tasks, utilization of resources, peak time, and cost.

Both Figs. 2 and 3 show the simulation result of resource allocation dynamically done using round-robin algorithm.

Similarly, to compare feasibility and performance of the heuristic search algorithm, gravitational search algorithm is evaluated using a cloud analyst, and Net Beans IDE is used for implementing the method of scheduling using gravitational search algorithm. The computer architecture of X86 architecture, Linux operating system, Xen hypervisor, and 1 GB of RAM has been set up for simulating the scheduling for the two different scenarios: requested for data center within the same region and requested for data center within the different regions; both of the simulations have been scheduled using gravitational search algorithm. The following figure is the implementation result of the gravitational search algorithm, which shows a number of tasks, utilization of resources, peak time and cost (Figs. 4 and 5).

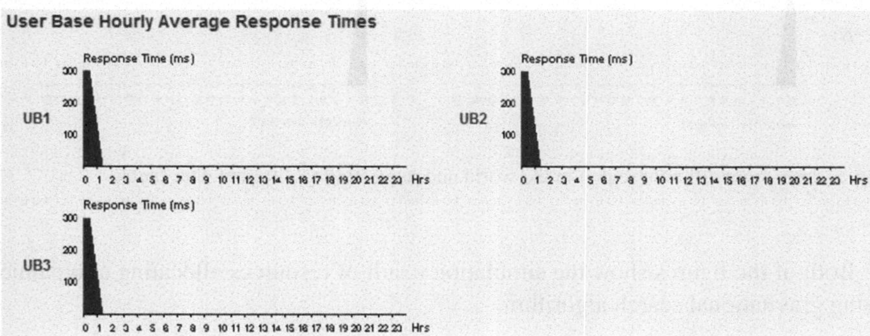

Fig. 2 User from same region in the world and requesting to same data center

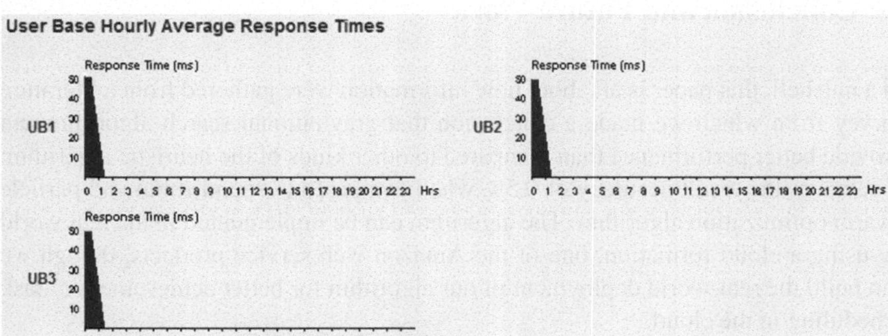

Fig. 3 User from different region in the world and requesting to same data center

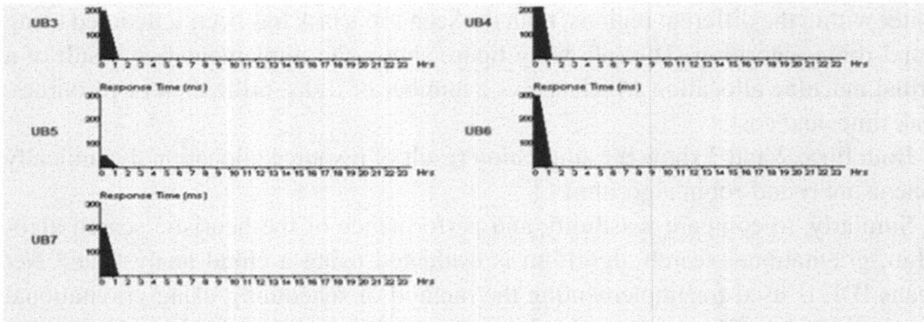

Data Center Request Servicing Times

Data Center	Avg (ms)	Min (ms)	Max (ms)
DC1	0.354	0.018	0.618
DC2	0.248	0.019	0.881
DC3	0	0	0

Fig. 4 User from same region in the world and requesting to different data center

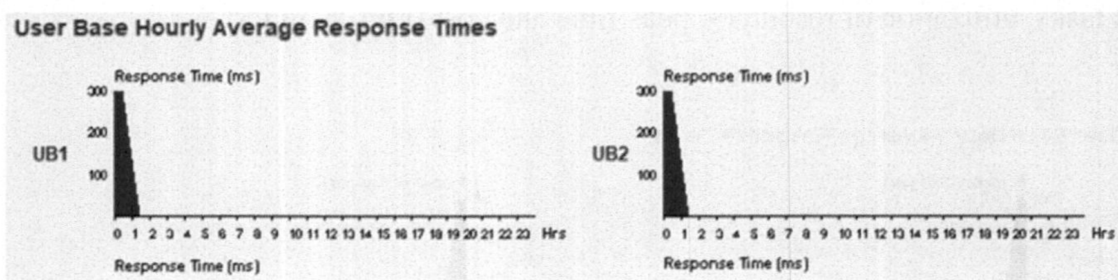

Fig. 5 User from different region in the world and requesting to different data center

Both of the figures show the simulation result of resources allocating a dynamic using gravitational search algorithm.

7 Conclusion and Future Work

In a nutshell, this paper is all about how information were gathered from a literature survey from which we made a conclusion that gravitational search algorithm can provide better performance than compared to other kinds of the heuristic algorithm. It achieves the result accuracy of 0.5% when compared to round-robin and particle swarm optimization algorithm. The algorithm can be implemented in the real world by using a cloud formation, one of the Amazon web service products, though we can build the real-world deployment of our algorithm for better achievement of task scheduling in the cloud.

References

1. Liu C, Han J, Shang Y, Liu C, Cheng B, Chen J (2017) Predicting of job failure in compute cloud based on online extreme learning machine: a comparative study. IEEE Trans J 5:1011–1025
2. Gulati A, Chopra RK (2014) Dynamic round Robin for load balancing in a cloud computing. IJCSMC 2:274–278
3. Mao C, Lin R, Xu C, He R (2017) Towards a trust prediction framework for cloud services based on PSO-driven neural networks. IEEE Trans J 79:88–115
4. Gautam P, Bansal R (2014) Extended round Robin load balancing cloud computing. IJEC 3:27–39
5. Farrag AAS, Mahmoud SA, El Sayed M, EI-Horbaty (2015) IEEE Conference on Intelligent Cloud Algorithms for Load Balancing problems: A Survey, pp 231–242, vol 3
6. Zarrabi A, Samsudin K (2014) Task scheduling on computational grids using gravitational search algorithm. J Clust Comput 17(3):1001–1011. ACM
7. Rastkhadiv F, Zamanifar K (2016) Task scheduling based on load balancing using artificial bee colony in cloud computing environment. IJBR 7(Special Issue-No 5):1058–1069
8. Ghanbari S, Othman M (2012) A priority based job scheduling algorithm in cloud computing 03(01):778–785. Elsevier conference
9. Zhong Z, Chen K, Zhai X, Zhou S (2016) Virtual machine-based task scheduling algorithm in a cloud computing environment. IEEE 21(6):660–667
10. Rashedi E, Nezamabadi-pour H, Saryazdi S (2009) GSA: gravitational search algorithm. Inf Sci 179(13):2232–2248. Elsevier
11. Liu X-F, Zhan Z-H, Deng JD, Li Y, Gu T, Zhang J (2018) Ant colony optimization for VM to savings of energy and efficient use of different resources. IEEE Trans 22(1)
12. de Moura Oliveira PB, Oliveira J, Cunha JB (2017) Trends in gravitational search algorithm. In: 14th international conference on distributed computing and artificial intelligence, DCAI 2017. Advances in intelligent systems and computing, vol 620. Springer, Berlin
13. Liu L, Qiu Z (2016) A survey on VM scheduling in cloud computing. In: 2016 2nd IEEE international conference in computer and communications (ICCC)
14. Zhang W-Z, Xie H-C, Hsu C-H (2017) Automatic memory control of multiple virtual machines on a consolidated server. IEEE Trans Cloud Comput 5(1):2–14

A New Cluster-Based Test Case Prioritization Using Cat Swarm Optimization Technique

Dharmveer Kumar Yadav and Sandip Dutta

Abstract Software regression testing is very expensive and time-consuming process. It is also an essential part of the software development life cycle. The main objective of regression testing is to run all possible combinations of test cases in a test suite. This process requires a large amount of efforts as well as time. Hence, to overcome the human effort and time, there is need for prioritization of test cases. The objective of this research is to execute those test cases have high priority before low priority of test case. We have proposed a new test case prioritization technique using cat swarm optimization (CSO) with clustering approach. CSO is a latest meta-heuristic algorithm based on the behavior of cats. In this paper, we present a clustering-based test case prioritization technique. The method consists of clustering the test cases based on fault detected by test cases. To measure the effectiveness of this algorithm, we used average percentage fault detection (APFD) metric. Moreover, experimental results show that CSO algorithm is an effective and efficient method to prioritize the test cases.

Keywords Cat swarm optimization · Test case prioritization · Software testing · Regression testing

1 Introduction

Regression testing is a process of retest parts of the modified code which was affected by the modified so that effort and time can be reduced. In regression testing, rerunning all the test cases would be very expensive process. To minimize the cost of regression testing select those test case which tests modified code using some defined factor

D. K. Yadav (✉) · S. Dutta
Department of Computer Science & Engineering, Birla Institute of Technology, Mesra, Ranchi, India
e-mail: Kumar.dharmveer@gmail.com

S. Dutta
e-mail: sandipdutta@bitmesra.ac.in

© Springer Nature Singapore Pte Ltd. 2019

V. Nath and J. K. Mandal (eds.), *Proceedings of the Third International Conference on Microelectronics, Computing and Communication Systems*,
Lecture Notes in Electrical Engineering 556, https://doi.org/10.1007/978-981-13-7091-5_36

[1]. Prioritization in regression testing is the scheduling of test cases based on some criteria, such as higher coverage, fault [2]. In the present era, there is an incremental growth in the field of software. Day by day, a variety of software are developed by the various software developers. Among this, in the present scenario, the user requirements are also changing dynamically. Both of these increased the work of the software developers as well as software testers and make the software testing process challenging. As user requirements are changed dynamically, then the test cases are also changed frequently. Hence, the primary task of a software tester is to save the available test suites such that if any modifications will be done, then he is able to reuse test cases. According to Arafeen et al. [3], software includes the subroutines, functions, and statements of an application system and software organizations believe that if a software can be developed independently, then its operation will have better testing and security features. Hence, the primary task of the testing is to avoid bugs in a software system by designing the suitable test cases and also to improve the quality of software. But to design the test cases is not the simple one, and it is a time-consuming and effort taking process. But, it cannot be judged easily whether a program that is undergoing a test case will be passed to the specified test. Testing process should be well planned, scheduled, designed, and has to be prioritized [4].

It also provides all possible faults present in the code and design and this process is also highlighted or proved the failure of a program. The main goal of the prioritization of test cases is to minimize the test suites and maximization of the objective function. So, it is tester task to schedule test cases so that it covers maximum code with minimum time and results in a reduction of the test suite cost. In literature, different types of prioritization categories are reported for test cases. Basically, these categorizations have been done on the basis of removal of bugs, execution of test cases, etc. The main objective of this research work is to investigate the capability of the cat swarm optimization algorithm for test cases prioritization. CSO is the latest meta-heuristic algorithm [5] based on the cat's behavior. It is applied to solve a large number of optimization problems and provides remarkable results [6–12]. The results reveal that CSO algorithm is more accurate, effective, and efficient method for prioritizing the test cases.

2 Related Work

In this section, we describe the various works in the domain of test cases prioritization. Harrold et al. developed a minimization technique for test suite by choosing a characteristic set of test cases from a test suite which provides the same coverage as the complete test suite for reducing the cost of regression testing [13]. The author has proposed a new methodology for assessing regression test selection techniques in terms of inclusiveness, precision, efficiency, and generality [14]. The authors also suggested criteria for selection of safe and unsafe regression test. A technique for test selection is also reported in [15]. Rothermel et al. have proposed some other techniques for prioritizing test cases and assess their performance on the basis of improving rate of fault detection [16]. They have observed the abilities to improve the rate of fault detection using test case prioritization of test suites [17]. The authors in [18] proposed a prioritization technique which improves the rate of fault detection in regression testing. They have presented a number of techniques for test cases prioritization. The author proposed another technique called version-specific prioritization [19] for the same purpose. A new technique [20] has presented a new method and claimed that neither of earlier studies observed that how the magnitude of changes affects the cost-effectiveness of regression testing nor do they study various change features with respect to test case prioritization. Researchers presented a cost-effective prioritization method [21], and presented a study of prioritization techniques, focused on an object-oriented programming language, Java, that is rapidly gaining popularity among the programmer [22]. Researcher proposed a general method for regression testing using black box testing [23]. The main focus of these algorithms was to assemble all used test cases by the fault types they have discovered, and then adjust their priorities dynamically according to the results of related test cases. A historical value-based approach was presented that is based on historical value to calculate the fault severity and cost for a cost-cognizant test case prioritization [24]. Researchers have studied the efficiency of test case minimization and prioritization on software testing using previous studies [25].

3 Proposed Approached Based on Clustering

The motivation of this research work is to cluster the test cases according to its characteristics in a test suite. A test suite consists of different types of test cases. The test cases consist of several types of characteristics. Hence, in our proposed approach, the test cases are differentiated on basis of characteristics. Test cases having similar characteristic are grouped in one cluster and test cases occurred heterogeneity (in terms of characteristics) are grouped in different clusters. The aim of the proposed approach is to select a set of test cases randomly from each cluster for testing purpose rather than to apply all test cases. It will reduce the total number of comparisons as well as effort and time.

A. *Cat Swarm Optimization*

CSO algorithm is an evolutionary optimization technique which is applied for searching the best solution by simulating the behavior of cats. In CSO algorithm, positions of the cats represent the potential solution for optimization problems. The objective of CSO algorithm is to derive the best possible direction of the optimal solution by balancing its two modes—seeking mode and tracing mode. The seeking mode describes the curiosity of cats toward the moving objects while tracing mode describes the outstanding hunting skills. A mathematical model is formed by combining these two modes to solve the optimization problems and cats are represented using the position and velocity vectors in a random search space [26]. A problem-specific fitness function is used to direct the next step of search. In CSO algorithm, position of cats represents the possible solution set and a flag is used to determine the type of mode of the cat.

B. Pseudocode of CSO Algorithm

This section describes the steps of CSO algorithm for clustering. The main steps of the CSO algorithm are given as follows:

Step 1: Load the dataset, initialize the parameters of the proposed algorithm and number of cats.

Step 2: Initialize the positions of cats in random fashion and evaluate the velocities of every cat.

Step 3: Compute the value of objective function (Euclidean distance) for each data to each cluster center and group the data into different clusters using the minimum values of objective function.

Step 4: Determine the fitness function ($Fit_G(X)$) of each cat, keep the positions of cats in a variable X_G, and determine the global best position (X_{gbest}) of the cat.

Step 5: If flag is on, goto step 6 and start the seeking mode process, else goto step 7 and start tracing mode process

Step 6: For each cat k, start seeking mode process:

- Make the j copies of each cat i where j = SMP.
- Calculate the objective function value and group the data into clusters.
- Evaluate the fitness function for each set of cat's positions (cluster centers).
- Compare the fitness function value ($Fit_S(X_S)$) of test case and the minimum fitness value serves as the best fitness value and compute the corresponding positions of cats and keep the best position in a variable X_S.
- If $Fit_G(X_G) < Fit_S(X_S)$, then $Fit_S(X_S) \leftarrow Fit_G(X)$ and $X_S \leftarrow X_G$ else, $Fit_S(X_S) \leftarrow Fit_S(X_S)$ and $X_{S,new} \leftarrow X_S$. $Fit_G(X_G)$ represents the global fitness value of test case, whereas $Fit_S(X_S)$ represents the fitness value of test case in seeking mode. $X_{S,new}$ represents the new test case in seeking mode process.

Step 7: For each cat k, start tracing mode process

- Update the velocity of cat k using Eq. 1.

$$V_{i,new} = w*V_i(t) + r*(X_{best}(t) - X_i(t)) \quad (1)$$

In Eq. 1, w denotes inertia weight, $V_i(t)$ represents the velocity of *i*th test case (cluster) in *t*th iteration, r is a random number in the range of (0, 1). $X_{best}(t)$ denotes best test case and $X_i(t)$ denotes current test case.

– Update the position of cat k using Eq. 2.

$$X_{i,d}(t+1) = X_{i,d}(t) + V_{i,d\,new}(t+1) \quad (2)$$

Calculate the objective function value and group the data.
Evaluate the fitness function value ($Fit_T(X_T)$) and keep the best positions of cats in a variable X_T.

Step 8 If $Fit_T(X_T) < Fit_S(X_S)$, then $Fit_G(X) \leftarrow Fit_T(X_T)$ and $X_G \leftarrow X_T$, else $Fit_G(X) \leftarrow Fit_S(X_S)$ and $X_G \leftarrow X_S$; and compute the global best position (X_{gbest}) of the cat. $Fit_T(X_T)$ represents the fitness value of test case in tracing mode, whereas $Fit_S(X_S)$ represents the fitness value of test case in seeking mode. $Fit_G(X)$ denotes global best test case.

Step 9: If maximum iterations are not met, go to step 5, else obtain the final solutions.

In the above, Fit_S, Fit_T, and Fit_G represent the value of fitness function in seeking mode, tracing mode, and global fitness of the CSO algorithm; $X_S$and X_T denote the best position achieved by cats in seeking mode and tracing mode, X_G represents the global positions of cats and X_{gbest} represents the global best position of the cat.

4 Analysis of Algorithm

The initialization of an algorithm begins with some population. The formation of population is a particle or group of particles. For a given problem population are selected randomly. From the PSO technique population has assigned position and velocity. In case of our test case prioritization algorithm, here we consider number of faults as a position of the particle covered by test cases and velocity is the total execution time of the particle to cover the faults. In our algorithm, total fault covered in less time. The stopping condition is required for an algorithm to stop. The stopping condition may be decided as number of iterations may be maximum, fully optimize solution, or population does not generate good solution. We have considered solution is fully optimized as the stopping condition. When stopping condition does not meet then mutation [25] approach will be performed on. During mutation, particles positions and velocities are changed. The test suite which consists of maximum fault in very less time will be the global best solution. Each particle's position and velocity is compared with their old position and velocity, if there are some improvements then position and velocity will be updated. This process will continue till stopping condition is met.

A. *APFD Metric*

In this paper, average percentage faults detected (APFD) metric is used to measure the performance of proposed algorithm. The average of percentage of faults detection [19] metric measures the average number of fault detected by test cases. The range of APFD values varies from 0 to 100 in which higher values indicate fastest fault detection rates. Let us consider T is the test suite containing n number of test cases and let F be a set of m faults covered by test case T. Let TF_i be the position of first test case in T' of T that exposes fault i. The APFD for test suite T' can be calculated using the equation as given below.

$$\text{APFD} = 1 - \frac{\text{TF1} + \text{TF2} + \text{TF3} + \cdots + \text{TFm}}{\text{n*m}} + \frac{1}{2n}$$

For example, we have considered test suite in which ten test cases are present as shown in Table 1. We can prioritize the test cases in order like T1- > T2- > T3- > T4- > T5- > T6- > T7- > T8- > T9- > T10, T3- > T1- > T2- > T6- > T5- > T4- > T7- > T8- > T10-T9. Ordering of the test cases is determined with the help of AFPD equation. We select those test cases from test suite for prioritization which gives highest value of APFD. This ensures that the maximum possible fault coverage will have been achieved by the test case. The test case prioritization is performed based on maximum number of faults exposed by running minimum number of test cases. Here, execution time is analyzed for every test case. The fault measuring approach is used. There are some test cases which forms Test Suite = {T1, T2, T3, T4, T5, T6, T7, T8, T9, T10} and the faults are represented as faults covered by test case are {F1, F2, F3, F4, F5, F6, F7, F8, F9, F10}. It has been represented in Table 1. Execution time of test case to detect the faults is shown in Table 2.

In Table 2, total number of faults exposed by the test cases and time taken to execute those test cases are shown. In this algorithm, we are considering maximum number of fault covered by test cases from test suite, and objective functions reach maximum number of fault covered in very less time.

Table 1 Information of ten test cases with fault

Test case	Fault detected by test cases									
	1	2	3	4	5	6	7	8	9	10
T1	*									
T2				*				*		
T3		*								
T4			*			*		*		
T5					*	*				
T6	*									*
T7		*			*					
T8		*								*
T9			*		*					
T10			*						*	

Table 2 Test case with execution time

Test case	Faults	Execution time
T1	6	12
T2	4	10
T3	3	13.5
T4	5	11.5
T5	2	15
T6	1	10
T7	4	14
T8	3	9
T9	2	11
T10	1	10

5 Experimental Results and Discussion

In this study, CSO algorithm is used to prioritize test cases. The test cases prioritization is done on the basis of the maximum number of faults covered by using minimum number of test cases. In this study, the execution time of each test case is also noted. The results of our study are taken on a test suite (TS) that comprises 38 test cases. The attributes of test cases are software reliability, use of software tools, lines of codes, number of faults, and efforts in hours. The proposed algorithm is implemented in MATLAB tool using windows operating system and 4 GB RAM. As discussed in proposed approach, first, we cluster our test suite by using the CSO algorithm into five different clusters. Here, it is also noted that a dynamic clustering approach is used to identify the clusters automatically from the given test suite. After performing the clustering on test suite, the next step is to select the test cases from each cluster randomly. In this study, three test cases are selected from each cluster randomly and apply these test cases to detect faults into a given program. Finally, we prioritize the test cases according to the maximum number of faults detected which is shown in Table 3. Two software applications are used for this work written in object-oriented language. These applications are used to validate the results. Fault is seeded into the application. A number of test cases are 50 and a number of fault seeded are 10. Next step is to perform clustering and prioritization. Data collected are execution time and test case failure rate. From Table 3, it is noted that the test cases selected from cluster 1 detects the maximum number of faults in comparison to other clusters, whereas test cases of cluster 5 detect minimum number of faults. Hence, the prioritization of test cases based on maximum number of faults is detected. So, we can schedule the test cases from cluster 1 first and after that schedule the test cases from clusters 3, 2, 4, and 5, respectively. Finally, we define the ordering of test cases from clusters as 1, 3, 2, 4, and 4. The execution time of each test case is also mentioned in Table 3. Success rate of this technique is compared with different techniques shown in Table 4.

Table 3 Performance of different test cases using fault detection and execution time

Cluster	Test cases	Fault detected	Execution time
1	1	6	10.4
	2	4	11.2
	4	5	12.33
2	3	3	15.75
	8	3	13.6
	6	1	10.2
3	2	4	10.88
	10	1	14.75
	7	3	11.2
4	5	2	9.55
	10	1	12
	9	1	10.33
5	7	2	18.55
	4	1	11.6
	9	0	10.28

Table 4 Comparison of the success rate

Cluster	Test cases	Success rate		
		CSO (%)	K-means (%)	Agglomerative (%)
1	1	65	45.00	50
	2	85	60	80
	4	80	75	50
2	3	83	50	75
	8	70	66.60	50
	6	85	60	80
3	2	65	60	60
	7	88.00	50	87
	10	70	65.00	66.00
4	5	80	40	64
	10	66.00	45	56.00
	9	58	46.00	66.60
5	7	87	75	50
	4	100	100	50
	9	78	69.00	67.00

6 Conclusion

This paper presents a new approach based on cat swarm optimization for prioritizing the test cases. The proposed approach prioritizes the test cases based on the fault detection rate in early life cycle of software development. The proposed approach is worked in two steps—first, the CSO algorithm is applied on test suite to cluster the test cases according to its attributes. In second, we randomly select three test cases from each cluster and evaluate the performance of test cases based on fault detection rate and execution time parameters. From results, it is concluded that the proposed approach is an efficient and effective method for prioritizing the test cases, also called as clustering-based prioritization technique. In future, we can apply other swarm-based approaches for prioritization of test cases. Proposed work is compared with K-means and agglomerative technique. We found from the experimental result that fault detection rate in CSO technique is more than K-means and agglomerative techniques. The analysis of our algorithm represents the percentage of faults detected using APFD metric for test case prioritization.

References

1. Yadav DK, Dutta S (2016,March) Test case prioritization technique based on early fault detection using fuzzy logic. In: 2016 3rd international conference on computing for sustainable global development (INDIACom), pp 1033–1036. IEEE
2. Yadav DK, Dutta S (2017) Regression test case prioritization technique using genetic algorithm. In: Advances in computational intelligence, pp 33–140. Springer, Singapore
3. Arafeen MJ, Do H (2013) Test case prioritization using requirements-based clustering. In: Proceedings of the IEEE 6th international conference on software testing, verification and validation, (ICST) March 18–22, pp 312–321. IEEE Xplore Press, Luxembourg
4. El-Koka A, Cha KH, Kang DK (2013) Regularization parameter tuning optimization approach in logistic regression. In: Proceedings of the 15th international conference on advanced communication technology (ICACT), January 27–30, pp 13–18. IEEE Xplore Press, PyeongChang
5. Tsai PW, Chu S-C, Pan J-S (2006) Cat swarm optimization. PRICAI Trends in Artificial Intelligence., Springer, pp 854–858
6. Tsai PW, Pan J-S, Chen S-M, Liao B-Y (2012) Enhanced parallel cat swarm optimization based on the taguchi method. Exp Sys Appl 39(7):6309–6319
7. Panda G, Pradhan PM, Majhi B (2011) IIR system identification using cat swarm optimization. Exp Syst Appl 38(10):12671–12683
8. Pradhan PM, Ganapati G (2012) Solving multi objective problems using cat swarm optimization'. Exp Syst Appl 39(3):2956–2964
9. Santosa B, Ningrum MK (2009) Cat swarm optimization for clustering. In: IEEE international conference of soft computing and pattern recognition (SOCPAR'09), pp 54–59
10. Kumar Y, Sahoo G (2015) An improved cat swarm optimization algorithm for clustering. Comput Intell Data Min 1:187–197
11. Kumar Y, Sahoo G (2015) A hybrid data clustering approach based on improved cat swarm optimization and k-harmonic mean algorithm, AI Commun 1–14
12. Kumar Y, Sahoo G (2014) A hybridize approach for data clustering based on cat swarm optimization. Int J Inf Commun Technol
13. Harrold M, Gupta R, Soffa M (1993) A methodology for controlling the size of a test suite. ACM Trans Softw Eng Methodol 2(3):270–285

14. Rothermel G, Untch RH, Chu C, Harrold MJ (1999) Test case prioritization: an empirical study. In: Proceedings of international conference of software maintenance, pp 179–188
15. Wong WE, Horgan JR, London S, Agrawal H (1997). A study of effective regression testing in practice. In: Proceedings of 8th IEEE international symposium on software reliability engineering (ISSRE' 97), Albuquerque, NM, pp 264–274
16. Rothermel G, Untch RH, Chu C, Harrold MJ (2001) Prioritizing test cases for regression testing. IEEE Trans Softw Eng 27(10):929–948
17. Elbaum S, Malishevsky A, Rothermel G (2000) Prioritizing test cases for regression testing. In: Proceedings of international symposium on software testing and analysis, pp 102–112
18. Rothermel G, Untch RH, Chu C, Harrold MJ (2001) Prioritizing test cases for regression testing. IEEE Trans Softw Eng 27(10):929–948
19. Elbaum S, Malishevsky A, Rothermel G (2002) Test case prioritization: a family of empirical studies. IEEE Trans Softw Eng 28(2):159–182
20. Elbaum S, Kallakuri P, Malishevsky A, Rothermel G, Kanduri S (2003) Understanding the effects of changes on the costeffectiveness of regression testing techniques. J Softw Verif Reliab 12(2):65–83
21. Elbaum S, Rothermel G, Kanduri S, Malishevsky AG (2004) Selecting a cost-effective test case prioritization technique. Softw Qual J 12(3):185–210
22. Do H, Rothermel G, Kinneer A (2006) Prioritizing Junit test cases: an empirical assessment and cost-benefits analysis. Empir Softw Eng 11:33–70
23. Qu B, Nie C, Xu B, Zhang X (2007) Test case prioritization for black box testing. In: The proceedings of 31st annual international computer software and applications conference. IEEECS press, Beijing
24. Park H, Ryu H, Baik J (2008) Historical value-based approach for cost-cognizant test case prioritization to improve the effectiveness of regression testing. In: The proceedings 2nd international conference on secure system integration and reliability improvement. IEEECS press, Washington, pp 39–46
25. Khan SR, Rehman I, Malik S (2009). The impact of test case reduction and prioritization on software testing technologies, pp 416–421
26. Andrews S (2006) An investigation into mutation operators for particle swarm optimization. In: Proceedings of IEEE congress on evolutionary computation, pp 1044–1051

Phase Portrait Analysis of Small Satellites: With and Without Torque

Shilpee Kumar and Sarbani Chakraborty

Abstract The qualitative behavior analysis of attitude dynamics of small satellite, especially nanosatellite, with lesser weight and smaller size is presented in this paper. Nanosatellites are high dimensional, nonlinear small satellite systems sensitive toward external disturbances and internal variations. The objective of the work is to ascertain the variation in angular velocity when the dynamical system is affected with and without torque. At first, the satellite behavior is inspected with Euler's equation of motion: torque-free motion of an axisymmetric satellite body and for the generalized torque-free motion. Next, the behavior is examined for the attitude control of satellite in space with torque. Any change in the initial condition is reflected along with the changes in the angular velocity at the respective axis. It is validated with the simulation result of the mentioned categories: time series, two-dimensional and three-dimensional phase portraits of torque-free and torque-imposed motions.

Keywords Small satellites · Phase portrait · Torque-free motion · Angular velocity · Equilibrium point · Euler's equation

1 Introduction

The rigid body satellite system in motion is complex with or without the influence of uncertainties and disturbances. Attitude control of such vehicle in space is very challenging. The compound dynamical behavior is highly sensitive toward small variations in the initial conditions [1]. This behavioral study, applying the method of phase portrait, [2] of satellite system in motion is done to acknowledge the region of instability and thus helps to discover the region of control associated with the trivial error.

S. Kumar (✉) · S. Chakraborty
EEE Department, BIT, Mesra, Ranchi 835215, India
e-mail: shilpeeprasad@gmail.com

S. Chakraborty
e-mail: schakraborty@bitmesra.ac.in

© Springer Nature Singapore Pte Ltd. 2019

V. Nath and J. K. Mandal (eds.), *Proceedings of the Third International Conference on Microelectronics, Computing and Communication Systems*,
Lecture Notes in Electrical Engineering 556, https://doi.org/10.1007/978-981-13-7091-5_37

The torque-free motion of a small satellite is the absence of external torques in attitude dynamics [3]. It is considered as axisymmetric rotational motion about a principle axis. The angular velocity remains unchanged as there is no changeable disturbance torque applied to the satellite body. The stability of the system is of practical importance and is established with the help of Euler's equation of motion. In case of unsymmetrical torque-free motion, the angular velocity remains unaffected with any change in the initial condition. A detailed study is elaborated in [4] to control the error and defined trajectory of the satellite with an active control law [5, 1].

With the diminishing size of satellite, attaining improved and stable system is a difficult task. We are focused on the analysis of satellite body to check the instability in the dynamical equation. It has been done for other complex machines such as robotic bodies, spacecraft, and underwater vehicles for the pictorial representation and understanding of the system [6, 7].

In the case of motion with external torques, the satellite no longer rotates entirely about the principle axis. For better picture, the two- and three-dimensional studies are emphasized in parts. In the first category, the graphical display of the angular velocity is two dimensional assuming one of the state as constant and the latter is portrayed in varying three dimensions. The fluctuation in response from the desired range along a certain axis can be avoided by detailed examination and applying appropriate control law. Internal variations and external disturbances are inevitable and a state of disorder. The slightest variation in the initial condition results in a large difference in the later state of a nonlinear dynamical system. The work done here is categorized into two and three axes response. It is termed as chaos and illustrated with the butterfly effect [8]. A detailed graphical study of this case shows the responsiveness of angular velocities and the tendency of the system to wobble uncertainly. The Jacobian method and equilibrium point technique helps to decide the focus point and the unstable region of the satellite dynamics.

In this paper, we present the fundamental approach of phase portrait in two and three dimensions for the satellite system with and without torque [9]. We calculate the equilibrium points and obtain the eigenvalues at each point by Jacobian matrix. It is used to demonstrate the region of stability and vice versa.

The remaining paper is methodized as follows. The Euler's equation of motion is formulated for all the three cases: axisymmetric torque-free motion, asymmetric torque-free motion is presented in Sect. 2. The portrait of dynamic equation of motion with flywheel, gravitational, and disturbance torque in two and three dimensions with varied condition is presented in Sect. 3. Simulation results demonstrating the effectiveness of the study is put together in Sect. 4. The result is concluded in Sect. 5 of the paper with the validated region of instability.

2 Mathematical Modeling of Satellite Body

The standard Euler's rotational equation of motion [9] in inertial coordinate system is

$$I\dot{\omega} + \omega \times I\omega = M \tag{1}$$

where M is the external moment governing on the body about its mass center, which is equal to the inertial momentum change of the system.

A. *Mathematical Formulation for Torque-free motion*

This section is dedicated for rotational motion of satellite without the influence of the external torques, and therefore, $M = 0$. The satellite system in motion rotates about one of the principle axes with a body-fixed reference frame Ω and has its origin at the center of mass as in Fig. (1).

The satellite equations of motion are

$$I_1\dot{\omega}_1 = \omega_2\omega_3(I_2 - I_3) \tag{2}$$

$$I_2\dot{\omega}_2 = \omega_3\omega_1(I_3 - I_1) \tag{3}$$

$$I_3\dot{\omega}_3 = \omega_1\omega_2(I_1 - I_2) \tag{4}$$

where I_1, I_2, and I_3 are the principle moment of inertia and ω_1, ω_2, and ω_3 are the angular velocities of satellite.

B. *Case I: Axisymmetric motion*

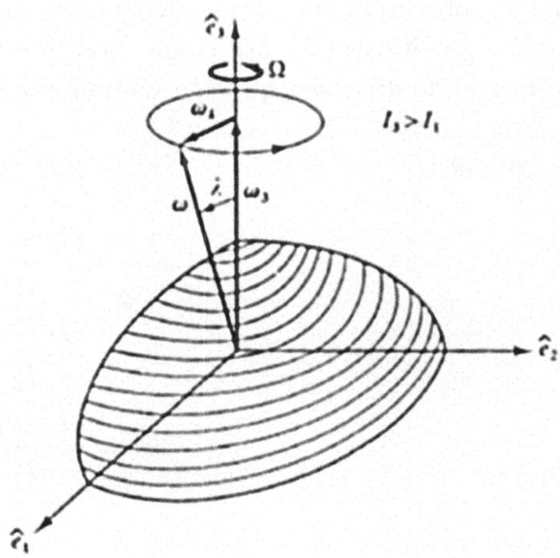

Fig. 1 Torque-free motion of axisymmetric satellite

In this case, the reference frame is in sync with set of principle axes, and the axis of rotation precesses about the symmetry axis with angular velocity Ω as in Fig. 1, and thus $I_1 = I_2$. Therefore, Euler's rotational equation of motion of a torque-free, axisymmetric satellite with $I_1 = I_2 = I$ becomes

$$I\dot{\omega}_1 - (I - I_3)\omega_3\omega_2 = 0 \tag{5}$$

$$I\dot{\omega}_2 - (I - I_3)\omega_3\omega_1 = 0 \tag{6}$$

$$I_3\dot{\omega}_3 = 0 \tag{7}$$

Equation (6) gives the spin rate of satellite $\omega_3 = \text{const} = \text{n}$ about the symmetry axis. Since the rotation is about Ω axis, and therefore the relative rate is

$$\Omega = \omega_3 \frac{(I_3 - I)}{I} \tag{8}$$

Rewriting Eqs. (4)–(6) as below, the reference for Model-I is obtained.
Model-I

$$\dot{\omega}_1 - \Omega\omega_2 = 0 \tag{9}$$

$$\dot{\omega}_2 - \Omega\omega_1 = 0 \tag{10}$$

$$\dot{\omega}_3 = 0 \tag{11}$$

C. *Case II: Unsymmetrical motion*

In this subsection, the torque-free motion of satellite is unsymmetrical. The satellite rotates about two of its principle axis those with the largest and smallest moment of inertia. In this case, $I_1 \rangle I_2 \rangle I_3$, without the loss of generality. Equations (2)–(4) are rearranged as
Model-II

$$\dot{\omega}_1 = \frac{(I_2 - I_3)}{I_1}\omega_2\omega_3 \tag{12}$$

$$\dot{\omega}_2 = \frac{(I_3 - I_1)}{I_2}\omega_3\omega_1 \tag{13}$$

$$\dot{\omega}_3 = \frac{(I_1 - I_2)}{I_3}\omega_1\omega_2 \tag{14}$$

3 Modeling of Satellite Dynamics with Torque

A satellite body keeps rotating about one of the principle axes until no external torque is imposed on the body. Let us consider the rotation about the third axis, that is, $\omega_3 =$ constant and $\omega_1 = \omega_2 = 0$. The condition fails as soon a disturbance torque is applied and the satellite no longer rotates purely about the said axis. The dynamic equation of motion of small satellite from (1) is

$$\dot{M} = H + U \tag{15}$$

where H is angular momentum and U is the applied control. The work done in this paper is about the study of dynamical behavior, and therefore $U = 0$. The angular momentum is equal to the sum of perturbation torques: flywheel torque, gravitational torque, and disturbance torque.

$$H = T_a + T_b + T_c \tag{16}$$

A. *Flywheel torque*

Flywheel torque is the effect of flywheel friction torque and other interfering factors which influence the satellite in motion. It arises in two work modes: current momentum mode and speed momentum mode.

B. *Gravitational torque*

A satellite body, especially in the case of nanosatellites, is considered as a point mass for analysis. It is due to its very small size relative to the Earth. The gravity potential present on one side of a satellite differs from another because of the nonuniform sphere size of the Earth. Thus, the facing part of the satellite toward the Earth observes more gravitational pull than the remaining. Therefore, the orientation of the satellite may result in one end of the satellite being closer to the Earth.

The gravitational disturbance torque is a constant torque felt by a satellite in the Earth's orbit. In order to determine the magnitude of the gravitational torque, a few listed assumptions are made.

1. The gravitational force of the Earth is only considered neglecting any third body effects.
2. Spherical size of the Earth is considered.
3. Mass of the satellite is assumed very less than the mass of the Earth.
4. A satellite is a rigid non-flexible body.

C. *Disturbance torque*

A satellite in space generates magnetic field torques due to the interaction between the Earth's magnetic field and satellite's residual magnetic dipole. The current running through the various components and wiring harness of the satellite body creates residual magnetic dipole. The switching of power between various components and subsystems activates transient and periodic fluctuations because of the residual magnetic dipole.

The low Earth orbiting satellite experiences drag when particles are surrounding the Earth collides with them, at or above the upper limits of the atmosphere. It then exhibits aerodynamic disturbance torque. Drag varies with the orbital distance and thus the coefficient of drag for the satellite varies. This difference is in proportion to the orientation of the satellite with respect to the velocity vector. It also depends on the atmospheric variations of thermal expansion, thermal contraction, solar flares, and so on.

The values of torque are chosen such that the satellite motion is enforced with uncertainties and disturbances.

$$H = \begin{bmatrix} T_{ax} + T_{bx} + T_{cx} \\ T_{ay} + T_{by} + T_{cy} \\ T_{az} + T_{bz} + T_{cz} \end{bmatrix} = \begin{bmatrix} -1.2 & 0 & \frac{\sqrt{6}}{2} \\ 0 & 0.35 & 0 \\ -\sqrt{6} & 0 & -0.4 \end{bmatrix} \begin{bmatrix} \omega_1 \\ \omega_2 \\ \omega_3 \end{bmatrix} \tag{17}$$

Back substituting the value of H in (16) and (1), the satellite equation of motion with external torque are

$$\dot{\omega}_1 = \frac{(I_2 - I_3)}{I_1}\omega_2\omega_3 - \frac{1.2}{I_1}\omega_1 + \frac{\sqrt{6}}{2}\omega_3 \tag{18}$$

$$\dot{\omega}_2 = \frac{(I_3 - I_1)}{I_2}\omega_3\omega_1 + \frac{0.35}{I_2}\omega_2 \tag{19}$$

$$\dot{\omega}_3 = \frac{(I_1 - I_2)}{I_3}\omega_1\omega_2 - \frac{\sqrt{6}}{I_3}\omega_1 - \frac{0.4}{I_3}\omega_3 \tag{20}$$

where $I_1 = 3$, $I_2 = 2$, and $I_3 = 1$.

In order to isolate the region of instability, equilibrium points of satellite system are calculated by solving the above system of (18)–(20). The five sets of equilibrium points are $P_0 = \begin{bmatrix}0\ 0\ 0\end{bmatrix}^T$, $P_1 = \begin{bmatrix}0.9728\ 2.3142\ 0.4163\end{bmatrix}^T$, $P_2 = \begin{bmatrix}-0.9728\ 2.3142\ -0.4163\end{bmatrix}^T$, $P_3 = \begin{bmatrix}0.1468\ -1.0897\ -1.2990\end{bmatrix}^T$, and $P_4 = \begin{bmatrix}-0.1468\ -1.0897\ 1.2990\end{bmatrix}^T$. The Jacobian matrix of the satellite system is used to evaluate the eigenvalues and thus the unstable region is determined. The generalized Jacobian matrix can be written as

$$J(X) = \begin{bmatrix} -0.4 & 0.33 * \omega_3 & (0.33\omega_2 + \frac{1}{\sqrt{6}}) \\ -\omega_3 & 0.175 & -\omega_1 \\ \omega_2 - \sqrt{6} & \omega_1 & -0.4 \end{bmatrix} \tag{21}$$

At each equilibrium point P_0, P_1, P_2, P_3, and P_4, we get three eigenvalues as $E_1 = (-0.4 \pm 1.000i), 0.1750$, $E_2 = (-0.0195 \pm 1.0989i), -0.6641$, $E_3 = (-0.0195 \pm 1.0989i), -0.6641$, $E_4 = (-0.0449 \pm 0.8488i), -0.5352$, and $E_5 = (-0.0449 \pm$

0.8488i), −0.5352, respectively. The eigenvalues at each equilibrium point are a pair of complex conjugates and a real number. It helps to conclude that the equilibrium points are saddle focus and unstable.

4 Simulation

In this section, the variable step Runge–Kutta method is used to solve differential equations numerically in MATLAB tool. In Model-I, the initial condition for torque-free satellite system is $\left[\, \omega_1(0)\ \omega_2(0)\ \omega_3(0) \,\right]^T = \left[\, 2\ 1\ 0.5 \,\right]^T$. In Model-I and II, the phase portrait of angular velocity in time series is insensitive toward any small change in the initial condition which is shown in Figs. 2 and 4. The initial condition remains the same, with no equation of angular velocity equal to zero. The elliptical trajectory of the satellite system without torque is stable and is shown in three-dimensional graph in Fig. 3. The parabolic and elliptical trajectories define the saddle and stable point, respectively, in Fig. 5.

The latter should not be confused with asymptotically stable. In Model-III, the saddle points are plotted for two varying angular velocities assuming third angular velocity as constants: 0.0522, 0.0174, and 0.0522 one each at a time while simulation. The saddle point portrait is shown graphically in Fig. 6. A small change in the initial conditions affects the satellite system on a large extent which can be easily identified in Fig. 7.

The phase portrait of the satellite system with torque is unpredictable and shown graphically with the butterfly effect in Fig. 8. Solution of the dynamic satellite system is projected in time series, and two-dimensional and three-dimensional phase portraits. It is summarized in Table 1.

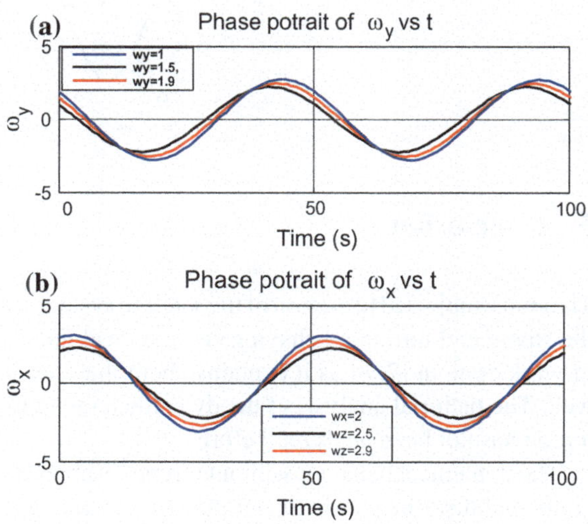

Fig. 2 Time series graph of torque-free motion of axisymmetric satellite

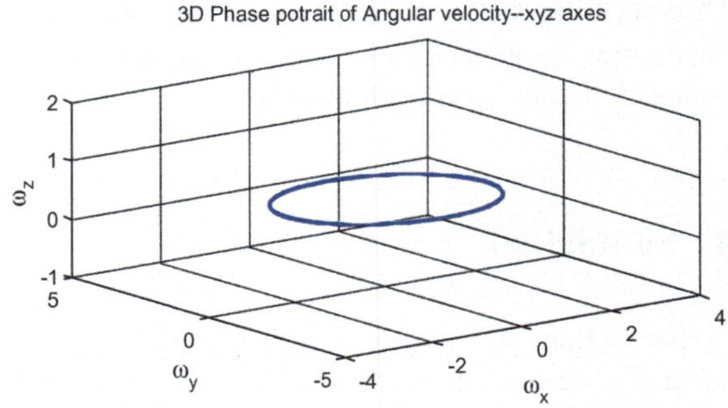

Fig. 3 Three-dimensional phase portrait of torque-free motion

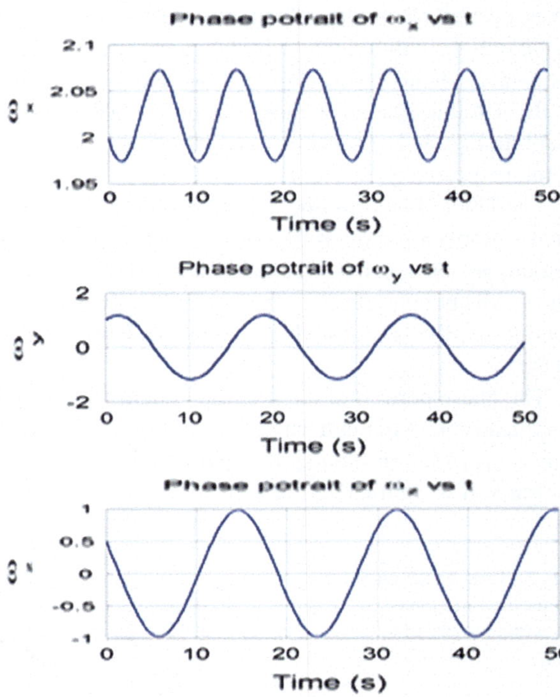

Fig. 4 Time series graph of unsymmetric motion of satellite

5 Conclusion

The study subjected to motion of the satellite system with and without torque indicates the stable and unstable behavior via graphical analysis. This study is an extension of work done in [7, 8] as it explains about the satellite in motion without torque as well. The pictorial analysis of the dynamic body in motion without any influence of torque has not been analyzed before.

The dynamical behavior without torque is sable and poise, i.e., without any uncertainty and disturbances in the output. An external torque or disturbances makes the

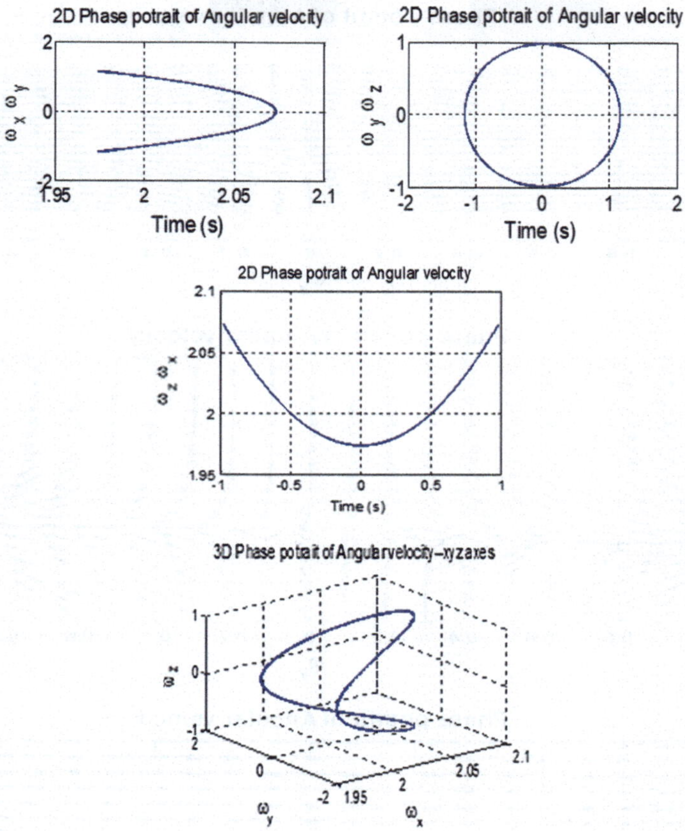

Fig. 5 Two- and three-dimensional portraits of generalized torque-free satellite motion

system wobble and tends toward uncertain movement. It is ascertained with a tiny change at the input corresponding to vast and uncertain variations at the output. In other words, we can conclude that the satellite system is chaotic. The simulation result verifies this dynamic behavior of the satellite in the two stated conditions. In order to follow the defined trajectory, an active and adaptive control law should be designed for the satellite system. The future work is to design a control law, such that the system should be insensitive toward the external disturbances, internal variations, and should be able to follow the mentioned trajectory [10].

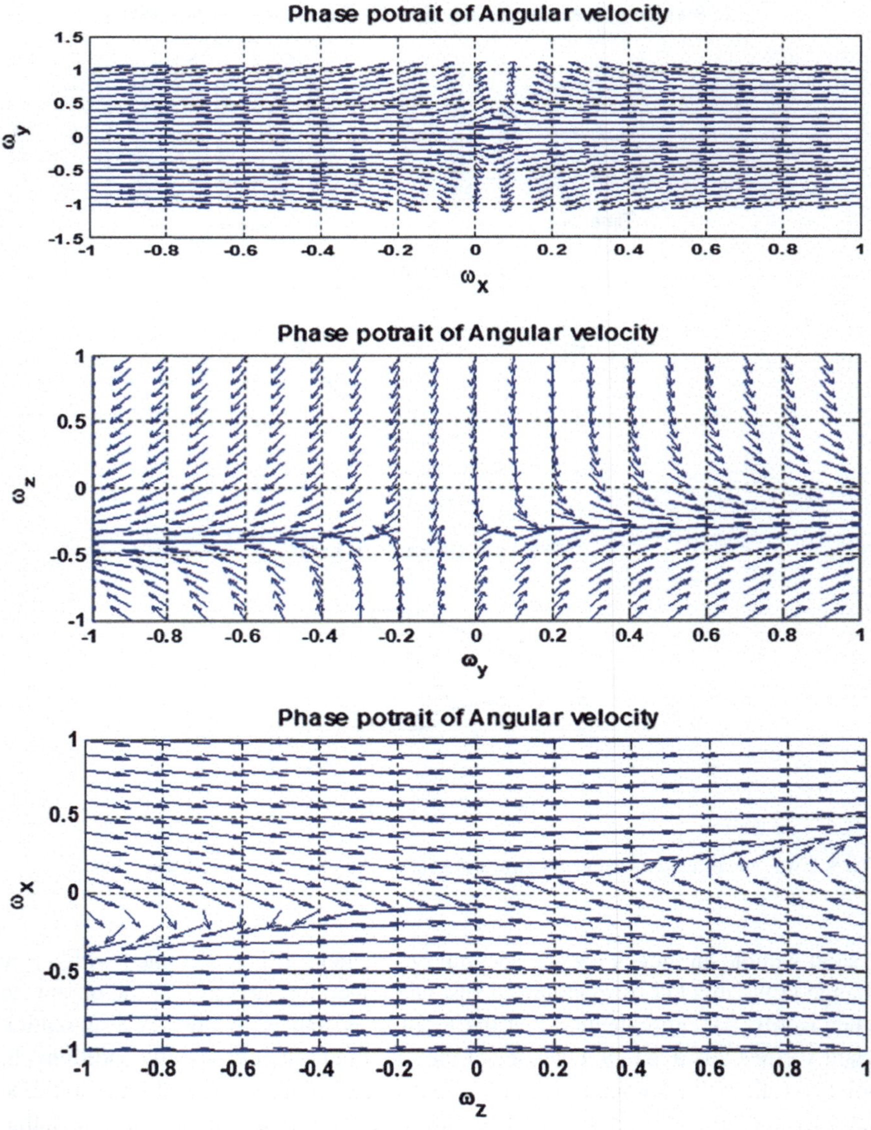

Fig. 6 Eigen values portrait of the angular velocities

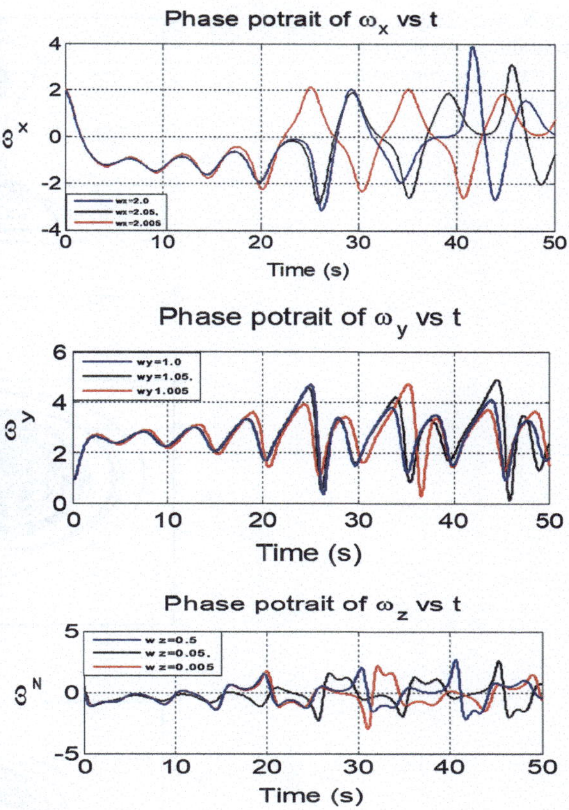

Fig. 7 Time series graph of satellite system with external torque

Table 1 Simulation summary

Figure number	Simulation graphs		
	Condition of simulation	Result expected	Result determined
1	No torque, ω_L varies as 1, 1.6, 1.9	Non-varying angular velocity	Same as expected, Fig. 2
2	Three-dimensional plot of torque-free satellite system	Stable phase trajectory of the satellite in motion	Same as expected, Fig. 3
3	Time series plot of satellite in motion with all angular velocities varying	Varying velocities at all the three axes	Same as expected, Fig. 4
4	Two- and three-dimensional plot of the satellite system, unsymmetric motion	Unstable and stable regions as calculated and plotted	Same as expected, Fig. 5
5	Saddle point and other plots of satellite with external torques	Stable ideally	Unstable and Chaotic as in Fig. 6, 7, 8

Fig. 8 Two- and three-dimensional phase portraits of satellite with torques

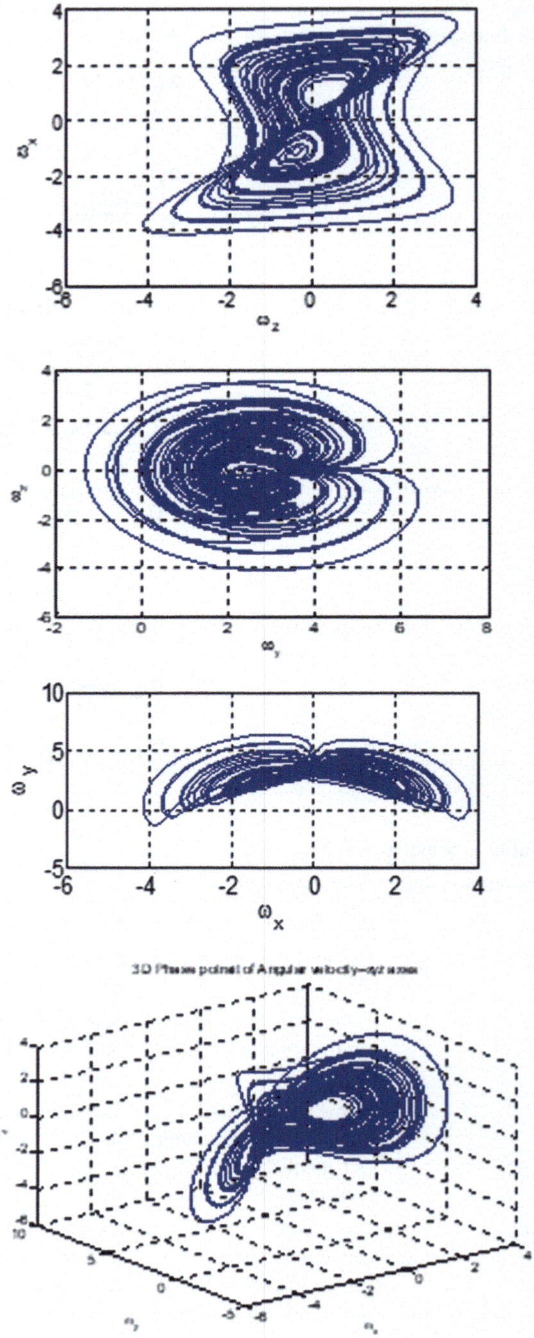

References

1. Yi X, Anvar A (2013) Small-satellite magnetorquer attitude control system modelling and simulation. In: International Congress on Modelling and Simulation, Adelaide, Australia, pp 984–990
2. Butkovskiy AG (1992) Phase potraits of control dynamical systems, pp 48–53. Springer, Berlin
3. Sofyali A, Caferov E Computational phase portrait analysis of two nonlinear small satellite models. In: IEEE conference, pp 491–496
4. Hamidzadeh SM, Esmaelzadeh R (2014) Control and synchronization chaotic satellite using active control. Int J Comput Appl 94:29–33
5. Sidi MJ (1997) Spacecraft dynamics and control: a practical engineering aproach. Cambridge University Press, Cambridge
6. Choi D, Atkeson CG, Cho SJ, Kim JY (2008) Phase place control of a humanoid. In: IEEE-RAS international conference on humanoid robot, vol 8, pp 145–150
7. Doroshin AV, Krikunov MM (2014) Attitude dynamics of a spacecraft with variable structure at presence of harmonics perturbations. Appl Math Model 38:2073–2089
8. Khan A, Kumar S (2018) Study of chaos in chaotic satellite systems. Pramana-J. Phys 13
9. Wie B (1998) Space vehicle dynamics and control. AIAA Education Series. AIAA, Reston, pp 331–357
10. Kose E (2017) Controller design ny using non-linear control methods for satellite chaotic systems. Electr Eng 763–773. Springer

Hardware Design of SEPIC Converter for Battery Charging Application Through PV Using P&O MPPT Algorithm

Mriduwani Verma, S. Shiva Kumar, Samarjeet Singh, Niteesha Kumari and V. Laxmi

Abstract This paper aims at design and hardware implementation of SEPIC converter with photovoltaic source in order to charge a battery. Also, in order to extract maximum power from the photovoltaic module, the Perturb and observe algorithm is being used. This algorithm examines the slope (DP/DV) of the PV and according to the slope the voltage is perturbed to track maximum power. The SEPIC converter is being used as it provides with non-inverted and a constant output voltage. The paper contains the simulation in MATLAB and the simulation results as well as hardware implementation of SEPIC for battery charging.

Keywords SEPIC · Perturb and observe · MATLAB · Battery charger PV module

1 Introduction

With photovoltaic source of energy being a renewable and also cleaner source is becoming a major power generation technique. The PV generation is done using a PV module that contains fundamental component of a solar cell and a series connection forms an array and a combination of series and parallel connections forms a PV module.

The efficiency of the PV module depends on operating point of the PV characteristics of it. The PV curve of the PV module is extracted from the I–V characteristics of the module. Also, for getting the best efficiency from the module, the maximum power of the module is tracked. There are various maximum power point tracking algorithms: perturb and observe, incremental conductance, fractional open-circuit

M. Verma (✉) · S. Shiva Kumar · S. Singh · N. Kumari · V. Laxmi
Department of Electrical and Electronics Engineering, Birla Institute of Technology, Mesra, Ranchi, Jharkhand, India
e-mail: mriduwani9204@gmail.com

S. Shiva Kumar
e-mail: shivkumar.ee@gmail.com

© Springer Nature Singapore Pte Ltd. 2019

V. Nath and J. K. Mandal (eds.), *Proceedings of the Third International Conference on Microelectronics, Computing and Communication Systems*, Lecture Notes in Electrical Engineering 556, https://doi.org/10.1007/978-981-13-7091-5_38

Table 1 Comparison of SEPIC and Cuk converters [5]

Converter	Power at maximum power point	Input ripple voltage	Input ripple current	Output voltage
Cuk converter	Lesser stable	More	More	Inverted
SEPIC	More stable with lesser ripple	Less	Less	Non-inverted

voltage, fractional short-circuit current, incremental resistance, and particle swarm optimization is some of the methods that have been developed for PV arrays [1].

In this paper, perturb and observe algorithm has been implemented as owing to its major advantage of its PV array independent and easier to implement. The P&O algorithm works on the PV characteristics and its slope [1] (i.e., DP/DV). Depending on the sign of the slope, if the slope is negative the voltage is perturbed toward MPP which is on the right, whereas it has to be perturbed toward MPP in case of positive slope of DP/DV which is on the left. The perturb and observe technique can be implemented by using voltage reference control with the help of using PI controller or using duty ratio control [2]. Certain applications in which these converters find their use are renewable energy sources like solar panel which provides DC voltage that is very fluctuating and in order to provide a constant voltage at the load end these converters will be needed [3]. Similarly, battery voltage decreases as it discharges, in order to maintain the voltage DC–DC converters are used. The main DC–DC converters are buck converter, a boost converter, buck–boost converter, Cuk converter, SEPIC [4]. SEPIC converter is being designed in this paper which is basically a boost–buck topology. Unlike Cuk converter, it provides non-inverted output and since it has low input current ripple (Table 1).

2 Proposed System Circuit Design

Figure 1 consists of PV array which is given as the input voltage source of SEPIC converter and at the load a battery is connected that has to be charged. The sensing circuit will provide with the PV array voltage and current to the MPPT controller which will in turn adjust the duty cycle for the switch accordingly.

Proposed circuit is explained in detail below:

1. Modeling of PV,
2. Perturb and observe MPPT algorithm,
3. Mathematical modeling of SEPIC, and
4. Battery modeling.

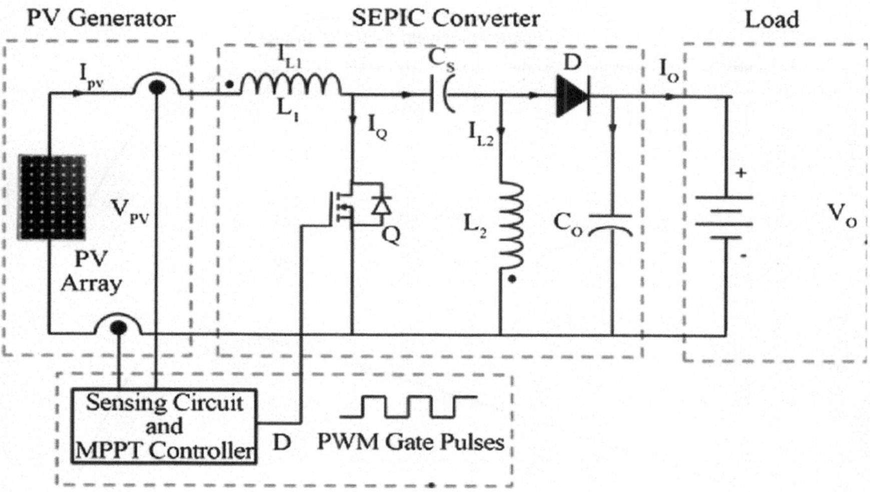

Fig. 1 Proposed system circuit diagram

2.1 Modeling of PV

In the modeling of PV, the single diode equivalent circuit of solar cell is being considered. This circuit consists of the ideal part that is the equivalent current source and diode, as well as the practical part that is the shunt resistance R_{sh} and the series resistance R_s [1].

The modeling of the solar cell is being defined by the voltage–current relationship of PV system as follows [1]:

$$I = I_{pv} - I_s(\exp\left(q\left(\frac{(V + R_s I)}{N_s kTa}\right) - 1\right) - \frac{V + R_s I}{R_{sh}} \quad (1)$$

where I_{pv} = Current (A), I_s = saturation current (A), q = electron charge (1.60217×10^{19} °C), k = Boltzmann constant (1.38065×10^{-23} J/K), a—diode ideality constants and R_{sh}—series and parallel resistances (Ω), N_s = no. of cells in series, and T = temperature (K).

The relation between power and voltage and the relation between current and voltage for variation in irradiation keeping the irradiation constant is shown in Fig. 2, and similarly variation in temperature with keeping the irradiation constant is shown in Fig. 3.

The PV panel used for the analysis is from HBL private Ltd model no. HB1275 (Table 2).

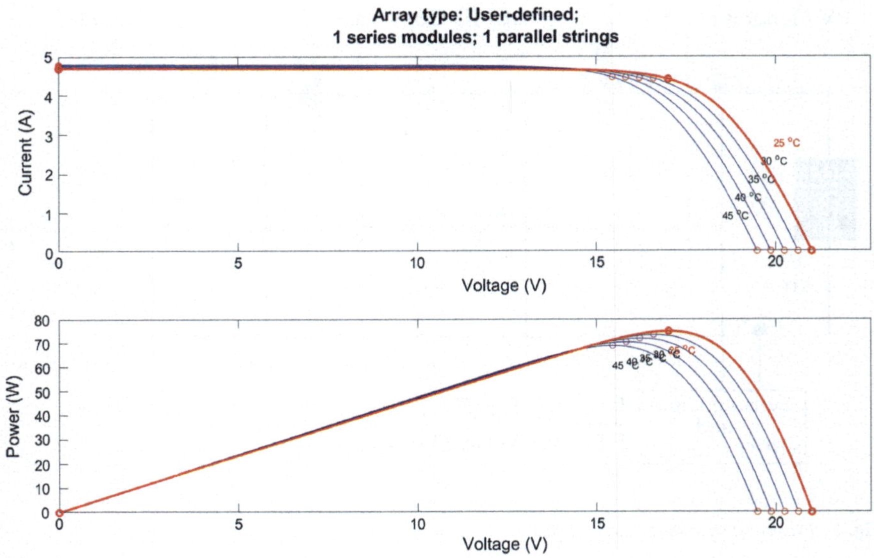

Fig. 2 The I–V curve and the P–V curve for constant temperature and varying irradiation

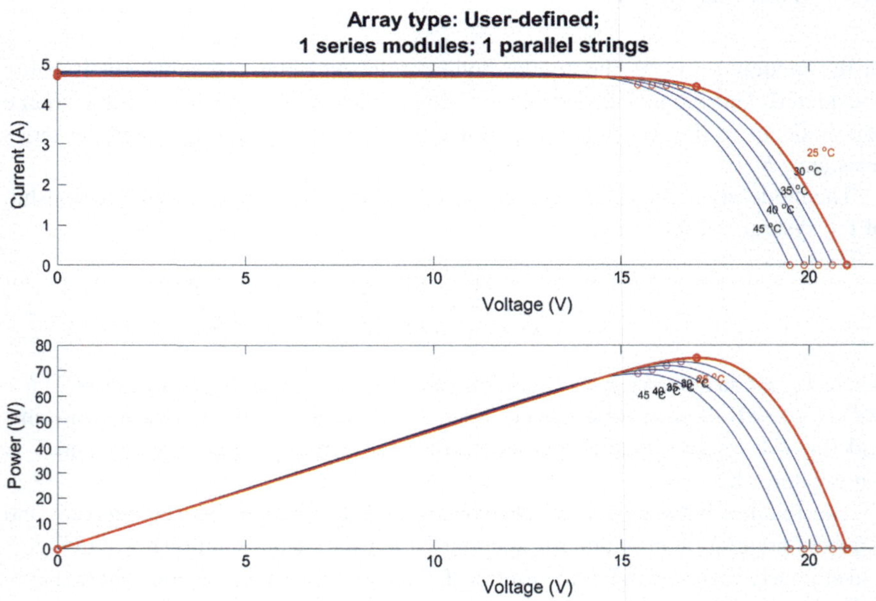

Fig. 3 The I–V curve and the P–V curve for constant irradiation and varying temperature

Table 2 PV panel parameter

S. no	Parameter	Value
1	Maximum power (Pm)	75 W
2	Open-circuit voltage (Voc)	21 V
3	Short-circuit current (Isc)	4.7 A
4	Voltage at max power (Vmp)	17 V
5	Current at max power (Imp)	4.41 A

2.2 *Perturb and Observe Method*

In this method, the PV voltage and current values are measured by the sensor and then the controller accordingly will choose the perturbation size in order to track the PV to its maximum power. Also, the duty cycle is set according to the voltage and power relationship [6] (Fig. 4.)

$$\frac{dP}{dV_{pv}}(n) = P(n) - \frac{P(n-1)}{V_{pv}(n) - V_{pv}(n-1)} \tag{2}$$

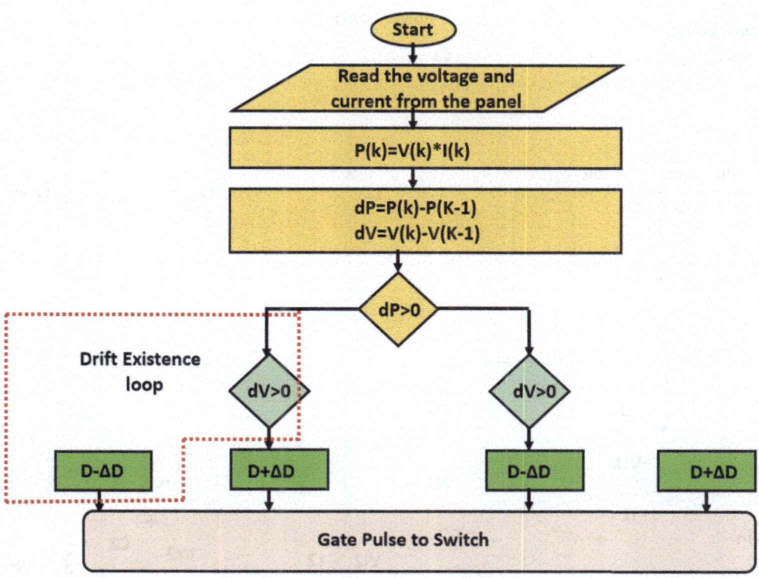

Fig. 4 Flowchart for perturb and observe algorithm

2.3 Mathematical Modeling of SEPIC

See Fig. 5.

Mode 1: When switch S is ON
When switch S is ON, then the inductor L1 gets charged by the source voltage Vs and since assuming that the coupling capacitor Cp is initially charged to source voltage Vin, in S being on the coupling capacitor discharges through inductor L2 and since as the polarity can be seen in Fig. 2, the diode is reverse biased and output capacitor discharges through load [7] (Fig. 6).

$$V_{c1} = L_2 \frac{di_2}{dt} \tag{3}$$

$$C_2 \frac{dV_{c2}}{dt} = -\frac{V_0}{R} \tag{4}$$

$$i_2 = i_{c1} = C_1 \frac{dV_{c1}}{dt} \tag{5}$$

$$V_i = L_1 \frac{di_1}{dt} \tag{6}$$

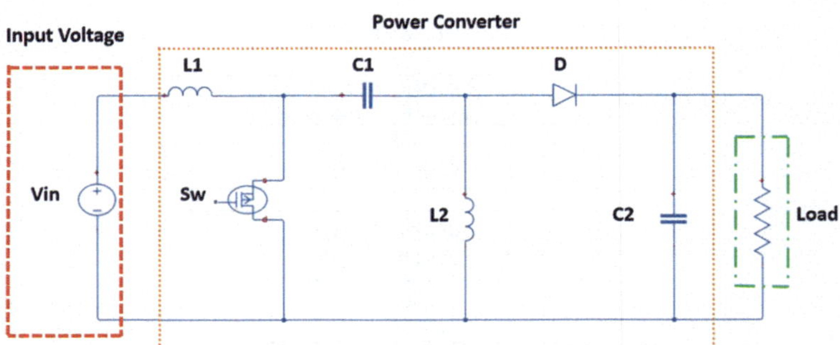

Fig. 5 SEPIC converter circuit diagram

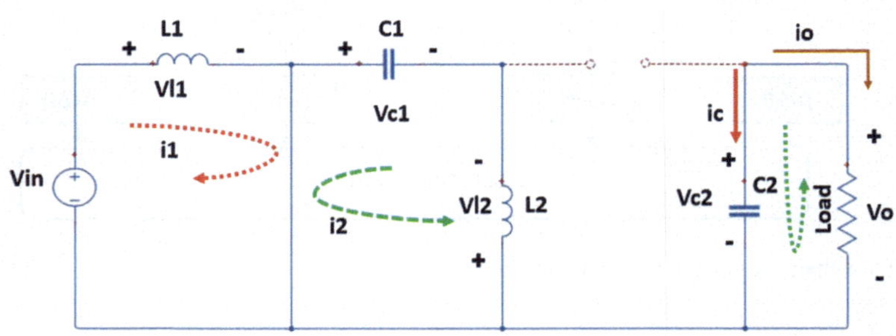

Fig. 6 The equivalent circuit SEPIC when switch is ON

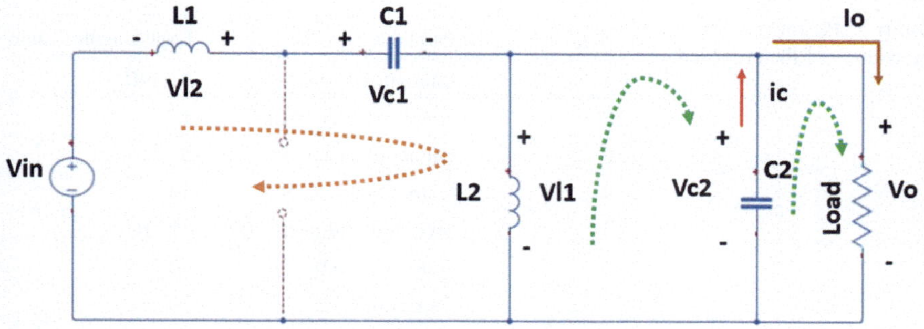

Fig. 7 Equivalent circuit when the switch is OFF

where

Vi	is the input voltage,
Vc1	is the voltage across the coupling capacitor,
Vo	is the output voltage across the load,
I1	is the input current, and
I2	is the current flowing through inductor L2.

Mode 2: When switch S is OFF:
When switch S is OFF, then the circuit is shown Fig. 3. The inductor L1 discharges through coupling capacitor Cp, thereby charging it. Since the inductor L2 changes polarity to oppose the change in direction of current, the diode D turns ON. The inductor L2 demagnetizes through output capacitor and also through the load (Fig. 7).

$$V_i = -L_1\frac{di_1}{dt} + V_{c1} + V_{c1} \tag{6}$$

$$V_{c1} = L_2\frac{di_2}{dt} \tag{7}$$

$$i_2 = \frac{V_{c2}}{R} + C_2\frac{dV_{c2}}{dt} \tag{8}$$

2.4 Design Parameters for SEPIC

The ideality constraints are considered while calculating the inductor and capacitor values [3].
Input inductance:
The inductor value is calculated based on the amount of ripple current the designer wants to allow (Table 3).

$$L_1 = \frac{V_0(1-D)}{f_s\Delta i_1} = L_2 \tag{9}$$

Table 3 Parameters for simulation of the proposed model

S. no	Parameter	Experimental value
1	Inductor L1 = L2	7.5 mH
2	Output voltage (Vo)	12
3	Output current (Io)	2
4	Output power (Po)	30
5	Switching frequency (fs)	10 kHz
6	Duty cycle (D)	60%
7	Capacitor C2	200e−6
8	Capacitor C1	47e−6
9	Irradiation	1000 W/m2
10	Temperature	25 °C

Input ripple current:

$$\Delta i_1 = 10\% \; of \; i_1 \tag{10}$$

Output capacitance:

$$C_2 = \frac{V_0(D)}{R f_s \Delta V_{c2}} \tag{11}$$

Output voltage ripple:

$$\Delta V_{c2} = 2\% \; of \; V_0 \tag{12}$$

2.5 *Battery as Load*

The lead–acid battery is being used as load. The lead acid battery is a secondary storage device. When a source is connected to the battery, the electrical energy is converted to chemical energy, and in this condition the battery is charging.

In the lead–acid battery, each cell contains lead metal and lead oxide in an electrolyte of 37% w/w of sulphuric acid. Figure 8 shows the simple battery model which consists of a voltage source connected in series with the impedance.

3 Simulation and Result

The simulation of the SEPIC is done in the MATLAB environment and keeping the runtime. 1 s and using solver Ode23 Bogacki-Shampine.

Fig. 8 Simple battery model

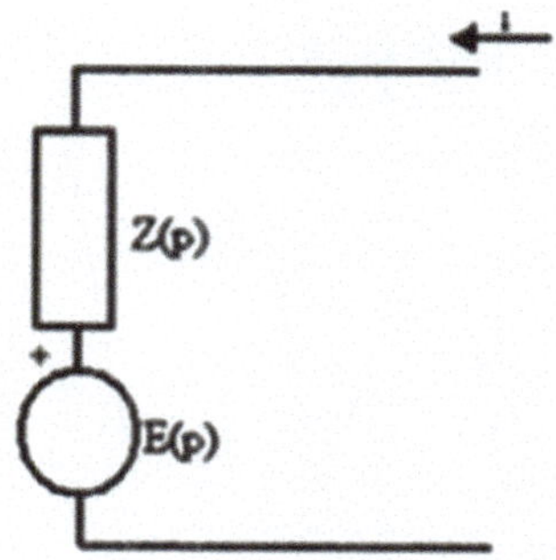

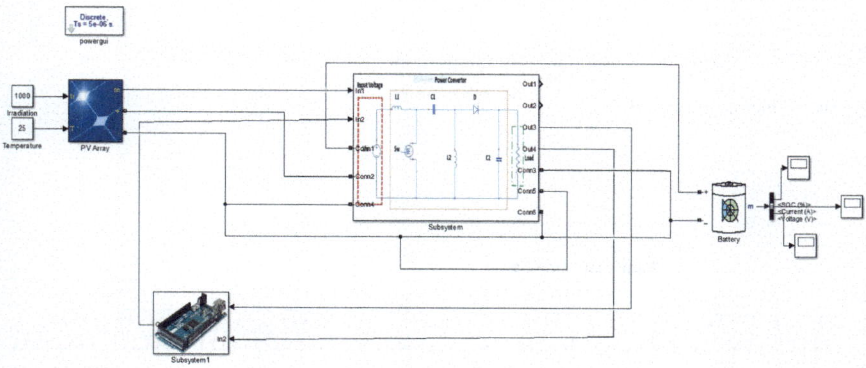

Fig. 9 Simulation diagram of proposed system

Result:
The result of the simulation is shown in Figs. 8, 9, 10, 11, 12 and 13. The current across the inductor rises and falls due to the magnetizing and demagnetizing of the core. The triangle current waveform gives ripple current in the inductor.

4 Hardware Design for Proposed Circuit

Gate Driver Circuit: The TLP250 optocoupler is used here and it provides isolation between the Power MOSFET bridge and the Arduino microcontroller.

Another function of TLP250 optocoupler is to provide amplified gate control signals to IRFP460 Power MOSFETs.

The TLP250 consists of a light-emitting diode and an integrated photodetector. This unit is 8-lead DIP package. TLP250 is suitable for gate driving circuit of IGBT or power MOSFET. Switching speed of this optocoupler is higher than others and it provides high voltage isolation.

Arduino: Arduino Uno is a microcontroller based on the ATmega328P. It has 14 digital input/output pins, 6 analog inputs, a 16 MHz quartz crystal, a USB connection,

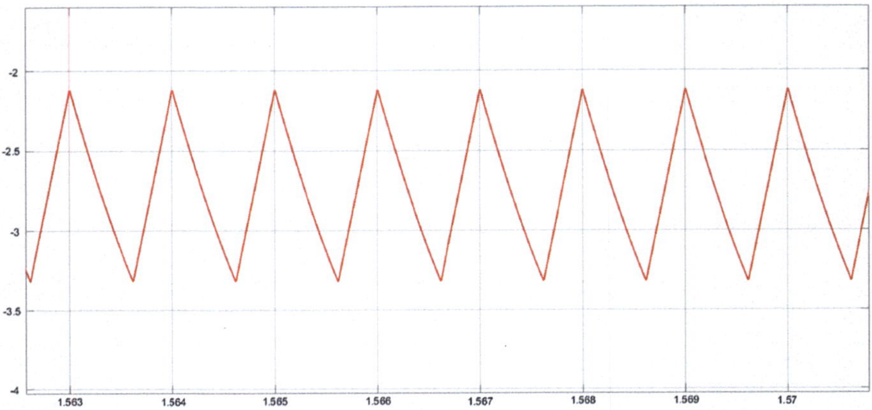

Fig. 10 Inductor current (IL1)

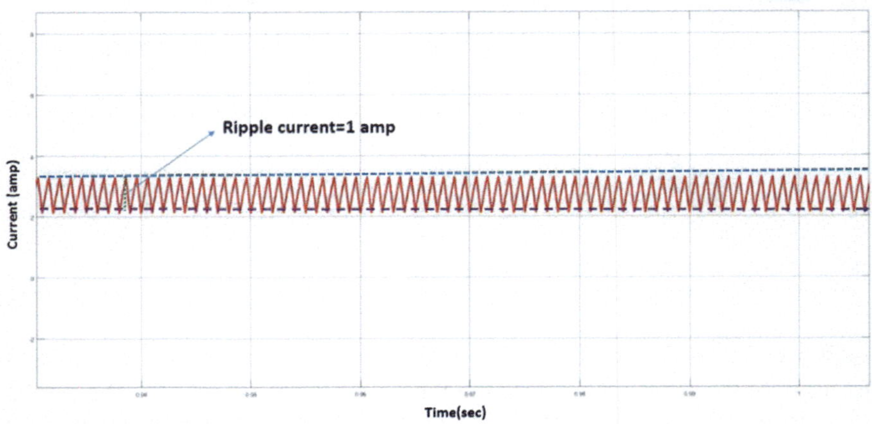

Fig. 11 Output voltage waveform

a power jack, an ICSP header, and a reset button. It contains everything needed to support the microcontroller, simply connect it to a computer with a USB cable, or power it with a AC-to-DC adapter. 5 V DC supply as input supply is provided to an Arduino.

Power circuit board: Power board consists of passive elements which include two coupled inductors and two coupling capacitors. The power switched used is IRF460P. The power board comes with a snubber circuit which reduces the voltage stress, the snubber circuit consists of 40 W 4.5-Ω resistor, 220 nf 2000 V capacitor, and ultrafast recovery diode.

PV panel: The maximum power that can be extracted is 75 W, at Vmax = 17 V and Imax = 4.41 A, Voc = 21 V, and Isc = 4.7 A PV module HB-1275 by HBL power systems Ltd. The PV panel is the input source to the SEPIC converter. The

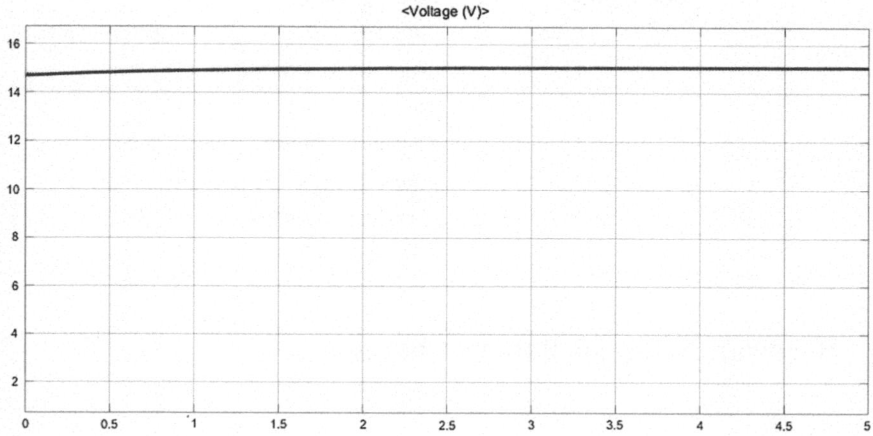

Fig. 12 Inductor current (IL1)

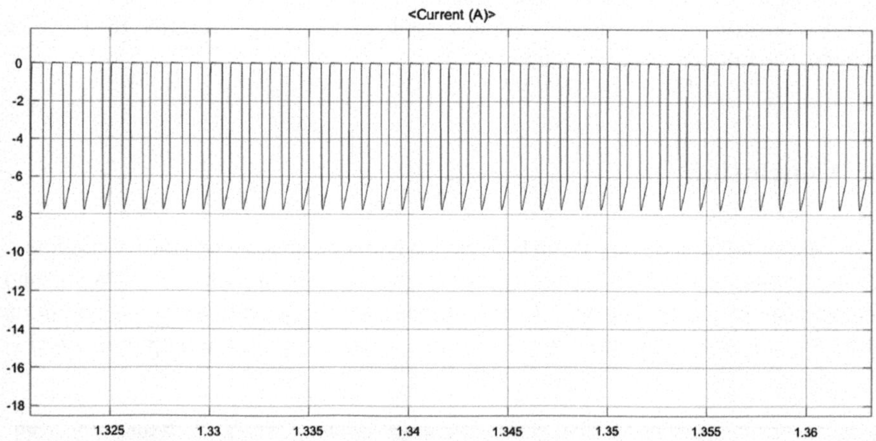

Fig. 13 Output current waveform

above specifications are at standard temperature and irradiations that is 25 °C and irradiation of 1 KW/m^2.

Battery: The battery used is a lead–acid battery. It has two lead plates as electrodes to avoid short-circuit through contact mostly through physical contact. The battery used here is of nominal voltage of 12 V and ampere rating of 7 Ah (Table 4).

Table 4 Hardware design parameters

S. no	Parameter	Experimental value
1	Input voltage (Vin)	18.92
2	Output voltage (Vo)	12.45 V
3	Output current (Io)	2.29 A
4	Output power (Po)	W
5	Switching frequency (fs)	10 kHz
6	Duty cycle (D)	60%

5 Hardware Setup for Battery Charging

See Fig. 14.

Result validation:
Hardware results are shown in Figs. 15, 16, 17, and 18. The output wave form of SEPIC converter was having less voltage ripple and less current ripple [8–13]. The battery was fully charged to a rated volt of 12 V.

6 Conclusion

From the simulation results, it can be seen that battery charges when the PV module is giving a voltage of 18.92 V at the input of SEPIC with irradiation of 1000 W/m2 and temperature of 25 °C. The MPPT controller sets the duty cycle for above condition at 58.8%. Implementing these results in the practical circuit, the output obtained is

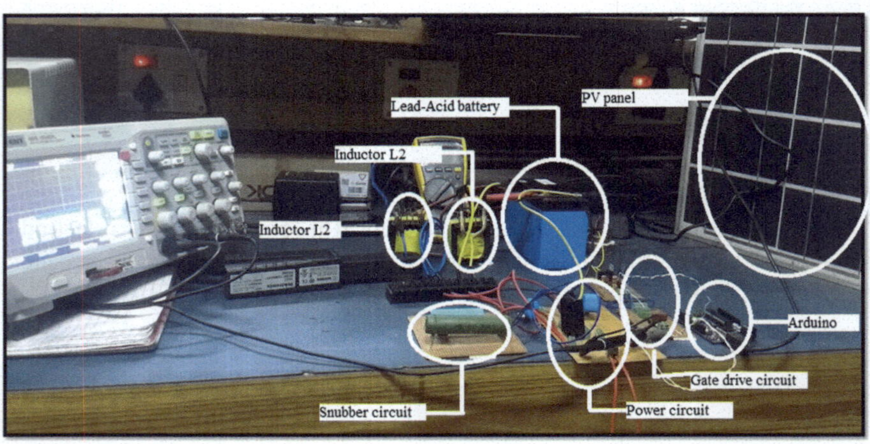

Fig. 14 Hardware setup of SEPIC converter with PV as source and battery as load

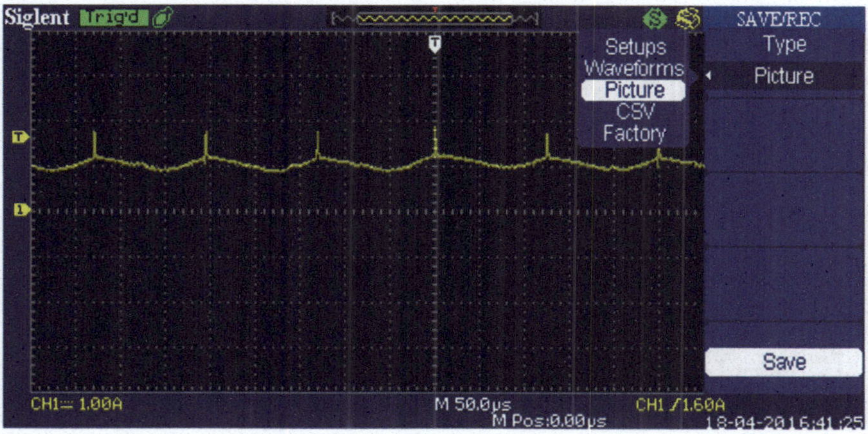

Fig. 15 Inductors current IL1

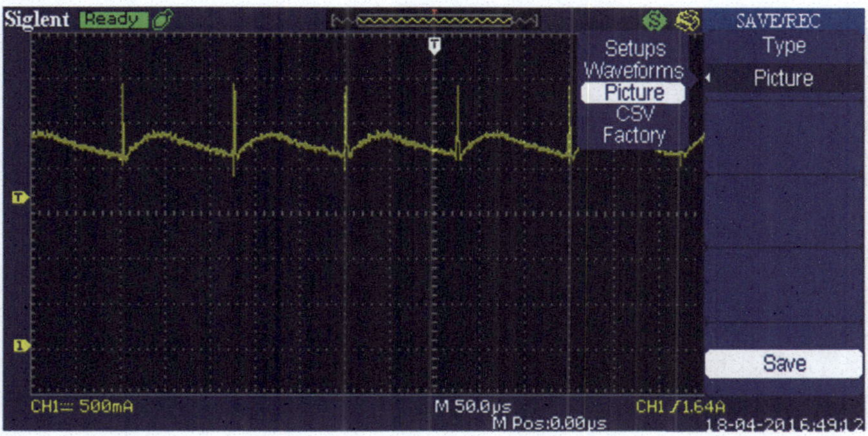

Fig. 16 Inductors current IL2

12.45 V, which is the required voltage for battery charging, i.e., the nominal voltage (i.e., 12 V) [14].

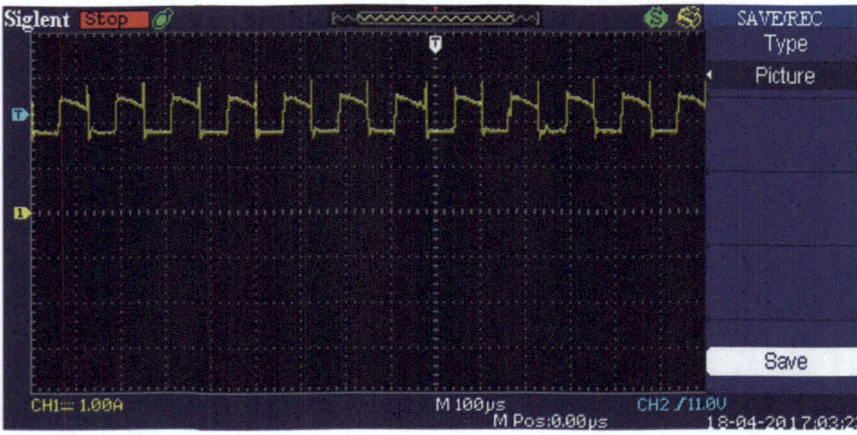

Fig. 17 Output current

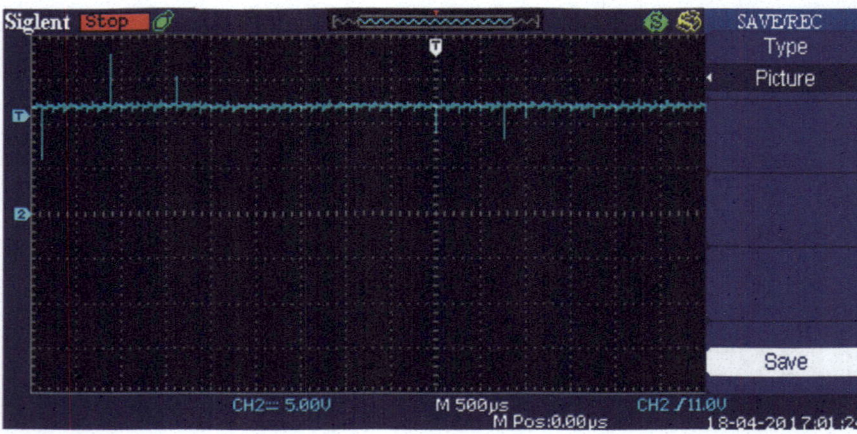

Fig. 18 Output voltage

References

1. Killi M, Samanta S (2015) An adaptive voltage-sensor-based MPPT for photovoltaic systems with SEPIC converter including steady-state and drift analysis. IEEE Trans Industr Electron 23(12)
2. Moradpour R, Ardi H, Tavakoli A (2018) Design and implementation of a new SEPIC-based high step-up DC/DC converter for renewable energy applications. IEEE Trans Industr Electron 65(2)
3. Falin J Senior Designing DC/DC converters based on SEPIC topology. Appl Eng Texas Instrum Incorp Analog Appl J
4. Rani PU, Saravanan S Simulation and modeling of SEPIC converter with high static gain for renewable application. In: 2017 asian conference on energy, power and transportation electrification (ACEPT)

5. Cabral HG, Marques AR, Pedrollo GR, de Faria PF, dos Reis FS buck-boost family converters for driving LED lamps, Power electronics laboratory of PUCRS Pontifical Catholic University of Rio Grande do Sul Porto Alegre, Brazil
6. Shihong Q, Min W, Chen T, Yao X Comparative analysis of incremental conductance and perturb- and-observation methods to implement MPPT in photovoltaic system. In: International conference on 2011 performance comparison of electrical and control engineering (ICECE)
7. Application Report SNVA168E–May 2006–Revised April 2013. AN-1484 Designing A SEPIC Converter by Texas instruments
8. Lee S-W, Do H-L (2018) Isolated SEPIC DC–DC converter withripple-free input current and lossless snubber. IEEE Trans Industr Electron 65(2)
9. Repetitive Avalanche And Dv/Dt Rated Irf460 Hexfet® Transistors Thru-Hole (To-204aa/Ae) Www.Irf.Com 1 500v, N-Channel
10. 8-bit AVR Microcontrollers ATmega328/P Datasheet co
11. Esram T, Chapman PL (2007) Comparison of photovoltaic array maximum power point tracking techniques. IEEE Trans Energy Convers 22(2):439–449
12. Villalva MG, Gazoli JR, Filho ER (2009) Comprehensive approach to modelling and simulation of photovoltaic arrays. IEEE Trans Power Electron 24(5):1198–1208
13. Haddad R, El Shahat A, Kalaani Y Lead acid battery modeling for photovoltiac applications, Department of Electrical Engineering, Georgia Southern University, Statesboro, GA, USA
14. Zhang D AN-1484 designing a SEPIC converter. Application report SNVA168E–May 2006–Revised April 2013, Texas instruments

Study and Design of Electro-Pneumatic Shifting System

Chaitanya Tyagi, Shantum Verma, Abhishek Pandey, Deepak Prasad and Vijay Nath

Abstract This paper describes the methodology of associate degree algorithmic program development toward the automatic external gear-shifting and clutch-actuation system for a successive case with easier and an economical suggests. Mechanically motivated manual gear mechanism bridges the gap between automatic and manual transmissions that provide the benefits of each form of transmissions. This could ideally result in quicker shifting time and supply important advantages within the variety of electronic aids like launch management and traction management. Removal of mechanical clutching would scale back fatigue and result in engineering profit. A complete shift time of concerning 100 ms for upshifts and 150 ms for downshifts while not acceleration flatlining is that the desired output. The developed system ought to be usable on any similar engine with a successive case by creating little modifications to the GCU code and mechanism force. The system may even be employed in autocross vehicles for reduced gear-shift times and exaggerated driver concentration, resulting in higher lap times.

1 Introduction

Gasoline engines develop helpful torsion over a restricted engine speed which varies. To be ready to use, the obtainable torsion over the varying of auto speed, gears are required to cut back or increase the engine speed consequently. The standard manual transmission uses a driver-operated clutch, generally operated by a pedal or lever, for regulation torsion transfer from the inner combustion engine to the transmission, and a gear stick, either operated by hand (as during a car) or by foot (as on a motorcycle). It allows the motive force to pick any gear quantitative relation ("gear") at any time. However, just in case of the motorcycles and a few forms of sports cars, it solely permits the motive force to pick the next higher or next lower gear [1]. This kind of

C. Tyagi · S. Verma · A. Pandey (✉) · D. Prasad · V. Nath
VLSI Design Group, Department of ECE, B.I.T. Mesra, Ranchi 835215, JH, India
e-mail: a.p.bitmesra@gmail.com

© Springer Nature Singapore Pte Ltd. 2019
V. Nath and J. K. Mandal (eds.), *Proceedings of the Third International Conference on Microelectronics, Computing and Communication Systems*, Lecture Notes in Electrical Engineering 556, https://doi.org/10.1007/978-981-13-7091-5_39

transmission is named an ordered manual transmission. Ordered transmissions are employed in racing for his or her ability to create fast shifts.

Modern industrial gearboxes are system sort that uses ratcheting drums and shift forks to pick totally different gears. Associate degree automatic case is one variety of car transmission which will automatically modify gear ratios because the vehicle moves, releasing the motive force from having to shift gears manually. Most automatic transmissions have an outlined set of substance ranges. Besides automatic, there are also alternative forms of machine-driven transmissions like Continuous Variable Transmissions (CVTs) and semiautomatic transmissions that liberate the motive force from having to shift gears manually by victimization the transmission's laptop to alter gear [2]. Turbo boost is often lost between gear changes manual, whereas in an automatic the accelerator will stay totally depressed. This but remains mostly dependent upon the quantity and optimum spacing of substances ratios for every unit. A replacement technology referred to as mechanically motivated manual transmission bridges the gap between automatic and manual transmissions and provides advantages of each. There are vital benefits in shift times, torque, engine rpm, clutch effort, and vehicle acceleration over ancient manual effort. Methods of actuation vary from coil actuators to gas or hydraulic systems [3]. All transmission styles had one goal in common to create shifting easier. The motive force gear-shifting strategy influences significantly within the vehicle dynamic behavior, performance, and fuel consumption and as a result of it changes the gear mechanism inertia and the engine speed. This result becomes even additional pronounced in performance or competition driving. The time throughout that the facility delivery from the engine to the wheels is interrupted thanks to a gear modification being motivated is noted because the "shift time" [4]. Longer a vehicle loses in finishing gear shifts, larger is that the final lap time [5].

The typical autocross vehicle incorporates a six speed case, yet reaches a (course limited) prime speed in competition of solely concerning 110 kmph. Choosing a final drive for this prime speed would lead to five gear shifts in but 4 s. As a result, final drive quantitative relation is extremely sensitive to shift delay time. Though vehicle mass, engine performance and traction still play a serious role, and typical vehicle acceleration is considerably restricted by the time it takes to finish a gear shift. The time taken to finish a shifter depends on the following parameters [6]:

1. Time of driver's reaction,
2. Time of clutch operation,
3. Time of shifter mechanism operation, and
4. Time supported case style.

Automated mechanical gearboxes are the logical selection for improving the shift time of a fixed-ratio case. Integration of the clutching and also the gear-shifting mechanisms seamlessly is the major advantage of this kind of gearboxes. When the effort commands by the motive force, the Gear Management Unit (GCU) cuts engine power (through ignition and fuel cut), disengages the clutch, actuates the shifter mechanism and reengages the clutch, dead a couple of hundreds of a second. Also, the shift rate will be simply controlled and call the engine rate throughout a gear shift will be matched with the shift time to make the engine stay within the most torsion band at all times. What is more is an automatic manual case which will manage the "launch RPM" and might shift mechanically for drag strip or acceleration events. Thus, the engine will be unbroken within the power band for nearly all of the time.

An automated gear-shifting mechanism will be faster than the manual throughout associate degree autocross, since the motive force now not must take care of the clutch or the throttle, whereas shifting, permitting him to tackle the track additional with confidence and quickly. Gas shifting systems are designed antecedent; however, typically involve serious parts, complicated physics, and restricted practicality [7].

The planned system aims to be a performance-oriented, easy-to-customize replacement for multiple forms of vehicles.

2 Development Methodology

The steps involved in the development of the shift system are as follows:

1. Choice of actuation mechanism from available alternatives.
2. Calculation of required actuation forces and procurement of appropriate parts.
3. Development of algorithm and incorporation of driver aids.
4. Integration of the control system with hardware.
5. Testing of system for shift times and stability.

3 Benchmarking and Choice of Shifting Mechanism

The shifting system that was benchmarked was manufactured by Translogic (Model no-TLS-PS1) and was an electric servomotor-type shifter that was readily available as an aftermarket alternative for use with bike engines. It had some flaws. The shifter failed to work properly often stalling the engine in the process of switching between gears. The shifting time lag caused delay between shifts leading to slow acceleration times. The high amperage demand by the shifter drained vital battery power. Hence, a new shifting system is to be developed that would allow for faster shifts, automatic clutch actuation, and incorporation of electronic driver aids. The choice of actuating mechanism must meet the following criteria:

1. It must allow the driver to shift at any time while accelerating, decelerating, and taking corners. This can be achieved by steering wheel mounted shifting paddles or buttons, and hence eliminates a traditional side mounted linkage based mechanical actuator.
2. It must be reliable, lightweight, and cost-effective.
3. It should minimize the time lag between shifts.
4. It must not drain large amount of battery power.

Shifting systems can be classified into four categories: mechanical, electrical, hydraulic, and electro-pneumatic.

Mechanical shifting employs a system of cables, levers, and springs to shift gears—much like throttle or clutch actuation we currently employ. It is extremely reliable and provides us with infinite number of shifts without draining the battery of its power. Also, it is the most economical option. However, this system is not capable of automatic upshifting since it is independent of electronic control [8].

Electrical shifting employs a servomotor-type or a magnetic-type shifter to change gears when actuated by a button. It is lightweight and reliable and is compatible with the automatic upshift system. However, it consumes much of the battery power.

Hydraulic shifting utilizes hydraulic pressure to actuate gears. The required pressure is created with the use of a pump. This method is fairly reliable. However, the biggest drawback is the means to power the pump. The pump requires a fairly large value of current to operate. Also, it is extremely bulky and any leakage will cause a potential fire hazard.

Electro-pneumatic shifting works on the same principle as that of hydraulic shifting, but uses compressed air instead of a liquid. This method is very reliable, and is capable of shifting with almost no time lag and also uses very low current (~0.4 A). The only drawback this system has is that the number of shifts is limited. However, the number of shifts can be controlled by appropriating the size of the pressure cylinder.

Comparing and contrasting the abovementioned alternatives, Electro-pneumatic shifting comes out as the best alternative and is hence recommended.

4 Components of Pneumatic Shifting

The following are the main components of the developed pneumatic shifting assembly:

1. Storage tank and regulator.
2. Manifold and solenoid assembly.
3. Pneumatic actuators.
4. Electric control.

A **storage tank** is responsible for storing compressed gas and releasing it at a high pressure for the actuators. A paintball-style tank can be used to store the working gas. However, paintball tanks have a storage pressure of about 1800 psi. Hence, a pressure regulator is attached to the tank to reduce and maintain the output pressure to the required working value. The output pressure can be varied through an adjustment screw on the top of the regulator. The gage on the pressure regulator allows correct output pressure to beset. For the gas to be used in the storage tank, CO_2 and compressed air are considered to be viable options. But compressed air requires pressure pumps which are not very economical and are not readily available. CO_2 is chosen as it is cheaply available commercially and can be stored in paintball tanks. The solenoid valves are responsible to control the flow of gas to the appropriate actuators on demand. These valves work on 12 V from the vehicle battery and are controlled by a microcontroller. The routing from the tank to the valves and from valves to the actuators is done through polyurethane tubing connected using push connectors.

A pneumatic actuator is a piston–cylinder arrangement which generates the required force for gear shifts. It can be a double-acting or a single-acting design. Double-acting actuators allow powered extension and retraction at the same time as opposed to single-acting which can either pull or push. The gear-shift actuator connects to the ratcheting drum via an adjustable splined lever. The amount of torque on the drum can be adjusted by adjusting the position of the actuator on the lever. The clutch actuator is connected to the pull lever via a clutch cable. This allows freedom for the lever to rotate while keeping the stroke constant.

5 Number of Shift Calculations

The number of shifts for a particular gas cylinder is determined by the amount of gas required for each shift. For this, the dimensions of the actuators need to be calculated. For shifting the gears on the Honda CBR600RR, a force of 120 N needs to be applied onto the shift lever, over a travel distance of 25 mm for either upshift or downshift. Thus, a double-acting (push–pull type) actuator with a bore of 16 mm and stroke of 50 mm was selected. A pressure of 100 psi on this actuator is enough to shift the gears. To actuate the clutch on this engine, a force of about 200 N is required. The travel distance of the clutch is about 25 mm. Hence, a single-acting (pull type) actuator is selected with a bore of 25 mm and stroke of 25 mm. The same pressure of 100 psi is supplied to the clutch as well as the shifter. From these dimensions, we can calculate that 0.4 g of air is required for an upshift and 0.6 g of air is required for a downshift. Since the number of upshifts and downshifts are equal for a race, we can assume that the total air for a "shift couple" is 1 g. A sample autocross track was needed to be selected to carry out the design simulations. From lap time simulations, a gear track map was obtained. This map shows the engaged gear at all the positions of the track. A driver is expected to shift about 27 times per lap on this track. For a 20 lap race, the total number of shifts comes to 540, i.e., 270 g of air. Keeping in mind pressure fluctuations and leakage losses, a tank capacity of 550 g was chosen to be enough for an endurance race.

6 Algorithms for the Different Sequences

i. **Upshift Sequence**

When the test car was run with a solenoid-based Translogic power shift gear actuator, an upshift time of 550 ms was recorded. Moreover, the driver had to manually actuate the clutch, adding to his effort. With the pneumatic shifter design, the driver was able to shift without manually actuating the clutch, also known as flat shifting. The advantage of flat shifting is that the driver can shift without letting go of the throttle and the engine constantly remains in the power band. This is possible by cutting the fuel and ignition for the engine appropriately. When ignition and fuel are cut for the engine, the rpm of the engine falls enough for the next gear to be engaged without the use of clutch. It was manually calculated that for a successful upshift the time needed was 50 ms. Along with the valve actuation time of 8.75 ms and including 10 ms buffer before and after the valve actuation, a total of 87.5 ms was calculated for the Ignition and Fuel Cut (IFC).

ii. **Downshift Sequence**

By making use of the Translogic power shift gear actuator, the downshift time was recorded to be 600 ms. Since the engine RPM increases on downshifting, IFC is not possible. Therefore, clutch actuation is very much needed. When the driver presses the downshift paddle, the GCU first disengages the clutch and then shifts to the lower gear. Here, after a successful shift the clutch is engaged instantaneously, bypassing the flow control valve and dumping straight to air. This dump is possible only when the car is in motion, i.e., it cannot be used to start the car from standstill. The total downshift time was 140 ms for the pneumatic system.

iii. **Launch Control**

Using a driver-operated manual clutch for launch starts was mediocre as both time and torque were lost in wheel slip. Slipping of the wheels is inefficient as the tires operate out of their traction limit and the extra torque is wasted, limiting the acceleration capability. Also, if the drivers are amateurs, launching the car at exact clutch bite point is not always possible. The basic idea behind the launch control system is to hold the engine at a particular RPM prior to launch, quickly releasing the clutch, and finally avoiding the wheel slip by arming the traction control. The primary rev limit for the test car was set to 13,500 RPM. The power band of the engine was from 8000 to 11,000 rpm, as found from dynamometer testing. The objective was to launch the car at the beginning of this band, i.e., 8000 RPM and to shift into the next gear at 11,000 RPM. Testing data showed that the rpm of the engine falls by around 2500 upon the engagement of the clutch during a launch. Therefore, a Secondary Rev Limit of 10,750 was set by means of a digital input to the ECU. The Secondary Rev Limit (SRL) held the engine at this RPM till the time the launch input was given. Once the clutch was engaged, the RPM fell to 8000, starting the power band. The timing for the SRL as well as clutch release was controlled by the microcontroller. The engagement of the clutch too is a major limiting factor on the efficiency of the launch. Too slow a release will eat up precious time while too fast release might stall the engine. Ideally, the clutch should be dumped at the bite point (the point when the clutch just starts to engage). To achieve this, the gas from the clutch actuator was released slowly via the flow control valve till the bite point was reached and then bypassed straight to the air. The microcontroller waited for a present time before bypassing the valves. This engaged the clutch instantaneously, delivering maximum torque without stalling. The bite point of the clutch was found by changing the actuation time of the slow release valve. The time when the bite point was hit was found to be 1950 ms.

iv. **Clutch Override**

The CBR600RR has a sequential gearbox with the gear sequence of 1-N-2-3-4-5-6. So, an instant jump to the neutral is not possible. This function overrides the microcontroller operation and actuates the clutch valve to disengage the clutch. The clutch is held disengaged as long as the button is kept pressed. Upon release, the clutch engages gradually through the flow control valve for a smooth roll of the car. This function is used when instant launch is not desired or possible.

v. **Testing and Data Analysis**

A standard test was carried out to ensure the stability of the shifting system under extreme conditions. The conditions were simulated as the worst-case scenario on an actual race track. This test required the driver to accelerate from 25 to 75 km/h and decelerate back to 25 km/h. The car would be in first gear at the start of the test. During the acceleration phase, the driver would upshift twice, thus reaching third gear and 75 km/h at the end of this phase. The brakes would be applied and the car would be decelerated back to 25 km/h and first gear, thereby requiring two downshifts. A shift pair was achieved by performing this exercise and performance of the shifting system while upshifting and downshifting was evaluated. During the acceleration phase the drop in RPM while the velocity is still increasing denotes an upshift. At this point, the plot of vehicle velocity was checked for any flatlining or drop which would suggest as low shift or lack of "seamless" power delivery while shifting through gears. As evident from the graph, the velocity continued to rise without any sign of flatlining. During the deceleration phase, the vehicle was observed for any signs of instability that may be caused due to sudden engagement of the clutch after the downshift. Also, the time of the downshift was measured by measuring the duration of flatlining of RPM as that is the time for which the clutch was disengaged. Another test was conducted wherein the vehicle accelerated slowly on partial throttle from 25 to 60 km/h. This test was done primarily to test the stability of the vehicle while upshifting while slowly accelerating on partial throttle and while slowly decelerating with partial brake application. The speed of shift was controlled by the GCU code, and hence remained the same even on partial throttle and braking and could induce some instability. The car was observed to be stable and drivable with the only characterizing difference being the small spike in RPM while downshifting as opposed to the test where full brakes were applied.

7 Conclusion and Future Scope

The pneumatic shifting system is chosen from a set of available alternatives which has been developed for the sequential gear box. The required performance parameters in shift times and ease of operation have been achieved. Upshifts and downshifts can be performed in 87.5 ms and 140 ms, respectively, using the developed system. The gear control unit has been programmed and this system is ready for implementation on any similar sequential gearbox with modifications in the valve actuation times, piston diameters, and piston stroke. Shift speed during heavy acceleration and braking and stability during slow acceleration and deceleration were found to be satisfactory with the system giving almost seamless power delivery during acceleration as evidenced by no flatlining or decrease of velocity between upshifts.

References

1. Eckert JJ, Corrêa Fernandac C, Santiciolli Fabio M, Costa Eduardo S, Dionísio Heron J, Dedini Franco G (2015) Gear shifting strategies co-simulations to optimize vehicle performance and fuel consumption. In: Multibody mechatronic systems, pp 143–152. Springer International Publishing
2. Ping C (2004) Shift-time limited acceleration: final drive ratios in formula SAE. SAE technical paper 2004-01-3554. https://doi.org/10.4271/2004-01-3554
3. Meng F, Gang T, Huiyan C (2015) Smooth shift control of an automatic transmission for heavy-duty vehicles. Neurocomputing 159:197–206
4. Goel V, Riya KP, Shikha P, Tanushree PD, Nath V (2019) Design of smartphone controlled robot using bluetooth. In: Nath V, Mandal J (eds) Nanoelectronics, circuits and communication systems. Lecture notes in electrical engineering, vol 511, pp 557–563. Springer, Singapore. https://doi.org/10.1007/978-981-13-0776-8_52
5. Juels A, Rivest RL, Szydlo M (2003) The blocker tag: selective blocking of RFID tags for consumer privacy. In: Proceedings of 8th ACM conference computer and communication security, pp 103–111
6. Engels DW, Sarma SE The reader collision problem, Nov 2001
7. Want R (2004) Enabling ubiquitous sensing with RFID. Computer 37(4):84–86
8. Beneli G, Pozzebon A (2013) RFID underwater: technical issue and applications. Intech, pp 379–395

Performance Analysis of Tunable Third- and Fifth-Order BandPass Filter Using an Active Inductor

Suman Nehra, P. K. Ghosh and Meena Singh

Abstract CMOS innovation has turned out to be overwhelming over bipolar innovation for analog circuits design in mixed signal system. CMOS technology has become dominant over bipolar technology for analog circuits. So proposed active floating inductor and active load resistor is used for the implementation of third- and fifth-order bandpass filter. Proposed active floating inductor is designed with CMOS and current sources. The tunability can obtain by varying the transconductance of inductor which in turn is controlled by the externally controlled bias current and bias voltage. Simulation results show that proposed bandpass filter will provide a large bandwidth and less power consumption than the filter designed with conventional inductor. In this paper, the simulations are obtained by using 65 nm CMOS technology on Tanner EDA tool 13.0.

Keywords Complementary metal oxide semiconductor (CMOS) · Bandpass filter (BPF) · Low-pass filter (LPF) · High-pass filter (HPF) · Active resistor · Power–delay product (PDP)

1 Introduction

Wireless communication system is on great demand nowadays; due to this, it increases the integrated CMOS modules. Portable systems should be of low cost, low power consumption, and high frequency range. Improvement of VLSI technology, associated with the demand for signal processing integrated into a single chip,

S. Nehra · P. K. Ghosh · M. Singh (✉)
Department of Electronics and Communication Engineering, College of Engineering and Technology, Mody University of Science and Technoloy, Lakshmangarh, India
e-mail: meena71singh@rediffmail.com

S. Nehra
e-mail: nehra.sumanvlsi@gmail.com

P. K. Ghosh
e-mail: pkghosh.ece@gmail.com

© Springer Nature Singapore Pte Ltd. 2019
V. Nath and J. K. Mandal (eds.), *Proceedings of the Third International Conference on Microelectronics, Computing and Communication Systems*,
Lecture Notes in Electrical Engineering 556, https://doi.org/10.1007/978-981-13-7091-5_40

has extremely good ability for design of analog circuits. Mostly, the VLSI circuits consist of amplifiers, filters, oscillators, low-noise amplifier, ADC, and DAC [1, 2]. The key driving factor filters are high gain, high packing density, low power consumption, easy in designing, etc. The most important part in analog circuit that should be integrated is analog filters [3, 4].

The performance of general-purpose processor and digital signal processing unit has been increased by scaling of CMOS technologies. Advanced and scalable CMOS technologies provide low cost, high integration, and good reliability which form digital and analog circuits in a single chip [5, 6]. Filters especially analog filters have a wide range of applications in different areas such as in control system, communication system, military, radar, medical instruments, and industrial electronics [7, 8]. Filter is a circuit that transforms an electrical signal at input in such a manner so that the output signal has specified features which may be in terms of frequency or in time domain depending upon the application. Filter is the circuit that works on signal in a frequency-dependent manner [9, 10].

Filters can be made from passive components and from active components [11]. Filters that are fabricated from active components have large number of advantages over filters fabricated with passive components [12, 13]. Passive inductor is large in size, and unable to work at moderate frequency range; standard values are not very close to each other [14, 15]. So active inductor is preferred for the designing of filters.

The organization of this paper is described below: In Sect. 2, we describe the proposed filter design using active inductor; Sect. 3 gives the experimental results. Section 4 concludes the paper.

2 Proposed Filter Design

Filters can be classified in many different ways, for example, according to the function they are to perform, and in terms of range of frequencies, as passband and stopband. An ideal filter has amplitude response unity for a given range of frequencies known as passband and zero for other range of frequency is known as stopband.

A. *Proposed Active Inductor*

Proposed differential floating active inductor was designed with the help of CMOS and current sources that is shown in Fig. 1. It was designed with seven CMOS which include four NMOS from M1 to M4 and three PMOS from M5 to M7. Cross-coupled pair was formed by M1 and M2, from DC point of view and transistors M3 and M4 are in common drain configuration [16].

The circuit function is an inductor at V_{in}. At the quiescent bias point, transistors M1 to M4 are saturated. Due to control voltages V_b and V_{bb} at the gates transistors M5, M6, and M7, they work in linear region or in saturation region. Therefore, at desired bias point, M5, M6, and M7 are modeled as gds5, gds5′ gds6, and gds7 representing the drain conductance. In this paper, conventional inductor is the inductor that was proposed by Lu et al. [17].

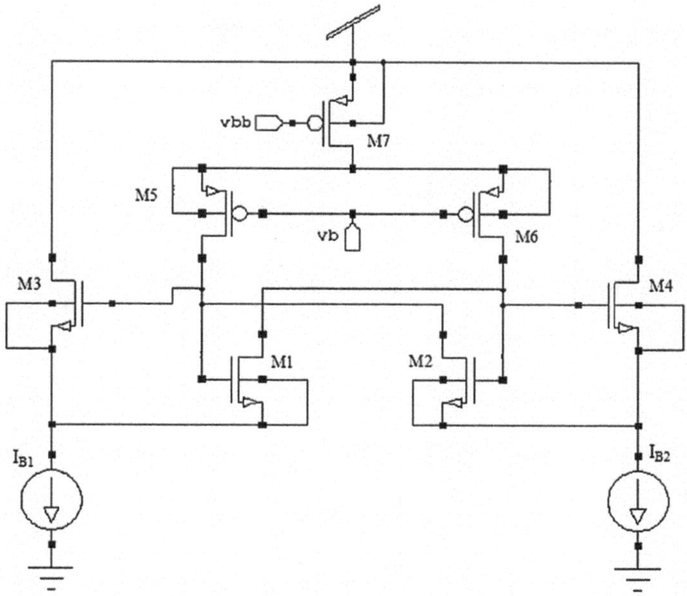

Fig. 1 Proposed active inductor

In simplified manner, small-signal model for the proposed active inductor is shown in Fig. 2a, and equivalent impedance for active inductor is shown in Fig. 2b. This analysis is done for the approximate analysis of a circuit.

For an input current I_{in}, the input voltage is V_{in}; for the differential port, input impedance Z_{in} is given below:

$$Z_{in} = \frac{v_{in}}{i_{in}} = \frac{\left[j\omega\left(C_{gs1} + C_{gs3}\right) - g_{m1} + g'_{ds5}\right]}{g'_{ds5}\left[g_{m1} + g_{m3} + j\omega\left(C_{gs1} + C_{gs3}\right)\right]} \tag{1}$$

By solving it with approximation, the value of inductance L_{eq} can be expressed as

$$L_{eq} = \frac{\left(C_{gs1} + C_{gs3}\right)}{g'_{ds5}\left(2g_{m1} + g_{m3} - g'_{ds5}\right)} \tag{2}$$

where g'_{ds5} is expressed as

$$g'_{ds5} = \frac{(g_{ds7} + g_{ds6}M_1)g_{ds5}}{g_{ds7} + g_{ds6}M_1 + g_{ds5}} \tag{3}$$

where g_{ds6}M1 and g_{ds6}M2 are given as

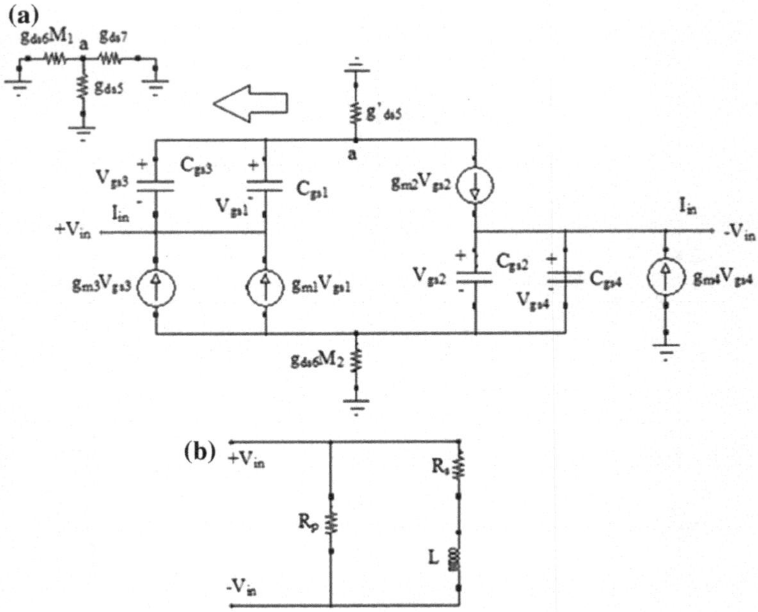

Fig. 2 **a** Proposed active inductor with small-signal model. **b** Proposed active inductor with equivalent circuit

$$g_{ds6}M_1 = \frac{g_{ds6}}{1-K} \tag{4}$$

and

$$g_{ds6}M_2 = g_{ds6}\frac{K}{1-K} \tag{5}$$

The Miller constant K is expressed as

$$K = \frac{v_2}{v_1} \tag{6}$$

The equivalent inductance depends on C_{gs1}, C_{gs3}, g_{m1}, g_{m3}, and g'_{ds5}; this was observed from Eq. (2). By manipulating the drain conductance by gate voltage, tuning of inductance can be achieved. Therefore, for tuning of active inductor, V_b and V_{bb} can be used as the controlling element.

B. *Bandpass Filter*

Bandpass filter passes a band of frequencies and rejects the frequencies outside that band. Cascade combination of a high-pass filter and a low-pass filter results into bandpass filter. Filters can be designed up to any limit their designing will depend on its application in which it was used. Higher order filter is complex in designing

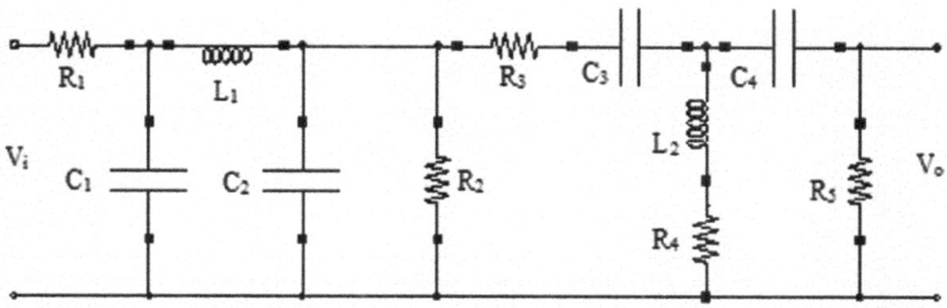

Fig. 3 Third-order bandpass filter topology

and very expensive so as we increase the order it will also increase the number of components used in it; due to this, size of the filter will also increase [18, 19].

a. *Third-order bandpass filter*

Figure 3 shows the circuit for third-order bandpass filter using passive components. It is created by cascade combination of third-order LPF and third-order HPF.

Transfer function of third-order bandpass filter G(s) is

$$G(s) = G_1(s)G_2(s) \tag{7}$$

where $G_1(s)$ is the transfer function of third-order low-pass filter and $G_2(s)$ is the transfer function of third-order high-pass filter.

Transfer function of third-order LPF is

$$G_1(s) = \frac{V_O(s)}{V_i(s)} = \frac{R_2}{(s^2R_2LC + sL + R_2)(1 + sR_1C)} \tag{8}$$

Transfer function of third-order HPF is

$$G_2(s) = \frac{s^2C^2R_2(sL + R_2)}{(sCR_2 + 1)\left[s^2LC + sC(R_1 + R_2) + 1\right]} \tag{9}$$

Transfer function of third-order BPF according to Eq. 1 is

$$\frac{s^2C^2R_2^2(sL + R_2)}{(sCR_2 + 1)(1 + sR_1C)(s^2R_2LC + sL + R_2)\left[s^2LC + sC(R_1 + R_2) + 1\right]} \tag{10}$$

The schematic for third-order bandpass filter using active components is shown in Fig. 4.

b. *Fifth-order bandpass filter*

To show the flexibility of active inductor in filters, fifth-order bandpass filter has been designed. Figure 5 shows the circuit for fifth-order bandpass filter using passive components.

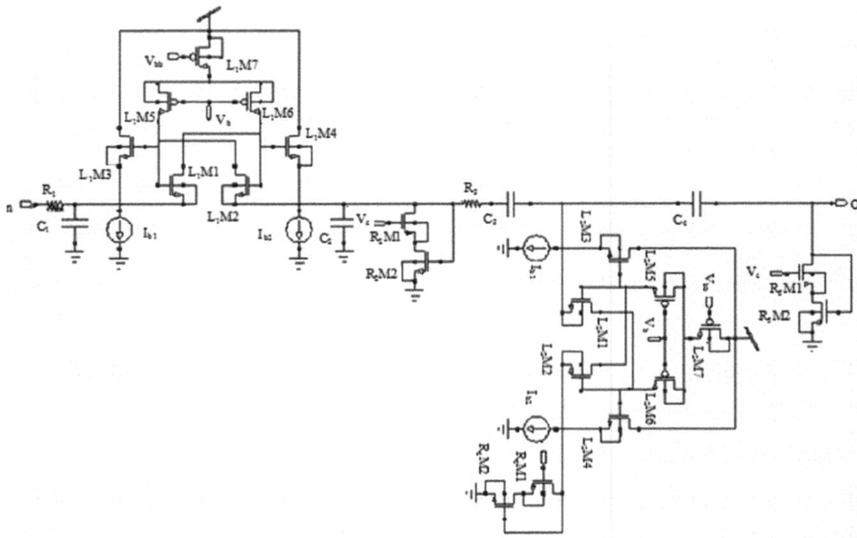

Fig. 4 Third-order BPF using active inductor

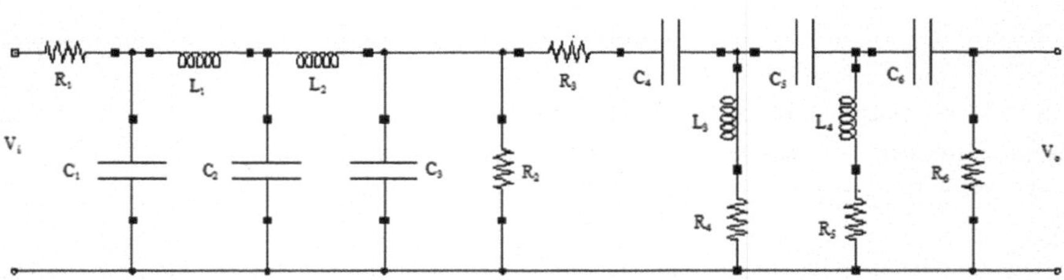

Fig. 5 Fifth-order BPF topology

Transfer function of fifth-order low-pass filter is

$$G_1(s) = \frac{R_2}{(s^2R_2LC + sL + R_2)(1 + s^2LC)(1 + sR_1C)} \quad (11)$$

Transfer function of fifth-order high-pass filter is

$$G_2(s) = \frac{s^2C^2R_2(sL + R_2)^2}{(s^2LC + sCR_2 + 1)[s^2LC + sC(R_1 + R_2) + 1](sCR_2 + 1)} \quad (12)$$

Transfer function of fifth-order bandpass filter according to Eq. 1 is

$$\frac{s^2C^2R_2^2(sL + R_2)^2}{(s^2LC + sCR_2 + 1)[s^2LC + sC(R_1 + R_2) + 1](sCR_2 + 1)(s^2R_2LC + sL + R_2)(1 + s^2LC)(1 + sR_1C)} \quad (13)$$

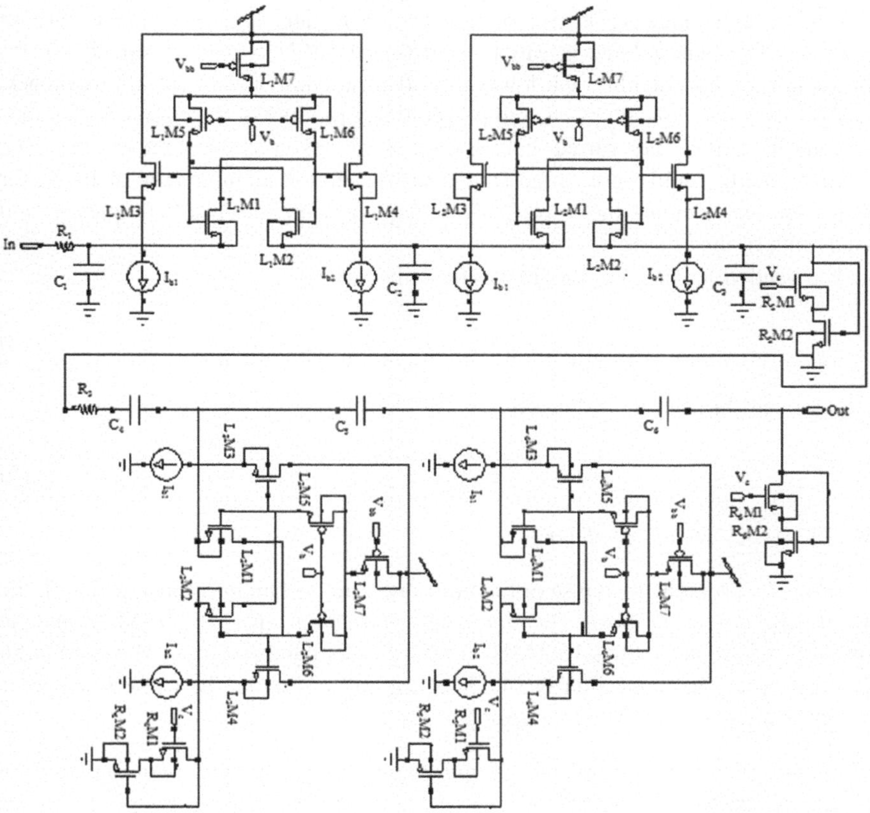

Fig. 6 Fifth-order BPF using active components

The schematic for fifth-order bandpass filter using active components is shown in Fig. 6.

3 Experimental Results

This segment talks about the outcomes got by the schematic-level simulations, and the simulations are done with 65 nm CMOS technology on Tanner EDA tool. In the design of active filters, the value of all the capacitances used in the circuit is taken as 1 p, source resistance is of 1 KΩ, supply voltage is of 1.0 V, and the bias voltage V_{bb} is of 0.5 V. In order to demonstrate the electronically tunable property of the inductor, bias voltage and bias current are varied to study the variation of transconductance of the CMOS.

When bias voltage Vb is varied, bias current I_{b1} and I_{b2} are taken as constant (10 μA). The bias voltage is varied from 0.7 to 0.9 V with the step of 0.05 V; for each step, we calculate the lower cut-off frequency, higher cut-off frequency, bandwidth, power consumption, and power–delay product of the circuit. When bias currents I_{b1} and I_{b2} are varied, bias voltage Vb is taken as constant (0.7 V). The bias currents I_{b1} and I_{b2} are varied from 5 to 15 μA with an increment of 1 μA; for each step, we calculate the lower cut-off frequency, higher cut-off frequency, and bandwidth of the circuit.

The bandwidth (BW) of the filter is described as

$$BW = f_h - f_l \tag{14}$$

f_o on a logarithmic scale is defined as

$$f_o = \sqrt{f_h f_l} \tag{15}$$

A. *Third-order bandpass filter*

Figure 7 shows the response of the third-order active bandpass designed with the specifications as cited above. The higher cut-off frequency f_h is 163.96 MHz, and the lower cut-off frequency f_l is 16.2 MHz for third-order bandpass filter. The bandwidth for third-order bandpass filter is 147.34 according to Eq. 7. The f_o for third-order bandpass filter is 52.20 MHz according to Eq. 8.

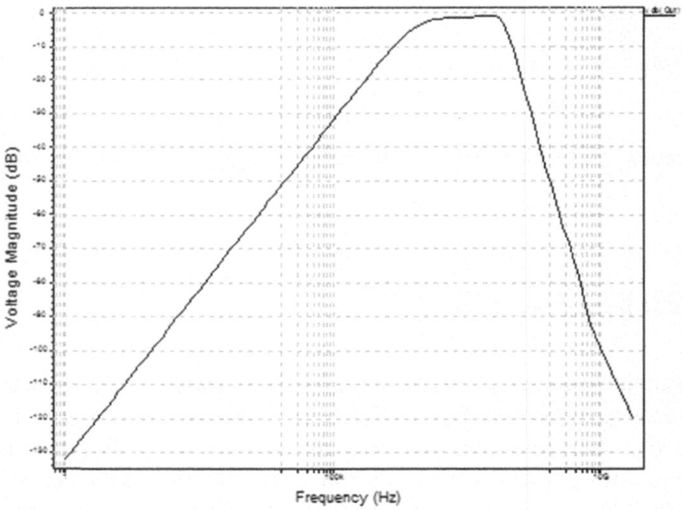

Fig. 7 Response of third-order BPF

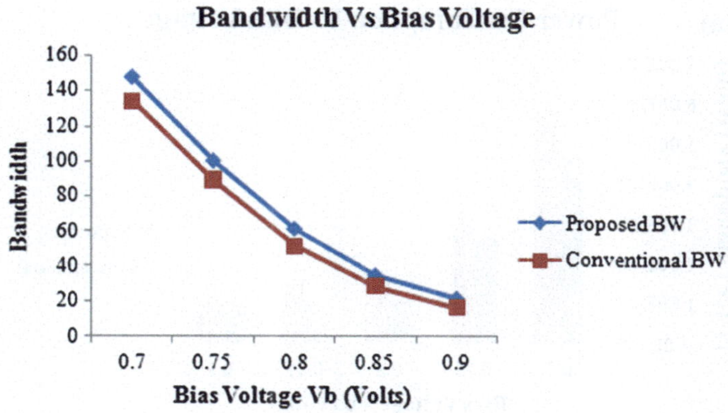

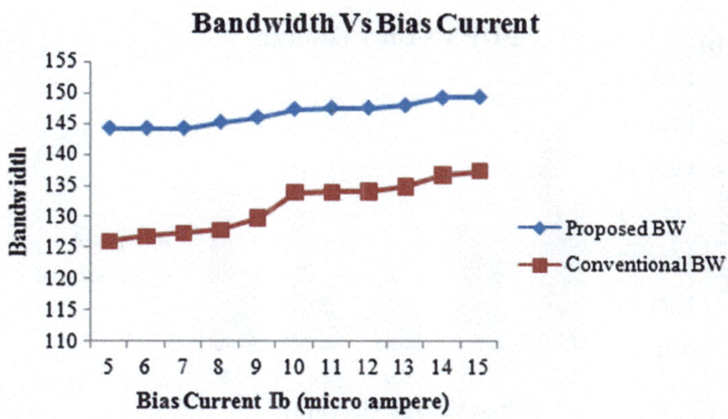

Fig. 8 Variation in bandwidth for third-order bandpass filter with **a** bias voltage, **b** bias current

The dependencies of the bandwidth of the filters on the bias voltage V_b and bias current I_b are shown in Fig. 8a and 8b, respectively. The same third-order conventional bandpass filter is also plotted on the same graph.

Figure 9a and Fig. 9b shows the variation in power consumption and power–delay product with bias voltage, respectively, for third-order proposed and conventional bandwidth pass filter at 65 nm technology. Filter designed with proposed active inductor shows the better result than the conventional one.

The comparative analyses of third-order proposed bandpass filter and conventional bandpass filter are given in Table 1.

B. *Fifth-order bandpass filter*

Figure 10 shows the frequency response of the fifth-order BPF designed with the specification as cited above. The higher cut-off frequency f_h is 128.34 MHz, and the lower cut-off frequency f_l is 52.23 MHz for fifth-order BPF. The bandwidth is 76.11 MHz, and the f_o is 81.87 MHz for fifth-order BPF.

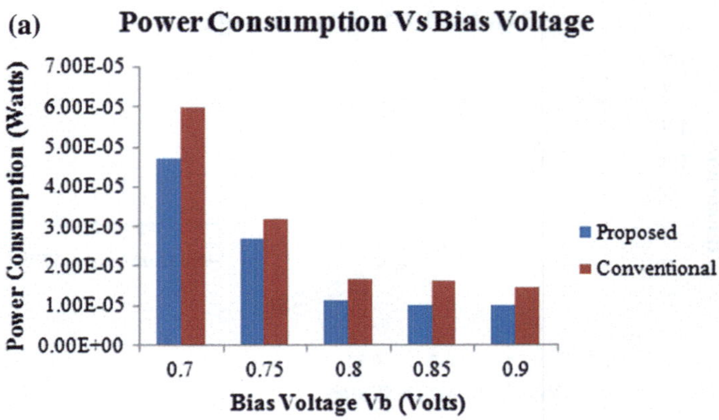

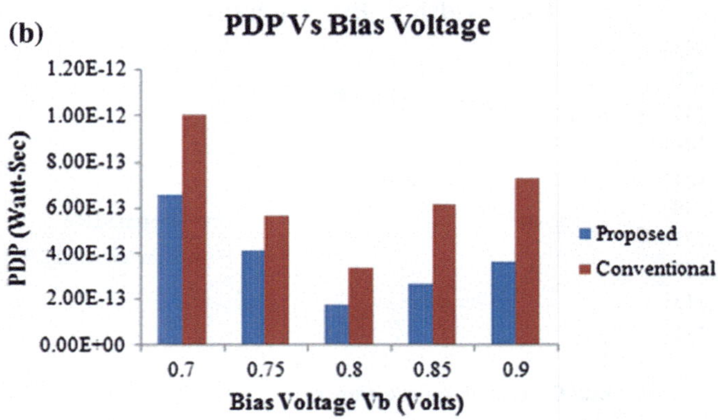

Fig. 9 **a** Variation in power consumption with bias voltage. **b** Variation in PDP with bias voltage for third-order BPF

Table 1 Comparative analysis of third-order proposed and conventional BPF

Parameters	Proposed bandpass filter	Conventional bandpass filter
Cut-off frequency f_h (MHz)	163.96	153.95
Cut-off frequency f_l (MHz)	16.62	20.02
Center frequency f_o (MHz)	52.20	55.52
Bandwidth (MHz)	147.34	133.93
Power consumption (W)	4.70E−05	5.99E−05
Delay (s)	1.41E−08	1.67E−08
Power–delay product (W-s)	6.63E−13	1.00E−12

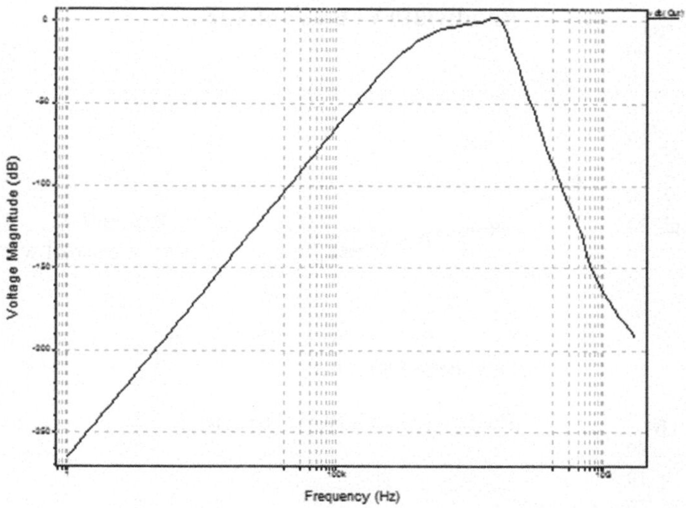

Fig. 10 Response of fifth-order BPF

Table 2 Simulation results of fifth-order proposed and conventional BPF

Parameters	Proposed bandpass filter	Conventional bandpass filter
Cut-off frequency f_h (MHz)	128.34	99.67
Cut-off frequency f_l (MHz)	52.23	58.45
Center frequency f_o (MHz)	81.87	76.32
Bandwidth (MHz)	76.11	41.22
Power consumption (W)	5.02E−05	6.72E−05
Delay (s)	2.06E−08	2.09E−08
Power–delay product (W-s)	1.04E−12	1.40E−12

The dependencies of the bandwidth of the filters on the bias voltage V_b and bias current I_b are shown in Fig. 11a and b, respectively. The same fifth-order conventional bandpass filter is also plotted on the same graph.

Figure 12a and Fig. 12b shows the variation in power consumption and power–delay product with bias voltage, respectively, for fifth-order proposed and conventional bandwidth pass filter. Filter designed with proposed active inductor shows the better result than the conventional one.

The simulation results of fifth-order proposed and conventional BPF are given in Table 2.

C. *Bandpass filter as a function of filter order*

The passband frequency decreases as we increase the order of the filter. So the stopband for the bandpass filter approaches to the ideal curve that is brick wall curve as we increase the order of the filter. Now in this section, we will give the comparison

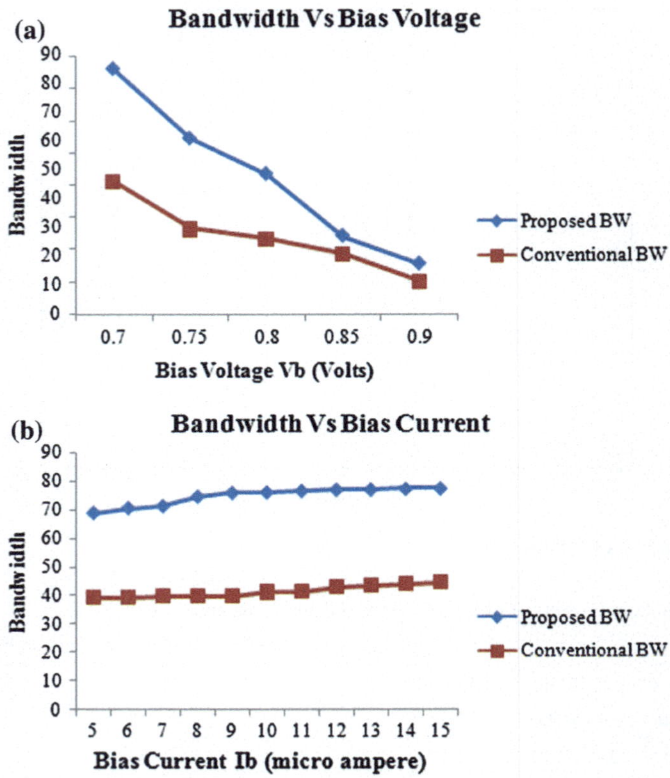

Fig. 11 Variation in bandwidth for fifth BPF with **a** bias voltage, **b** bias current

results according to the order of the filter a 65 nm technology. Figure 13 shows the frequency response of the filters according to the order of the filter. Figure shows that curve moves towards the ideal response according to the order of the filter.

Variation in bandwidth with bias voltage and bias current according to the order of filter is shown in Fig. 14a and Fig. 14b, respectively. The comparison result shows that bandwidth decreases as we increase the order of filter.

Figure 15a shows the variation in power consumption with bias voltage, and variation in power–delay product with bias voltage is shown in Fig. 15b, according to the order of filter.

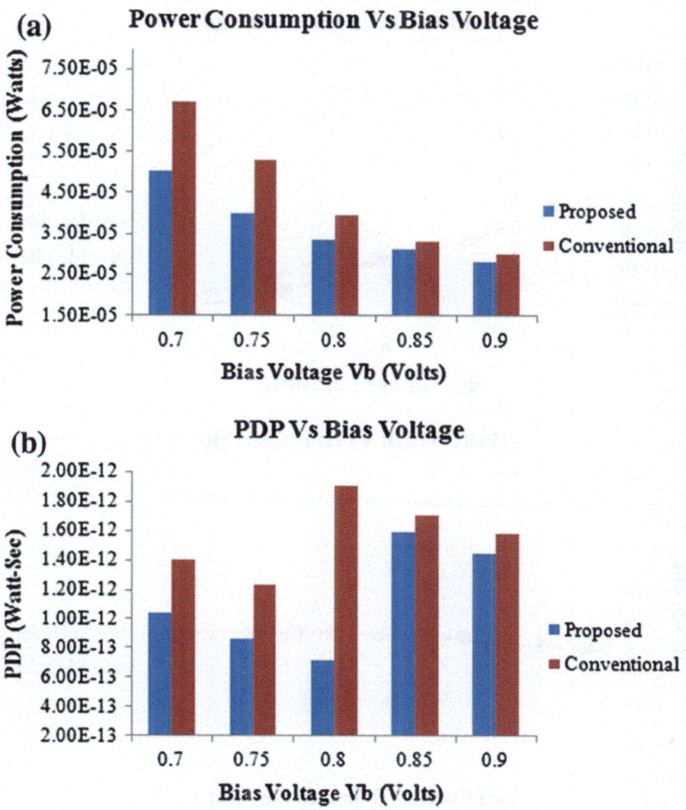

Fig. 12 **a** Variation in power consumption with bias voltage, **b** variation in PDP with bias voltage for fifth-order BPF

Fig. 13 Response of BPF as a function of filter order

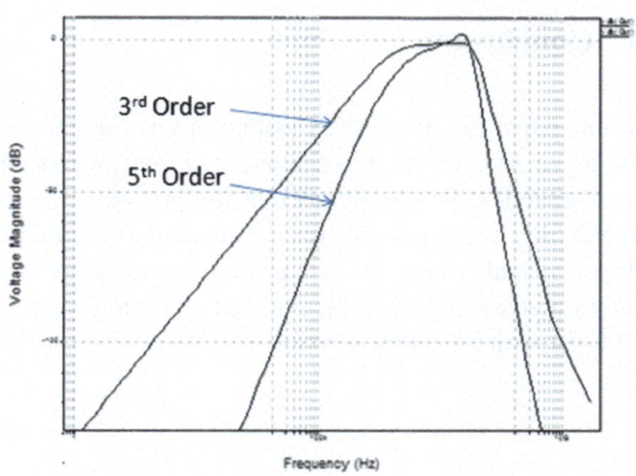

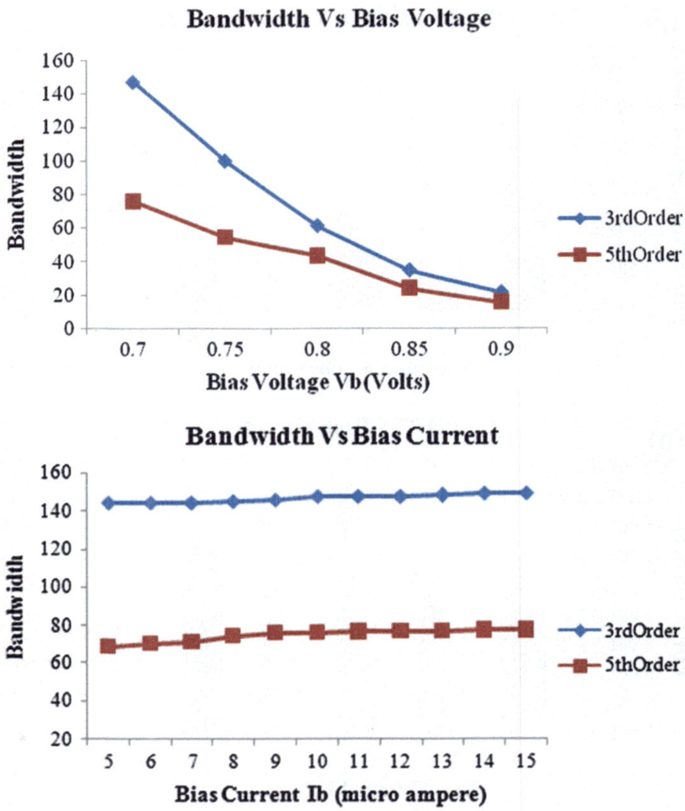

Fig. 14 Variation in bandwidth according to the order of filter with **a** bias voltage, **b** bias current

4 Conclusion

In this paper, 65 nm CMOS technology is used for the simulation of circuit. With the assist of external bias voltage and bias current, the inductance of an inductor is regulated. It was analyzed that the proposed circuit provides enough accuracy; due to this, it is applicable for the designing of third- and fifth-order bandpass filter. Experimental results show that the filter designed with proposed floating inductor shows better results than the convention filter. Proposed floating inductor uses seven CMOSs with two current sources.

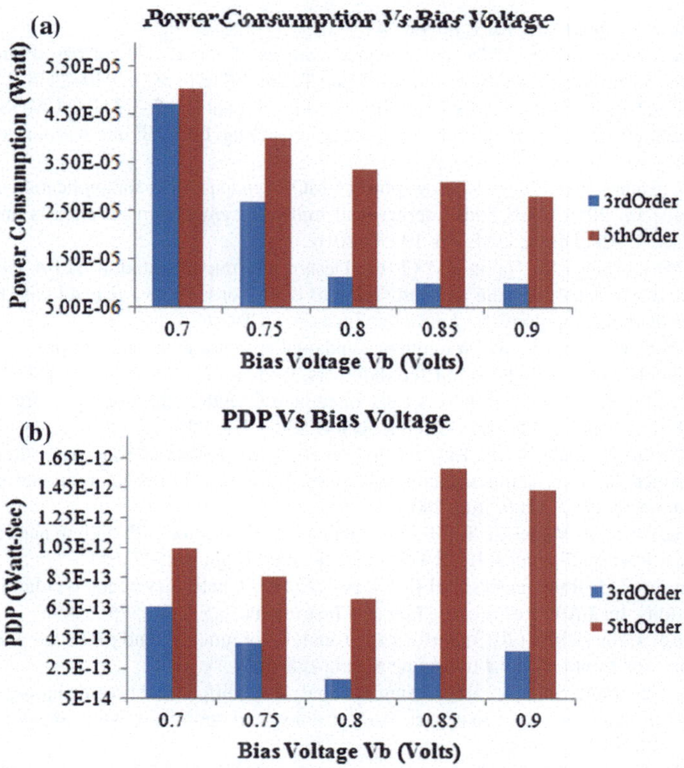

Fig. 15 According to the order of filter variation in **a** power consumption, **b** PDP with bias voltage

Acknowledgements The authors Suman Nehra and P. K. Ghosh thankfully acknowledge the authority of College of Engineering and Technology, Mody University of Science and Technology, Lakshmangarh for providing the opportunity to use the resources of the Institute.

References

1. Tomar A, Pokharel R, Kanaya H, Yoshida K Design of digitally controlled LC oscillator with wide tuning range in 0.18 μm TSMC CMOS technology. In: Microwave conference, 2008. APMC 2008. Asia-Pacific, pp 1–4, Dec 2008
2. Ferri G, Stornelli V, di Simone A (2011) A CCII-based high impedance input stage for biomedical applications. J Circuits Syst Comput 20(8):1441–1447
3. Heydarzadeh S, Torkzadeh P (2013) 1 GHz CMOS band-pass filter design using an active inductor and capacitor. Amer J Electric Electron Eng 1(3):37–41
4. Krishnamurthy SV, El-Sankary K, El-Masry E (2010) Noise-cancelling CMOS active inductor and its application in RF band-pass filter design. Hindawi Publ Corp Int J Microw Sci Technol, Article ID 980957
5. Thanachayanont A, Payne A (1998) A 3-V RF CMOS bandpass amplifier using an active inductor. In: Circuits and systems, 1998. ISCAS '98. Proceedings of the 1998 IEEE international

symposium on monterey, vol 1, pp 440–443
6. Manjula J, Malarvizhi S (2013) Performance analysis of active inductor based tunable band pass filter for multiband RF front end. Int J Eng Technol (IJET) 5(3):2930–2938
7. Stornelli V, Ferri G, Leuzzi G, De Marcellis A (2006) A tunable 0.5–1.3 GHz CMOS 2nd order bandpass filter with 50 Ω input-output impedance matching. In: IEEE international symposium on circuits and systems (ISCAS), pp 1–4
8. Chu H-C, Huang Y-H, Hung C-C Low-power CMOS bandpass filter for application of cochlear prosthesis. In: 2016 IEEE 59th international midwest symposium on circuits and systems (MWSCAS), Abu Dhabi, UAE, 16–19 Oct 2016
9. Myo ZMM, Aung ZM, Naing ZM (2017) Design and implementation of BA Atsuwe, GA Igwue and AN Amah, Design and simulation of eighth-order active-R band pass filter using multiple feedback topology. Int J Adv Res Innov 5(1):39–41
10. Andriesei C, Goras L (2008) On frequency and quality factor independent tuning possibilities for RF band-pass filters with simulated inductors. Roman J Inf Sci Technol 11(4):367–382
11. Gao Z, Ma J, Yu M, Ye Y (2008) A fully integrated CMOS active band pass filter for multi band RF front-ends. IEEE Trans Circuits Syst II 55(8):718–722
12. Frye RC, Liu K, Badakere G, Lin Y A hybrid coupled—resonator band pass filter topology implemented on lossy semiconductor substrates. In: IEEE/MTT-S international microwave symposium, pp 1757–1760, June 2007
13. Leuzzi G, Stornelli V, Del Re S (2011) A tunable active inductor with high dynamic range for band-pass filter applications. IEEE Trans Circuits Syst II 58(10):647–651
14. Thangasamy V, Duraikannan S (2014) Λ-type 5.75 GHz Chebyshev bandpass filter for WLAN applications. Int J Adv Res Electric Electron Instrument Eng 3(1):6576–6589
15. Manjula J, Malarvizhi S (2013) Performance analysis of tunable band pass filter and VCO for multiband RF front end. Int J Intell Eng Syst 6(4):8–16
16. Nehra S, Ghosh PK, Soni P (2016) Design and analysis of differential floating inductor for filter. In: International conference on electrical, electronics, and optimization techniques (ICEEOT), pp 885–888. IEEE Explore
17. Lu L, Hsieh H, Liao Y (2006) A wide tuning-range CMOS VCO with a differential tunable active inductor. IEEE Trans Microw Theory Tech 54(9):3462–3468
18. Nehra S, Ghosh PK (2017) Design of second order low pass filter with differential floating inductor. Int J Control Theory Appl 10(33):121–128
19. Nehra S, Ghosh PK (2017) Design of 7th and 9th order low pass filter with higher cut-off frequency using 65 nm CMOS technology. Int J Control Theory Appl 10(33):107–119

Analysis of Impact of Circular Cut on Microstrip Patch Antenna at 2.49 GHz for S Band Application

Mitchell S. Prajapati, Abhishek Rawat, Ravi Parikh and Priyank Joshi

Abstract In this paper, the impact of circular cut on the geometry of microstrip patch antenna is analyzed. Initially, the geometry of microstrip antenna is prepared for the 2.49 GHz central frequency with the bandwidth of 50.4 MHz which is one of the operational bands of IRNSS receiver. In this band, it provides very low return loss and RHCP around 2.0 which is suitable for IRNSS receiver second band of operation 2.49 GHz. This antenna is fed by coaxial probe with proper impedance matching. Different circular cuts alter the key properties of this antenna. So, the comparative analysis is done by changing the radius of cuts at two diagonal angles at patch geometry to analyze the variation in important parameters of microstrip patch antenna.

Keywords Microstrip patch antenna · Ansoft HFSS 15

1 Introduction

Microstrip patch antenna is widely used in wireless communication application due to its several advantages as it is small in size, low in weight, easy to design, fabricate and install, and low in cost. In this paper, microstrip patch antenna is designed for S band from 2.4654 to 2.5158 GHz. This microstrip patch antenna can be used for satellite communication, amateur radio, amateur satellite, amateur television, and wireless networking applications.

The IRNSS is the Indian Regional Navigation Satellite System. It is an independently developed navigation system established and controlled by the Indian Space Research Organization (ISRO). IRNSS receiver is capable of receiving standard positioning service signal at L5 and S band frequency. Microstrip patch antenna can have different shapes of patch according to design and appropriate applications. A

M. S. Prajapati (✉) · A. Rawat · R. Parikh · P. Joshi
Institute of Infrastructure Technology Research and Management, Ahmedabad, India
e-mail: mishelprajapati@gmail.com

© Springer Nature Singapore Pte Ltd. 2019

V. Nath and J. K. Mandal (eds.), *Proceedings of the Third International Conference on Microelectronics, Computing and Communication Systems*, Lecture Notes in Electrical Engineering 556, https://doi.org/10.1007/978-981-13-7091-5_41

dielectric substrate is used between ground and conducting patches. The selection of dielectric constant of the substrate depends on the application of the antenna and parameter requirements. Low dielectric constant substrates enhance the antenna bandwidth and efficiency. High dielectric constant substrates cause surface wave excitation and provide low bandwidth. The loss tangent is chosen very small as it affects the efficiency of antenna [1].

Radiation of patch depends on its geometry and dimensions. The radiating patch is photo etched on dielectric substrate. So, in such a way a compact, lightweight microstrip patch antenna can be designed with good radiated power [2]. This paper contains various sections about microstrip patch antenna designing. Section 2 shows antenna geometry and its detailed structure. Section 3 shows comparison of various antenna parameters by changing the radius of cut in mm on patch.

2 Microstrip Patch Antenna Design

In this paper, the proposed antenna is having ground plane, Roger RT/duroid 5880(tm) substrate material, and a conducting patch with coaxial feed line. The basic geometry of microstrip patch antenna is given in Fig. 1.

Initially, the dimensions of microstrip patch antenna can be calculated by following equations:

The width of patch is given by Eq. (1) as given below:

$$w = \frac{c}{2f0\sqrt{(\frac{\varepsilon r+1}{2})}} \tag{1}$$

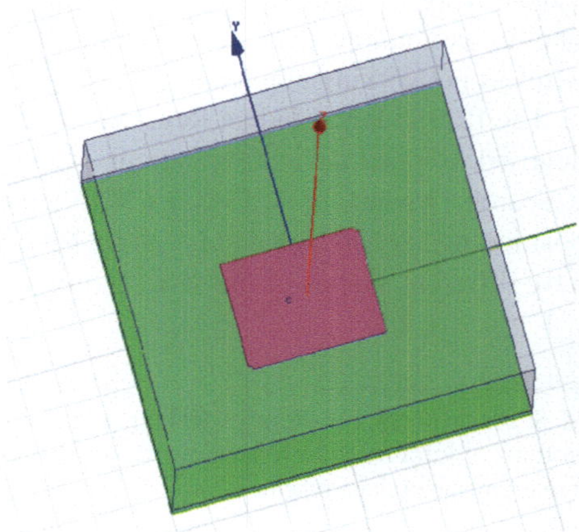

Fig. 1 Model of proposed microstrip patch antenna in HFSS

The length of patch is given by Eq. (2) as given below:

$$L = \frac{c}{2f0\sqrt{\varepsilon eff}} - 0.82\left(\frac{(\varepsilon eff + 0.3)(\frac{w}{h} + 0.264)}{(\varepsilon eff - 0.258)(\frac{w}{h} + 0.8)}\right) \tag{2}$$

Here, w is width of patch, C is speed of light, and εr is dielectric constant for substrate material, and its value is 2.2 and f0 is resonant frequency of antenna, L is the length of patch, εeff is the effective dielectric constant, and h is the height of dielectric substrate. There are four basic types of feeding techniques for microstrip patch antenna which are microstrip feed, coaxial feed, aperture coupling, and proximity coupling. Here, the patch is fed by a coaxial feed line, with the matching of impedance. The coaxial feed is simple to design, easy to fabricate, and easy to match impedance with impedance of antenna. The patch is cut at two diagonal angles to set the axial ratio around 2. The proposed antenna is designed using HFSS. Here, rectangular patch is used with the dimensions of 38 mm × 29 mm, and dimension of substrate is chosen as 100 mm × 90 mm with height of 3.2 mm to maximize the radiation from the patch. The length and width of ground plane remain same as substrate. Here, Roger RT/duroid 5880(tm) is used for substrate material, its relative permittivity or dielectric constant is 2.2, and loss tangent is 0.0009.

There are two ways to get circular or elliptical polarization for microstrip patch antennas. One is that two orthogonal patch modes are simultaneously excited with equal amplitude and ±90° out of phase using dual feed method and second is to cut the corners of patch in such a way to get the axial ratio as per our requirement by using single feed design [3, 4]. Single fed microstrip antenna is simple easy to fabricate and gives a simple structure. In proposed design, the cut in patch method is used to get axial ratio less than two with single feed technique. The following structure of microstrip patch antenna is designed and simulated in Ansoft HFSS 15. The dimensions and specification of proposed antenna are given in Table 1.

Table 1 Dimensions and specification of proposed antenna

Sr. No.	Antenna parameters	Value
1	Operating frequency	2.49 GHz
2	Substrate material	Roger/RT Duroid
3	Dielectric constant	2.2
4	Loss tangent	0.0009
5	Patch dimension	38 mm × 29 mm
6	Substrate dimension	100 mm × 90 mm × 3.2 mm
7	Cut radius on patch	0.25, 0.5, 1.0, 1.5, 2.0, 2.5, 3, 3.5 mm

Table 2 Comparison of different simulated results

Sr. No.	Cut radius (R) in mm	Return loss	Gain	VSWR	BW in MHz	Resonant frequency of antenna in GHz
1	0.25	−21.8737	7.4113	1.4031	49.8	2.4800
2	0.5	−20.5455	7.4145	1.6363	49.5	2.4900
3	1	−22.9331	7.4317	1.2414	50	2.4900
4	1.5	−20.9398	7.4134	1.5632	50.4	2.4900
5	2	−23.2232	7.3956	1.2005	51.3	2.5000
6	2.5	−22.7697	7.3941	1.2651	51.7	2.5100
7	3	−21.9807	7.4044	1.3859	51.6	2.5100
8	3.5	−23.0324	7.4438	1.2273	52.5	2.5200

3 Result and Discussion

The basic parameters for any antenna are its gain, bandwidth, radiation pattern, return loss, SWR, efficiency, beam width, directivity, input impedance, etc. which decides the performance of antenna in its real-world application. Here, proposed antenna is simulated in Ansoft HFSS 15, in which the parameters of antenna are calculated such as return loss, gain, bandwidth, axial ratio, and VSWR. The HFSS15-based model is proposed to measure different parameters for different radii of cuts on patch. Here, the radius of cut is varied from 0.25 to 3.5 mm and according to it, various plots are given below (Figs. 2, 3, 4 and 5).

Ansoft High-Frequency Structure Simulator (HFSS) is a software for simulating high-frequency electromagnetic fields. HFSS delivers 3-D full-wave accuracy for components to enable RF and high-speed design. Here, the following parameters are found by using HFSS software. Table 2 shows the value of various parameters according to change in radius of cut by using HFSS.

Above parameters are designed for microstrip patch antenna for S band application. The antenna operates in s band with the center frequency of 2.49 GHz. Comparison by changing the value of cut radius in mm is done and the results are shown. Here, it observed that as the cut radius increases bandwidth, return loss, gain, and VSWR vary nonlinearly and resonant frequency of antenna varies linearly.

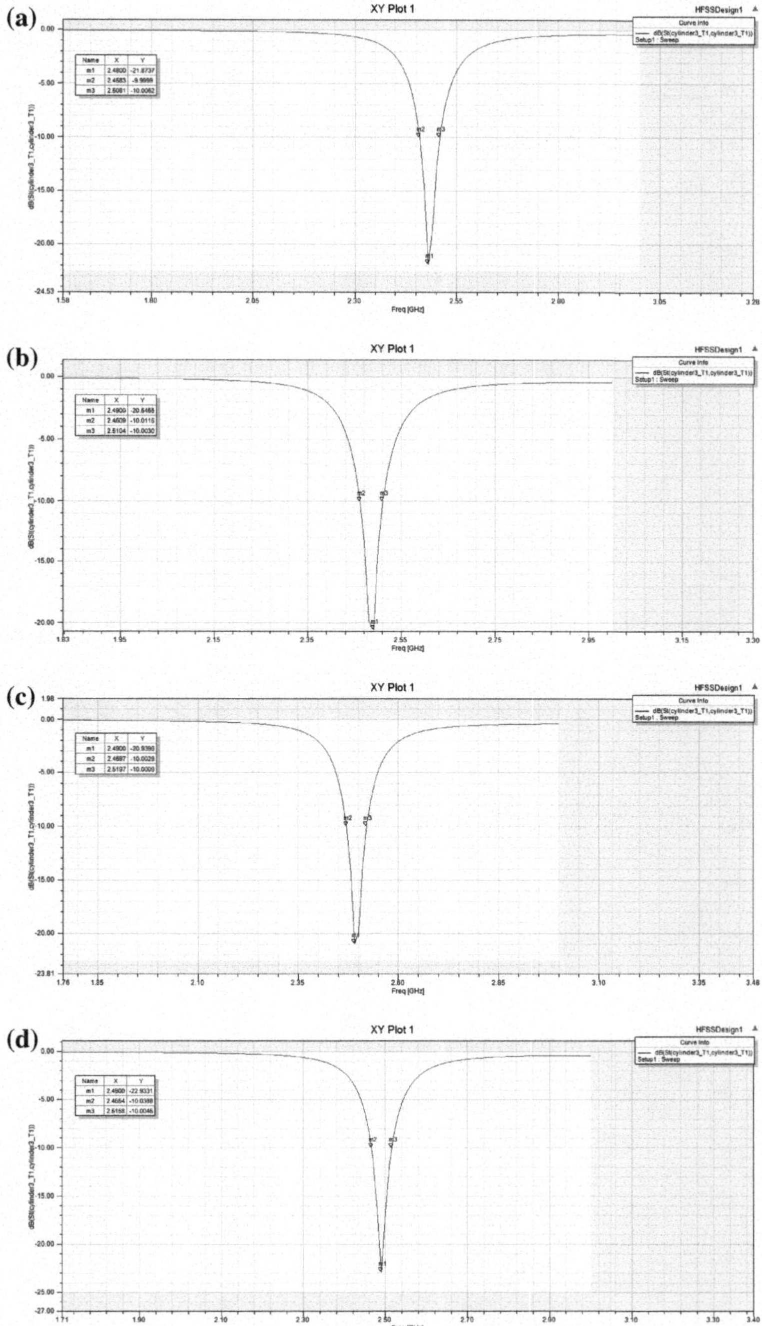

Fig. 2 XY plot by varying cut radius on patch (figure shows cut radius = 0.25 mm, 0.5 mm, 1 mm, 1.5 mm, 2 mm, 2.5 mm, 3 mm, and 3.5 mm, these are named as **a**, **b**, **c**, **d**, **e**, **f**, **g**, and **h**)

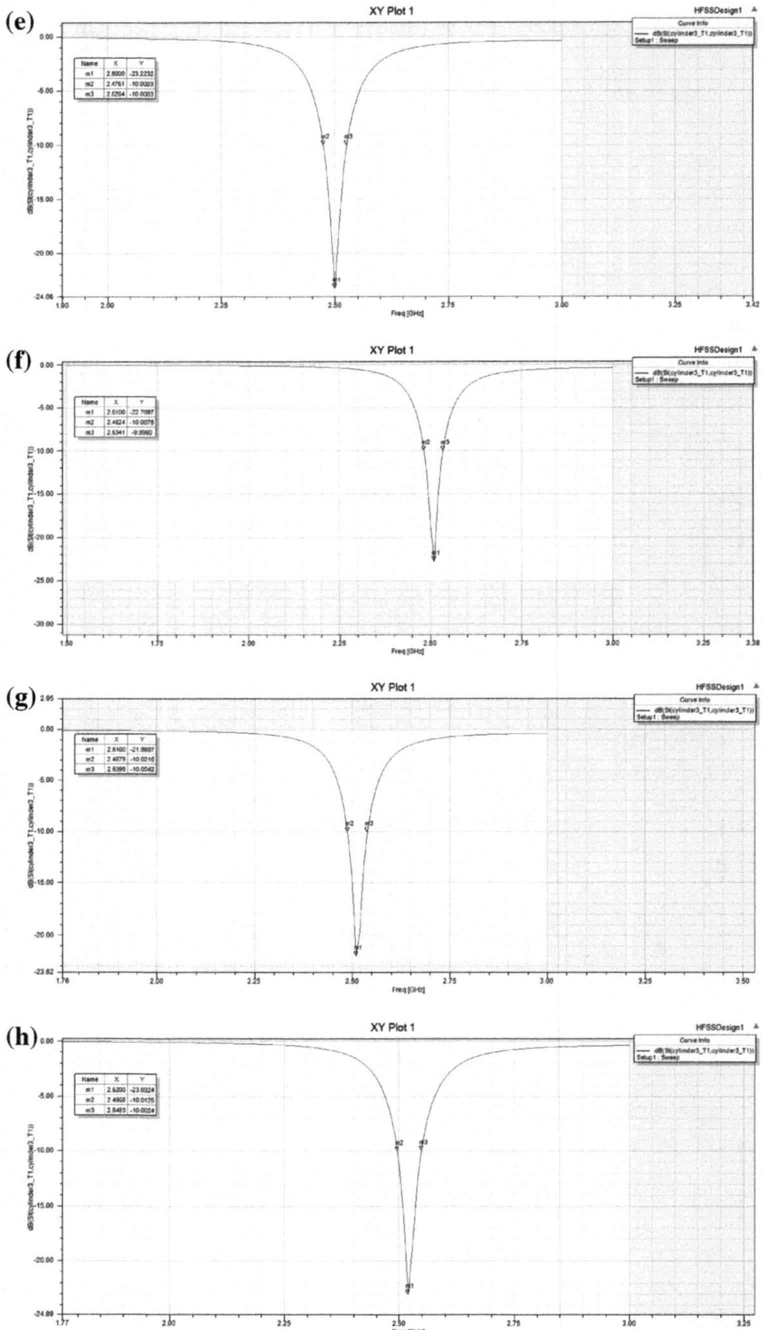

Fig. 2 (continued)

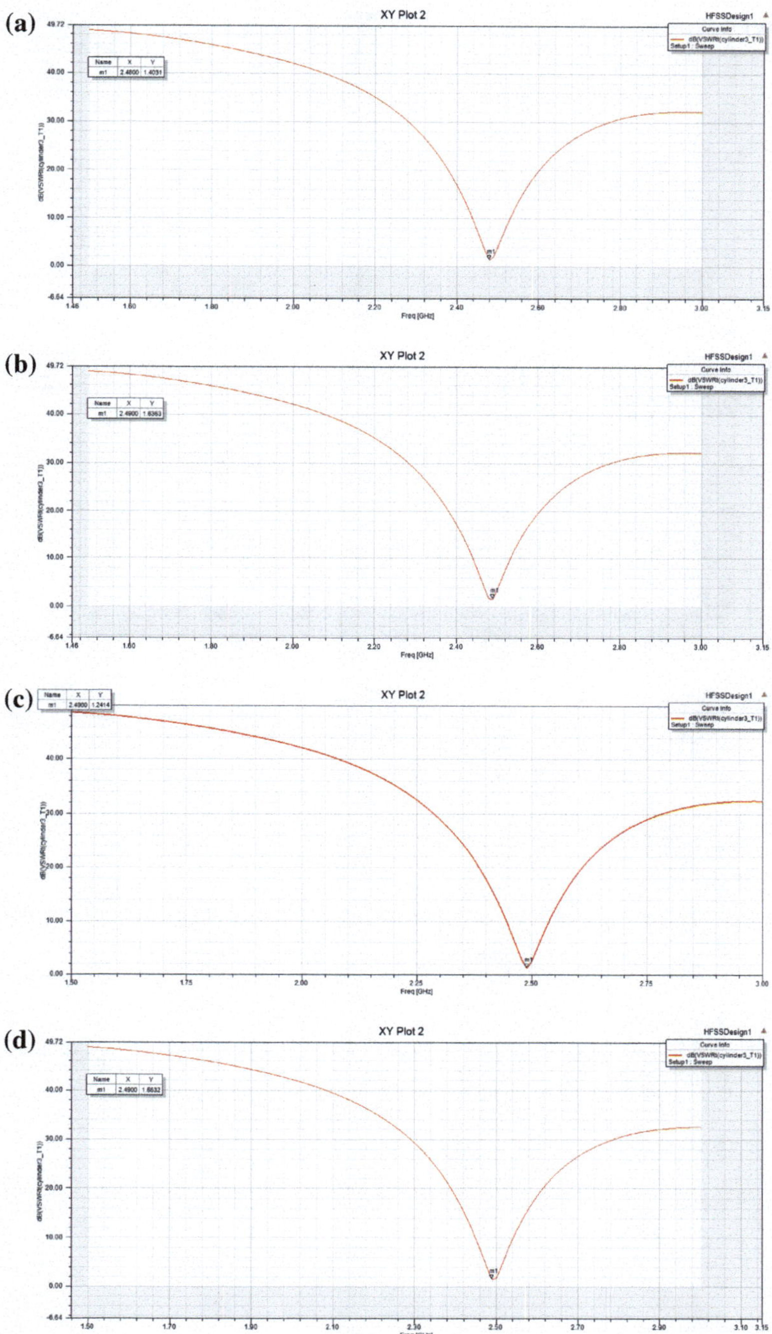

Fig. 3 VSWR plot by varying cut radius on patch (figure shows cut radius = 0.25 mm, 0.5 mm, 1 mm, 1.5 mm, 2 mm, 2.5 mm, 3 mm, and 3.5 mm, these are named as **a**, **b**, **c**, **d**, **e**, **f**, **g**, and **h**)

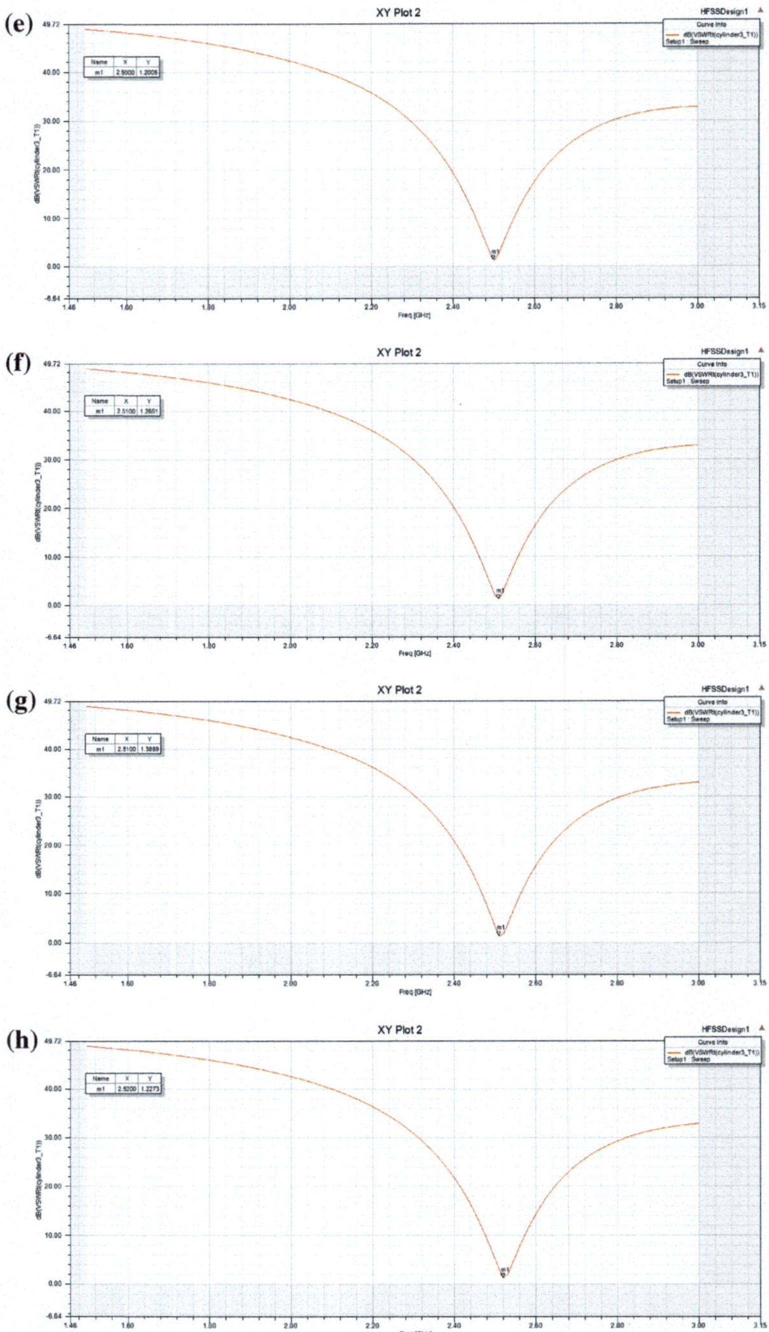

Fig. 3 (continued)

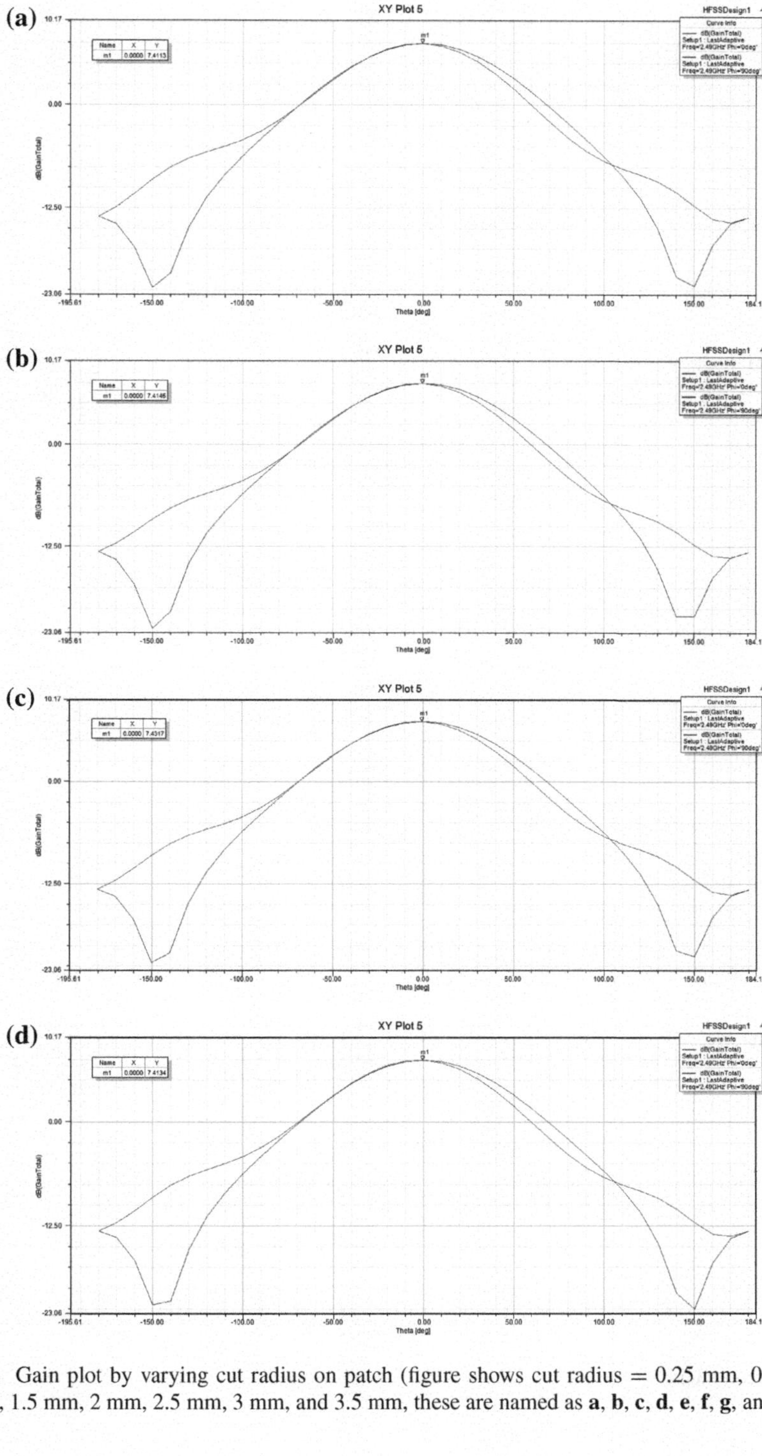

Fig. 4 Gain plot by varying cut radius on patch (figure shows cut radius = 0.25 mm, 0.5 mm, 1 mm, 1.5 mm, 2 mm, 2.5 mm, 3 mm, and 3.5 mm, these are named as **a**, **b**, **c**, **d**, **e**, **f**, **g**, and **h**)

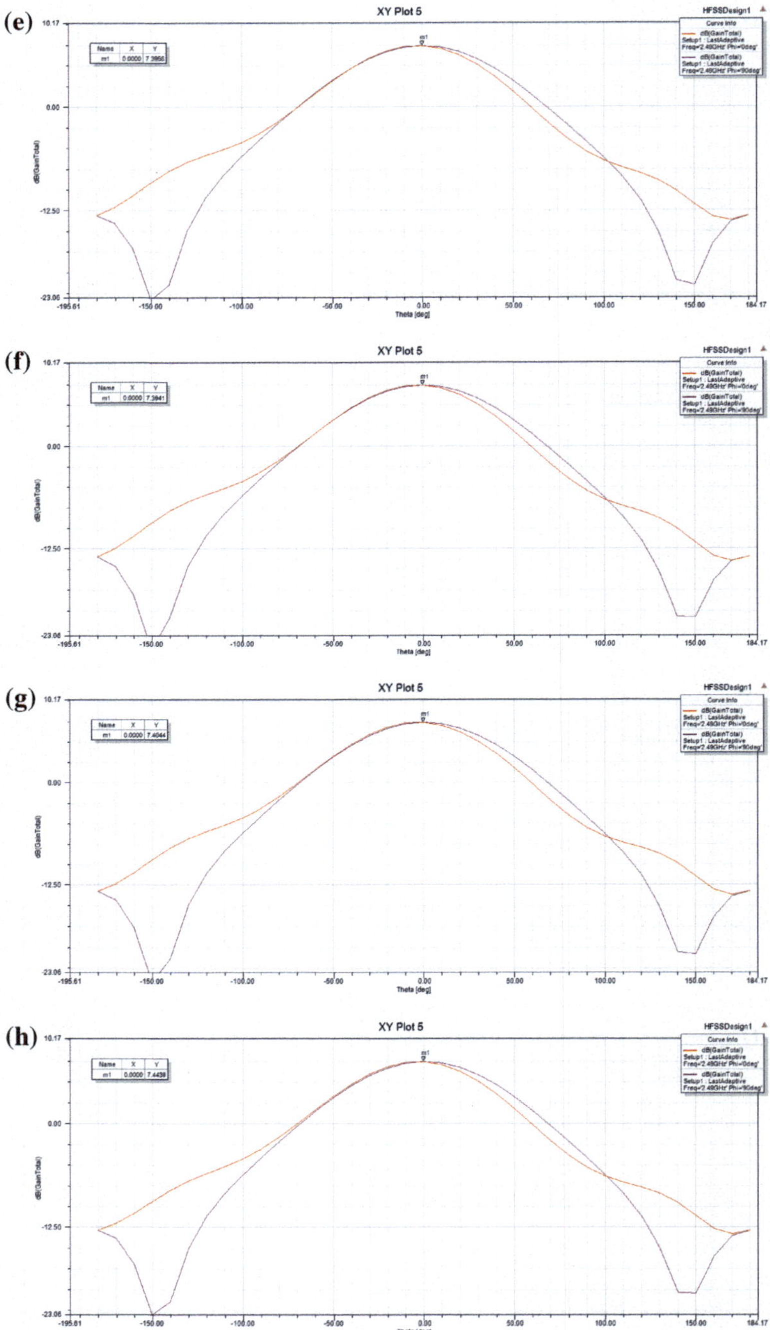

Fig. 4 (continued)

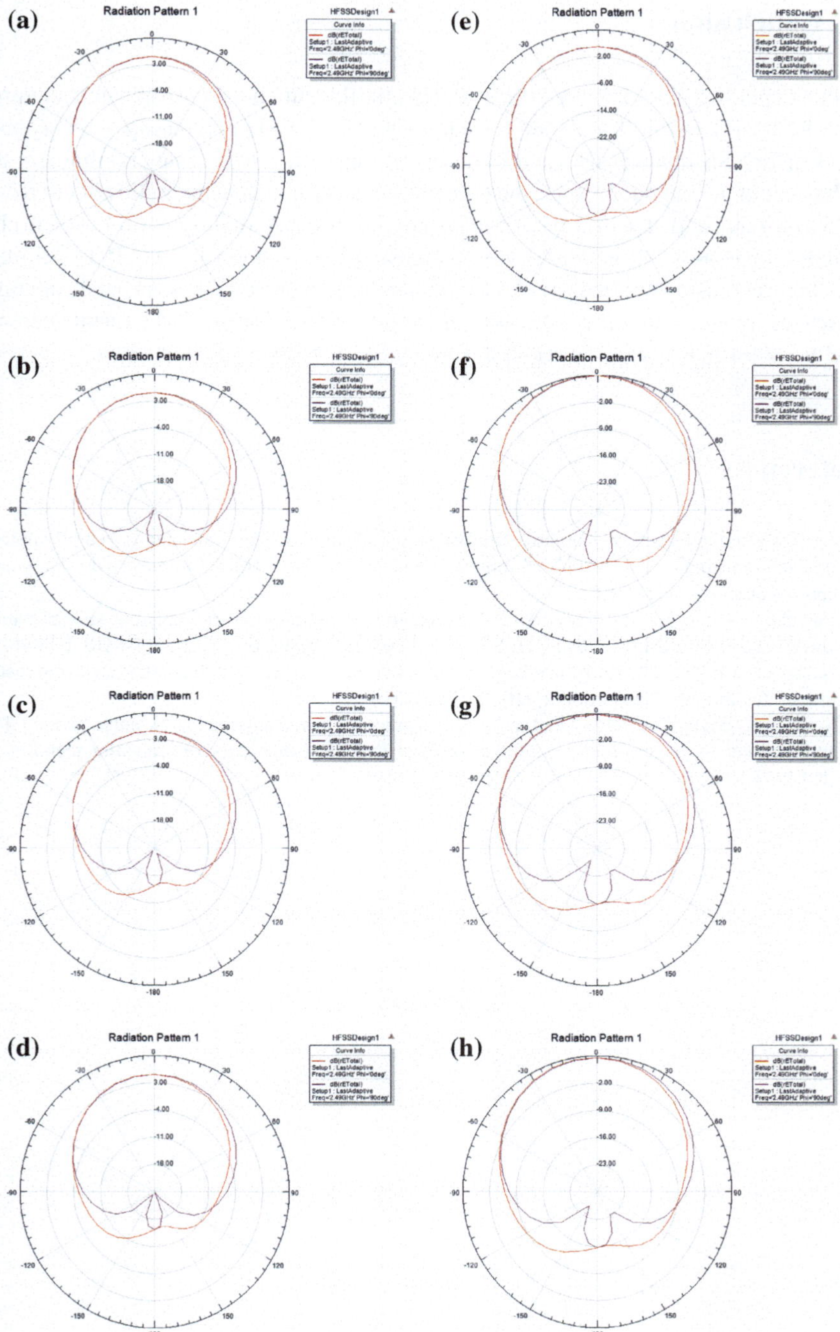

Fig. 5 Radiation pattern by varying cut radius on patch (figure shows cut radius = 0.25 mm, 0.5 mm, 1 mm, 1.5 mm, 2 mm, 2.5 mm, 3 mm, and 3.5 mm, these are named as **a**, **b**, **c**, **d**, **e**, **f**, **g**, and **h**)

4 Conclusion

In this paper, we have analyzed the effect of circular cut on microstrip patch antenna with Roger/RT duroid for resonant frequency of 2.49 GHz compatible with second band of IRNSS receiver. Here, it is observed that return loss varies nonlinearly as radius of cut on patch increases, the gain increases with increasing radius of cut from 0.25 to 1.5 mm and then decreases nonlinearly. Gain varies slightly with change in cut radius. The bandwidth increases with increasing radius of cut from 0.25 to 2.5 mm and then decreasing nonlinearly. And, resonant frequency of microstrip patch antenna increases by increasing the radius of cut from 0.25 to 3.5 mm. These results can be useful for further research work in designing of microstrip patch antenna.

References

1. Sandhya Reddy B, Senthil Kumar V, Srinivasan VV, Mehta Y (2015) Dual band circularly polarized microstrip antenna for IRNSS reference receiver. In: 2015 IEEE International microwave and RF conference (IMaRC)
2. Vayada MG, Vayada MG, Patel A (2015) Design and simulation of rectangular shaped patch antenna used for ISM band using HFSS. Int J Adv Res Comput Eng Technol (IJARCET) 4(4)
3. Narbudowicz AZ (2013) Advanced circularly polarised microstrip patch antennas. Dublin Institute of Technology. https://doi.org/10.21427/d7f31n
4. Sidhu AK, Sivia JS (2016) Double U slotted microstrip rectangular patch antenna for GPS applications. In: International conference on electrical, electronics, and optimization techniques (ICEEOT)—2016

FPGA Implementation of Adaptive Filtering Algorithms for Noise Cancellation—A Technical Survey

Pankaj Goel and Mahesh Chandra

Abstract In an environment where noise is our constant companion, quality of speech signal degrades considerably. Over a period of time, researchers have developed and implemented many algorithms to enhance the quality of speech signal. In this article, an attempt is made to present the significant work of researchers in the area of Field-Programmable Gate Arrays (FPGA) implementation of noise cancellation algorithms using adaptive filters. FPGA based hardware platforms have been commonly used to implement prototypes for active noise canceller (ANC) systems. The commonly used software tools for FPGA based implementations include MATLAB/Simulink, Modelsim, Xilinx System Generator, Xilinx ISE, Vivado design suite, and Altera Quartux II development platform. The performance of the algorithms achieved using FPGA based platforms give an insight of what may be achieved from the final commercial implementation. In this study, it has been observed that the performance of various adaptive filter structures varies considerably in terms of rate of convergence, throughput rate, improvement in signal-to-noise ratio (SNR), maximum clock speed, hardware complexity, power consumption and cost.

Keywords Active noise cancellation · Speech enhancement · Adaptive algorithm · DSP · FPGA · Hardware implementation

1 Introduction

The unwanted signal present in the information bearing signal is referred to as noise which degrades the signal strength in terms of quality and intelligibility. Noise can be broadly classified into two main categories, namely White Noise or Non-White

P. Goel (✉) · M. Chandra
Electronics and Communication Engineering, BIT Mesra, Mesra, Ranchi 835215, Jharkhand, India
e-mail: pankajgoel80@gmail.com

M. Chandra
e-mail: shrotriya@bitmesra.ac.in

© Springer Nature Singapore Pte Ltd. 2019

V. Nath and J. K. Mandal (eds.), *Proceedings of the Third International Conference on Microelectronics, Computing and Communication Systems*, Lecture Notes in Electrical Engineering 556, https://doi.org/10.1007/978-981-13-7091-5_42

Noise. White noise is spectrally flat but non-white is not. Furthermore, noise can either be stationary like computer's fan noise or non-stationary like Machine gun noise.

Various commonly used adaptive filters used to diminish the noise are Least Mean Square (LMS) algorithm, Normalized LMS algorithm (NLMS), Variable Step Size LMS algorithm (VSSLMS), Filtered-X LMS algorithm (FxLMS), Hybrid LMS algorithm and Recursive Least Square (RLS) algorithm.

The hardware implementation can be done using either Digital Signal Processors (DSPs) or Field-Programmable Gate Arrays (FPGAs). DSPs have advantages such as the use of C programming language and therefore more user-friendly, and relatively simple design flow, whereas FPGAs offer advantages such as high degree of parallelism and low power consumption.

In this article, the important work of researchers is presented in the field of FPGA-based hardware implementations of noise cancellation algorithms using adaptive filters. FPGA implementations of different algorithms are compared based on their maximum clock speed, resources required, power consumption and cost, and in the end, a conclusion is drawn.

2 Adaptive Filtering Algorithms

An adaptive filter is a filter capable of tracking the variations in the statistical characteristics of input data and capable of self-adjusting its parameters. In non-stationary environments where characteristics of the environment are unknown or changing with time, a fixed filter doesn't give good results. In such situations, an adaptive filter gives efficient results as an adaptive filter can track the changes in the environment and update the filter weight vector coefficients. Adaptive filters are usually more complex and difficult to implement as compared to non-adaptive systems, but such type of systems offers increased system performance whenever the input signal is time varying or its characteristics are unknown [1].

An active noise cancellation (ANC) system is one which is used to suppress the noise from the signal of interest. One of the popular algorithms used for ANC is Least Mean Square (LMS) algorithm devised by Widrow and Hoff in 1959. It is a widely used algorithm because of its simplicity. This algorithm belongs to the family of stochastic gradient algorithms involving filtering and adaptive process.

The ANC system consists of two types of inputs, one is a primary input also called as source signal $s(n)$ and the other is reference input or noise signal $x(n)$. The source signal gets distorted by a noise signal $x_1(n)$ which is usually correlated with noise signal $x(n)$. The reference signal can vary in amplitude, phase or time with respect to noise portion of the primary signal [2, 3].

The desired signal $d(n)$ is the result of addition of primary signal $s(n)$ to correlated noise signal $x_1(n)$. The reference signal $x(n)$ will be given to adaptive filter such that its output $y(n)$ can be subtracted from desired signal $d(n)$ to yield error signal $e(n)$.

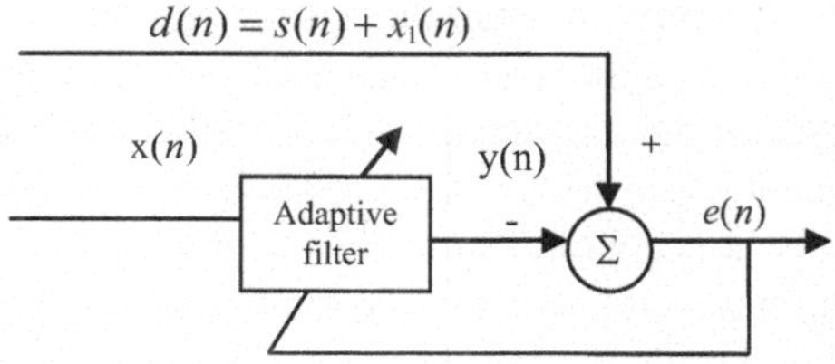

Fig. 1 Active noise cancellation system using LMS adaptive filter algorithm

The output of the summer block, $e(n) = d(n) - y(n)$ is then fed back to adaptive filter in order to update filter coefficients [4, 5] (Fig. 1).

One of the primary disadvantages of the LMS algorithm is slow convergence rate due to the fixed step size for each tap weight in every iteration. To improve the performance of active noise cancellation system, a variant of LMS algorithm known as Variable Step Size Least Mean Square (VSSLMS) algorithm was developed in which step size is usually expressed as a vector $\mu(n)$ for each iteration. Each element of vector $\mu(n)$ can have different step size value corresponding to an element of filter tap weight vector. During the adaptation process if $w(n)$ is far from its optimum value, the step-size parameter will be relatively high to achieve fast convergence and whenever the filter coefficients $w(n)$ approaches the steady-state solution, the step-size parameter will decrease so as to reduce the excess Mean Square Error (MSE). There are many other variants of LMS such as Normalized LMS (NLMS), Block LMS (BLMS), Filtered-x LMS (FxLMS), Sign-error LMS algorithm that has been developed by researchers to improve the performance.

3 FPGA Implementation of Adaptive Filter Algorithms

As the integrated circuit technology has made rapid progress in the past few decades, multiple hardware platforms have come into existence for implementation of computationally intensive signal processing applications. Using these hardware platforms, designers can implement the prototype hardware for an algorithm in a short period of time. Therefore, hardware implementation of adaptive filtering algorithms has become an interesting area of research lately. As adaptive filtering algorithms find applications in diverse fields, choosing an appropriate hardware for an application is a key for successful commercial implementation. A particular hardware should be chosen keeping in mind the performance/area/cost tradeoff.

DSPs and FPGAs have emerged as predominant hardware platforms for adaptive filtering algorithms implementation. In the past, the use of DSPs was ubiquitous, but with the need of many applications outstripping the processing capabilities of DSPs, the use of FPGAs is growing rapidly. Apart from higher processing capabilities compared to DSPs, FPGAs offer many other advantages such as high degree of parallelism, low power consumption and low cost.

Adaptive filter architectures have been implemented on FPGA platform using different approaches. One approach is to implement filter structures using hardware description languages such as VHDL or Verilog and the second approach is to implement using MATLAB/Simulink and Xilinx System Generator development tools. The concept of retiming has been used to implement highly pipelined adaptive filter architectures and the maximum clock speed of retimed adaptive filter architectures thus obtained is found to be superior to conventional architectures [6].

Three different architectures have been proposed for implementing a least mean square (LMS) adaptive filtering algorithm to enhance signal-to-noise ratio (SNR) of speech signal, and then the comparison is done based on the performance and area for different filter lengths. To implement the architecture, Xilinx Multimedia board is used. The audio is captured/playback using on-board AC97 audio codec and the three architectures are implemented using Xilinx Virtex-II FPGA platform. The three proposed architectures are pure software architecture, HW/SW Co-design architecture and pure HW architecture [7].

Adaptive noise cancellation systems based on LMS and NLMS algorithms are implemented to remove the 60 Hz interference from ECG signal. MicroBlaze soft processor core is used to implement the two architectures. The first one uses 16 bit fixed-point fractional format based on LMS algorithm whereas the second architecture uses 16-bit fixed-point integer format based on NLMS algorithm [8].

An adaptive filter based on NLMS algorithm is implemented on FPGA platform. In the proposed implementation, to reduce the computational complexity and improve the speed of the architecture, bit-shift technique is used. The results obtained confirm that when the normalized frequency is set to 0.04375π offset from the centre frequency, the attenuation value obtained is 99.21 dB [9].

Delayed LMS adaptive filter algorithm is implemented on FPGA platform using VHDL hardware description language. Instead of a single large block, a number of sub-blocks are used with efficient adder tree in the implemented structure which results in less area and power consumption, and more efficient in terms of area-delay product (ADP) and energy-delay product (EDP) [10].

A Sign-error LMS algorithm based pipelined adaptive filter architecture is implemented with an aim to reduce the computational complexity. The proposed architecture has low latency, high throughput, a simple structure but at the same time, its convergence rate is lesser compared to DLMS. The calculations required for sign-error LMS algorithm implemented using 4:2 adder are less than half of the conventional architectures. Also, this design provides faster convergence and higher throughput than the delayed LMS (DLMS) algorithm [11].

High throughput fast block LMS (FBLMS) adaptive algorithm is implemented on FPGA platform. Concept of distributed arithmetic is used to design the filter structure. The advantage of Block LMS algorithm compared to LMS algorithm is that the throughput rate is higher and area consumption is reduced. In Block LMS, error calculations are not done for every sample but once per block of data. Similarly, weights are not updated for every sample rather over a block of data. The resultant structure is highly computation and area efficient. FPGA implementation of the proposed structure confirms that it takes 45% less area than that of the existing FBLMS algorithm based adaptive filter [12].

High speed highly pipelined adaptive filter structures based on Delayed LMS algorithm has been implemented on Virtex-II Pro FPGA platform. To achieve fully pipelined circuits, the delays in DLMS have been redistributed using the concept of retiming. The delay reduction is 66.7% and convergence time is five times faster in the resultant retimed delayed architectures. The throughput rate of 182 Msample/s is achieved using retimed architectures implemented using virtex-II FPGA technology [13].

High speed highly pipelined architectures for adaptive noise cancellation was implemented on Virtex-II Pro XC2VP30 FPGA device. To achieve highly pipelined parallel architectures for better performance, retiming technique was used resulting in higher performance than conventional architectures. Retiming is a technique to relocate delay elements in a circuit such that its input/output characteristics are not affected. The retimed highly pipelined 64-tap direct form adaptive filter structure has maximum operating frequency of 187.614 MHz compared to the conventional structure which has maximum operating frequency of 10.576 MHz. It has also been observed that the hardware resource consumption varies considerably for different structures. For direct form retimed delayed LMS 64-tap structure (DF-RDLMS), 2464 delay elements are required whereas only 397 and 325 delay elements are required for binary tree form retimed delayed LMS 64-tap structure (TDF-RDLMS) and transpose form retimed delayed LMS 64-tap structure (TF-RDLMS), respectively. Sequential architecture has the advantage that it require less hardware compared to parallel architectures, and therefore suitable especially for large size filters. Simulation results show an SNR improvement of approximately 8.8 dB on speech signal taken from TIMIT database for input SNR of 6.57 dB and approximately 28 dB on ECG signals taken from MIT-BIH arrhythmia database for input SNR of 1.2436 dB [14–16].

Retimed adaptive filter structures based on delayed LMS algorithm have been implemented for speech enhancement application and synthesis results obtained for different FPGA platforms. It has been observed from the results that the implementation on different FPGA platforms leads to considerable variation in the performance of various filter structures. For instance, the Maximum operating frequency and power consumption of TF-RDLMS architecture for Virtex-4 series device are 90.901 MHz and 448.16 mW respectively whereas for Spartan-6 series device is 96.899 MHz and 20.45 mW respectively. Therefore careful examination of the performance parameters should be done while selecting an FPGA device for an application. Simulation has been performed to check the improvement in SNR. For a Hindi speech signal taken from Hindi speech database and white noise from Noisex 92 database, output SNR of approximately 38.75 dB is achieved for input SNR of 20.41 dB [17, 18].

High-speed parallel structures are implemented using Delayed block FxLMS (DBFxLMS) and delayed block FsLMS (DBFsLMS) algorithms. To design an active noise cancellation system for linear noise processes, Filtered-X LMS-based algorithms are suitable. The performance of the FxLMS algorithm diminishes if the noise source or medium introduces nonlinearity. The throughput rate for DBFxLMS and DBFsLMS structures is higher (nearly L times) and hardware consumption is less than the DFxLMS and DFsLMS structures. The DBFxLMS structure has more than L times higher throughput rate, consumes L times more multipliers and adders, and (3Q + 2)(3Q + 2) less registers (where Q is the secondary path filter length) as compared to the existing FxLMS-based structure. The proposed DFxLMS and DBFxLMS structures have 55 and 58% less ADP, 30 and 41% less EPS and significantly higher throughput than those of the existing FxLMS-based structures [19].

Wavelet-based filters are very effective to remove noise from the information bearing signals. To remove power line interference from ECG signal, a real-time denoising technique using wavelet has been implemented on FPGA. In this work, Donoho denoising algorithm based architecture and soft-thresholding technique is used. The concept of this algorithm is based on the assumption that the wavelet coefficients with smaller magnitudes are caused by noise and wavelet coefficients with larger magnitudes are caused by information bearing signal. Using the soft-thresholding technique, the smaller wavelet coefficients are replaced by zero and larger wavelet coefficients are retained. Noise is estimated using the second level wavelet coefficients because power line interference is a narrow band signal [20].

Cutset retiming and clock period minimization techniques are used to obtain optimized retimed filter structures. A lattice filter structure has been implemented using Spartan3E series FPGA platform. The filter structure obtained after cutset retiming technique consumes 190 slices whereas the filter structure obtained after clock period minimization technique consumes 220 slices. To optimize the performance, various optimization techniques such as supply voltage scaling, parallel prefix tree adders, etc. is used at structural level [21].

An LMS adaptive filter using conventional multiplier adder has been implemented for active noise control application. It has been observed that the LMS adaptive filter is able to successfully track the unknown system with the convergence speed of 1.46 ms at $1/2^{10}$ step size [22].

LMS adaptive filter based on the transposed direct form has been implemented on Xilinx Spartan-3 FPGA platform. Different types of noises were considered such as fan noise, car noise, and sinusoidal noise. ModelSim SE 6.2c simulator and Xilinx ISE 9.1i development environment was used for the simulation and synthesis of the VHDL design, respectively. Noise reduction is achieved sufficiently for stationary noise in a frequency range between 188 and 4000 Hz. The results obtained in the work have proved that the FPGA platform is well suited for the complex real-time audio processing tasks in augmented reality audio systems [23] (Table 1).

Table 1 Comparison of various adaptive noise cancellation algorithms implemented on FPGA Platform

S. No.	Algorithm	Development Tool /Implementation Platform	Significant Outcomes
1	LMS adaptive filtering implementation [7]	Xilinx MicroBlaze development tool and Virtex II FPGA board	System speed increases but with an extra area and lower flexibility
2	Adaptive noise cancellation systems based on LMS and NLMS algorithms [8]	Xilinx MicroBlaze development tool and Xilinx FPGA implementation	60 Hz interference suppression from ECG signal in real time is achieved
3	Normalized LMS algorithm implementation using bit-shift technique [9]	FPGA platform	Attenuation of 99.21 dB is obtained with the proposed design
4	DLMS filter structure implementation using optimized pipelining technique [10]	VHDL hardware description language and FPGA platform	Proposed design results in better ADP & EDP, and less area and power consumption
5	Pipelined Sign-error LMS adaptive filter architecture [11]	MATLAB	Faster convergence and higher throughput compared to conventional adaptive filter structures
6	Fast block adaptive filter implementation using distributed arithmetic [12]	FPGA platform	Proposed algorithm takes 45% less area than existing Fast block LMS algorithm
7	High speed highly pipelined Delayed LMS filter structures FPGA implementation [13]	MATLAB/Simulink and Xilinx System Generator development tools and Virtex-II Pro FPGA board	RDLMS architectures are superior to conventional LMS architectures in terms of maximum operating frequency. The delay reduction is 66.7%, throughput rate is 182 Msample/s and convergence time is five times faster in the resultant retimed delayed architectures
8	LMS based Parallel and Sequential architectures implementation based on Retiming concept [14]	MATLAB/Simulink and Xilinx System Generator development tools and Virtex-II Pro FPGA board	Sequential architecture requires less hardware compared to parallel architectures, and therefore suitable especially for large size filters. SNR improvement of approx. 8.8 dB and 28 dB is achieved for speech signal and ECG signal with input SNR of 6.57 dB and 1.2436 dB respectively

(continued)

Table 1 (continued)

S. No.	Algorithm	Development Tool /Implementation Platform	Significant Outcomes
9	FPGA implementation of High speed retimed delayed LMS algorithm based adaptive filter structures [17]	MATLAB/Simulink and Xilinx System Generator development tools and Spartan 6 FPGA board	Implementation on different FPGA platforms leads to considerable variation in the performance of various filter structures. Therefore careful examination of the performance parameters should be done while selecting an FPGA device an application
10	Delayed block FxLMS (DBFxLMS) and delayed block FsLMS (DBFsLMS) algorithms implementation [19]	Hardware/ASIC implementation	The proposed DFxLMS and DBFxLMS structures have 55 and 58% less ADP, 30 and 41% less EPS and significantly higher throughput than those of the existing FxLMS-based structures
11	Donoho denoising algorithm based wavelet filter architecture [20]	MATLAB/Simulink, Xilinx system generator development tool and Virtex II FPGA board	Efficiently removes power line interference from ECG signal
12	Lattice filter implementation based on cutset retiming and clock period minimization technique [21]	VHDL hardware description language and Xilinx FPGA Sparten3E XC3S100E device	Hardware consumption is less than conventional filter structures
13	conventional multiplier adder-based LMS adaptive filter implementation [22]	Altera Quartux II development platform and Altera DE2 development board	LMS adaptive filter successfully track the unknown system with the convergence speed of 1.46 ms at $1/2^{10}$ step size
14	Transposed direct form LMS adaptive filter implementation [23]	MATLAB/Simulink, Modelsim SE6.2c, Xilinx ISE 9.1i development environment, Xilinx Spartan-3 FPGA board	Noise reduction is achieved sufficiently for stationary noise in a frequency range between 188 and 4000 Hz

4 Conclusion

Efficient noise cancellation techniques have a wide range of applications in wireless and mobile communication systems. FPGA implementations of several adaptive filter algorithms having different characteristics have been reviewed in this paper. The parallel and sequential retimed adaptive filter structures have better maximum operating frequency as compared to conventional adaptive filters but at the cost of more hardware resources consumption. For instance, the retimed highly pipelined 64-tap direct form adaptive filter structure has maximum operating frequency of 187.614 MHz compared to the conventional structure which has maximum operating frequency of 10.576 MHz. The sign-error LMS algorithm and fast block LMS (FBLMS) algorithms have relatively less computational complexity. For instance, the proposed FBLMS algorithm implemented using distributed arithmetic takes 45% less area than existing Fast block LMS algorithms. The DBFxLMS and DBFsLMS algorithms have 55 and 58% less area-delay product (ADP), 20 and 30% less energy per sample (EPS) than those of existing FxLMS-based algorithms. Hence this article is an attempt to give an insight into the research work done in the field of FPGA implementation of various adaptive filter algorithms for noise cancellation applications.

References

1. Diniz PSR Adaptive filtering—algorithms and practical implementation, 3rd edn. Kluwer Academic Publishers, Springer
2. Poularikas AD, Ramadan ZM (2006) Adaptive filtering primer with MATLAB. CRC Press
3. Haykin Simon (2008) Adaptive filter theory, 4th edn. Pearson, India
4. Gupta VK, Chandra M, Sharan SN (2012) Real time implementation of adaptive noise canceller. In: International conference on systemics, cybernetics and informatics, pp 24–28
5. Dhal M, Ghosh M, Goel P, Kar A, Mohapatra S, Chandra M (2015) A unique adaptive noise canceller with advanced variable-step BLMS algorithm. In: 2015 International conference on advances in computing communications and informatics (ICACCI)
6. Parhi KK (2010) VLSI digital signal processing systems—design and implementation. Wiley
7. Elhossini, Areibi, and Dony, An FPGA implementation of the LMS adaptive filter for audio processing. In: IEEE International conference on reconfigurable computing and FPGA's, 2006, San Luis Potosi, Mexico, pp 1–8
8. Bahoura M, Ezzaidi H (2015) Real-time implementation of an adaptive noise canceller based on MicroBlaze soft processor. In: IEEE 28th Canadian conference on electrical and computer engineering (CCECE) 2015, Halifax, NS, Canada, pp 1180–1183
9. Dai J, Wang Y (2010) NLMS adaptive algorithm implement based on FPGA. In: IEEE Third international conference on intelligent networks and intelligent systems (ICINIS 2010), Shenyeng, China, pp 366–369
10. Priya P, Babu P (2014) An efficient architecture for the adaptive filter using delayed LMS algorithm. In: IEEE International conference on information communication and embedded systems (ICICES2014), Chennai, India, pp 1–6
11. Liu M, Wang M, Liu D (2016) A pipelined sign-error LMS adaptive filter architecture with low computational complexity. In: IEEE International conference on anti-counterfeiting, security, and identification (ASID) 2016, pp 86–90
12. Baghel S, Shaik R (2011) FPGA implementation of fast block LMS adaptive filter using distributed arithmetic for high throughput. In: IEEE International conference on communications and signal processing, 2011, Calicut, India, pp 443–447
13. Yi Y, Woods R (2005) High speed FPGA based implementations of delayed—LMS filters. J VLSI Signal Process 113–131
14. Bahoura M, Ezzaidi H (2011) FPGA implementation of parallel and sequential architectures for adaptive noise cancellation. In: Circuits, systems and signal processing, Springer, Boston, pp 1521–1548

15. Garofolo JS, Lamel LF, Fisher WM, Fiscus JG, Pallett DS, Dahlgren NL (1993) DARPA TIMIT acoustic-phonetic continuous speech corpus, CD-ROM, NTIS edition
16. Goldberger AL, Amaral LAN, Glass L, Hausdorff JM, Ivanov PCh, Mark RG, Mietus JE, Moody GB, Peng C-K, Stanley HE (2000) PhysioBank, PhysioToolkit, and PhysioNet: components of a new research resource for complex physiologic signals. Circulation 101(23):e215–e220
17. Goel P, Chandra M (2018) VLSI implementations of retimed high speed adaptive filter structures for speech enhancement. In: Microsystem technologies. Springer, GmbH Germany, pp 1–8
18. Samudravijaya K, Rao PVS, Agrawal SS (2000) Hindi speech database. In: Proceedings of international conference on spoken language processing, ICSLP-2000, Beijing, China, Oct, vol 4, pp 456–459
19. Mohanty BK, Singh G, Panda G (2017) Hardware design for VLSI implementation of FxLMS- and FsLMS-based active noise controllers. In: Circuits, systems, and signal processing. Springer, vol 36, Issue 2, pp 447–473 (first online Apr 2016)
20. Bahoura M, Ezzaidi H (2010) FPGA implementation of wavelet-based denoising technique to remove power-line interference from ECG signal. In: Proceedings of the 10th IEEE international conference on information technology and applications in biomedicine, Corfu, Greece, pp 1–4
21. Yagain D, Vijaya Krishna A, Chennapnoor S (2012) Design optimization platform for synthesizable high speed digital filters using retiming technique. In: 10th IEEE International conference on semiconductor electronics (ICSE), Kuala Lumpur, Malaysia, pp 551–555
22. Mustafa R, Umat C, Ali MAM, Al-asady AD (2009) Design and implementation of least mean square adaptive filter on Altera Cyclone II field programmable gate array for active noise control. In: IEEE Symposium on industrial electronics and applications (ISIEA 2009), Kuala Lumpur, Malaysia, vol 1, pp 479–484
23. Fohl W, Matthies J (2009) A FPGA based adaptive noise cancelling system. In: Proceedings of the 12th international conference on digital audio effects (DAFX-09), Como, Italy, 01–04 Sept 2009

A New Neighborhood-Based Outlier Detection Technique

Umang Gupta, Vandana Bhattacharjee and Partha Sarathi Bishnu

Abstract Outlier detection is one of the most vital and essential issues in data mining tasks. We propose a new method to detect and analyze outliers. We apply neighborhood-based outlier detection technique to detect and analyze the outliers. Using weights of the neighbors of each data and a unique parameter OBN is used to identify the outlier. Our proposed algorithm is tested on real datasets and compared with the existing technique and the results are presented.

Keywords Data mining · Outlier · Neighborhood

1 Introduction

The set of data objects that are significantly dissimilar with the remaining data objects are called outliers and outlier mining means detection and analysis of outliers [1]. Many outlier detection techniques have been developed [2–6] which can be classified into four classes: the statistical technique, the distance-based technique, the density-based local outliers technique, and the deviation-based technique [1]. Outliers detection technique has been applied to many domains, such as data cleaning, financial fraud detection [7], medical analysis, marketing, electronic commerce, network intrusion detection, and biology [1, 8–11]. Neighborhood-based technique is one of the most important techniques that find outliers based on dissimilarity measure. Neighborhood-based outliers are those data that do not have enough neighbors, where neighbors are defined based on closeness from the given object [1]. This

U. Gupta · V. Bhattacharjee (✉) · P. S. Bishnu
Department of Computer Science and Engineering, Birla Institute of Technology, Mesra, Ranchi 835215, Jharkhand, India
e-mail: vbhattacharya@bitmesra.ac.in

U. Gupta
e-mail: umanggupta7@gmail.com

P. S. Bishnu
e-mail: psbishnu@bitmesra.ac.in

© Springer Nature Singapore Pte Ltd. 2019

V. Nath and J. K. Mandal (eds.), *Proceedings of the Third International Conference on Microelectronics, Computing and Communication Systems*, Lecture Notes in Electrical Engineering 556, https://doi.org/10.1007/978-981-13-7091-5_43

approach does not require any a priori knowledge of data distributions as the statistics method do [10]. Density-based outlier detection uses the concept of density to find the Level of Outlierness (LON) and ranks the data points accordingly [11].

In this paper, we present a new technique using neighborhood to detect and analyze the outliers. Here, first, we identify number of neighbors associated to each other, and then we calculate weights of the neighborhood of all the data. Finally, using a special parameter called Outlierness Based on Neighborhood (OBN) is used to identify the outliers. To validate the performance of our proposed algorithm, we have compared with the existing technique [12] using real datasets from the UCI machine learning datasets.

The organization of the paper is as follows: in Sect. 2, we illustrate a review of the related work. Section 3 presents our proposed outlier-based algorithm. In Sect. 4 we discuss all the experiments and conclusions in Sect. 5.

2 Related Work

In this section, we discuss various outliers detection and prediction techniques proposed by several researchers [9, 13–15]. The idea of distance-based method has been first introduced by Knorr and Ng. According to Knorr and Ng, "A point o in a data set is an outlier with respect to the parameter k and d, if no more than k points in the data set are at a distance d or less from o" [13]. Ramaswamy et al. [15] revised the definition of outliers introduced by Knorr and Ng. "Given a k and n, a point o is an outlier if no more than $n - 1$ other points in the data set have a higher value for D_k than o". Angiulli et al. [16] presented another method to detect top n outliers of a large and high dimension dataset. Angiulli and Pizzuti [14] introduced a new method which improves the distance-based outlier detection. To find the highest n distance-based outliers, Bay and Schewabacher suggested a simple pruning rule and randomization on large and high dimensional datasets [17]. Li et al. [9] suggested another distance-based technique which applies a neighbor-based distance function to identify the outliers from categorical data.

Zhu et al. [18] suggested an outlier detection technique based on k-neighborhood using minimum spanning tree, by spotting the local outliers and local outlying clusters efficiently. Moreover, the authors proposed a new dissimilarity measure based on k-neighborhood to form minimum cost spanning tree to identify outliers can be found. Recently, Bhattacharya et al. [6], suggested a new outlier detection technique using informative distance measure. Chen et al. [12] proposed rough set-based outlier detection on neighborhood for continues and discrete data. The granular ability of the proposed algorithm helps to detect the outliers effectively based on the majority class in the neighborhood of the data.

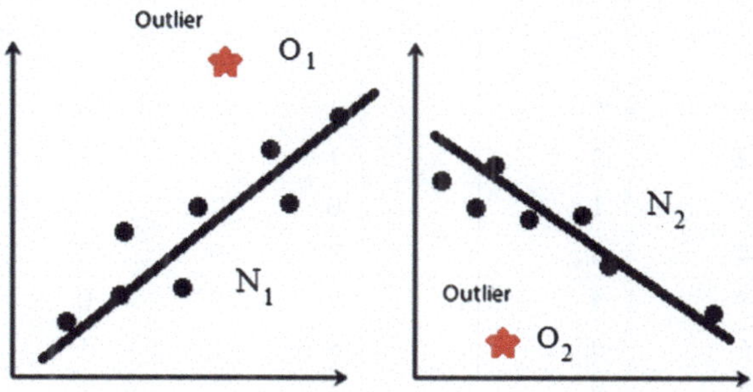

Fig. 1 Outliers

3 Proposed Algorithm for Outlier Detection

Before presenting our algorithm, we discuss some basic definitions.

An *outlier* can be defined as an observation which is different from the others in the sample. It is considered as the deviation from the normal obtained pattern of data. Outliers may be a result of malicious activity on the network. An example is credit card fraud, etc. The outliers can be represented in Fig. 1: N_1 and N_2 are normal data O_1 and O_2 are two outliers to the respective dataset. Since the outlier data (O_1 and O_2) are far from the rest of the data or they do not follow the patterns of the other data (N_1 and N_2). It is observed that the numbers of neighbors of the outlier data (O_1 and O_2) are less than the non-outlier data.

Outlier detection techniques using nearest neighbor based needs a dissimilarity measure between data. The dissimilarity between two data can be computed in various techniques. Euclidean distance and Manhattan distances are used for continues data. A simple matching dissimilarity is used for categorical data. The distance or similarity is combined and computed for multivariate data [19, 20].

Let a set of data $D = \{x_1, x_2, \ldots, x_n\}$, contains n number of data and d is the dimension of the data. For each x_i, $1 \leq i \leq n$ number of r-neighbors are $N_r(x_i) = \left\{ y_j \middle| x_i, y_j \in D, dist(x_i, y_j) \leq r \right\}$, where $dist(x_i, y_j) = \frac{\sum_{f=1}^{p} \delta^{f}_{(x_i,y_j)} d^{f}_{(x_i,y_j)}}{\sum_{f=1}^{p} \delta^{f}_{(x_i,y_j)}}$, $1 \leq j \leq i$.

For the calculation of our parameter, OBN, we define *weight of neighborhood* as follows:

The weight of the neighborhood of each data point x_i, $1 \leq i \leq n$, is defined as $W[N_r(x_i)] = \sum_{j=1}^{N_r(x_i)} dist(x_i, y_j)$. Finally, for each x_i, $1 \leq i \leq n$, we calculate OBN as follows:

$$OBN(x_i) = \frac{W[N_r(x_i)]}{\sum_{j=1}^{N_r(x_i)} W[N_r(y_j)]}.$$

If $OBN(x_i > \mu)$ then x_i is an outlier.

Example 1 Let the data be $\{(1, 2), (2, 1), (2, 2), (5, 6), (6, 5), (6, 6), (5, 5), (10, 10)\}$. The Manhattan distances are as follows:

	1	2	3	4	5	6	7	8
1	0	2	1	8	8	9	7	17
2	2	0	1	8	8	9	7	17
3	1	1	0	7	7	8	6	16
4	8	8	7	0	2	1	1	9
5	8	8	7	2	0	1	1	9
6	9	9	8	1	1	0	2	8
7	7	7	6	1	1	2	0	10
8	17	17	16	9	9	8	10	0

The neighbor matrix are:

	1	2	3	4	5	6	7	8
1	1	1	1	0	0	0	0	0
2	1	1	1	0	0	0	0	0
3	1	1	1	0	0	0	0	0
4	0	0	0	1	0	1	1	0
5	0	0	0	0	1	1	1	0
6	0	0	0	1	1	1	0	0
7	0	0	0	1	1	0	1	0
8	0	0	0	1	1	1	0	1

The neighbor of first data are {1, 2, 3} and the neighbor of fourth data are {4, 6, 7}. Similarly the neighbor of eighth data is {4, 5, 6, 8}.

The $W[N_r(x_i)]$ values are {3, 3, 2, 2, 2, 2, 2, 26} and $\sum_{j=1}^{N_r(x_i)} W[N_r(y_j)]$ values are {8, 8, 8, 6, 6, 6, 6, 32}. The OBN values are {0.3750, 0.3750, 0.25, 0.3333, 0.3333, 0.3333, 0.3333, 0.8125} respectively. From the OBN values, it is observed that the OBN value of the x_8 data is quite high (0.8125) from the rest of the OBN values. Hence, we can say that the x_8 data is an outlier data.

Next, we present our proposed neighborhood-based outlier detection algorithm as follows:

Algorithm 1: Neighborhood-based outlier
Input: Data D, r, μ;
Output: Set of outliers;

Step 1: for each data x_i, $1 \leq i \leq n$, calculate $N_r(x_i)$;
Step 2: for each data x_i, $1 \leq i \leq n$, calculate $W[N_r(x_i)]$;
Step 3: for each data x_i, $1 \leq i \leq n$, calculate $OBN(x_i)$;
Step 4: return set of outliers;

Description of the algorithm

In this technique, we calculate the number of neighbors present for each data point. Next, we calculate weight of the neighborhood of each data to calculate OBN. The OBN value is compared with the threshold value μ and if the OBN value of the data is less than the threshold value, then the data is not outlier. Here, r and μ are user-defined constants.

4 Experiments

4.1 Data Set

Four real-life datasets are used to test our proposed algorithm. These data sets are *E. Coli*, Glass, Wine, and Yeast (UCI machine learning repository: http://archive.ics.uci.edu/ml/datasets.html). The actual number of outliers present in the datasets is 9, 9, 10, and 8, respectively. Our goal is to identify the outliers correctly using neighborhood-based outlier detection algorithm.

The description of the datasets are as follows (Table 1).

4.2 Experimental Setup

In this experiment, all the programs were computed using a computer with 4 GB of RAM, Intel Pentium 4, 2.3 GHz CPU, Windows 10. The proposed algorithm was compared with the existing algorithm [12]. Both the algorithms were executed on all the four datasets under the same computing environment. The time taken to detect more than 90% outliers in each case was noted. The μ values for *E. Coli*, Glass, Wine, and Yeast datasets are more than 0.5, 0.57, 0.55, and 0.6, respectively. And, the VDM values for *E. Coli*, Glass, Wine, and Yeast datasets are more than 0.9, 0.85, 1.2, and 2.5, respectively. For r value, we take the distance of nearest third data. For distance calculation, we use Manhattan distance.

Table 1 Description of real datasets

Sr. No	Dataset	No. of data	Dimension	Actual No. of outliers
1	*E. Coli*	336	7	9
2	Glass	214	9	9
3	Wine	214	19	10
4	Yeast	1484	8	8

Table 2 Comparison of time efficiency

Sr. No.	Dataset	Proposed technique	Chen et al. [12]
		Time	Time
1	*E. Coli*	5.7655	6.5807
2	Glass	2.2972	8.7654
3	Wine	2.2987	3.2519
4	Yeast	142.5100	153.5619

Table 3 Comparison for outliers detection

Sr. No.	Dataset	Actual	Proposed algorithm	Chen et al. [12]
1	*E. Coli*	9	9	12
2	Glass	9	8	12
3	Wine	10	11	15
4	Yeast	8	8	6

All the algorithms have been coded in Octave-3.2.4. GNU Octave (http://www.gnu.org/software/octave) is a high-level programming language for statistical and numerical computation.

4.3 Result and Analysis

In the second rows of the table (Table 2), it is seen that our proposed algorithm takes 5.7655 ms as compared to 6.5807 ms (*E. Coli* dataset) of algorithm of Chen et al. Similar results can be observed for Glass, Wine and Yeast data sets. Thus, it is seen that our algorithm performs very well compared to Chen et al. [12], algorithm.

In Table 3, we display the comparison of power of outlier detection. It is seen that for *E. Coli*, and Yeast dataset perfectly, we can identify the outliers by our proposed algorithm and number of outliers just one less and one more in cases of Glass and Wine datasets, respectively, i.e., our proposed technique identified more than 90% numbers of outliers correctly which is not true for Chen et al. [12] technique.

5 Conclusion

This paper discusses a very interesting topic that of outlier detection. In this paper, a new outlier analysis model is introduced with a uniform framework, which provides the basic idea for understanding and implementation of the outlier detection approach. This paper proposes a neighborhood-based metric for outlier detection scheme, where the OBN is measured to detect the outliers, if any. The proposed algorithm is validated on a real-time platform to detect outliers. The result shows the correctness of the approach. The basic idea of the proposed algorithm is that an outlier data point will have a sparse neighborhood, leading to a high value of OBN. Thus, it will get marked as an outlier. Further validation on a bigger data set and an enhancement of the proposed approach is a part of the ongoing research. Feature selection may also be helpful in improving the results.

References

1. Han J, Kamber M (2000) Data mining: concepts and techniques, 2nd edn. Morgan Kaumann Publishers, An imprint of Elsevier
2. Arning A, Agarwal R, Raghaban P (1996) A linear method for deviation detection in large databases. In: Proceedings 1996 international conference data mining and knowledge discovery, Portland, Oregon, pp 164–169
3. Knorr E, Ng R (1997) A unified notion of outlier: properties and computation. In: Proceedings of 1998 international conference knowledge discovery and datamining, New Port Beach, pp 219–222
4. Breuning MM, Kriegel HP, Ng R, Sander J (2000) LOF: identifying density based local outliers. In: Proceedings of ACM-SIGMOD international conference on management of data, Dallas, TX, May 2000, pp 93–104
5. Shi Y (2008) Detecting clusters and outliers for multi-dimensional data. In: 2008 IEEE international conference on multimedia and ubiquitous engineering, pp 429–432
6. Bhattacharya G, Ghosh K, Chowdhury AS (2017) kNN classification with an outlier informative distance measure. In: PReMI 2017, LNCS 10597, pp 21–27
7. Yue D, Wu X, Wang Y, Li Y, Chu C-H (2007) A review of data mining based financial fraud detection research. In: Proceedings of 2007 international conference on wireless communications, networking and mobile computing, Shanghai, P. R. China, 21–25 Sept, pp 5514–5517
8. Jian M, Cheng L, Xiang W (2008) Research of gravity-based outliers detection. In: International conference on intelligent information hiding and multimedia signal processing
9. Li S, Lee R, Lang S-D (2007) Mining distance-based outliers from categorical data. In: Seventh IEEE international conference on data mining—workshops
10. Cherednichenko S (2005) Outlier detection in clustering. University of Joensuu, Department of Computer Science; Barnett V, Lewis T (1994) Outliers in statistical data. Wiley
11. Gupta S, Bhattacharjee V (2017) Intrusion detection system using outlier analysis. In: Proceedings of ICEMIT 2017, in Book series, Advances in soft computing. Springer
12. Chen Y, Miao D, Zhang H (2010) Neighborhood outlier detection. Expert Syst Appl 37:8745–8749
13. Knorr EM, Ng RT (1998) Algorithm for mining distance based outliers in large data sets. In: Proceedings of the 24th VLDB conference, NY, USA
14. Angiulli F, Pizzuti C (2005) Outlier mining in large high-dimensional data sets. IEEE Trans Knowl Data Eng 17(2)
15. Ramaswamy S, Rastogi R, Shim K (2000) Efficient algorithms for mining outliers from large data sets. In: Proceedings of international conference management of data (SIGMOD '00), pp 427–438
16. Angiulli F, Basta S, Pizzuti C (2006) Distance-based detection and prediction of outliers. IEEE Trans Knowl Data Eng 18(2)
17. Bay SD, Schewabacher M (2003) Mining distance-based outliers in near linear time with randomization and a simple pruning rule. In: Proceedings of international conference on knowledge discovery and data mining (KDD '03)

18. Zhu Q, Fan X, Feng J (2014) Outlier detection based on K-neighborhood MST. In: 2014, IEEE IRI 2014, 13–15 Aug 2014, San Francisco, California, USA, pp 718–724
19. Tan P-N, Steinbach M, Kumar V (2005) Introduction to data mining. Addison-Wesley
20. Bishnu PS, Bhattacharjee V (2013) A modified K-modes clustering algorithm. In: PReMI 2013. Lecture notes in computer science, vol 8251. Springer, Berlin, Heidelberg

Structural Testing of Multichip Module SoC Components

Amruta Hosur and N. Shylashree

Abstract In modern VLSI, the testing of manufactured VLSI chips is as important as the whole process of designing and manufacturing. The structural testing plays an important role as the functional testing. Structural testing, unlike functional testing, involves testing of all the physical connections, macros, other functional blocks in the SoC (System on chip). In the structural testing of SoC boundary Scan standards (IEEE 1149.1, IEEE 1687, and IEEE 1500) are used. The complete testing of SoC can be achieved by testing the individual components of the SoC. In case of multichip module SoC, the testing can be done at the complete SoC level as well as at the chip level. In this paper, the structural testing of one of the unique component of the SoC "Fuse Macro" is explained.

Keywords Structural testing · Multichip module · Automatic test equipment (ATE)

1 Introduction

In modern VLSI, area, speed, and the power are of a major concern; the concept of SoC is helping in reducing the area of complex designs. The SoC can have multiple cores inside it along with other IPs (hard macro)/soft macros/blocks inside it. One of the unique components of the SoC is fuse macro. The role of fuse macro is different in different scenarios. In some cases, fuse macro is used to have the repairable rows and columns for memory and in some fuses, macro is used to fuse only the required IPs of the SoC. The multichip module SoCs includes the multiple chips having different functionality/different IPs or same functionality chips on the same SoC.

A. Hosur
VLSI Design and Embedded Systems, Rashtreeya Vidyalaya College of Engineering, Bengaluru 560059, India

N. Shylashree (✉)
Department of ECE, Rashtreeya Vidyalaya College of Engineering, Bengaluru 560059, India
e-mail: shylashreen@rvce.edu.in

© Springer Nature Singapore Pte Ltd. 2019

V. Nath and J. K. Mandal (eds.), *Proceedings of the Third International Conference on Microelectronics, Computing and Communication Systems*, Lecture Notes in Electrical Engineering 556, https://doi.org/10.1007/978-981-13-7091-5_44

The structural testing of any macros or functional blocks needs to be done at the chip level to ensure the proper interconnections between the IPs in the chip as well as at the SoC level to ensure the testing of interconnections between the multiple chips. IEEE 1149.1 [1] gives the boundary scan protocols at the board level, IEEE 1687 [2] gives the standards at the chip level and IEEE 1500 [3] gives the structural testing standard at the chip level. Design for testability (DFT) is the design of the actual RTL along with the added features for increasing the testability of the design after the manufacture of chip. Testability means the controllability and observability for all the internal nodes and nets in the design.

DFT includes boundary scan standards, built-in self-test, memory built-in self-test for testing memories (MBIST), etc.

While carrying out the structural testing, one should reach the test data registers (TDRs) in the RTL to give the test data as an input and to check the response of the RTL or design for the given test data input. To access the test, data registers boundary scan standards help at different levels.

2 Fuse Macro

Fuse macro is the unique component or the functional block of the SoC it is meant for different applications one like storing the repair bits of the memory [4] one for fusing the IPs of the SoC according to the requirement etc.

The cost for the manufacturing of VLSI circuits is huge and it is recommended to manufacture the chips with the same configurations so that it should be error-prone easy to manufacture. Thus manufactured chip can be configured according to the need of customers by incorporating the concept of fusing only required IPs in the chip or SoC.

Here, the structural testing is carried out for one such fuse macro having the bits from different IPs of the chips on SoC. The SoC contains many components inside like memory (SRAM and DRAM), CPU, clocking block, Phase-locked loop (PLL), Chip Low Dropout Regulators (CLDO), fuse macro, etc. For the structural testing of such SoC, it is necessary to generate the test patterns for all the components of the SoC. Since Fuse macro is associated with all other IPs in the SoC, it is recommended to generate the test patterns for fuse macro with the higher priority.

The different IPs in the SoC has the fuse bit associated with each of it and the fuse macro contains the fuse bit from different IPs of the SoC. The SoC is having the fuse SRAM which is having one to one correspondence with the bits in the fuse macro. Each IP in the SoC is having the fuse parameter file, which contains the information about the fuse bit of the respective IP. The Fuse macro in the SoC can be represented as shown in Fig. 1. Fuse macro is the nonvolatile memory, storing the information about the programmed fuse bits.

The IPs in the SoC may be in the different chips of the SoC. The fuse macro contains the fuse bits from different IPs located at different chips of the SoC. There is always one to correspondence between the fuse macro and the fuse bits.

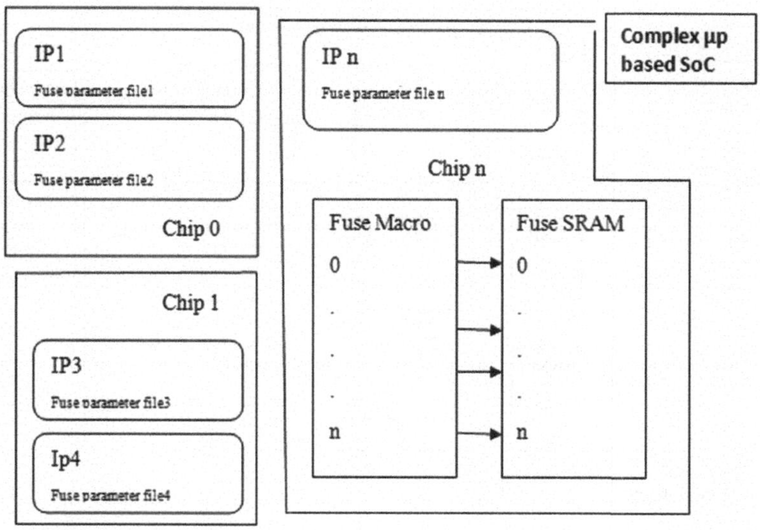

Fig. 1 Conceptual diagram of the fuse macro in SoC

It is recommended to access the fuse macro via the fuse SRAM for giving the test data input to the fuse macro and for observing the response of the actual RTL for the given test data input.

It should be designed in such a way that when the Fuse bit of any particular IP or functional block is programmed, it should go to the secure state, if the customer returns, the device because of some error hit, and to fix that error, it should be refused. For refusing, the device should be booted in nonsecure state.

3 Test Data Register Access

In DFT the controllability and the observability for each of the intermediate nets and the nodes is are increased by incorporating the boundary scan standards and the scan paths along with the actual RTL. There are test data registers in the design which will take the test data input and captures the response of the actual RTL for the given test data input.

IEEE 1149.1 is the IEEE standard, which is also known as JTAG standard for the boundary scan at the board level which has the Test Access Port (TAP) controller and Test Mode Signal (TMS) here the test data is captured and shifted according to the TMS signal [1].

IEEE 1687 which is also known as IJTAG, it is the same as the IEEE 11149.1 but this standard is applicable at the chip level.

For accessing the embedded cores at the SoC IEEE 1500 standard is used. According to this standard, there will be wrapper signals for accessing embedded cores.

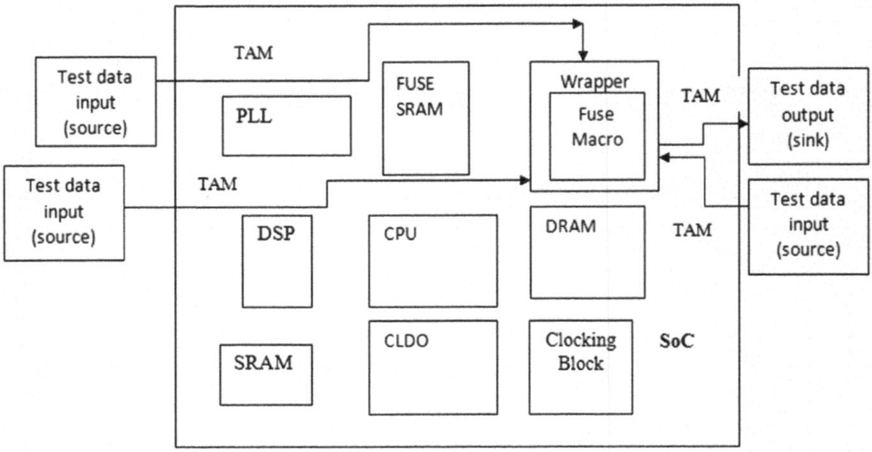

Fig. 2 Test data access

These IEEE standards will help in accessing the test data registers down the hierarchy in the SoC. The test data access can be represented as shown in Fig. 2.

4 Structural Testing for Fuse Macro

For the functional testing of the SoC, Automatic Test Pattern Generator (ATPG) will generate the test patterns for structural testing test patterns, which needs to be designed according to the requirement of particular macro/functional block by checking the various functionalities of the block.

Since it is not possible to communicate with the fuse macro directly, fuse SRAM acts as an intermediate media to communicate fuse macro and Fuse macro.

Fuse macro and the Fuse SRAM are connected with the bridge having the mechanism to read and write to and from the Fuse Macro and Fuse SRAM.

To ensure there are no structural defects, one should ensure the proper data shifting between the Fuse macro and the Fuse SRAM. Once the test patterns are written for the Fuse macro, it needs to dump on to the ATE for the post-silicon testing. Since fuse macro is associated with other IPs in the SoC, the test pattern generation for fuse macro is of highest priority.

5 Test Pattern Description

The patterns for any IP/functional block in a SoC need to be decided based on the functionality of that particular IP and the interconnections to that IP with other blocks in SoC.

The test patterns for fuse macro includes Fuse sense SRAM, Fuse programming, Fuse read, Fuse write, Fuse SRAM check, Fuse margin read, Fuse blank check, Refuse Key unlock, and Power shorts check. These test cases include the testing of data shifting in Fuse SRAM as well as in Fuse Macro.

Fuse read, Fuse sense, Fuse write, Fuse programming test cases has the variation for single row read, write, sense, and programming also. These test sequences need to be separately written for single rows too because when the device is booted in different modes sometimes, there is a need to shift the data to the single row only. Accessing of the complete Fuse macro in such cases may harm the already programmed fuse bits.

5.1 Fuse Sense SRAM

This test will take the fuse data from Fuse box and place it into Fuse SRAM, and then it will the Fuse SRAM to ensure the data sensing.

This test case has to be written for single row variation also, where the specific row number of the Fuse SRAM needs to be specified from where the data needs to be sensed. This test should be done at the chip level as well as at the SoC level. Before starting any test, the device should be reset. The flow of the test is shown in Fig. 3.

The RTL will be having the signal for ensuring sensing, the respective signal needs to be checked to assure the sensing.

5.2 Fuse Write

This pattern will write the data into the entire SRAM. This test should be done separately for single row variation also. The signal in RTL corresponding to data writes needs to be checked to assure the test pass. The flow for the test is shown in Fig. 4.

For writing to the single row the information about the row number needs to be given separately.

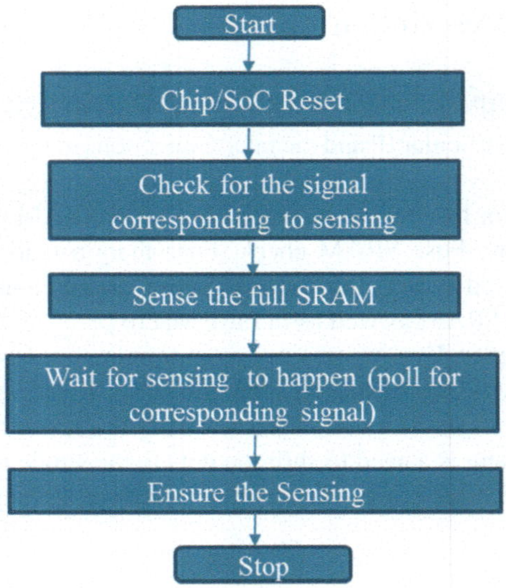

Fig. 3 Fuse sense SRAM

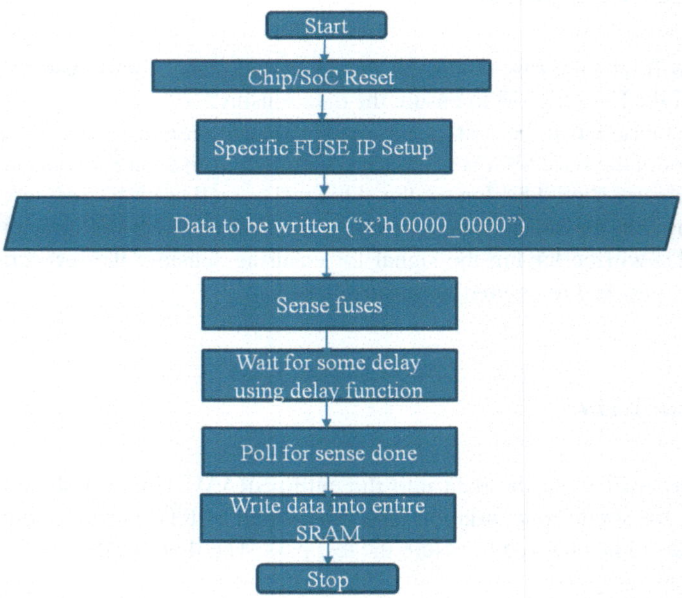

Fig. 4 Fuse write

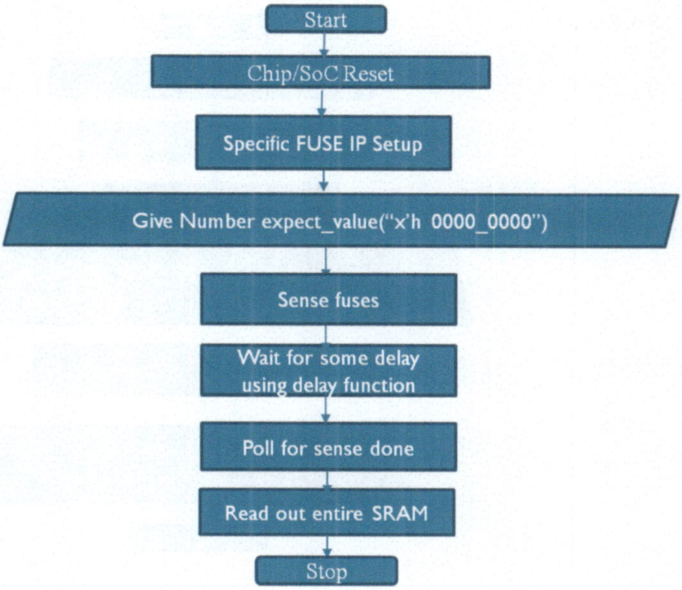

Fig. 5 Fuse read

5.3 Fuse Read

This test sequence will read the data from the Fuse SRAM. Since the Fuse SRAM acts as an intermediate to shift the data to and from the Fuse macro to ensure the proper shifting of the data to the Fuse macro, the Fuse read test should be carried out. The flow of the test sequence is shown in Fig. 5.

Fuse read should be done for single row variation also; the only difference is in that test sequence the row number should be given. Here, the read data is compared with the expected value o check the data in Fuse SRAM.

5.4 Fuse SRAM Check

In this test sequence, the data shifting of the Fuse SRAM is checked, Unlike the Fuse read and write, where the data is read from the SRAM and written into the SRAM separately here, the data is written and read back in the same sequence. The flow of the Fuse SRAM Check is shown in Fig. 6.

In this test sequence, the length of the Fuse SRAM for which the data shifting needs to be done should be specified. The corresponding signals for assuring the test pass in the RTL should be checked.

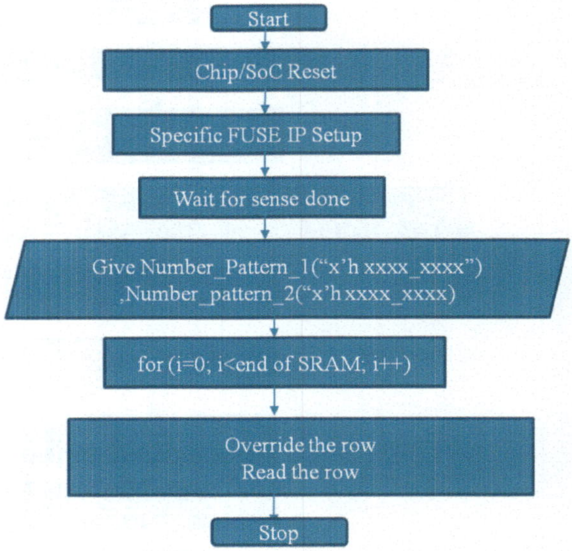

Fig. 6 Fuse SRAM check

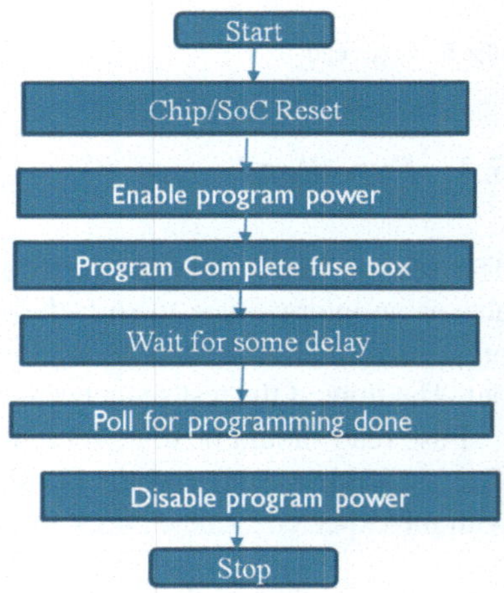

Fig. 7 Fuse programming

5.5 Fuse Programming

Fuse programming is like hard fusing or fuse blowing. In this test case, the actual programming to the fuse bits in the Fuse box/fuse macro is done. Unlike fuse write where the data is written only on the Fuse SRAM.

Sometimes, Fuse programming has to be done for single row also, i.e., programming to the single bit of the fuse macro. Hence, fuse programming test sequence has to be written for single row variation also. The flow of the Fuse programming test sequence is shown in Fig. 7.

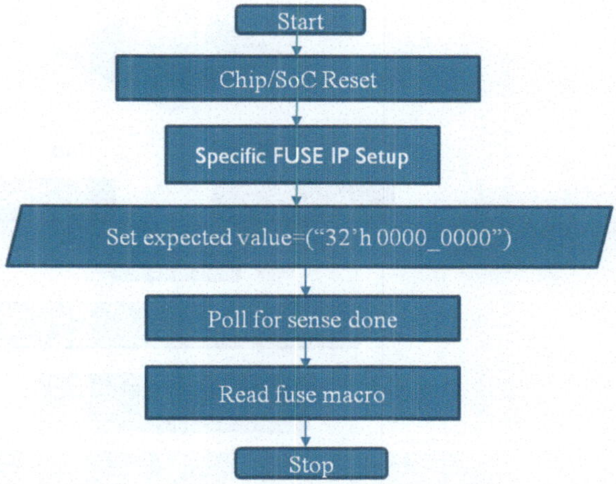

Fig. 8 Fuse blank check

The corresponding signals in the RTL to assure the Fuse programming have to check so ensure the test pass.

5.6 *Fuse Blank Check*

These test cases ensure all values in the Fuse Macro as zeros, which gives the access to the all on chip structure and debug functionality. The flow of the test sequence is as shown in Fig. 8.

In this test sequence set, the expected value as zero and read the fuse macro/fuse box. The corresponding signals in the RTL which ensures the blank check needs to checked to assure the blank check.

5.7 *Fuse Margin Read*

In margin read mode, the fuses are sensed from fuse box are compared to the reference resistance (Rref). This test case helps to check how strongly the fuse box is programmed.

The flow of the fuse margin read test case is as shown in Fig. 9.

In this, the test sequence needs to set some expected value and need to be checked whether it is matching with the expected value 0 assure the programming of fuse macro as strongly as expected.

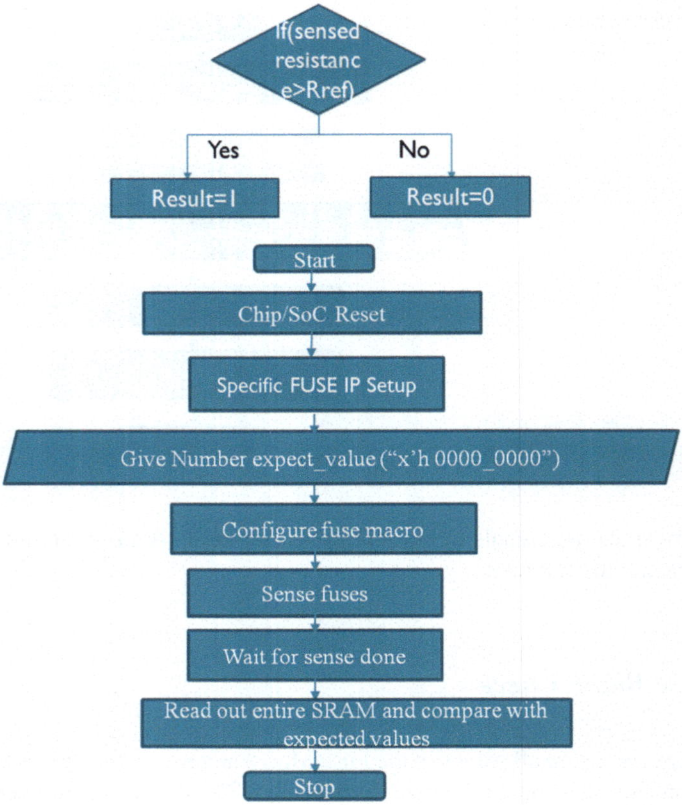

Fig. 9 Fuse margin read

5.8 *Refuse Key Unlock*

When the programming to the Fuse macro is done, it should be moved to the secure state. Sometimes if any error hits, it needs to be refused and need to be reprogrammed again according to fix the error. To refuse the already programmed fuse, the device should be moved to nonsecure state. Hence it is important to test the refuse key unlock of the fuse. The flow for this test sequence is as shown in Fig. 10.

The password for refusing should match with the refuse key, then only it is possible to program the fuse. The corresponding signals in the RTL to ensure the refusing needs to be checked to assure the test pass.

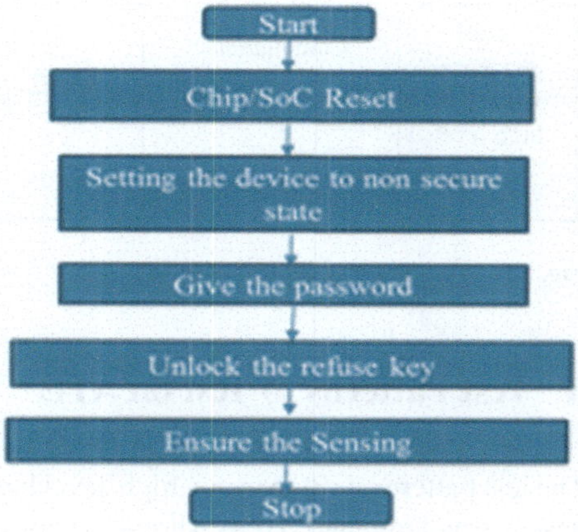

Fig. 10 Refuse key unlock

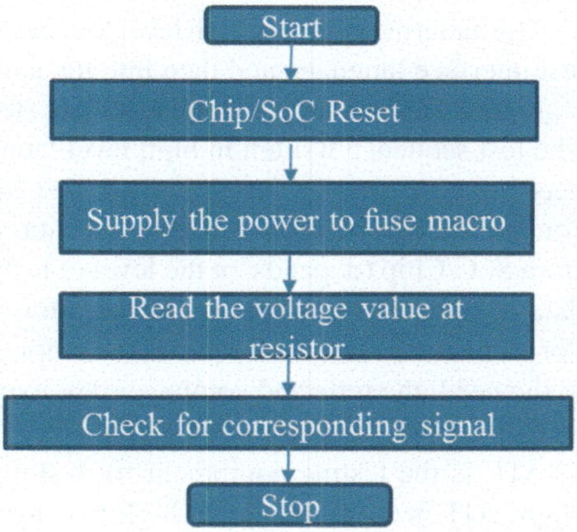

Fig. 11 Power shorts check

5.9 Power Shorts Check

This pattern ensures for no power shorts within the fuse macro. This test turns on the power for fuse macro. A test automatically starts when the signal corresponding this is enabled. A voltage value is read from here. This voltage value must be some specified value above 0. If the value is above the threshold the read of the corresponding signal ensuring for no shorts goes high and if the read voltage value is below the specified threshold the signal ensuring for no shorts goes low.

The flow of this test sequence is shown in Fig. 11.

The testing for any shorts between the power signals is also as important as checking of the test data shifting between the fuse SRAM and fuse Macro. The signals corresponding to the power shorts check present in the RTL needs to be checked.

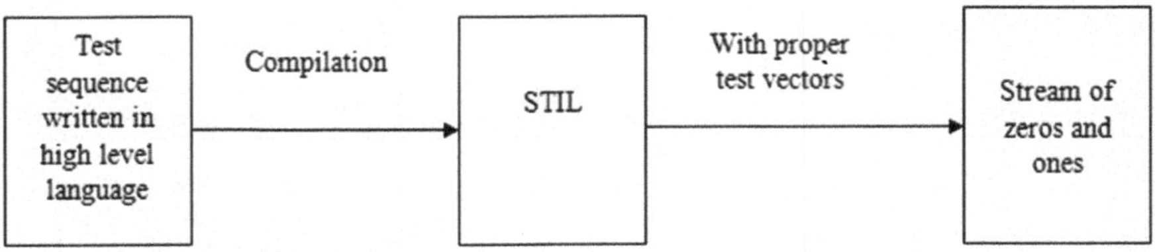

Fig. 12 Test pattern generation for ATE

6 Test Patterns to Test on ATE

The test patterns written in any high-level language or sometimes written in hardware description language (HDL) should be converted to the stream of zeros and ones to dump on to the ATE.

The patterns written in high level language needs to be converted first into Standard test interface language and then into the stream of zeros and ones.

Figure 12 shows the test intermediate steps to generate the test patterns for ATE. The test sequence written in high level language should be compiled clean and this needs to be converted into Standard Test Interface Language (STIL) [5]. STIL file contains the information about the test data capturing and shifting down the hierarchy from SoC/Chip (depends on the level of testing) till it reaches the corresponding test data register. STIL also contains the information about the number of test vectors. For some test cases, it is required to compress and expand the test vectors according to the need, the reflected compressed or expanded test vectors need to be verified in STIL file of the respective test sequence.

ATE is the testing equipment for testing the post silicon RTL or manufactured chips. ATE is the collection of high-performance computer controlled testing instruments. The test patterns required for functional testing (generated by ATPG), as well as structural testing need to be dumped on to the ATE in the form of stream of zeros and ones.

7 Visual Environment to Verify Simulation Results

Once the test patterns are made suitable to dump on ATE, before carrying out the ATE testing, test patterns should be simulated in EDA tools (like synopsys/cadence EDA tools) with debug platform, and results need to be verified in visual environment.

Once the status of test pattern (pass/fail) is verified in the visual environment by seeing the corresponding waveforms, the test pattern can be dumped on to the ATE. This increases the coverage of test pattern when the ATE based testing is done.

The debug platform of the EDA is helps when any error in the simulation of the test sequence hits. The point of occurrence of error is traced back to point of error and to fix it.

The test patterns for fuse macro explained in Sect. 5 are simulated in **Verdi a tool by synopsys** and the simulation results for those test sequences are listed in Sect. 8.

8 Results of Simulation

The test patterns for Fuse macro explained in Sect. 5 are simulated and the signals of respective test sequences are verified in the visual environment and are listed.

Figure 13 shows the result of Fuse sense SRAM. Here the signals ensuring the sensing is high.

Figure 14 shows the result of Fuse sense single row. The signals include the one for assuring the sensing of single row, and one showing the row number of the SRAM which is being sensed.

Figure 15 shows the result of Fuse programming. Here, the signal corresponding to the programming of the fuse macro is high and there is signal showing the programmed data.

Figure 16 shows the result of Fuse programming single row. Here, the signal assuring the programming of any row of the fuse macro is high.

Figure 17 shows the result of Fuse read SRAM. Here, the signal ensuring the reading of the SRAM is high. And there is a signal corresponding to the read data. Here the read data is zero.

Figure 18 shows the result of read of the single row of the SRAM. Here, the signal assuring the read of the single row of the SRAM is high and there is a signal corresponding to the row of the SRAM and read data of the row.

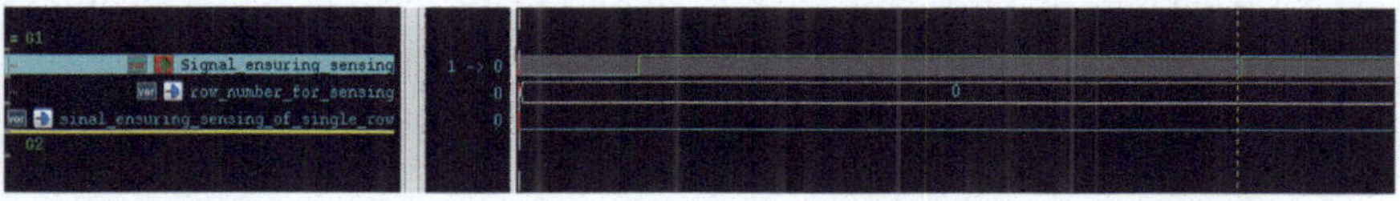

Fig. 13 Result of fuse sense SRAM

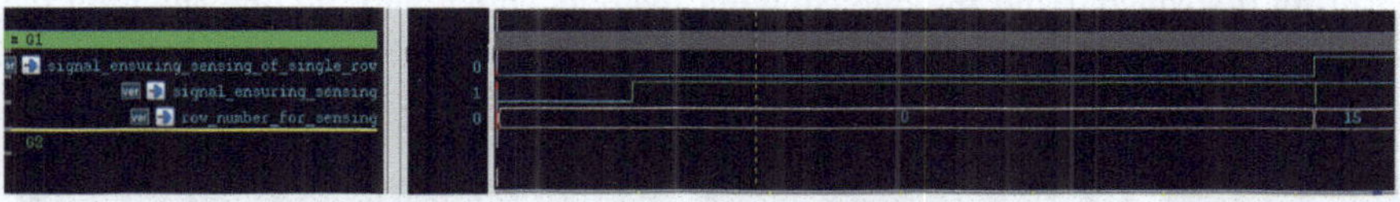

Fig. 14 Result of fuse sense single row

Fig. 15 Result of fuse programming

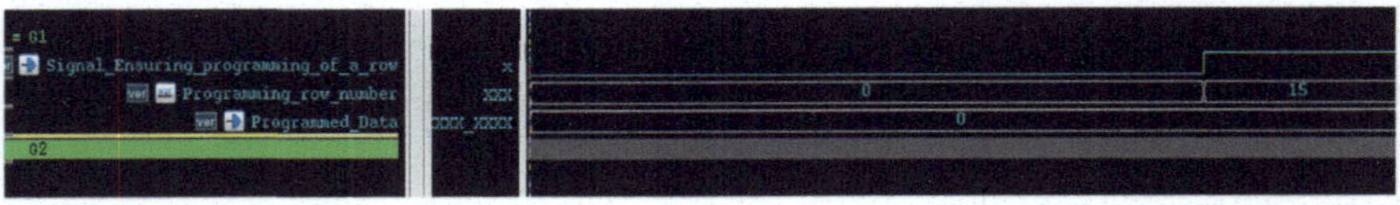

Fig. 16 Result of fuse programming single row

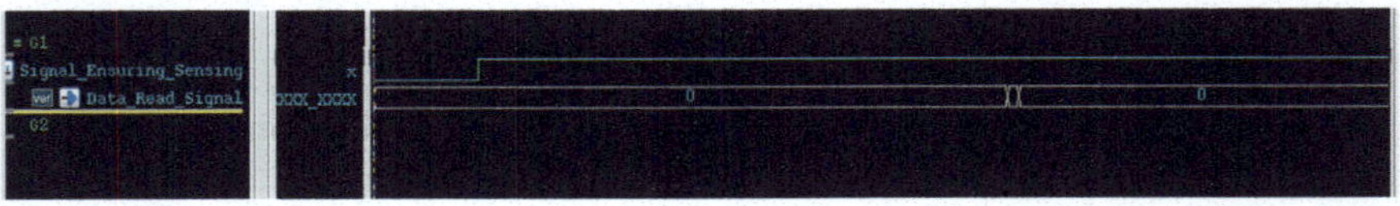

Fig. 17 Result of read SRAM

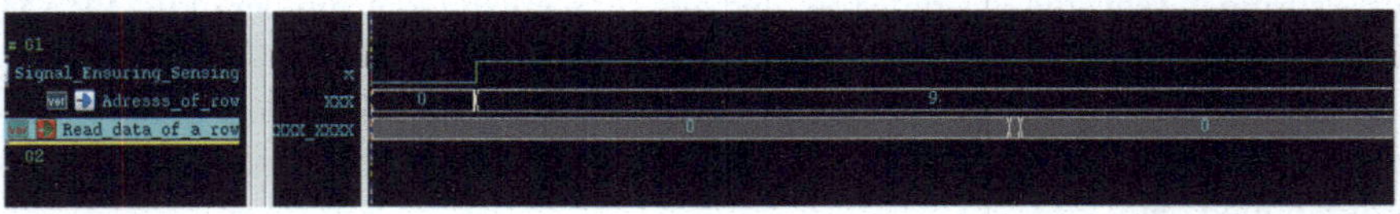

Fig. 18 Result of read SRAM single row

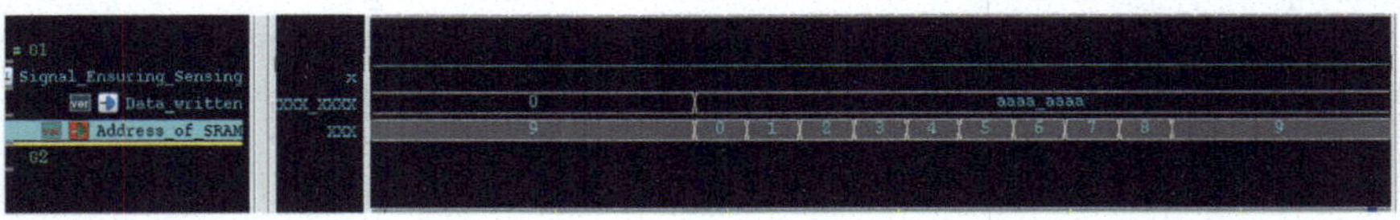

Fig. 19 Result of write SRAM

Figure 19 shows the result of the write to the SRAM. Here, the length of the SRAM to where the data is written is chosen as 9. There is signal showing the written data.

Figure 20 shows the result of the write to single row of the SRAM. There is a signal ensuring the write data and row number of the SRAM.

Figure 21 shows the result of the SRAM check. Here, the complete data shifting (writing and reading back) is tested in the same sequence. There are signals one ensuring the sensing and the one showing the data signal and the one showing the address of the SRAM.

Fig. 20 Result of write SRAM single row

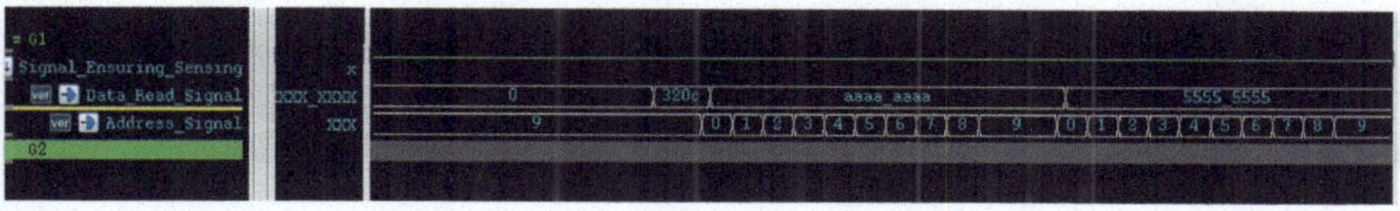

Fig. 21 Result of SRAM check

Fig. 22 Result of blank check

Fig. 23 Result of fuse margin read

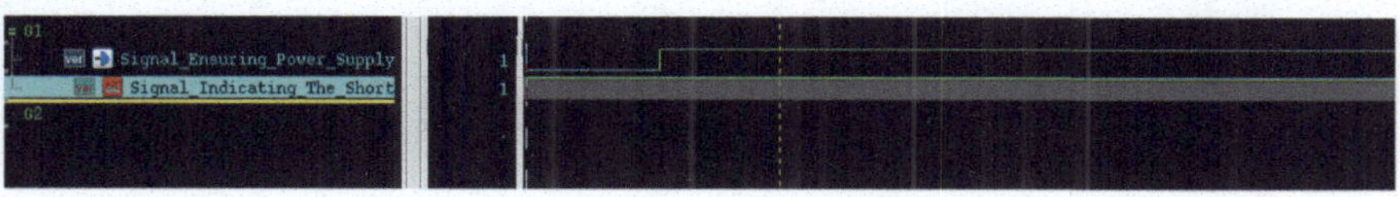

Fig. 24 Result of power short check

Figure 22 shows the result of the blank check, The signals in the figure includes one ensuring the sensing and one having the read data from the fuse macro. Since this is the blank check the data read from the fuse macro is zero.

Figure 23 shows the result of Fuse margin read. In this test case, fuses sensed from the fuse macro are compared with the reference resistance. If the margin read fail signal is high, then it implies that fuse is not programmed strongly. Here, the signal corresponding to the margin read fail is zero.

Figure 24 shows the result of power shorts check. To check for any power shorts first, the power to fuse macro need to be applied. The corresponding signals of the test sequence are included in Fig. 24.

Figure 25 shows the results of refuse key unlock. When the refuse password matches with refuse key, the key unlocks and in this nonsecured state, the fuse can be reprogrammed and can be fused again. In Fig. 25, the refuse password given is zero and hence result of the refuse key unlock is also zero.

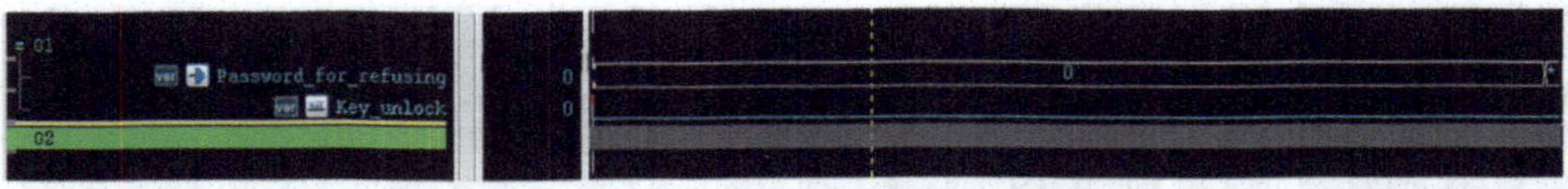

Fig. 25 Result of refuse key unlock

9 Conclusion

In this paper, the description for various test patterns required for structural testing of one of the unique components fuse macro of the SoC is given importance of testing the data shifting to and from fuse macro and fuse SRAM is described. Since in modern SoCs, the DFT concept is included, accessing the test data registers has become easy. If the structural testing of any functional block is proper at the chip level as well as SoC level in multichip module, there will be an assurance that functional testing data reaches the RTL correctly. Since the test sequences of every component of SoC differ, even though the DFT associated with RTL remains the same, they cannot be compared with one another.

Here, the description about only one component of the SoC is given, and in a similar way, the structural testing needs to be carried out for other blocks of the SoC depending on the functionality of the block.

References

1. Joint Test Action Group. In: IEEE standard for test access port and boundary-scan architecture
2. Joint Test Action Group. In: IEEE standard for access and control of instrumentation embedded within a semiconductor device
3. Joint Test Action Group. In: IEEE standard testability method for embedded core-based integrated circuits
4. Chauhan J, Panchal C, Suthar H (2017) Scan methodology and ATPG DFT techniques at lower technology node. In: International conference on computing methodologies and communication (ICCMC), pp 508–514, 18 July 2017
5. IEEE Standard for Extensions to Standard Test Interface Language (STIL) (IEEE Std 1450-1999) for Test Flow Specification Sponsor Test Technology Standards Committee of the IEEE Computer Society, Approved 6 Dec 2017, IEEE-SA Standards Board

Analysis of Growth Rate of Tikka Disease Using Image Processing

Meena Singh, B. P. Singh and Ekta Rewar

Abstract The assessment of growth rate of disease is considered as an important parameter for near accurate estimation of disease severity as well as plant life to maximize the yield. The selection of most appropriate method to identify the severity and growth rate of disease is considered as a necessity of agriculturist. The involvement of engineering presimulation technique is one form available in the current methodologies that allows the prediction and estimation of growth rate of disease in appropriate direction. The current study emphasizes on presimulation technique adopted for the assessment of growth rate of Tikka disease on Groundnut crop. The presimulation technique is based on the assortment and selection of artificial data sets. The assortment of artificial data sets is done based on the real data sets collected from the Groundut crop field. Image processing technique in C# language platform is used to create the artificial data sets. These data sets are prepared for both with and without fungicidal applications. A novel algorithm is proposed to effectively predict the rate of increase of Tikka disease using image processing techniques.

Keywords Groundnut leaf · Tikka disease · Early leaf spot · Late leaf spot · Fungicides

---

M. Singh (✉)
Department of ECE, University Polytechnic, Birla Institute of Technology, Mesra, Ranchi 835215, India
e-mail: meena71_singh@rediffmail.com

B. P. Singh · E. Rewar
Department of ECE, Manipal University Jaipur, Jaipur, India
e-mail: bpsinghgkp@gmail.com

E. Rewar
e-mail: ektarewar6@gmail.com

© Springer Nature Singapore Pte Ltd. 2019

V. Nath and J. K. Mandal (eds.), *Proceedings of the Third International Conference on Microelectronics, Computing and Communication Systems*,
Lecture Notes in Electrical Engineering 556, https://doi.org/10.1007/978-981-13-7091-5_45

1 Introduction

Plant diseases cause major economic losses every year to the farmers worldwide. The research in agriculture always aims to increase the productivity and quality of the field produce to achieve higher profits. Newer and newer methods are evolved to resolve this issue [1]. Earlier available technique is based on the identification of disease severity with the help of naked eyes. The observations for such analysis are under the hands of expert person in the field [2]. The visual observations for detecting disease in the large agricultural fields are always expensive, time consuming and highly labour intensive. The accuracy and precision of human vision approach totally depends on the eyesight of the person or the expert hired. The adeptness of this traditional method (naked eye observation) is still accepted in the fields by agriculturists.

The image processing technique has proved to be an efficient method that appropriately controls the parameters related to plant disease to achieve higher accuracy and lower financial outlay. Image processing techniques go into finer details of visual symptoms [3]. When leaf gets infected, it displays a range of symptoms like coloured spots, streaks, etc. The symptoms of disease show a continuous variation in colour, shape and size as it deepens [4]. These attributes are used in image processing for detecting and predicting the severity of disease to forecast its effect on the yield.

Many advanced computing techniques have been developed to help farmers to monitor proper growth of crops and to identify disease so that effective measures can be taken before yield gets too much affected. Digital image processing is one of the accurate techniques proposed by Lindow and Webb [5] in 1983. Other techniques based on thresholding [6–8], colour analysis [9, 10], fuzzy logic [11–15], colour transformation [16], Otsu method [17], k-means clustering [1, 18–20], fast computing algorithm [21–23], neural network [1, 24, 25], etc. have been proposed to detect diseases on different types of plants.

These literature surveys are yet to reveal the rate of increase of foliar disease. Hence, the proposed research work has been selected in this direction.

2 Methodology

The image processing-based methodology has been adopted for the predication and analysis of the growth of Tikka disease. The cluster of artificial data sets has been created in C# programming tool. The methodology consists of two steps; first is the creation of artificial data sets and the second is the determination of growth rate of disease by histogram analysis and colour processing. Figure 1 shows the block diagram of the algorithm to detect the rate of increase of disease severity using artificially created data set. The steps required in flowchart are discussed below:

Fig. 1 Block diagram of methodology designed

Fig. 2 Diseased samples of early leaf spots without fungicides on Groundnut leaves

Fig. 3 Diseased samples of late leaf spots without fungicides on Groundnut leaves

2.1 Creating Artificial Dataset

Data sets of Groundnut leaves were collected from the experimental field of Rajasthan Agricultural Research Institute, Durgapura, Jaipur (RARI). These data sets were collected in the form of images at regular intervals of 10 days in the duration from May 2016 to October 2016. Typical samples of Groundnut leaf infected by Tikka disease are shown in Figs. 2, 3, 4 and 5.

To start with the practical on the foliar disease of Groundnut crop, artificial data sets have been created to collaborate with the actual data sets. These data sets assumed approximately equivalent to the actually collected data sets (shown in Figs. 2, 3, 4 and 5) and depicted in Figs. 6, 7, 8 and 9.

Fig. 4 Diseased samples of early leaf spots with fungicides on Groundnut leaves

Fig. 5 Diseased samples of late leaf spots with fungicides on Groundnut leaves

Fig. 6 Artificial data sets of early leaf spot without fungicides

2.2 *Histogram Analysis*

Histogram is a graph or plot, showing number of pixels in an image at each intensity value. An 8-bit greyscale image has 256 (from 0 to 255) possible intensities. So, the histogram graphically displays 256 intensity lines or numbers showing the distribution of pixels corresponding to the greyscale values. The representation of histogram may be drawn for each colour, as well as, for the combination of Red, Green and Blue (RGB) channels.

Fig. 7 Artificial data sets of late leaf spot without fungicides

Fig. 8 Artificially data sets of early leaf spot with fungicides

Fig. 9 Artificially data sets of late leaf spot with fungicides

2.3 Growth Rate of Disease

The estimation of disease severity based on growth rate is predicted using a combination of pixels obtained from RGB values. The graph is drawn using those values.

3 Algorithm Designed

The human vision system is more sensitive to colour information. Colour is the most significant feature that humans perceive while viewing images in the real world. All the existing colours are the combinations of three primary colours, namely Red (R), Green (G) and Blue (B). These colours (RGB) are distributed in the individual layer. The change in colour of a leaf shows an inclination toward disease severity. A novel algorithm has been proposed to predict the severity of Tikka disease from histogram analysis by extracting the values of RGB pixels. The proposed algorithm consists of five different parts are the following:

- Creation of a healthy leaves,
- Creation of disease on leaves,
- Histogram analysis,
- Extracting the values of RGB pixels, and
- Estimating the rate of increase of disease using colour analysis.

Figure 10 shows the flow diagram of the algorithm designed for the growth rate of artificially created data sets of disease based on colour pixels.

The above proposed algorithm is coded in C# (programming language) for creating artificial data sets using unsafe command for the speed and quality of the program so that the program may not be affected. Bitmap data is used to read and process the RGB pixels in the image. To construct the histogram, first step is to "bin" the entire range of RGB pixel intensity into the series of the intervals. Thereafter, an array may be created to store the computed histogram of individual RGB channel. Further, the colour of each pixel in an image is obtained using a command of [picture.GetPixel(i,j)]. The array is presented by [histogram.drawhistogram(myvalues)]. A working principal of histogram analysis is based on artificial data sets created to find out the rate of increase of disease in the range of RGB from 0 to 255. Current problem suggest a colour combinations like $R = 0$, $G = 130$ ($\pm 5\%$) and $B = 0$ for the data sets in pixels of a healthy leaf. If the values of RGB are not within the prescribed limits (0, 130, 0), then increment in the values of RGB pixels has been considered by 4, 2 and 0.01 that declares the leaf to be diseased in %.

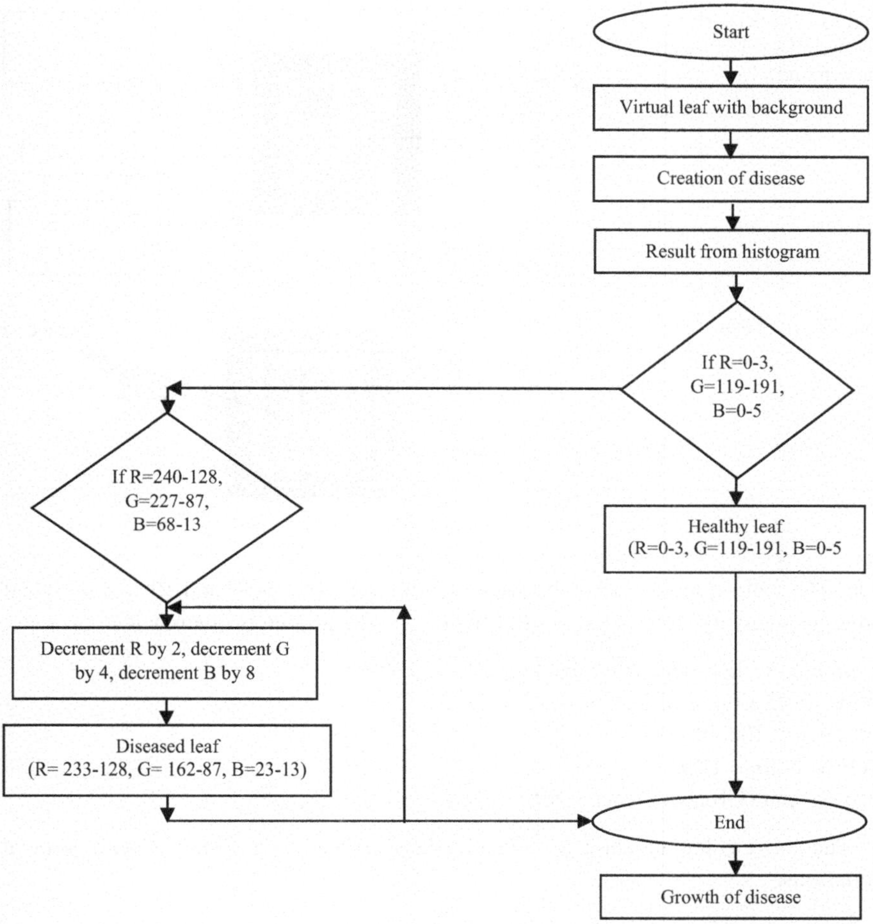

Fig. 10 Flowchart of algorithm designed

3.1 Mathematical Equation for Algorithm Designed

Mathematical equations have been formulated from the designed algorithm. These may suggest the exact solution for the data sets created for presimulation as well as for nearly exact solution of the real data sets or diseased leaf. Figure 15 shows artificially created diseased leaf with background for Tikka disease to formulate an appropriate equation.

As shown in Fig. 11, a square consisting of 100 * 100 pixels has been considered for the formation of equation. The whole size of 100 * 100 pixels is taken as the total area, i.e., 100%. Further, the area showing different colours in the same (leaf), i.e., brown, yellow, green, and white is calculated using software analysis (C#). The area considered in this case has diseased portion (brown pixels) of about 50% of total

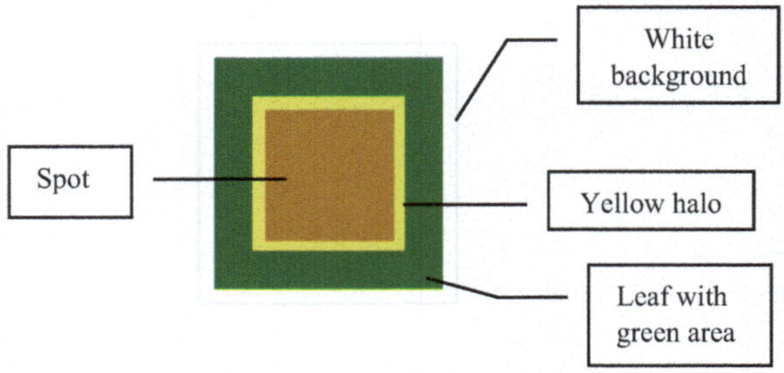

Fig. 11 Artificial Tikka disease infected leaf with background

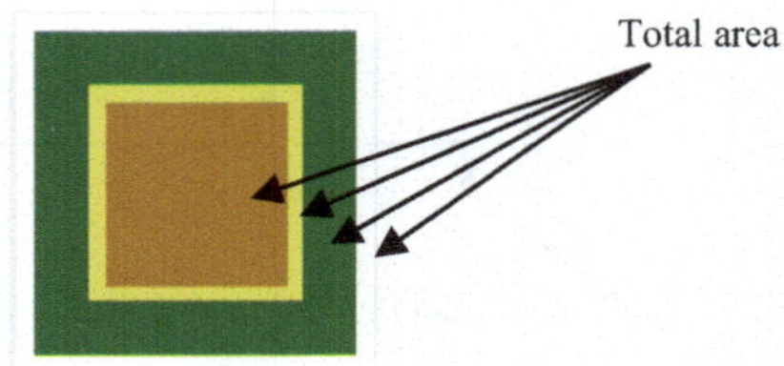

Fig. 12 Total leaf area

leaf. The yellow, green and white areas are about of 10%, 30% and 10% of total leaf area, respectively. The calculation of the diseased leaf area is given below:

Total area of leaf with white background = 100%
White background of leaf = 10%
Green area (healthy leaf) = 30%
Yellow halo = 10%
Brown spot (Diseased leaf) = 50%

The above calculated area has been further verified using designed mathematical equations.

- Equation for total area of a leaf

Equation for the total area of a leaf shown in Fig. 12 is given as

$$a_1 = \lim_{n \to 100} \sum_{i=0}^{n} \Delta i \left\{ \lim_{m \to 100} \sum_{j=0}^{m} \Delta j [f_0(i, j) + f_1(i, j) + \cdots + f_{mn}(i, j)] \right\} \quad (1)$$

where

$$\Delta i = \frac{n - i}{N} = \frac{100 - 0}{100} = 1$$
$$\Delta j = \frac{m - j}{M} = \frac{100 - 0}{100} = 1$$

a_1 total area of a leaf
i and j number of rows and column

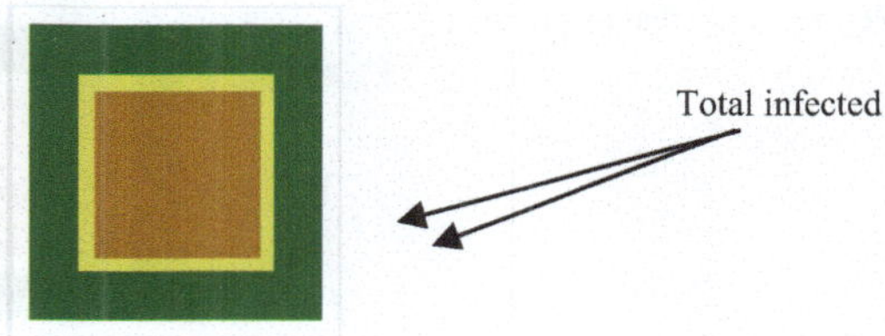

Fig. 13 Label representation of total infected area of a leaf

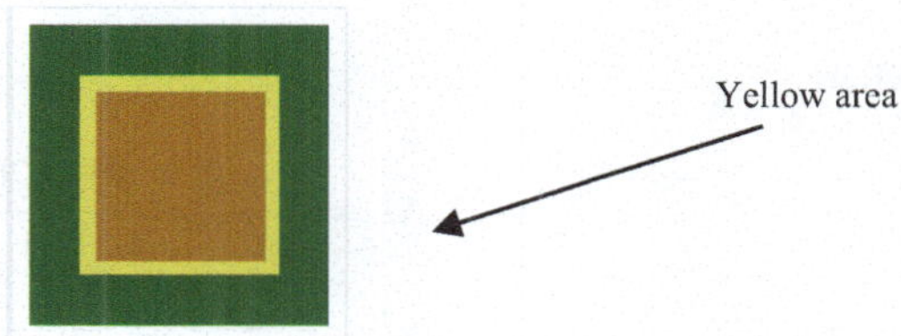

Fig. 14 Labelled yellow part of a leaf

n and m size of the square (100 * 100)
N and M height and width of square (in terms of pixels, i.e., 100 * 100).

- Equation of leaf for infected area

The total infected area as shown in Fig. 13 is expressed in terms of mathematical equation as

$$a_2 = \lim_{n \to 95} \sum_{i=5}^{n} \Delta i \left\{ \lim_{m \to 95} \sum_{j=5}^{m} \Delta j [f_0(i, j) + f_1(i, j) + \cdots + f_{mn}(i, j)] \right\} \quad (2)$$

where

$$\Delta i = \frac{n - i}{N} = \frac{95 - 5}{100} = 0.9$$
$$\Delta j = \frac{m - j}{M} = \frac{95 - 5}{100} = 0.9$$

a_2 infected area of a leaf
i and j number of rows and columns
n and m size of the square (100 * 100)
N and M height and width of square (in terms of pixels, i.e., 100 * 100).

- Equation of leaf for yellow area

The separate equation for yellow part for Fig. 14 is proposed as

$$a_3 = \lim_{n \to 80} \sum_{i=20}^{n} \Delta i \left\{ \lim_{m \to 80} \sum_{j=20}^{m} \Delta j [f_0(i, j) + f_1(i, j) + \cdots + f_{mn}(i, j)] \right\} \quad (3)$$

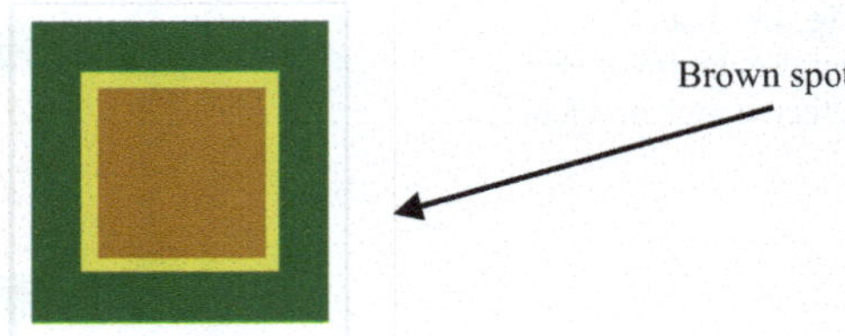

Fig. 15 Labelled brown part of a leaf

where

$$\Delta i = \frac{n - i}{N} = \frac{80 - 20}{100} = 0.6$$
$$\Delta j = \frac{m - j}{M} = \frac{80 - 20}{100} = 0.6$$

a_3 yellow area of a leaf
i and j number of rows and columns
n and m size of the square (100 * 100)
N and M height and width of square (in terms of pixels, i.e., 100 * 100).

- Equation of leaf for brown area

The separate equation for diseased part for Fig. 15 is proposed as below

$$a_4 = \lim_{n \to 75} \sum_{i=25}^{n} \Delta i \left\{ \lim_{m \to 75} \sum_{j=25}^{m} \Delta j [f_0(i, j) + f_1(i, j) + \cdots + f_{mn}(i, j)] \right\} \quad (4)$$

where

$$\Delta i = \frac{n - i}{N} = \frac{75 - 25}{100} = 0.5$$
$$\Delta j = \frac{m - j}{M} = \frac{75 - 25}{100} = 0.5$$

a_4 brown area of a leaf
i and j number of rows and column
n and m size of the square (100 * 100)
N and M height and width of square (in terms of pixels, i.e., 100 * 100).

Equations 1, 2, 3 and 4 suggest the values for the total area, total infected area, yellow area and brown area with the value 1, 0.9, 0.6, 0.5, respectively. The calculation for the same is given as

Total area of leaf with white background = 100%
White background of leaf = 100 – 90 = 10%
Green Area (healthy leaf) = 90 – 60= 30%

Yellow halo = 60 – 50 = 10%
Brown spot = 50%

3.2 Mathematical Equation for Healthy and Diseased Leaf

Mathematical equations have been designed for the leaf under different condition in the current section. The conditions for RGB values of diseased and healthy leaves are given as

$$Total\ leaf = \left\{\begin{array}{ll} 1 & if\ R = 0 - 3,\ 119 \le G \ge 191,\ B = 0 - 5 \\ 0 & if\ R = 240 - 128\ than\ R \downarrow 2 \\ & if\ G = 227 - 87\ than\ G \downarrow 4 \\ & if\ B = 68 - 13\ than\ B \downarrow 8 \\ NA\ otherwise & \end{array}\right\} \quad (5)$$

where 1 represents healthy leaf and 0 represents diseased leaf.

The elaborated form of equation has conditions for leaf as R lies between 0 and 3, G lies between 119 and 191 and B lies between 0 and 5. This indicates healthy condition of the leaf. Similarly, the diseased leaf shown has values of R in between 128 and 240 with an increment of 2; the values of G lie in between 87 and 227 with an increment of 4, and the values of B lie in between 13 and 68 with the increment of 8. Therefore, the individual equation has been formed for healthy and diseased leaf.

- Equation for healthy leaf

Equation for the healthy leaf for the above stated conditions (the values of R, G and B lying between 0 and 3, 119 nad 191 and 0 and 5) is expressed as

$$Healthy\ leaf = \left\{\begin{array}{l} \lim\limits_{n\to 3} \Delta i \sum\limits_{a=0}^{n} R(a) \\ \lim\limits_{n\to 191} \Delta i \sum\limits_{b=119}^{n} G(b) \\ \lim\limits_{n\to 5} \Delta i \sum\limits_{c=0}^{n} B(c) \end{array}\right\} \quad (6)$$

- Equation for diseased leaf

Equation for the healthy leaf for the above stated conditions (the values of R, G and B lying between 128 and 240, 87 and 227 and 13 and 68) is given as

$$Diseased\ leaf = \left\{ \begin{array}{l} \lim\limits_{n \to 240} \Delta i \sum\limits_{a=128}^{n} R(a) \\ \lim\limits_{n \to 227} \Delta i \sum\limits_{b=87}^{n} G(b) \\ \lim\limits_{n \to 68} \Delta i \sum\limits_{c=13}^{n} B(c) \end{array} \right\} \tag{7}$$

4 Result and Analysis

The graphs depicted in Figs. 16, 17, 18 and 19 show the growth of Tikka disease over time.

The growth rates of artificially created data sets of early and late leaf spots without fungicides with time are shown in Figs. 16 and 17. The increase of red colour along with the decrease in green colour and a slight increase in blue colour tend the image toward a brown colour, indicating the presence of disease on leaves. The contribution of each colour is justified from Figs. 16 and 17 and these show that the values of

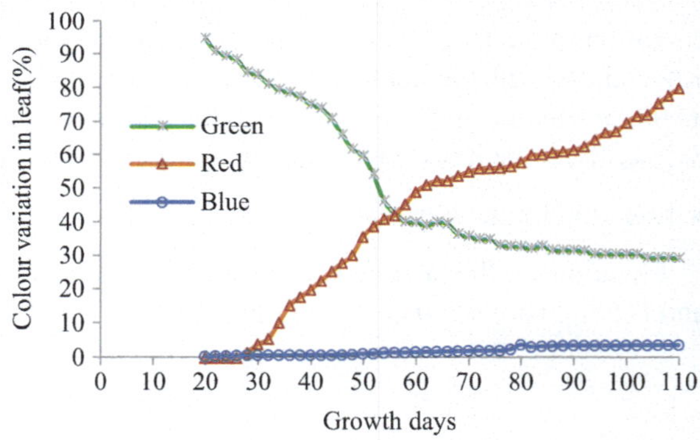

Fig. 16 Colour variation of early leaf spot without fungicide over time

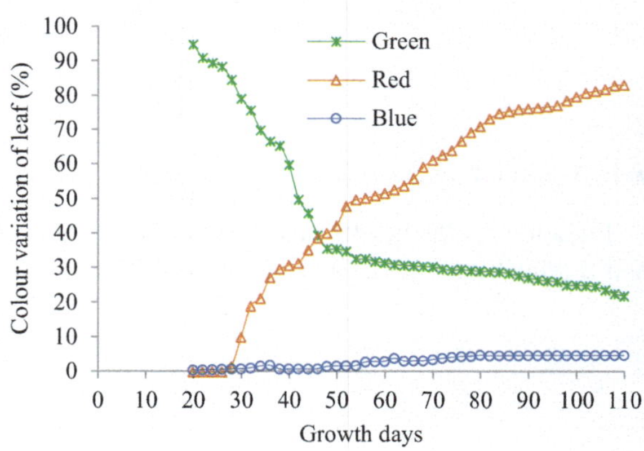

Fig. 17 Colour variation of late leaf spot without fungicide over time

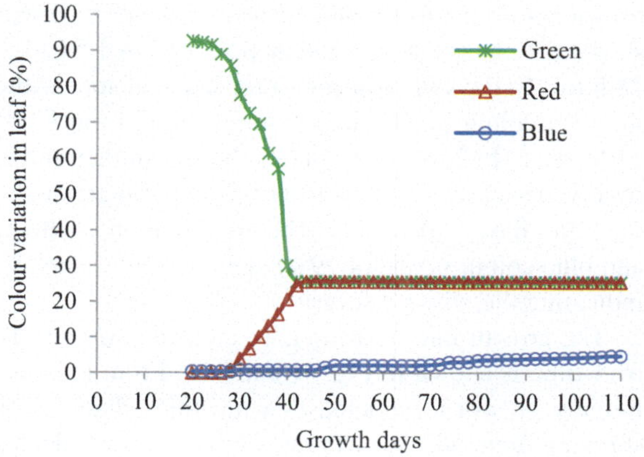

Fig. 18 Colour variation of early leaf spot with fungicide over time

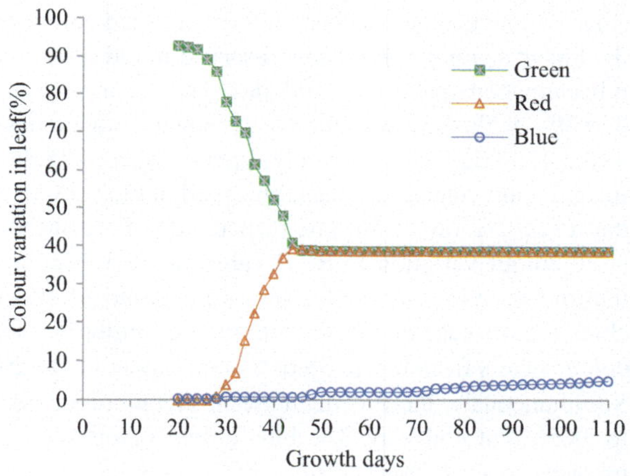

Fig. 19 Colour variation of late leaf spot with fungicide over time

green pixels have negligible contribution in formation and increment in brown colour to indicate the severity of growth rate of Tikka disease. The representation of healthy part in Groundnut leaf through artificial algorithm has been shown by green pixels. At the same time, it has been observed that the contribution of the red pixel value is more as Tikka disease symptom shows reddish brown lesions indicating an increase in disease severity. However, the blue pixels show cooperatively lesser involvement toward disease severity.

Initially in Fig. 16, no change has been observed in red colour pixels (0%) till 26th day and shows the linear increment in red colour up to 70th day and finally, the change reaches up to 79.93%. The occurrence of green colour pixels is 94.86% at 20th day after crop sown. A linear decrement has been observed up to the 70th day and decreased to 29.3% on 110th day, whereas relatively insignificant change has been observed in blue colour that may vary from 0.5% at 20th day to 3.83% on 110th day, respectively. The combination of green (29.3%), red (79.93%) and blue colour (3.83%) at 110th day shows the final variation in leaf colour in conjunction with brown colour pixels to indicate as a evidence of disease severity. Similarly, Fig. 13 shows the change in red colour pixels from 1.54 to 83.11% during 84 days (from 28th to 110 days of crop sown) for late leaf spot without fungicide. The decrement in green colour pixels shows similar trend from 94.86% at 20th day to 22.01% at 110th day. The blue colour pixels shows similar trend with different intensity level as observed in case of early leaf spot without fungicide, from 0.5% at 20th day to 4.79% on 110th day, respectively. The combination of the green (22.01%), red (83.11%) and blue colour pixels (4.79%) at 110th day shows the final pixels of brown colour indicating the, disease severity.

The growth rate of artificially created early and late leaf spots with fungicides over time is shown in Figs. 18 and 19. From the above graphs, it is observed that initially growth rate increases with time (change in colour of the pixel) but after spraying fungicides growth rate of disease starts decreasing and ultimately becomes constant at the age of nearly 40 days. Initially for early leaf spot with fungicides (Fig. 18), no change has been observed in red colour pixels (0%) till 26th day, after the linear increment has been observed in red colour up to 42nd day and thereafter, it became constant up to 110th day. The occurrence of green colour pixels has been 92.83% at 20th day after the crop is sown. A linear decrement has been observed up to the 42nd day and ultimately dipped down to 25.76% on 110th day. The relative insignificant change has been observed in blue colour, varying from 0.438% at 20th day to 5.02% on 110th day, respectively. It is observed that after 46th day, there is no colour variation in RGB values as they become constant. This may suggest that the spraying fungicides shows insignificant effect or no change in rate of the change of disease on Groundnut leaves. Similarly, Fig. 19 shows the change in red colour pixels from 0% to 38.68%, corresponding to 20th to 46th days, respectively. Simultaneously, there is decrease in green colour pixels from 92.83% at 20th day to 38.68% at 46th day. The blue colour pixels show a similar trend with different intensity levels in contrast to early leaf spot with fungicide from 0.438% at 20th day to 5.02% on 110th day, respectively. It is observed that after 50th day, there is no colour variation in RGB pixel values.

5 Conclusion

In this paper, presimulation methodology is adopted to simulate the growth rate of Tikka disease on groundnut leaves. An algorithm has been proposed for effective calculation of the growth rate of Tikka disease with and without fungicides by creating artificially data sets. The data sets have been generated in C# (programming language) using image processing technique and assumed to be equivalent to the real data sets collected from the experimental field of RARI. Thereafter, RGB pixel values have been extracted from histogram analysis to predict the growth of Tikka disease. Further, an equation has been proposed for the correct assessment of healthy as well as diseased leaf.

Present analysis using image processing indicates that the increase of red component, decrease of green component and slight increase of blue component in image tends toward brown colour to indicate the presence of disease on leaves. The analysis of growth of Tikka disease obtained through visual analysis and image processing are almost analogous.

References

1. Al Bashish D, Braik M, Bani-Ahmad S (2010) A framework for detection and classification of plant leaf and stem diseases. In: International conference on signal and image processing. IEEE, Chennai, pp 113–118
2. Ingram DS, Robertson NF (1999) Plant Disease. Harpercollins Publisher, pp 1–288
3. Deshmukh Kripali S (2012) Disease detection of crop using hybrid algorithm. Int J Eng Res Technol 1(10):1–5
4. Agrios GN (2005) Plant pathology. Elsevier Academic Press, pp 1–922
5. Lindow SE (1983) Webb RR quantification of foliar plant disease symptoms by microcomputer-digitized video image analysis. J Phytopathol 73(4):520–524
6. Tucker CC, Chakraborty S (1997) Quantitative assessment of lesion characteristics and disease severity using digital image processing. J Phytopathol 145(7):273–278
7. Skaloudova B, Krvan V, Zemek R (2006) Computer-assisted estimation of leaf damage caused by spider mites. Comput Electron Agric 53(2):81–91
8. Macedo-Cruz A, Pajares G, Santos M, Villegas-Romero I (2011) Digital image sensor-based assessment of the status of Oat (Avena sativa L.) crops after frost damage. Sensors 11(6):6015–6036
9. Wiwart M, Fordonski G, Zuk-Golaszewska K, Suchowilska E (2009) Early diagnostics of macronutrient deficiencies in three legume species by color image analysis. Comput Electron Agric 65:125–132
10. Pugoy RADL, Mariano VY (2011) Automated rice leaf disease detection using color image analysis. In: 3rd international conference on digital image processing. SPIE, Hengdu, vol 8009, pp F1–F7
11. Sannakki SS, Rajpurohit VS, Nargund VB, Kumar A (2011) Leaf disease grading by machine vision and fuzzy logic. Int J 2(5):1709–1716
12. Sekulska-Nalewajko J, Goclawski J (2011) A semi-automatic method for the discrimination of diseased regions in detached leaf images using fuzzy C-means clustering. In: VII international conference on perspective technologies and methods in MEMS design. IEEE, Polyana-Svalyava, pp 172–175
13. Zhou Z, Zang Y, Li Y, Zhang Y, Wang P, Luo X (2011) Rice plant-Hopper infestation detection and classification algorithms based on fractal dimension values and fuzzy C-means. Math Comput Model 58:701–709
14. Sannakki Sanjeev S, Rajpurohit Vijay S, Nargund VB, Kumar R Arun, Yallur Prema S (2011) Leaf disease grading by machine vision and fuzzy logic. Int J Comput Sci 2(5):1709–1716
15. Jagtap Sachin B, Hambarde Shailesh M (2014) Agricultural plant leaf disease detection and diagnosis using image processing based on morphological feature extraction. ISOR J VLSI Signal Process 4:24–30
16. Amoda Niket, Jadhav Bharat, Naikwadi Smeeta (2014) Detection and classification of plant disease by image processing. Int J Innov Sci Eng Technol 1(2):70–74
17. Badnakhe Mrunalini R, Deshmukh Prashant R (2012) Infected leaf analysis ad comparison by Ostu threshold and K-means clustering. Int J Adv Res Comput Sci Softw Eng 2(3):449–452
18. Kulkarni AH, Ashwin Patil RK (2012) Applying image processing technique to detect plant diseases. Int J Modern Eng Res 2(5):3661–3664

19. Renugambal K, Senthilraja B (2015) Application of image processing techniques in plant disease recognition. Int J Eng Res Technol 4(3):919–923
20. Bashir Sabah, Sharma Navdeep (2012) Remote area plant disease using image processing. IOSR J Electron Commun Eng 2(6):31–34
21. Camargo A, Smith JS (2009) Image pattern classification for the identification of disease causing agents in plants. Comput Electron Agric 66(2):121–125
22. Revathi P, Hemalatha P (2012) Classification of cotton leaf spot diseases using image processing edge detection techniques. In: International conference on emerging trends in science, engineering and technology, ISBN: 978-1-4673-5144-7
23. Anami Basavaraj S, Suvrna S Nandyal, Govardhan A (2010) A combined color, texture and edge feature based approach for identification and classification of indian medicinal plants. Int J Comput Appl 6(12):45–51
24. Huang KY (2007) Application of artificial neural network for detecting phalaenopsis seedling diseases using color and texture features. Comput Electron Agric 57:3–11
25. Kai S, Zhikun L, Hang S, C Hunhong G (2011) A research of maize disease image recognition of corn based on BP networks. In: Third international conference on measuring technology and mechatronics automation. IEEE, Shangshai, pp 246–249

Development of Autonomous Garbage Collector Robot

Anukriti Jha, Anshuman Singh, Roshan Kerketta, Deepak Prasad, Kumari Neelam and Vijay Nath

Abstract The problem of garbage collection and dumping in India is not a hidden fact. Due to rapid economic and technological growth, India faces many other problems like corrosive urban planning, poor civic amenities, overpopulation. Till now, no steps have been taken to manage the garbage properly has worked as per expectations. And, we are on a constant lookout for innovative ways of managing the mammoth task of garbage collection and segregation. This paper is based on the problem of garbage and it looks forward to provide a solution, namely the autonomous garbage collecting bot works with the Arduino controller. The robot is built on an aluminium base of size 30 × 40 cm which is powered by a battery of 12 V, 7.5 A. The robot movement is controlled by motors connected to the programmed Arduino. The robot can only be designed to move on hard, cemented surfaces. When it senses an obstruction, it follows the code and proceeds to lift the garbage as per the designed mechanism. All this can be achieved through Arduino which is the CPU of the bot.

Keywords Arduino · Swacch Bharat Abhiyaan · MSMWWs

1 Introduction

Refuse collection and segregation is the major problem not only in urban areas but also in villages and small towns of India. Indian cities alone generate more than 100 million tons of solid waste a year. In 2000, India's Supreme Court directed all Indian cities to implement a waste management programme that would include household collection of segregated waste, recycling and composting. These directions have had no effect till now. No major city runs a comprehensive programme of

A. Jha · A. Singh · R. Kerketta
Department of Computer Science and Engineering, Birla Institute of Technology, Mesra, Ranchi 835215, Jharkhand, India

D. Prasad (✉) · K. Neelam · V. Nath
Department of ECE, Birla Institute of Technology, Mesra, Ranchi 835215, Jharkhand, India
e-mail: prasaddeepak007@gmail.com

© Springer Nature Singapore Pte Ltd. 2019

V. Nath and J. K. Mandal (eds.), *Proceedings of the Third International Conference on Microelectronics, Computing and Communication Systems*, Lecture Notes in Electrical Engineering 556, https://doi.org/10.1007/978-981-13-7091-5_46

the kind as directed by the Supreme Court. The small towns and villages are far from implementing it. These robots which are in use have various cons like high cost, not user-friendly and can collect garbage only from cans that are full of refuse. Also, Municipal solid waste workers (MSWWs) or refuse collectors are exposed to health hazards like respiratory problems, or accidental hazards. They also have to face toxic waste which can prove to be very detrimental to their health. There are certainly many health hazards associated with mismanagement of garbage and rubbish as well as injuries caused by work-related accidents. In order to help in garbage collection and revolutionaries the whole process, the idea of an autonomous garbage collector came into existence which aims to provide cleaner surroundings and footpaths. This robot can sense obstacles through the two types of sensors attached to it. The Arduino is then programmed to move according to the set instructions along a predefined path. After the refuse bin is full, the sensors attached to it send the signal through the WiFi module to a web page. Hence this project even incorporates the principle of Internet of Things. The robot moves then to empty the trash bin in a designated place. Internet of Things/Everything: An overview [1] demonstrated the use of IoTs. Design Strategy for Smart Toll Gate Billing System [2] describe the smart networking setup. The design of Smart Embedded System for Agricultural Update Using Internet of Things [3] described the utility of wide applications of IoT in agriculture. The study and Design of Smart Embedded System for Train Track Monitoring Using IoTs [4] describes the utility of embedded systems in train track monitoring for safety and security. Study and Design of Smart Embedded System for Smart City Using the Internet of Things [5] describes the wide utility of the Internet of Things for developing the smart cities in the country. Study and Design of Smart Embedded System for Remote Health Monitoring Using the Internet of Things [6] describes the applications of IoTs in health sectors for diagnosing and cure of patients in remote village areas. The design of Smart Embedded System for Auto Toll Billing System Using IoTs [7] explains the plan of activities how to integrate all toll billing system together in the country and collect toll automatically without vehicle stop and delay. Different Aspects of Smart Grid: An Overview [8] describes the utility of IoTs to develop smart grid for power generation, storage, transmission, distribution, revenue collection with minimum power loss. Automated Toll Plaza Using Barcode Laser Scanning Technology [9] describes the better technique than RFID in new toll plaza. The design of All-Terrain Rover Quadcopter for Military Engineering Services [10] explains the monitoring of our military and giving information about security and safety. The design of Robot Monitoring System for Aviation [11] describes the monitoring strategy of all types of machinery in space. Electronic Toll Collection System Using Barcode Technology [12] demonstrate the new techniques in toll collection centre for collecting revenue without any delay from passing vehicles. Design of Smartphone Controlled Robot Using Bluetooth [13] describes the mobile strategy to control the robot working in remote and security areas. Design of Earthquake Indicator System Using ATmega328p and ADXL335 for Disaster Management [14] demonstrate the role of observation of pre-disaster activities. The study and Design of Smart Embedded System for Aviation System: A Review [15] explains the jest work done in these areas. Study and Design of Smart Industry: A Review [16] explains the components required to develop a smart city. Design of Smart Security Systems for Home Automation [17] demonstrate the operation, control and maintenance of home appliances through mobile. Auto-Train Track Fault Detection System [18] defines the different types of problem occurred to track fault.

2 Methodology

The said [2–18] references are giving good ideas for starting the new work as smart garbage collector systems using IoTs. In this paper, the design of the autonomous garbage collecting bot uses engineering method. First, the needs of Municipal authorities are analysed, and then the basic operations which would be performed by the bot are decided. After that, the circuitry, the input and output devices are decided upon, then a basic framework is built. The basic points to focus upon are illustrated in the figure below. The functioning of the bot can be classified into four main categories. They are motion control of the bot, obstacle detection, garbage collection and disposal of garbage.

2.1 Movement of the Robot

The bot travels on the predetermined path by using the Arduino which is linked with a combination of motors, drivers and sensors consisting of four geared motors of 35 rpm each, motor drivers, three infrared sensors and three ultrasonic sensors on every three sides. The infrared sensors give input to the Arduino Infrared sensors are more sensitive. As soon as garbage is detected or a light obstacle is encountered, the base of the robot rotates and the whole setup faces the direction of the garbage. Arduino is programmed to move in that direction. Then, when the robot actually touches the refuse, which is detected by the less sensitive ultrasonic sensors, Arduino is programmed to start the tilting of the shafts and the rotation of the armature.

2.2 Garbage Collection

The whole setup is on a rotating base. The geared motors move the armature which in turn rotates the rectangular shafts attached. This mechanism picks up the garbage. The robot does not always pick up garbage every time it encounters an obstacle, it only operates for a specific size of the obstacles and certain other predefined conditions as programmed in the Arduino after receiving input from the sensors. Whenever the bot will encounter an obstacle, the Arduino is programmed to perform specific operations. The garbage collecting plates are made of aluminium to suit outdoor usage as well as to be durable enough to last long. Aluminium is light, corrosion free and cheap. Hence, it is a good choice for making the body of the bot. The main motive or objective of our bot is to collect garbage of small sizes which are usually found on the streets and dry leaves, papers and plastic. The shape of the bot is cubical and rounded at the edges to move easily in narrow, confined places. The entire circuitry of the bot is set up above the base and proper ground clearance is provided so that it moves unaltered by minute particles or grains. The motors are installed on the base and connected directly to the already programmed Arduino for proper movement of the bot. The robot can move only on flat, tough cemented surfaces for a proper operation like footpaths. When the infrared sensors detect a stationary obstacle, the garbage collecting plates rotate and the garbage is put into a collecting bin provided at the rear of the bot. The robot keeps collecting the garbage until the bin is full. This is again determined through the infrared sensors attached to the bin which detect the depth of the bin. Once the garbage is picked up, it is dumped [19].

2.3 *Disposal of Garbage*

The collecting bin is built such that it can be tilted. Three infrared sensors are used which improve the functionality of the robot tremendously. They are attached to the three sides of the robot, front, left and right. Whenever the infrared sensors detect a small obstacle, the whole setup rotates in that direction. The robot's Arduino is programmed to move in that direction towards the refuse. Then, as it gets closer, less sensitive ultrasonic sensors detect the garbage and Arduino gives directions for shaft rotation and tilting action which is actually responsible for picking up the garbage. The circuitry containing the AVR microcontroller, when it gives direction that the bin is full, the robot moves so as to dispose the garbage at a proper designated place. The data that the bin is full, is sent through the WiFi module [20].

3 Block Diagram

Figure 1 shows the various blocks involved in the proposed architecture.

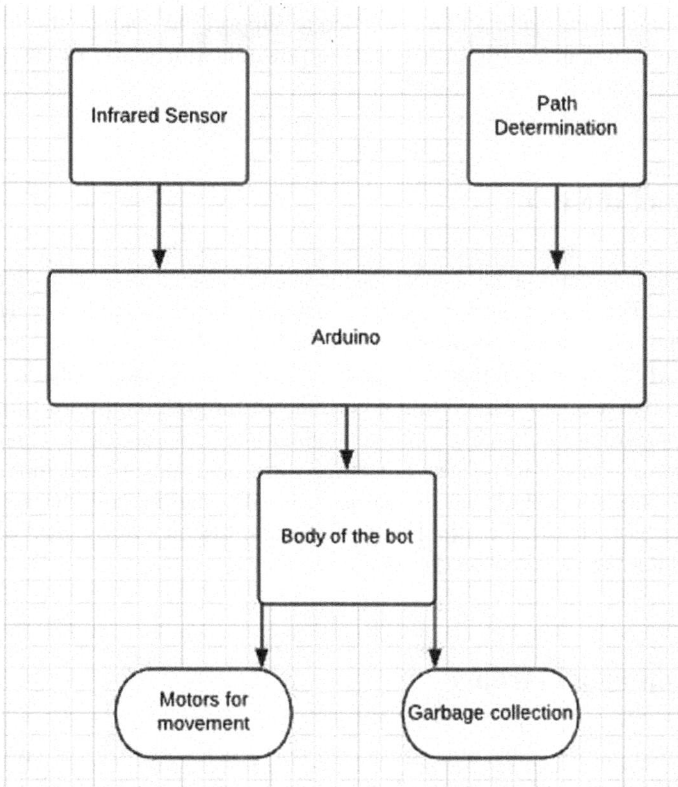

Fig. 1 Basic building block of autonomous garbage collector

4 Required Circuitry

The following components are required to complete the circuits:

i. For the movement of the bot, we have a system of motors which helps the bot move to accomplish its task.
ii. A power source so that the bot could extract energy and function.
iii. The bot is provided with a microcontroller i.e. Arduino Uno kit with adapter, so that we could predefine its path and response to obstruction.
iv. Autonomous and safe motion of the robot is obtained by using infrared sensors attached to the microcontroller.

Other components are added as per the needs of the circuits. Arduino Uno is the controller that governs the functionalities of the three stages to meet the major requirements. It is a microcontroller board based on the ATmega328. It has 14 digital input/output pins (of which 6 can be used as PWM outputs), 6 analog inputs, a 16-MHz ceramic resonator, a USB connection, a power jack, an ICSP header and a reset button. It contains everything that is needed to support the microcontroller; simply connect it to a computer with a USB cable or power it with an AC-to-DC adapter or battery to get started.

5 Setup of the Bot

The robot has a rectangular base on which the whole mechanism is set up. It measures 30 × 40 cm. The robot has four wheels which are rotated through a geared motor. Installation of the motors is such as to allow the bot to accurately detect the garbage. Above the wheel and motor, there is a flat base. This base is made so that it can rotate with the complete setup. This rotation is achieved through the power motor. Then there is a long armature on which flat, rectangular blades are attached. This armature can also rotate. On the top of the bot, three ultrasonic sensors and three infrared sensors are placed which detect the garbage. Finally, there is a garbage collecting bin near the base. This bin is tilted to dispose the refuse at a designated place. The circuitry of the bot is placed on the rotating base [21].

6 Collection of Garbage

There are two garbage bins attached to the bot which serve for the collection of garbage. Ultrasonic sensors are attached to the bins which calculate the depth of the bin every time and transmit that data over the web server through the WiFi module. Thus, the principle of IoT is also implemented through this research work. The sensor board has AVR microprocessor, a buzzer which acts as an output device when the bin is full and also an LED. The ultrasonic sensors calculate the depth of the bin and this information is processed by the microcontroller which is programmed to do further actions which are shown in Fig. 2.

7 Scope of Improvement

Basic microprocessor chip, Arduino can be upgraded to MyRIO. For better navigation of the bot, navigation cameras can be attached which can further improve the functionality. Segregation of waste if done at the time of collection can further improve its usage in municipal waste management. For segregation of waste, proper sensors can be utilized to segregate the waste into degradable and non-degradable waste, depending on dry and wet waste. Finally, an LED screen can be attached with bin to improve the user interface. WiFi models of setup are shown in Fig. 3 which collect the information of fill or empty collection pot.

Fig. 2 Bin sensor and buzzer with data transmitted through WiFi module

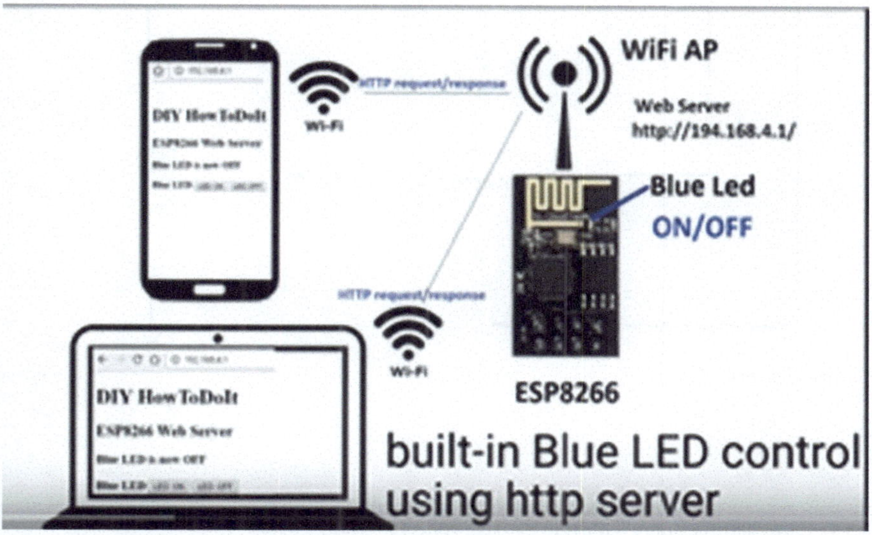

Fig. 3 A basic WiFi module showing transmittance of data

8 Result

This is just a proposal and the working model is yet to be prepared which is estimated to function according to the need. It can surely work at local municipal level for autonomous garbage collection and improve manpower usage which otherwise would have been wasted on collection and segregation of garbage. Improvement in garbage collector setup shown in Fig. 4.

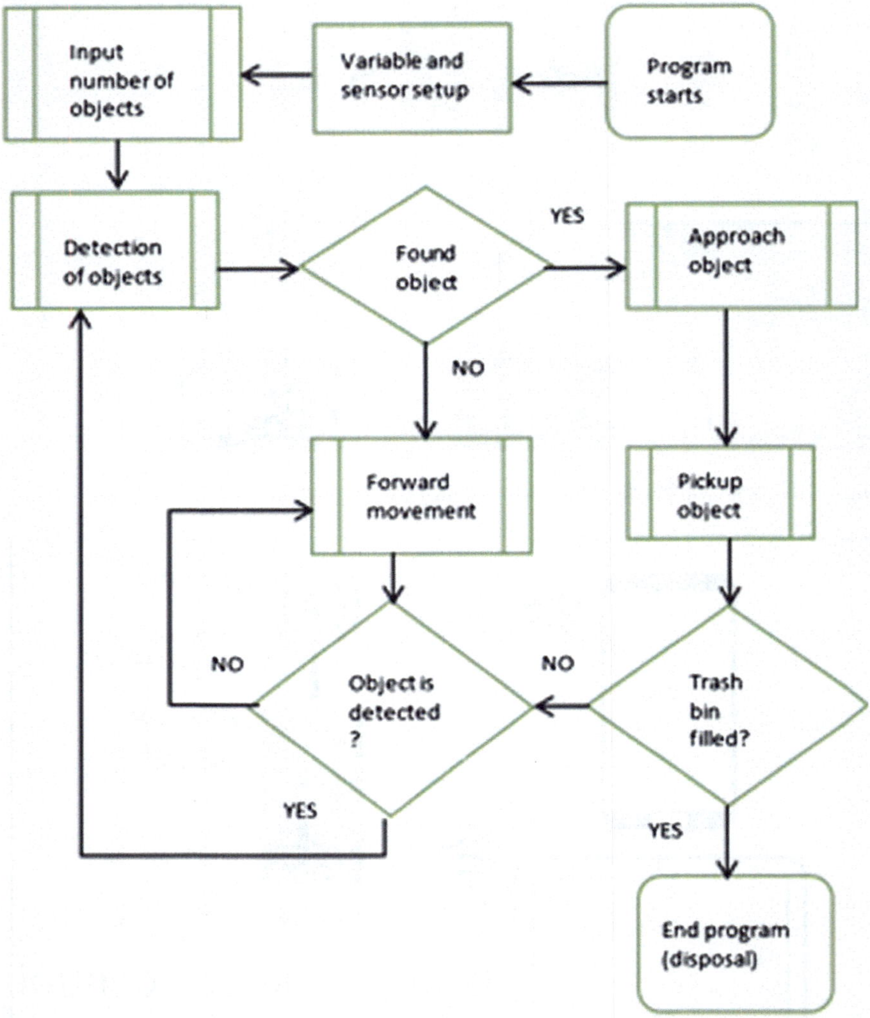

Fig. 4 Flowchart for improvement in garbage collector setup

9 Conclusion

Collection, processing and segregation of garbage is a major part of the municipal workload. Making this autonomous using a garbage collector robot can actually help in saving work power which would otherwise be wasted in this futile work. Majority of the work can be taken care of by this bot. It cannot only work for the collection of refuse but also with slight modification and with good sensors can also help in segregation of dry and wet garbage as illustrated in scope for improvement. Using and programming Arduino is much easier. The idea illustrated this project is very much simple and can be easily implemented.

References

1. Nidhi N, Prasad D, Nath V (2018) Internet of things/everything: an overview. Int J Pure Appl Math119(12). ISSN: 1314-3395
2. Bohra V, Prasad D, Nidhi N, Tiwari A, Nath V (2019) Design strategy for smart toll gate billing system. In: Nath V, Mandal J (eds) Proceeding of the second international conference on microelectronics, computing and communication systems (MCCS 2017). Lecture notes in electrical engineering, vol 476, pp 615–621. Springer, Singapore. https://doi.org/10.1007/978-981-10-8234-4_49. Online ISBN: 978-981-10-8234-4
3. Razi Q, Nath V (2019) Design of smart embedded system for agricultural update using internet of things. In: Nath V, Mandal J (eds) Nanoelectronics, circuits and communication systems. Lecture notes in electrical engineering, vol 511, pp 372–382. Springer, Singapore. https://doi.org/10.1007/978-981-13-0776-8_34. Online ISBN: 978-981-13-0776-8
4. Fatma S, Nath V (2019) Study and design of smart embedded system for train track monitoring using IoTs. In: Nath V, Mandal J (eds) Nanoelectronics, circuits and communication systems. Lecture notes in electrical engineering, vol 511, pp 385–395. Springer, Singapore. https://doi.org/10.1007/978-981-13-0776-8_35. Online ISBN: 978-981-13-0776-8
5. Kumar A, Nath V (2019) Study and design of smart embedded system for smart city using internet of things. In: Nath V, Mandal J (eds) Nanoelectronics, circuits and communication systems. Lecture notes in electrical engineering, vol 511, pp 397–408. Springer, Singapore. https://doi.org/10.1007/978-981-13-0776-8_36. Online ISBN: 978-981-13-0776-8
6. Anand S, Nath V (2019) Study and design of smart embedded system for remote health monitoring using internet of things. In: Nath V, Mandal J (eds) Nanoelectronics, circuits and communication systems. Lecture notes in electrical engineering, vol 511, pp 409–414. Springer, Singapore. https://doi.org/10.1007/978-981-13-0776-8_37. Online ISBN: 978-981-13-0776-8
7. Kumar C, Nath V (2019) Design of smart embedded system for auto toll billing system using IoTs. In: Nath V, Mandal J (eds) Nanoelectronics, circuits and communication systems. Lecture notes in electrical engineering, vol 511, pp 415–424. Springer, Singapore. https://doi.org/10.1007/978-981-13-0776-8_38. Online ISBN: 978-981-13-0776-8
8. Nidhi N, Prasad D, Nath V (2019) Different aspects of smart grid: an overview. In: Nath V, Mandal J (eds) Nanoelectronics, circuits and communication systems. Lecture notes in electrical engineering, vol 511, pp 451–456. Springer, Singapore. https://doi.org/10.1007/978-981-13-0776-8_41. Online ISBN: 978-981-13-0776-8
9. Raj U, Nidhi N, Nath V. (2019) Automated toll plaza using barcode-laser scanning technology. In: Nath V, Mandal J (eds) Nanoelectronics, circuits and communication systems. Lecture notes in electrical engineering, vol 511, pp 475–481. Springer, Singapore. https://doi.org/10.1007/978-981-13-0776-8_44. Online ISBN: 978-981-13-0776-8
10. Chaudhary S, Prava A, Nidhi N, Nath V (2019) Design of all-terrain rover quadcopter for military engineering services. In: Nath V, Mandal J (eds) Nanoelectronics, circuits and communication systems. Lecture notes in electrical engineering, vol 511, pp 507–513. Springer, Singapore. https://doi.org/10.1007/978-981-13-0776-8_47. Online ISBN: 978-981-13-0776-8
11. Mohan R, Suraj AK, Agarawal S, Majumdar S, Nath V. (2019) Design of robot monitoring system for aviation. In: Nath V, Mandal J (eds) Nanoelectronics, circuits and communication systems. Lecture notes in electrical engineering, vol 511, pp 535–547. Springer, Singapore. https://doi.org/10.1007/978-981-13-0776-8_50. Online ISBN: 978-981-13-0776-8
12. Hari Charan EVV, Pal I, Sinha A, Baro RKR, Nath V. (2019) Electronic toll collection system using barcode technology. In: Nath V, Mandal J (eds) Nanoelectronics, circuits and communica-

tion systems. Lecture notes in electrical engineering, vol 511, pp 549–556. Springer, Singapore. https://doi.org/10.1007/978-981-13-0776-8_51. Online ISBN: 978-981-13-0776-8
13. Goel V, Riya, Kumari P, Shikha P, Tanushree, Prasad D, Nath V. (2019) Design of smartphone controlled robot using Bluetooth. In: Nath V, Mandal J (eds) Nanoelectronics, circuits and communication systems. Lecture notes in electrical engineering, vol 511, pp 557–563. Springer, Singapore. https://doi.org/10.1007/978-981-13-0776-8_52. Online ISBN: 978-981-13-0776-8
14. Sinha PK, Saraiyan S, Ghosh M, Nath V. (2019) Design of earthquake indicator system using ATmega328p and ADXL335 for disaster management. In: Nath V, Mandal J (eds) Nanoelectronics, circuits and communication systems. Lecture notes in electrical engineering, vol 511, pp 565–572. Springer, Singapore. https://doi.org/10.1007/978-981-13-0776-8_53. Online ISBN: 978-981-13-0776-8
15. Raju D, Eleswarapu L, Saiv R, Nath V. (2019) Study and design of smart embedded system for aviation system: a review. In: Nath V, Mandal J (eds) Nanoelectronics, circuits and communication systems. Lecture notes in electrical engineering, vol 511, pp 573–590. Springer, Singapore. https://doi.org/10.1007/978-981-13-0776-8_54. Online ISBN: 978-981-13-0776-8
16. Maurya DK, Kumar A, Kaunoujiya S, Prasad D, Nath V. (2019) Study and design of smart industry: a review. In: Nath V, Mandal J (eds) Nanoelectronics, circuits and communication systems. Lecture notes in electrical engineering, vol 511, pp 591–598. Springer, Singapore. https://doi.org/10.1007/978-981-13-0776-8_55. Online ISBN: 978-981-13-0776-8
17. Sanjay Kumar S, Khalkho A, Agarwal S, Prakash S, Prasad D, Nath V. (2019) Design of smart security systems for home automation. In: Nath V, Mandal J (eds) Nanoelectronics, circuits and communication systems. Lecture notes in electrical engineering, vol 511, pp 599–604. Springer, Singapore. https://doi.org/10.1007/978-981-13-0776-8_56. Online ISBN: 978-981-13-0776-8
18. Goel V, Kumar S, Muralidharan A, Markham N, Prasad D, Nath V. (2019) Auto-train track fault detection system. In: Nath V, Mandal J (eds) Nanoelectronics, circuits and communication systems. Lecture notes in electrical engineering, vol 511, pp 605–610. Springer, Singapore. https://doi.org/10.1007/978-981-13-0776-8_57, Online ISBN: 978-981-13-0776-8
19. Saravana Kannan G, Sasi Kumar S, Ragavan R, Balakrishnan M (2016) Automatic garbage separation robot using image processing technique. Int J Sci Res Publ 6(4)
20. Alsahafi H, Almaleky M (2014) Design and implementation of metallic waste collection robot. In: SEE2014 Zone I conference, 3–5 April 2014. University of Bridgeport, Bridgpeort, CT, USA
21. Nurlansa O, Istiqomah DA, Pawitra MAS (2014) AGATOR (automatic garbage collector) a automatic garbage collector robot model. Int J Future Comput Commun 3(5)

Study and Design of Biometric Security Systems: Fingerprint and Speech Technology

Udit Anchalia, Konda Praneeth Reddy, Abhay Modi, Kumari Neelam, Deepak Prasad and Vijay Nath

Abstract In this research, the article demonstrates about the design of Biometric systems based on fingerprint and speech tech. In the initial stages, we are discussing about Fingerprint technology, and then we integrate both the technologies for the biometric security. So, we can say that it is a multimode biometric/security system, which integrates fingerprint verification and speaker verification. It can overcome some of the limitations of a single biometrics and it is more robust. A single feature is not sufficient to exact identification and it is not always reliable therefore multi model identifications are used to identify people with two different features such as voice and fingerprint. With the introduction of these two modalities, it achieves much better accuracy than single. It is also helpful in an emergency if one modality has some problems at that condition others modalities completing the task fairly. This research article describes two modalities together voice and fingerprint. This idea has simulated using Simulink of MATLAB.

Keywords Fingerprint · Speaker recognition · Biometric · Security

1 Introduction

Due to the growing importance of the technology, we should be able to restrict other uses that are to provide access restrictions to whatever info we have. There are many places where biometric security systems can be used. For example, in the field of LAW, this biometric security plays an important role to hide the best evidences.

U. Anchalia · K. P. Reddy · A. Modi · K. Neelam · D. Prasad (✉) · V. Nath
VLSI Design Group, Department of ECE, Birla Institute of Technology, Mesra, Ranchi 835215, Jharkhand, India
e-mail: prasaddeepak007@gmail.com

U. Anchalia
e-mail: uditanchalia7@gmail.com

V. Nath
e-mail: profvnath@gmail.com

© Springer Nature Singapore Pte Ltd. 2019
V. Nath and J. K. Mandal (eds.), *Proceedings of the Third International Conference on Microelectronics, Computing and Communication Systems*,
Lecture Notes in Electrical Engineering 556, https://doi.org/10.1007/978-981-13-7091-5_47

Mainly, this tech is based on the identification. There are many characteristics or attributes in an individual for the identification purposes. These attributes change from person to person that is they're unique. There are mainly two types of attributes in an individual namely Physical and Behavioral above all these attributes fingerprint can be named under most user-friendly attribute, because it doesn't need any training to know how fingerprint tech. Works and also because a finger has different patterns consisting of spiral's, loops and curves, which are almost different from person to person. For biometric identification, fingerprints are the best and fastest and unique technique. They are secure and easy to implement. It is unique for every person and can't duplicate in a lifetime. The speech technology can be kept under behavioral. Communication plays a key role in speech technology biometric system that is talking to a machine. This communication includes recognition of speech generalization of speech detection of speech. Speech-based person's identity which is popularly known as Speaker verification (SV). There mainly two types of speaker verification system dependent system have been trained from. One is the text-dependent system where the user while testing the system is restricted to the say the limited amount of text and the other kind of the system is text independent where the user is free to speak any desired words. It is our moral responsibility to enhance the security and identification system for execution of large database, e.g., institutes, universities, cities, states or country. For implementing this task some important algorithms have been used in this article such as gender estimation, key based one many matching, removing boundary minutiae. With the help of these techniques, a new identification systems have been created which are faster and secure than other existing models in the market. These biometric techniques are tested for college/universities students' identification purpose. These matching results are highly applicable where a large database has a country like India.

2 Literature Survey

2.1 For Fingerprint

Filip Orság and Martin Drahanský define the strategies of biometric security system using fingerprint and speech technology. In their book, the first chapter defines the biometric security system and concept of the integration of fingerprint and speech technology [1].

Adewole et al. developed tools using Microsoft Visual Basic .Net programming language which deals with the interaction with central database that contains all the official records of institute/universities teaching and non-teaching staffs as well as their monthly attendance. For implementation of this system, some standard criteria are followed. In these criteria, non-academic staffs of the faculty are also considered. In this criterion, no one can do the attendance of others. The entire candidate can see their attendance. It will also send to candidates/staffs for their verification of attendances. Monthly summary attendance can archive through ERP and it will also be used by the account department for their salary generation and other official records. In this, the system has the facility of grace period IN and OUT that can manage by any working days equivalent extra hours.

Sopan et al. explains the attendance and fingerprint acquisition module. This system relies on automatic phenomenon including information acquisition of fingerprint, wireless transmission, signal processing, and fingerprint matching for generating attendance report.

After taking the attendance systems to share the attendance information through message to students as well as their guardian's mobile via GSM and also store the information for official records for calculating the % of attendance in the class. Through this system, the student is always alert and bound to attend the class. For enabling the system on such important places, it has saved the stationary items as well as save the attendance count in the class.

2.2 For Voice Biometric

Jasneet Kaur and Sukhdeep Kaur has briefed different steps, which are required in voice-dependent attendance system [2]. We have briefed the different algorithms for feature extraction, i.e., MFCCs, LPCs etc. Along with this, various classification algorithms have been briefed which can be used for testing and training purposes. As of now, they have surveyed the pattern of speech recognition systems and the need of different steps in the proposed work to be done by them in future. After surveying the speech recognition related work, It has been found that MFCCs or GFCCs can be used for feature extraction process. For classification, different algorithms can be explored for better results, i.e., GMMs, ANN, SVM etc. This paper has been written after surveying various journals and key points have been elaborated for better understanding of voice dependent attendance system.

Prasanna and Sinha describe the enhancement techniques in speech biometric-based attendance system. The clients the system by matching a call from the pre-registered mobile phones [3]. An interactive voice response (IVR) unit guides a new client in the enrolment and enrolled clients in the verification process. This unit uses text-independent verification with MFCC() features and i-vector-based speaker modeling for authentication the clients. A normal cosine distance scoring done with score normalization is applied as the classifier and a fixed threshold is applied for making the decision. This developed system has been tested 110 students for 60 days on a regular basis. The system performance in terms recognition rate is about 94.2% and average time response is 26 s.

Design Strategy for Smart Toll Gate Billing System [4] can help to data recognition. Fingerprint-Based Attendance Management System with SMS alert to parents [3] gives the new development and cure in rural medical systems. Through this technique, the doctors are able to diagnose the patient in remote village area. The development of academic attendance monitoring system using fingerprint identification [5] demonstrates new setup of attendance in the classroom, manpower in factory, and corporate. Face recognition-based lecture attendance system [6] added new steps in attendance system without figure print just you move through the door and your attendance is over. It creates a healthy environment in the corporate sector for preparing the salary and monitoring the culture and output of employee. Study and Design of Smart Embedded System for Smart City Using Internet of Things [7] gives the new dimension in creating a positive environment in the city and properly monitor the existing sectors in side of the city including, hospital, banks, electric supply, schools, and colleges, etc. Fingerprint Image Enhancement and Recognition [8] demonstrate the improvement in fingerprint and recognize the human for attendance and other co-curricular activities. Intelligent Biometric Techniques in Fingerprint and Face Recognition [9] explains the smart fingerprint setup. Biometrics: Identity Verification in a Networked World [10] is utilize to recognize the worldwide network in banking and other corporate sectors where 24 $\times$ 7 h works are going on and three shift manpower working on same systems. Spoken Language Processing [11] provides helps to understand the voice of the human in any language and process as per needs. Speech and Audio Signal Processing [12] demonstrate the both feature for better understanding and processing.

3 Design of Biometric Security System

Biometric security systems come under the category of two-piece systems which consists of special and processing hardware. This special hardware contains a sensor which is in turn connected to processing hardware. The identification and recognition process are carried out in a processing hardware. Generally, this processing hardware is a personal computer. The separate parts are very dangerous and are a source to a great damage to security systems. To overcome this hazardous attack, we need to calculate and some info which is cryptographic and which in turn is combined with some private cryptographic keys.

The initial target is to set-up cryptographic key among all the applied biometric technology. Each biometric technology has in turn generates some amount of vectors, which are again considered as cryptographic keys. After obtaining these keys a hash function should be calculated this may be stored on storage. In the storage, it will not contain any secret information. As the feature of biometric attributes is not stored.

In this way, all the cryptographic keys are encrypted with all the biometric uses from a given biometric attribute, this information containing hashes and encrypted values will be stored in a database. The verification can be done through hash values. When a user locks a pin, he needs to provide biometric attribute which will generate cryptographic key and produces qualifiers and gives a certain hash value to it, and these hash values are compared with function columns stored in the disk. The hazards biometric access control systems are shown in Fig. 1.

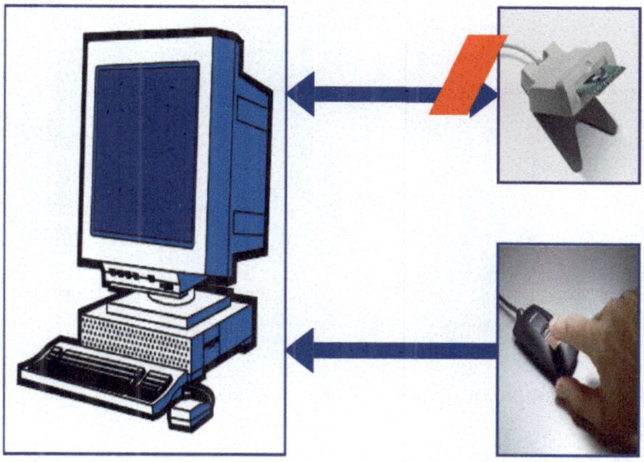

Fig. 1 Biometric access control system with attack hazards

4 Recognition System

4.1 Fingerprint Recognition

The main thing in fingerprint recognition is extracting the minutiae from the fingerprint. Each fingerprint has five types. The process of fingerprint analysis has six steps:

1. Fingerprint input: The quality of the image acquired by the machine should be high for the automatic identification. It should use the scanner which can identify different skin tones, finger injuries, wet and dryness of skin.
2. Complexity of Algorithm for the image quality improvement: Quality of image can be improved by recovering the original image from the damaged one. The use of some filters improves the structures of ridge and furrows in a damaged sector and prepare image to match the real one.
3. Performance of image preprocessing: It is the preparation step for the extraction of ridges of fingers thresholding and slimming is performed in this step.
4. Fingerprint classification: In this process, assigning the fingerprint to one of the classes take place. This process is named as the toughest process for both the human as well as the machine. In this process, PNN classifier is applied to classify.
5. Extraction of minutiae: In this process, we use an extractor which extracts only 3-types of fingerprint-bifurcation, continuous line and rejoining. After this some tests are performed, if minutiae lie on edge, they are deleted. The test consists of detecting the degree of bias of the papillary line.
6. Verifying: Two minutiae sets are compared and their comparison complexity strongly relies on reliability of minutiae obtaining process. Ratha method is used for comparison.

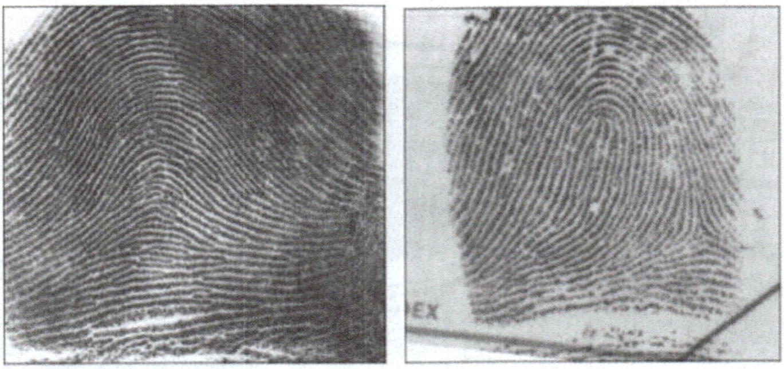

Fig. 2 Sample of fingerprint

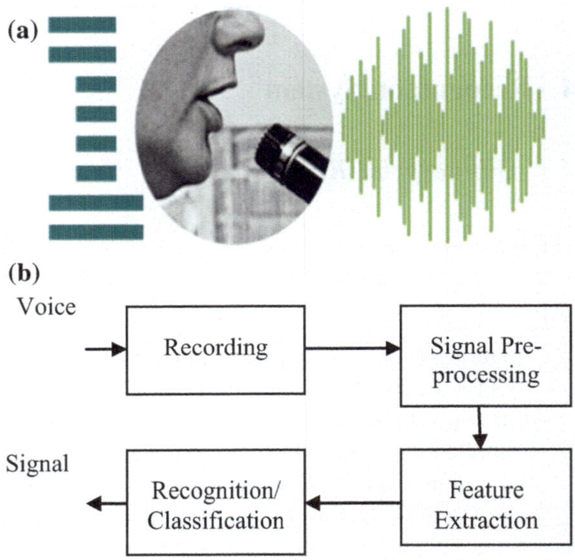

Fig. 3 **a** Speech recognition setup. **b** Steps of speaker recognition

4.2 Speaker Recognition Setup

The fingerprint recognition sample is shown in Fig. 2 and speech recognition setups are shown in Fig. 3.

Speaker recognition system contains mainly four steps:

A. **Recording**: The primary step is nothing other than recording a signal and this is done by sound hardware which is able to sample the sound of frequency 11,121 Hz, under precision of 2 bytes. The quality recording will affect the result of recognition.
B. **Preprocessing**: It is considered that the primary phase of other phases in speech recognition to distinguish the voiced or unvoiced signal and create feature vectors. It also adjusts or modifies the speech signal. There are some steps:

i. Windowing: It is for minimizing the disruption at the starting and at the end of the form, the form and window function is being multiplied.
ii. *Preemphasis*: It is, therefore, aimed at compensating for lip radiation and necessary attenuation of high frequencies in the sampling process.
iii. *Framing*: Here, the waveform is divided into small parts approximately 30 ms. This process continues until the whole signal is broken down into small forms.

C. ***Feature Extraction***: This step is nothing but extracting the frames. There are many possible frames ad the best frame is selected in this process. Many features are based on some parameters of speech signals these parameters are: MEL frequencies Ceptral coefficients, 0 processing rates, autocorrelation, and prediction coefficients.
D. ***Recognition***: The vector feature sometimes doesn't work perfectly so it requires neural networks to determine the speaker.

4.3 System Development Protocols

This developed biometric system has two-layer security model for the attendance system. We should collect some database for both identification systems. After the speech/fingerprint data recording/collecting Verification and resultant of attendance are marking. In the testing phase, the user is asked to key in their speaker ID and the system verifies the same by checking the presence of the speaker ID/thump impression in the enrolled ID list. On successful verification of the claim, the system marks the attendance of the client/user and a message is displayed "successful authentication". On rejection of the claim, the message is displayed "authentication failed". It develops a comparison algorithm and user interface. This will be a double security system more robust and not replicated.

5 Applications

There are several applications of these developed systems. As people already know that voice authentication is used by government intelligence agencies worldwide. There is no chance for proxy attendance as well as it also increases the accuracy of the system. Voice mail, security is used for the e-commerce. It is used in bank transactions, credit card activation, and payments. It is applicable for shopping by telephone, phone banking and trading, password resetting, accessing customer care. It is an automatic system. It is a very fast processing system, therefore, it consumes very less time, and it also saves paper and environment.

6 Conclusion

In older days, password and pincode played an important role in securing the data. Due to the advancements in technology, these means became the weakest method of security, so the biometric is the perfect replacement for this, which contains the password in the form of crypto key as it is very difficult to detect or decode, there is a point of securing the data, so with a piece of machine or instrument, we can leave our tension of loosing the privacy of our data.

References

1. Drahansky M (2001) Finger abdruckerkennung mittels neuronaler Netze. Diploma thesis
2. Adewole KS et al (2014) "z". Comput Eng Intell Syst 5(2). ISSN: 2222-1719 (paper), ISSN: 2222-2863 (Online)
3. Kaur J, Kaur S (2016) A brief review: voice biometric for speaker verification in attendance system. Imp J Interdiscip Res (IJIIR) 2(10). ISSN: 2454-1362
4. Bohra V, Prasad D, Nidhi N, Tiwari A, Nath V (2019) Design strategy for smart toll gate billing system. In: Nath V, Mandal J (eds) Proceeding of the second international conference on microelectronics, computing and communication systems (MCCS 2017). Lecture notes in electrical engineering, vol 476, pp 615–621. Springer, Singapore. https://doi.org/10.1007/978-981-10-8234-4_49. Online ISBN: 978-981-10-8234-4
5. Nawas T, Pervaiz S, Korrani A, ud din A (2009) Development of academic attendance monitoring system using fingerprint identification. Int J Comput Sci Netw Secur 9(5)
6. Kawaguchi Y, Shoji T, Lin W, Kakusho K, Minoh M (2009) Face recognition-based lecture attendance system. Department of Intelligence Science and Technology, Graduate School of Informatics, Kyoto University
7. Kumar A, Nath V (2019) Study and design of smart embedded system for smart city using internet of things. In: Nath V, Mandal J (eds) Nanoelectronics, circuits and communication systems. Lecture notes in electrical engineering, vol 511, pp 397–408. Springer, Singapore. https://doi.org/10.1007/978-981-13-0776-8_36. Online ISBN: 978-981-13-0776-8
8. Emyroglu Y (1997) Fingerprint image enhancement & recognition. Yldyz Technical University, Turkey
9. Jain LC, Halici U, Hayashi I, Lee SB, Tsutsui S (1999) Intelligent biometric techniques in fingerprint and face recognition. CRC Press LLC
10. Nanavati S, Thieme M, Nanavati R (2002) Biometrics: identity verification in a networked world. Wiley
11. Huang X, Acero A, Hon HW (2001) Spoken language processing. Prentice Hall, New Jersey, USA. ISBN: 0-13-022616-5
12. Gold B, Morgan N (2000) Speech and audio signal processing. Wiley, New York, USA. ISBN: 0-471-35154-7

Design of Low-Power 3-Bit CMOS Flash ADC for Aerospace Applications

N. Nidhi, M. Kumari, D. Prasad, A. Pandey, S. S. Solanki, A. Kumar, K. K. Thakur and V. Nath

Abstract CMOS Flash ADC is an important device of modern electronics and useful for aerospace applications. It is also frequently used in many other applications such as satellite communications, wireless communication, medical, education, and transportation, etc. Amongst all types of ADC available in market, flash ADC is the fastest ADC. In this paper, a low-power 3-bit flash ADC has been designed and verified. Flash ADC is a power-hungry device, it means the power consumptions is very high and it is a major issue of this device. So, here, a new technique is applied to reduce the power consumption so that it can utilize durable applications with high speed. This ADC rail-to-rail supply voltage is ± 1 V. The circuit is designed and simulated using Cadence analog and digital system design tools with 90 nm CMOS technology.

Keywords ADC · Comparator · CMOS (Complementary metal–oxide–semiconductor field effect transistor) · Multiplexer

1 Introduction

In modern era, people are surrounded by electronic devices used with different applications. An analog to digital converter (ADC) and digital-to-analog converter (DAC) are most commonly used in the electronic device. Meanwhile, many signals in this real world are analog signals, so by using these two interfaces, the analog signals are processed in digital electronic devices. ADCs consist of three important parameters such as resolution, speed, and power consumption which cannot be changed

N. Nidhi · M. Kumari · D. Prasad (✉) · A. Pandey · S. S. Solanki · V. Nath
Department of ECE, Birla Institute of Technology, Mesra, Ranchi, Jharkhand, India
e-mail: prasaddeepak007@gmail.com

V. Nath
e-mail: profvnath@gmail.com

A. Kumar · K. K. Thakur
ARTTC BSNL, Hazaribag Road, Ranchi, Jharkhand, India

© Springer Nature Singapore Pte Ltd. 2019

V. Nath and J. K. Mandal (eds.), *Proceedings of the Third International Conference on Microelectronics, Computing and Communication Systems*, Lecture Notes in Electrical Engineering 556, https://doi.org/10.1007/978-981-13-7091-5_48

after designing. These days ADC architecture needs a high-speed operation and less power consumption. There are different types of ADCs present like Sigma-Delta ADC, SAR ADC, Dual Slope ADC, and Flash ADC. Amongst all the ADC's flash, ADC are the fastest and most commonly used. Flash ADC has its applications where there is a need of low resolution and very high speed conversion. Due to its parallel architecture, it is also known as parallel ADC. Due to its low-resolution reason, Flash ADCs are also used as subblocks in some low-power ADCs such as pipelined ADC and sigma-delta ADC. Thus, a low-resolution flash-type ADC play an important role in decreasing the power consumption of another high-resolution-type ADC. Flash ADC consists of resistor ladder, comparator, thermometer code-to-binary code encoder, etc. Among all these, resistor ladder and comparator are the major parts and it consumes more power. So here, we include thermometer-to-binary code encoder, which consumes low power so that it can help to improve the complete system performance. Different architectures can be used for the conversion of thermometer to binary code, and most commonly used is direct conversion by using 2:1 multiplexers. This type of circuit is well known for its low power and high speed. Design of three-stage CMOS Comparator in 90 nm Technology demonstrates the low-power high-gain operational amplifier [1]. On improving the performance of Traff's Comparator defines the utility of comparator [2]. The design of low-power and high-speed multiplexer-based thermometer to gray encoder illustrates the new technique for gray encoder [3]. Fat tree encoder design for ultra-high-speed flash A/D converters demonstrates the new design model of ADC [4]. A High-Speed CMOS Current Comparator in 90 nm CMOS Process Technology explains the new design of current comparator [5]. A 200-MHz 6-bit folding and interpolating ADC in 0.5-μm CMOS demonstrate the new design technique of ADC [6]. Low-Power and High-Speed CMOS Comparator Design using 0.18 μm technology shows the design strategy of CMOS comparator [7]. A low-power, high-speed, and IF Range Flash-Type ADC designed with the concept of TMCC and binary counter defines the new core design of ADC using TMCC [8]. A 6 nW CMOS operational amplifier for biomedical and sensor applications shows the utility of operational amplifier in medical applications [9]. A current mirror is based on two stages of CMOS Cascade Op-Amp for high-frequency application and express the design strategy of two-stage operational amplifier using mirroring techniques [10]. Ultra-high gain CMOS op-amp design using self-cascading and positive feedback demonstrate the new design styles of op-amp [11]. Switched Capacitor Circuit Realization of Sigma-Delta ADC for Temperature Sensor expresses the utility of switched capacitor in sigma-delta ADC [12]. The design of 30-MHz CMOS Operational Amplifier defines the utility of op-amp for communication [13]. A 21 nW CMOS Operational Amplifier for Biomedical Application explains the utility of low-power op-amp in biomedical applications [14]. The design of Comparator in Sigma-Delta ADC Using 45 nm CMOS Technology demonstrates the utility of comparator in ADC [15]. The design of CMOS Integrator Circuit for Sigma-Delta ADC for aerospace application demonstrates the use of integrator circuit in the design of sigma-delta ADC [16]. Linearity improvement of gain-enhanced op-amp using cross-coupled architecture defines the new design techniques of op-amp and its applications in communications [17].

2 Methodology

Very Large-Scale Integration (VLSI) itself create a very good environment to improve the circuits with different optimization techniques and new technology adopted in the market. Due to advancement in VLSI design techniques and architecture able to give bulk production of electronics goods in the world market. In analog electronics, ADC and DAC are most required devices which frequently used in communications applied areas [18]. This paper includes several logic styles to improve the performance of the circuit and reduce the power consumptions. These are transmission gate logic, dual-pass transistor logic (DPL), static CMOS, swing restored pass transistor logic (SRPL) etc. Digital logic block such as 2 inputs XOR, 4:1 MUX, 2:1 MUX, etc., are implemented by using these logic styles. Among all these logic styles static CMOS logic gives better performance as compared to others. For the conversion of thermometer to binary code, various architectures are presented. Most commonly used architecture is a direct conversion method. Also, there is some other method like the conversion of thermometer code to a gray code, and then converts it to binary code [19]. Another method is Wallace tree architecture using full adders [20]. This architecture is implemented using full adder circuits. It contains a large number of hardware, so it is having high-power consumption and low speed. In this paper authors present an architecture in which all the AND gates and NAND gates are replaced by 2:1 multiplexers in the gray code based architectures presented [19] and proposed a FAT tree architecture in which thermometer code is first converted to 1 of N code after that it will be converted to binary code [4]. This is also implemented with logic gates. The transistor count is high as compared to other proposed architectures.

The circuit diagram of CMOS flash ADC is shown in Fig. 1. Due to its parallel architecture, 2^N resistors and $2^N - 1$ comparators are used for the conversion of analog signal to N bits of digital data. The Vin is applied to one input terminal of each comparator and this input voltage is compared with another reference voltages which is generated from the resistor ladder [3] which is shown in Fig. 1. For increment in the resolution of circuit, the number of comparators and resistors can be increased and decreased. Comparator symbolic diagram is shown in Fig. 2.

2.1 Comparators

The comparators compare the one analog signal (V_P) to another reference signal (V_N) and give digital output waveform based on the comparison. It is widely used in the conversion of ADC.

$$V_P < V_N \text{ Then } V_O = VSS = \text{logic } 0$$
$$V_P > V_N \text{ Then } V_O = VDD = \text{logic } 1$$

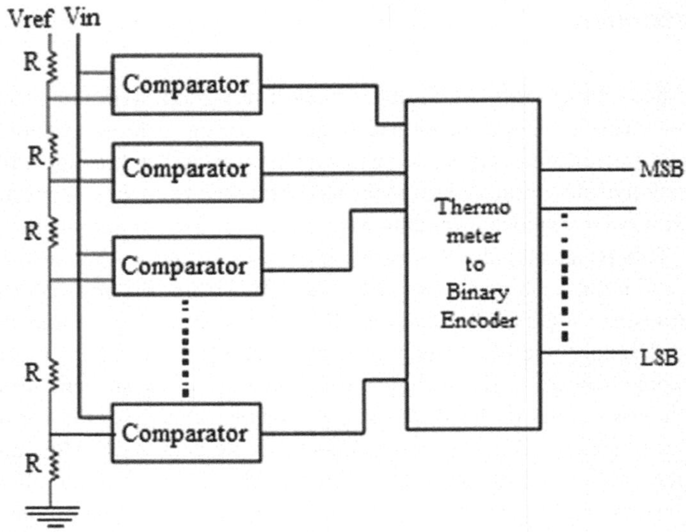

Fig. 1 Circuit diagram of CMOS flash ADC

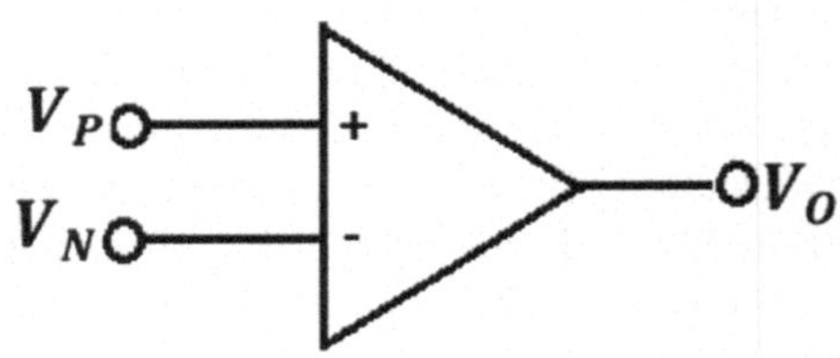

Fig. 2 Symbol of comparator

Comparator (Fig. 2) internal structure is shown in Fig. 3. This is known as CMOS two-stage comparator. It works on open loop mode, so that compensation in comparator does not require. Here larger bandwidth is achieving and giving faster response. The comparator circuit includes two stages, the first stage is a differential stage and the other stage is the output stage.

The advantage behind this type of design is that it requires less number of transistors and required small area.

The CMOS comparator is simulated using cadence virtuoso analog and digital system design tool of UMC PDK 90 nm CMOS technology as shown in Fig. 4 where, differential stage consist of transistors M1, M2, M3, and M4, transistors M3 and M4 are connected in current mirror mode transistor M5 and M8 connected in current mirror mode and create bias for this circuit. Transistors M6 and M7 create the buffer stage of the comparator. Here M1, M2, M5, M7, and M8 are the NMOS transistors; and M3, M4, and M6 are the PMOS transistors. The output of two-stage CMOS open loop comparator is connected with load capacitor C_L which is shown in Fig. 5 where the reference DC signal is applied from inverting terminal and input sinusoidal signal

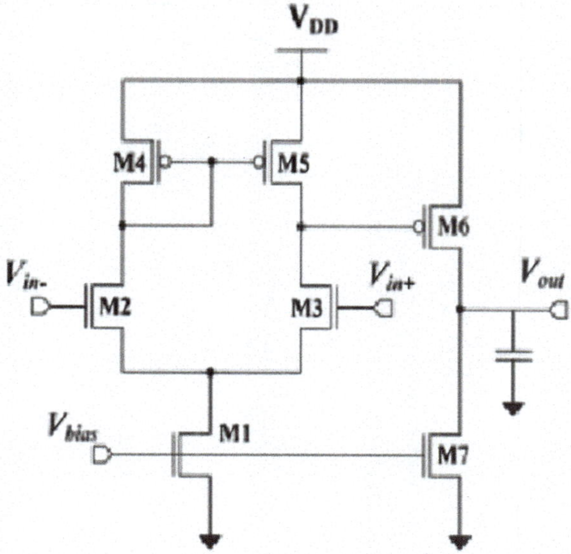

Fig. 3 Two-stage CMOS open-loop comparator

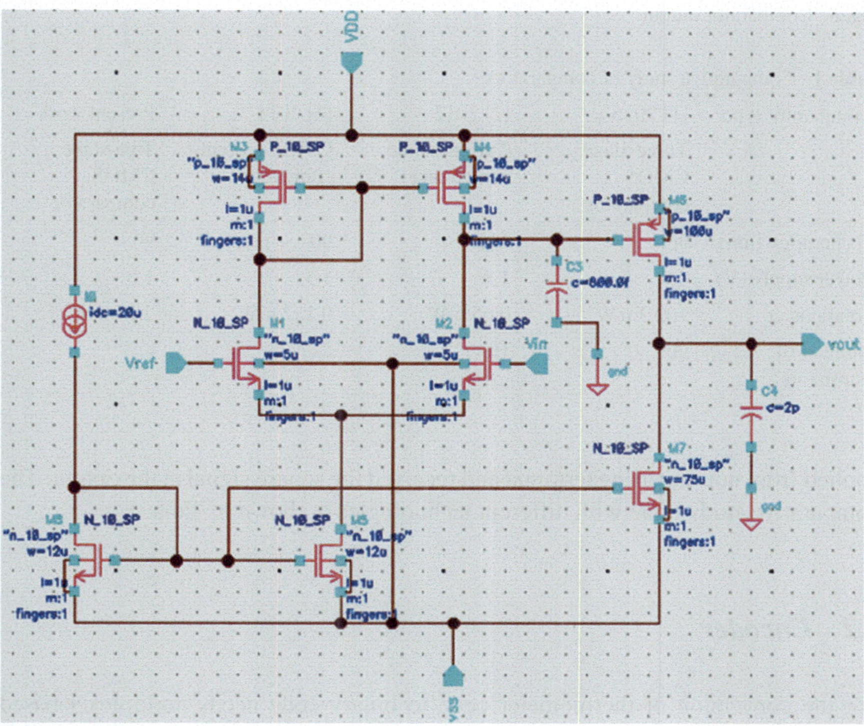

Fig. 4 Design of two-stage CMOS comparator

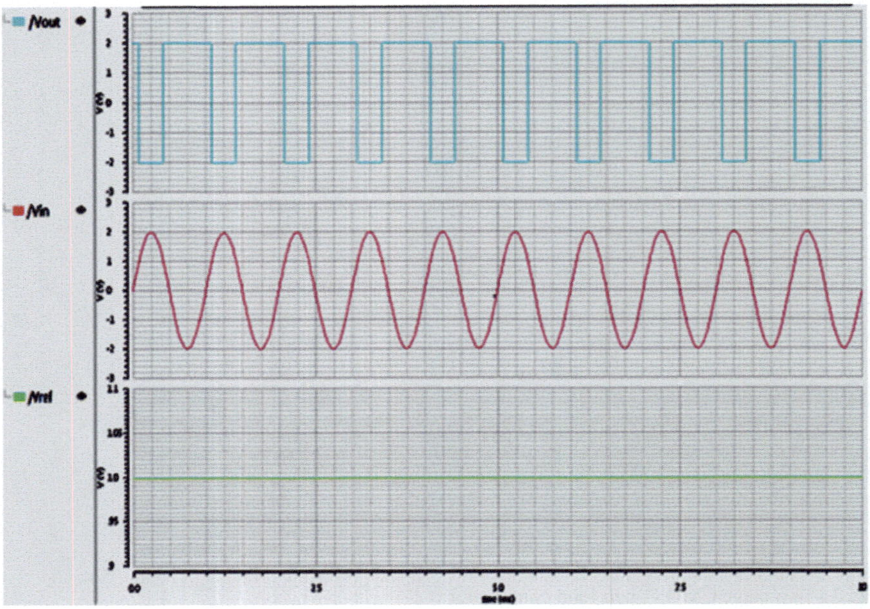

Fig. 5 Comparator output

Table 1 Comparative study of comparators

Comparator type	[1] 2015	[2] 2012	[5] 2014	Present work
	Three-stage CMOS comparator	Current mode comparator	CMOS current comparator	Two-stage CMOS comparator
Technology (nm)	90	180	90	90
Power supply (V)	1.8	1.8	1.5	1.0
Total power	102.3 μW	0.65 mW	0.122 mW	2.24 mW
Number of transistors	20	10	22	8

applied from non-inverting terminal and received the square signal at the output. The comparative study table with different technologies is shown in Table 1.

2.2 Encoder

For the conversion of thermometer code to binary code needs multiplexer-based encoder which is shown in Fig. 6. The CMOS logic gates are used to design multiplexers. The truth table for 3-bit multiplexer based encoder is shown in Table 2.

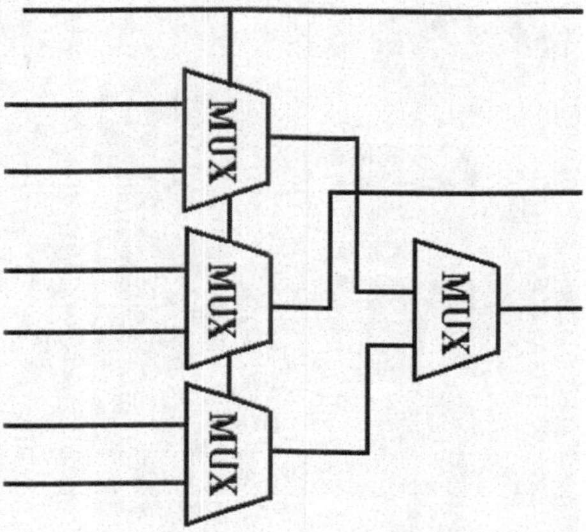

Fig. 6 Encoder for 3-bit ADC using multiplexer

Table 2 Truth table for 3-bit encoder

C7	C6	C5	C4	C3	C2	C1	X2	X1	X0
0	0	0	0	0	0	0	0	0	0
0	0	0	0	0	0	1	0	0	1
0	0	0	0	0	1	1	0	1	0
0	0	0	0	1	1	1	0	1	1
0	0	0	1	1	1	1	1	0	0
0	0	1	1	1	1	1	1	0	1
0	1	1	1	1	1	1	1	1	0
1	1	1	1	1	1	1	1	1	0

2.3 Flash ADC

The present design simulation of flash ADC has required 2^N resistors, i.e., for 3-bit flash ADC needs eight resistors and $2^N - 1$ comparators, i.e., seven comparators. An analog signal is given to one input of each comparator so that it can compare with the other reference voltage which is generated by the present used ladder resistor. The comparator outputs 0 or 1 depends upon the comparison of input voltage (V_{in}) and reference voltage (V_{ref}). If input voltage (V_{in}) is larger than reference voltage (V_{ref}), output will be 1 else 0. Then this output is served for the encoder for getting digitized output. Figure 7 shows the customized design of Flash ADC. The overall output waveform of flash ADC is shown in Fig. 8.

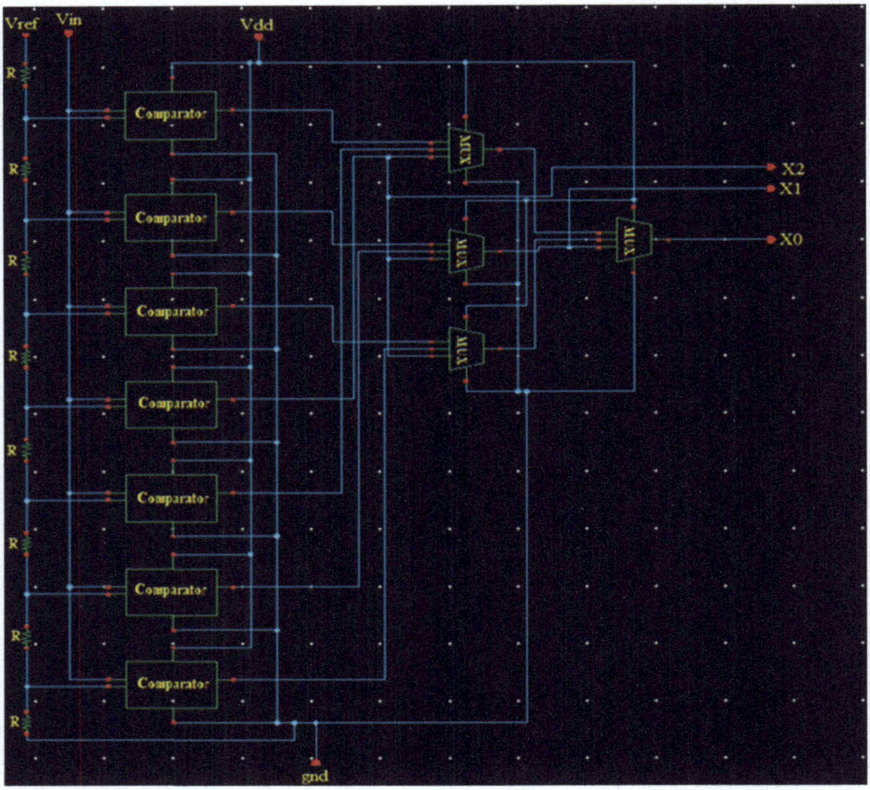

Fig. 7 Circuit diagram of flash ADC

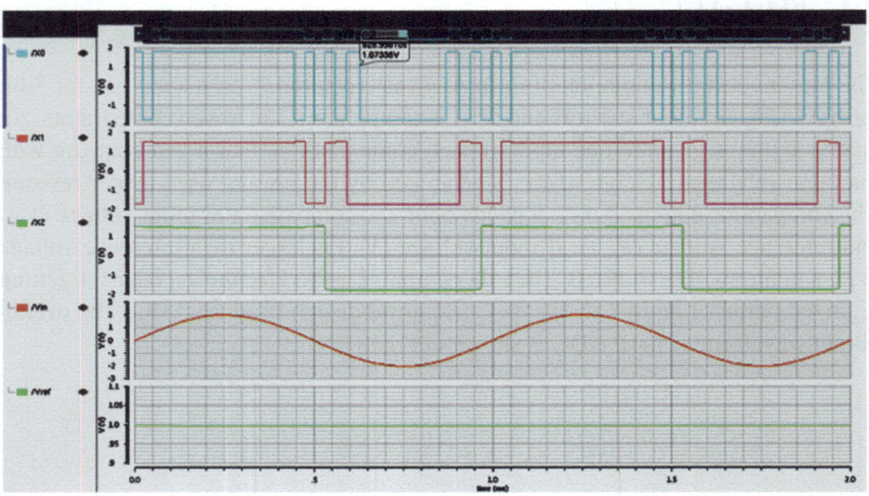

Fig. 8 Output waveform of flash ADC

3 Conclusion

In this paper, 3-bit CMOS Flash ADC is designed using UMC PDK 90 nm CMOS technology of cadence virtuoso analog and digital design tool. It includes two-stage CMOS open loop comparators, and multiplexer based encoders. The output response of CMOS flash ADC is shown in Fig. 8. Its simulation results are satisfying the Aerospace and other parallel applications. The power supply of flash ADC is ±1 V. It is very challenging to design a low-power high-speed flash CMOS ADC device because of it has power hungry property. Due to the adoption of new techniques able to successfully design the new challenging circuits. It is simple, smaller size, and functional for medium to high-resolution applications.

Acknowledgements The authors are heartily thankful to Dr. S. Pal, H.O.D of ECE Department and Dr. Mahesh Chandra, Dean Student Welfare for their constant encouragement and Prof. M. K. Mishra, VC of BIT Mesra for providing the suitable infrastructure/laboratory to carry out this research work.

References

1. Prathibha B, Jyothi H (2015) Design of three stage CMOS comparator in 90 nm technology. IJSTE, vol 1, June 2015
2. Sridhar R, Pandey N, Bhatia V, Bhattacharyya A (2012) On improving the performance of traff's comparator. In: 2012 IEEE 5th India international conference on power electronics (IICPE)
3. Gupta Y, Garg L, Khandelwal S, Gupta S, Saini S (2013) Design of low power and high speed multiplexer based thermometer to gray encoder. IEEE ISPACS, pp 501–504
4. Lee D, Yoo J, Choi K, Ghaznavi J (2002) Fat tree encoder design for ultra-high speed flash a/d converters. In: IEEE Midwest symposium on circuits and systems. IEEE
5. Rath A, Mandal SK, Das S, Dash SP (2014) A high speed CMOS current comparator in 90 nm CMOS process technology. IJCA
6. Jiang X, Wang Y (1998) A 200 MHz 6-bit folding and interpolating ADC in 0.5-μm CMOS. In: IEEE international conference on circuits and systems, vol 1, pp 5–8, June 1998
7. Panchore M, Gamad RS (2010) Low power and high speed CMOS comparator design using 0.18 μm technology. Int J Electron Eng Res 2(1):71–77
8. Mukherjee S, Saha D, Mostafa P, Saha D, Chatterjee S, Sarkar CK (2012) A low power, high speed, IF range flash type ADC designed with the concept of TMCC and binary counter. In: IEEE India conference (INDICON-2012)
9. Chakraborty A, Pandey SK, Saw, Nath V (2015) A 6nW CMOS operational amplifier for bio-medical and sensor applications. In: 2015 global conference on communication technologies (GCCT), Thuckalay, pp 242–245, 2015. https://doi.org/10.1109/gcct.2015.7342659, http://ieeexplore.ieee.org/stamp/stamp.jsp?tp=&arnumber=7342659&isnumber=7342608
10. Kundu R, Pandey A, Chakraborty S, Nath V (2017) A current mirror based two stage CMOS cascode op-amp for high frequency application. J Eng Sci & Technol (JESTEC) 12(3):686–700
11. Chakraborty S, Pandey A, Nath V (2017) Ultra high gain CMOS op-amp design using self-cascoding and positive feedback. Microsyst Technol 23(3):541–545. https://doi.org/10.1007/s00542-016-2971-7. Online ISSN 1432-1858
12. Pandey A, Javed KM, Prasad D, Nath, V, Solanki SS, Singh LK (2017) Switched capacitor circuit realization of sigma-delta adc for temperature sensor. In: Singh R, Choudhury S (eds)

Proceeding of international conference on intelligent communication, control and devices. Advances in intelligent systems and computing, vol 479. Springer, Singapore, pp 1129–1135. https://doi.org/10.1007/978-981-10-1708-7_134
13. Prasad D, Pranav A, Nimbargi A, Singh J, Ray MK, Mishra M, Kumar M, Nath V (2017) Design of 30 MHz CMOS operational amplifier. In: Singh R, Choudhury S (eds) Proceeding of international conference on intelligent communication, control and devices. Advances in intelligent systems and computing, vol 479. Springer, Singapore, pp 519–525. https://doi.org/10.1007/978-981-10-1708-7_59
14. Tyagi S, Saurav S, Pandey A, Priyadarshini P, Ray M, Pal BB, Nath V (2017) A 21nW CMOS operational amplifier for biomedical application. In: Nath V (eds) Proceedings of the international conference on nano-electronics, circuits & communication systems. Lecture notes in electrical engineering, vol 403. Springer, Singapore, pp 389–396. https://doi.org/10.1007/978-981-10-2999-8_33
15. Kumar V, Singh KK, Pandey A, Nath V (2017) Design of comparator in sigma-delta ADC using 45 nm CMOS technology. In: Nath V (eds) Proceedings of the international conference on nano-electronics, circuits & communication systems. Lecture notes in electrical engineering, vol 403. Springer, Singapore, pp 381–387. https://doi.org/10.1007/978-981-10-2999-8_32
16. Prasad D, Nath V (2018) Design of CMOS integrator circuit for sigma delta ADC for aerospace application. In: Bhattacharyya S, Sen S, Dutta M, Biswas P, Chattopadhyay H (eds) Industry interactive innovations in science, engineering and technology. Lecture notes in networks and systems, vol 11. Springer, Singapore, pp 377–383. https://doi.org/10.1007/978-981-10-3953-9_36
17. Chakraborty S, Pandey A, Prasad D, Vedam V, Nath V (2018) Linearity improvement of gain enhanced op-amp using cross-coupled architecture. Microsyst Technol 1–10. Online ISSN: 1432-1858. https://doi.org/10.1007/s00542-018-3885-3
18. Zimmermann R, Fichtner W (1997) Low-power logic styles: CMOS versus pass-transistor logic. IEEE J Solid-State Circuits 32:1079–1090
19. Sheikhaei S, Mirabbasi S, Ivanov A (2005) An encoder for a 5GS/s 4-bit flash ADC in 0.18/spl mu/m CMOS. In: Canadian conference IEEE electrical and computer engineering, 2005, pp 698–701, May 2005
20. Kaess F, Kanan R, Hochet B, Declercq M (1997) New encoding scheme for high-speed flash ADC's. In: IEEE international symposium on circuits and systems, vol 1, pp 5–8

Application of Burglary Alarm System to Avoid Railway Accidents

Rishabh Jain, Anvesh Ashu, Shivam Lal, Kumari Neelam, Deepak Prasad and Vijay Nath

Abstract This research article demonstrates a study result of the application of the burglary alarm system at railway and road intersections. This proposed system has many advantages over existing systems that are being used. The use of a laser alarm system is very common in homes and banks to prevent burglaries. But what if it could be applied to stop accidents at railway and road intersections! The only difference being that instead of a buzzer we will use a high-frequency source whose signals will be sent to the control room. By using the automatic burglaries alarm at the level crossing the trains detected 3 km before with IR detector and train will also start the siren. This proposed system will provide safety to pedestrian and vehicles users. It will be also be avoiding the accidents and is highly useful for normal public. Simultaneously crossing door will be closed and after passing the train 1 km and above, the door will be automatically open and stop the alarm. This system makes use of IR transmitter, fire sensor, zig Bee and other embedded systems. And, it will furnish an efficient, reliable, and secure alternative for the existing railway management systems.

Keywords Laser alarm · IR transmitter · LCD · Photo diode · Fire sensor · Embedded system

R. Jain · A. Ashu · S. Lal
Department of CSE, Birla Institute of Technology, Mesra, Ranchi 835215, Jharkhand, India

K. Neelam (✉) · D. Prasad (✉) · V. Nath
Department of ECE, Birla Institute of Technology, Mesra, Ranchi 835215, Jharkhand, India
e-mail: kumarineelam375@rediffmail.com

D. Prasad
e-mail: prasaddeepak007@gmail.com

© Springer Nature Singapore Pte Ltd. 2019

V. Nath and J. K. Mandal (eds.), *Proceedings of the Third International Conference on Microelectronics, Computing and Communication Systems*, Lecture Notes in Electrical Engineering 556, https://doi.org/10.1007/978-981-13-7091-5_49

1 Introduction

From the survey, it is found that more than 15,000 people are killed each year in rail accidents [1]. The numbers of cases of railway accidents are little; they can have very severe consequences. In countries like India having an extensive rail network, there are thousands of deaths every year because of rail accidents (as per [2] in the year 2014 with 2547 deaths) and will continue to increase in the recent years. In some countries, like Vietnam, Spain, and USA, the number of railway accidents have multiplied over the years and still tend to increase. Some causes include damaged train or light signals, improper maintenance of railway tracks, sudden appearance of animals like dog, cow, buffalo, etc., unprotected crossings, negligence by the conductor, etc. [3].

We have proposed a mechanism to reduce the rail accident. The proposal is simple. There will be a number of alarm circuits applied at the intersection. If someone tries to cross the railway crossing when the train is about to arrive, then they must pass through the lasers and as soon as they do so, the signal will be sent to the control room to alert the train driver of the approaching train. Obviously, the circuits have to be switched on only after the crossing gates have been closed. Even in places where railway tracks intersect and often trains have to cross the intersection one after the other, there is a risk of an accident. Thus, if these alarm systems are applied at these intersections, it can greatly help to reduce accidents at these places. The alarm would alert both the train drivers, thereby helping them to react accordingly. The alarm system comprises of a simple circuit which includes an LM358, IR transmitter, photodiode, resistors, and a high-frequency emitter source. At present, India, Italy, China, etc. [4, 5] have many suppliers of train alarm systems at rail–road intersections. However, due to their high cost, these systems are only installed at a few important intersections where traffic density is too high. Other intersections have high probability of accidents. Due to the specificity of the manufacturers, the replacement, maintenance, or repair of the systems is also very difficult.

2 Component

LM358 is a dual Operational Amplifier IC. It contains two independent Op-Amps that can be used separately. Op-amp is the heart part of analog electronics. It has numerous applications such as transducers, amplifiers, instrumentation amplifiers, multipliers, differentiator, integrator, active filters, general signal conditioning, and DC gain blocks. Infrared or IR is a range of light frequencies that have a longer wavelength than visible light. Hence, they are not visible to the human eye.

An IR transmitter and detector is a device that emits infrared light. And IR detector detect is used to detect the arrival of the train in the railway track which is installed below the rails. IR detector sends the signal to the LED so that it glows and the siren sound is dispatched from the speaker which act as the warning signal for a period of time set in the timer. An electronic device that converts light to electrical current is known as photodiode. It is a special type of PN junction diode that operates in reverse bias region. When light focus on the photodiode then reverse bias current flows in the junction. It is proportional to the intensity of light. Thus, this alarm system is economically viable and its installation and repairing processes are also easier than the other methods in use. LEDs are used as indicator and alarm or speaker.

3 Methodology

Circuit diagram of Burglary Alarm system is shown in Fig. 1. The IR transmitter is jointed to a current-limiting resistor of 150 Ω and to supply. It is installed in the range of photodiode. The cathode terminal of the photodiode is jointed to the power supply while anode terminal is jointed to a 10 KΩ resistor and the second end of the 10 KΩ resistor is grounded. The anode terminal of the photodiode is jointed to pin 5 (the non-inverting terminal) and Wiper of the 10 KΩ POT is jointed to pin 6 (the inverting terminal) of LM358 while the other two ends of the POT is jointed to the power supply Vcc and ground. Pins 4 and 8 of LM358 are power supply pins. They are jointed to ground and power supply Vcc respectively. From pin 7 of LM358 output is taken. One terminal of the high-frequency source is jointed to pin 7 of LM358 while the other terminal is jointed to a power supply Vcc.

In this research article LM358 Op-Amp is used as a comparator. When the proposed system power is ON, the IR transmitter or IR LED emits infrared light, which focuses on the surface of the photodiode. As it is jointed in reverse bias, when the light focus on it, it starts to conduct and current flows through it. Since it is jointed to pin 5 therefore, the output of the comparator will be high. As the high-frequency source is jointed between the power supply Vcc and output of LM358 (pin 7), no alarm signal is generated. When an intrusion is made between the IR transmitter and photodiode, the light focusing on the photodiode is interrupted and it doesn't conduct. Due to that, the input at pin 5 potential is less than the input at pin 6, therefore, the output of the comparator is low [6]. This will trigger the high-frequency source and a signal is sent to the control room and to the drivers of any approaching train. Working model of (Fig. 1) shown in Fig. 2.

Note: The circuit diagram shown was made in Multisim 2001. Due to the limitation of the software photodiode was not available. Hence, an optocoupler has been used instead of a combination of IR transmitter and photodiode, only to represent circuit but the actual circuit has been described above. An optocoupler performs the same function. It uses a "short optical transmission path" to transfer an electric signal between different parts of the circuit. Here, it helps to join the two isolated parts (Part A and Part B) of the circuit, thereby completing the circuit. However, we have not used optocoupler in the actual model of our paper as it does not meet our needs because the optocoupler requires the different parts (to be connected) to be close.

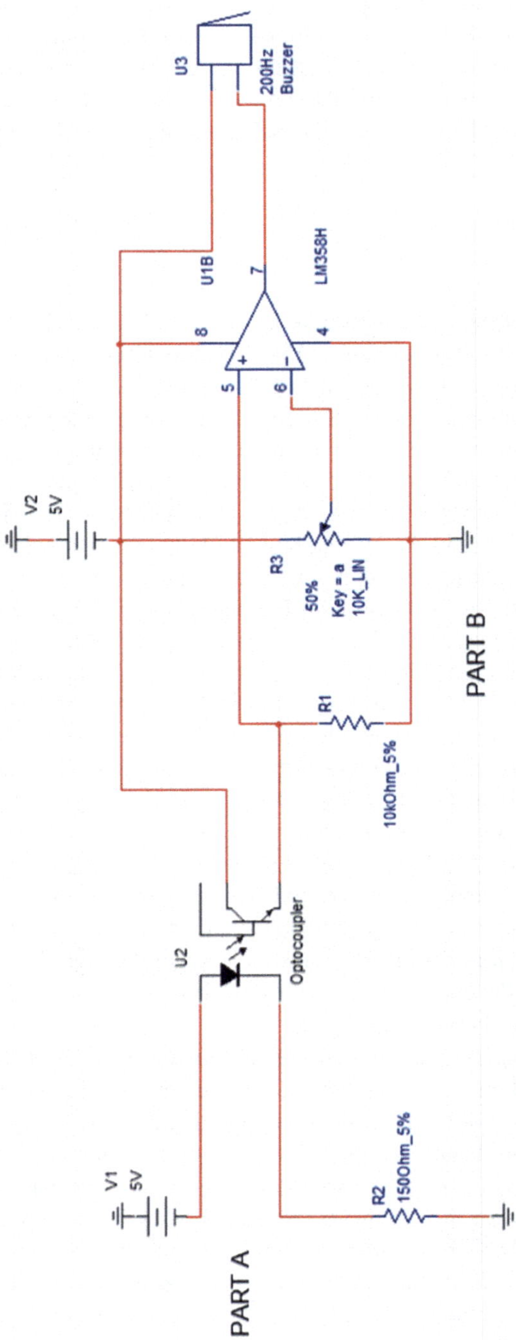

Fig. 1 Circuit diagram of burglary alarm system

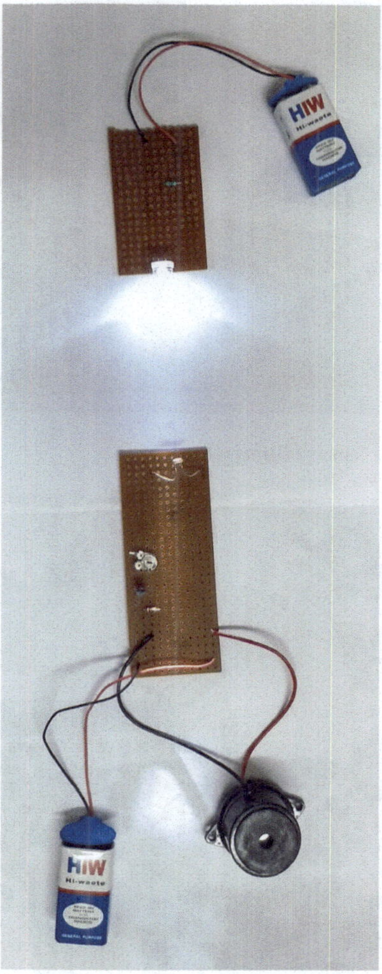

Fig. 2 Working model of burglary alarm system

4 Major Railway Accidents in India

Bihar Train Derailment, India
On June 6, 1981, a passenger train on its way from Mansi (Dharma Pul) and Saharsa (Bihar) [7, 8]. Train derailment near Sahrsa (Bihar) pictorial views is shown in Fig. 3.

Firozabad Rail Disaster, India
On August 20, 1995, on the Delhi–Kanpur section of India's Northern Railway near Firozabad, the Purushottam express (from Puri) collided with Kalindi express (from Kanpur [9, 10]. The train derailment near Firozabad is shown in Fig. 4.

Fig. 3 Train derailment near Saharsa (Bihar) [8]

Fig. 4 Train derailment near Firozabad [10]

Khanna Rail Disaster, India
On November 26, 1998, on the Khanna–Ludhiana section of India's Northern Railway in Punjab, the Jammu Tawi–Sealdah Express [11, 12]. The train derailment at Khanna near Ludhiana (Punjab) pictorial view is shown in Fig. 5.

Kanpur Rail Tragedy 2016, India
On November 10, 2016, 60 kms from Kanpur [13–15], 14 coaches of the Indore—Patna Express had derailed. 151 casualties were reported [16]. The train derailment near Kanpur (UP) is shown in Figs. 6 and 7.

Fig. 5 Train derailment at Khanna near Ludhiana (Punjab) [12]

Fig. 6 Train derailment near Kanpur (UP) [13]

5 Results and Discussion

The circuit can be split into two parts—Part A and Part B (as shown in the Fig. 1 and working model in Fig. 2). Part A constitutes the IR transmitter whereas part B consists of the photodiode and the rest of the circuit. To implement this idea, need to place Part A on one side of the tracks and part B on the other. This allows the IR light emitting from part A to reach part B while spreading throughout the tracks. Thus, if any intrusion is made between the tracks, it can easily be detected which is shown in Fig. 8.

This will also help to railway officer to detect if any crack develops in the tracks so that precautions can be taken. Also, it will also help us to identify if any unwanted changes have been made to the track or the track has been tampered with. For example,

Fig. 7 Train derailment near Kanpur (UP) [15]

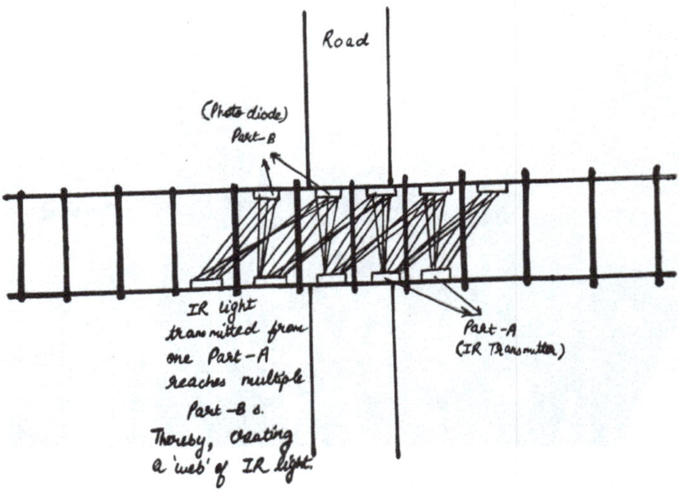

Fig. 8 Implementation of our proposed ideas

if a bomb has been planted on the track, it can be easily identified with the help of this circuit and a signal is sent to the control room. Moreover, since IR is invisible to the human eye, there is less chance of the bomber being alerted. The performance of the system can be improved by increasing the range of IR detection. This can be done by using a converging lens. Another way to extend the range is to implement a series of photodiodes. Instead of using IR transmission and detection, a high-intensity LASER light and an LDR can also be used to implement the same circuit. This proposed idea successfully implemented the use of burglar alarm system in the railway network. This research article is significant in the lowering of accidents in the railways.

6 Conclusion

In countries like India, which have extensive and dense railway network, automatic barrier systems are important. The application and dissemination of these systems will help us to reduce the cost of the railway sector as well as contribute to the reduction of traffic accidents. The proposed topic uses simple electronic devices, thus saving both money and time required in maintenance. Moreover, it will be highly effective in India, as people are well aware of the fact that many people on two wheelers try to cross the railway crossing even after the gates have been closed. This system would alert the guard of any such violator so that appropriate action can be taken to avoid harm to anyone. Also, strict actions can be taken against the violator so that the frequency of such violations can be reduced. People must understand that these rules are made for their safety and it would be best for them to obey it with complete sincerity.

Acknowledgements We would like to specially thank the Vice Chancellor Dr. M. K. Mishra as well as the Head of Department of ECE Dr. Srikanta Pal and the Head of Department of CSE Dr. Vandana Bhattacharjee who gave us this golden chance to do this paper.

References

1. http://timesofindia.indiatimes.com. Accessed on 01 Mar 2018
2. Borale SD (2016) Fingerprint based attendance management system with SMS alert to parents. Int J Res Adv Technol (IJRAT) (E-ISSN: 2321-9637) Special issue national conference convergence 2016, 06–07 Apr 2016
3. http://www.slideshare.net/dorismmahoney/common-causes-of-train-accidents-22193684. Accessed 04 Apr 2018
4. http://www.forbixindia.com/electronics/products/traffic-light-control/unmanned-railway-crossing-alarm. Accessed 23 Mar 2018
5. http://www.bombardier.com/en/transportation/products-services/rail-control-solutions.html. Accessed 02 Apr 2018
6. Anand S, Nath V (2019) Study and design of smart embedded system for remote health monitoring using internet of things. In: Nath V, Mandal J (eds) Nanoelectronics, circuits and communication systems. Lecture notes in electrical engineering, vol 511, pp 409–414. Springer, Singapore. https://doi.org/10.1007/978-981-13-0776-8_37
7. REUTERS (1981) AROUND THE WORLD; Toll From Train Crash Reaches 215 in India. New York Times. Retrieved 6 June 2010
8. https://crazycrashes.wordpress.com/all-the-worst-train-wrecks/. Accessed on 06 Apr 2018
9. The leading emergency management site on the net. Emergency-management.net. Retrieved 01Sept 2012
10. https://www.pinterest.com/pin/189291990560013402/. Accessed on 06 Apr 2018
11. https://en.wikipedia.org/wiki/Khanna_rail_disaster. Accessed 06 Apr 2018
12. http://shreejiexpress.com/. Accessed on 06 Apr 2018
13. https://www.indiatoday.in/india/story/indore-patna-train-derailment-kanpur-isi-link-railways-bihar-police-motihari-955565–2017-01-17. Accessed 06 Apr 2018
14. http://www.xinhuanet.com/english/2016/11/21/c_135846185.htm. Accessed 06 Apr 2018
15. https://mashable.com/2016/11/20/india-train-crash-death-toll/. Accessed 06 Apr 2018
16. https://en.wikipedia.org/wiki/2016_Kanpur_train_accident. Accessed on 06 Apr 2018

Design of Password-Based Door Locking System

Pratik Roopchandka, Saiba Khanam, Ratnesh Dhan, Kumari Neelam, Deepak Prasad and Vijay Nath

Abstract This paper validates the use of password-based door locking system which can protect our houses, offices and factories. 8051 microcontroller is the important device which is used in the circuit to control, monitor, and process. In this system, the user will have a unique password. The checking is done by matching the entered password with the password already fed into the system. If an appropriate password is entered, then the door motor is rotated by the system and the status is exhibited on the LCD, whereas if the password is incorrectly matched, then the door doesn't open and a statement of "WRONG" is shown in LCD. A keypad is used to enter the password and if we will enter the wrong password three times, then the buzzer will be activated to indicate for the wrong password.

Keywords Motor · Microcontroller · Keypad · LCD · Motor driver board

1 Introduction

In today's modern world, the technology is so advanced that we have unlocked different ways to cope up with our everyday problems. Such technological example is the DOOR LOCKING BY PASSWORD. It is a system which uses a password to unlock and lock doors in certain areas. It uses an 8051 Microcontroller to carry out the whole process smoothly. The previous locking system where one uses a metallic lock and key has become too mainstream. So the system proves to be the

---

P. Roopchandka · S. Khanam · R. Dhan
Department of CSE, Birla Institute of Technology, Mesra, Ranchi 835215, Jharkhand, India

K. Neelam · D. Prasad (✉) · V. Nath
Department of ECE, Birla Institute of Technology, Mesra, Ranchi 835215, Jharkhand, India
e-mail: prasaddeepak007@gmail.com

K. Neelam
e-mail: kumarineelam375@rediffmail.com

V. Nath
e-mail: vijaynath@bitmesra.ac.in

© Springer Nature Singapore Pte Ltd. 2019

V. Nath and J. K. Mandal (eds.), *Proceedings of the Third International Conference on Microelectronics, Computing and Communication Systems*, Lecture Notes in Electrical Engineering 556, https://doi.org/10.1007/978-981-13-7091-5_50

best replacement for it. This system is the result of the highest intelligence. These security systems are quite simple and highly efficient. This advanced lock system has an electronically controlled assembly that has the capability to control the output by the password. The best examples of such loads would be mechanical and electrical loads. In this system, the 8051 Microcontroller acts as a control provider for the load actuation. The system also has a keyboard which is used for the input whereas the output gets operated correspondingly. It is the most accurate demonstration of a precise door locking system which grants access to an individual who enters the correct password (password saved in memory) and stops the trespasser who doesn't know the password to the secured area.

2 Design Requirements

Following hardware's is *required* to make the password based door locking system:-

(a) 8051 microcontroller
(b) 8051 development board
(c) 8051 programmer
(d) Keypad (4 × 4)
(e) 16 × 2 LCD
(f) L293D motor driver board
(g) DC motor
(h) 10 KΩ potentiometer
(i) Connecting wires
(j) Power supply.

A. ***Microcontroller***: It is an important part of this system. It executes the following functions:-

i. Entering the digital input from keypad.
ii. After processing the data it will display on LCD.
iii. Machine will sense the password entered by keyboard and verify it is correct or not. If it is correct instruct to stepper motor for rotation.
iv. It will send the data to the computer through serial port.

B. ***LCD***: Here, two sets of LCD display is used one for numeric display and another for alphabet display.
C. ***Buzzer***: It is used to indicate when the wrong password opens the door.
D. ***Keypad***: It is used to enter the different character as password and other required characters such as (i) 0–9; (ii) Enter; and (iii) Escape.
E. ***Motor driver IC*** (***L293D***): Motor is a device which converts electrical energy to mechanical energy. Actuators are a device which actually gives the acceleration to the mechanical arms. There are different types of motors are available in the market which work on different voltages. So, the motor driver is required which

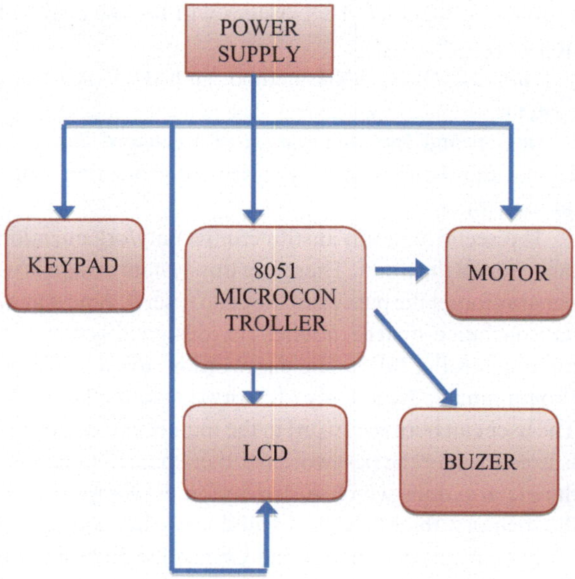

Fig. 1 Block diagram door locking system

control by microcontroller itself. Motor and microcontroller will be connected by a bus and run by driver software it will create the interface between hardware and software [1].

In this research article, L293 motor driver is used. It is monolithic and used for high voltage. It is best suited up to 5 kHz switching frequency. L293 IC packaged in plastic with 16 pins in which 4 centre pins is connected together and used for heat sink. L293DD is assembled in 20 lead surface mount in which 8 centre pins is connected together and used for heat sinking. The basic building block of Door Locking System is shown in Fig. 1.

Software Requirements: The following software is required to execute to the systems:-

- Kiel μvision IDE for assembly and C/C++ coding
- Willar programmer
- Proteus software for circuit diagram and simulation.

This research article has written on the different study of recently available journals, manuals and books to enable this idea in reality. Ideas came from the study of different literature and enable with the help of coding from books and manuals. As per the demand of current market started the research work on the said topic and ultimately achieved the goal. This consists of major five blocks keypad, microcontroller, motor, LCD, and buzzer. All the devices are interfaced with the microcontroller and central supply provided to all components. In future, this system will become more advanced and robust after implementing the biometric with fingerprint and speech technology together. This system will be low cost and useful in different places in home security.

They use BPN, ANN method, and MATLAB use to effectively implement this system.

Design and Implementation of Password Based Security Lock System by M. Hymavathi demonstrate the password protection will be in hand of programmer or administrator.

The security of any institute/university/organization or house is of prime importance for all the times. Therefore this system design is user-friendly. If any time house persons forget the password they can reset it. A new password will be forwarded to the user on their e-mail or mobile. This designed system is fully operational through 8-bit microcontroller 8051. The password is stored in EEPROM (Electronically Erasable Programmable Read Only Memory) and directly interfaced with the microcontroller. The user can burn password in the memory or can change as per their need. A keypad is used to enter the password or other character and relay are used to lock and unlock the electric door which is display on the lamp. If any wrong information enters in the memory, then WRONG will display in another LCD.

The design of Smartphone-Controlled Robot Using Bluetooth [2] is helpful to design control system of open and closed door system. Different Aspects of Smart Grid: An Overview [3] demonstrates the concept for a complete setup of automation. Wireless Fingerprint-Based Security System using Zigbee [4] demonstrate the new feature to open and close door system to design more stronger and robust systems. Fingerprint Identification in Biometric Security Systems [5] demonstrates another solution for design the security key. A design of real-time control robot system using Android Smartphone [6] demonstrates how to utilize your mobile to open and close your door remotely or physically.

3 Methodology

The microcontroller 8051 is the heart part of this proposed embedded password based door locking system. It is connected with 4 × 4 keypads which are used to enter the password. When entered password is matched with predefined password then microcontroller gives the instruction to motor for open the door otherwise door will not be opened and display the WRONG password on LCD panel [2, 7]. Now, in current market, old/traditional locking systems are being replaced by new embedded locking systems [6, 8, 9]. These techniques are a bridge between mechanical and electronic devices and it is highly intelligent and highly required in the digital world. These locking systems are simple, low cost, reliable and highly efficient. The system consists of certain ports like PORT1, PORT2 etc. PORT1 is connected to the data pins of the LCD and also to the control pins (RS and E) to P3.0 and P3.2. That's

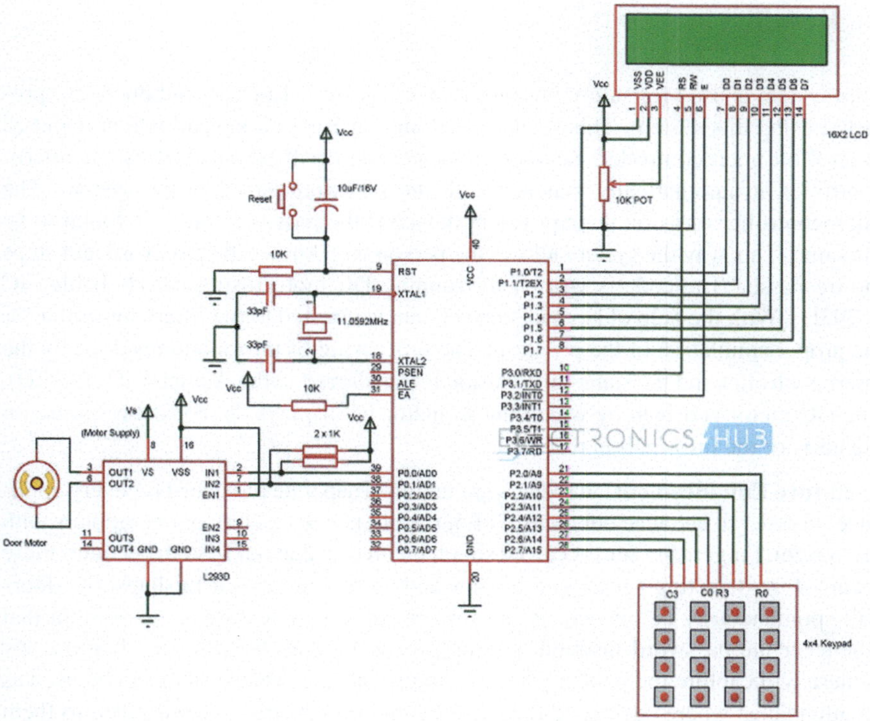

Fig. 2 Circuit diagram

the fixed place for PORT1 whereas PORT2 needs to be asserted to the keypad. P0.0 and P0.1 should be used for Motor Drivers. After these declarations, a message pops up showing “PASSWORD TO BE ENTERED” on the LCD. The password is generally of 5-digits which are to be inserted by the operator. The entry continues to the comparison between the entered and the saved password. The correct password turns the P0.0 pin HIGH and P0.1 pin LOW and the door opens. The door opening is indicated on the LCD by the pop-up status “OPENING OF DOOR”. The closing of the door takes place when P0.0 pin is in LOW state and P0.1 pin is in HIGH state. The indicator is the pop up stating “CLOSING OF DOOR”. If the door is opened and the user wants to change their password they can press 0 or 1 key and enter their user name, and then the option will come to change the password [10, 11]. If the user wants to include more users then press# key and enters the user’s ID, and then the machine will generate different passwords for all the users. The circuit diagram of the open and close door system is shown in Fig. 2.

4 Result

This system is used for safety purpose in many sectors. Lockers and doors are protected using this system. This security system consists of a keypad which is linked to an 8-bit microcontroller. Keypad password entry will be checked by the microcontroller, it automatically matches with the password saved in the system. The microcontroller turns on the equivalent device if the password entry is found to be the same. This way the system allows the person who knows the password and stops the trespasser. The code is written for running DC motor using the H- Bridge IC (L293D). With the help of keypad, correct data is inserted to the Microcontroller for the proper simulation of the program. The processing of information is done by the microcontroller and the same information is transferred to the Actuator IC (L293D). The DC motor is driven by which the Actuator IC displays the correct response on the basis of the information received.

Future Enhancement: In today's era require a separate password for every appliance. In order to enhance our control of our appliance, it can link our computers with the system. Fingerprint sensor can be supplemented so that our system becomes more secure. Fire detecting sensor can be attached so that in case of mishaps, the doors will spontaneously be opened. It can further add a new feature to our system that whenever the password inserted is wrong for more than three times, then the spy camera will capture the person who is trying to intrude. This system can be used as an attendance where each candidate will be entering their password given to them by the institution.

5 Discussion

There are certain advantages and disadvantages of the password based door locking system.

MERITS: Security is the prime concern nowadays which is provided successfully by this system. Acute power is consumed and such security is achieved easily and the system is well organized too.

USES: The simplicity of this circuit can be plus point for better security in the domestic places. It can be quite useful in the administrations, as it would prove to be the best access granter to highly secured places. Load switching control through password is also an application of this project.

DEMERITS: The lack of feasibility to run the circuit remotely is its limitation. If the persons forget the password the door remains closed.

6 Conclusion and Future Scope

This research articles clearly indicate the advancement in technology and need of society. This system is highly useful for office, smart homes, and hospitals, etc. It adds the GSM module with microcontroller 8051, then it can be checked locking and unlocking of their doors of assets from a remote place through their mobiles/internet. Its software is open source, then it is easy to access, design, burn, and implement. This proposed system has been successfully designed and implemented in door locking systems. In future, security key will be more stronger if it is added to the biometric fingerprint and speech technology together.

References

1. Lalanne T, Lempereur C (1998) Color recognition with a camera: a supervised algorithm for classification. In: IEEE southwest symposium on image analysis and interpretation, Tucson-Arizona, pp 198–204
2. Goel V, Kumari P, Shikha P, Prasad D, Nath V (2019) Design of smartphone controlled robot using bluetooth. In: Nath V, Mandal J (eds) Nanoelectronics, circuits and communication systems. Lecture notes in electrical engineering, vol 511, pp 557–563. Springer, Singapore. https://doi.org/10.1007/978-981-13-0776-8_52
3. Nidhi N, Prasad D, Nath V (2019) Different aspects of smart grid: an overview. In: Nath V, Mandal J (eds) Nanoelectronics, circuits and communication systems. Lecture notes in electrical engineering, vol 511, pp 451–456. Springer, Singapore. https://doi.org/10.1007/978-981-13-0776-8_41
4. Thakur MK, Kumar RS, Kumar M, Kumar R (2013) Wireless fingerprint based security system using Zigbee. Int J Inventive Eng Sci (IJIES) 1(5). ISSN:2319-9598
5. Mary Lourde R, Khosla D (2010) Fingerprint identification in biometric security systems. Int J Comput Electr Eng 2(5)
6. Song M, Kim B, Ryu Y, Kim Y, Kim S (2010) A design of real time control robot system using android smartphone. In: The 7th international conference on ubiquitous robots and ambient intelligence (URAI), Busan Korea
7. Signals, Systems and Computers (2004) Conference record of the thirty-eighth Asilomar conference on publication 7 Nov 2004, vol 1, pp 577–581
8. Shilpa V, Pradeep HS, Kurian MZ (2012) The symbian robot. Int J Comput Sci Inf (IJCSI) 1(3)
9. Klingmann M (2009) Accelerometer-based gesture recognition with the iPhone, Goldsmiths University, MSc in Cognitive Computing, London, Sept 2009
10. Azid SI (2012)Technology intelligent home: SMS based home security system with immediate feedback. Int J Adv Res Sci Eng 2(5). http://www.ijarse.com. ISSN-2319-8354(E)
11. Liu T, Guo H, Wang Y (2008) A new approach for colour-based object recognition with fusion of colour models. In: Congress on image and signal processing conference, Sanya-China, vol 3, pp 456–460

Development of Wireless Power Transfer System with Internet of Things

Suvid Sahay, Nitika Sharma, Shubham Raj, Kumari Neelam, Deepak Prasad and Vijay Nath

Abstract The main objective of the article is to develop a device for wireless power transfer. The concept of WPT was earlier realized by Nikola Tesla. This article can be used for charging batteries which cannot be connected physically such as pacemakers implanted in the body that runs on a battery. The battery is replaced every year and hence the body is operated. This article is written to recharge a battery wirelessly. This technique can also be used to charge smartphones, iPhone, iPad, laptops, etc. The risk of getting an electrical shock is also reduced as the device is galvanically isolated. The research in this article is still in progress as it is an emerging technology.

Index Terms Electric vehicles · Green house gas · Wireless power transfer

1 Introduction

The world is getting technologically advanced day by day. For making our lives simpler, new technologies are being developed day by day. But in all these technological advancements, the conventional wire system makes our life a mess like charging of mobiles, laptops, and cameras which involves the use of a lot of wires. It also takes up a lot of electric sockets and not to mention the fact that each device has its own design for the charging port. But there is a possibility of charging these devices simultaneously without the use of wires and not creating a mess in this process. This can be done with the process of inductive coupling which is a simple and effective way of power.

S. Sahay · N. Sharma · S. Raj
Department of CSE, Birla Institute of Technology, Mesra, Ranchi 835215, Jharkhand, India
e-mail: rajshubham10@gmail.com

K. Neelam · D. Prasad (✉) · V. Nath
Department of ECE, Birla Institute of Technology, Mesra, Ranchi 835215, Jharkhand, India
e-mail: prasaddeepak007@gmail.com

V. Nath
e-mail: vijaynath@bitmesra.ac.in

© Springer Nature Singapore Pte Ltd. 2019

V. Nath and J. K. Mandal (eds.), *Proceedings of the Third International Conference on Microelectronics, Computing and Communication Systems*, Lecture Notes in Electrical Engineering 556, https://doi.org/10.1007/978-981-13-7091-5_51

The efficient transmission of electric power between two different points to through vacuum, as well as an atmosphere without using wire or any other substance is Wireless Power Transmission Transfer (WPT). This system is used for the applications where either an instantaneous amount or a continuous delivery of energy is needed, but where conventional wires are unaffordable, inconvenient, hazardous, or impossible. There are different ways through which power can be transmitted by the use of short-range Inductive coupling, Resonant Induction for mid-range and EM wave power transfer for high range. With the help of WPT, we can transport power to locations, which aren't possible or impractical to reach. The next big thing in this era could be charging devices through WPT.

The aim of this article is to design and construct a method to transmit wireless electrical power through space and charge a designated low power device. The transmission of power from an AC line to a resistive load can be done by using resonant coils. Investigation of various geometrical and physical form factors evaluated in order to increase coupling between transmitter and receiver.

The charging process would be made simpler and it would be easier to charge a low power device with the help of the WPT system. Also, it ensures the safety of the device since it the risk of short circuit is eliminated through this.

The objective also includes the prospect of charging multiple low power devices simultaneously using a single source which would use a single power outlet.

The concept of electronics and integrated circuits utilizes from [1–3]. The concept of the Internet of Things [4, 5] demonstrates the software-based multi-switch connectivity. Being Researches on the Propagation of Electric Action with Finite Velocity through Space [6] generate the concept of power. Modeling and control of series–series-compensated inductive power transfer system [7] demonstrate the concept of remotely device charging and discharging capability. Controlling the renewable microgrid using semidefinite programming technique [8] demonstrate the concept generation, transmission and distribution of power. Modeling and design of dynamic wireless power transfer system for EV applications [9] generate the concept of power for wireless transmission. Energy big data: A survey [10] demonstrate the generation of highly efficient power. A comparative study of power supply architectures in wireless EV charging systems [11] is the best concept of PWT. A feasible cellular internet of things: Enabling edge computing and the IoT in dense futuristic cellular networks [12] defines the utility of power and lending toward the consumers or customers. High-efficiency contactless power transfer system for electric vehicle battery charging application [13] demonstrate the e-charging device by the server. Simplified mathematical modeling of phase-shift-controlled series–series-compensated inductive power transfer system [14] demonstrate the concept of power transmission remotely. Different Aspects of Smart Grid: An Overview [15] explain the appropriate setup of electric supply generation from solar, thermal, hydraulic, wind etc. and transmission by low lossless medium and distribution to different setups as per requirement of consumers and billing and collection of the charges of energy. Design of Smartphone Controlled Robot Using Bluetooth [16] will help to understand the charging of devices through WPT.

2 Methodology

The main aim of this research article is to charge a battery using inductive methodology, the time to charge the battery is too short. The energy provided should be sufficient to power an EV. Also, it should power devices over larger distances. The Multisim software is used to simulate the circuit for power transfer system.

Wireless power transfer is discussed in many other articles published by many other authors. One thing which is common in all the articles is that the EV is the next big thing in this world, which reduces the emission of GHG and hence reduces global warming. Thus, with the help of WPT, the driving range of the electric vehicles can be increased by charging the battery continuously by the method of inductive coupling.

3 Circuit Setup and Working Principle

(1) ***Inductive Coupling***: *Magnetic* coupling or even inductive coupling would work by following the principle of electromagnetism. When a wire is at proximity to a magnetic field, it automatically generates a magnetic field in the specific wire. Transfer of energy between one to another wire through magnetic fields has been termed as inductive coupling. This transfer of energy is performed by the transfer of the common magnetic field present in both the circuits.

In the field of electrical engineering, two conductors are said to be magnetically coupled or mutual-inductively coupled in case if they are configured such that a voltage is induced across the end of the other wire due to the change in the flow of current in the first wire by electromagnetic induction. Measurement of the amount of inductive coupling present between both the conductors is done with the help of their mutual inductance.

Increase in efficiency of Power transfer of inductive coupling can be obtained if we increase the number of turns present in the coil, the current strength, the radial magnetic field strength and the cross-sectional area of the coil. Magnetic fields decay in a very small period of time, resulting in a very short range of effectiveness of inductive coupling.

(2) *Inductive Charging*: The electromagnetic field is used by inductive charging, for transferring energy between the two objects. Energy is sent by a charging station with the help of inductive coupling to an electrical device, which contains batteries in which the energy is stored. Because the two coils are placed at a very small gap between each other, Inductive charging is termed as a kind of very short- distance wireless energy transfer.

An induction coil is typically used by induction chargers forth the creation of an alternating electromagnetic field present within a charging base station, and also a second induction coil present in the portable device absorbs all the power from the electromagnetic field and converts it again in the electrical current for charging the

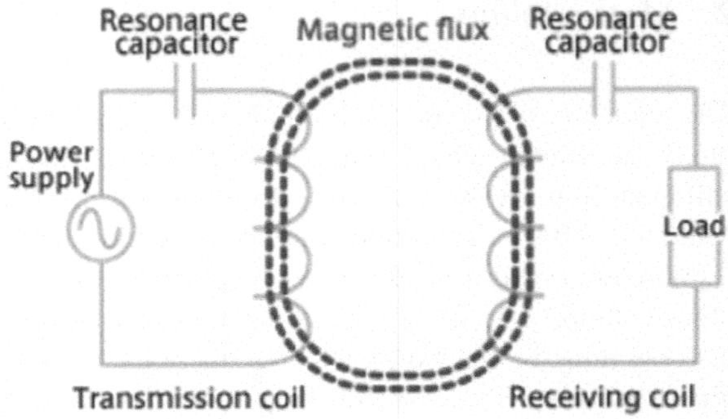

Fig. 1 Circuit setup for wireless power transfer

battery. An electrical transformer is formed by the combination of the two induction coils which are in proximity.

Resonant inductive coupling is used by the inductive charging system to achieve greater distances. The basic circuit of the wireless transfer circuit is shown in Fig. 1.

Software Simulation:
The Multisim software is used for the simulation of wireless power transfer. Multisim is an industry-standard SPICE simulation and circuit design software for analog, digital, and power electronics in education and research. And some equipment name and quantities are discussed below in Table 1.

Equipment:
This paper proposes a wireless power transfer system that consists of a transmitter with a resonator and a receiver without a resonator. This system utilizes magnetic

Table 1 List of component and quantities

Component name	Quantity
Voltage source, Vdc 15 V	1
Resistor 39 Ω 5 W	4
Resistor 5.6 kΩ	2
Diode 1N4148	2
Diode OA79	4
MOSFET IRF540	2
Radio frequency choke 120 μH	2
Capacitor 10 nf	1
Capacitor, C2 100 μF	1
Voltage regulator IC LM 7812	1
12 V DC FAN	1
Power cord	1
Coil, L 8 μH	2
Soldering lead (50 gm)	1

coupling and resonance. The receiver is simple, but this system has longer transfer distance than electromagnetic induction and requires no resonant frequency synchronization between the transmitter and the receiver that is required for the magnetic resonance method. In general, receivers have a main function other than power receiving function and it is favorable that receivers have less restriction for placement and a simple adjustment mechanism. The proposed system offers solutions for these issues. The construction of this circuit comprises of a transmitter and a receiver. The wireless power transmission can be defined as the energy can be transmitted from the transmitter to a receiver through an oscillating magnetic field.

Major Problems due to the use of wired technology:

(1) Cost is one of the major reason which separates wired technology from the wireless technology. Cost of the wired technology is constant for a long time whereas the price of power transmission has been decreasing day by day in the case of wireless technology. This gives wireless technology an edge over other wired technology.
(2) It can be clearly seen that wired technology is not easily portable. The equipment must be connected to a power source for it to function. Moving these units may take time and energy. In the case of a wireless system, portability is not an issue. It can be easily transported from places to places. As shown in Fig. 4, a Samsung wireless charger is used to charge the phone. This charger is easily portable and also compact. The transmitter and receiver circuit for wireless power transfer is shown in Fig. 2. The wireless charging setup has developed by Samsung shown in Fig. 3.
(3) Wireless technologies eliminate the case of space usage. It can be easily noted that wired technologies takes a lot of space. Also having a lot of wires at a table creates a mess and the wires get entangled between them as clearly seen in Fig. 4.
(4) Another instance in which wired technology is much less reliable is due to the fact that the wires can be damaged like cuts or accidentally breaking any wire. Also, naked wires can give electrical shocks to the person. These risks are also eliminated in the case of wireless systems.

An ideal room powered completely by the wireless system is shown in Fig. 5. A source coil is used to power all the electrical system in the room. Although this kind of concept is still in progress, there will come a time not too soon when this becomes a reality.

4 Results and Discussion

The WPT systems may find its use in gadgets like Smart Phones, iPod, DSLRs, PCs, laptops, and other electronic devices. WPT can be used in any device containing a battery or a cell inside it. This contains devices like vacuum cleaners, iPods, laptops,

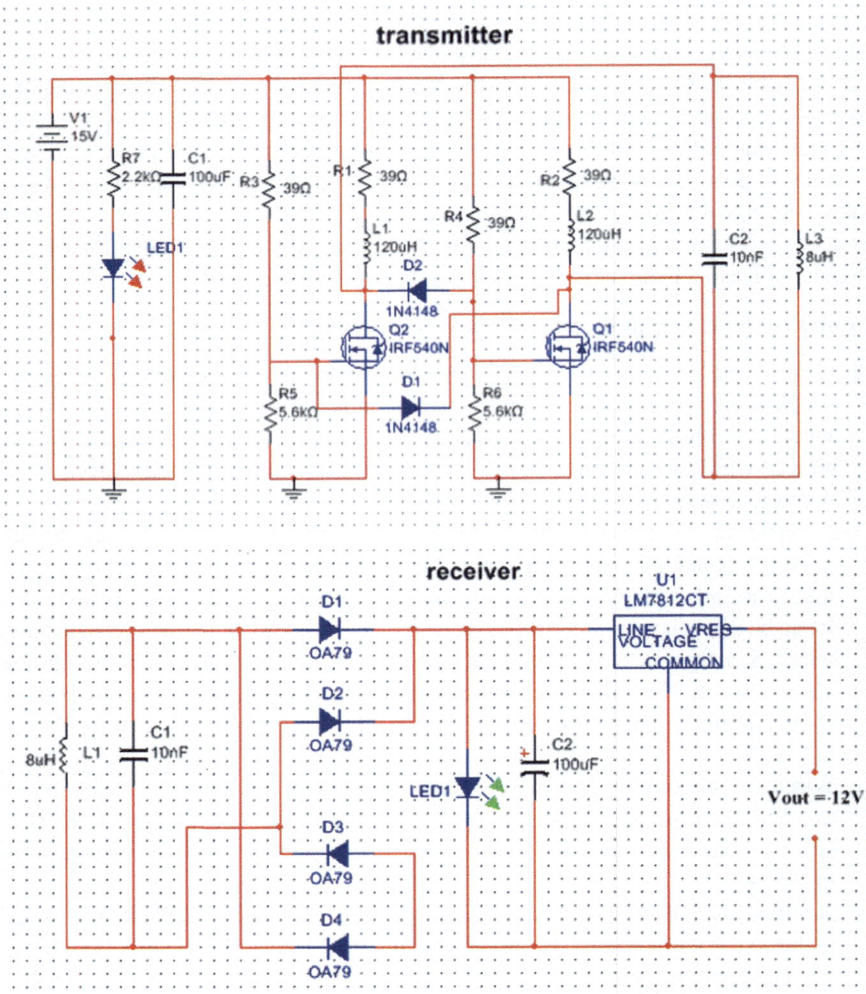

Fig. 2 The transmitter and receiver circuit for wireless power transfer

personal computers, and tablets. Wireless POWER System can also charge an electronic device like super caps (viz. supercapacitor). Devices which consume power with a low voltage cable wire can also be based on WPT. Also, they find application at the railway station. When there is only a couple of switchboards and we got too many users.

This article also discusses the advantages and disadvantages of wireless power transfer by the method of inductive coupling.

Advantage: (1) Improve range issues, (2) usage is simple and easy, (3) guarantees safety as cables are not used.

Fig. 3 A wireless charger developed by Samsung

Fig. 4 A wired mess consisting of charging wires

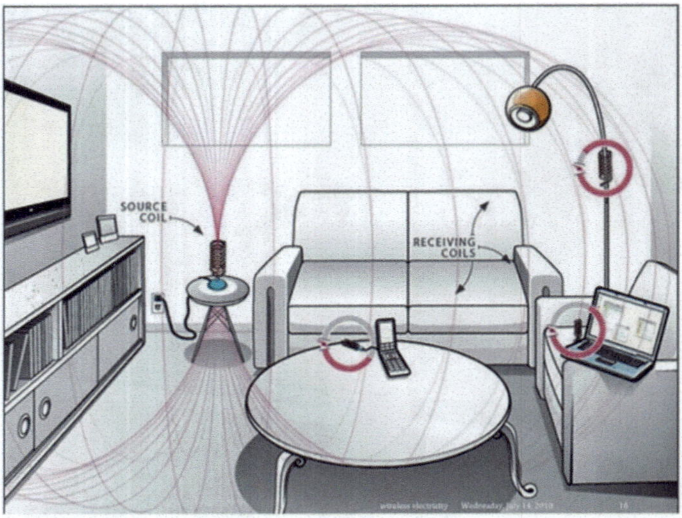

Fig. 5 A room completely powered by wireless power transfer

Disadvantage: (1) It is costly but prices will reduce in future, (2) Efficiency is not that much.

5 Conclusion

Nikola Tesla is the godfather of transfer of wireless energy. He is a fascinating figure and a brilliant personality in the history of electrical engineering. EM energy can be transferred wirelessly, was found by Tesla long in past in the 1800s. Another thing that Tesla stated that in the future all our electronics product will be able to receive energy from the space.

There are still many aspects in which wireless power transfer can be improved. The size of the coils can be reduced significantly as for greater extent of power transfer, larger coils are required. Also, the coils should be placed at close proximity to each other. As for today, the distance between the coils should be not more than 3–5 m. The efficiency of power transfer is also not much great. Few researchers have transferred the power to around 45%, whereas the other few have transferred it to 80% in their respective projects. Hence unless and until the efficiency in power transfer is not increased to more than 95% this project cannot be utilized in practical life as the energy provided is not freely available.

6 Future Scopes

From the past century, Electronic revolution has always been a tangled affair. Remember that time, when telephones were always wired in the wall of our houses? And when the internet needed one extra cord attached from the backside of our computer system? Wireless is a very big development in the future, which has given us very much ease to our lives and it is also good for the transfer of information. But the question arises is about the energy itself?

We all live in a world which is wireless. Mobile phones, Wi-Fi network, Bluetooth, and also networks like 2G, 3G, and 4G. By the magic of wireless, we get a connected network which is always on, and always connected with digital devices. Even after this, all those devices including mobiles to our PCs, still they depend on wires and plugs to be charged. When the battery of our mobile or laptops gets discharged and there is no option to charge it, they become completely useless for the whole day, Product ads always use this problem of charging batteries as a major factor, by featuring various lists and issues. Developed and heavy powerful devices such as mobiles, iPod, PCs, and laptop are completely worthless without an outlet when that battery bar hits 0%, (that same thing which everybody present in the cyber cafe or at the airport wants to implement).

If all the effort we have to do to give life to our dead devices by charging it is set them at the right place. That would be the real achievement of the aim of the wireless power.

Acknowledgements We would like to express our special thanks and gratitude to the Vice Chancellor Mr. M. K. Mishra as well as the Head of Department of ECE, Dr. Srikanta Pal and Head of Department of CSE, Dr. Vandana Bhattacharjee for giving me this golden opportunity to perform a wonderful research work on the topic: Design of Wireless Power Transfer System, the completion of this research work helped me in doing a lot of research and development and I came to know about many new things.

References

1. Millman J, Halkias CC, Integrated electronics: analog and digital circuits and systems
2. Rashid MH, Power electronics: circuits, devices, and applications
3. Boylestad RL, Nashelsky L, Electronic devices and circuit theory
4. Internet of Things' Connected Devices to Almost Triple to Over 38 Billion Units by 2020. Juniper Research, 28 July 2015. http://www.juniperresearch.com/press/press-releases/iot-connected-devices-to-triple-to-38-bn-by-2020, 24 Oct 2015
5. M2 M world of connected services: the internet of things. an infographic prepared by Beecham research. 4 November 2015
6. Hertz Heinrich, Waves Electric (1893) Being researches on the propagation of electric action with finite velocity through space. Macmillan & Co., London, p 152
7. Zahid ZU, Dalala ZM, Zheng C, Chen R, Faraci WE, Lai J-SJ, Lisi G, Anderson D (2015) Modeling and control of series-series compensated inductive power transfer system. IEEE J Emerg Sel Top Power Electron 3(1):111–123

8. Rana MM, Li L, Su SW (2017) Controlling the renewable microgrid using semidefinite programming technique. Int J Electr Power Energy Syst 84:225–231
9. Song K, Zhu C, Koh K-E, Kobayashi D, Imura T, Hori Y (2015) Modeling and design of dynamic wireless power transfer system for EV applications. In: Proceedings of the annual conference of the industrial electronics society, pp 005229–005234
10. Jiang H, Wang K, Wang Y, Gao M, Zhang Y (2016) Energy big data: a survey. IEEE Access 4:3844–3861
11. Esteban B, Sid-Ahmed M, Kar NC (2015) A comparative study of power supply architectures in wireless EV charging systems. IEEE Trans Power Electron 30(11):6408–6422
12. Dama S, Sathya V, Kuchi K, Pasca TV (2017) A feasible cellular internet of things: enabling edge computing and the IoT in dense futuristic cellular networks. IEEE Consum Electron Mag 6(1):66–72
13. Zheng C, Lai J-S, Chen R, Faraci WE, Zahid ZU, Gu B, Zhang L, Lisi G, Anderson D (2015) High-efficiency contactless power transfer system for electric vehicle battery charging application. IEEE J Emerg Sel Top Power Electron 3(1):65–74
14. Aditya K, Williamson SS (2016) Simplified mathematical modelling of phase-shift controlled series-series compensated inductive power transfer system. In: Proceedings of the international symposium on industrial electronics, pp. 842–848
15. Nidhi N, Prasad D, Nath V (2019) Different aspects of smart grid: an overview. In: Nath V, Mandal J (eds) Nanoelectronics, circuits and communication systems (Lecture notes in electrical engineering), vol 511. Springer, Singapore, pp 451–456. https://doi.org/10.1007/978-981-13-0776-8_41, ISBN: 978-981-13-0776-8
16. Goel V, Riya, Kumari P, Shikha P, Tanushree, Prasad D, Nath V (2019) Design of smartphone controlled robot using bluetooth. In: Nath V, Mandal J (eds) Nanoelectronics, circuits and communication systems (Lecture notes in electrical engineering), vol 511. Springer, Singapore, pp 557–563. https://doi.org/10.1007/978-981-13-0776-8_52, ISBN: 978-981-13-0776-8

Design of Water Overflow Indicator Alarm and Controller

Gurpreet Singh, Kumari Nivedita, Sachin Sanjay Minz, Kumari Neelam, Deepak Prasad and Vijay Nath

Abstract It is very common nowadays that the water reservoirs often overflow, as there is only one pump operator who has to manage filling up several water tanks in many blocks all by itself and often the pumps are switched off after much delay. This causes a lot of wastage of very precious water needed by the human race to survive. Even the underground water basins would run out of water in some years in the near future. The main concern is that roughly all the households face the problem of wastage of water because of the water overflow tank. According to a study, we waste around 45% of the water because of the tank overflow. To remove this problem, companies have come up with an overflow tank. Countries like the USA and some European countries are on it and arousing water level indicator and controller alarm for preventing water to be wasted. India along with some other Asian countries is contributing quite high in wasting of Water. As the water level rises or falls, different circuits in the controller send different signals. These signals are used to switch ON or switch OFF the motor pump as per our requirements.

Keywords LED · Transistor · Buzzer · Resistor · Tank overflow

1 Introduction

Nowadays, water crisis is one of the major global risks. One drop of water waste can be a huge concern for our future survival. Nearly 80% of the world population lack access to safe drinking water [1]. Thus, solving this water wastage problem is of huge concern. The major aim is to reduce this problem. This water indicator tank not

G. Singh · K. Nivedita · S. S. Minz
Department of CSE, Birla Institute of Technology, Mesra, Ranchi 835215, Jharkhand, India

K. Neelam · D. Prasad (✉) · V. Nath
Department of ECE, Birla Institute of Technology, Mesra, Ranchi 835215, Jharkhand, India
e-mail: prasaddeepak007@gmail.com

V. Nath
e-mail: vijaynath@bitmesra.ac.in

© Springer Nature Singapore Pte Ltd. 2019

V. Nath and J. K. Mandal (eds.), *Proceedings of the Third International Conference on Microelectronics, Computing and Communication Systems*, Lecture Notes in Electrical Engineering 556, https://doi.org/10.1007/978-981-13-7091-5_52

only indicates the level of water present in the water tank but also gives an indication when the tank is full. When the water tank is empty the wires in the tank are switched off as the circuit is incomplete and the resistor pulls down the switch thus opens the switch. As the water rises up in the first water tank its fills percentage of water filled is displayed on the screen. Nowadays in the world, the majority of the developing nations are using this to save water [2]. These tanks are perfectly designed to prevent any sort of water wastage by reducing the overflow of water. The crisis for drinking water shortage in India is at an alarming level. It may soon become a global crisis. Hence, it is our primary duty to preserve water. Water tanks used for the domestic purpose have a major problem that is overflowing of water tanks in day to day life [3]. It is known to all that water is very precious to mankind, so for solving this water wastage problem we use a simple circuit known as "Water Level Indicator". It works on the principle that water conducts electricity hence water itself can be used to switch on and off the main water supply leading water to the water tank. This project demonstrates a simple circuit to the water level indicator [4]. Design Strategy for Smart Toll Gate Billing System [6] demonstrates the concept of the automation system in the water supply. A Literate review of RFID-enabled Healthcare Applications and Issues [7] shows the concept of consumer utility and billing. Study and Design of Smart Embedded System for Smart City Using Internet of Things [8] demonstrate the concept to add the water tank and users tanks and see the utility and requirement on a single server. Study and Design of Smart Embedded System for Remote Health Monitoring Using Internet of Things [9] define the concept of how to utilize the water properly, purity of water, consumptions of water, tank clinking etc. An Ultra-Low-Power Internet-Controlled Home Automation System [10] demonstrate the important factor and utility of water in the house in the city. Design and Implementation of a Reaction Timer Using CMOS Logic [11] demonstrate the concept delivery of water to the consumer as per their needs without any wired communication. Design of Smart Embedded System for Agricultural Update Using Internet of Things [12] demonstrate the needs of water in agriculture and its applications. Study and Design of Smart Embedded System for Train Track Monitoring Using IoTs [13] demonstrate the fully use of water in railways for drinking, clinging, etc. Study and Design of Smart Embedded System for Smart City Using the Internet of Things [8] demonstrate the concept for the use of water and arrangement of clean water in the city level. Different Aspects of Smart Grid: An Overview [14] explain the utility of water for power generation in the smart grid.

2 Literature Survey

The simplest way to set up water level indicator can be made with the help of resistor, buzzer, transistor, and LEDs. We can utilize this project to measure water levels and also control the wastes of water [5]. We can consider some sections to measure various levels of filling up of water in water basins with help of conducting wires. Ends of these conducting wires are torn down some apart to each other in the water basins. We can use three to four number of wires to represent different levels of water rise in tanks.

(1) Level 1: It is the lowest level of water tank which indicates us that 25% of the water basin is now filled up by lighting up LED 1. It also shows that water has started filling.
(2) Level 2: It is the middle level of water tank which indicates us that 50% of the water basin is filled up by lighting up LED 2.
(3) Level 3: It is the last level of water saving indicator that lights up LED 3 when 90% of the tank is filled. It is the last sensor before buzzer which tells it's time to switch off the water supply.
(4) Level 4: Buzzer—It is the last stage of water level indicator system which will be active only when pump is not closed at level 3. It will produce a sound so that water supply can be turned of before a lot of precious water is saved [6].

Water level indicator works on circuit diagram listed below. It is connected to a 9 V D.C. voltage source. The positive terminal is connected to the upper head of water basins. Whereas, negative end of D.C. source with LEDs or Buzzer accordingly as required. They are connected to 330 Ω resistors and other ends with upper part of water basins.

3 Equipment Required

(1) **Input Voltage**: 9 V D.C. power supply is used to give the required current for the circuit operation.
(2) **Water Sensors**: Conducting wires can be placed in the water basin for water sensor purposes. As the current that will pass through these wires is in the range of nanoamperes. These wires are placed at different levels such that at every level LEDs will light up.
(3) **Transistors**: Any transistor can be used but BC548C will be preferable for its low price and long lifetime [7].
(4) **LEDs**: Led with red color light and made up of aluminum gallium arsenide is used as it is easily available in the market.
(5) **Resistors**: Resistors of 330 kΩ is used such that the current is in nanoamperes.
(6) **Buzzer**: Simple Buzzer is used to alarm about water overflow.

4 Circuit Setup

Due to the limitation of software in which we created our project circuit model, it did not have any LED, so we used Bulb in our diagram to represent LED in our circuit diagram as shown in Fig. 1.

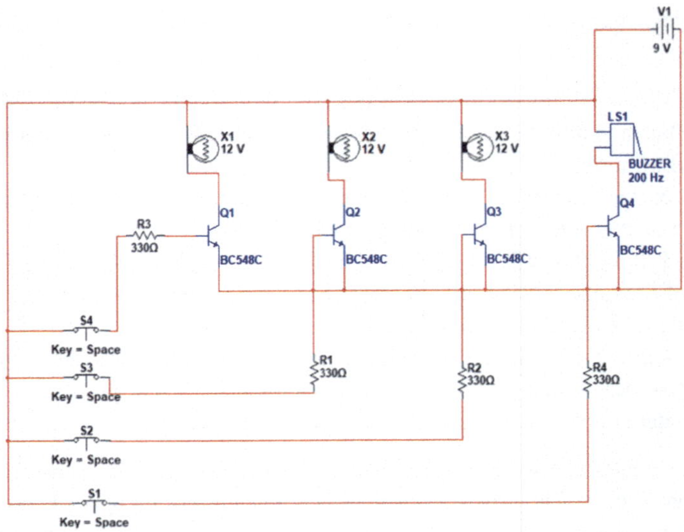

Fig. 1 Schematic diagram of the water level indicator and controller

5 Working Principle

Case I: Water Not Available
The Water Level Indicator employs a simple mechanism to detect and indicate the water level in an overhead tank or any other water container. When the water basin/tank is vacant there is no path for the flow of current between any of the conducting wires/water sensors and thus no current will pass through the wires and therefore no LED will be in active state or simply not glow.

Case II: Intermediate Level II
Now as the water starts rising up in the basin/tank, a conductive path is developed between the conducting wire 1 and the common conducting wire and the corresponding transistors get sufficient biasing at their base, and they start conducting and now the outputs will be V. Thus, this will make LED 1 to glow up at a level of 25% of the water basin and thus we will get to know water has risen to 25% of water basin/tank.

Case III: Intermediate Level II
When the water level is at the half stage of filling it. Conducting wire 2 and common wire will be in active state. Also, at 25% conducting wire 1 was also active so therefore Led 1 and Led 2 will glow up indicating us 50% of the water tank is filled up.

Case IV: Intermediate Level III
This is the last LED stage, here when the water level is at 90% stage of the water basin conducting wire 3 and common conducting wire will be now active. Here also conducting wire 1 and 2 were active beforehand thus this will lead to LED 3 to glow

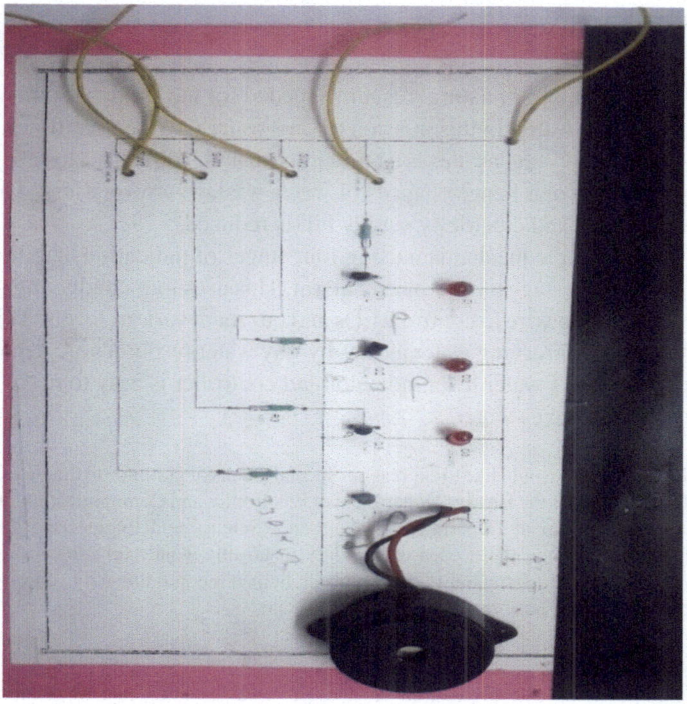

Fig. 2 Complete set up of the designed system

up with LED 2 and LED 1. This is the last stage before filling up of tank and buzzer to create an alarm for wasting of water.

Case V: Water Tank Full

When the water basin is completely filled with water, the top level conducting wire 4 and common conducting wire will be in active state and the corresponding transistor gets into conduction mode and makes buzzer to beep and thus alarming us that water tank is full, and we need to close the water supply to save water. The complete set up of the designed system is shown in Fig. 2.

6 Result

Water level indicator and controller is widely applicable for use in domestic as well as commercial and industrial purposes. These are hugely capable to save a lot of precious water and energy efficient. With the help of this water indicator, we can easily control the amount of water to be filled in a water basin. They can be made at a cheap rate. Everyone can easily save water as on filling of water basin a buzzer will be heard and thus water supply can be stopped.

7 Conclusion

Water is the most important natural resource needed for the survival of mankind and it is our duty to save water. In this mission to save water it, water overflow alarm and controller is very important as it uses a buzzer alarm and it directs us to save water by closing water supply and hence chances of water wastage are very less. The overall cost of water supply and electricity supply bill is reduced.

It uses a quite simple mechanism using four stages of indication and three stages having LEDs and last stage having buzzer alarm. This uses four conducting wires and a common conducting wire to control LEDs and buzzer at various levels. This system is superefficient in conserving water as day-by-day water resources are depleting and every drop counts, this water level indicator and controller is easy to make and very cheap. Moreover, it saves water.

Acknowledgements We would like to express our special thanks of gratitude to the Vice Chancellor Dr. M. K. Mishra as well as the Head of Department of Electronics and Communication Engineering Dr. Srikanta Pal and Head of Department of Computer Science and Engineering Dr. Vandana Bhattacharjee, who gave me the golden opportunity to do this wonderful project on the topic: Design of Water Overflow Alarm and Indicator, which helped me in doing a lot of Research and I came to know about many new things.

References

1. https://circuitdigest.com/electronic-circuits/water-level-indicator-alarm-circuit. Accessed 4 Mar 2018
2. Ruiz-Garcia L, Lunadei L (2011) The role of RFID in agriculture: applications, limitations and challenges. Comput Electron Agric 79:42–50
3. Abbasi AZ, Islam N, Shaikh ZA (2014) A review of wireless sensors and networks' applications in agriculture. Comput Stand Interfaces 36:263–270
4. Nidhi N, Prasad D, Nath V (2019) Internet of things/everything: an overview. Int J Pure Appl Math 119(12). ISSN: 1314-3395
5. Srinivasan A (2006) Handbook of precision agriculture: principles and applications. Food Products Press, Binghamton, NY
6. Bohra V, Prasad D, Nidhi N, Tiwari A, Nath V (2019) Design strategy for smart toll gate billing system. In: Nath V, Mandal J (eds) Proceeding of the second international conference on microelectronics, computing & communication systems (MCCS 2017). Lecture notes in electrical engineering, vol 476. Springer, Singapore, pp 615–621. https://doi.org/10.1007/978-981-10-8234-4_49
7. Wamba SF, Anand A, Carter L (2013) A Literate review of RFID enabled healthcare applications and issues. Int J Inf Manag 875–891
8. Kumar A, Nath V (2019) Study and design of smart embedded system for smart city using internet of things. In: Nath V, Mandal J (eds) Nanoelectronics, circuits and communication systems. Lecture notes in electrical engineering, vol 511. Springer, Singapore, pp 397–408. https://doi.org/10.1007/978-981-13-0776-8_36. ISBN: 978-981-13-0776-8
9. Anand S, Nath V (2019) Study and design of smart embedded system for remote health monitoring using internet of things. In: Nath V, Mandal J (eds) Nanoelectronics, circuits and communication systems. Lecture notes in electrical engineering, vol 511. Springer, Singapore, pp 409–414. https://doi.org/10.1007/978-981-13-0776-8_37. ISBN: 978-981-13-0776-8

10. Rajiv P, Raj R, Singh R, Nagarkar R, Chaurasia AK, Agarwal S, Nath V (2018) An ultra-low-power internet-controlled home automation system. In: Nath V (eds) Proceedings of the international conference on microelectronics, computing & communication systems. Lecture notes in electrical engineering, vol 453. Springer, Singapore, pp 271–280. https://doi.org/10.1007/978-981-10-5565-2_24. Online ISBN: 978-981-10-5565-2
11. Bohra V, Nidhi N, Singh S, Prasad D, Thakur A, Kumar A, Nath V (2019) Design and implementation of a reaction timer using CMOS logic. In: Nath V, Mandal J (eds) Proceeding of the second international conference on microelectronics, computing & communication systems (MCCS 2017). Lecture notes in electrical engineering, vol 476. Springer, Singapore, pp 533–543. https://doi.org/10.1007/978-981-10-8234-4_43. Online ISBN: 978-981-10-8234-4
12. Razi Q, Nath V (2019) Design of smart embedded system for agricultural update using internet of things. In: Nath V, Mandal J (eds) Nanoelectronics, circuits and communication systems. Lecture notes in electrical engineering, vol 511. Springer, Singapore, pp 372–382. https://doi.org/10.1007/978-981-13-0776-8_34. Online ISBN: 978-981-13-0776-8
13. Fatma S, Nath V (2019) Study and design of smart embedded system for train track monitoring using IoTs. In: Nath V, Mandal J (eds) Nanoelectronics, circuits and communication systems. Lecture notes in electrical engineering, vol 511. Springer, Singapore, pp 385–395. https://doi.org/10.1007/978-981-13-0776-8_35. Online ISBN: 978-981-13-0776-8
14. Nidhi N, Prasad D, Nath V (2019) Different aspects of smart grid: an overview. In: Nath V, Mandal J (eds) Nanoelectronics, circuits and communication systems. Lecture notes in electrical engineering, vol 511. Springer, Singapore, pp 451–456. https://doi.org/10.1007/978-981-13-0776-8_41. Online ISBN: 978-981-13-0776-8

An Assessment of Advanced Transportation Research Opportunities

Rudraksh Agrawal, Deepak Prasad, Kumari Neelam, Abhishek Pandey and Vijay Nath

Abstract These days, more and more aerial vehicles like the airplanes and jets are constructed from metal despite that diamond has a weight-to-strength ratio over 60 times of that of the currently being used aerospace aluminium, Diamond is costlier for sure but that's not the real issue, the problems we are not technically equipped to shape in the desired form and it shatters easily. Nanotechnology is where it comes handy to make inexpensive shatter resistant diamond (structure resembling the diamond fibre) in the exactly desired shape as we want. This would allow us to make and replicate any airline like the Boeing 747 which would be stronger and lighter than 50 times. As we know to travel in space is expensive and for a few privileged class. Nanotechnology will effectively cut off the costs and increase the utility of space flight and space travel. With nanotechnology: the weight-to-strength ratio and the high cost of the components would critically reduce and raise the standards of performance, both of the mentioned parameters will be benefited equally. In addition to inexpensive and giving high efficiency with its properties of strength and lightweight for space ships, it will also provide an unbelievable powerful and fats supercomputers with applications beyond the reach of the human brain.

Keywords Nano technology · Nano gear · Human brain

1 Introduction

Transporting goods and people from one point to another is becoming an increasingly critical aspect of our society. We are social animals and move and commute a lot may it be for work or to home; goods are commuted all over the world and even to the next door. Mobility, in layman terms is the measure of distance a person or a

R. Agrawal · D. Prasad · K. Neelam · A. Pandey (✉) · V. Nath
VLSI Design Group, Department of ECE, B.I.T. Mesra, Ranchi 835215, JH, India
e-mail: a.p.bitmesra@gmail.com

V. Nath
e-mail: vijaynath@bitmesra.ac.in

© Springer Nature Singapore Pte Ltd. 2019

V. Nath and J. K. Mandal (eds.), *Proceedings of the Third International Conference on Microelectronics, Computing and Communication Systems*,
Lecture Notes in Electrical Engineering 556, https://doi.org/10.1007/978-981-13-7091-5_53

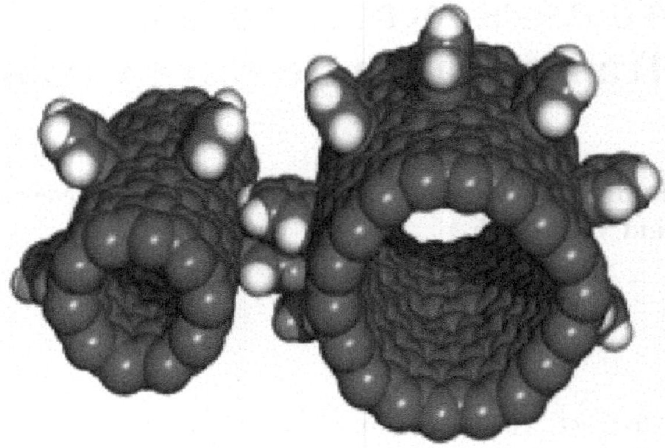

Fig. 1 The nano gear

goods is commuted over a given time period. If the time period is less the efficiency is more. As mobility is connected to wealth and development, the growth in this field is associated to the developing countries as population trends have shown rapid changes in industry and economy from the developing countries as depicted in Fig. 1. The amount of time a person gives to travel or commute from one place to other is almost equal in all countries with similar income levels, but there is a correlation of distance and the income of a person. This leads to an increase in travel speed along with wealth. Demand is getting high for speedy transportation to decrease the waste of time and increase one's efficiency this trend is constantly there for some decades now and experts believe it will continue in the future.

Nanotechnology is important in the world of science as it has the power to make almost any manufactural product lighter, stronger, faster, smarter, cleaner and safer. We have seen some of its capabilities and yet to discover the full potential of it. The scientists are continuously thriving to gain the full power of this gift of science. New products that are being solved by nanotechnology will make our life more simple and solve the other problems. That is righteously the main aim of science. Nanotechnology History: A non-technical primer, the Nanotechnology Revolution [1] demonstrate the innovation in nanotechnology. Nanocrystals Production and Fabrication [2] demonstrate the light device fabrication and production of low power devices. Modifying light [3] explain low power consumptions. Engine of Creation: The Coming of Era of Nanotechnology [4] illustrate the new generation of electronics goods and applicable in medical, aerospace etc. There is plenty of room at the bottom [5] demonstrate the creation of the bottom surface. Assembly required [6] demonstrate the typical procedure of assembly. The History of liquid crystal displays [7] defines the new generation display device. Domain divided vertical alignment mode with optimized fringe field effect [8] defines the different types of fringes. The advance of PVA technology for multimedia applications [9] demonstrate the solar panel preparation and generation of solar energy. The Integrated amorphous silicon colour sensor on ICT panel for LED backlight feedback control system [10] demonstrate the preparation of LED devices and applications in the market. Materials Development and

Exploitation: The Determining Factor for the Pace of Technological Advancement [11] defines material development. Research Challenges in Nanoscience and Nanotechnology in Developing Counties [12] demonstrate the innovation in the new world. Nanoscience and Nanotechnology [13] demonstrate the recent innovation in science and technology. Nanoparticles in Cancer Treatment [14] defines the recent technique to use in the cure of cancer. Megabash on Nanomaterials [15] demonstrates the utility of nanoscience and nanotechnology in the current market.

2 Carbon Nanotubes

A prominent member of the structural fullerene family which includes bucket balls is the nanotube (also referred to as Buckytube). As we know bucket balls are more of a sphere, whereas a nanotube resembles more to a long cylinder or maybe a wire with one of the ends covered with a bucket ball hemisphere. The etymology of the name is because of the fact of their size the diameter of a nanotube is just about a few nanometers while mentioned earlier can length up to a few centimetres as desired of the function (closely resembling the human hair structure but 50,000 times smaller). The implementation of CMOS 5-stage ring oscillator on single carbon nanotube is shown in Fig. 2. An optimistic view of earth is demonstrated in Fig. 3.

3 Space Elevator: A Technical Reality

With advances towards ultra-strong nanofibres, the realm which until this point seemed science fiction is standing towards a reality where the concept of building the elevator which can carry cargo up to 60–70 k miles.

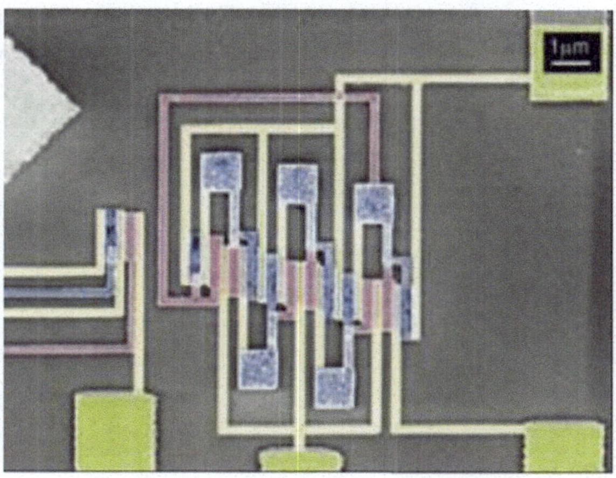

Fig. 2 Implementation of a CMOS-type 5 stage ring oscillator on a single carbon nanotube

Fig. 3 An optimistic view (not to scale)

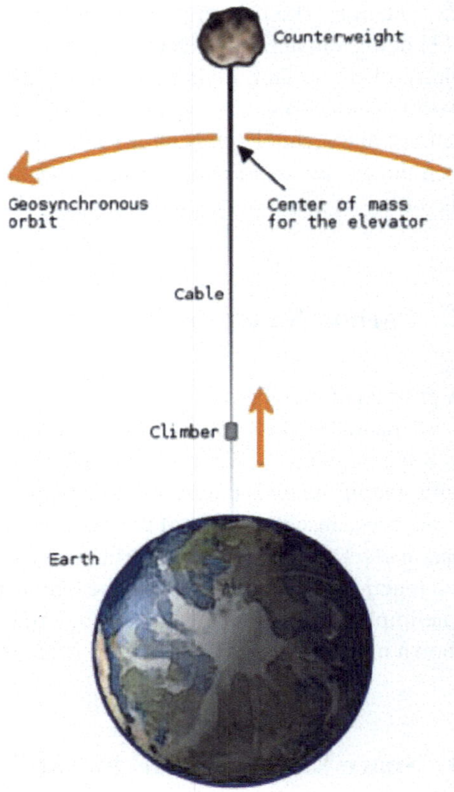

The materials can be transported from earth to space using a space elevator. This concept requires the need for a long cable and a deadpool. The hydraulic pumped electromagnetic vehicles to transport payloads and people between different planets.

The space elevator comprises of an extra long cable hinged to the surface of earth carrying a counterweight at the other end ensuring the centrifugal force is acting its part of making the string taut. This phenomenon ensures the elevator remains geosynchronous.

Planets rotation would be further accelerated by the carriage of moving the gravitational midpoint.

4 Current Research

(a) **Nano-Materials**

The nanomaterials field is a wide and a diverse field comprising of a number of subfields which develop or study a lot of distinctive properties of a lot of distinctive nanoscale dimensioned materials. Colloidal and interface science has arisen a numerous materials which are quite useful in this field of science, such as fullerenes, carbon nanotubes and various nanorods. Nanoelectronics and nano Ionics associates to fast ion transport. Flavours are currently the widely used commercial applications and bulk application. As mentioned already nanotechnology has many applications, it has also contributed towards the field of medicine by using Nao medicines and making appliances using nanomaterials. Nanopillars a type of nanoscale material is used in making of the solar cells which can save up the high cost of the traditionally used silicon solar cell.

(b) **Bottom-up approaches**

Smaller components can be arranged in tortious assemblies. Watson Crick method is used to base pair DNA and other nucleic acids. Atomic force can be used to deposit in a covet pattern using nanoscale write head upon a surface this process is known as pen nanolithography. The notable assemblies of materials using semiconductor materials that allows computing applications stacks gating and lasers.

(c) **Top-down approaches**

Smaller versions made into their bigger versions using this technique:-NEMS can be created using solid chain state drives relating to advancement in microelectromechanical systems. The methods of eradicating directly pre-occur gasses are applied to the same time. This phenomenon is greatly seen in the making of sub-100 nm sections of materials in transmission microscopy using electron speed gun.

(d) **Functional approaches**

Components of the desired usage can be assembled with the advancement of this field. Functionality without the regard of usage is pointless. Present magnetic nano chains are derived and synthesized by the process of artificial magnetic assembly of superparamagnetic nano-materials. Electronic properties of these nanomaterials also contribute in molecular-scale electronics. These could be used in single molecular components and nanodevices like the rotaxane. Synthetic molecular motors, generally called the nanocar in the scientific world can be created by synthetic chemical methodology.

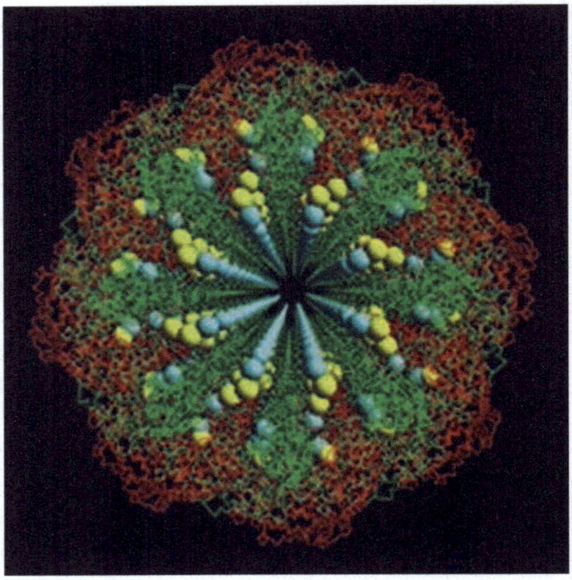

Fig. 4 Ames research centre conducted a series of experiment using nanoscale assembly to create sensing elements and many other things [2]

5 Nanotechnology in Transportation

Areas of Vehicle Nanoscience [1] Research Windows materials that are glare-free and dustproof, Super-grip tires which are wear resistant and have a long life, Anti-corrosive in nature, Wear-proof parts, Digital electronic safety sensors, Concept of a new drive, Lightweight materials [2] and Colour differing individually.

Ames Research Centre conducted a series of experiment using nanoscale assembly to create sensing elements is shown in Fig. 4.

(a) **Bike and Nanotechnology**
A bicycle which is light in weight is always easier to pedal. Light and strong bicycle frames can be possible and bicycle manufacturers are starting to make them. Carbon nanotubes are derivatives of carbon shaped in cylinders of carbon atoms that are 100 times stronger than steel and weight much lesser. Nanotechnology has been beneficial to produce extra durable and feather weighted bicycle frames. Carbon nanotubes could be helpful in bicycles which are shown in Fig. 5.

(b) **More Efficient Batteries with Nanoscience**
A123 System is often considered as one of the top 10 most futuristic and innovative companies in transportation in 2011 by the Fast Company Magazine, is developing electric car batteries using the nanoscience. The battery technology uses the principles of nanotechnology and the nano-phosphate lithium-ion battery developed at the Massachusetts Institute of Technology. Nano powered battery streamliner set a world record of 307.66 mpH shown in Fig. 6.

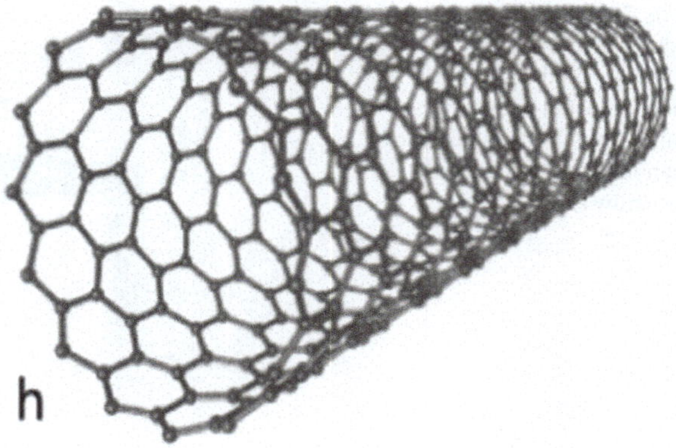

Fig. 5 Carbon nanotube could be helpful in bicycles

Fig. 6 Nanopowered battery streamliner set a world record of 307.66 mpH

(c) **Fuel Cells Using Nanotechnology**
With the help of enzyme associated electrically to a carbon nanotube, a cheaper fuel cell powered vehicles can possibly exist. The National Renewable Energy Lab (NREL) in Texas is mutating hydrogenase with carbon nanotube, a micro-bacterial enzyme that can produce electric current with virtually no harmful or producing pollutant by-products. These "bio hybrid" catalysts eliminate the need of the raw material platinum, a not so abundant and expensive component

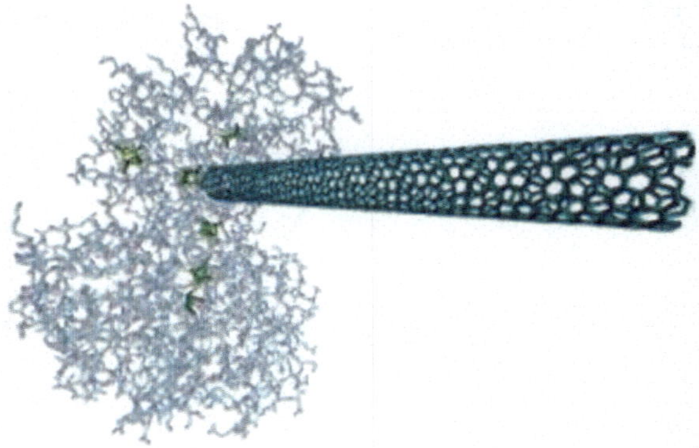

Fig. 7 Graphic representation of a single-walled carbon nanotube and hydrogenase an example of a widely used nanomaterial (elongated structure)

which is used in fuel cell catalysts. Graphic representation of a single-walled carbon nanotube and hydrogenase is shown in Fig. 7.

(d) **Lighter Nanotech Cars Save Fuel**
Scientists at Ford are thriving continuously in the field of nanotechnology for designing auto motives (mainly cars) 300–700 lb less heavy without altering the structural integrity or reducing the performance of the vehicle. Ford predicted by the inception of 2016, about 65% of the automotive industry will be redefined by the science of nanotechnology. Reducing the weight of the vehicles will make cars with better gas mileage.

(e) **A Better Catalytic Converter**
Mazda's converters have their applications connected with nanomaterials for catalytic metal; each less than mere six nanometres in diameter, fixed upon the surface of a miniature ceramic sphere using the technology used behind embedded systems in advance computers. Mazda has created a catalytic converter using the principles of nanotechnology. With advancement in the field of nanoscience, these catalytic converters use 65–85% less valuable metals like rhodium, palladium and platinum, which are in the reaction as catalysts to reduce the time of chemical reactions of pollutants such as carbon monoxide, hydrocarbons and nitrogen oxide.

6 Conclusion

Having the ability to change the physical and chemical properties of materials this technological gift, nanotechnology, can shape the future of material science and can be practically applied in various fields as mentioned like in medical technology, military applications, solar energy, electronics and fuels cells. Few scientists and connoisseur of this field believe nanotechnology is capable to elongate the human lifetime or produce replicator-like devices that can create anything and everything from simple input materials. Current research trend in nanotechnology is focused on nanomaterials like quantum dots for biological imaging, tissue engineering and drug delivery and nanorobotics where machines are created using nanomaterials with increased functionality and eco-friendly. We will mainly focus on how nanotechnology and science can help in transportation, electronics and computer science more precisely atom computers. Nano-Electromechanically Systems (NEMS) has the capacity to create mirrors, sensor elements, electrical circuitry and also mechanical gears in silicon surfaces permits the producer of mini-sized sensors like those which are useful to initiate the airbags of your car. The NEMS technique results in the near mutation of the much needed electronic circuitry on a single silicon chip with the mechanical mechanism, similar to the method used to produce digital based computerized chips used in motherboards. NEMS products are being used more and more these days causing manufacturers to produce more, this process is slow as the standard capital investment in low. Once this process is looked upon they will the flow of nanosized electronic equipment in the technologic market. Few nanomaterials already developed by scientists include quantum dots, nanowires and Buckyballs. Coming to its wide application in the electronics sector—mainstream electronics companies like IBM, Kodak and Idec are developing nanophotonic, opto-electrical materials and CMOS for IC engines. Memory devices which are dense and low power are build using transistors without the usage of p–n junction and nanowires; Using carbon nanotubes to direct electrons to illuminate pixels, resulting in a lightweight, millimetre thick "nanomedicine" display panel and using magnetic random access memory (MRAM), computers will be able to "boot" almost instantly. MRAM is enabled by manometer scale magnetic tunnel junctions and can quickly and effectively save data during a system shutdown or enable resume play features. Latest in the field of computing nanoelectronics have taken huge steps by making a functional computer using five atoms only. Quantum computing is dependent on the atomic-scale units, familiarly called "qubits," that can be simultaneously 0 and 1—a state known as a superposition. Most on the quantum computing depends on the functioning of nanomaterials and making a similar functional part to replace the original parts of a computer like the RAM. Concluding to it the next big thing in the world of technological advancement is actually small (one-billionth of a metre).

Acknowledgements We would like to thanks Dr. M. K. Mishra, Vice Chancellor BIT Mesra Ranchi for providing such great infrastructure. A piece of thanks also goes to Dr. S. Pal for their helping hand.

References

1. Keiper A (2005) Nanotechnology history: a non-technical primer, the nanotechnology revolution. New Atlantis
2. Bullen C (2004) Nanocrystals production and fabrication. Glasgow, Scotland
3. Castellanos AJ Modifying light, American Scientist, vol 95, no 5, Sept–Oct 2006
4. Eric D (1974) Engine of creation: the coming of era of nanotechnology. Springer, New York

5. Richard F (1959) There is plenty of room at the bottom. In: American physical society, Caltech. U.S.A
6. Manoharam H (2003) Assembly required, Stanford Engineering and Science Institute, Nanoscience and Nanotechnology
7. Kawamoto H (2002) The history of liquid crystal displays. Proc IEEE 90:260
8. Kim KH et al (1998) Domain divided vertical alignment mode with optimized fringe field effect. Proc Asia Display
9. Kim KH, Kim SS (2003) Advance of PVA technology for multi-media applications. In: SD international symposium digest 208
10. Lee KC et al Integrated amorphous silicon colour sensor on ICT panel for LED backlight feedback control system. In: International symposium 2005 digest of technical papers, p 1376
11. Obikwelu DON (2010) Materials development and exploitation: the determining factor for the pace of technological advancement: key note address delivered at the international conference at the raw materials research and development Council, Abuja on new and advanced materials
12. Obikwelu DON (2008) Research challenges in nanoscience and nanotechnology in developing counties. Int J Eng 2(3):261–270
13. Obikwelu DON (2009) Nanoscience and nanotechnology. Paper Delivered at the University of Nigeria Teaching Hospital, Enugu
14. Obikwelu DON (2008) Nanoparticles in cancer treatment. Paper Delivered at the University of Nigeria Teaching Hospital, Enugu
15. Trevor O Megabashon nanomaterials. In: The occupational hygiene newsletter, vol 17, no 4, Nov 2004

Preemption of Traffic Signal Using Global Positioning System (GPS)

Ankur Shrivastava, Shiksha Rawat, Harsh Kumar Singh, Kumari Neelam, Deepak Prasad and Vijay Nath

Abstract The manipulation of an automated traffic signal is triggered by a Traffic Signal Preemption system. The signals are generally a preconception end to support the emergency vehicles like ambulances and fire truck. This results in their counter times being shortened and the right of-way is offered in an organized and smooth way. The paper offers a pioneering and practical server-centric model to enable preemption using an elementary mobile phone application which uses Global Positioning System (GPS) and a microcontroller which regulates traffic signals.

Keywords Traffic signal preemption · Global positioning system (GPS) · Mobile application · Arduino Uno · Arduino microcontroller · Emergency vehicle · Road traffic · Traffic management

1 Introduction

Human entity is an appreciated strength for any country. The occurrence of casualties such as fire, mishap, cardiac arrest etc. is very common. It is very important for emergency teams to reach the accident spot on time. It is very important to reduce response time. However, the growth of population in cities has caused in high traffic densities. This is an obstacle to emergency vehicles [3].

On hearing the vehicle's siren, drivers give way to emergency vehicles, which is expected from them. An emergency Vehicle passing a crossroads poses danger to traffic upcoming the signal from other roads since most traffic signals today are robotic [4]. Thus, in order to pass the intersection, an emergency vehicle has to wait for the entire traffic signal cycle to complete. This increases response time considerably [2].

A. Shrivastava · S. Rawat · H. K. Singh
Department of CSE, Birla Institute of Technology, Mesra, Ranchi 835215, Jharkhand, India

K. Neelam · D. Prasad (✉) · V. Nath
Department of ECE, Birla Institute of Technology, Mesra, Ranchi 835215, Jharkhand, India
e-mail: prasaddeepak007@gmail.com

© Springer Nature Singapore Pte Ltd. 2019

V. Nath and J. K. Mandal (eds.), *Proceedings of the Third International Conference on Microelectronics, Computing and Communication Systems*,
Lecture Notes in Electrical Engineering 556, https://doi.org/10.1007/978-981-13-7091-5_54

Thus, the need of the hour is Traffic Signal Preemption, in cities where traffic densities are growing continuously [1]. Preemption of Traffic signals assists in reducing response time and increases road safety with minimized cost. Stand-alone hardware is used by numerous prevailing technologies to accomplish this goal [3]. In this paper, we present cost-effective solution in which the preciseness of Global Positioning System (GPS) and the pros of a server-centric structured system are being utilized.

In Sect. 2 of this paper, we have discussed prevailing applications, the machinery, and hardware used and their faults. In Sect. 3, the design of this proposed system is discussed briefly. Benefits of the proposed system are discussed in Sect. 4 and in Sect. 5 discussion of the future possibility and possible enhancements are done.

2 Review of Current Operations

Traffic preemption devices are useful in a variety of ways such as Line of Sight (LOS), an acoustic device, and localized radio signals. The pattern of waves which are transmitted by the emergency vehicles is sensed b the acoustic system Installed. Figure 1 depicts an audio sensor. This type of system can either be used self-sufficiently or can be jointly used with another system. This system also has some disadvantages. Traffic preemption can be easily triggered in the wrong direction or intersection as the sound waves are reflected and may traverse in all directions [4]. Nonhomogeneity of signals produced from different sirens is another drawback. A malicious illegitimate preemption can easily be triggered by imitating the audio of a siren. An additional worry is a hardware for sensors at each intersection [8].

Fig. 1 Traffic intersection installed with acoustic sensors

A beam of infrared light toward traffic signals is released by Line-of-Sight traffic signal preemption systems in the face of the vehicle. Sensors at the intersection sense the beam and change traffic lights appropriately and provide correct direction to the emergency vehicle accordingly. In some circumstances, a visible range strobe light may also be used. Causes of poor preemption are barriers to the line of sight, and atmospheric conditions [2]. The system which is used in preemption can be copied illegally [3].

Traffic preemption systems based on radio waves use radio signals to provoke preemption. Majority of the shortcomings of LOS and Acoustic systems are then overpowered. Radio signals aren't obstructed visually, by lighting or weather conditions. Supplementary hardware is a high expense. All the above restrictions will be removed by use of GPS technology in this paper [6]. Section 4 lists the advantages of using such a system.

3 Proposed Model

Global Positioning System (GPS) hardware, which is prevailing in most smartphones, receives and transmits GPS signals which will be used in the projected model [1]. The traffic signal controllers which are interconnected to a server are used to transmit these signals through the internet. The model consists of three main components as portrayed in Fig. 2. An Arduino microcontroller and an Ethernet Shield are used at each traffic intersection to permit it to connect to the internet. The staff of each ambulance will have to carry an Android smartphone with a particular Android Application installed beforehand. This app receives GPS directs from the vehicle and transmits them to a web server. A Web Server acquires the coordinates of all emergency vehicles by the GPS. Prevention signals to traffic signal controllers are then sent accordingly [4].

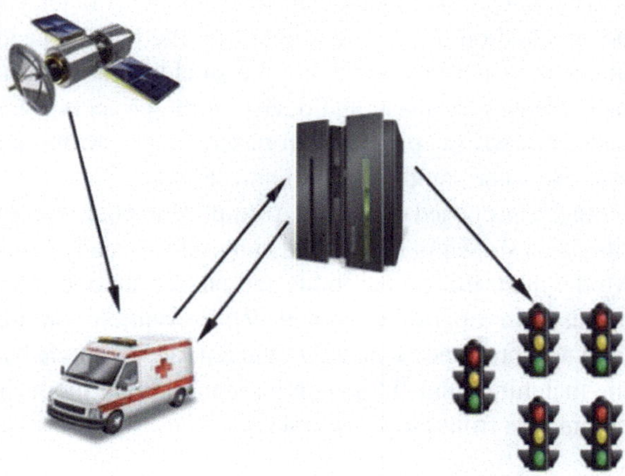

Fig. 2 Main components as portrayed in the model

Fig. 3 Arduino Uno microcontroller

The GPS coordinates of the satellites are obtained by an emergency vehicle with an Android device installed, and then transmits the coordinates to the web server. Position, speed, and other factors are then calculated by the web server and then pre-emption instructions are sent to the corresponding traffic light where the ambulance is about to reach.

Arduino Microcontroller: Open-Source Hardware platform is an Arduino. A microcontroller development stage which can be programmed is present in it and a development environment for creating custom software for the board is also there. Add-on boards can be used to deliver further capabilities. All the components incorporating Circuit diagrams and other requirement are open source [5].

A number of connection sockets are present in Arduino. These can be connected to exterior electronic components, such as microphones, relays, light sensors, motors, laser diodes, buzzers etc. A computer by a USB cable or from a 9 V DC power basis can be used to power an Arduino board [3]. The Arduino board can be detached from the computer once the program is uploaded [8] which is shown in Fig. 3.

An Arduino board can be added with additional competency by enhancing auxiliary boards. Shield have pins in a similar arrangement as the, thus they can simply be attached on the Arduino board by the help of headers. The Arduino board regulates the shield by retrieving the shields pins using the pins in the Arduino. Easy incorporate of sensors and device in the projects is permitted to users by Arduino, instead of writing programs from scratch [5]. Arduino Ethernet Shield, Wi-Fi Shield etc., are some samples of Arduino shields.

In this proposed model, an Arduino Uno microcontroller with an Arduino Ethernet Shield as shown in Fig. 4, that can perform many functions will be used. In Normal Mode, the traffic signal lights and buzzer are operated normally, pre-stored timings are used in a periodic manner [7]. Prevention commands from the web server are then accepted and switched to the emergency mode by manipulating signal lights to the matching road. To switch back-to-Normal mode, accept return-to-normal mode commands from the web server.

Fig. 4 An Arduino Uno with an ethernet shield mounted over it

4 Android App Using Microcontroller

Android contains an operating system, key applications and middleware [9]. Developers are offered an open developed platform which helps them to build tremendously advanced and rich applications. Developers can access the same framework Application Programming Interfaces (APIs) completely which the fundamental applications use. The components can be reused to streamline the design of architecture. Applications can use advantages offered by any other applications [10]. The same mechanism can be used to replace constituents by the user. The Android Software Development Kit (SDK) provides the tools and Application Programming interfaces (APIs) essential to start developing applications on the Android platform. Location and time report using GPS satellite system can be obtained during all weathers, anywhere on earth, when four or more satellites have an unhindered line of sight. Satellites use High precision atomic clocks to measure by time [1]. A GPS receiver owns information concerning the position of a specified satellite at any given time, the time taken by the signals to navigate the distance between the receiver and satellite can be used to determine the receiver's accurate location [8]. A ground headset can calculate its 3D position that is latitude, longitude, and altitude by diagramming the data from three different satellites. A fourth satellite is used to correct mistakes in timing offsets which helps in expanding the accuracy of the location of the emergency vehicle. The more satellites there are, To benefit developers, Location provider, Location manager, Google maps, and geocoding are offered by Android in the above four classes.

In this model, an Android application will be used on a device such as smartphone or a tablet, installed in every emergency vehicle [6]. The shortest path to the destination will be displayed when the staff of the emergency vehicle selects their destination using Google maps, which can only be accessed when the login is authenticated. Once the vehicle is identified as an emergency vehicle, the staff can request an emergency status by selecting a priority level. The app regularly updates the geo-coordinates of the vehicle, obtained using the Global Positioning System (GPS) [8].

Web Server: By using the Hypertext Transfer Protocol, a web server can accomplish its chief function, which is to deliver files and web pages, on client's demand [2]. The web server will monitor Geo-coordinates of the emergency vehicle once its login is authenticated from the Android device application. For each emergency vehicle, corresponding traffic controller receives prevention commands from the closest upcoming traffic signal. The return-to-normal mode commands are sent to the traffic controller once the emergency vehicle has passed the intersection [6].

5 Advantages of This System

This proposed system possesses many benefits in economic terms and the ease with which it can be implemented. Using an Android-based phone reduces the cost of implementing the hardware at each and every traffic intersection and on each and every emergency vehicle [7]. With newer models arriving in the market, the cost of a smartphone is reducing considerably [6]. It is requisite that the Arduino Microcontroller and Ethernet Shield are installed at each traffic intersection. And they are open source and come at a market price of around $60. This cost can be dropped further since the hardware is open source if personalized manufacturing is done in mass. Upgradation in GPS technology in recent years helps in the retrieval of accurate locations [8]. In GPS satellites, a precise atomic clock delivers a method to compute location to the adjacent millimeter. Due to the accuracy of the Global Positioning System, here is a minute chance of false activating of the system. Preemption in opposite directions can be easily triggered in acoustic and radio-based systems [9]. This entire system will be like a centralized secure web server, which will have the same security benefits as the server. The central server will not recognize any illegal hardware used to pre-empt the signals. In contrast, other systems do not provide such kind of security over illegal access [10]. In heavily populated cities, the shortcoming that tall buildings avoid satellite signals to reach the GPS receivers is overshadowed by the mobile tower Triangulation [4] (Table 1).

6 Circuit Diagram of Simple Traffic Signal Block Diagram

The circuit diagram of traffic signal system is shown in Fig. 5.

Table 1 Tabulates the various benefits and demerits of the proposed model

Parameter	Acoustic system	Line-of-sights system	Radio system	Proposed model (GPS)
Dedicated emitter required	No	Yes	Yes	No
Electronics noise interference	No	No	Yes	No
Line of sight	Yes	No	No	No
Affected by weather	No	Yes	No	No
Preemphasis of other approaches	Yes	No	Yes	No
Illegal triggering of preemption	High	High	High	Low
Centralized traffic signal monitoring and log Statistics	No	No	No	Yes

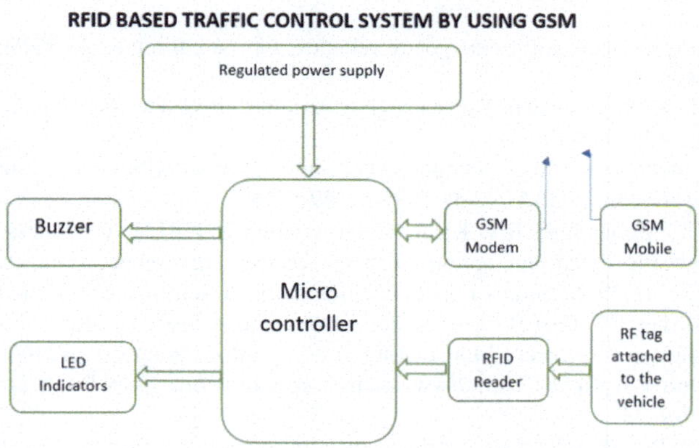

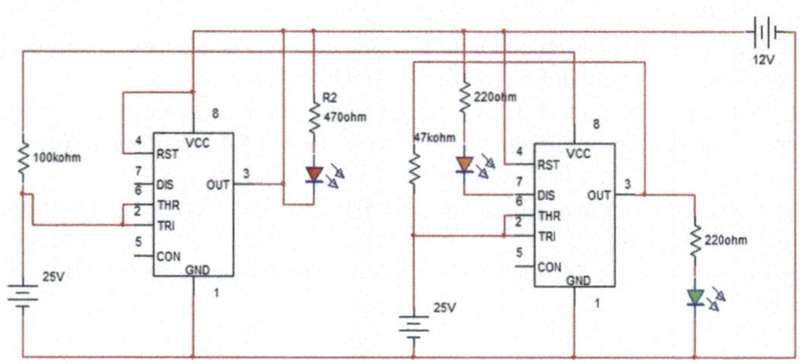

Fig. 5 Circuit diagram of traffic signal system

7 Conclusion

The Arduino once installed can also be used for other purposes. Traffic updates on LED screens can also be displayed using the same microcontroller, so congested routes can be avoided by the commuters [7]. Smart Traffic Management: Traffic signal timings can be varied slightly using the same microcontroller and obtaining traffic data from ground controllers and Google traffic and congested roads can be eased a bit [1]. Violations to traffic rules can automatically be viewed at the central office by using, cameras and other speed detecting sensors connected to the same controller [6, 8, 10].

The Arduino microcontroller can be used for similar applications like dynamic display of bus timings and arrival at bus stops, Anti-collision system in railways, automated closing of railway gates using the same microcontroller and connecting it with different types of hardware [4].

References

1. Traffic signal pre-emption for emergency vehicles: a cross-cutting study, FHWAJPO-05-010, January 2006
2. A report by U.S. Department of Transportation. Arduino. http://www.arduino.cc. Global Traffic Technologies, LLC (GTT)
3. Mahajan K, Mahajan M (2013) Navigating the current location without the availability of GPS. Int J Eng Adv Technol (IJEAT) 2(3). ISSN: 2249-8958
4. Shatnawi SZ, Mohammad BA, Kalaf AI, Al-zoubi HR (2012) A wireless mobile-phone approach to traffic signal preemption for faster service of emergency vehicles. Int J Comput Appl 46(3):35–41. Published by Foundation of Computer Science, New York, USA
5. Gould PJ, Fisher DK (2012) Open-source hardware is a low-cost alternative for scientific instrumentation and research. Mod Instrum 1:8–20. https://doi.org/10.4236/mi.2012.12002. Published Online April 2012 http://www.SciRP.org/journal/mi. ISSN Print: 2165-9257. ISSN Online: 2165-9273
6. Dong L, Chen W (2010) Real-time traffic signal timing for urban road multi-intersection. Intell Inf Manag J 21605912
7. Singhal M, Shukla A (2012) Implementation of location based services in Android using GPS and web services. Int J Comput Sci 16940784–16940814
8. Collura J, Willhaus EW (2001) Traffic signal preemption and priority: technologies, past deployments, and system requirements. In: Proceedings of the 11th ITS America, Florida, USA
9. Vreeken J, Wiering M, Veenen, Koopman A (2004) Intelligent traffic light control. Technical report. UU-CS-2004-029
10. Mendizabal J, Berenguer R, Melendez J (2009) GPS and Galileo. McGraw Hill. ISBN 978-0-07-159869-9

Evolution of Pacemaker: A Review

Yash Goyal, Sanskar Agarwal, Subrata Barman, Ashutosh Pranav, Deepak Prasad and Vijay Nath

Abstract An artificial pacemaker is an electronic device in which rhythmic electric stimulus is used which helps in maintaining an appropriate heart rhythm for a long time of the heart muscle which consequently regulates the pumping of the heart. The components of the pacemaker are: pulse generator, circuitry package, a battery (wirelessly charged) and lead system. Indication for installation of an everlasting pacemaker and the opposite pacemaker mode selection are created usually by keeping cardiac diseases in mind.

Keywords Arteries · Muscle · Ventricle · Fabric

1 Introduction

An output pulse is defined as the magnitude of electrical charge provided by the pulse generator during the stimulus and for stimulating the heart we need a lot of output pulses [1]. But for effective pacing, we need to have appropriate width and energy of the output pulse to depolarize the myocardial cells. A pacemaker evolved from time to time according to the needs. As it changed, the complications associated with it decreased, implantation methods were relieved and the demand for the pacemaker increased [2]. The following pacemakers will be discussed in detail: (a) Hyman's Pacemaker (b) Dual-Chamber Pacemaker. Hyman's pacemaker is considered to be one of the earliest artificial pacemakers. He constructed the pacemaker with the help of his brother Charles. In both the chambers of Dual-chamber pacemaker, there are pacing electrodes. It helps in maintaining a physiological relationship between atrial and ventricle contraction [3]. This paper is constructed in the following manner: A detailed explanation of the conduction system in heart and how electric pulses play their role in mediating the contraction is demonstrated in Sect. 2. Hyman's Pacemaker is explained and its design is described in Sect. 3. Dual-chamber pacemaker

Y. Goyal · S. Agarwal · S. Barman · A. Pranav · D. Prasad · V. Nath (✉)
Department of CSE, Birla Institute of Technology, Mesra, Ranchi 835215, Jharkhand, India
e-mail: vijaynath@bitmesra.ac.in

© Springer Nature Singapore Pte Ltd. 2019
V. Nath and J. K. Mandal (eds.), *Proceedings of the Third International Conference on Microelectronics, Computing and Communication Systems*,
Lecture Notes in Electrical Engineering 556, https://doi.org/10.1007/978-981-13-7091-5_55

is demonstrated in Sect. 4. Various modern advancements are discussed in Sect. 5. This consists of all the important data such as heart rate analysis, circuitry of various pacemaker models, etc. Finally, the conclusion is enunciated in Sect. 6.

2 Conduction System in Heart and Problems Associated

Before discussing various pacemaker and their advancements, one should know excitation and conduction system in heart. Our heart consists of three kinds of cardiac muscles [4]:

(i) Atrial Muscles
(ii) Ventricle Muscles
(iii) Specialized Fibres (excitation and conduction fibres)

For the control of the heart, i.e. regular pumping, we need to have proper excitation and conduction system, which has been illustrated in the in Fig. 1.

It consists of the

1. The atrioventricular node (A-V node)
2. Internodal tracks
3. Sinoatrial node (SA node)
4. His' Bundle
5. Purkinje fibres.

The natural pacemaker exhibits an interesting property named automaticity which generates electrical impulses via self-activation and delivers them in such a way that heartbeats and muscle contracts to the muscles of the heart. The SA Node controls the

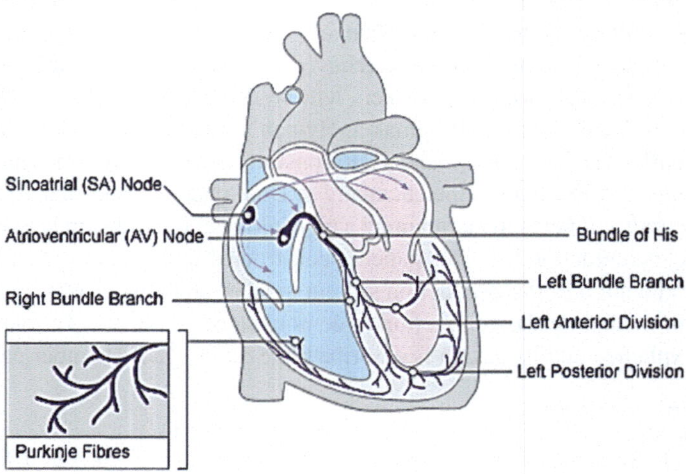

Fig. 1 Cardiac conduction system

usual rhythm of the heart, in the range of 60–100 beats per minute. The two tracks, namely Bachmann's bundle and internodal tracks start contraction of the atrial walls by transmitting external excitation throughout the atria. Simultaneously, the A-V node receives the impulse, as it's the only electrical pathway present between atria and ventricles. As a result, some amount of delay is there in the AV node, due to which the atria contracts before contraction in ventricle begin. As the His bundle is activated, the signal separates into the other branch, which is transmitted to the right ventricle, as well as the left bundle branch leading to the left ventricle. Both of these bundle branches mentioned above are terminated in Purkinje fibres. The fibres which are responsible for distributing the excitation throughout the two ventricles are Purkinje fibres themselves. They also cause a coordinated ventricular contraction [5].

A disease which is associated with cardiac problems resulting in irregular heartbeats is called Arrhythmias. At large, arrhythmias can affect the proper flow of blood, including times where the natural pacemaker creates some irregular rate. When it interrupts the conventional conduction pattern and a different part of the heart manipulates the rhythm. This disease can involve a nonstandard rhythm and classified as follows:

(a) When the rate is <60 bpm: Bradycardia
(b) When the rate is >100 bpm: Tachycardia

It can also be characterized by an irregular heart rate as shown in the figure below. An "artificial pacemaker" plays a key role in such situations.

3 Hyman's Pacemaker

Albert S. Hyman himself designed the primary heart pacemaker in the year 1933. According to the inventor of Hyman Pacemaker, "the introduced electrical impulse by DC power solely provides a manageable petulant point from that a wave of excitation might arise usually and bend over the heart on its familiarized pathways." Figure 2 shows the model of Hyman's pacemaker [6].

It uses a hand wound along with a generator (spring-driven) which provides 6 min of pace making in one winding. The operation of Hyman's model is as mentioned as: The crank wounded the spring motor that drove the generator at a defined speed and rotated the device disc. The generator equips the current to a surface of contact on the interrupter disc. The magnet parts along with them deliver the magnetic flux essential to come up with current in the generator [4]. After the current is produced by the device at the desired rate (in bpm), which is regulated by the impulse controller. It is represented by the periodic pacing waveform transferred to the electrode needle. The induction bulb is on when a stimulus is distorted. In Fig. 3, an appropriately labelled block diagram of the Hyman's pacemaker is illustrated.

Fig. 2 Representation of Hyman's pacemaker

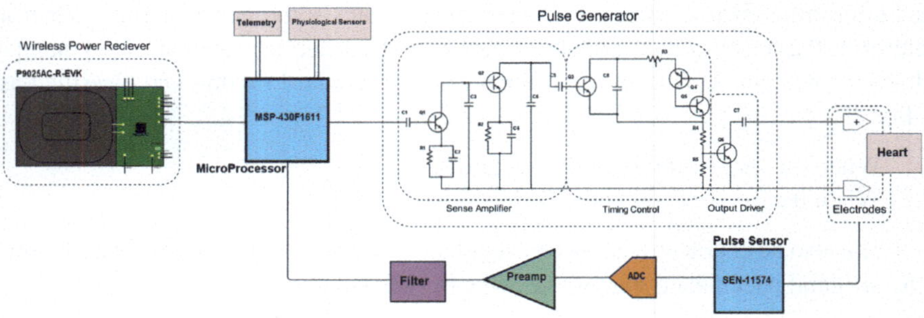

Fig. 3 Wireless pacemaker circuitry

4 Dual-Chamber Pacemaker

It generally needs two pacing electrodes, in the right ventricle and in the right atrium. The elementary work is to first sense any electrical activity around or in the atrium and then check if pacing is required. When dual-chamber predicts about the pacing, it starts to send pacing pulses in the region of atrium and ventricle and matches with the natural way of heart's pacing. The dual chamber was proposed by Berkovits in the 1970s, it was a bifocal pacemaker which could sense only the ventricle but still could pace both chambers. The input of the dual chamber consists of an atrial standstill provided with A-V block (or sinus node syndrome), and output is taken from bifocal lens to the atrium. After some time, the output is taken to the ventricle and it completes the whole cycle. Berkovits improvised his original design to a demand pacemaker implemented in the dual chamber [7].

The basic working of the dual chamber includes timing control circuits which are provided by sense amplifiers which determined each timeout stimulating period of atrial and ventricular cycles. However, once a planned quantity (200 ms) of atrial-stimulating impulse is reached, the generation of ventricular-stimulating impulse takes place. 3 electrodes were provided for the implemented pacemaker which are:

1. Atrial stimulation electrode
2. Neutral electrode
3. Pacing of Ventricle and sensing electrode.

The transistor-cum-switch (SFET), in the feedback path, was inserted to the ventricular electrode because the atrial contraction may lead to erroneous detection. The negative pulse was transferred through the diode D_a and charging the capacitor C_a produced at the atrial electrode and opening the circuit. In this way, during each Atrial stimulation sense amplifier was disabled for a short interval of time [8].

5 New Features

Well, we may say that pacemaker has saved countless lives since its invention in the 1930s, but the complications caused by normal pacemakers could sometimes be fatal [9]. So, in order to reduce costly and invasive operations, the world is moving to make a wireless and battery-less pacemaker. This new pacemaker is very cheap and is powered by microwaves, allowing it to be implemented directly into a patient's heart [5]. While the original pacemakers require wires for its operation, which could cause infection and bleeding, this new pacemaker takes its energy directly from radio waves allowing it to prevent any type of surgery. This extremely powered pacemaker also helps doctors to adjust the pacing of the signals sent to the device and of course, could be done by increasing or decreasing the amount of power sent to the receiver. This device is designed cleverly to store any excess energy, keeping the extra power as a reserve once it reaches or crosses its predefined threshold.

These pacemakers can be easily implemented in patient's heart (inside or outside), because of their small size, allowing them both to keep an identical pace, which would ultimately help the patients to face minor heart attacks [10].

6 Conclusion

Pacemakers have evolved tremendously since their invention, Hymann and dual-chamber pacemakers were the two most important pacemakers of the era. Their design, integrated circuit styles have all been progressively improved so as to provide better treatment. The new improvements such as wireless pacemaker and lead-free pacemaker have extended the use of pacemakers not only for the control of the rhythm of heart but also for cardiac resynchronization therapy and cardiomyoplasty.

Acknowledgements We would like to thanks Dr. M. K. Mishra, Vice Chancellor BIT Mesra Ranchi for providing such great infrastructure. A piece of thanks also goes to Dr. S. pal for their helping hand.

References

1. https://www.medicinenet.com. Accessed on 01 Mar 2018
2. Woollons DJ (1995) To beat or not to beat: the history and development of heart pacemakers. Eng Sci Educ J 4:259–268; Berkovits BV (1971) Atrial and ventricular demand pacer, US Patent 3,595,242, July 1971
3. Greatbatch W, Holmes CF (1991) History of implantable devices. IEEE Eng Med Biol 10:38–49
4. Geddes LA (1990) Historical highlights in cardiac pacing. IEEE Eng Med Biol 9:12–18
5. Mond HG (1999) Recent advantages in pacemaker lead technology. Cardiac Electrophys Rev 3:5–9
6. Sanders RS, Lee MT (1996) Implantable pacemakers. Proc IEEE 84(3):480–486
7. Guyton AC, Hall JE (1996) Textbook of Medical Physiology, 9th edn. W. B. Saunders, Toronto
8. Bronzino JD (2000) The biomedical engineering handbook, 2nd edn, vol. 1. CRC Press and IEEE Press, Boca Raton, FL
9. Schaldach M, Furman S (1975) Advances in pacemaker technology. Springer-Verlag, Berlin
10. Sutton R, Bourgeois I (1991) The foundations of cardiac pacing. Part I. Futura Publishing Company, Mount Kisco, NY

Study and Development of Solar-Powered Water Pumping System

Adity, Alwish Lakra, Kajal Gupta, Kamal Murmu, Deepak Prasad and Vijay Nath

Abstract Solar energy is the radiant light from the sun and can be harnessed from different technologies. It is an important renewable energy source which is generally used for generating electricity. It can be harnessed very easily as the only condition is availability of sunlight and landmass. But it is mainly used in rural areas for their development. Solar powered pumps can be used in farmsfields, villages and animal herds. It is normal power pump with an electric motor which uses electricity produced by solar panel at the site. The motor can us DC as well as AC power but due to reduced efficiency and cost of AC and DC motor is generally preferred. This can be used for pumping water. Basically, two elements are used in water pumping system. These are PV panels and pumps and the smallest one is solar pump. When exposed to light, each solar cell produce a direct electric current using the specially designed layers of semiconducting material. The wiring in the panel helps to collect this direct current and send it to the DC pump, which pumps it according to the sun or is stored for future use.

Keywords Energy · Renewable · Fuel · Semiconductor · Solar cell · Pump · PV panel

1 Introduction

The energy obtained from the sun in the form of heat and light is solar energy. It is one of the most renewable and readily available energy sources on earth. The fact that this it is present in abundant quantity and is free makes it the most important source of energy [1]. Use of solar energy in agricultural fields has already increased productivity across the world. Figure 1 shows the solar module [2].

Adity · A. Lakra · K. Gupta · K. Murmu · D. Prasad (✉) · V. Nath
VLSI Design Group, Department of ECE, Birla Institute of Technology, Mesra, Ranchi 835215, Jharkhand, India
e-mail: prasaddeepak007@gmail.com

© Springer Nature Singapore Pte Ltd. 2019

V. Nath and J. K. Mandal (eds.), *Proceedings of the Third International Conference on Microelectronics, Computing and Communication Systems*,
Lecture Notes in Electrical Engineering 556, https://doi.org/10.1007/978-981-13-7091-5_56

Fig. 1 A solar module

Fig. 2 A typical assembly of solar cell

Using PV cells solar energy can be converted into electricity as shown in Fig. 2 means "light" and "electricity", using a photovoltaic effect. When the sunlight falls on PV cells, due to the collisions with photons electrons get excited and start to move generating electricity [3].

According to scientific research, the earth receives 173 trillion terawatts of solar, which is more than the world consumption of energy. This verifies that the sun is the most useful and reliable resource on earth [4]. 174 peta watts (PW) of incident energy is received by the earth at the upper atmosphere. Earth receives 174 peta watts energy at the top layer of atmosphere out of which almost 30% is reflected back into space.

2 Working of Solar Cell

It is the electronic device which converts solar energy into electricity. The solar cell is sandwiched between two layers of silicon which are doped such that lower layer is p-type and upper layer is n-type [5].

The photons from sunlight carry energy and when it falls on the upper layer and reach to the lower level they give their energy to electrons. They use this energy to jump to upper layer and escape into the circuit. The amount of energy we get from sun, 1.5×10^{18} kW/year, is approximately 10,000 times more than the consumption of entire world. The density of power radiated from sun is 1.373 kW/m^2. The amount of electricity produced depends on the size and efficiency of cell as well as intensity of sunlight striking the surface. Actually, most of the cells can convert only 10–20% of energy into electricity but a cell can have maximum efficiency of 30% according to Shockley–Queisser limit. This cell has a single junction and is made up of silicon. There are some cells which has 46% efficiency having multiple junctions so that they can catch photons of different energies [6]. The photovoltaic is more effective than any other equipment like electric grid or generators. PV systems are much cheaper than that of installation of other sources of energy either for livestock watering or crop irrigation.

3 Solar-Powered Water Pumping System Configuration

Solar-powered pumping system can be divided into two major parts that are Battery coupled powered pump and direct-coupled powered pump. So many factors are needed to be considered to find which system is best according to the given condition.

3.1 Battery Coupled Water Pumping System

The requirements of battery-coupled water pumping system are PV cells, tank and pressure switch. As the name suggests, battery is also needed along with control regulator shown in Fig. 3. During day hour's electricity generated by the PV panels charge, the batteries and the power is supplied to the pump anytime when it is required. A steady-operating voltage is also supplied to the pumps DC motor functioning for a longer period of time. Hence, during nights, the system can still act as a constant source of power for the running of the pumps. But it comes with its drawbacks attached to it. First, the overall efficiency could be affected by the battery-operating voltage. On changing, the temperature voltage produced by the batteries can be a few volts lessen than that obtained inside panels. A pump controller helps to increase voltage as well as helps to minimize this effect, and the pump efficiency will drop off even more due to continuous changing weather [7].

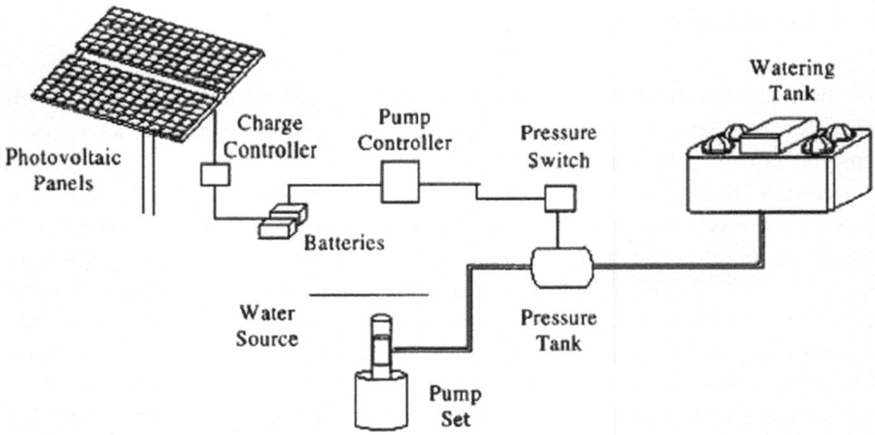

Fig. 3 Battery-coupled water pumping system

3.2 Direct-Coupled Water Pumping System

Direct-coupled pumping systems are made as large as they can store an extra amount of water for nights and the other days when the sun does not appear. The capacity to store water should be high as depending on climatic conditions and pattern of water usage, anytime water may be required. Storage of water in tanks has its own disadvantages as if we store water in open tank then due to evaporation process, loss of water can be considerable and on the other hand, if we use closed tanks, then the cost will increase. During the cold weather, water stored may freeze or purity of water may be affected [8]. In this system, electricity from the PV modules is directly sent to the pump, which sends the water to required places as shown in Fig. 4. The amount of water pumped depends on the intensity of sunlight incident on the PV cells and the type of pump used. Thus, it can only be used in day time. The intensity and angle of incidence of sunlight alter the quantity of water pumped. During the optimum sunlight periods, the pump efficiency is almost 100%.

4 Main Solar-Powered Stock Watering System Components

The major components of a solar power water system are solar module, pump, controller, storage tank. Solar electric systems are also called as photovoltaic systems. Direct current is produced by most of the solar panels and modules. An array is a collection of modules. There are two major types of mounting structures which are fixed and tracking structures. Fixed mounts are cheaper and can endure larger wind intensity but they must face true south direction (nonmagnetic south). For portability, it can be mounted on trailer. The tracker will increase the cost but the pumping

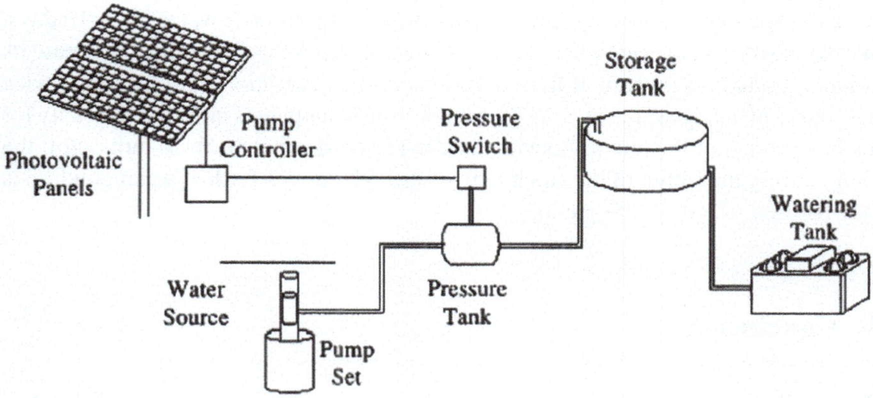

Fig. 4 Direct-coupled system

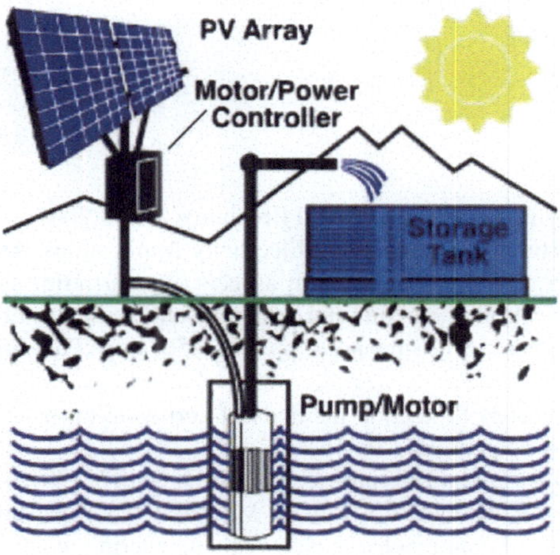

Fig. 5 Solar-powered based stock water system

efficiency will also be increased compared to fixed arrays. There are two types of pumps: DC and AC. DC pumps use 1/3rd to ½ the energy of AC pumps. DC pumps are of many types, namely centrifugal or displacement or submersible or surface pumps [9] The solar powered based stock water system in Fig. 5.

They use diaphragms, vanes or pistons to send the water out of the discharge outlet forcefully. The spinning impeller is used by centrifugal pumps to add energy to the water. Submersible pumps are placed down the wells and because of their important properties like not exposed to freezing temperatures and can exist without priming, they are highly reliable. Surface pumps are used mainly to move water through a pipeline, and as the name suggests they are located at or near surfaces. Supply of water to high elevations or long distances can be achieved through surface pumps. Batteries may often reduce the overall efficiency of the system. Therefore, it increases the maintenance cost. Some may even prefer to store water for 3–10 days, instead of storing electricity in batteries. The controller helps the pump to increase its efficiency and also protects it from high/low voltage conditions. An inverter device may come in handy as it converts DC current from solar panels into AC electricity for the functioning of the pump. A switch is also required which helps to turn on/off the pump during the filling of the stock tank. Cutoff electrodes for low water conditions may be used to protect the pump.

5 Conclusion

Photovoltaic power is more cost-effective than an electric system, gas, or diesel. For this reason, photovoltaic power is more useful for today world. Photovoltaic power is more useful for irrigation systems, street light, household appliances, etc.

References

1. Unisolar Solar Energy Produces Catalog and Brochures. USA (2001)
2. Solar Cell EIE, Department of Research on Electricity Applications, Ankara Turkey (1992)
3. Nidhi N, Prasad D, Nath V (2019) Different aspects of smart grid: an overview. In: Nath V, Mandal J (eds) Nanoelectronics, circuits and communication systems. Lecture notes in electrical engineering, vol 511. Springer, Singapore, pp 451–456. https://doi.org/10.1007/978-981-13-0776-8_41
4. Eker B, Akdogan A (2005) Protection methods of corrosion on solar systems. In: TMMOB machinery engineering society, Mersin, Turkey
5. Goel V, Riya Kumari P, Shikha P, Tanushree Prasad D, Nath V (2019) Design of Smartphone controlled robot using Bluetooth. In: Nath V, Mandal J (eds) Nanoelectronics, circuits and communication systems. Lecture notes in electrical engineering, vol 511. Springer, Singapore, pp 557–563. https://doi.org/10.1007/978-981-13-0776-8_52
6. Strang G, Nguyen T (1996) Wavelets and filter banks. Wellesley Cambridge Press, Wellesley, MA
7. Guide to Solar Powered Water Pumping Systems in New York State. New York State Energy Research and Development Authority (NYSERDA), Jan 2005
8. A Green and Prosperous Middle East, 24 Jan 2013. http://greenandprosperousmiddleeast.blogspot.com/2013/01/solar-panels-irrigation-and-local.html
9. Savco Solar Energy Clean Renewable Energy for the 21st Century Solar Water Pumping (2008). http://www.savcosolar.com/pumping.html

Study and Design of Obstacle Detection Mechanism

Pratik Kumar, Abhishek Kalra, KotniJyothi Prakash, Deepak Prasad and Vijay Nath

Abstract In today's world, robotics is a fast growing and interesting platform. The Robot has sufficient intelligence to cover maximum area of provided space. It has the infrared sensor which is used to sense the obstacles coming in between the path of the robot. It would move in the specified direction and avoided the obstacle which was coming in its path. Autonomous-intelligent robot could perform desired tasks without continuous human guidance. The minimum number of gear motor allows the walking robot to lower the power consumption while constructing a program that can produce coordination of multi-degree of freedom for sufficient to produce the basic walking robot and one voltage movement of the robot. It was found that two gear motors regulators are in need to control the load where it was capable of supplying enough current to drive two gear motors for each wheel.

Keywords Wheeled robot · Autonomous intelligent · Atmel microcontroller · Gear motor

1 Introduction

Nowadays, Robotics was part of today's communication and communication was apart of advancement of technology, so it is decided to work on robotics field, and design something which will make human life today's aspect [1]. There are different types of mobile robots which can be divided into several categories based on the wheeled robot, crawling robot, and legged robot. This project deals with a wheeled autonomous robot. It was the part of automation; robot has sufficient intelligence to cover the maximum area [2]. This robot used infrared sensor to detect the obstacles in between the path and then avoid them to complete its objective. The infrared

P. Kumar (✉) · A. Kalra · K. Prakash · D. Prasad · V. Nath
Department of ECE, Birla Institute of Technology, Mesra, Ranchi 835215, Jharkhand, India
e-mail: pratikdalmia34@gmail.com

D. Prasad
e-mail: prasaddeepak007@gmail.com

© Springer Nature Singapore Pte Ltd. 2019

V. Nath and J. K. Mandal (eds.), *Proceedings of the Third International Conference on Microelectronics, Computing and Communication Systems*, Lecture Notes in Electrical Engineering 556, https://doi.org/10.1007/978-981-13-7091-5_57

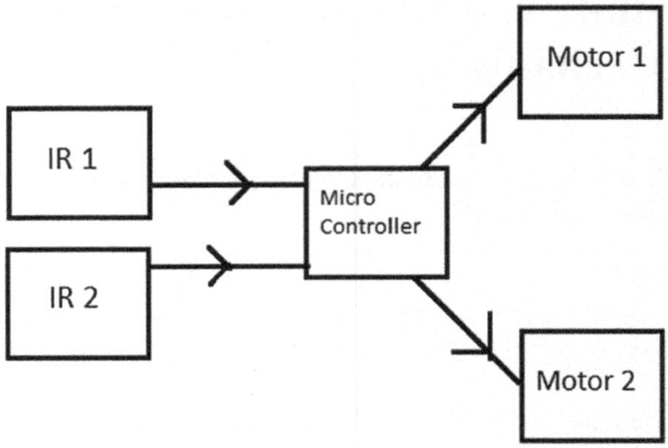

Fig. 1 Flowchart of algorithm

transmitter continuously generates an infrared signal, when an obstacle comes in the path the in IR signal reflected back from the object and was received by the IR sensor and then generate a positive signal with the help of the receiver's circuit that was there was an *o*bstacle in the path. In such a way, the robot was able to detect obstacles of provided space and is able to avoid obstacles coming in between a path of robot with the help of microcontroller board and complete the journey. The main motto is to design a type of robot or the technology in which this technology can be used in today's very fast transportation to avoid the accident that generally happens in congested or the metropolitan areas by applying an emergency break [3, 4]. If they use this technology in the car or any vehicle, it will automatically sense the obstacles, and then it will take a side to the available free space. An obstacle may be a living thing or any object. Autonomous-intelligent robot is robots could perform a given task in unstructured environments without continuous human guidance. Thus by using this technology in vehicles, we made the drive safe [5, 6] (Fig. 1).

2 Methodology

As the robot was switched ON, first, it will check either start signal was received or not, if not then the program counter will not go to the corresponding address it will remain on the previous address until it gets a negative low signal Then, the robot does a continuous check any obstacle in path, if there was no obstacle then the robot will go straight. If any obstacle is found on the left side, then the Atmel controller sends a command to the motor drive to stop the right motor and move the left motor and just opposite as an obstacle is found in the right side.

Fig. 2 Line follower

3 Application

Obstacle avoiding technique has a very useful in real life, this technique can also use as a vwasion belt of blind people by changing the IR sensor by a kinetic sensor, which was on type of microwave sensor whose sensing range was very high and the output of this sensor varies according to the object position changes. This technique makes a blind people able to navigate the obstacle easily by placing three vibratos in left, right and the center of a belt named as vwasion belt and makes a blind people able to walk anywhere. On top of obstacle avoiding robot temperature/pressure sensors can be added to monitor the atmospheric conditions around. This was useful in places where the environment was not suitable for humans. The same technology can be used in various applications by modifying the microcontroller program [7–9] (Fig. 2).

4 Working Principle of an IR Sensor

An infrared sensor circuit was one of the basic and popular sensor modules in an electronic device. This sensor was analogous to human's vwasionary senses, which can be used to detect obstacles and it was one of the common applications in real time. This circuit comprises of the following components (Figs. 3 and 4).

- LM358IC 2 IR transmitter and receiver pair
- Resistors of the range of kilo ohms.
- Variable resistors.
- LED (Light-Emitting Diode).

In this project, the transmitter section includes an IR sensor, which transmits continuous IR rays to be received by an IR receiver module. An IR output terminal of the receiver varies depending upon its receiving of IR rays. Since this variation cannot be analysed as such, therefore this output can be fed to a comparator circuit. Here an operational amplifier (op-amp) of LM 339 was used as comparator circuit.

Fig. 3 Automatic vacuum cleaner

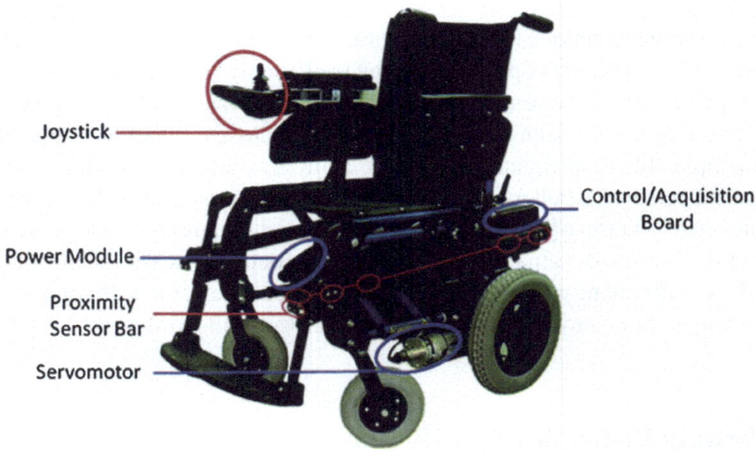

Fig. 4 IR automated wheelchair

When the IR receiver does not receive a signal, the potential at the inverting input goes higher than that non-inverting input of the comparator IC (LM339) as a result output of the comparator goes low, but the LED does not glow. When the IR receiver module receives a signal, the potential at the inverting input goes low as a result output of the comparator (LM338) goes high and the LED starts glowing and resistor R1 (1000), R2 (1 k) and R3 (33) are used to ensure that minimum 10 mA current passes through the IR LED devices like Photodiodes and normal LED, respectively. Resistor VR2 (preset = 5000) was used to adjust the output terminals. Resistor VR1 (preset = 10000) was used to set the sensitivity of the circuit (Fig. 5).

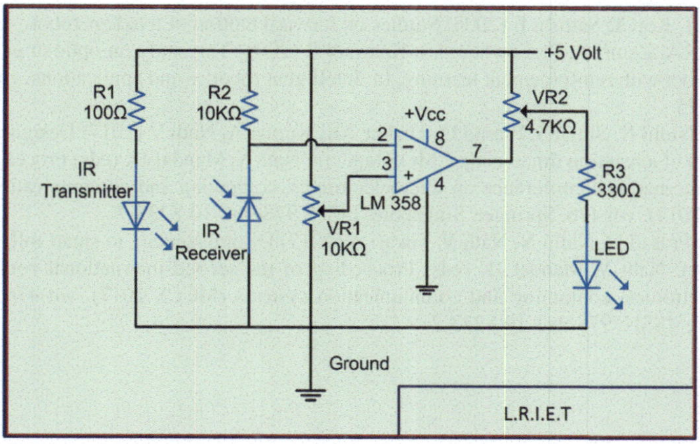

Fig. 5 Circuit diagram of IR sensor

5 Conclusion

From this study, a walking robot that achieved the stated objectives had been developed. This robot was able to produce the basic walking movements using two gear motors. We developed the robot with a very good intelligence which was easily capable to sense the obstacle and by processing the signal coming from the sensor it was perfectly avoiding the obstacle coming in between the path. Robot takes the left or right or the forward movement in according to the sensing signal with the help of the two gear motors, which makes the movement of the robot smooth. In future, the sensing range can be increased by increasing the sensor quality with the help of ultrasonic sensor or the infrared signal spread all over the provided area.

Acknowledgements Authors would like to thanks Prof M. K. Mishra, Vice Chancellor, BIT Mesra for providing such great infrastructure to carry out this work. A piece of thanks also goes to HOD ECE, Dr. S. Pal for their constant support and encouragement.

References

1. Senthil Kumar N, Saravanan M, Jeebananthan S (2012) Microprocessors and microcontrollers, 4th edn. Oxford University Press. ISBN: 978-0- 19-806647-7
2. Ram B, Fundamental of microprocessors and microcontrollers, 7th edn. Dhanpat Rai Publication. ISBN: 978-81-89928-60-5
3. Lam MC, Prabuwono AS, Arshad H, Chan CS (2011) A real-time vwasion-based framework for human-robot interaction. Lecture notes in computer science (part 1) pp 257–267
4. Prasad D, Nidhi N, Nath V (2018) Internet of things/everything: an overview. Int J Pure Appl Math
5. Washida T, Kuroki Y (2004) Sensor system of a small biped entertainment robot. Adv Robot VSP Robot Soc Jpn 18(10):1039–1052

6. Maikoto T, Koji Y, Satoshi E (2005) Studies on forward motion of five legs robot
7. Besari ARA, Zamri R, Prabuwono AS, Kuswadi S (2009) The study on optimal gait for five-legged robot with reinforcement learning. In: Intelligent robotics and applications, vol 5928, pp 1170–1175
8. Vohra V, Nidhi N, Singh S, Prasad D, Thakur AK, Kumar A, Nath V (2017) Design and implementation of a reaction timer using CMOS logic In: Nath V, Mandal JK (eds) Proceeding of the second international conference on microelectronics, computing and communication systems (MCCS 2017), vol 476. Springer, Singapore. ISSN: 978-981-10-8233-7
9. Bohra V, Prasad D, Nidhi N, Nath V, Tiwari A (2017) Design strategy to smart toll and billing system. In: Nath V, Mandal JK (eds) Proceeding of the second international conference on microelectronics, computing and communication systems (MCCS 2017), vol 476. Springer, Singapore. ISSN: 978-981-10-8233-7

Author Index

© Springer Nature Singapore Pte Ltd. 2019

V. Nath and J. K. Mandal (eds.), *Proceedings of the Third International Conference on Microelectronics, Computing and Communication Systems*,
Lecture Notes in Electrical Engineering 556, https://doi.org/10.1007/978-981-13-7091-5

The manufacturer's authorised representative in the EU is Springer Nature Customer Service Centre GmbH, Europaplatz 3, 69115 Heidelberg, Germany. If you have any concerns regarding our products, please contact ProductSafety@springernature.com

Printed and bound by CPI Group (UK) Ltd, Croydon, CR0 4YY
25/09/2026
02222884-0017